Problems Workbook for

Organic Chemistry

Paris Svoronos

and

Ed Sarlo

Department of Chemistry
Queensborough Community College of CUNY
Bayside, New York

WCB Wm. C. Brown Publishers
Dubuque, Iowa • Melbourne, Australia • Oxford, England

Book Team

Editor *Craig Marty*
Developmental Editor *Elizabeth M. Sievers*
Production Editorial Manager *Julie A. Kennedy*
Visuals/Design Freelance Specialist *Barbara J. Hodgson*

Wm. C. Brown Publishers
A Division of Wm. C. Brown Communications, Inc.

Vice President and General Manager *George Bergquist*
National Sales Manager *Vincent R. Di Blasi*
Assistant Vice President, Editor-in-Chief *Edward G. Jaffe*
Marketing Manager *John W. Calhoun*
Advertising Manager *Amy Schmitz*
Managing Editor, Production *Colleen A. Yonda*
Manager of Visuals and Design *Faye M. Schilling*
Publishing Services Manager *Karen J. Slaght*
Permissions/Records Manager *Connie Allendorf*

Wm. C. Brown Communications, Inc.

Chairman Emeritus *Wm. C. Brown*
Chairman and Chief Executive Officer *Mark C. Falb*
President and Chief Operating Officer *G. Franklin Lewis*
Corporate Vice President, Operations *Beverly Kolz*
Corporate Vice President, President of WCB Manufacturing *Roger Meyer*

Copyediting and Production by *Shepherd, Inc.*
Cover and Interior Design by *Shepherd, Inc.*

Library of Congress Catalog Card Number: 92–70067

ISBN 0–697–14551–4

Printed in the United States of America by Wm. C. Brown Communications, Inc., 2460 Kerper Boulevard, Dubuque, IA 52001

10 9 8 7 6 5 4 3 2 1

Contents

Preface

Every year the challenge of Organic Chemistry is faced by college sophomores pursuing careers in science. With the reputation of being one of the most difficult undergraduate courses, Organic Chemistry has been a stumbling block for many. But it does not necessarily have to be that way. Many students, could do well in the course provided they have confidence in their academic abilities. The study of organic chemistry requires many hours of work on a continuous and regular basis. A student must set aside several hours a day for study and must avoid the temptation to put off serious studying until immediately before an exam. Cramming for an exam almost never works and is a bad habit to get into.

But what should be studied? Organic Chemistry is so comprehensive that it is not surprising that students face it with trepidation. Although study habits vary with individuals, the following points are important.

First, lecture notes should be reviewed thoroughly. It is in the lecture that the instructor focuses on the material that he or she thinks is the most important. A good idea for the student is to re-write lecture notes: this is an opportunity to organize thoughts about the material presented and also an opportunity to formulate questions about what is not fully understood.

Second, the text should be studied, not merely read, paying particular attention to the topics covered in lecture.

Third, and absolutely essential, the problems must be worked in detail. It is a very common experience to read something and believe we understand it, only to find out that we cannot answer someone's question about what we have just read. The proof of our understanding is the ability to successfully solve problems. The more problems, the better.

This workbook presents a variety of problems which are common to all undergraduate courses in Organic Chemistry, but with an emphasis on reaction mechanisms. Why mechanisms? Well, because this is the thread that binds together a variety of topics and organic reactions. For example, condensations (such as the Aldol and Claisen), aromatic substitutions, and free radical additions to alkenes are only truly understood and remembered by a knowledge of their mechanisms. And mechanisms are learned only by working many problems which require the student to write them out. Repetition reinforces and is essential to the learning process.

This workbook also contains problems dealing with spectroscopy and organic synthesis. The problems vary in degree of difficulty and are suitable for all levels of learning, from junior college to pre-graduate school. Of course, answers to problems are included at the end of each chapter, but they should be referred to *only* after an attempt is made to work out the solutions. Even if the student arrives at an incorrect answer, he or she can learn from this mistake.

Each chapter provides a sample multiple choice test after the section on problems, and the answers can be found in Appendix I. These tests provide a means of testing overall understanding of material covered in each chapter.

There are recurring themes in Organic Chemistry—resonance, the formation and relative stabilities of reactive intermediates, and aromatic behavior, for example. Very often by working a problem, even incorrectly, the student gains insight into fundamentals which help in solving other problems—problems which at first appear to be unrelated but really aren't.

Organic Chemistry is probably not for the timid. It is certainly not for those who are academically lazy. But for the thousands of students who successfully navigate their way through the sometimes stormy waters, it offers a profound sense of accomplishment.

At the beginning of every organic chemistry course most students' conception of the subject agrees in many respects with Friedrich Wohler's quotation (1835) that stated:

"Organic chemistry nowadays almost drives me mad. To me it appears like a primeval tropical forest full of the most remarkable things, a dreadful endless jungle into which one does not dare to enter for there seems to be no way out."

Hopefully by the end of the course, and with the aid of this manual, the initial conception will be replaced by the 1970 quotation of the 1957 Nobel laureate Sir Alexander Todd.

"At one time organic chemistry appeared to the student as a vast body of facts, often apparently unconnected, which simply had to be learned, but the more recent developments in theory have changed all this so that organic chemistry is now a much more ordered body of knowledge in which a logical pattern can be clearly seen."

We, the authors, hope that you will agree with Sir Todd's quotation at the completion of the course.

Acknowledgements

Wayne B. Counts
Georgia Southwestern College

Seth Elsheimer
University of Cental Florida

James P. Hagen
University of Nebraska at Omaha

Christopher T. Waidner
Georgetown University

Helen S. Hauer
Delaware Technical & Community College

James H. Hutchinson
Middle Tennessee State University

Jack H. Stocker
University of New Orleans

1 General Principles

Questions

1.1 Carbon's orbitals hybridize during covalent bond formation.
a. What is meant by *hybridization of orbitals?*
b. Describe *sp*³, *sp*² and *sp* hybridization. What bond angles and geometry are associated with each type of hybridization?
c. Describe the formation of sigma (σ) and pi (π) bonds.
d. What are bonding and antibonding orbitals?

1.2 Define
a. octet rule
b. bond length
c. bond dissociation energy
d. hydrogen bond
e. Lewis base
f. Lewis acid

1.3 Compare the following compounds:

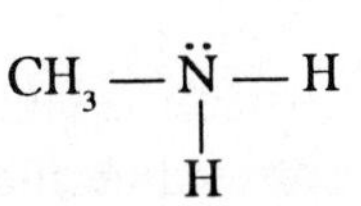

and

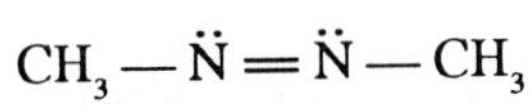

a. What orbital do the lone electron pairs occupy?
b. Which compound is more basic?

1.4 BF_3 has a dipole moment of zero. Propose a structure for BF_3 that is consistent with this fact.

1.5 Indicate whether the structures of each pair represent the same compound, structural isomers, or different non-isomeric compounds.

a. 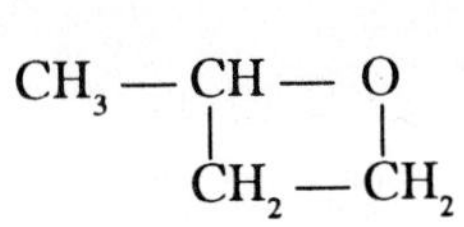

and

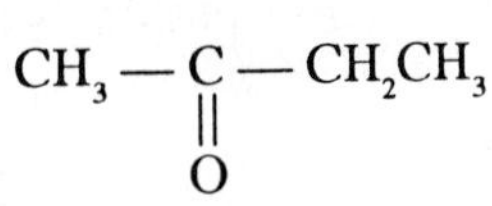

b. 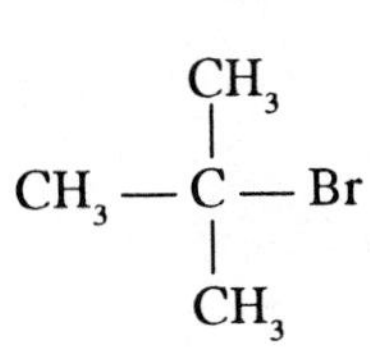

and

$(CH_3)_3C—Br$

c. 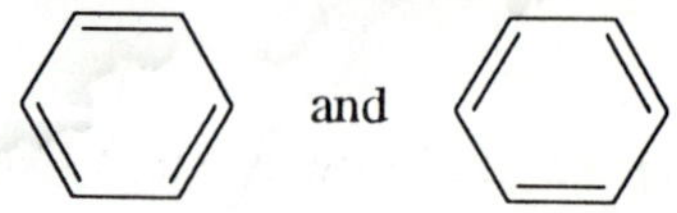

d. 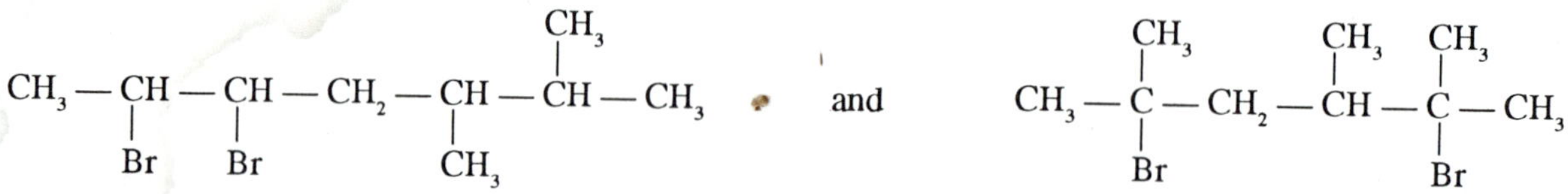

e. 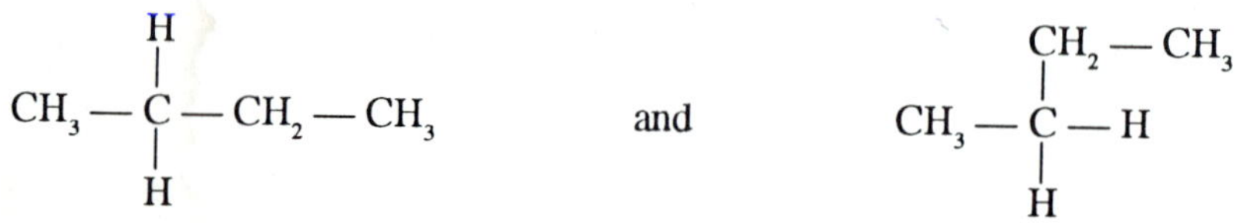

f. 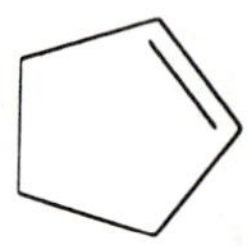 and $CH_3—CH=CH—CH_2—CH_3$

g. $CH_3—C≡C—CH_2—CH_3$ and $CH_2=CH—CH=CH—CH_3$

h. 

i. 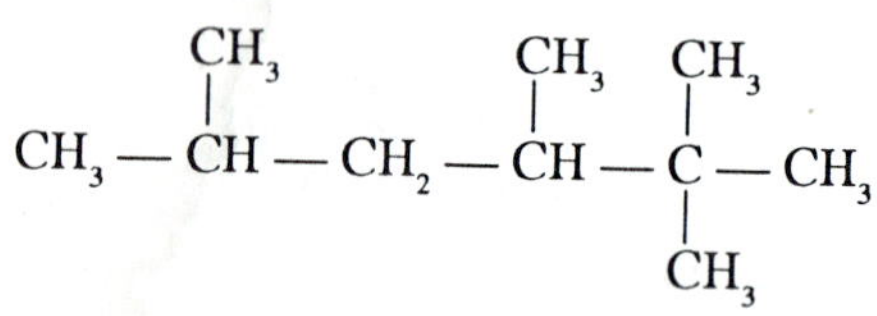 and $(CH_3)_2CH—CH_2CH(CH_3)—C(CH_3)_3$

j. 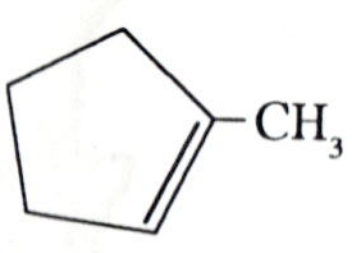 and 

k. $O=\overset{+}{N}—O^-$ and $^-O—\overset{+}{N}=O$

l. 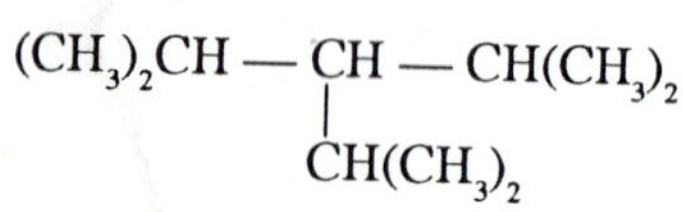 and $[(CH_3)_2CH]_3CH$

m. 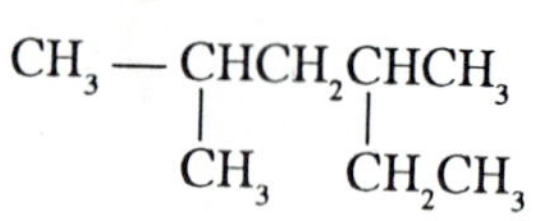 and 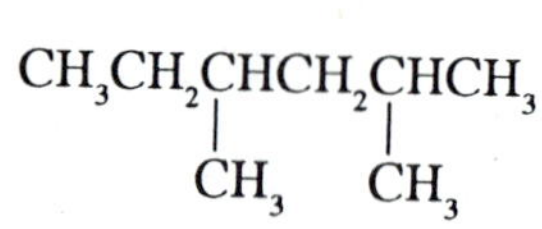

1.6 Indicate the direction of the dipole moment (if any) in each compound:
a. CH_3Cl c. CO_2 e. $(CH_3)_3N$ g. CBr_4
b. H_2O d. CH_3OH f. $BrCCl_3$

1.7 The dipole moment of NH_3 is 1.46 D.
a. In which direction is the dipole?
b. The dipole moment of NCl_3 is smaller than that of NH_3. Does this fact support your answer to (a)? Explain.

1.8 a. For the hypothetical molecule A:B, explain the meaning of the curved arrows:

Reaction 1 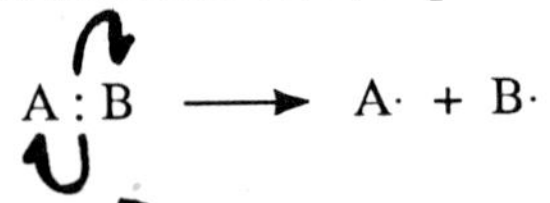

Reaction 2 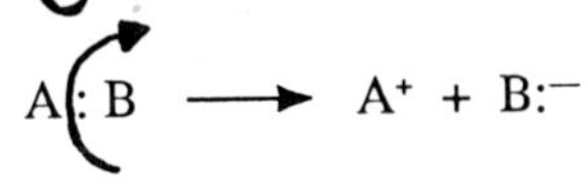

b. For the following two Lewis structures A and B, explain the meaning of the double-headed arrow:

Reaction 3 A $\longleftrightarrow$ B

1.9 What is meant by *resonance*? Write resonance structures for the following:

a. $CH_3SO_3^-$

b. $CH_3-\ddot{O}-\overset{+}{C}H_2$

c. $CH_2=CH-\dot{C}H-CH_3$

1.10 Where possible, draw contributing resonance structures for the following:

a. $\overset{-}{C}H_2-\overset{+}{N}\equiv\ddot{N}$

b. $\overset{+}{C}H_2-\ddot{O}-H$

c. $CH_3-\overset{+}{C}=\ddot{N}-H$

d. 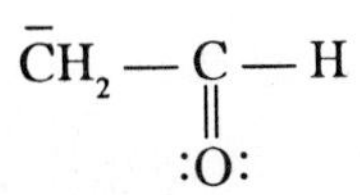

e. $CH_2=CH-\overset{+}{C}H-CH_3$

f. 

g. $H_2\ddot{N}-CH=CH_2$

h. $CH_3-\overset{+}{C}=\ddot{O}:$

i. 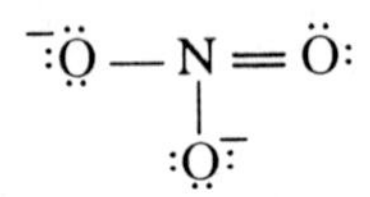

j. 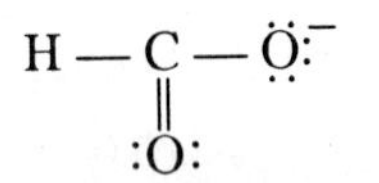

k. $CH_3-\ddot{O}:^-$

l. $CH_3-\overset{-}{C}=\overset{+}{N}=\ddot{O}:$

1.11 For each of the examples in question 1.10, where resonance forms are possible, which structure contributes most to the composition of the hybrid?

1.12 Identify the following pairs as two *resonance structures*, *constitutional isomers*, or *different (non-isomeric) compounds*.

a. 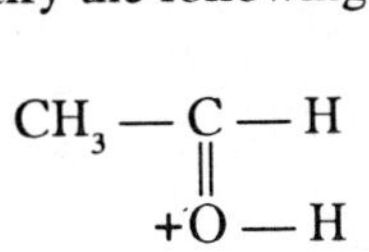 and 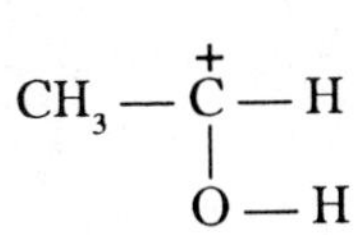

b. 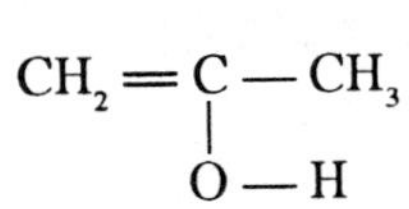 and 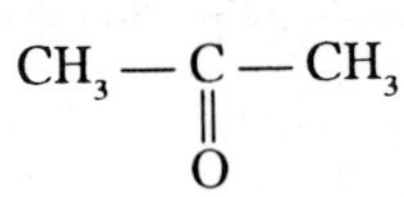

c. 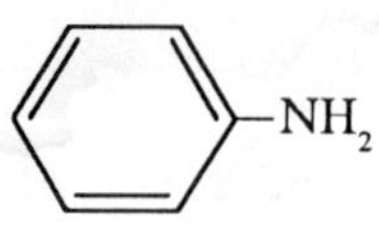 and 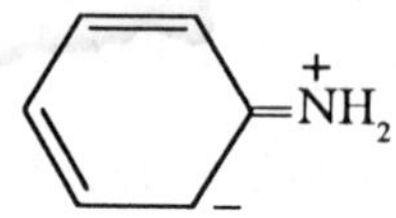

d. 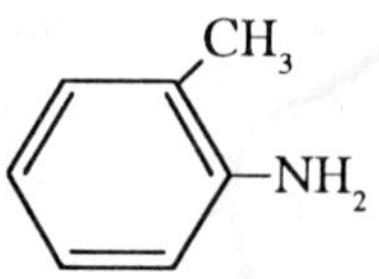 and 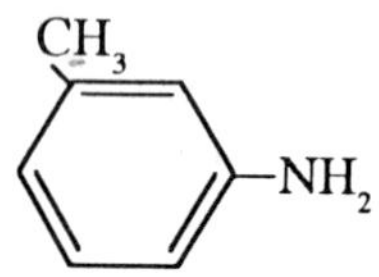

e. 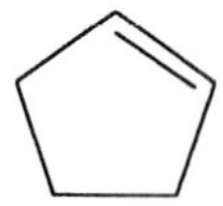 and 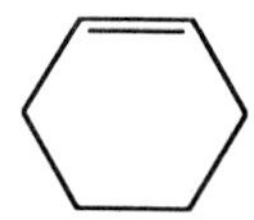

f. $CH_3-\bar{N}-\overset{+}{N}\equiv N$ and $CH_3-N=\overset{+}{N}=N^-$

g. $\bar{C}H_2-\overset{+}{N}\equiv N$ and $CH_2=\overset{+}{N}=\bar{N}$

h. $CH_3-\underset{CH_3}{\underset{|}{CH}}-CH_2-\underset{CH_3}{\underset{|}{CH}}-CH_3$ and $CH_3-CH_2-\underset{CH_3}{\underset{|}{CH}}-CH_2-\underset{CH_3}{\underset{|}{CH}}-CH_3$

i. $CH_3-\overset{CH_3}{\overset{|}{\underset{CH_2-CH_3}{\underset{|}{C}}}}-CH_2-CH_3$ and $CH_3C(CH_2CH_3)_3$

1.13 a. In the nitric acid molecule, HNO_3, there are two nitrogen-oxygen bond distances of 1.20 Å and one nitrogen-oxygen bond distance of 1.44 Å. Write a Lewis structure for nitric acid and match the bond distances.
b. Explain why two of the bond distances are equal.
c. In the nitrate ion, NO_3^-, all nitrogen-oxygen bond distances are equal. Explain.

1.14 Describe van der Waals attractions and repulsions.

1.15 Explain why
a. boron trifluoride, BF_3, has no dipole moment.
b. trimethylamine, $(CH_3)_3N$, has a considerably lower boiling point than *n*-propylamine, $CH_3CH_2CH_2NH_2$, even though both compounds have the same molecular weight.
c. NF_3 is a weaker base than ammonia.
d. acetic acid, $CH_3-\overset{O}{\overset{\|}{C}}-OH$, is a stronger acid than water.
e. in the following resonance structures, the one on the right is a more important contributor than the one on the left:

$$CH_3-\overset{H}{\overset{|}{\underset{+}{C}}}-\ddot{O}-H \quad \text{and} \quad CH_3-\overset{H}{\overset{|}{C}}=\overset{+}{\underset{\cdot\cdot}{O}}-H.$$

f. propionic acid, CH_3CH_2COOH, has a higher boiling point than methyl acetate, $CH_3-\underset{O}{\underset{\|}{C}}-OCH_3$.

g. $AlCl_3$ is a Lewis acid.

h. ethanol, C_2H_5OH has a higher boiling point than dimethyl ether, $CH_3—O—CH_3$, even though they have the same molecular weight.
i. organic compounds containing oxygen or nitrogen are soluble in concentrated sulfuric acid.

1.16 Write Lewis structures for the following species and calculate the formal charges, if any, in each structure:

a. phosphorous acid, H_3PO_3
b. thionyl chloride, $SOCl_2$
c. cyanate ion, NCO^-
d. nitrous acid, HNO_2
e. carbon monoxide, CO
f. diazomethane, CH_2N_2
g. hydrazoic acid, HN_3
h. methyl isocyanate, CH_3NCO

1.17 Assign formal charges where appropriate to each of the atoms in the following structures:

a. 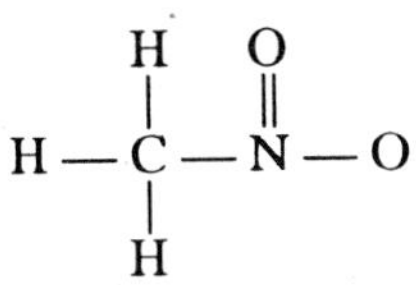

f. 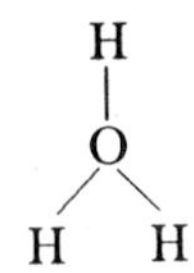

b. 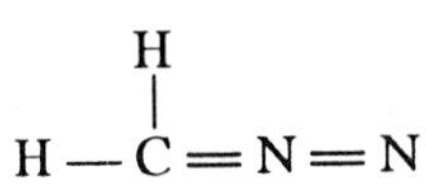

g. 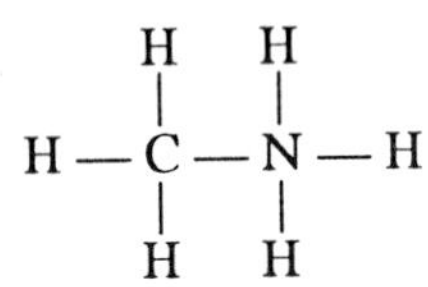

c. 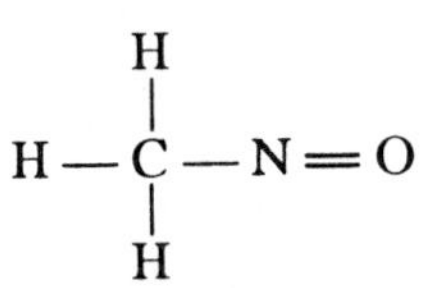

h. 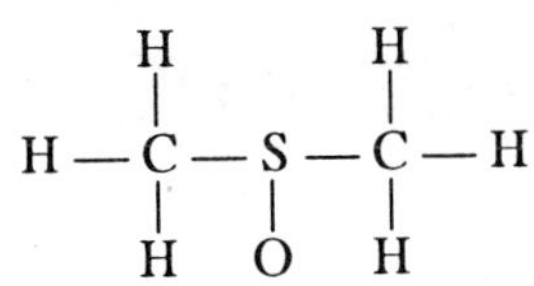

d. 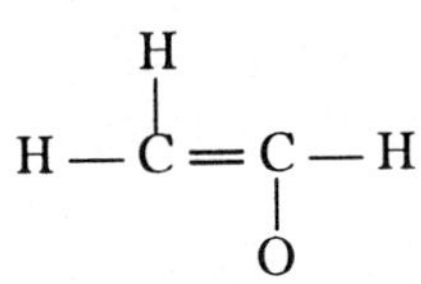

i. 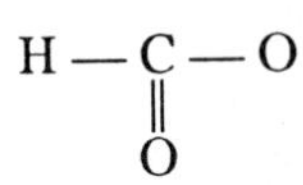

e. 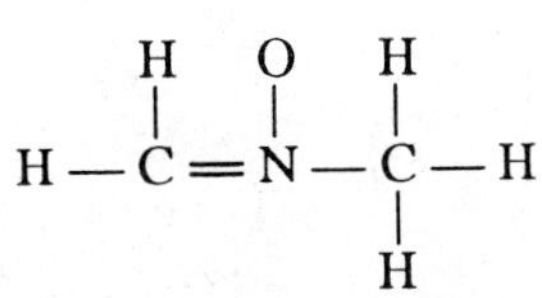

1.18 Identify the shape, hybridization type, and bond angles of each of the following chemcials:

a. H_2O
b. NH_3
c. CO_2
d. CH_3^+
e. SO_2

1.19 Match the compounds in the first column with the families in the second column.

1.	CH_3CH_2OH	a.	carboxylic acid
2.	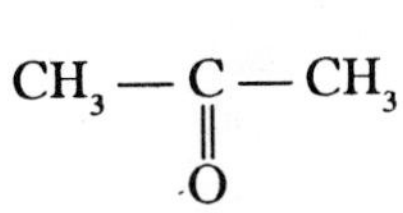	b.	ether
3.	$CH_3CH_2—S—H$	c.	1° alcohol
4.	$CH_3—NH—CH_3$	d.	2° alcohol
5.	$CH_2=CH—CH_3$	e.	3° alcohol

6. $CH_3\underset{\underset{OH}{\vert}}{C}HCH_2CH_3$	f. anhydride
7. $CH_3(CH_2)_8CH_3$	g. aldehyde
8. $CH_3-\underset{\underset{O}{\Vert}}{C}-OCH_3$	h. ketone
9. $(CH_3)_3N$	i. thiol
10. $CH_3CH_2CH_2Cl$	j. sulfide
11. $CH_3-\underset{\underset{O}{\Vert}}{C}-H$	k. 1° amine
12. $CH_3-\underset{\underset{O}{\Vert}}{C}-Cl$	l. 2° amine
13. CH_3-COOH	m. 3° amine
14. $CH_3-\underset{\underset{O}{\Vert}}{C}-O-\underset{\underset{O}{\Vert}}{C}-CH_3$	n. amide
15. $CH_3-C\equiv C-H$	o. ester
16. CH_3-S-CH_3	p. alkane
17. $CH_3-O-CH_2CH_3$	q. alkene
18. $CH_3-\underset{\underset{O}{\Vert}}{C}-NH_2$	r. alkyne
19. $CH_3CH_2NH_2$	s. alkyl halide
20. $CH_3-\overset{\overset{OH}{\vert}}{\underset{\underset{CH_3}{\vert}}{C}}-CH_3$	t. acyl halide

1.20 Identify the functional groups that are indicated by an arrow:

a.

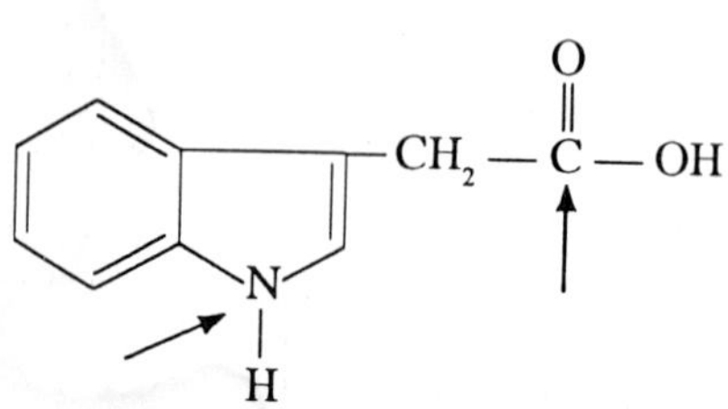

Indole-3-acetic acid (Auxin)

Growth hormone in plants

b.

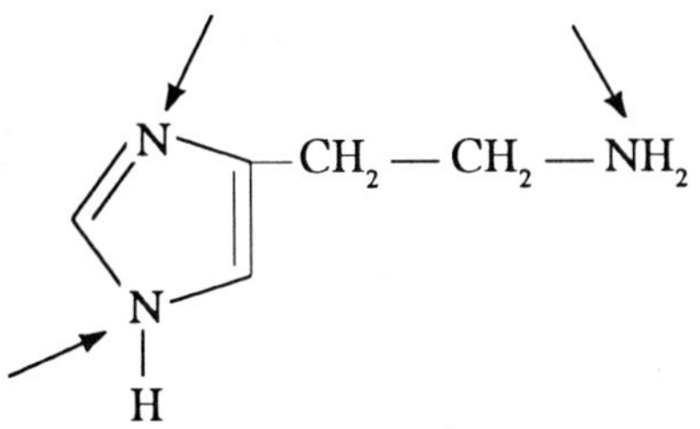

Histamine

Its release by the body causes common allergic reactions

c.

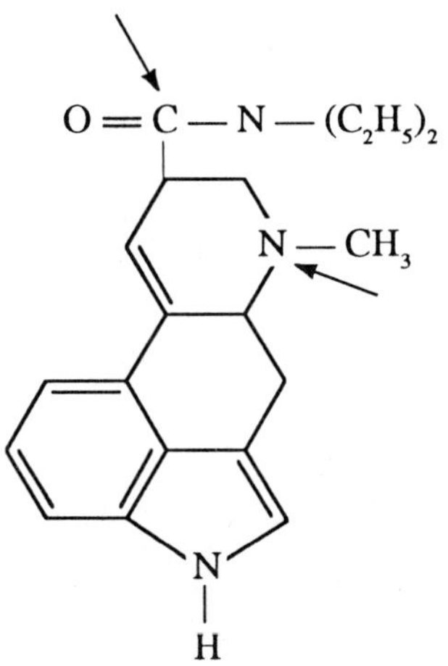

Lysergic acid diethylamide (LSD)

Hallucinogen

d.

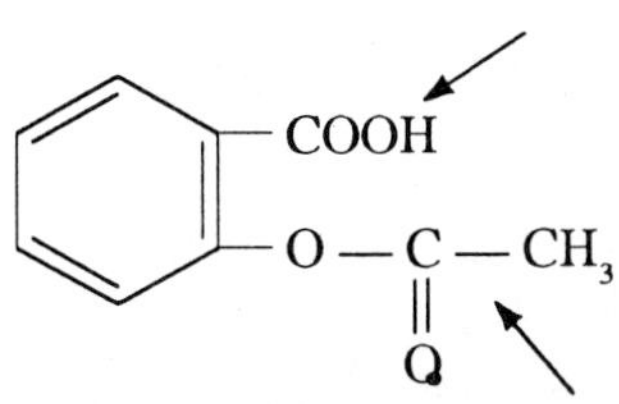

Acetylsalicylic acid (Aspirin)

Analgesic, nonsteroidal anti-inflammatory

e.

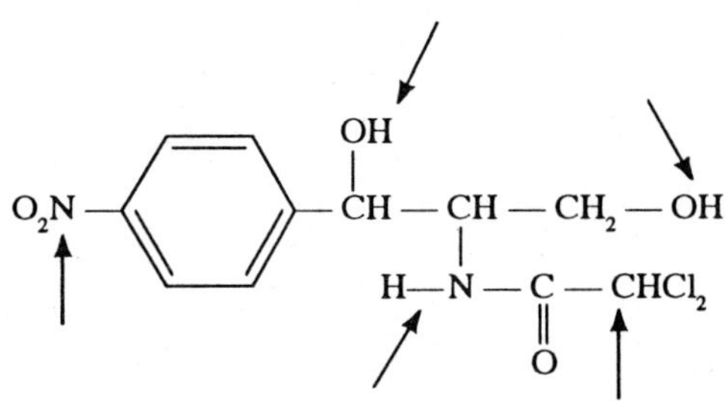

Chloramphenicol

Antibiotic

f.

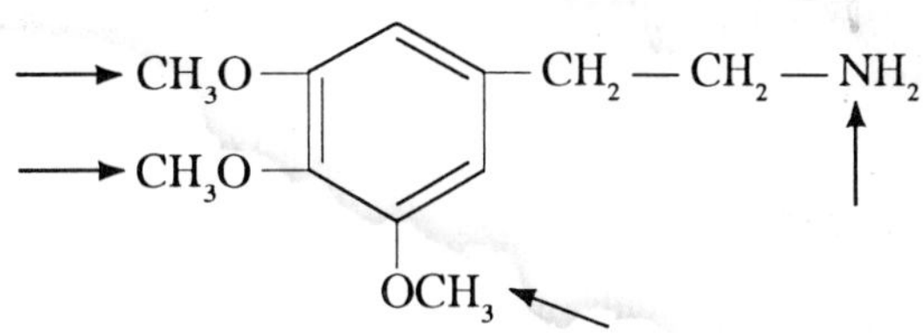

Mescaline

Hallucinogen

g.

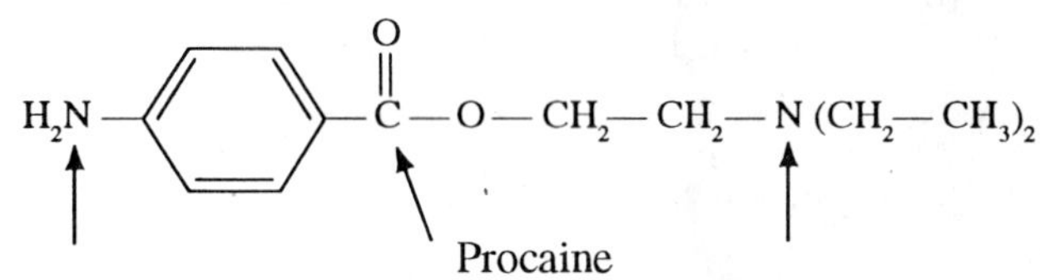

Procaine

Local anesthetic

h.

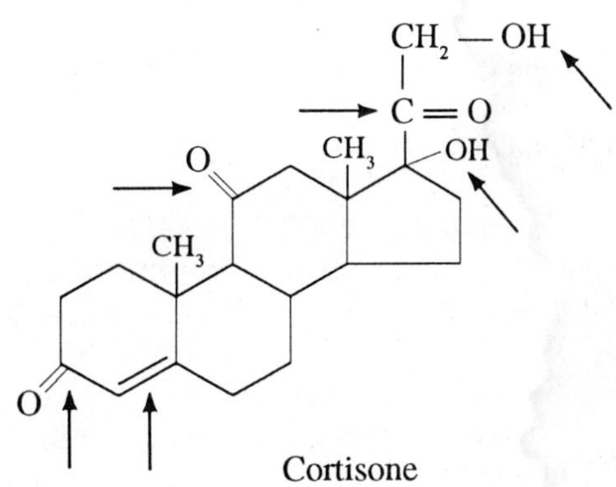

Cortisone

Anti-inflammatory, used in treatment of arthritis and other inflammatory conditions

i.

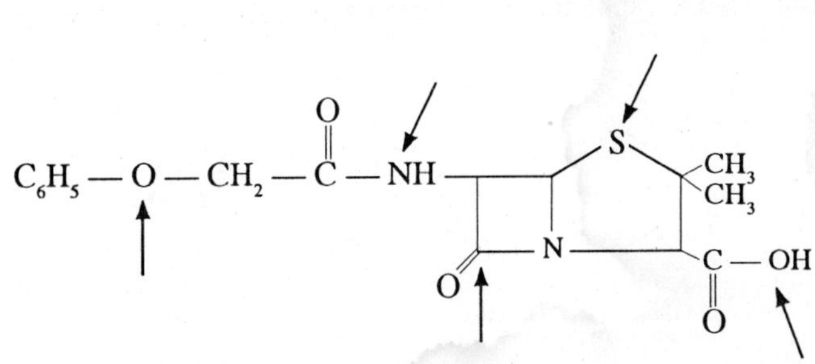

Penicillin V

Antibiotic

1.21 Give structures for
a. all ethers of molecular formula $C_4H_{10}O$.
b. all esters of molecular formula $C_4H_8O_2$.
c. all 1° alcohols of molecular formula $C_6H_{14}O$.
d. all ketones of molecular formula $C_5H_{10}O$.
e. all carboxylic acids of molecular formula $C_5H_{10}O_2$.
f. all aldehydes of molecular formula $C_5H_{10}O$.
g. all ketones of molecular weight 100.
h. all unsymmetrical anhydrides of molecular formula $C_6H_{10}O_3$.
i. an ester of molecular weight 74.
j. all 1° amines of molecular formula $C_4H_{11}N$.
k. all aldehydes of molecular weight 72.
l. all alkynes of molecular weight 68.
m. all 3° amines of molecular formula $C_5H_{13}N$.
n. a carboxylic acid that also carries a double bond and has a molecular formula $C_5H_8O_2$.

1.22 Calculate the percent composition of each element in the compounds below:
a. C_3H_6O
b. C_4H_9Cl
c. CH_2Cl_2
d. $C_6H_8O_2N_2S$
e. $C_{16}H_{26}O_4N_2S$

1.23 Given below are the percentage compositions of the elements composing four different compounds. Determine the empirical formula of each original chemical.

	Composition percentages			
	C	H	N	Other
Compound *a*	55.77	11.70	32.53	
Compound *b*	56.49	4.74	10.99	Cl 27.79
Compound *c*	12.60	3.20		Br 84.1
Compound *d*	76.57	6.43		

1.24 A compound contains carbon, hydrogen, bromine, and possibly oxygen. Combustion of a 0.2001g sample affords 0.1902 g of CO_2 and 0.0907g of H_2O. A second sample weighing 0.1523 g yields 0.2058 g of AgBr when treated with $AgNO_3$. What is the empirical formula of this unknown compound?

1.25 Combustion of 6.51 mg of a compound yielded 20.46 mg of carbon dioxide and 8.36 mg of water. At 100°C and 760 torr (760 mm Hg) pressure, 100 mL of the compound vapor weighted 275 mg. Calculate
a. the molecular weight
b. the empirical formula
c. the molecular formula

1.26 Combustion of 0.1908 g of a substance gave 0.2895 g CO_2 and 0.1192 g H_2O. Nitrogen analysis on a 0.1825 g sample yielded 40.2 mL of N_2 collected over water at 25°C (vapor pressure of water at 25°C is 24 mm Hg) and 755 mm Hg atmospheric pressure. When 1.082 g of the compound was dissolved in 28 g of benzene, the freezing point was lowered by 1.71°C. Calculate the molecular weight and molecular formula of the compound. (K_f of benzene = 5.12).

1.27 A 0.1 *M* solution of an acid HA is found to ionize to an extent of 1.2%. Calculate
a. the K_a
b. the pK_a
c. the pH of the solution.

1.28 Which value represents the stronger acid in the sets below.

a. $K_a = 3.1 \times 10^{-4}$	or	$K_a = 7.0 \times 10^{-6}$
b. $pK_a = -2.0$	or	$pK_a = 2.0$
c. $pK_a = 4.8$	or	$pK_a = 1.2$

1.29 In the following equilibria, indicate the direction that is favored (pK_a values are given in parentheses).

a.	HSO_4^- (– 1.7)	+	H_3O^+	$\rightleftharpoons$	H_2SO_4	+	H_2O (– 9.0)
b.	$(CH_3)_3C—OH$ (19.0)	+	OH^-	$\rightleftharpoons$	$(CH_3)_3C—O^-$	+	H_2O (17.0)
c.	$HC\equiv C—H$ (26.0)	+	NH_2	$\rightleftharpoons$	$HC\equiv C^-$	+	NH_3 (36.0)
d.	$C_6H_5S—H$ (7.8)	+	HCO_3^-	$\rightleftharpoons$	$C_6H_5S^-$	+	H_2CO_3 (6.5)
e.	NH_4^+ (9.4)	+	HCO_3^-	$\rightleftharpoons$	NH_3	+	H_2CO_3 (6.5)
f.	CH_4 (47.0)	+	$HC\equiv C^-$	$\rightleftharpoons$	CH_3^-	+	$HC\equiv C—H$ (26.0)

1.30 Given below are a set of acids (pK_a values in parentheses). Arrange the conjugate bases in order of strongest to weakest.
a. NH_4^+ (9.4)
b. C_6H_5OH (10)
c. C_2H_5SH (10.5)
d. $HC\equiv CH$ (26)
e. CH_4 (47)
f. HCO_2H (3.7)
g. H_3O^+ (–1.7)

1.31 Using the homolytic bond dissociation energy values in the table below, calculate the net energy change for each reaction. Predict whether the reactions are exothermic or endothermic.

Homolytic Bond Dissociation Energies

$$A:B \longrightarrow A\cdot + B:$$

Bond	ΔH (kcal/mole)	Bond	ΔH (kcal/mole)
H—H	104	CH_3CH_2—H	98
D—D	106	CH_3CH_2—F	106
F—F	38	CH_3CH_2—Cl	81.5
Cl—Cl	58	CH_3CH_2—Br	69
Br—Br	46	CH_3CH_2—I	53.5
I—I	36	CH_3CH_2—OH	91.5
H—F	136	CH_3CH_2—OCH_3	80
H—Cl	103	$CH_3CH_2CH_2$—H	98
H—Br	87.5	$CH_3CH_2CH_2$—F	106
H—I	71	$CH_3CH_2CH_2$—Cl	81.5
CH_3—H	104	$CH_3CH_2CH_2$—Br	69
CH_3—F	108	$CH_3CH_2CH_2$—I	53.5
CH_3—Cl	83.5	$CH_3CH_2CH_2$—OH	91.5
CH_3—Br	70	$CH_3CH_2CH_2$—OCH_3	80
CH_3—I	56	CH_3CH_2O—H	103
CH_3—OH	91.5	HO—H	119
CH_3—OCH_3	80	CH_2=CH—H	108
CH_3—CH_3	88	CH_2=$CHCH_2$—H	85
CH_3CH_2—CH_3	85	H_2C=$CHCH_2$—Br	47
$CH_3CH_2CH_2$—CH_3	85		

a. $I_2 \longrightarrow 2I\cdot$

b. $CH_4 + Br_2 \longrightarrow CH_3Br + HBr$

c. $CH_3Br + CH_3\text{—}CH_3 \longrightarrow CH_3CH_2Br + CH_4$

d. $CH_3F + H_2 \longrightarrow CH_4 + HF$

e. $CH_3\text{—}CH_2\text{—}CH_3 + CH_4 \longrightarrow CH_3\text{—}CH_2\text{—}CH_2\text{—}CH_3 + H_2$

f. $CH_3\text{—}CH_2\text{—}CH_3 + CH_4 \longrightarrow 2\,CH_3\text{—}CH_3$

g. $CH_3CH{=}CH_2 + Br_2 \longrightarrow Br\text{—}CH_2CH{=}CH_2 + HBr$

1.32 Using the heterolytic bond dissociation values in the table below, calculate the net energy change for each reaction. Predict whether the reactions are exothermic or endothermic.

Heterolytic Bond Dissociation Energies

$$A:B \longrightarrow A^+ + B:^-$$

Bond	ΔH (kcal/mole)	Bond	ΔH (kcal/mole)
$CH_3—Cl$	227	$CH_3CH_2—Br$	184
$CH_3—OH$	274	$CH_3CH_2—I$	176
$H—OH$	390	$CH_2{=}CHCH_2—Br$	165
$H—I$	315	$CH_2{=}CHCH_2—OH$	223
$H—Br$	324	$C_6H_5CH_2—OH$	215
$H—Cl$	334	$C_6H_5CH_2—I$	149

a. $CH_3—Cl + H—OH \longrightarrow CH_3—OH + H—Cl$

b. $CH_3CH_2—Br + H—I \longrightarrow CH_3CH_2—I + H—Br$

c. $CH_3—OH + H—Cl \longrightarrow CH_3—Cl + H—OH$

d. $H_2C{=}CHCH_2—Br + H—OH \longrightarrow H_2C{=}CHCH_2—OH + H—Br$

e. $C_6H_5CH_2—OH + H—I \longrightarrow C_6H_5CH_2—I + H—OH$

1.33 Consider the following reaction at 25°C:

$$C_2H_4 + HCl \longrightarrow C_2H_5Cl$$

in which $\Delta H° = -16$ kcal/mole and

$\Delta S° = -30$ cal/mole.K

a. What is the reason for the negative value of $\Delta S°$?
b. What effect does the negative sign have on the value of $\Delta G°$?
c. Calculate $\Delta G°$ for the reaction.
d. Calculate K_{eq}.
e. If the reaction is run at a higher temperature, how will this affect the value of K_{eq}? What will be the effect on the value of K_{eq} as the temperature increases?

1.34 Balance the following equations in acid medium

a. $ReO_2 + Cl_2 \longrightarrow HReO_4 + Cl^-$

b. $Cr_2O_7^{-2} + CH_3CH_2OH \longrightarrow CH_3COOH + Cr^{+3}$

c. $C_6H_5CH_3 + MnO_4^- \longrightarrow C_6H_5COOH + Mn^{+2}$

d. $C_8H_{10} + MnO_4^- \longrightarrow C_8H_6O_4 + Mn^{+2}$

1.35 Balance the following equations in basic medium

a. $Cr(OH)_4^- + BrO^- \longrightarrow CrO_4^{-2} + Br^-$

b. $C_6H_5CH_3 + MnO_4^- \longrightarrow C_6H_5CO_2^- + MnO_2$

c. $C_9H_{12} + MnO_4^- \longrightarrow C_9H_5O_6^{-3} + MnO_2$

d. $C_6H_{10}O + MnO_4^- \longrightarrow C_6H_8O_4^{-2} + MnO_2$

Multiple Choice Test

1.1 What is the charge on each of the species below?

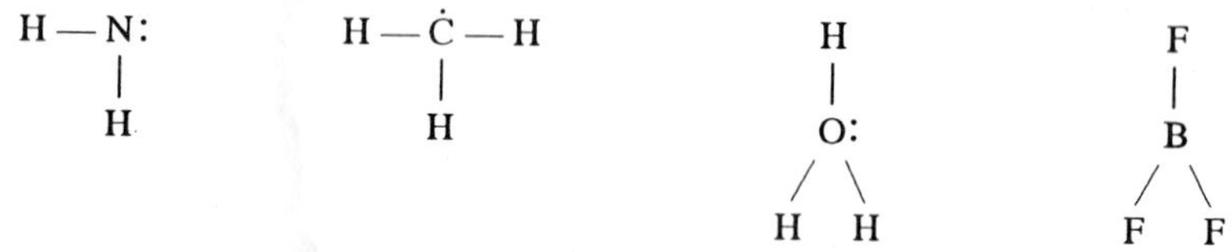

a. +1, −1, −1, 0
b. −1, 0, +1, 0
c. 0, +1, +1, 0
d. +1, 0, +1, 0

1.2 What is the correct descending order of electronegativity of the following elements?
a. F > N > O > C > Na
b. F > C > O > N > Na
c. F > O > N > C > Na
d. F > O > N > Na > C

1.3 Which pair of elements would form an ionic bond?
a. H and F
b. H and C
c. S and N
d. Li and F

1.4 For each of the following three bonds, specify whether the *negative* end of the dipole is towards the left or the right.

C—Br	Cl—H	H—H
A	**B**	**C**

a. A–left, B–left, C-no dipole
b. A–left, B-right, C-no dipole
c. A–right, B–left, C-no dipole
d. one cannot decide unless all other attached elements are given

1.5 The correct electronic configuration of nitrogen is
a. $1s^22s^22p_x^{2}2p_y^{1}$
b. $1s^22s^12p_x^{1}2p_y^{1}2p_z^{1}$
c. $1s^22s^22p_x^{1}2p_y^{1}2p_z^{1}$
d. $1s^22s^22p_x^{2}2p_y^{1}2p_z^{1}$

1.6 Listed below are the electron-dot formulas for several simple molecules and ions, including formal charges where necessary. Which of them are correct Lewis structures?

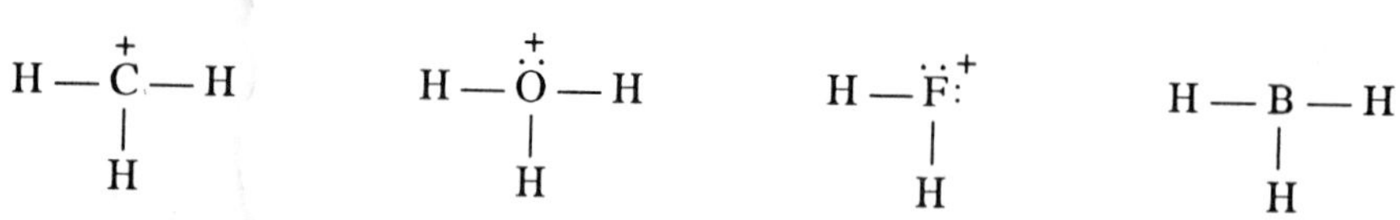

A **B** **C** **D**

a. A, B
b. A, B, D
c. B, C
d. A, D

1.7 Which of the following species have no formal charge on their carbon atom?

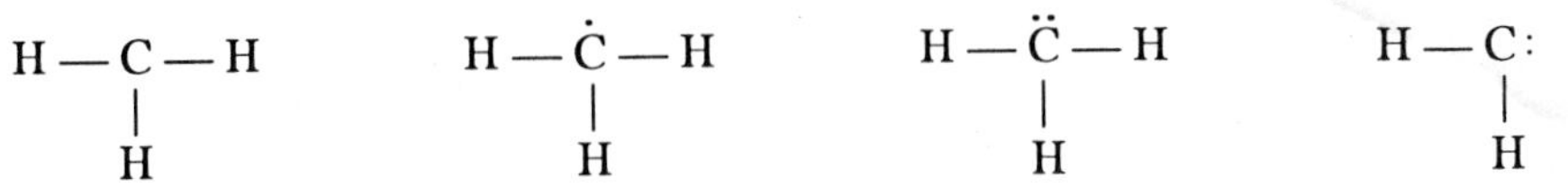

A B C D

a. A, C c. B, C
b. A, D d. B, D

1.8 Which of the following structures have one or more atoms with trigonal geometry? (charges have been omitted)

$H_3C{:}$	H_4N	$H_2C{=}CH_2$	BF_3
A	**B**	**C**	**D**

a. A, C c. A, C, D
b. B, D d. C, D

1.9 Which of the following molecules would have a dipole moment?

CH_2Cl_2	CCl_4	CH_3Cl	$H_2C{=}CH_2$	$H{-}C{\equiv}C{-}H$	H_3O^+
A	**B**	**C**	**D**	**E**	**F**

a. all but B c. C only
b. A, C and F e. C and F

1.10 The formal charges on nitrogen and the starred oxygen atom of the compound shown below are

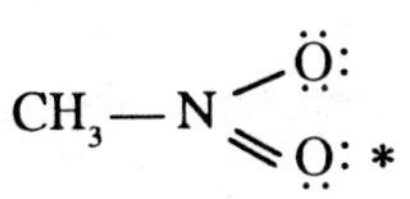

a. N +1; O* 0 c. N +1; O* +1
b. N +1; O* –1 d. N –1; O* 0

1.11 Which of the following species have tetrahedral bond angles?

NH_3	BF_3	CH_3^+	CH_3^-	H_2O	H_3O^+
A	**B**	**C**	**D**	**E**	**F**

a. A, D and E c. A, D, E, F
b. A, E d. only D

1.12 Which structure and formal charge distribution is correct for diazomethane, CH_2N_2?

a. 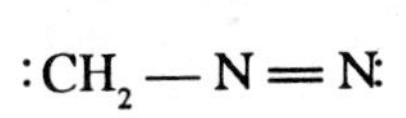 c. 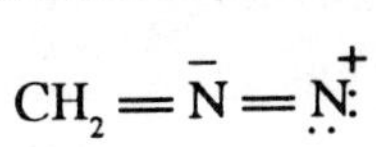

b. 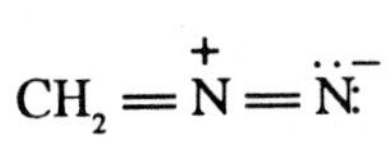 d. 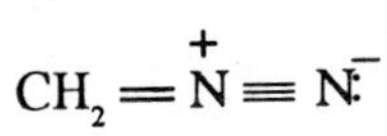

1.13 The highly toxic compound *dioxin*, a byproduct of herbicide industrial production, has an elemental analysis of 44.8% carbon, 1.25% hydrogen, and 44.0% chlorine. Its molecular weight is 320 ± 10. What is the molecular formula?

a. $C_{12}H_4O_2Cl_4$ c. $C_6H_2OCl_2$
b. $C_{10}H_8O_4Cl_4$ d. $C_{14}H_4OCl_4$

1.14 Which of the following species is planar?

CH_3^+	BF_3	H_3O^+	NH_4^+	$H_2C{=}CH_2$	$AlCl_3$
A	**B**	**C**	**D**	**E**	**F**

a. all but C and D c. all but C
b. A and B d. B, E, F

1.15 How many constitutional isomers are possible for $C_4H_{10}O$?

a. three c. five
b. four d. seven

1.16 Which of the isomers in question 1.15 are alcohols?

a. two c. four
b. three d. all

1.17 Which of the following compounds will have the highest boiling point?

a. $CH_3CH_2CH_2CH_2CH_3$ c. $CH_3CH_2CH_2CH_2CH_2OH$

b. $CH_3CH_2OCH_2CH_3$ d. $CH_3CH_2CH_2CH{=}CH_2$

1.18 Which compound would show the highest miscibility with water?

a. $CH_3{-}CH_2{-}CH_2{-}CH_2{-}CH_3$ c. $CH_3{-}CH_2{-}O{-}CH_2{-}CH_2{-}CH_3$

b. $CH_3{-}CH_2{-}CH_2{-}CH_2{-}CH_2Cl$ d. $CH_3{-}CH{=}CH{-}CH_2{-}CH_3$

1.19 In the reaction below, the strong base is

$$H_2O + NH_2^- \rightleftharpoons HO^- + NH_3$$

a. H_2O c. HO^-
b. NH_2^- d. NH_3

1.20 Match the hybridization of carbon with the correct shape

A. sp^3 X. trigonal

B. sp^2 Y. linear

C. sp Z. tetrahedral

a. AX, BY, CZ c. AY, BZ, CX
b. AZ, BX, CY d. AX, BZ, CY

Answers for Questions

1.1 a. **Hybridization** refers to the process whereby atomic orbitals with dissimilar energies "mix", interacting to produce a new set of orbitals each having the same energy (degenerate orbitals).

b. Carbon hybridizes its orbitals in three ways to form hybrid sp^3, sp^2, and sp orbitals. In the hybridized states, carbon's outer electrons have the following distributions:

sp^3: four equivalent orbitals each with 1/4 *s*-character and 3/4 *p*-character; bond angles 109°; tetrahedral shape. Example: CH_4.

sp^2: three equivalent orbitals each with 1/3 *s*-character and 2/3 *p*-character; bond angles 120°; trigonal, planar shape. The remaining electron is in a *p* orbital that is set perpendicular to the three sp^2 orbitals. Example: $CH_2=CH_2$.

sp: two equivalent orbitals with 1/2 *s*-character and 1/2 *p*-character; bond angles 180°; linear geometry. The remaining two electrons are in *p* orbitals that are perpendicular to each other and to the *sp* orbitals. Example: $HC\equiv CH$.

c. Atomic orbitals that overlap along a *common axis* generate a molecular orbital whose cross section is circularly symmetrical. These are **sigma (σ) bonds**. All single bonds are of this type.

Atomic orbitals that overlap along *parallel axes* are **pi (π) bonds**. A carbon-carbon π bond is not as strong as a carbon-carbon σ bond.

d. **Bonding molecular orbitals** arise from a combination of atomic orbitals of *like phase*. Electrons occupying these orbitals spend most of their time between the two nuclei and thereby strengthen the bond.

Sigma bonding molecular orbital (σ) resulting from overlap of two *p* orbitals that lie along a common axis. Overlap can occur between *s*, *p*, *d* or *f* orbitals.

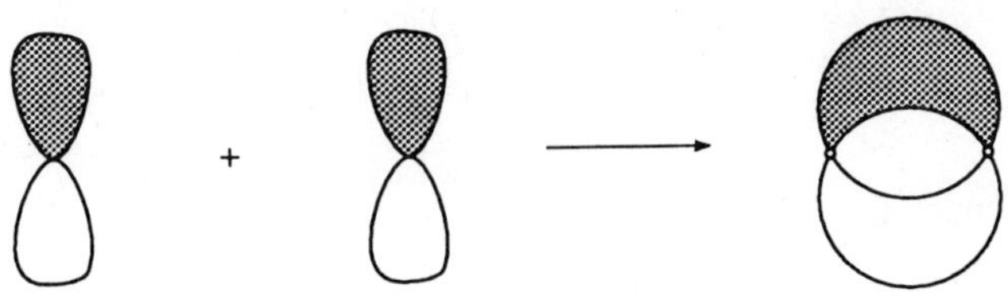

Pi bonding molecular orbital (π) resulting from overlap of two *p* orbitals that lie along parallel axes.

Sigma antibonding molecular orbital (σ^*)

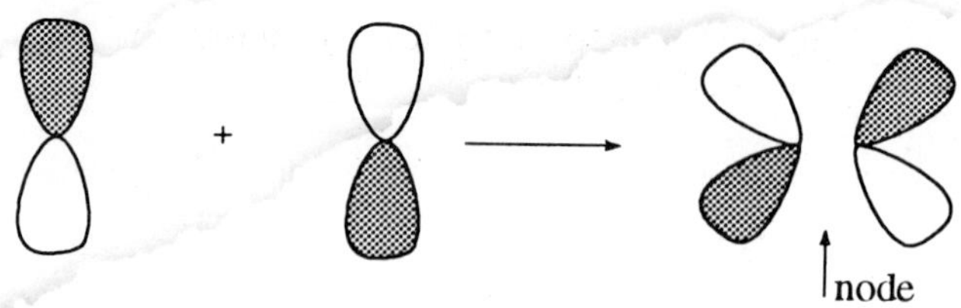

Pi antibonding molecular orbital (π^*)

1.2 a. **Octet rule:** the Lewis structure of any molecule or ion containing elements of the second row of the periodic table cannot have more than eight elecrons placed around a second-row element. Molecules or ions in which each atom has an octet of electrons are particularly stable.

b. **Bond length:** the average distance between two bonded nuclei.

c. **Bond dissociation energy:** the energy needed to homolytically separate two bonded atoms. In general, the shorter the bond, the higher the energy.

d. **Hydrogen bonding:** the attraction between a hydrogen atom bonded to a strongly electronegative atom such as O, N, or F and the nonbonding electron pair on a nearby O, N, or F atom.

e. **Lewis base:** a compound that possesses an available pair of electrons.

f. **Lewis acid:** a compound that is an electron pair acceptor.

1.3 a. *sp*³; and *sp*²

b. The first compound is more basic because *sp*² orbitals have more *s*-character; thus their electrons are more strongly held.

1.4 BF_3 is trigonal, planar in structure (*sp*² hybridized) and dipoles cancel:

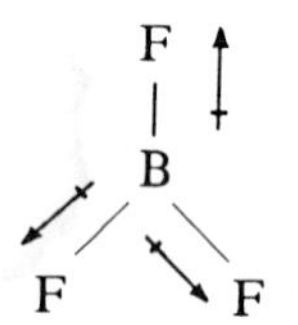

1.5 Same compound: b, c, e, i, k, l, m.
Structural isomers: a, d, g, h, j.
Different, non-isomeric compounds: f.

1.6 a.

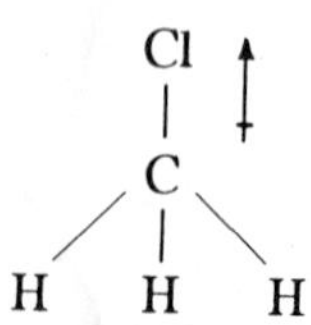

Direction of dipole moment is toward chlorine.
Molecule has a tetrahedral structure.

b.

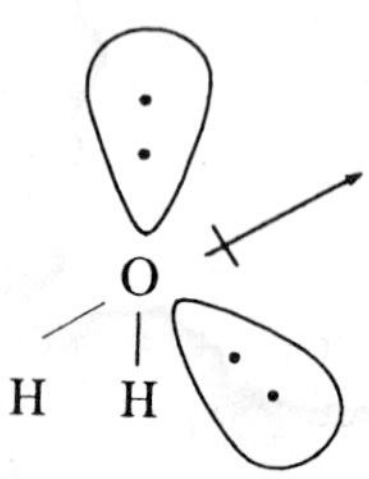

Direction of dipole moment is toward oxygen. It bisects the two unshaired electron pairs in the *sp*³ orbitals. The water molecule shows a bent, or angular, geometry because the bonding and nonbonding electron pairs are positioned toward the corners of a tetrahedron.

c. O=C=O — This molecule has a dipole moment of zero. It is a linear molecule.

d. 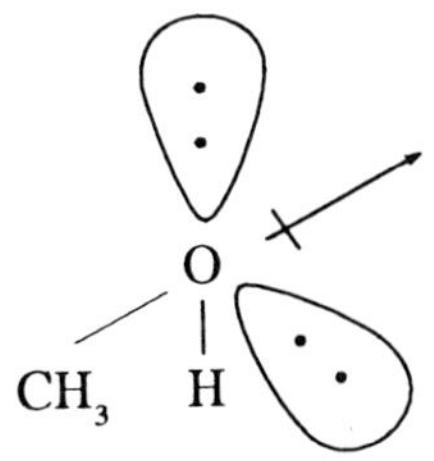

Direction of dipole moment is toward oxygen. This molecule is similar to water; it too exhibits an angular geometry.

e. 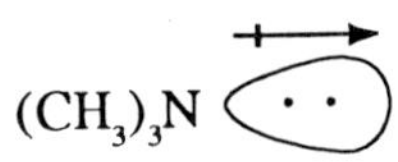

Direction of dipole moment is toward the electron pair. This molecule has a pyramidal structure, and its bonding and nonbonding electron pairs are oriented toward the corners of a tetrahedron.

f. 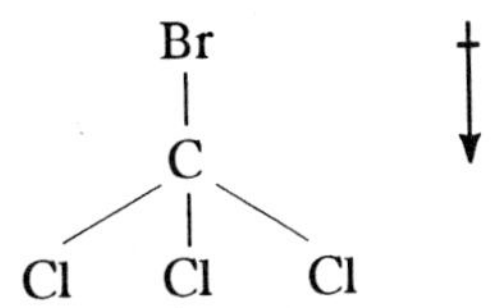

In this case, the direction of the dipole moment bisects the three chlorine atoms. The net dipole moment is the vector sum of the individual dipoles of the C — Cl bonds minus the dipole of the C — Br bond. This molecule exhibits tetrahedral geometry.

g. CBr_4 — This molecule has a dipole moment of zero. It has a tetrahedral structure.

1.7 a. 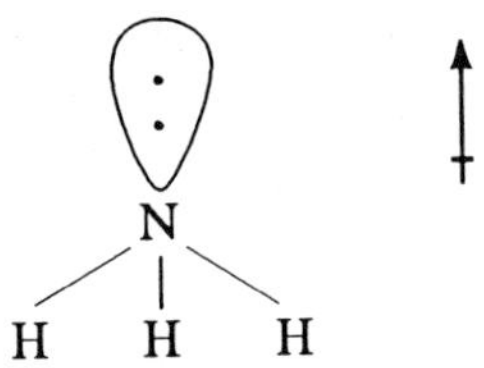

Direction of dipole moment is toward the unshaired electron pair.

b. 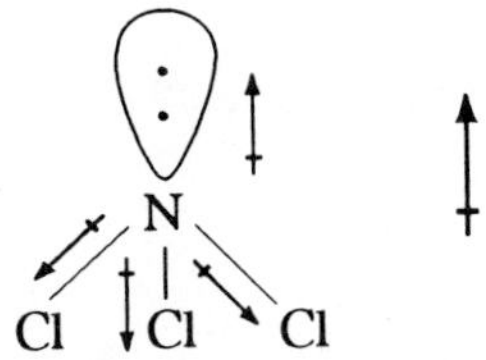

The vector sum of opposing dipoles results in a smaller overall dipole moment than that of the preceding molecule.

1.8 a. The curved arrow indicate election movement, singly or in pairs.
Reaction 1: Homolytic cleavage to form A· and B· radicals.
Reaction 2: Heterolytic cleavage to form ions A^+ and $:B^-$.

b. The double-headed arrow signifies *resonance*, where A and B are contributing structures.

1.9 **Resonance** occurs when a molecular orbital is associated with more than two (positively charged) nuclei. In other words, the electrons encompass more than two nuclei and, in this sense, are said to be *delocalized*. The involvement of more than two nuclei allows longer electronic waves to move through larger volumes of space. Since longer wavelengths are associated with lower energies (i.e., $E = hc/\lambda$), the result is that the resonance structure has lower energy overall.

The resonance phenomenon is indicated by writing more than one Lewis structure for the compound. The structures differ *only* in the arrangement of the electrons, not in the position of the atoms. The compound's "true" structure is a hybrid of the contributing forms which shows characteristics of these forms to a greater or lesser extent, depending on the degree of their contribution.

a. 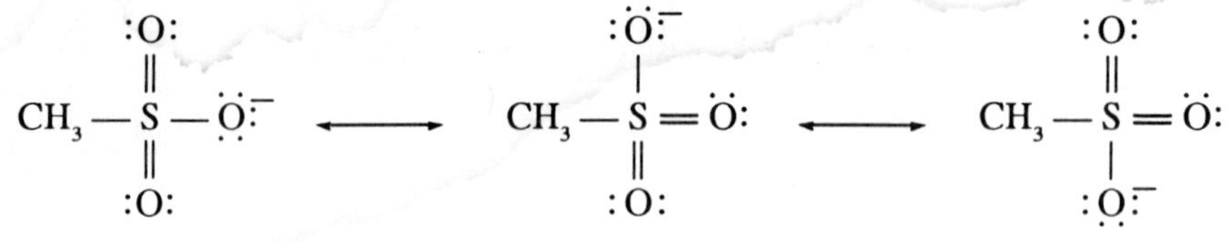

Note: S is a third-row periodic table element and can accommodate more than 8 valence electrons.

b. 

c. 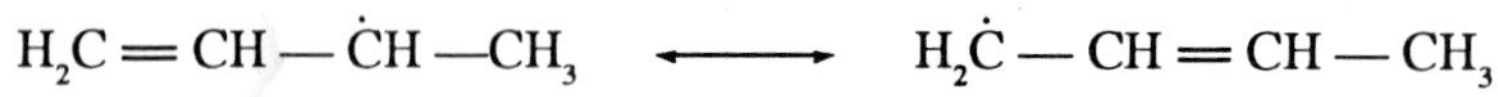

1.10 a. 

b. 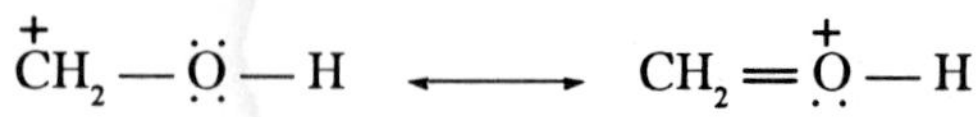

c. 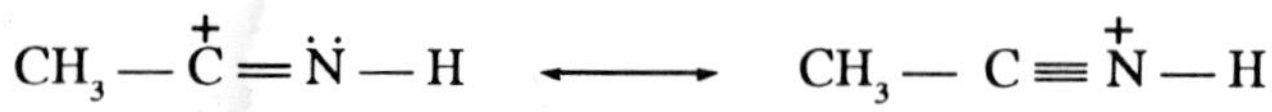

d. 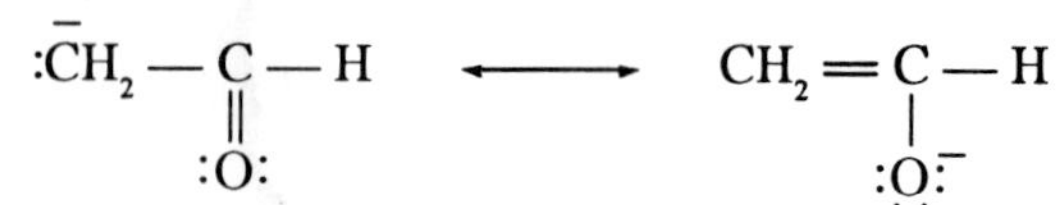

e. 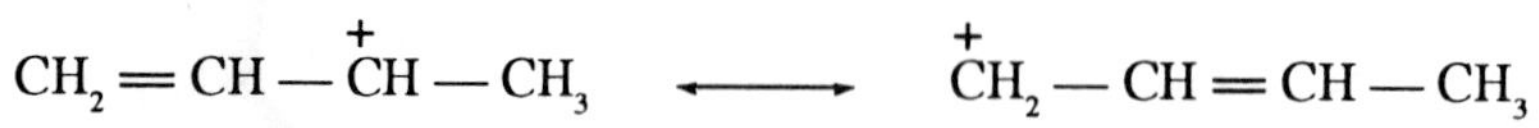

f. 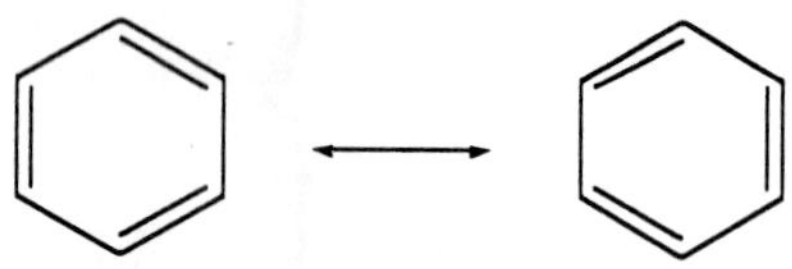

g. 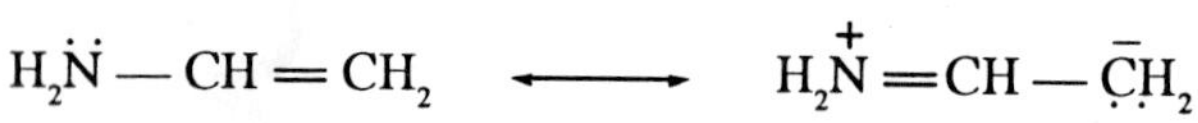

h. 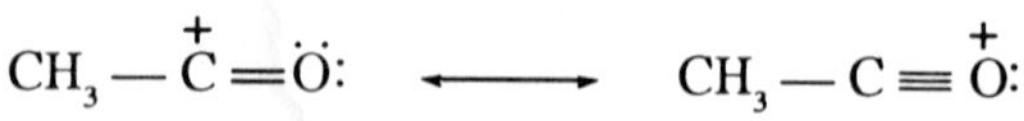

i. 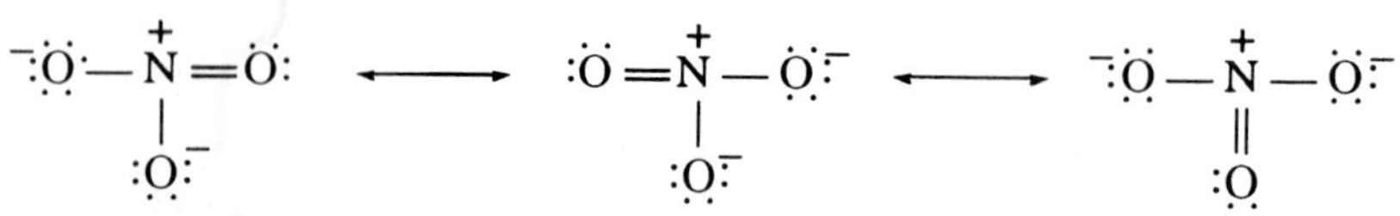

j. 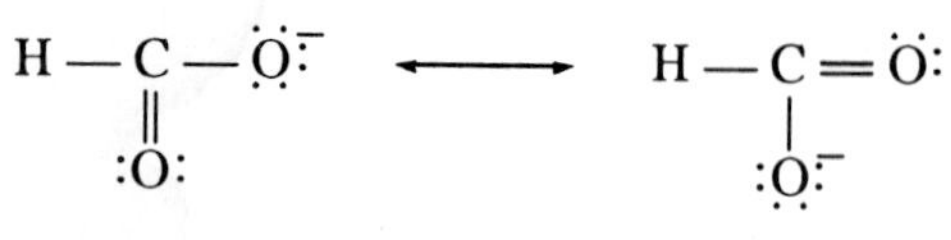

k. No other structure possible

l. 

1.11 a. 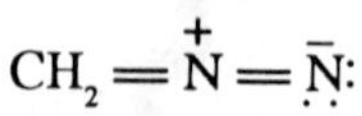

The negative charge is on the most electronegative atom.

b. 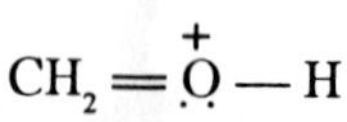

Carbon obeys the octet rule (not followed in $\overset{+}{C}H_2-\ddot{O}H$).

c. 

Carbon obeys the octet rule (not followed in $CH_3-\overset{+}{C}=\ddot{N}-H$).

d. 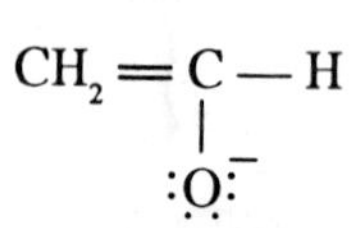

The negative charge is on the most electronegative atom.

e. $CH_2{=}CH{-}\overset{+}{C}H{-}CH_3$ The carbocation is allylic and secondary.

f. Equal contribution.

g. $H_2\ddot{N}{-}CH{=}CH_2$ Neutral molecule.

h. $CH_3{-}C{\equiv}\overset{+}{O}:$ Carbon obeys the octet rule (not followed in $CH_3{-}\overset{+}{C}{=}\ddot{O}:$).

i. Equal contribution.

j. Equal contribution.

k. No other structure possible

l. $CH_3{-}C{\equiv}\overset{+}{N}{-}\ddot{O}:^-$ The negative charge is on the most electronegative atom.

1.12 Resonance structures: a, c, f, g.
Isomers: b (keto-enol), d.
Different: e, h, i.

1.13 a.

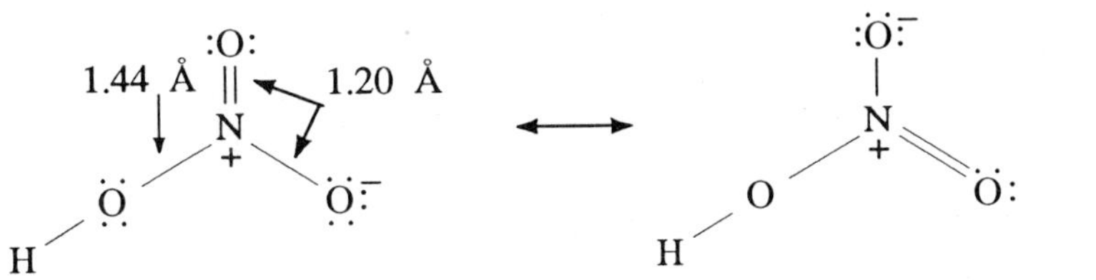

b. The two N — O bonds are equivalent as indicated by the above resonance structures.

c. The nitrate anion is a resonance hybrid of three contributing structures:

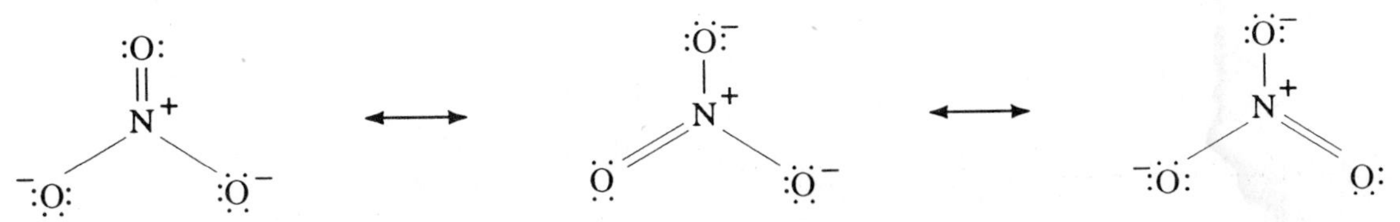

These structures are equivalent; therefore, all N — O bond lengths are the same.

1.14 Van der Waals *attractions* among nonpolar atoms or groups arise from instantaneous dipole-induced dipole attractions. When molecules approach each other a small fleeting dipole in one molecule occurs, which induces a dipole in the second molecule. This attraction is maximal when the two molecules (or atoms) are a particular distance apart. If they get closer, *repulsion* (sometimes called *nonbonded repulsion*) occurs because two molecules are trying to occupy the same space at the same time. Negatively charged electron clouds will repel each other as will positively charged nuclei.

1.15 a. BF_3 is an sp^2 hybridized, trigonal, flat molecule, with each F—B—F angle equal to 120°. Thus all B—F dipoles cancel each other.

b. Intermolecular hydrogen bonding is possible for the compound *n*-propylamine because of its –N—H bonds; trimethylamine lacks this group.

c. Fluorine has an electron-withdrawing effect that reduces the availability of nitrogen's lone pair of electrons; thus the basicity of NF_3 is less than that of NH_3.

d. The conjugate base of acetic acid, $CH_3-C(=O)-O^-$, can be written as two different contributing resonance forms:

Thus because its negative charge is delocalized, the acetate anion is a much weaker *base* than the hydroxide anion. Consequently, its conjugate acid (acetic acid) is a stronger acid than water.

e. In the structure on the right, all atoms have a complete valence shell of electrons. The left-hand structure contains a positively charged carbon atom with only six valence electrons. The greater stability of the right-hand structure results in its greater contribution to the hybrid.

f. Hydrogen bonding can occur with propionic acid via the O—H group, which increases the boiling point:

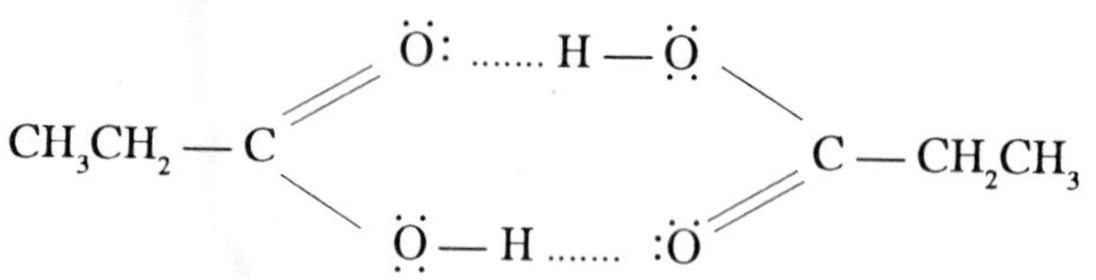

Hydrogen bonding is not possible for methyl acetate.

g. A Lewis acid is an electron pair acceptor. Aluminum has an empty *p* orbital that can accomodate the pair of electrons provided by a Lewis base.

h. Ethanol exhibits intermolecular hydrogen bonding; dimethyl ether does not.

i. Strongly electronegative elements like oxygen or nitrogen enable proton acceptance from H_2SO_4; for example, the oxygen in ethanol:

ethanol

1.16 a.

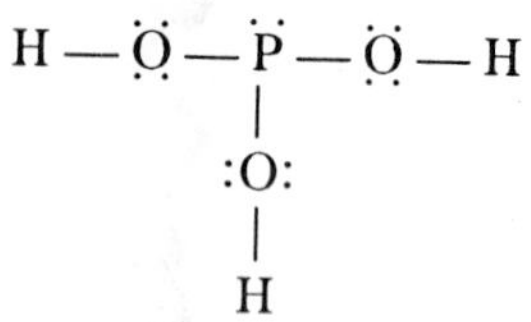

b.

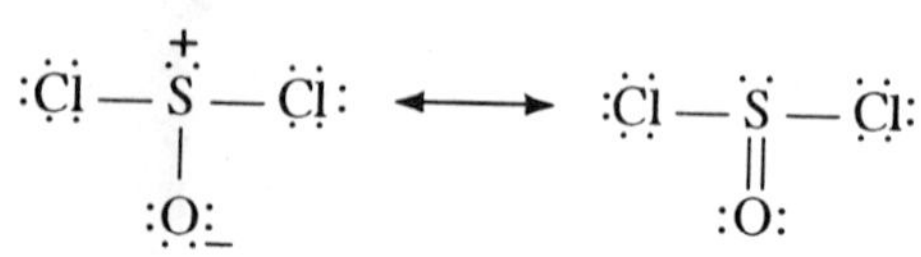

c.

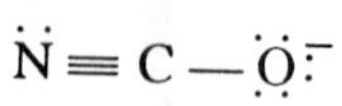

d.

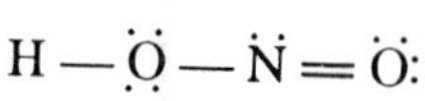

e.

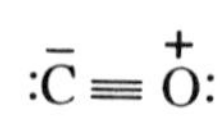

f.

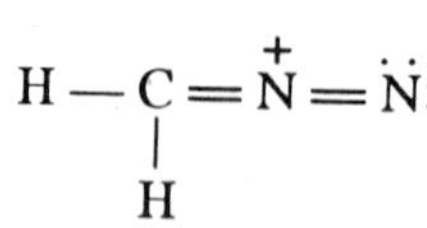

g.

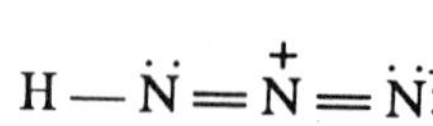

h.

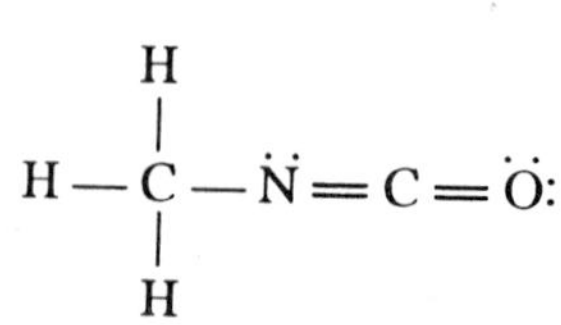

1.17 a. $CH_3-\overset{+}{N}(-O^-)=O$ d. $CH_2=C(-O^-)-H$ g. $CH_3-\overset{+}{N}H_3$

b. $CH_2=\overset{+}{N}=N^-$ e. $CH_2=\overset{+}{N}(-O^-)-CH_3$ h. $CH_3-\overset{+}{S}(-O^-)-CH_3$

c. $CH_3-N=O$ f. H_3O^+ i. $H-C(=O)-O^-$

1.18 a.

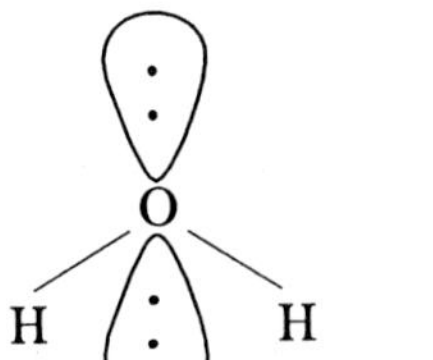

Angular, *sp*³, 105°.

d.

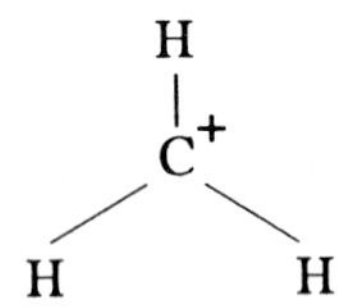

Trigonal, *sp*², 120°.

b.

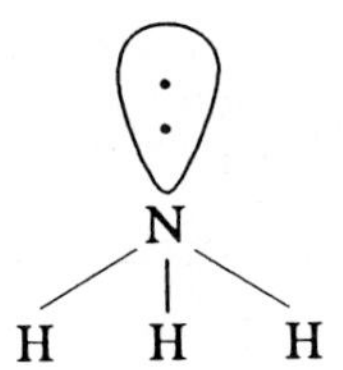

Trigonal pyramidal, *sp*³, 105°.

e.

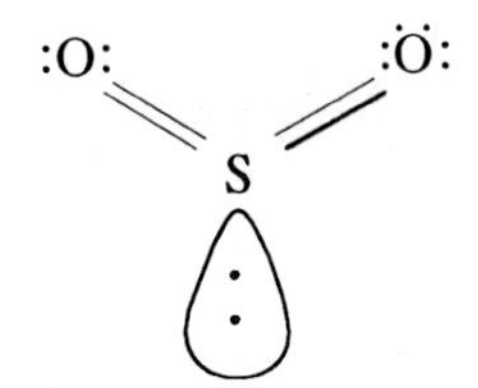

Angular, *sp*², 120°.

c. $:\ddot{O}=C=\ddot{O}:$
Linear; C: *sp*, O: *sp*²; 180°.

1.19

1.	c	5.	q	9.	m	13.	a	17.	b
2.	h	6.	d	10.	s	14.	f	18.	n
3.	i	7.	p	11.	g	15.	r	19.	k
4.	l	8.	o	12.	t	16.	j	20.	e

1.20 a.

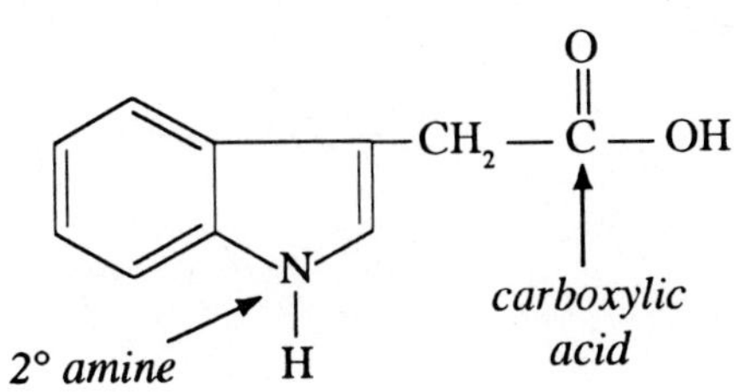

Indole-3-acetic acid (Auxin)

b.

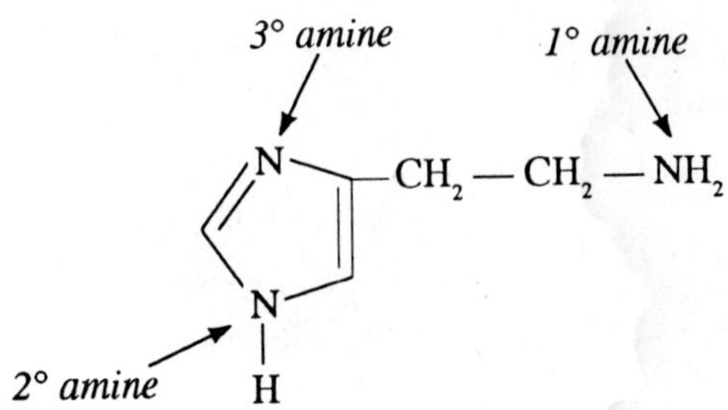

Histamine

c.

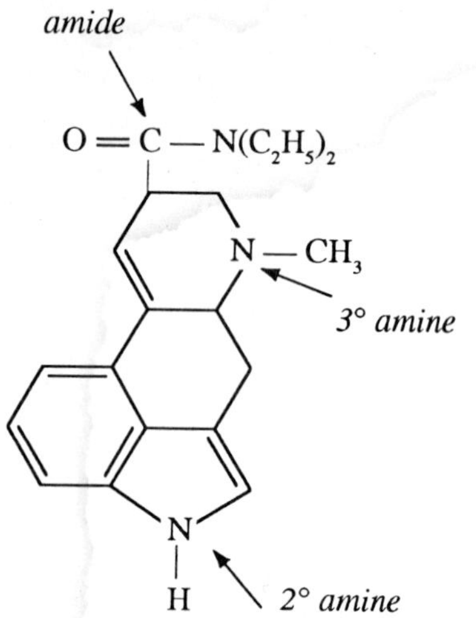

Lysergic acid diethylamide (LSD)

d.

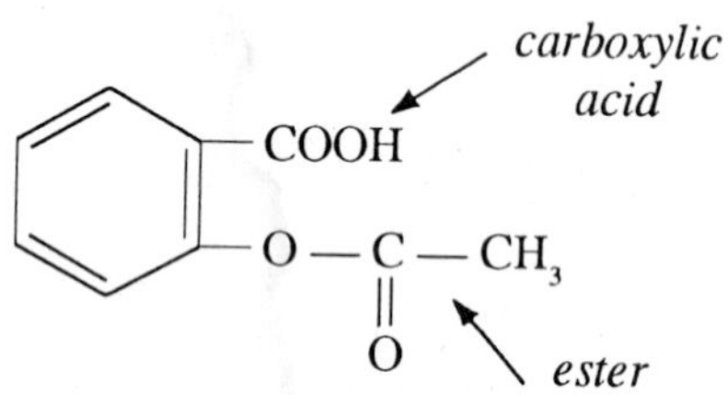

Acetylsalicylic acid (Aspirin)

e.

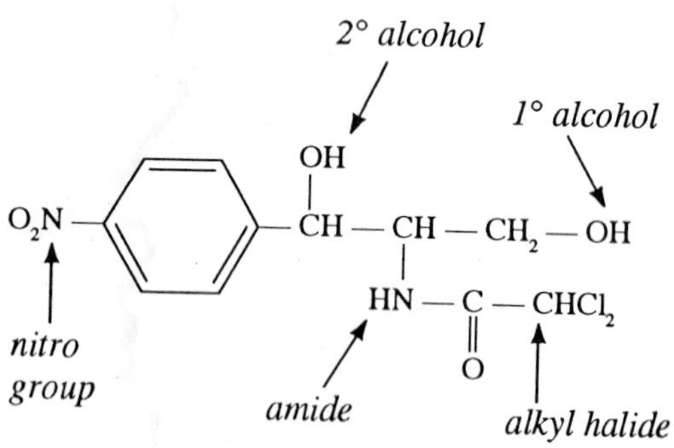

Chloramphenicol

f.

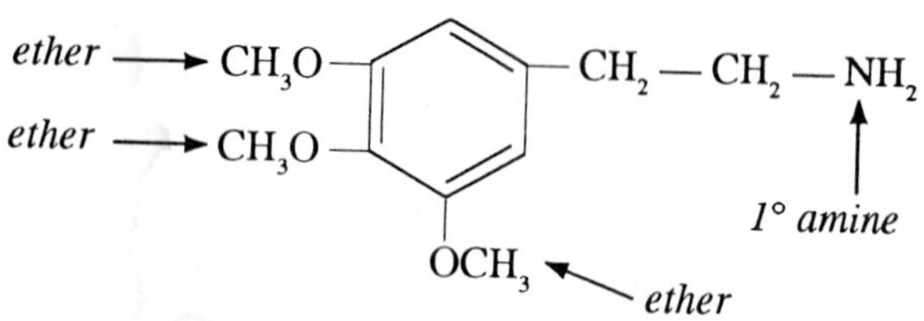

Mescaline

g.

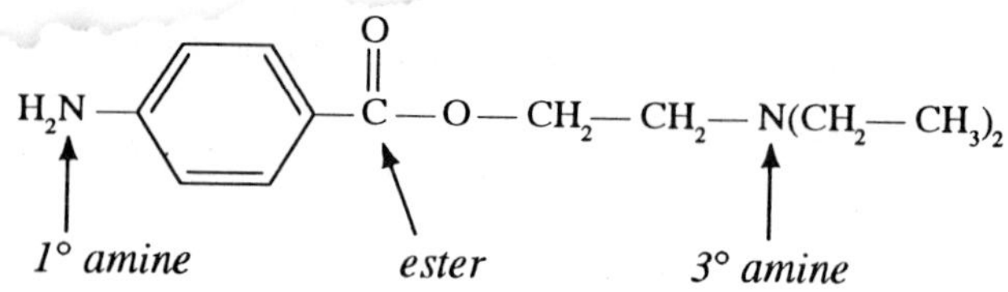

Procaine

h.

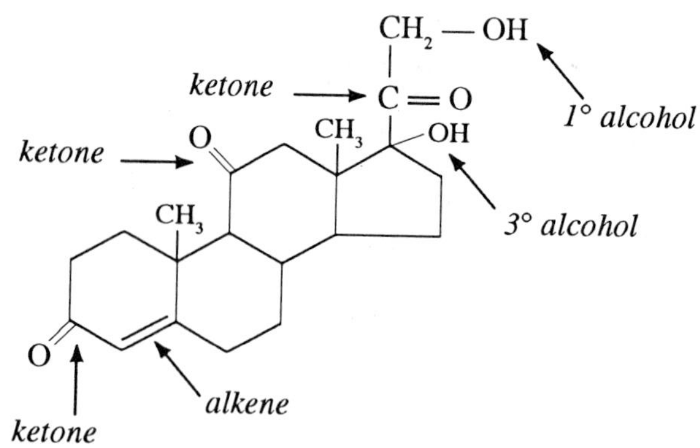

Cortisone

i.

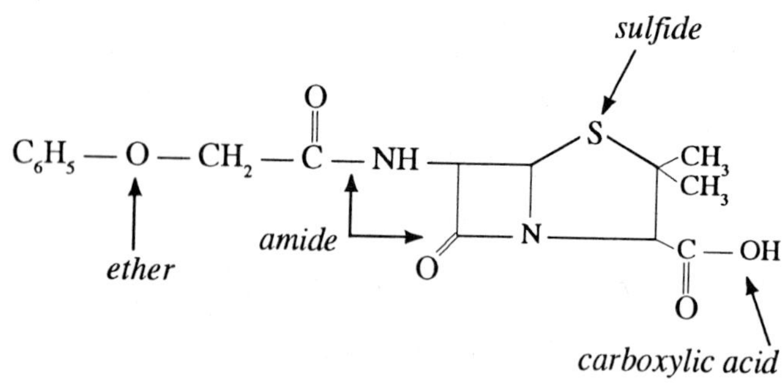

Penicillin V

1.21 a. Three ethers of molecular formula C_4H_{10}:

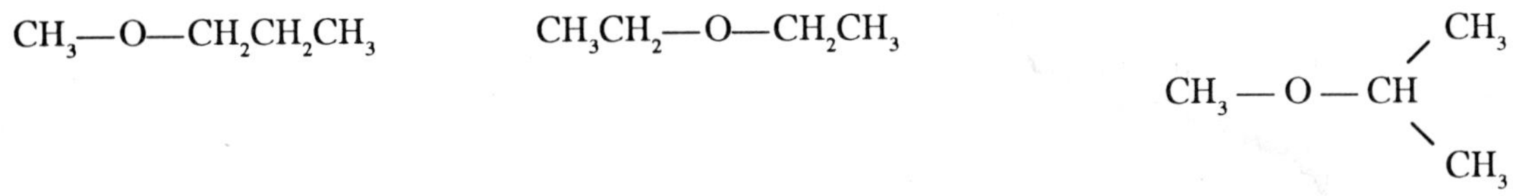

b. Four esters of molecular formula $C_4H_8O_2$:

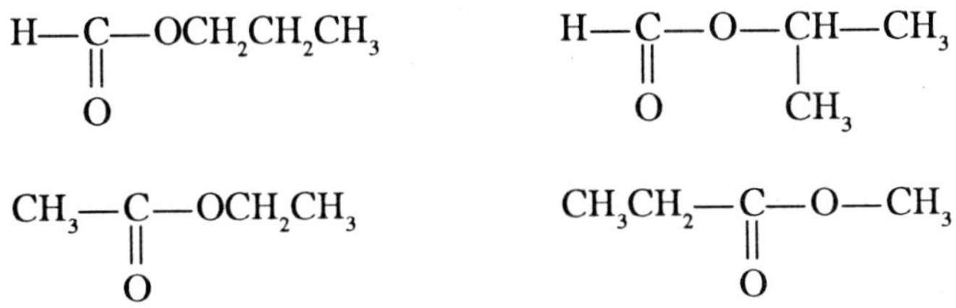

c. Seven 1° alcohols of molecular formula $C_6H_{14}O$:

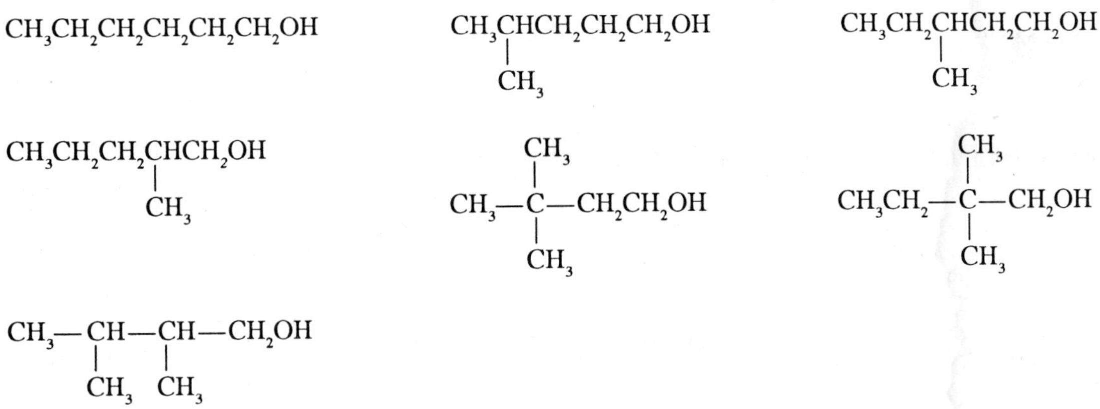

d. Three ketones of molecular formula $C_5H_{10}O$:

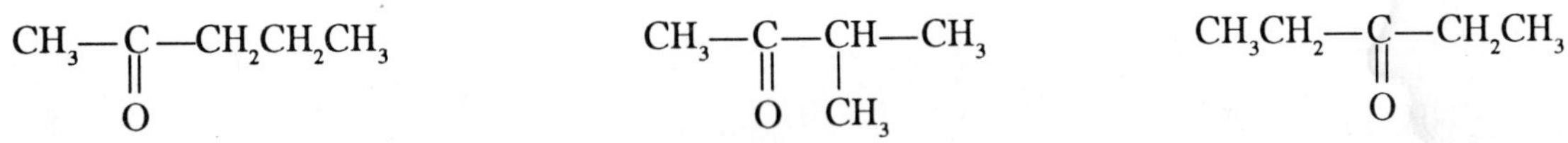

e. Four carboxylic acids of molecular formula $C_5H_{10}O_2$:

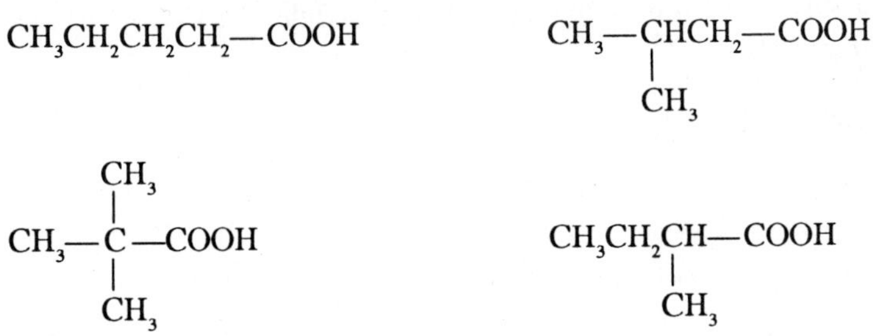

f. Four aldehydes of molecular formula $C_5H_{10}O$:

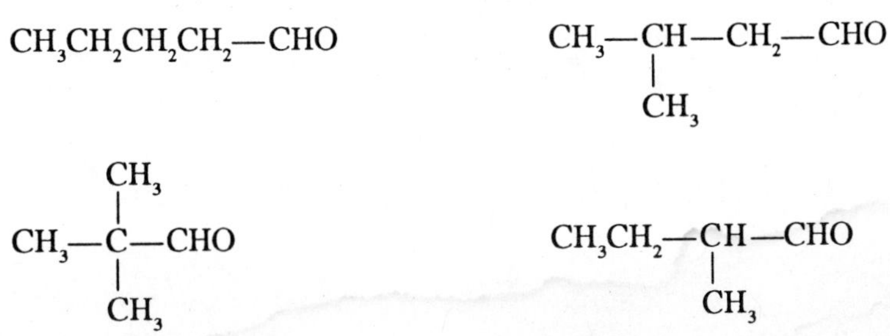

g. There are six ketones of molecular weight 100. The general formula of ketones is $R-C(=O)-R$, thus the two R groups must have a combined molecular weight of 72 (100 minus 28, the molecular weight of C=O). This means that the molecular formula is $C_6H_{12}O$:

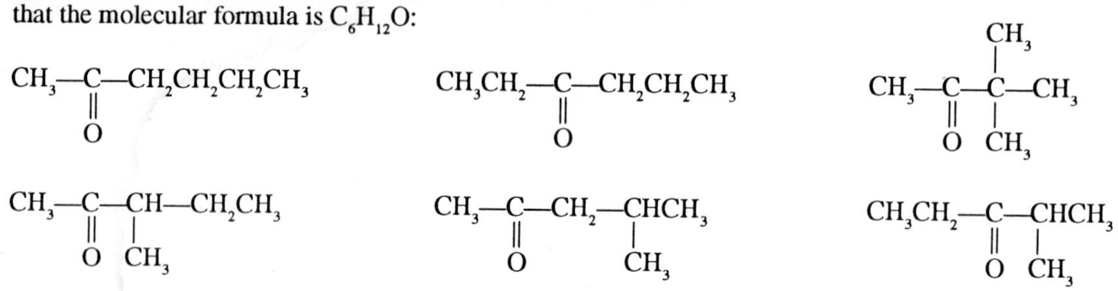

h. Unsymmetrical anhydrides of molecular formula $C_6H_{10}O_3$:

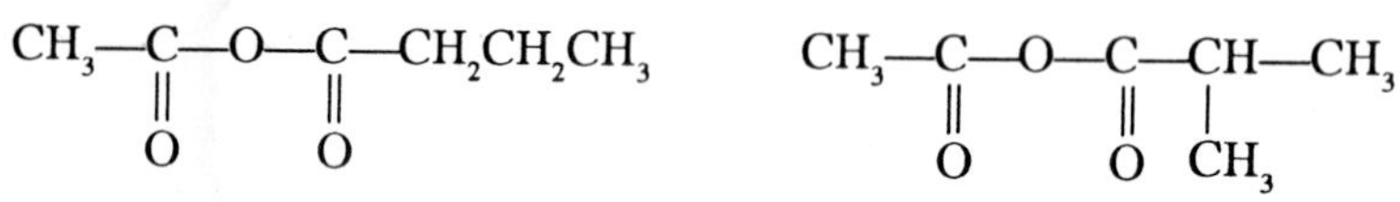

i. The general formula of an ester is $R-C(=O)-OR'$. Since the combined molecular weight of R and R′ is 30 (74 minus 44, the molecular weight of CO_2), the structure should be $CH_3-C(=O)-O-CH_3$ or $H-C(=O)-O-CH_2CH_3$

j. Four 1° amines of molecular formula $C_4H_{11}N$:

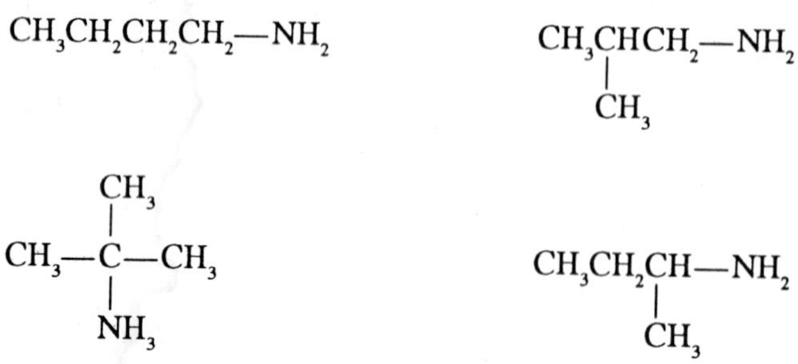

k. There are two aldehydes of molecular weight 72. The general formula for aldehydes is R—CHO, which suggests that the molecular weight of R is 43 (72 minus 29, the molecular weight of CHO). This means that the molecular formula is C_4H_8O and the R is C_3H_7:

$CH_3CH_2CH_2-CHO$ and

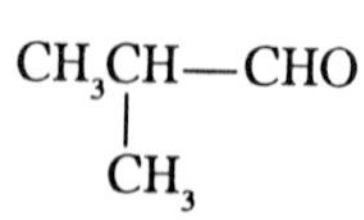

l. The general formula for an alkyne is R—C≡C—R′ or R—C≡CH. Since the combined molecular weight of R and R′ is 44 (68 minus 24, the molecular weight of the two carbon atoms), the three possible structures are:

$CH_3CH_2CH_2C\equiv CH$

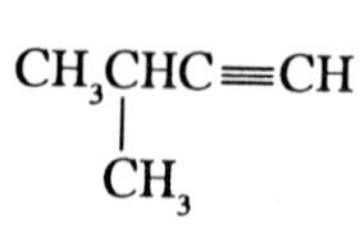

$CH_3CH_2C\equiv CCH_3$

m. Three 3° amines of molecular formula $C_5H_{13}N$:

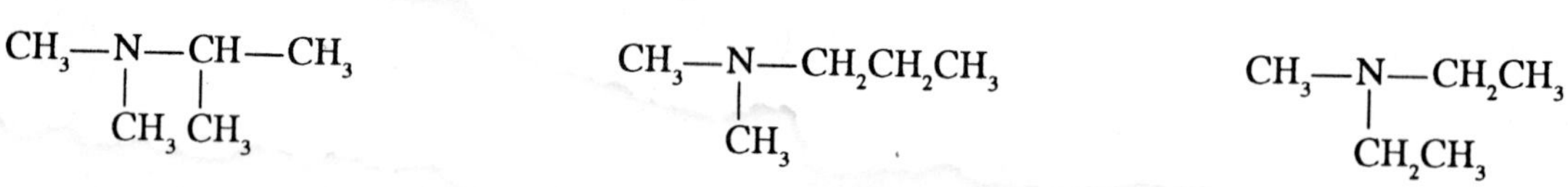

n. Eight unsaturated carboxylic acids of molecular formula $C_5H_8O_2$:

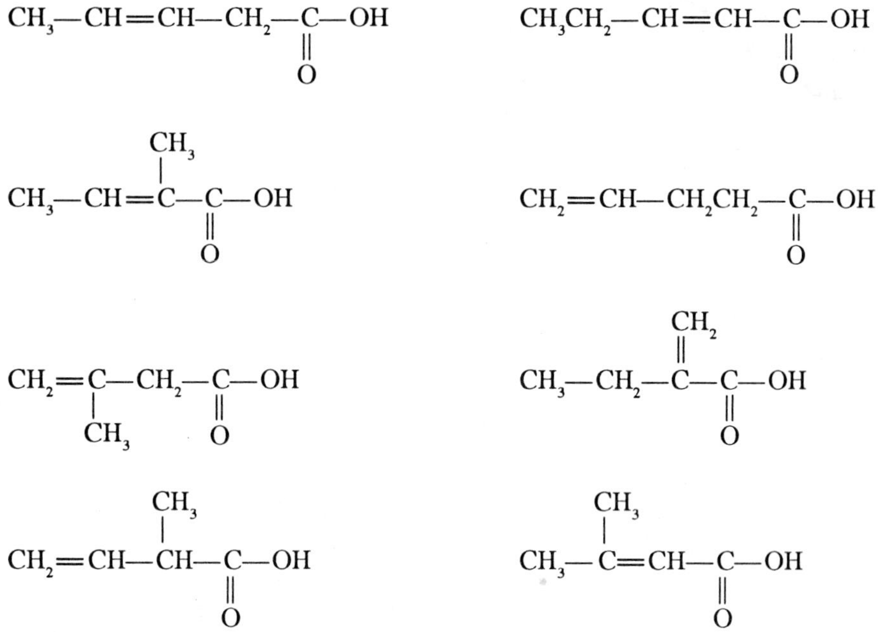

1.22 Problem (a) is solved. The solutions of all other problems can be found by following the same procedure, and therefore, only the final answer is given.

a. C_3H_6O

Molecular weight $= (3 \times 12) + (6 \times 1) + (1 \times 16)$
$= 36 + 6 + 16$
$= 58.$

Carbon 36/58 × 100 = 62.1%
Hydrogen 6/58 × 100 = 10.3%
Oxygen 16/58 × 100 = 27.6%

Total 100.0%

	Formula	Mol. wt.	%C	%H	%O	%N	%Other
b.	C_4H_9Cl	92.5	51.9	9.7			38.4 (Cl)
c.	CH_2Cl_2	85.0	14.1	2.4			83.5 (Cl)
d.	$C_6H_8O_2N_2S$	172.0	41.9	4.7	18.6	16.3	18.6 (S)
e.	$C_{16}H_{26}O_4N_2S$	342.0	56.1	7.6	18.7	8.2	9.4 (S)

1.23 Problem (a) is solved. All other problems follow the same procedure, and therefore, only the final answer is given.

a. 55.77% carbon; 11.70% hydrogen; 32.53% nitrogen. Since 55.77 + 11.70 + 32.53 = 100, there was no oxygen present in the original chemical.

C =(55.77/12) = 4.65 gram atoms   4.65/2.32 = 2
H = (11.70/1) = 11.70 gram atoms   11.7/2.32 = 5
N = (32.53/14) = 2.32 *gram atoms   2.32/2.32 = 1

* *Use as denominator being the smallest value.*

Empirical formula: C_2H_5N

Note: Always make sure that the sum is 100%. If it is not, the rest is assumed to be oxygen.

b. C_6H_6ClN

c. CH_3Br

d. C_6H_6O

1.24 In 44 g of CO_2 there are 12 g of carbon.

$$\frac{44}{0.1902} = \frac{12}{y}\ ;\ \text{thus } y = \frac{(12)(0.1902)}{44} = 0.0519 \text{ g carbon.}$$

In 18 g of H_2O there are 2 g of hydrogen.

$$\frac{18}{0.0907} = \frac{2}{z}\ ;\ \text{thus } z = \frac{(2)(0.0907)}{18} = 0.0101 \text{ g hydrogen.}$$

In 187.8 g of AgBr there are 79.9 g of bromine.

$$\frac{187.8}{0.2058} = \frac{79.9}{w}\ ;\ \text{thus } w = \frac{(79.9)(0.2058)}{187.8} = 0.0876 \text{ g bromine.}$$

In a 0.2001 g sample of the unknown compound, the bromine content would be

$$\frac{0.0876}{0.1523} \times 0.2001 = 0.1151 \text{ g bromine.}$$

The content of oxygen is calculated by subtracting the sum of the weights of the known elements from the weight of the sample: 0.2001 g – (0.1151 g + 0.0519 g + 0.0101 g) = 0.0230 g of oxygen.

Element	Gram atoms	Ratio (Gram atoms divided by smallest gram atom value)
Carbon	0.0519/12 = 0.0043	0.0043/0.0014 = 4
Hydrogen	0.0101/1 = 0.0101	0.0101/0.0014 = 7
Bromine	0.1151/80 = 0.0014	0.0014/0.0014 = 1
Oxygen	0.0230/16 = 0.0014	0.0014/0.0014 = 1

Thus the empirical formula is C_4H_7BrO.

1.25 a. The molecular weight (MW) can be calculated using the ideal gas law:

PV = nRT ; where n = sample mass/MW.

P = 760 torr = 1 atm

V = 100 ml = 0.1 L

T = 100°C = 373°K

R = 0.082 atm L/°K mole

sample mass = 275 mg = 0.275 g

$$MW = \frac{(\text{sample mass})RT}{PV}$$

$$= \frac{(0.275 \text{ g})(0.082 \text{ atm L/°K mole})(373°\text{K})}{(1 \text{ atm})(0.1 \text{ L})}$$

$$= 84.$$

b. In 6.51 mg of the compound, there are 5.58 mg of carbon, 0.93 mg of hydrogen, and 0 mg of oxygen. (Calculated from the amounts of CO_2 and H_2O produced from combustion; for procedure, see problem 1.24.) Thus the empirical formula is CH_2.

c. From the MW and the empirical formula, the molecular formula is C_6H_{12}.

1.26 In 44 g of CO_2 there are 12 g of carbon.

$$\frac{44}{0.2895} = \frac{12}{y} \;;\; \text{thus } y = \frac{(12)\,(0.2895)}{44} = 0.0790 \text{ g carbon in } 0.1908 \text{ g sample.}$$

In 18 g of H_2O there are 2 g of hydrogen.

$$\frac{18}{0.1192} = \frac{2}{z} \;;\; \text{thus } z = \frac{(2)\,(0.1192)}{18} = 0.0132 \text{ g hydrogen in } 0.1908 \text{ g sample.}$$

$P = 755$ mm Hg – 24 mm Hg = 731 mm Hg (731/760 = 0.962 atm)

$V = 40.2$ mL = 4.02×10^{-2} mL

$T = 25°C + 273° = 298K$

MW of N_2 = 28

$R = 0.082$ atm L /K mole

$PV = nRT = (\text{mass}/MW)RT$

$$\text{mass} = \frac{PV(MW)}{RT} = \frac{(0.962 \text{ atm})\,(4.02 \times 10^{-2} \text{ L})\,(28)}{(0.082 \text{ atm L /K mole})\,(298K)} = 0.0443 \text{ g nitrogen in } 0.1825 \text{ g sample.}$$

Mass of nitrogen in a 0.1908 g sample = (0.0443/0.1825) (0.1908)
= 0.0424 g nitrogen.

Mass of oxygen in a 0.1908 g sample = 0.1908 g – (0.0790 g + 0.0132 g + 0.0463 g) = 0.0523 g oxygen.

Element	Gram atoms	Ratio (Gram atoms divided by smallest gram atom value)
Carbon	0.0790/12 = 0.0066	0.0066/0.0033 = 2
Hydrogen	0.0132/1 = 0.0132	0.0132/0.0033 = 4
Oxygen	0.0523/16 = 0.0033	0.0033/0.0033 = 1
Nitrogen	0.0463/14 = 0.0033	0.0033/0.0033 = 1

Thus the empirical formula is C_2H_4NO.

Molality = moles of solute per 1000 g solvent.

Number of moles of solute = 1.082 g /MW.

Depression in freezing point (ΔT): $\Delta T = K_f \times$ molality.

$1.71°C = (5.12)\,(1.082 \text{ g}/MW)\,(1000)/28$

$$\text{Thus } MW = \frac{(5.12)\,(1.082)\,(1000)}{(28)\,(1.71)} = 115.7$$

MW of empirical formula C_2H_4NO is 58; MW of molecular formula is 115.7. Therefore, the molecular formula is $C_4H_8N_2O_2$.

1.27 At equilibrium: $HA + H_2O \rightleftharpoons H_3O^+ + A^-$

$[H_3O^+] = [A^-] = (0.1M)(0.012) = 1.2 \times 10^{-3}M$ and $[HA] = 0.1000 - 0.0012 = 0.0988M$

$$K_a = \frac{[H_3O^+][A^-]}{[HA]} = \frac{(1.2 \times 10^{-3})^2}{(0.0988)} = 1.5 \times 10^{-5}$$

Thus, $pK_a = -\log K_a = -\log(1.5 \times 10^{-5}) = 4.8$ and $pH = -\log[H_3O^+] = -\log[1.2 \times 10^{-3}] = 2.9$

1.28 a. $K_a = 3.1 \times 10^{-4}$ b. $pK_a = -2.0$ c. $pK_a = 1.2$

1.29 A stronger acid or base reacts to form a weaker acid or base.

a. to left
b. to left
c. to right
d. to left
e. to left
f. to left

1.30 A weaker acid will have a stronger conjugate base:

$CH_3^- > HC{\equiv}C^- > C_2H_5S^- > C_6H_5O^- > NH_3 > HCO_2^- > H_2O$

1.31 a. Bonds broken: I_2 requires 36 kcal/mole
Bonds formed: None
ΔH = 36 kcal /mole (endothermic)

b. $CH_3{-}H + Br{-}Br \longrightarrow CH_3{-}Br + H{-}Br$

Bonds broken: +104 +46 = 150 kcal/mole
Bonds formed: −70 −87.5 = −157.5 kcal/mole

ΔH = −7.5 kcal/mole (exothermic)

c. $CH_3{-}Br + CH_3{-}CH_3 \longrightarrow CH_3CH_2{-}Br + CH_3{-}H$

+70 +88 = 158 kcal/mole; −69 −104 = −173 kcal/mole

ΔH = −15 kcal/mole (exothermic)

d. $CH_3{-}F + H{-}H \longrightarrow CH_3{-}H + H{-}F$

+108 +104 = +212 kcal/mole; −104 −136 = −240 kcal/mole

ΔH = −28 kcal/mole (exothermic)

e. $CH_3CH_2CH_2{-}H + CH_3{-}H \longrightarrow CH_3CH_2CH_2{-}CH_3 + H{-}H$

+98 +104 = +202 kcal/mole; −85 −104 = −189 kcal/mole

ΔH = +13 kcal/mole (endothermic)

f. $CH_3CH_2{-}CH_3 + CH_3{-}H \longrightarrow 2CH_3CH_2{-}H$

+85 +104 = +189 kcal/mole; 2(−98) = −196 kcal/mole

ΔH = −7 kcal/mole (exothermic)

g. $H{-}CH_2CH{=}CH_2 + Br{-}Br \longrightarrow Br{-}CH_2CH{=}CH_2 + HBr$

+85 +46 = +131 kcal/mole; −47 −87.5 = −134.5 kcal/mole

ΔH = −3.5 kcal/mole (exothermic)

1.32 a. $CH_3{-}Cl + H{-}OH \longrightarrow CH_3{-}OH + H{-}Cl$

Bonds broken: +227 +390 = +617 kcal/mole

Bonds formed: –274 –334 = –608 kcal/mole

ΔH = +9 kcal/mole (endothermic)

b. $CH_3CH_2{-}Br + H{-}I \longrightarrow CH_3CH_2{-}I + H{-}Br$

+184 +315 = +499 kcal/mole

–176 –324 = –500 kcal/mole

ΔH = –1 kcal/mole (exothermic)

c. $CH_3{-}OH + H{-}Cl \longrightarrow CH_3{-}Cl + H{-}OH$

+274 +334 = +608 kcal/mole

–227 –390 = –617 kcal/mole

ΔH = –9 kcal/mole (exothermic)

d. $H_2C{=}CHCH_2{-}Br + H{-}OH \longrightarrow H_2C{=}CHCH_2{-}OH + H{-}Br$

+165 +390 = +555 kcal/mole

–223 –324 = –547 kcal/mole

ΔH = +8 kcal/mole (endothermic)

e. $C_6H_5CH_2{-}OH + H{-}I \longrightarrow C_6H_5CH_2{-}I + H{-}OH$

+215 +315 = +530 kcal/mole

–149 –390 = –539 kcal/mole

ΔH = –9 kcal/mole (exothermic)

1.33 a. Two molecules combine to form only one (fewer degrees of freedom in product).

b. Since $\Delta G° = \Delta H° - T\Delta S°$, a negative value of $\Delta S°$ tends to make $\Delta G°$ more positive (or less negative).

c. $\Delta G°$ = –16000 cal/mole – [(298°K) (–30 cal/mole)] = –7060 cal/mole.

d. $$\log K_{eq} = \frac{-\Delta G°}{(2.303)\,RT} = \frac{-(-7{,}060)}{(2.303)\,(1.98)\,(298)} = 5.196$$

$$K_{eq} = 1.57 \times 10^5$$

e. The term $T\Delta S°$ becomes increasingly more positive at higher temperatures. Considering $\Delta H°$ as a constant, this results in a less negative value of $\Delta G°$ and a smaller K_{eq}. At 533.3K, $\Delta G°$ becomes zero.

$$\Delta G° = \Delta H° - T\Delta S°$$
$$0 = -16{,}000 - (T)(-30)$$
$$\text{and } T = \frac{16{,}000}{30} = 533.3\,K$$

1.34 The first equation is balanced in detail. All others follow the same pattern and, therefore, only the final equation is given.

a. $ReO_2 + Cl_2 \longrightarrow HReO_4 + Cl^-$

Step 1: Separate the equation into two half-equations:

$ReO_2 \longrightarrow HReO_4$ $Cl_2 \longrightarrow Cl^-$

Step 2: Balance all elements except hydrogen, oxygen, and the charges:

$ReO_2 \longrightarrow HReO_4$ $Cl_2 \longrightarrow 2Cl^-$

Step 3: Balance the charges by adding H^+

$ReO_2 \longrightarrow HReO_4$ $\qquad$ $Cl_2 \longrightarrow 2Cl^- + 2H^+$

Step 4: Balance oxygens by adding H_2O:

$ReO_2 + 2H_2O \longrightarrow HReO_4$ $\qquad$ $Cl_2 \longrightarrow 2Cl^- + 2H^+$

Step 5: Balance hydrogens with **H** (uncharged hydrogen):

$ReO_2 + 2H_2O \longrightarrow HReO_4 + 3\mathbf{H}$ $\qquad$ $Cl_2 \longrightarrow 2\mathbf{H} + 2Cl^- + 2H^+$

Step 6: Multiply each half equation by a factor so that all **H**'s are cancelled (in this case, by a factor of 2 or 3):

$2ReO_2 + 4H_2O \longrightarrow 2HReO_4 + 6\mathbf{H}$ $\qquad$ $3Cl_2 + 6\mathbf{H} \longrightarrow 6Cl^- + 6H^+$

Step 7: Add the two half-equations:

$2ReO_2 + 4H_2O + 3Cl_2 \longrightarrow 2HReO_4 + 6Cl^- + 6H^+$

b. $2Cr_2O_7^{-2} + 16H^+ + 3C_2H_6O \longrightarrow 4Cr^{+3} + 3C_2H_4O_2 + 11H_2O$

c. $5C_7H_8 + 18H^+ + 6MnO_4^- \longrightarrow 5C_7H_6O_2 + 6Mn^{+2} + 14H_2O$

d. $5C_8H_{10} + 12MnO_4^- + 36H^+ \longrightarrow 5C_8H_6O_4 + 12Mn^{+2} + 28H_2O$

1.35 The first equation is balanced in detail. All others follow the same pattern and, therefore, only the final equation is given.

a. $Cr(OH)_4^- + BrO^- \longrightarrow CrO_4^{-2} + Br^-$

Step 1: Separate the equation into two half-equations:

$Cr(OH)_4^- \longrightarrow CrO_4^{-2}$ $\qquad$ $BrO^- \longrightarrow Br^-$

Step 2: Balance all elements except hydrogen, oxygen, and the charges:

$Cr(OH)_4^- \longrightarrow CrO_4^{-2}$ $\qquad$ $BrO^- \longrightarrow Br^-$

Step 3: Balance the charges by adding OH^-

$OH^- + Cr(OH)_4^- \longrightarrow CrO_4^{-2}$ $\qquad$ $BrO^- \longrightarrow Br^-$

Step 4: Balance oxygens by adding H_2O:

$Cr(OH)_4^- + OH^- \longrightarrow CrO_4^{-2} + H_2O$ $\qquad$ $BrO^- \longrightarrow Br^- + H_2O$

Step 5: Balance hydrogens with **H**:

$Cr(OH)_4^- + OH^- \longrightarrow CrO_4^{-2} + H_2O + 3\mathbf{H}$

$BrO^- + 2\mathbf{H} \longrightarrow Br^- + H_2O$

Step 6: Multiply each half equation by a factor so that all **H**'s are cancelled (in this case, by a factor of 2 or 3):

$2Cr(OH)_4^- + 2OH^- \longrightarrow 2CrO_4^{-2} + 2H_2O + 6\mathbf{H}$

$3BrO^- + 6\mathbf{H} \longrightarrow 3Br^- + 3H_2O$

Step 7: Add the two half-equations:

$2Cr(OH)_4^- + 2OH^- + 3BrO^- \longrightarrow 2CrO_4^{-2} + 5H_2O + 3Br^-$

b. $C_7H_8 + 2MnO_4^- \longrightarrow C_7H_5O_2^- + 2MnO_2 + OH^- + H_2O$

c. $3C_9H_{12} + 16MnO_4^- \longrightarrow 3C_9H_5O_6^{-3} + 16MnO_2 + 7OH^- + 7H_2O$

d. $C_6H_{10}O + 2MnO_4^- \longrightarrow C_6H_8O_4^{-2} + H_2O + 2MnO_2$

Answers for Multiple Choice Test

1.1 d	**1.6** b	**1.11** c	**1.16** c
1.2 c	**1.7** d	**1.12** b	**1.17** c
1.3 d	**1.8** d	**1.13** a	**1.18** c
1.4 c	**1.9** b	**1.14** a	**1.19** b
1.5 c	**1.10** a	**1.15** d	**1.20** b

2 Alkanes and Cycloalkanes

Questions

2.1 Write structures for the following alkyl groups:

a. ethyl
b. *n*-propyl
c. isopropyl
d. *n*-butyl
e. isobutyl
f. *s*-butyl
g. *t*-butyl
h. neopentyl

2.2 Give structures and names for

a. all pentanes of molecular formula C_5H_{12}.
b. all hexanes of molecular formula C_6H_{14}.
c. all heptanes of molecular formula C_7H_{16}.
d. all monochloropentanes of molecular formula $C_5H_{11}Cl$.
e. all dibromobutanes of molecular formula $C_4H_8Br_2$.
f. all monobromochloroderivatives of *n*–butane.
g. all monochlorination products of 2–methylpentane.
h. all monochlorination products of 3–ethyl–2–methylpentane.
i. all dichlorinated derivatives of propane.

2.3 Write the structure of

a. an alkane (molecular formula C_5H_{12}) that has an isopropyl group.
b. an alkane (molecular formula C_6H_{14}) that has a *t*–butyl group.
c. an alkane (molecular formula C_8H_{18}) that has two isopropyl groups.
d. an alkane (molecular formula $C_{10}H_{22}$) that has two ethyl and one t-butyl groups.
e. an alkane (molecular formula C_6H_{14}) that has no 2° hydrogens.
f. an alkane (molecular weight 72) that would yield only one monochloro derivative.
g. an alkane of molecular weight 86 that would yield two monochloro derivatives.
h. an alkane of molecular weight 72 that would yield three different monochloro derivatives.
i. a compound whose 2° hydrogens equal half the total number of 1° hydrogens.
j. the intermediate that is formed most rapidly when 2,2,3–trimethylbutane reacts with bromine in the presence of light.
k. a $C_2H_4Br_2$ isomer that gives only one $C_2H_3Br_3$ derivative upon reaction with Br_2/UV.

2.4 Name the following compounds:

a. $(CH_3)_4C$

b. $BrCH_2CH_2CH_2CH_2Br$

c. $CCl_3CH_2CCl_3$

d. 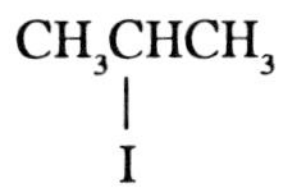

e. 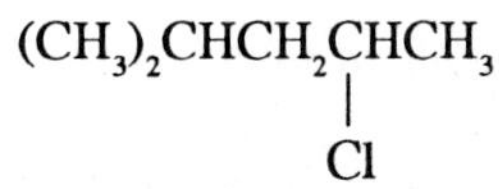

f. $(CH_3)_3CCH_2CH_2C(CH_3)_3$

g. 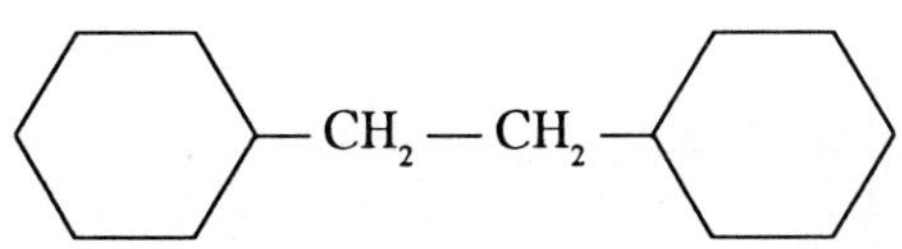

h. 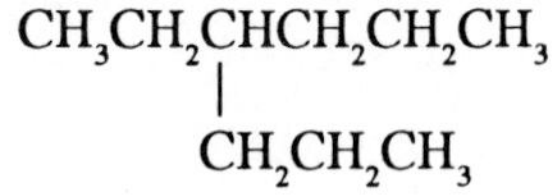

i. 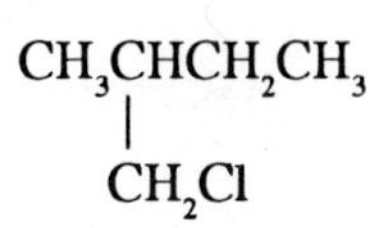

j. 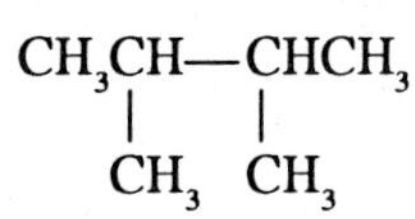

k. 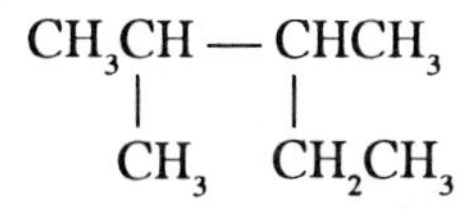

l. 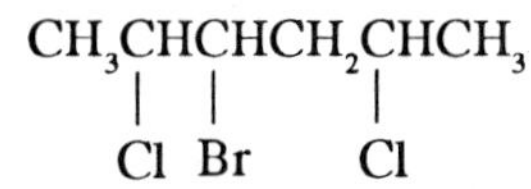

m. 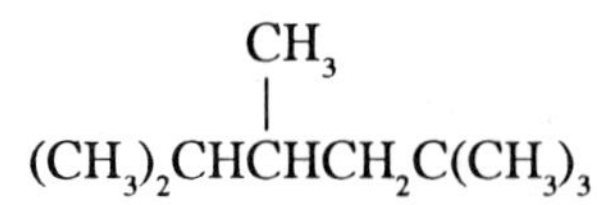

n. 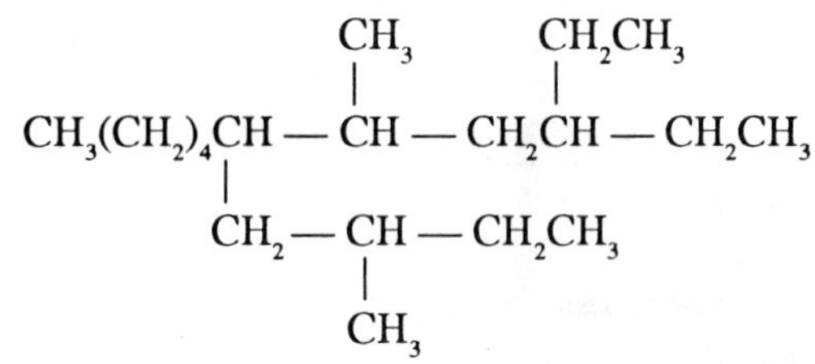

o. 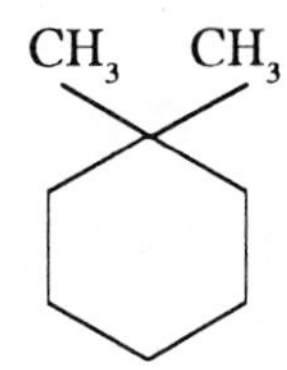

p. 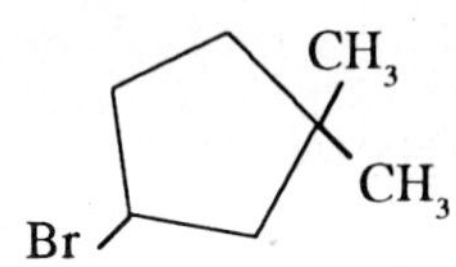

q. 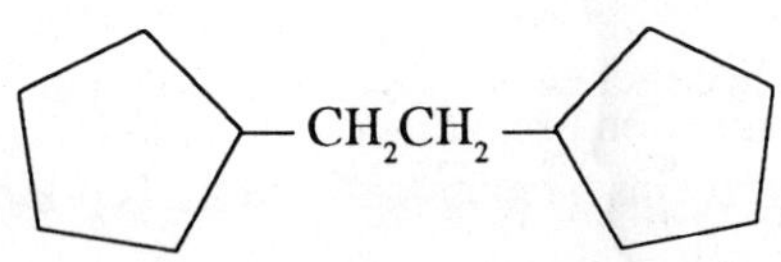

r. 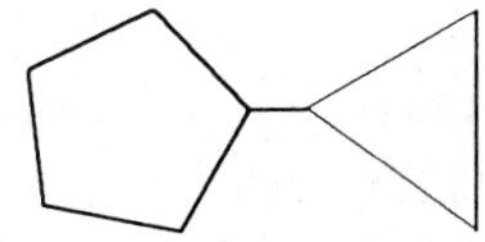

s. 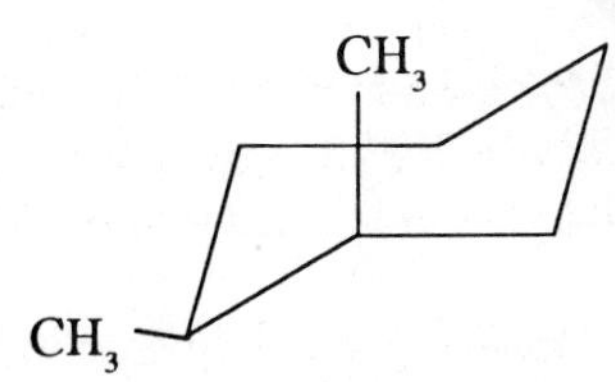

t. 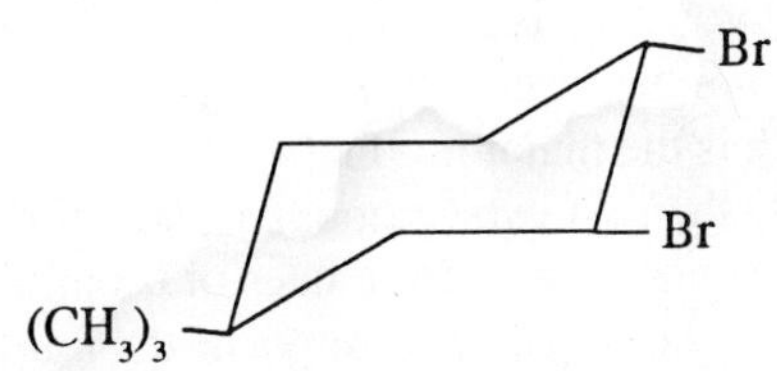

u. 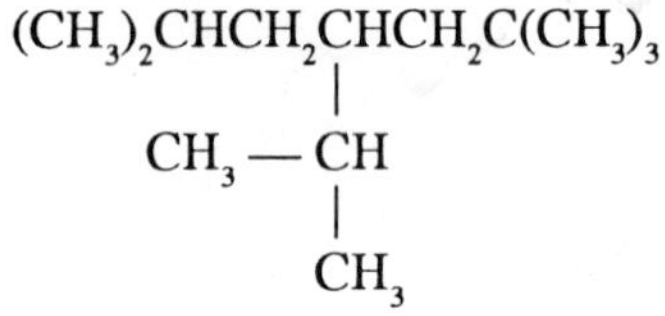

v. $(CH_3)_3CCH_2CH_3$

w. 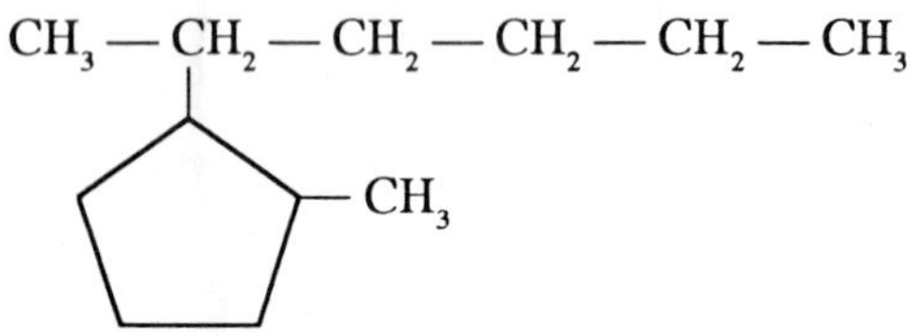

x. 

y. 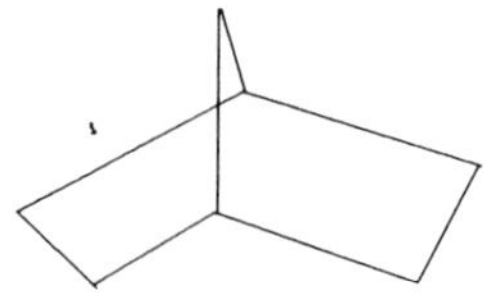

z. 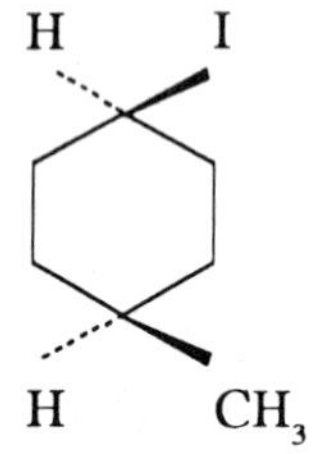

2.5 Draw a structure for the following compounds:

a. 3–ethylpentane
b. 2,3–dimethylbutane
c. isobutylbromide
d. 2–methyl–6–(2–methylbutyl)dodecane
e. 2–bromo–2–methylpropane
f. 1–bromo–2,2–dimethylpropane
g. 2–chloro–4–methylpentane
h. chloroform
i. 3–(3,4–dimethylcyclohexyl) heptane
j. 2–bromo–2–methylbutane
k. 6–ethyl–5–isopropyl–2,3–dimethyloctane
l. 2–bromo–3–chloro–3–ethyl–2–methylpentane
m. *trans*–1,3–dibromocyclohexane
n. 1–cyclobutylpentane
o. *trans*–2–*cis*–3–dimethyl–*t*–butylcyclohexane
p. *cis*–1–chloro–2–methylcyclopropane
q. 1,1,2–trimethyl–3–isopropylcyclohexane
r. *cis*–1,3–dichlorocyclohexane
s. dicyclohexylmethane
t. bicyclo[3.1.1]heptane
u. 7–chlorobicyclo[4.3.1]decane
v. bicyclo[2.2.1]heptane
w. 1–methylbicyclo[4.1.0]heptane

2.6 The bromination of methane with Br_2/UV is slowed by addition of small amounts of HBr gas. Explain mechanistically (not by just stating Le Chatelier's principle) why this happens. (*Hint:* Start by writing the individual steps of the mechanism).

2.7 Write

a. the two propagation steps in the reaction of dichloromethane with Cl_2/UV.
b. the three terminating steps in the above reaction.
c. the initiating step in the reaction of propane with Br_2/UV.

2.8 Explain mechanistically the formation of small amounts of ethane and chlorinated ethane observed in the UV chlorination of methane.

2.9 Careful analysis of the monobromination products of the reaction of propane with bromine in a 1:1 Br_2 to C_3H_8 molar ratio reveals the production of a third compund in addition to the expected products. Combustion analysis of 0.808 g of this compound gives 0.528 g of CO_2 and 0.216 g of H_20. Chlorination of this third compound gives only one monochloroderivative.

a. What are the formulas of the expected two products?
b. Which is the major product?
c. If only the first two compounds were formed, what would be the expected percentage of each compound? (See problem 2.11 for relative rates of bromination.)
d. What is the empirical formula of the third compound?
e. What is the structural formula of the third compound?
f. What is the structural formula of the monochloro derivative of the third compound?

2.10 Identify the reagents and/or products missing in the reactions below:

a. 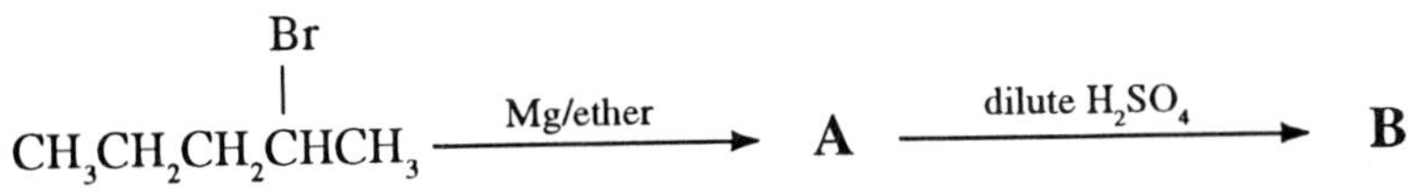

b. 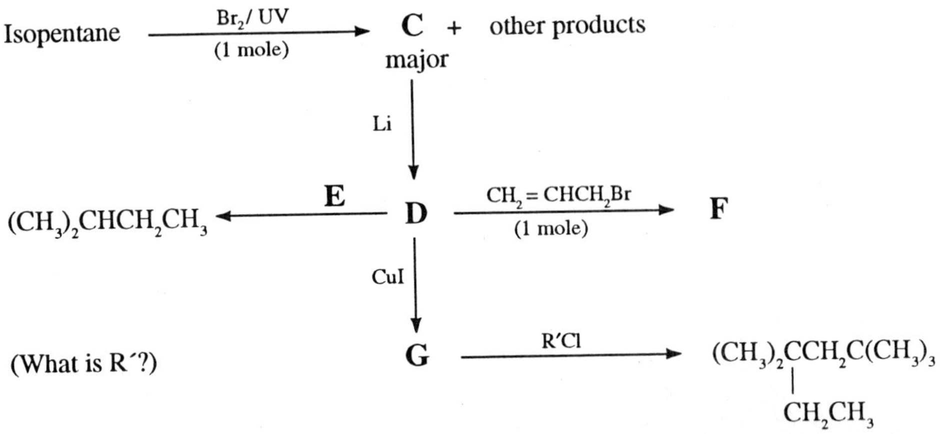

c. 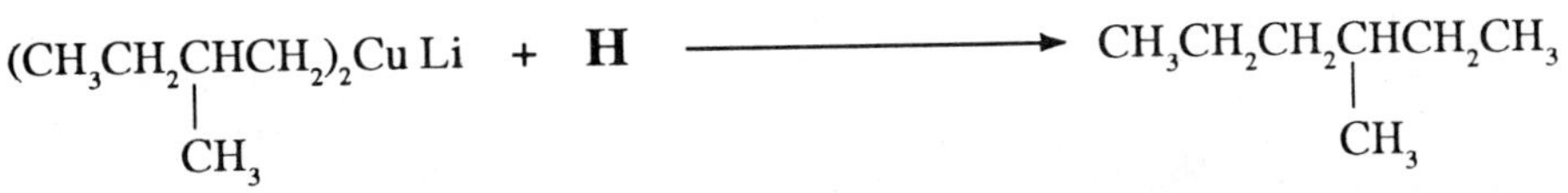

d. 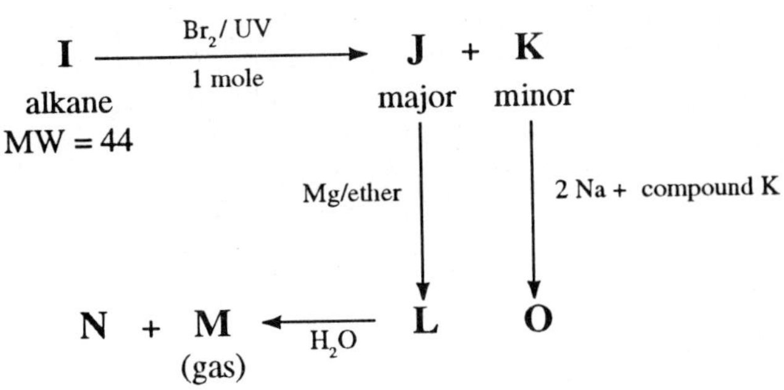

e. 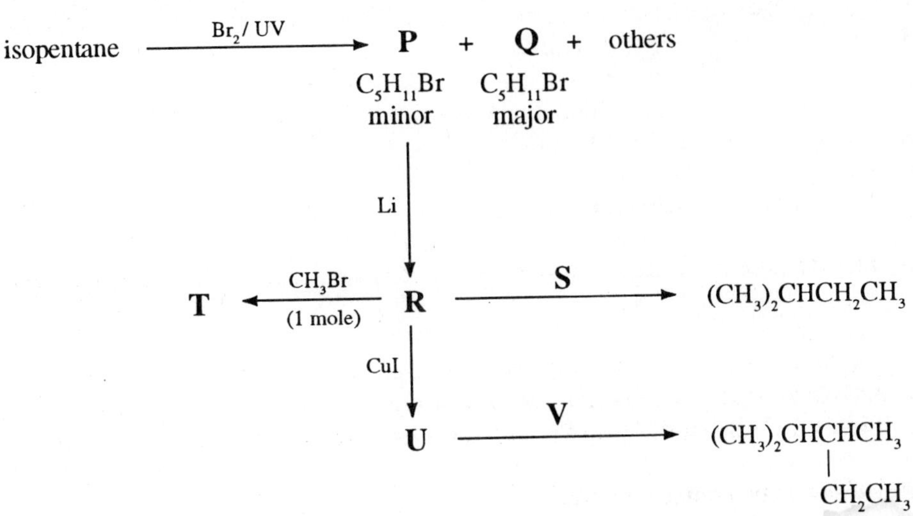

2.11 Given the relative reactivities of chlorination and bromination below, predict the percentage of each product formed when the following undergo those reactions. Name all products

Types of hydrogen	1°		2°		3°
Relative rates of chlorination	1	:	3.8	:	5.0
Relative rates of bromination	1	:	82	:	1600

a. $CH_3CH_2CH_3$

b. $CH_3CH_2CH_2CH_3$

c. $(CH_3)_2CHCH_3$

d. $CH_3CH_2CH_2CH_2CH_3$

e. 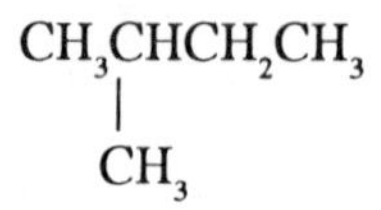

2.12 Suggest syntheses for

a. 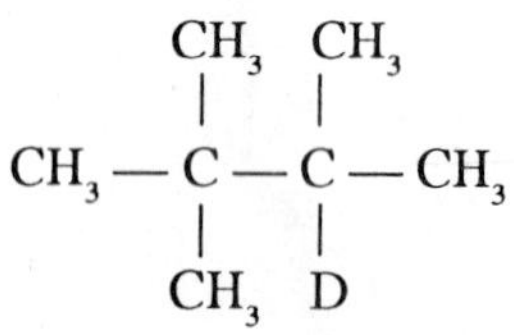 from 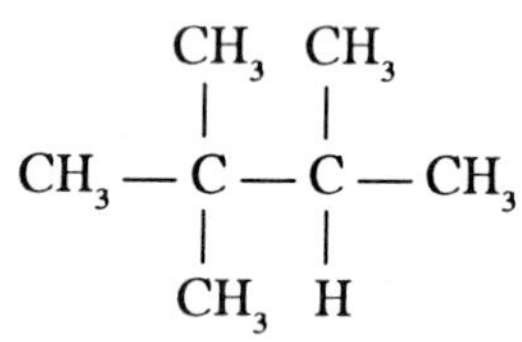

b. 2–methylpentane from any monobrominated propane.
c. 2,5–dimethylhexane from any monobrominated butane.
d. 2–methylbutane from ethane and/or propane.

2.13 Use the bond dissociation energies from the table in question 1.31 to answer the following questions.

a. Is the following reaction exothermic or endothermic?

$$Cl_2 + CH_3CH_3 \longrightarrow 2CH_3Cl$$

b. Is the bromination reaction exothermic or endothermic?

$$Br_2 + CH_3CH_3 \longrightarrow 2CH_3Br$$

c. There are two possible mechanisms in the chlorination process:

Mechanism A

Initiating step — Step 1. $Cl{-}Cl \longrightarrow 2\,Cl^{\cdot}$

Propagating steps — Step 2. $Cl^{\cdot} + CH_3CH_2{-}H \longrightarrow CH_3CH_2^{\cdot} + HCl$

Step 3. $CH_3CH_2^{\cdot} + Cl{-}Cl \longrightarrow CH_3CH_2Cl + Cl^{\cdot}$

Mechanism B

Initiating step — Step 1. $Cl{-}Cl \longrightarrow 2\,Cl^{\cdot}$

Propagating steps — Step 2. $Cl^{\cdot} + CH_3CH_3 \longrightarrow CH_3^{\cdot} + CH_3Cl$

Step 3. $CH_3^{\cdot} + Cl{-}Cl \longrightarrow CH_3{-}Cl + Cl^{\cdot}$

Calculate ΔH° for each step to determine which mechanism is more likely.

d. Which mechanism would be more likely in a bromination reaction?

2.14 Using the data from problem 2.13, draw the energy vs. progress of reaction graphs for the following processes:

a. $2\ Cl^{\cdot} \longrightarrow Cl_2$

b. $Cl_2 \longrightarrow 2\ Cl^{\cdot}$

c. $CH_3{-}H\ +\ Cl^{\cdot} \longrightarrow CH_3^{\cdot}\ +\ HCl$

d. $CH_3^{\cdot}\ +\ Cl{-}Cl \longrightarrow CH_3{-}Cl\ +\ Cl^{\cdot}$

e. $CH_3Cl\ +\ Cl^{\cdot} \longrightarrow {}^{\cdot}CH_2Cl\ +\ HCl$

2.15 Sketch the Newman projections of both chair and boat conformations of cyclohexane, sighting down the $C_2{-}C_3$ bonds.

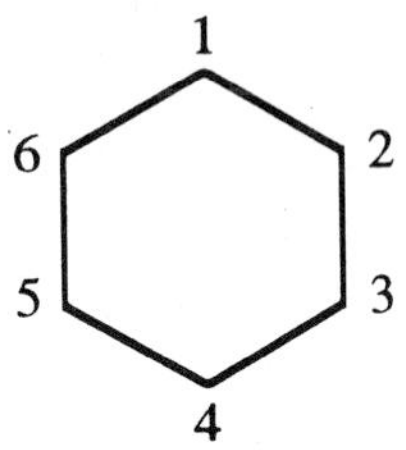

2.16 Cyclobutane is not flat but puckered into the shape shown. The bond angle in this conformation is reduced to 88° from 90°. How does this affect angle strain? Why then does cyclobutane adopt this shape?

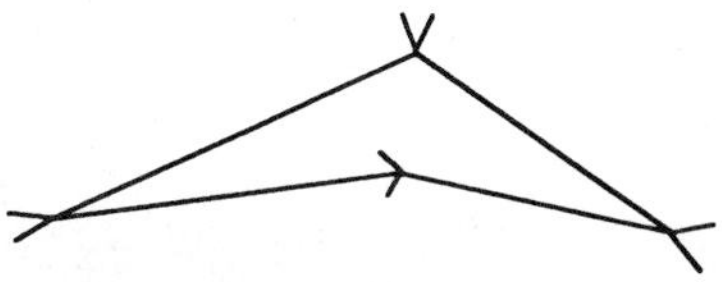

2.17 a. The heat of combustion (ΔH_c°) of *trans*–1,3–dimethylcyclohexane is greater than that of *cis*–1,3–dimethylcyclohexane. Draw the chair structure for each compound and explain the reason for the difference in ΔH_c°.

b. The heats of combustion of *n*–butane and isobutane are respectively –635 kcal/mole and –633 kcal/mole. Assuming that at 25°C the difference in ΔH_c° approximates the value of ΔG° for the equilibrium *n*–butane $\rightleftharpoons$ isobutane, calculate the value of K_{eq}. (R = 2 cal/mole K)

2.18 Compare the relative stabilities of *cis*– and *trans*–1,3–di-*tert*–butylcyclohexane. Explain why the *trans* isomer adopts the twist-boat conformation rather than the chair.

2.19 *Trans*–3–*cis*–4–dibromo–*t*–butylcyclohexane has a smaller dipole moment than its diastereomer, *cis*–3–*trans*–4–dibromo–*t*–butylcyclohexane. Write the chair structure for each and offer an explanation for the difference in dipole moments.

2.20 a. Compare conformations I and II and calculate the potential energy difference between them using the following values for the energy differences between axial and equatorial conformations.

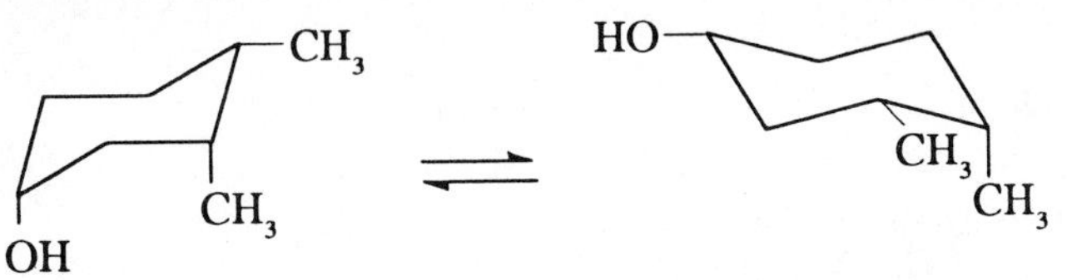

I II

CH_3/CH_3	gauche interaction	0.9 kcal/mole
CH_3/H	1,3–diaxial interaction	0.9 kcal/mole
CH_3/OH	1,3–diaxial interaction	2.3 kcal/mole
H/OH	1,3–diaxial interaction	0.25 kcal/mole

b. Assuming that this potential energy difference approximately equals ΔG° for the equilibrium, calculate the value of K_{eq} at 25°C. (given R = 2.0 cal/mole · K)

c. Using this value for K_{eq}, calculate the percentage of molecules in conformation II.

Multiple Choice Test

2.1 How many constitutional isomers are there for C_6H_{14}?
a. Three b. four c. five d. six e. seven

2.2 Which of the following has the highest boiling point?

a. $CH_3CH_2CH_2CH_2CH_2CH_3$

c. 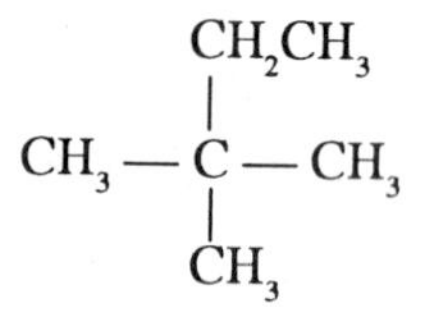

b. 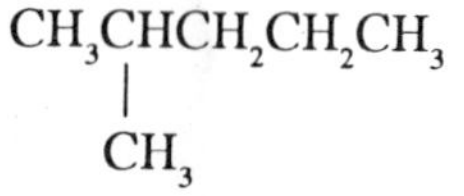

d. 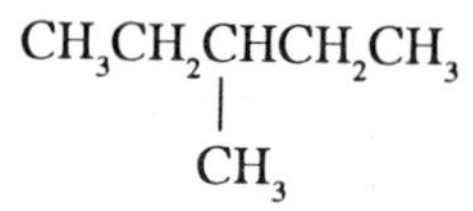

2.3 The hydrocarbon that has only 1° and 4° carbons is:
a. hexane b. 2-methylpentane c. 2, 2-dimethylpropane d. 2, 3-dimethylbutane

2.4 What is the correct IUPAC name of the compund shown below?

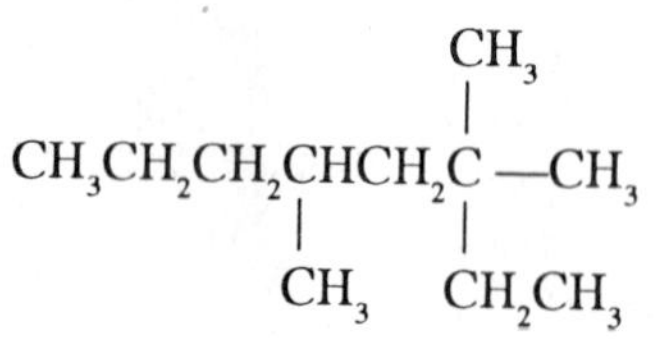

a. 4,6-dimethyl-6-ethylpentane
b. 2,4-dimethyl-2-ethylheptane
c. 3,3,5-trimethyloctane
d. 4,6-dimethyl-6-isobutylpentane

2.5 The eclipsed and staggered forms of ethane are said to differ in
a. configuration b. conformation c. resonance d. constitution

2.6 Which is the most stable radical?

a. 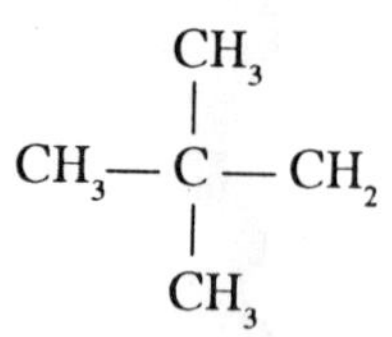

b. 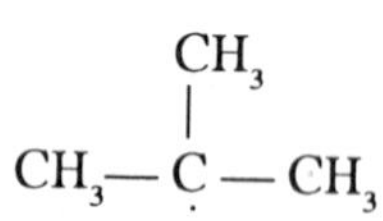

c. $CH_3\text{—}\dot{C}HCH_3$

d. 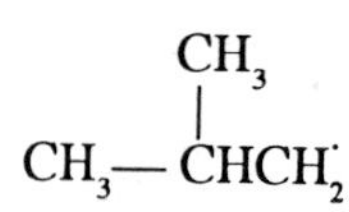

Consider the following steps for the bromination of methane in answering questions 2.7–2.9.

A. $Br^{\cdot} + CH_4 \longrightarrow CH_3^{\cdot} + HBr$
B. $Br^{\cdot} + CH_4 \longrightarrow CH_3Br + H^{\cdot}$
C. $CH_3^{\cdot} + Br_2 \longrightarrow CH_3Br + Br^{\cdot}$
D. $CH_3^{\cdot} + HBr \longrightarrow CH_3Br + H^{\cdot}$
E. $CH_3^{\cdot} + Br^{\cdot} \longrightarrow CH_3Br$
F. $Br_2 + UV \longrightarrow 2\,Br^{\cdot}$
G. $Br^{\cdot} + Br^{\cdot} \longrightarrow Br_2$
H. $CH_3^{\cdot} + CH_3^{\cdot} \longrightarrow CH_3CH_3$

2.7 Which of the above reactions is/are the initiation step(s)?
a. A b. G c. F and E d. F

2.8 Which of the above can be considered as termination step(s)?
a. E b. G c. H d. E, G, H

2.9 Which of the above are the acceptable chain-propagation steps?
a. A and B b. A and C c. B and C d. A, B, C and D

2.10 The C_6H_{14} alkane(s) with only 1° and 2° hydrogens include(s)
A. *n*-hexane **C.** 2,2-dimethylbutane
B. 2-methylpentane **D.** 2,3-dimethylbutane

a. only A b. A and C c. all but D d. all but B

2.11 Which reagent will yield 3-methylhexane when reacted with ethyl bromide?

a. $(CH_3CH_2CH_2)_2CuLi$

b. $((CH_3)_2CHCH_2)_2CuLi$

c. $(CH_3CH_2CHCH_2)_2CuLi$ (with CH_3 on the CH carbon)

d. $(CH_3CH_2)(CH_3CH_2CH_2)CuLi$

2.12 Compound C_5H_{12} forms four structurally different monochloroderivatives. C_5H_{12} is, therefore,
a. *n*-pentane b. 2-methylbutane c. 2, 2-dimethylpropane d. 2, 3-dimethylpropane

2.13 Cyclobutane is an example of molecule that shows
a. angle strain. b. 109.5° bond angles. c. angle and torsional strain. d. a stable ring structure.

2.14 UV bromination of isopentane yields ______ as the major product.
a. 1-bromo-3-methylbutane c. 2-bromo-2-methylbutane
b. 3-bromo-2-metrhylbutane d. 1-bromo-2-methylbutane

2.15 The stability order in the conformation of the central C-C bond of butane (most stable conformation first) is
a. eclipsed > gauche > anti c. gauche > eclipsed > anti
b. eclipsed > anti > gauche d. anti > gauche > eclipsed

2.16 Which of the following statement(s) for dimethylcyclohexane is/are correct?
A. *cis*-1,2-is more stable than *trans*-1,2.
B. *cis*-1,3-is more stable than *trans*-1,3.
C. *cis*-1,4-is more stable than *trans*-1,4.

a. B b. A and C c. C only d. all are correct

2.17 The compound shown is called
a. bicyclo[3.2.1]heptane c. bicyclo[2.1.1]heptane
b. tricyclo[2.1.1]octane d. bicyclo[3.2.1]octane

2.18 Cyclopentane exhibits more torsional strain than cyclohexane because
a. cyclopentane exists predominantly in the boat form.
b. its atoms are in a partially eclipsed conformation.
c. its atoms are in the staggered conformation.
d. its bond angles deviate greatly from 109°.

2.19 Consider the potential energy diagram for the various conformations of cyclohexane. The correct matching of each conformation with the proper letter in the diagram is

a. A-1, B-2, C-3

b. A-1, B-3, C-2

c. A-3, B-1, C-2

d. A-2, B-3, C-1

1 (halfchair)

2 (boat)

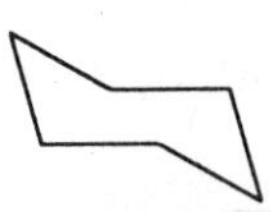

3 (chair)

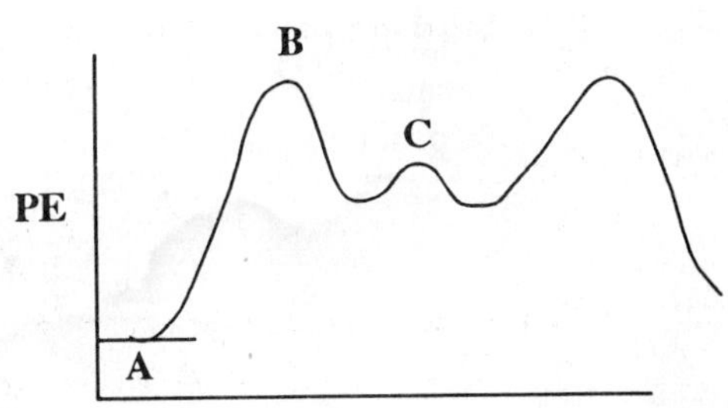

2.20 Which compound below is *sec*-butylcyclohexane?

a. 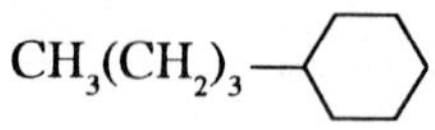

c. 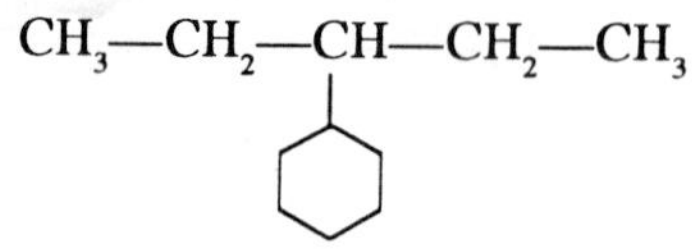

b. 

d. 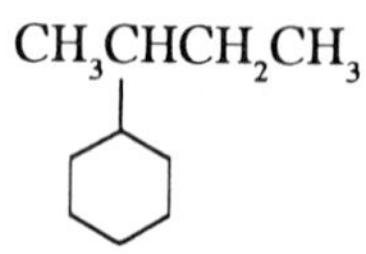

Answers for Questions

2.1 a. $CH_3CH_2—$

b. $CH_3CH_2CH_2—$

c. $(CH_3)_2CH—$ or

$$\begin{array}{l} CH_3—CH— \\ \quad\quad\;\; | \\ \quad\quad CH_3 \end{array}$$

d. $CH_3CH_2CH_2CH_2—$

e. $(CH_3)_2CH—CH_2—$ or

$$\begin{array}{l} CH_3—CH—CH_2— \\ \quad\quad\;\; | \\ \quad\quad CH_3 \end{array}$$

f.

$$\begin{array}{l} CH_3—CH—CH_2CH_3 \\ \quad\quad\;\; | \end{array}$$

g.

$$\begin{array}{l} \quad\quad CH_3 \\ \quad\quad\;\; | \\ CH_3—C— \\ \quad\quad\;\; | \\ \quad\quad CH_3 \end{array}$$

h.

$$\begin{array}{l} \quad\quad CH_3 \\ \quad\quad\;\; | \\ CH_3—C—CH_2— \\ \quad\quad\;\; | \\ \quad\quad CH_3 \end{array}$$

2.2 a. Three pentanes of molecular formula C_5H_{12}:

$CH_3CH_2CH_2CH_2CH_3$
pentane
(*n*-pentane)

$$\begin{array}{l} \quad\quad CH_3 \\ \quad\quad\;\; | \\ CH_3—C—CH_3 \\ \quad\quad\;\; | \\ \quad\quad CH_3 \end{array}$$
2,2-dimethylpropane
(neopentane)

$$\begin{array}{l} CH_3—CH—CH_2CH_3 \\ \quad\quad\;\; | \\ \quad\quad CH_3 \end{array}$$
2-methylbutane
(isopentane)

b. Five hexanes of molecular formula C_6H_{14}:

$CH_3CH_2CH_2CH_2CH_2CH_3$

hexane
(*n*-hexane)

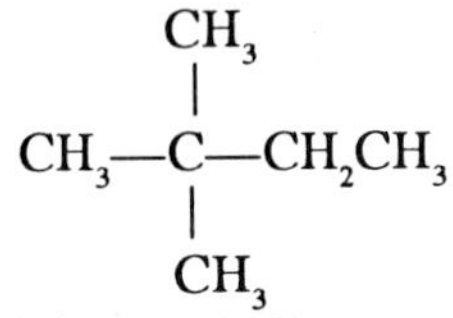

2,2-dimethylbutane

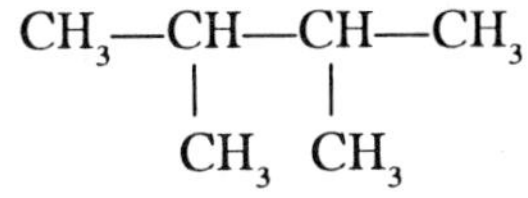

2,3-dimethylbutane

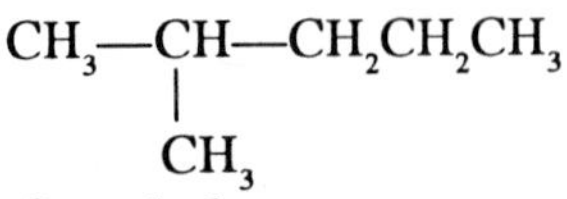

2-methylpentane
(isohexane)

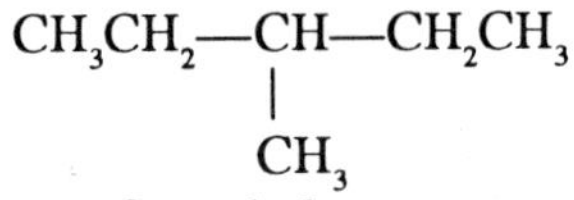

3-methylpentane

c. Nine heptanes of molecular formula C_7H_{16}:

$CH_3CH_2CH_2CH_2CH_2CH_2CH_3$

heptane
(*n*-heptane)

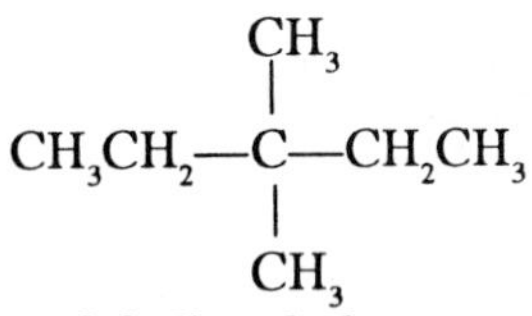

3,3-dimethylpentane

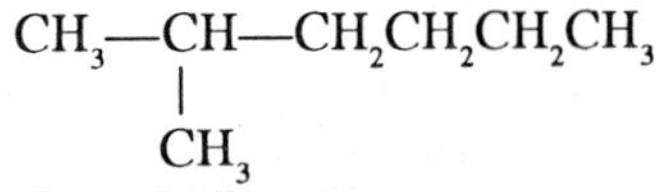

2-methylhexane
(isoheptane)

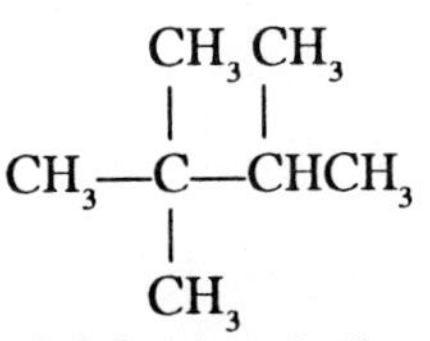

2,2,3-trimethylbutane

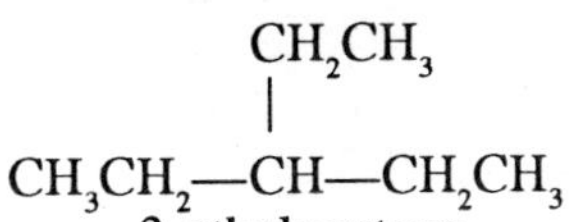

3-ethylpentane

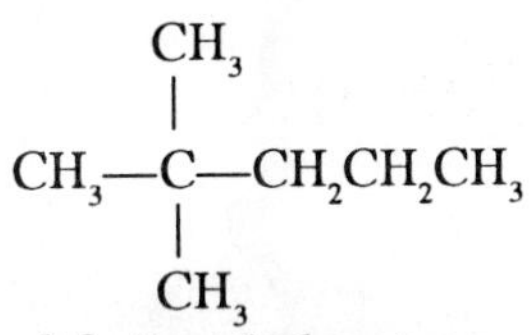

2,2-dimethylpentane

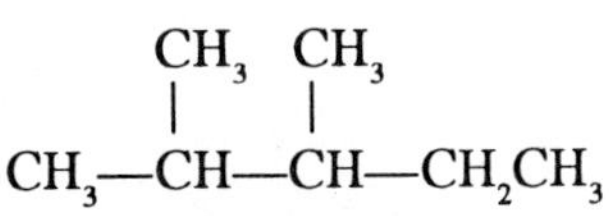

2,3-dimethylpentane

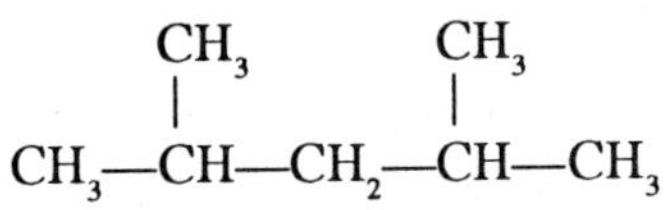

2,4-dimethylpentane

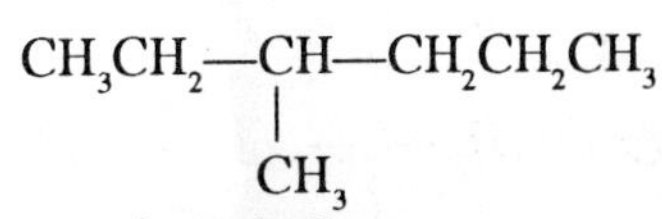

3-methylhexane

d. Eight monochloropentanes of molecular formula $C_5H_{11}Cl$:

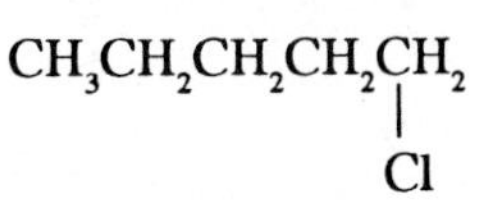

1-chloropentane

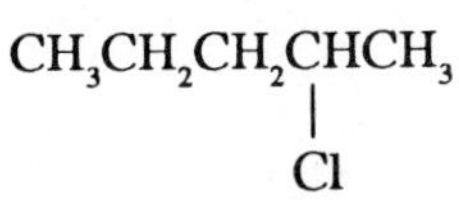

2-chloropentane

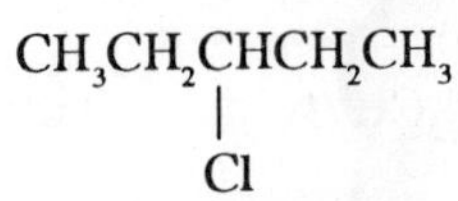

3-chloropentane

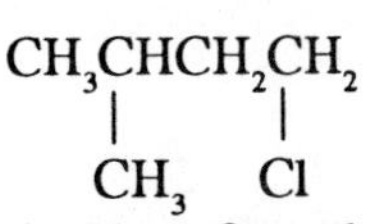

1-chloro-3-methylbutane

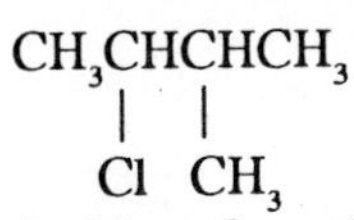

2-chloro-3-methylbutane

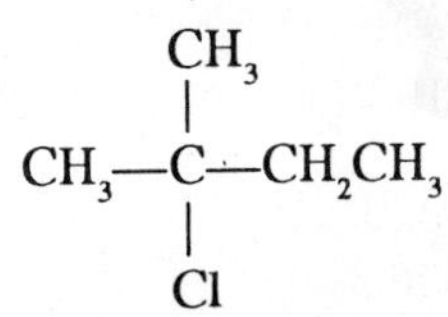

2-chloro-2-methylbutane

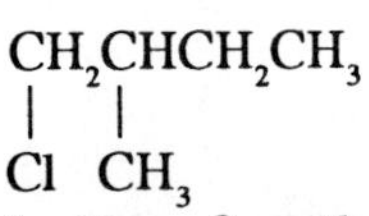

1-chloro-2-methylbutane

(note that the alternate structure

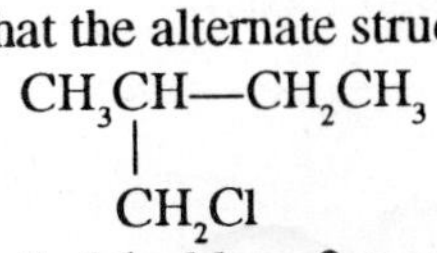

is also called 1-chloro-2-methylbutane)

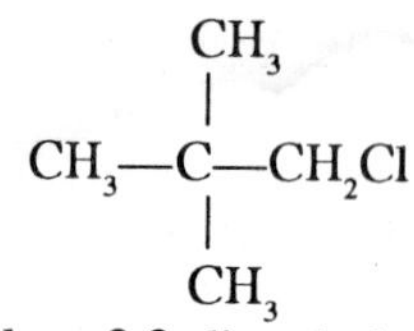

1-chloro-2,2-dimethylpropane

Which is equivalent to

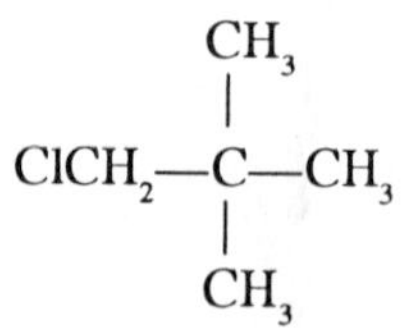 or 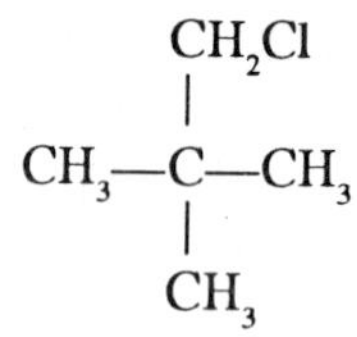 or 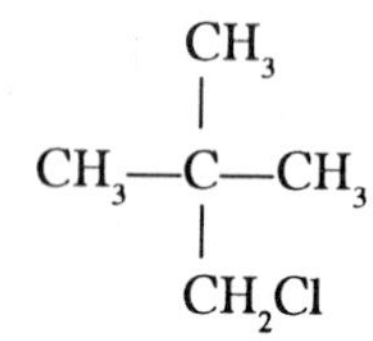

e. Nine dibromobutanes of molecular formula $C_4H_8Br_2$:

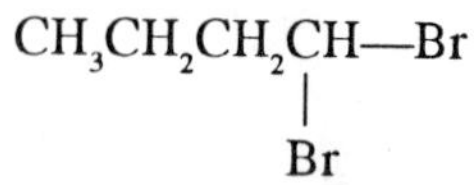

1,1-dibromobutane

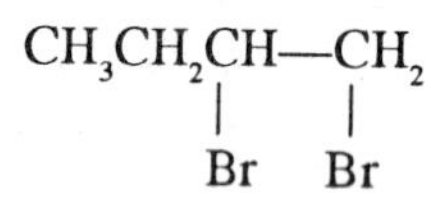

1,2-dibromobutane

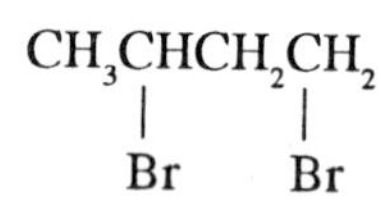

1,3-dibromobutane

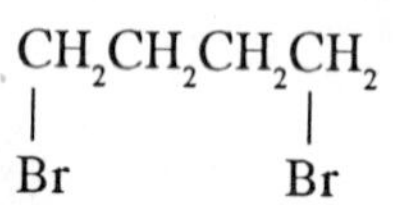

1,4-dibromobutane

$CH_3—CH_2—CH(Br)_2—CH_3$

2,2-dibromobutane

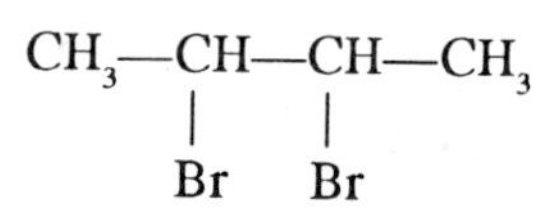

2,3-dibromobutane

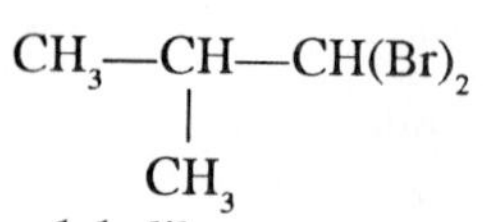

1,1-dibromo
2-methylpropane

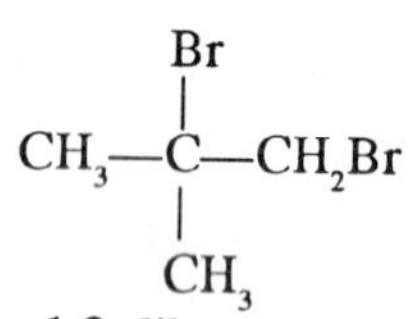

1,2-dibromo
2-methylpropane

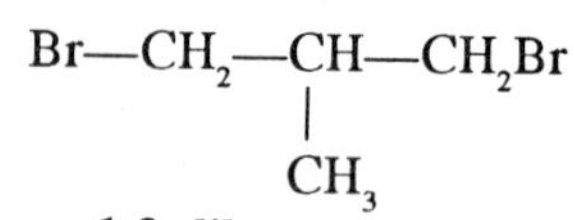

1,3-dibromo
2-methylpropane

f. Eight monobromochloroderivatives of *n*-butane:

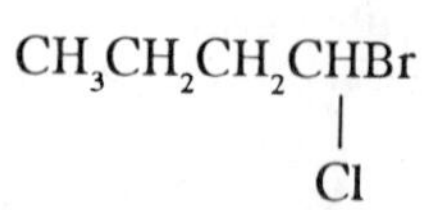

1-bromo-1-chlorobutane

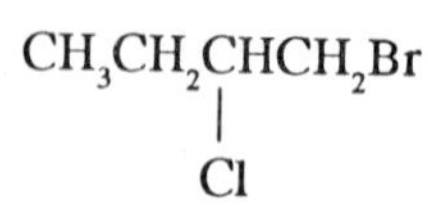

1-bromo-2-chlorobutane

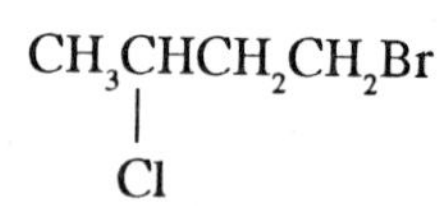

1-bromo-3-chlorobutane

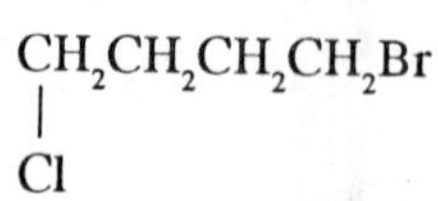

1-bromo-4-chlorobutane

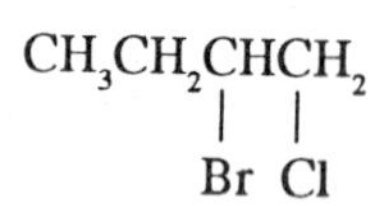

2-bromo-1-chlorobutane

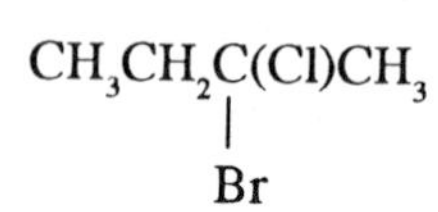

2-bromo-2-chlorobutane

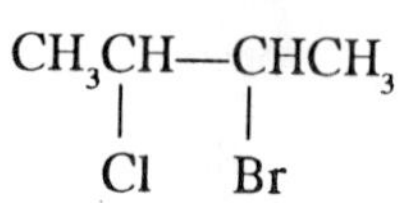

2-bromo-3-chlorobutane

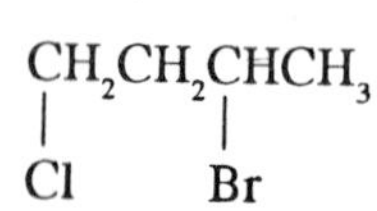

3-bromo-1-chlorobutane

g. Five monochlorination products of 2-methylpentane:

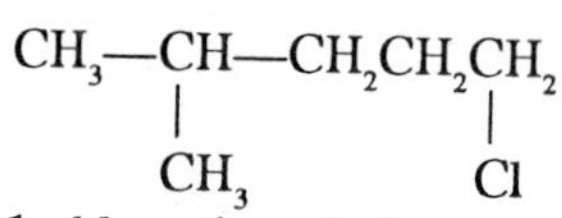

1-chloro-4-methylpentane

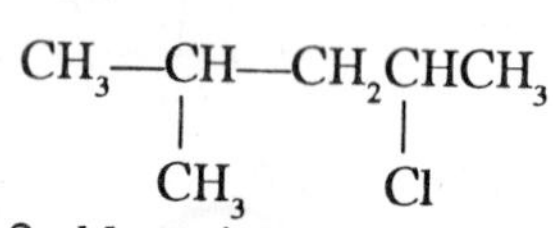

2-chloro-4-methylpentane

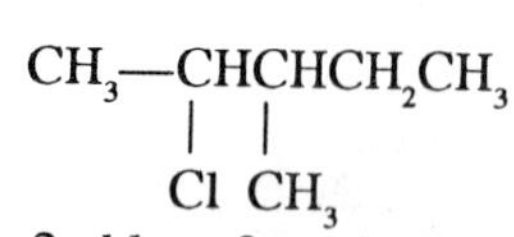

2-chloro-3-methylpentane

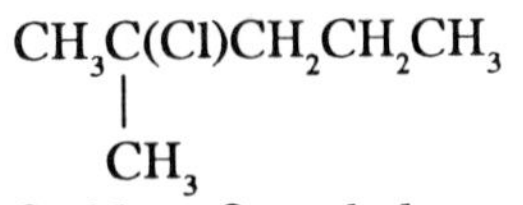
2-chloro-2-methylpentane

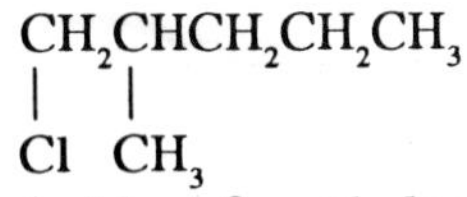
1-chloro-2-methylpentane

h. Five monochlorination products of 3-ethyl-2-methylpentane:

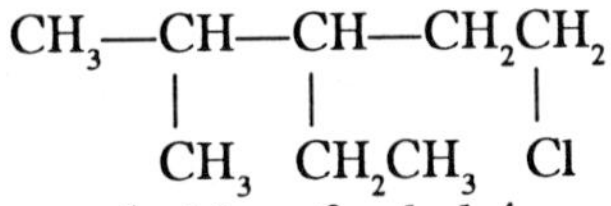
1-chloro-3-ethyl-4-methylpentane

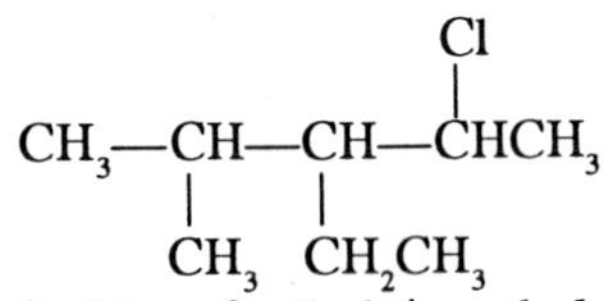
2-chloro-3-ethyl-4-methylpentane

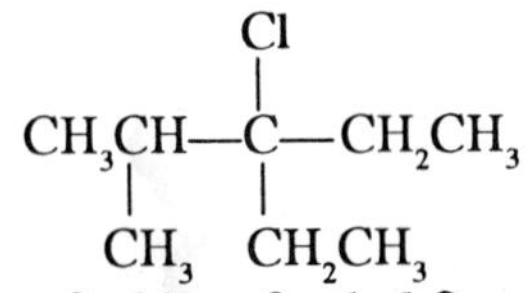
3-chloro-3-ethyl-2-methylpentane

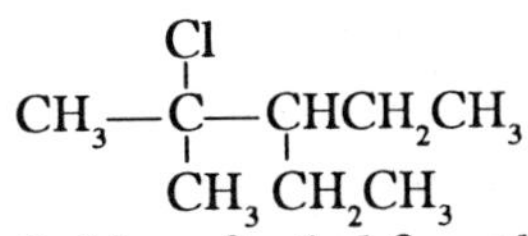
2-chloro-3-ethyl-2-methylpentane

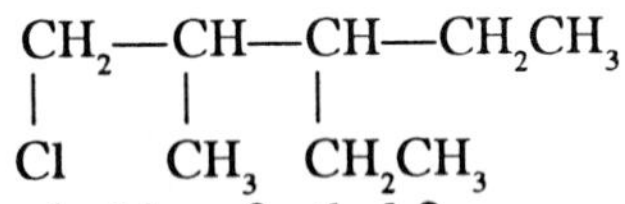
1-chloro-3-ethyl-2-methylpentane

i. Four dichlorinated derivatives of molecular formula $C_3H_6Cl_2$:

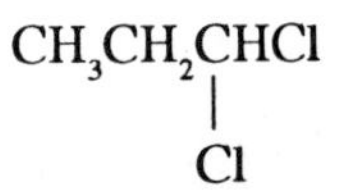
1,1-dichloropropane

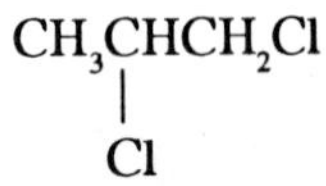
1,2-dichloropropane

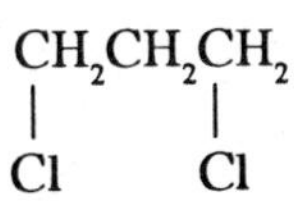
1,3-dichloropropane

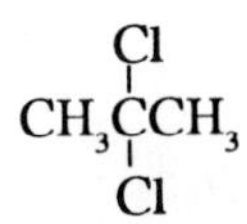
2,2-dichloropropane

2.3 a.

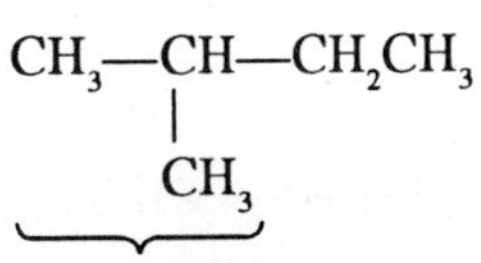
isopropyl group

b.

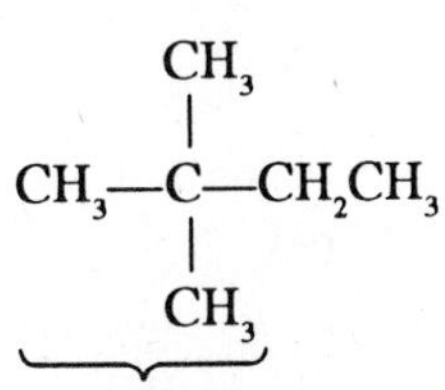
t-butyl group

c.

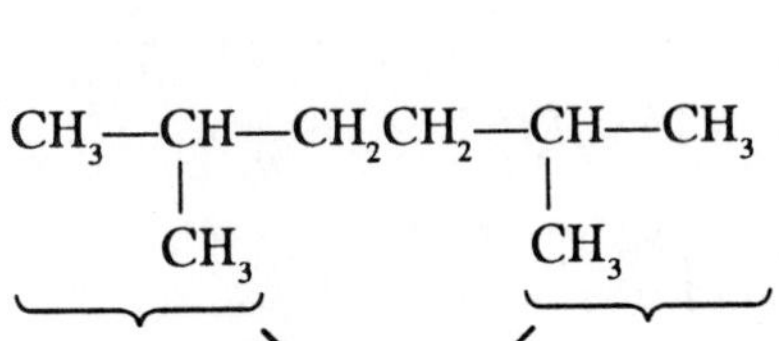
isopropyl groups

d.

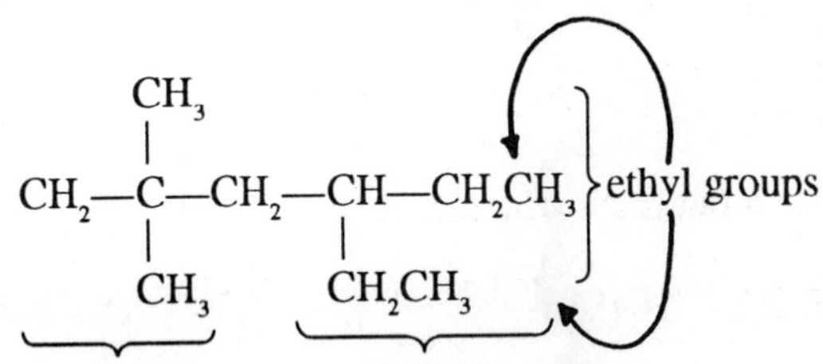

t-butyl group ethyl groups

e.

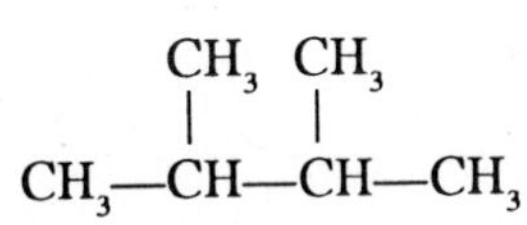

f.

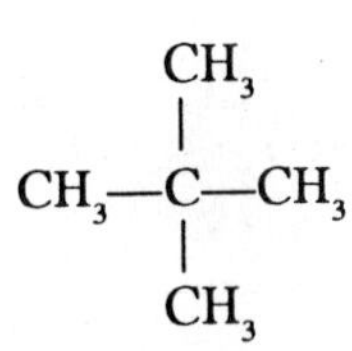

g. Molecular formula is C_6H_{14} and the structure is

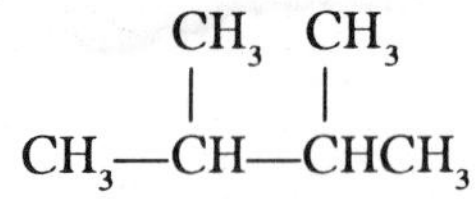

h. Molecular formula is C_5H_{12}, and the structure is $CH_3CH_2CH_2CH_2CH_3$.

i. $(CH_3)_3CCH_2CH_2CH_2CH_3$

j. $(CH_3)_3C—\dot{C}(CH_3)_2$

k. $BrCH_2CH_2Br$

2.4
a. neopentane (or 2,2-dimethylpropane)
b. 1,4-dibromobutane
c. 1,1,1,3,3,3-hexachloropropane
d. 2-iodopropane (or isopropyl iodide)
e. 2-chloro-4-methylpentane
f. 2,2,5,5-tetramethylhexane
g. 1,2-dicyclohexylethane
h. 4-ethylheptane
i. 1-chloro-2-methylbutane
j. 2,3-dimethylbutane
k. 2,3-dimethylpentane
l. 3-bromo-2,5-dichlorohexane
m. 2,2,4,5-tetraethylhexane
n. 3-ethyl-5-methyl-6-(2-methylbutyl) undecane
o. 1,1-dimethylcyclohexane
p. 3-bromo-1,1-dimethylcyclopentane
q. 1,2-dicyclopentylethane
r. cyclopropylcyclopentane
s. *cis*-1,2-dimethylcyclohexane
t. *cis*-3-*trans*-4-dibromo-1-*t*-butylcyclohexane
u. 4-isopropyl-2,2,6-trimethylheptane
v. 2,2-dimethylbutane
w. 2-(2-methylcyclopentyl) hexane
x. bicyclo[4.3.0]nonane
y. bicyclo[2.2.1]heptane
z. cis-1-iodo-4-methylcyclohexane

2.5
a. 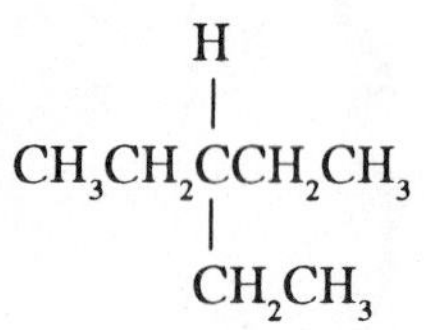

b. 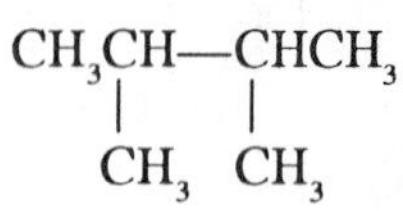

c. 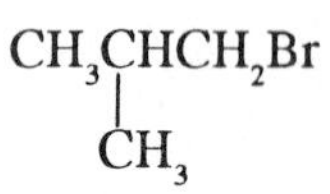

d. 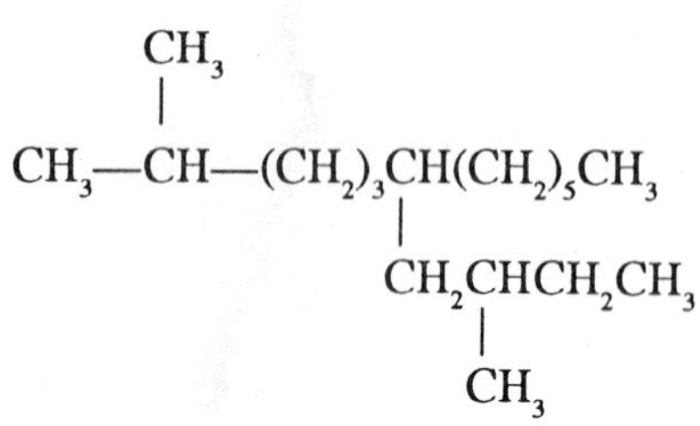

e. $(CH_3)_3C—Br$

f. $(CH_3)_3CCH_2Br$

g. 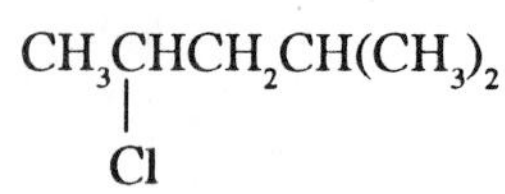

h. $HCCl_3$

i. 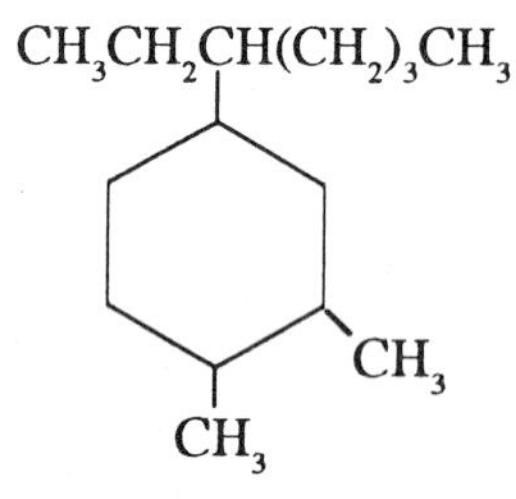

j. 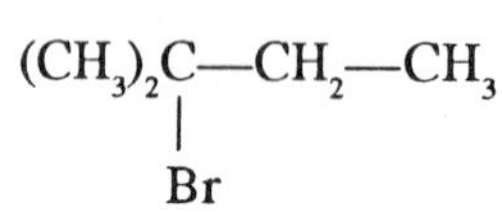

k. 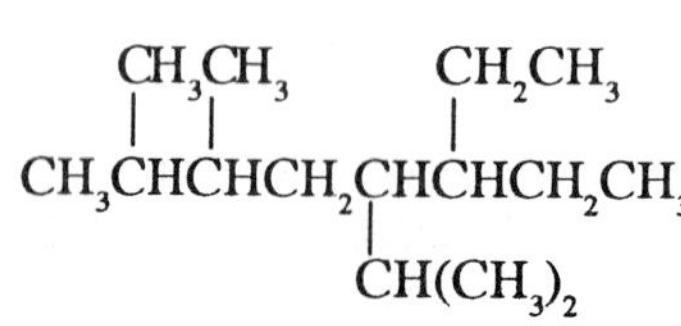

l. 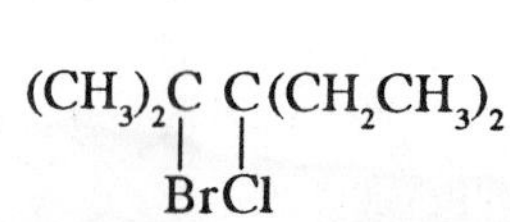

m. 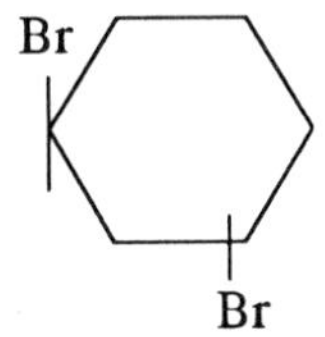

or 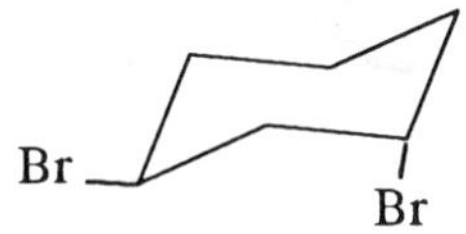

n. 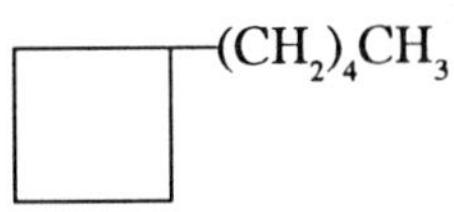

o. 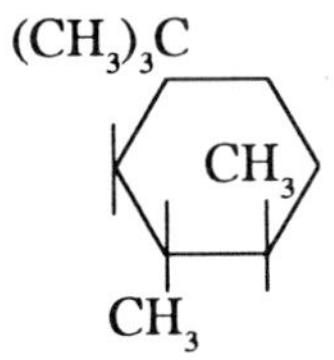

or 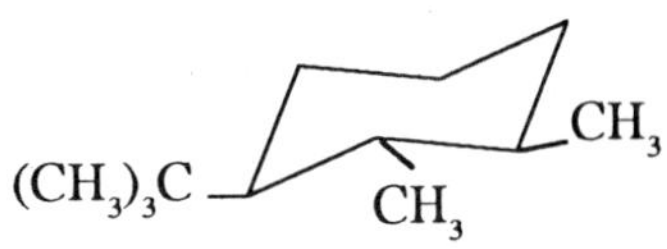

p. 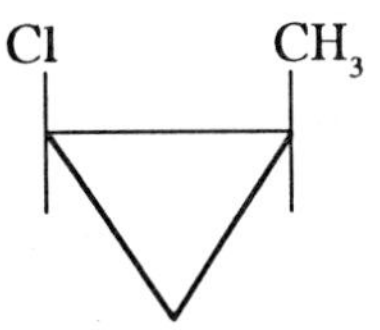

q. 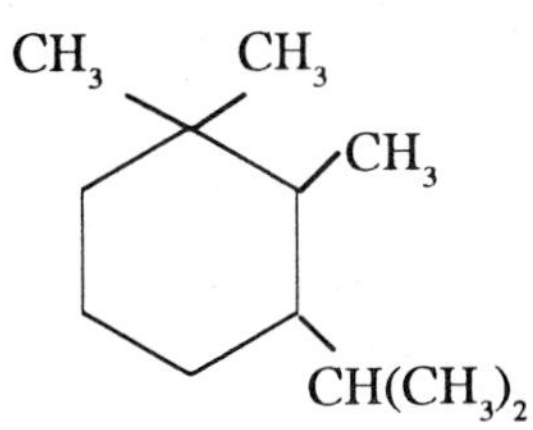

r. 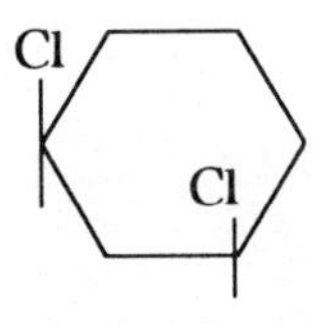

or 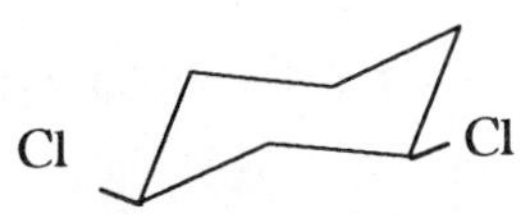

s. 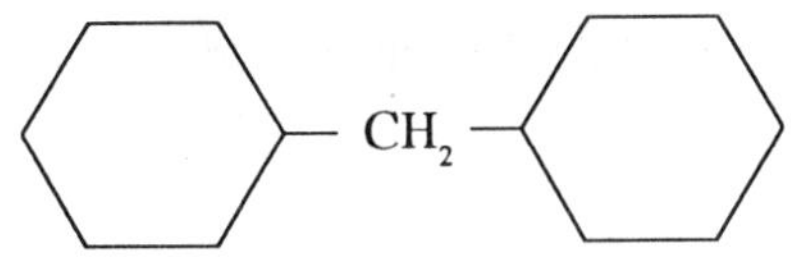

t. 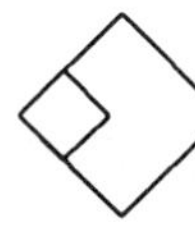

u. 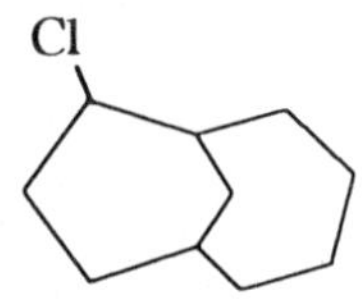

v. 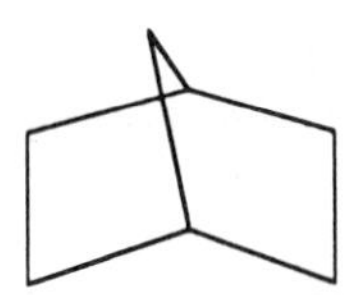
or 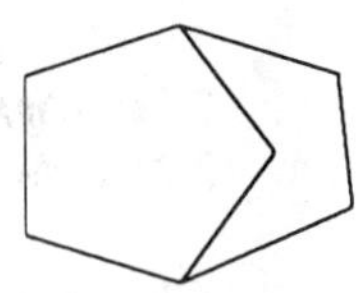

w. 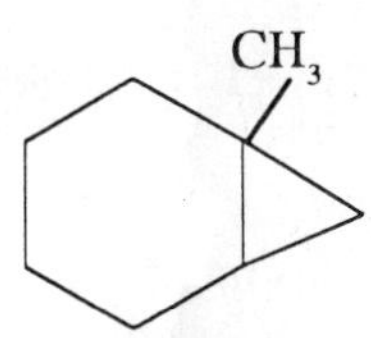

2.6 The general mechanism is:

initiation step: $Br_2 \xrightarrow{UV} 2\,Br^{\cdot}$

propragation steps: $Br^{\cdot} + CH_3{-}H \longrightarrow CH_3^{\cdot} + HBr$

$CH_3^{\cdot} + Br{-}Br \longrightarrow CH_3{-}Br + Br^{\cdot}$

When HBr is added, the following step can occur:

$CH_3^{\cdot} + H{-}Br \longrightarrow CH_3{-}H + Br^{\cdot}$ (reversal of first propagation step)

Thus, the second propagation step does not occur and no CH_3—Br is formed if enough HBr is available to react with all $CH_3^{\cdot}$ radicals.

2.7 a. $Cl^{\cdot} + CH_2Cl_2 \longrightarrow \dot{C}HCl_2 + HCl$

$^{\cdot}CHCl_2 + Cl{-}Cl \longrightarrow CHCl_3 + Cl^{\cdot}$

b. $Cl^{\cdot} + Cl^{\cdot} \longrightarrow Cl_2$

$Cl^{\cdot} + CHCl_2^{\cdot} \longrightarrow CHCl_3$

$CHCl_2^{\cdot} + CHCl_2^{\cdot} \longrightarrow CHCl_2{-}CHCl_2$

c. $Br_2 \xrightarrow{UV} 2\,Br^{\cdot}$

2.8 $Cl_2 \xrightarrow{UV} 2\,Cl^{\cdot}$

$Cl^{\cdot} + CH_3{-}H \longrightarrow CH_3^{\cdot} + HCl$

$CH_3^{\cdot} + Cl{-}Cl \longrightarrow CH_3{-}Cl + Cl^{\cdot}$

} normal initiation and propagation steps

$CH_3^{\cdot} + CH_3^{\cdot} \longrightarrow CH_3{-}CH_3$
ethane

} terminating step

$CH_3{-}CH_3 + Cl^{\cdot} \longrightarrow CH_3{-}CH_2^{\cdot} + HCl$

$CH_3{-}CH_2^{\cdot} + Cl{-}Cl \longrightarrow CH_3{-}CH_2Cl + Cl^{\cdot}$
Ethyl chloride

2.9 a. $CH_3CH_2CH_2Br$ and $CH_3CH\,(Br)\,CH_3$
b. $CH_3CH\,(Br)\,CH_3$
c. $CH_3CH_2CH_2Br$ (1°H abstracted) $6 \times 1 = 6$ $CH_3CH(Br)CH_3$ (2°H abstracted) $2 \times 82 = 164$

The relative rates (per hydrogen atom) of hydrogen abstraction by bromine are 1600 : 82 : 1 for 3° : 2° : 1° hydrogen atoms respectively.

$CH_3CH_2CH_2Br \quad \dfrac{6}{(164+6)} \times 100 = \dfrac{6}{170} \times 100 = 3.53\%$

CH_3CHCH_3 (with Br on the central carbon) $\quad \dfrac{164}{(164+6)} \times 100 = \dfrac{164}{170} \times 100 = 96.47\%$

d. Ratio of C in CO_2 = (12/44)
Mass of C in 0.528 g of the third compound = (12/44) × 0.528 = 0.144 g of C

Ratio of H in H_2O = (2/18)
Mass of H in 0.216 g of the third compound = (2/18) × 0.216 = 0.024 g of H

Mass of Br in 0.808 g of compound = 0.808 – (0.144 + 0.024) = 0.640 g

gram atoms C = 0.144/12 = 0.012
gram atoms H = 0.024/1 = 0.024
gram atoms Br = 0.640/80 = 0.008

Dividing by the smallest number (0.008) leads to the empirical formula of $C_3H_6Br_2$

e. Since only one monochloroderivative is formed, the structural formula is $CH_3CBr_2CH_3$.
f. The structural formula of the monochloro derivative is $CH_3CBr_2CH_2Cl$.

2.10 a. **A** = $CH_3CH_2CH_2CHMgBr$ (with CH_3 on the CH carbon) **B** = $CH_3(CH_2)_3CH_3$

b. **C** = $(CH_3)_2CCH_2CH_3$ (Br on the central C)
$\quad\quad\quad\quad\; |$
$\quad\quad\quad\quad Br$

F = $(CH_3)_2CCH_2CH{=}CH_2$
$\quad\quad\quad\quad |$
$\quad\quad\quad CH_2CH_3$

D = $(CH_3)_2CCH_2CH_3$
$\quad\quad\quad\quad |$
$\quad\quad\quad\quad Li$

$\quad\quad\quad\quad CH_3$
$\quad\quad\quad\quad |$
G = $(CH_3CH_2C)_2Cu\,Li$
$\quad\quad\quad\quad |$
$\quad\quad\quad\quad CH_3$

E = H_2O (or any other compound with an acidic H)

$R' = (CH_3)_3CCH_2—$

c. **H** = $CH_3CH_2—I$ (—Cl or —Br will also work)

d. **I** = $CH_3—CH_2—CH_3$

M = $CH_3CH_2CH_3$

J = $CH_3—CH—CH_3$
$\quad\quad\quad\quad |$
$\quad\quad\quad\quad Br$

N = MgBr(OH)

K = $CH_3—CH_2—CH_2—Br$

O = $CH_3(CH_2)_4CH_3$

L = $(CH_3)_2CHMgBr$

e. **P** = $(CH_3)_2CHCHCH_3$
$\quad\quad\quad\quad\quad |$
$\quad\quad\quad\quad\quad Br$

T = $(CH_3)_2CHCHCH_3$
$\quad\quad\quad\quad\quad |$
$\quad\quad\quad\quad\quad CH_3$

Q = $(CH_3)_2CCH_2CH_3$
$\quad\quad\quad\quad |$
$\quad\quad\quad\quad Br$

U = $((CH_3)_2CHCH)_2Cu\,Li$
$\quad\quad\quad\quad\quad\quad |$
$\quad\quad\quad\quad\quad\quad CH_3$

R = $(CH_3)_2CHCHLi$
$\quad\quad\quad\quad\quad |$
$\quad\quad\quad\quad\quad CH_3$

V = $CH_3CH_2—I$ (—Cl or —Br will also work)

S = H^+/H_2O

2.11 The problem is solved for the first compound. All others follow the same pattern and only the final answers are given.

a. $CH_3—CH_2—CH_3 + X_2 \xrightarrow{h\nu} \underset{\textit{n}\text{-propyl halide}}{CH_3—CH_2—\underset{X}{\underset{|}{CH_2}}} + \underset{\text{isopropyl halide}}{CH_3—\underset{X}{\underset{|}{CH}}—CH_3}$

The ratio of products is found from the number of hydrogens available for abstraction and the relative rates at which they are abstracted. There are six 1° hydrogens available for the formation of the *n*-propyl halide and two 2° hydrogens available for the isopropyl halide.

Chlorination

$$\frac{n\text{-propyl chloride}}{\text{isopropyl chloride}} = \frac{\#\text{ of }1^\circ\text{ H}}{\#\text{ of }2^\circ\text{ H}} \times \frac{\text{rel. abstraction rate }1^\circ\text{ H}}{\text{rel. abstraction rate }2^\circ\text{ H}} = \frac{6}{2} \times \frac{1}{3.8} = \frac{6}{7.6}$$

n-propyl chloride: $^{6}/_{13.6} \times 100 = 44.1\%$

isopropyl chloride: $^{7.6}/_{13.6} \times 100 = 55.9\%$

Bromination

$$\frac{n\text{-propyl bromide}}{\text{isopropyl bromide}} = \frac{6}{2} \times \frac{1}{82} = \frac{6}{164}$$

n-propyl bromide: $\frac{6}{170} \times 100 = 3.5\%$

isopropyl bromide: $\frac{164}{170} \times 100 = 96.5\%$

b. 28.3% 1-chlorobutane; 1.8% 1-bromobutane
71.7% 2-chlorobutane; 98.2% 2-bromobutane
c. 64.3% isobutyl chloride; 0.6% isobutyl bromide
35.7% *t*-butyl chloride; 99.4% *t*-butyl bromide
d. 21.0% 1-chloropentane; 1.2% 1-bromopentane
53.2% 2-chloropentane; 65.9% 2-bromopentane
26.6% 3-chloropentane; 32.9% 3-bromopentane
e. 13.9% 1-chloro-3-methylbutane; 0.2% 1-bromo-3-methylbutane
35.2% 2-chloro-3-methylbutane; 9.2% 2-bromo-3-methylbutane
23.1% 2-chloro-2-methylbutane; 90.2% 2-bromo-2-methylbutane
27.8% 1-chloro-2-methylbutane; 0.3% 1-bromo-2-methylbutane

2.12 a.

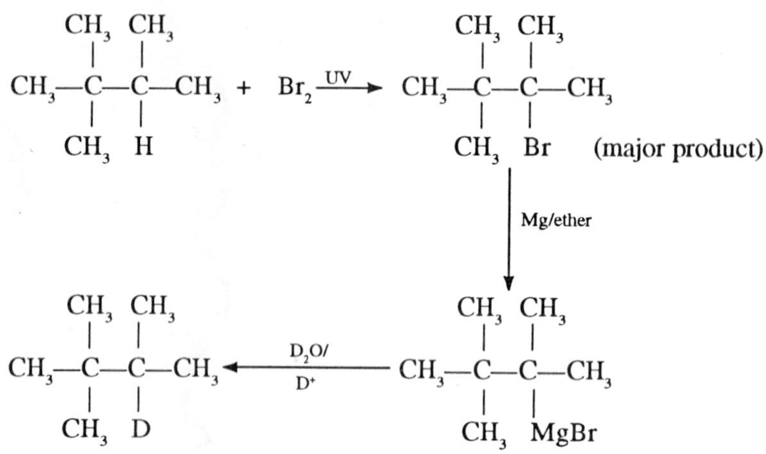

Note that chlorination will yield a greater percentage of 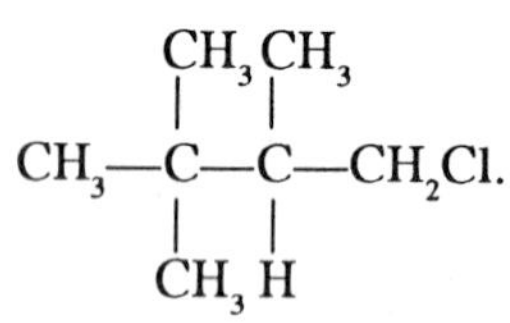

Chlorine is less selective than bromine in abstracting a 3° hydrogen. Thus chlorination is not recommended.

b. The structure of 2-methylpentane is 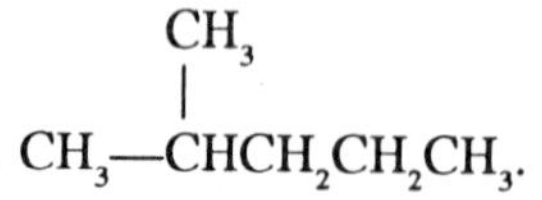

Since the final product is a six-carbon alkane one should follow the Corey-House synthesis:

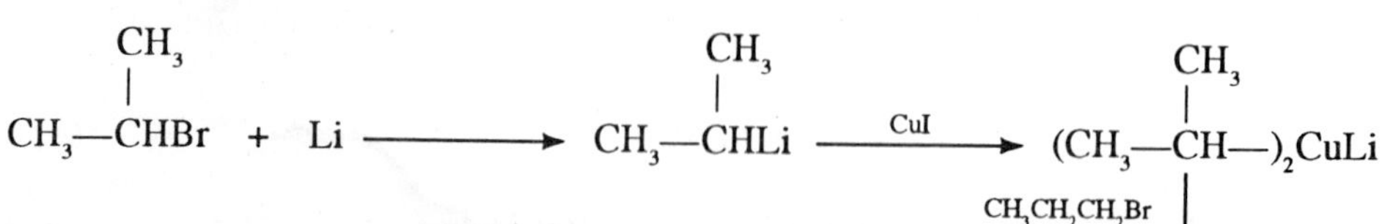

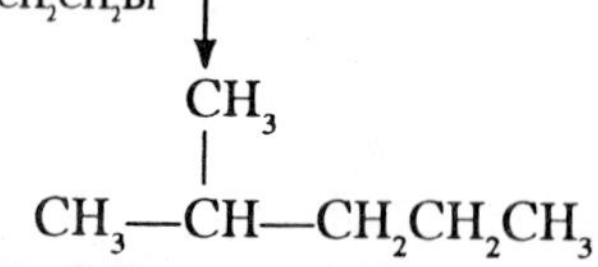

Note that the final step must use a 1° alkyl halide (in this case $CH_3CH_2CH_2Br$).

c. The structure of 2,5-dimethylhexane is $CH_3CH(CH_3)CH_2CH_2CH(CH_3)CH_3$

The molecule is symmetrical, which suggests isobutyl bromide as a starting compound in the Corey-House synthesis.

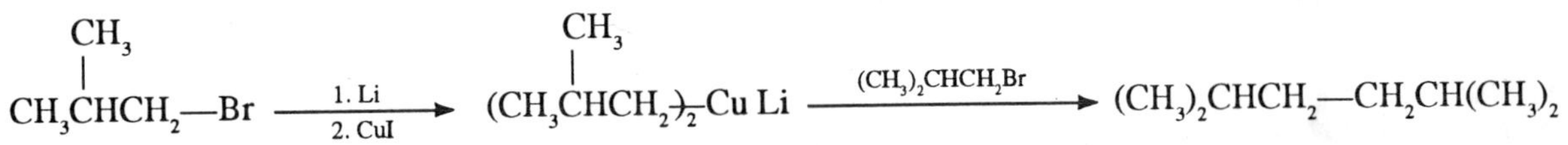

Note: although the yields are lower, the Wurtz reaction can also be used here since the molecule is symmetrical.

$$2(CH_3)_2CHCH_2Br \xrightarrow{2\,Na} (CH_3)_2CHCH_2{-}CH_2CH(CH_3)_2 + 2\,NaBr$$

d. The structure of 2-methylbutane is $(CH_3)_2CH{-}CH_2{-}CH_3$.

The versatile Corey-House synthesis is again the method of choice:

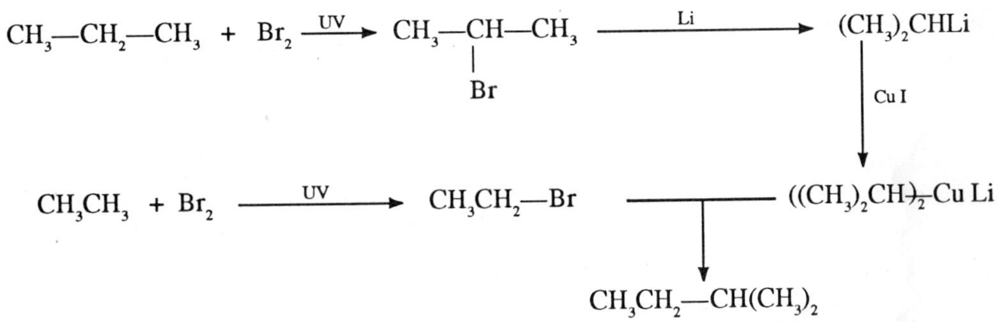

2.13 a.

	Cl_2	+	CH_3CH_3	→	$2CH_3Cl$
ΔH° (kcal/mole)	58		88		2 × 84

$\Delta H^\circ_{reaction}$ = −2(84) + 58 + 88 = −22 kcal/mole. Exothermic.

b.

	Br_2	+	CH_3CH_3	→	$2\,CH_3Br$
ΔH° (kcal/mole)	46		88		2 × 70

$\Delta H^\circ_{reaction}$ = −2(70) + 46 + 88 = −6 kcal/mole. Exothermic.

c. Since both mechanisms have the same initiating step, the factor that determines the most likely mechanism is the ΔH° value of the first *propagating* step (Step 2):

Mechanism A

Step 2.	$Cl^{\cdot}$	+	$CH_3CH_2{-}H$	→	$CH_3CH_2^{\cdot}$	+	H—Cl
ΔH° (kcal/mole)			98				−103

$\Delta H^\circ_{reaction}$ = −103 + 98 = −5 kcal/mole. Exothermic.

Mechanism B

Step 2.	$Cl^{\cdot}$	+	$CH_3{-}CH_3$	→	$CH_3^{\cdot}$	+	$CH_3{-}Cl$
ΔH° (kcal/mole)			88				−84

$\Delta H^\circ_{reaction}$ = −84 + 88 = +4 kcal/mole. Endothermic.

Thus, Mechanism A would be more likely because ΔH° of the determining step is less than zero (i.e., this step is exothermic).

d. Since both mechanisms have the same initiating step, the value of ΔH° for Step 2 will again be the determining factor:

Mechanism A

Step 2.	Br· +	CH_3CH_2—H ⟶	CH_3CH_2·	+	H—Br
ΔH° (kcal/mole)		98			−87

$\Delta H^\circ_{reaction}$ = −87 + 98 = +11 kcal/mole. Endothermic.

Mechanism B

Step 2.	Br· +	CH_3—CH_3 ⟶	CH_3· +	CH_3—Br
ΔH°		88		−70

$\Delta H^\circ_{reaction}$ = −70 + 88 = +18 kcal/mole. Endothermic.

Both are endothermic steps; the first one is a little less endothermic and, therefore, a little more likely.

2.14 a.

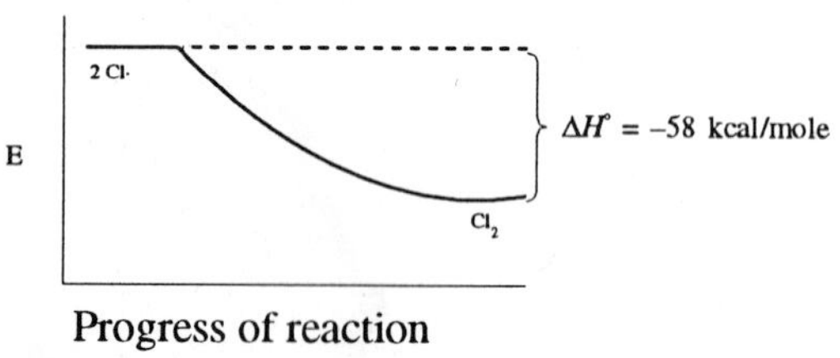

d.

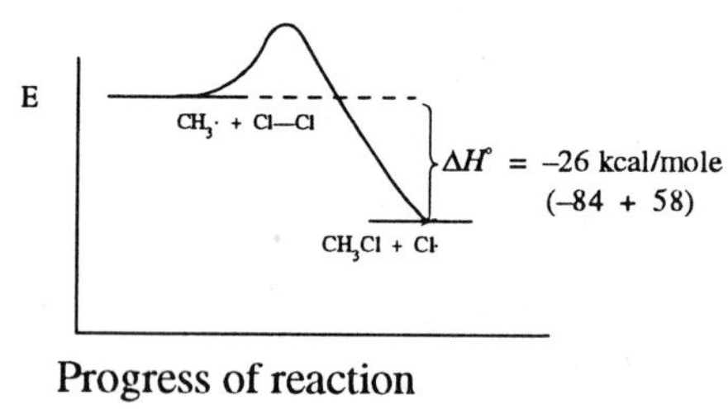

b.

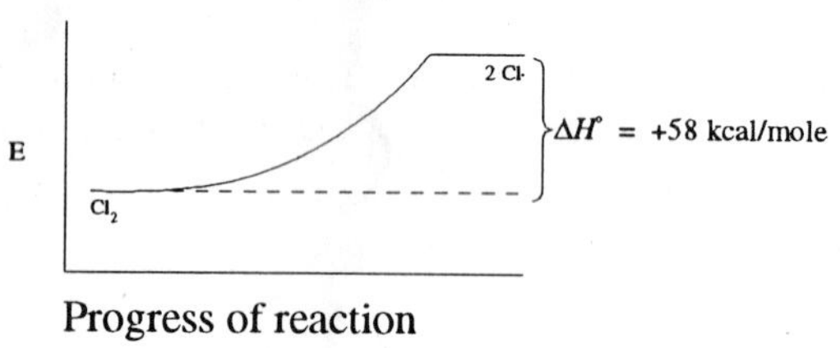

e.

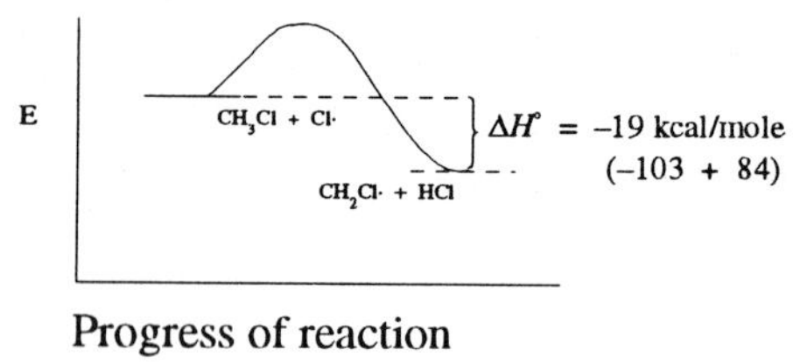

c.

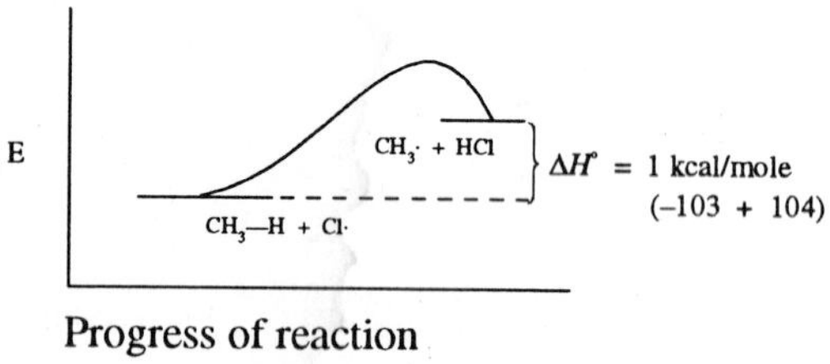

2.15 chair

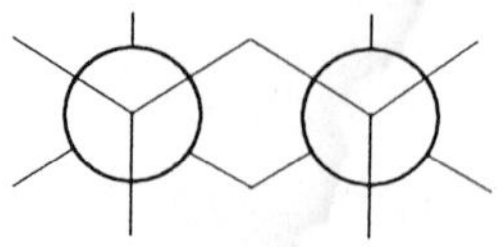

no eclipsed conformations

boat

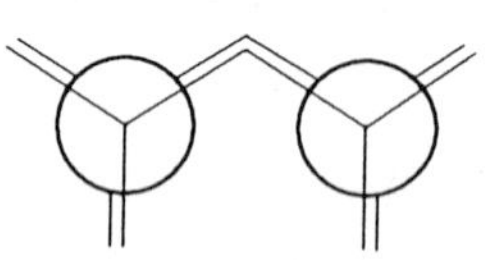

eclipsed conformations

2.16 The puckered ring increases angle strain, but reduces torsional strain and some steric strain by removing eclipsed conformations. Overall this conformation results in a lower potential energy.

2.17 a.

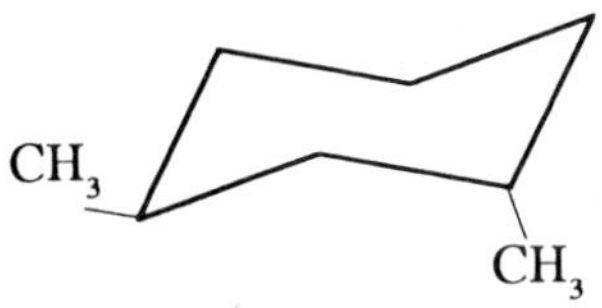

trans-1,3-dimethylcyclohexane

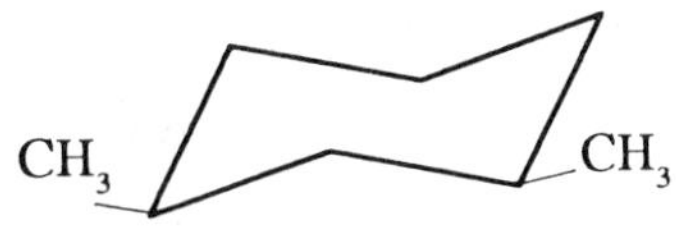

cis-1,3-dimethylcyclohexane

In both cases the following combustion reaction occurs: C_8H_{16} + $12O_2$ $\longrightarrow$ $8CO_2$ + $8H_2O$. Since the *trans* isomer gives off more heat it must contain more energy. In the *cis* isomer both CH_3 groups are equatorial, but in the *trans* isomer one CH_3 group is axial and it encounters 1,3-diaxial steric interactions, which increase its energy content.

b. For the equilibrium *n*-butane $\rightleftharpoons$ isobutane at 25°C
$\Delta G° \approx \Delta H° = -2$ kcal/mole = –2000 kcal/mole. Thus,

$$\ln K_{eq} = \frac{-\Delta G°}{RT} = \frac{2000 \text{ cal}}{(2 \text{ cal/mole K})(298 \text{ K})} = 3.36$$

$$K_{eq} = 28.7$$

2.18 *cis*-1,3-di-*t*-butylcyclohexane *trans*-1,3-di-*t*-butylcyclohexane

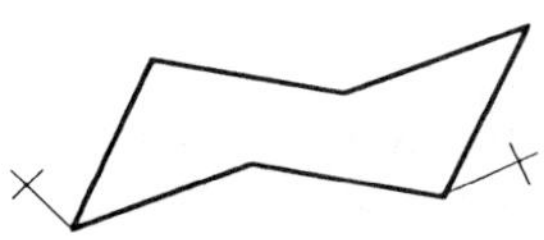

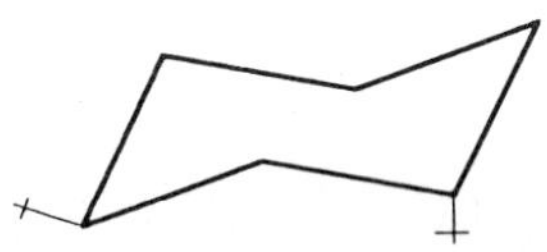

chair

as

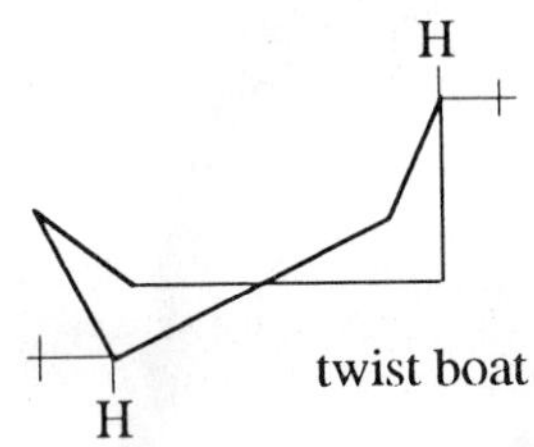

twist boat

The bulky *t*-butyl group would cause a large increase in potential energy if it occupied an axial position. For this reason, the *trans* isomer does not exist in the chair form but adopts the twist-boat conformation, where both *t*-butyl groups are nearly equatorial.

2.19

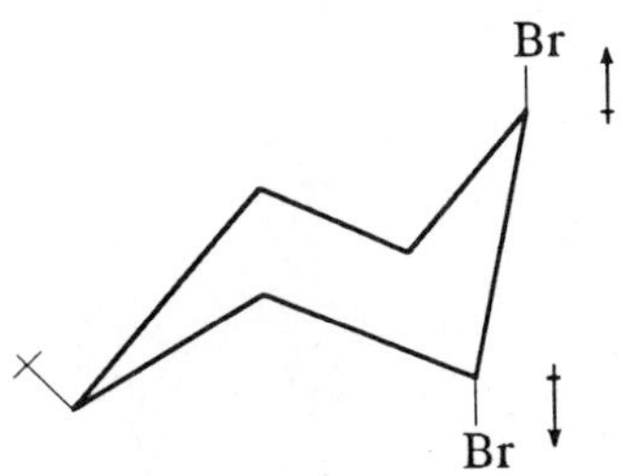

Trans-3-*cis*-4-dibromo-*t*-butyl-cyclohexane

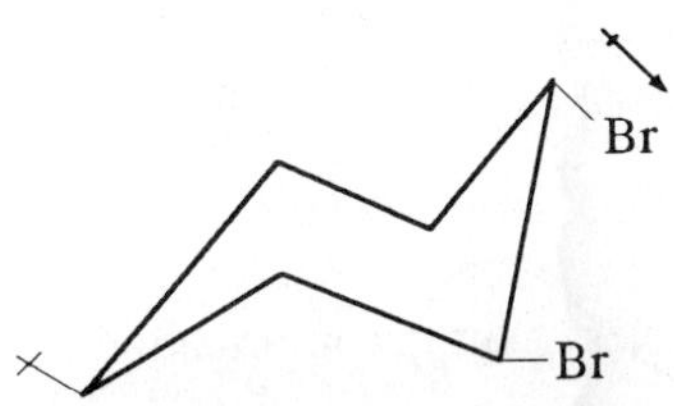

cis-3-*trans*-4-dibromo-*t*-butyl-cyclohexane

The *diaxial* bromine atoms are *anti* to each other. This arrangement reduces the dipole moment of this molecule.

2.20 The following interaction energies appear in conformations I and II.

a.

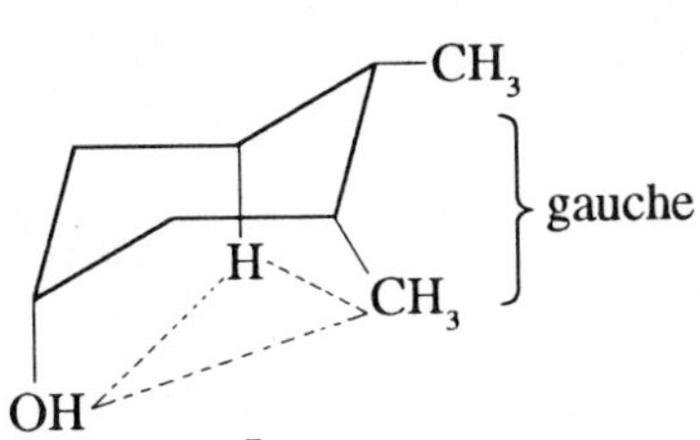

I

CH_3/CH_3	gauche	0.9	kcal/mole
CH_3/H	1,3-diaxial	0.9	kcal/mole
CH_3/OH	1,3-diaxial	2.3	kcal/mole
H/OH	1,3-diaxial	0.25	kcal/mole
		4.35	kcal/mole

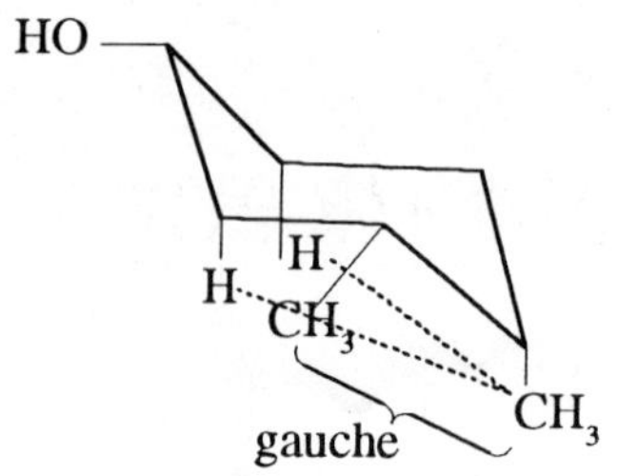

II

CH_3/CH_3	gauche	0.9	kcal/mole
two CH_3/H	1,3-diaxial	1.8	kcal/mole
		2.7	kcal/mole

Conformation II is more stable by 1.65 kcal/mole (4.35 – 2.7).

b.

$$\Delta G^\circ \approx -1650 \text{ cal/mole}$$

$$R = 2 \text{ cal/mole K}$$

$$T = 298 \text{ K}$$

$$\Delta G^\circ = -RT \ln K_{eq}$$

$$\ln K_{eq} = \frac{-\Delta G^\circ}{RT}$$

$$\ln K_{eq} = \frac{1650 \text{ cal/mole}}{(2)(298)} = 2.77 \quad \text{and} \quad K_{eq} = 16$$

c.

$$K_{eq} = \frac{[\,II\,]}{[\,I\,]} \quad \text{if } [\,I\,]_i = 1 \text{ mole/L; then } [\,II\,]_{eq} = x; \text{ and } [\,I\,]_{eq} = 1 - x$$

$$16 = \frac{x}{1-x}$$

$$x = 0.94$$

94% of molecules occur in conformation II.

Answers for Multiple Choice Test

2.1 c	**2.6** b	**2.11** c	**2.16** a
2.2 a	**2.7** d	**2.12** b	**2.17** d
2.3 c	**2.8** d	**2.13** c	**2.18** b
2.4 c	**2.9** b	**2.14** c	**2.19** c
2.5 b	**2.10** b	**2.15** d	**2.20** d

3 Alkenes and Cycloalkenes

Questions

3.1 Identify or give definitions for the following:

a. alkene
b. π-bond
c. unsaturated hydrocarbon
d. geometrical isomerism
e. catalytic hydrogenation
f. Markovnikov's rule

3.2 Give structures and names for the following:

a. all alkenes of molecular formula C_4H_8
b. all alkenes of molecular formula C_5H_{10}
c. all monochloro derivatives of 2-methyl-2-butene
d. an alkene (C_6H_{12}) whose hydrogens are all equivalent

3.3 Name the following compounds:

a. 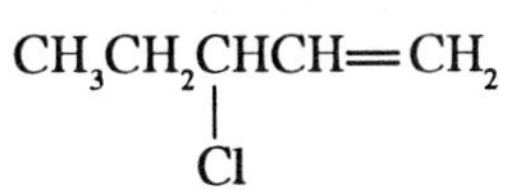

b. $(CH_3)_2CHCH{=}CH_2$

c. 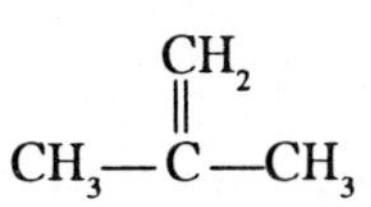

d. $(CH_3)_3CCH{=}CH_2$

e. 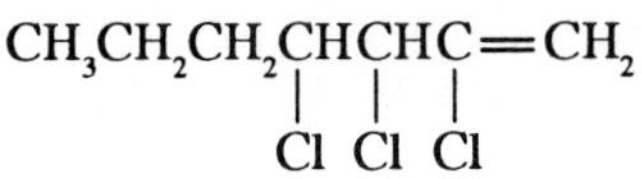

f. $CH_2{=}CHCl$

g. 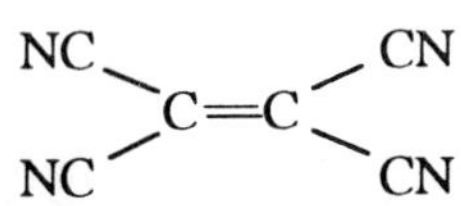

h. $(CH_3)_2C{=}C(CH_3)_2$

i. $CH_2{=}CH(CH_2)_5CH_3$

j. 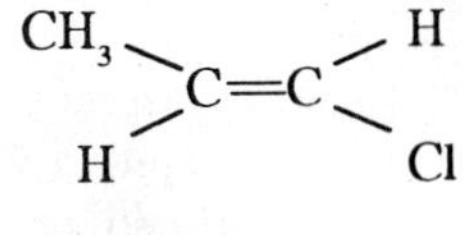

k. 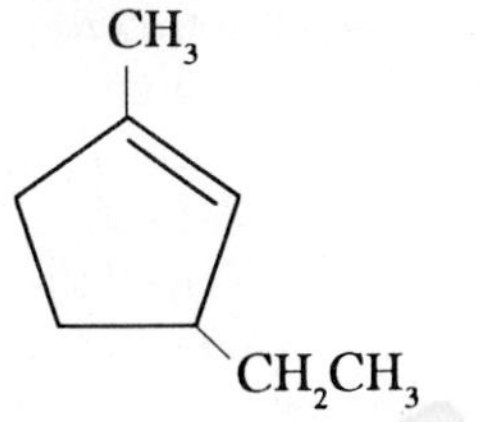

l. 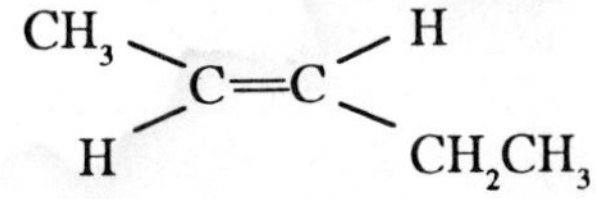

m.
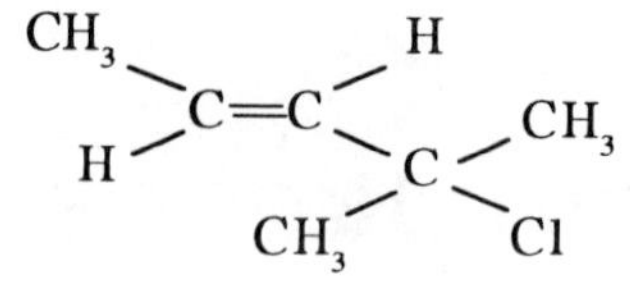

n.
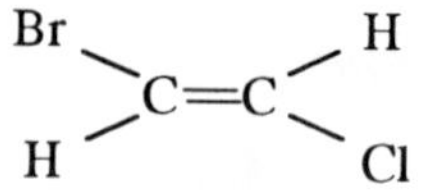

o.
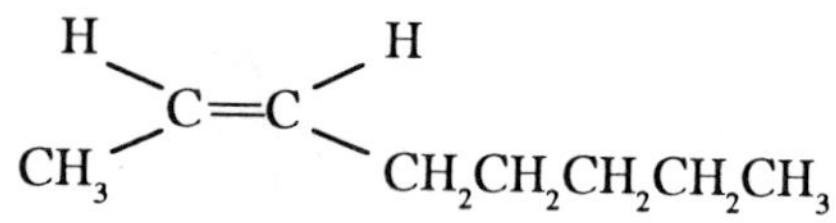

p.
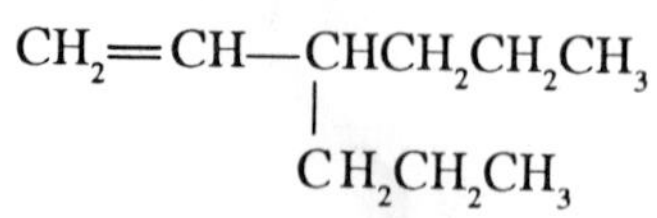

q.
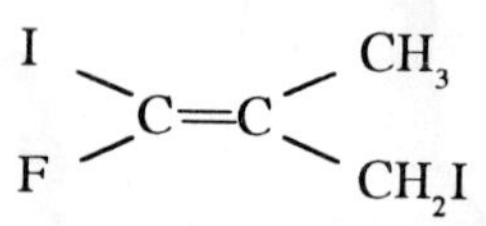

r. $(CH_3CH_2)_2C{=}CH_2$

s. $CCl_2{=}CH_2$

t.
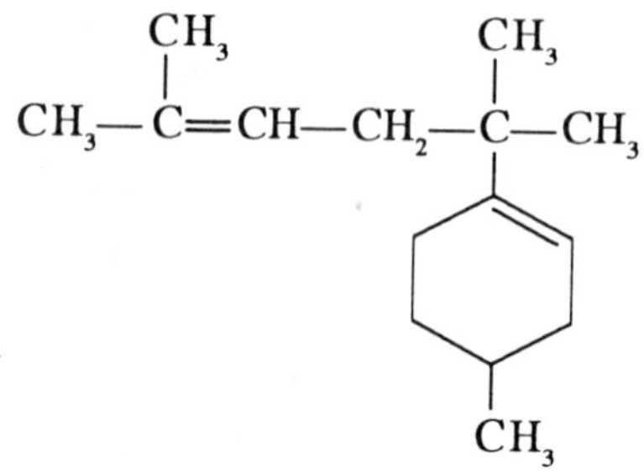

u.
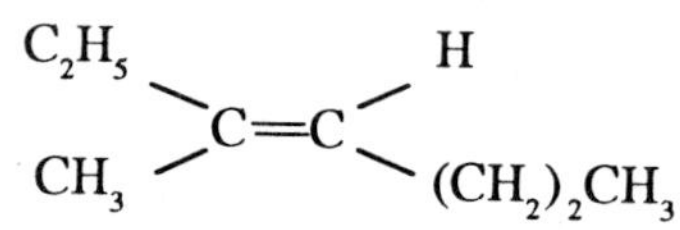

v.
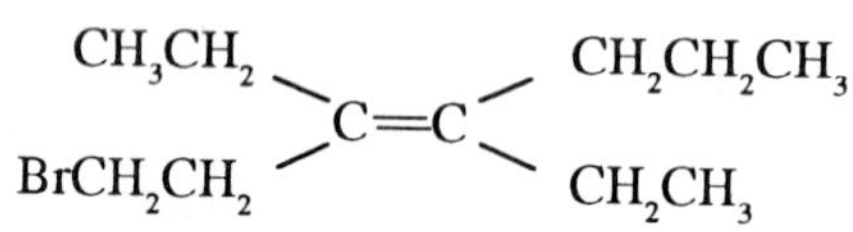

w. $CH_2{=}CH{-}CH_2CH_2NO_2$

3.4 Write structures for:

a. 1-butene
b. tetrafluoroethylene
c. allyl iodide
d. *cis*-1,2-dibromoethene
e. 3,4-dimethyl-1-pentene
f. vinyl chloride
g. *trans*-1,2-dicyanoethene
h. isobutylene
i. 2,3-dimethyl-2-butene
j. 2-methyl-1-butene
k. 3,5-dinitrocyclohexene
l. 3-isopropylcyclopentene
m. 1-cyclohexylcyclohexene
n. allyl chloride
o. (*E*)-5-methyl-3-heptene
p. (*E*)-3-methyl-2-hexene
q. 1,2-dichloro-1-pentene
r. tetravinylmethane
s. (*Z*)-di-t-butylethylene
t. (*E*)-1,2-dichloroethene
u. (*Z*)-3-methyl-3-heptene
v. (*Z*)-1-chloro-2-hexene

3.5 Which of the following exhibit *cis*/*trans* (*Z*/*E*) isomerism?

a. 2-butene
b. 1,2-dibromopropene
c. allyl chloride
d. 3-cyano-2-pentene
e. 1,4-dibromo-2-butene
f. 2-methyl-1-butene
g. 1,1-dibromo-1-pentene
h. 2,3-dimethyl-2-butene
i. 2-bromo-2-butene
j. 3-methyl-2-pentene
k. trimethylethylene
l. tetrafluoroethylene
m. 3,3-dimethyl-1-butene
n. 1-chloropropene
o. 3-methyl-1-butene
p. 2-methyl-2-butene
q. 1,2,3-triiodopropene
r. 2,4,4-trimethyl-1-pentene

3.6 Identify the missing chemicals in the reactions below:

a. $\underset{\displaystyle Br}{\underset{|}{C}H_2}CH{=}CH_2 + Br_2/CCl_4 \longrightarrow$ **A**

b. $\underset{\displaystyle Br}{\underset{|}{C}H_2}CH{=}\underset{\displaystyle Br}{\underset{|}{C}H} + HBr \longrightarrow$ **B**

c. $CH_3(CH_2)_7CH{=}CH(CH_2)_7CO_2H + Br_2/CCl_4 \longrightarrow$ **C**

d. 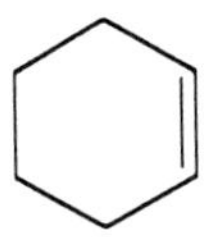 $+ \quad Br_2/CCl_4 \longrightarrow$ **D**

e. **E** $+ Br_2 \longrightarrow HO_2C{-}\underset{\displaystyle Br}{\underset{|}{C}H}\underset{\displaystyle Br}{\underset{|}{C}H}{-}CO_2H$

f. 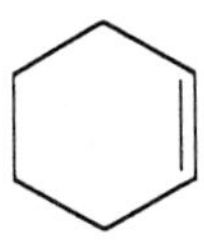 $+ \quad HOCl \longrightarrow$ **F**

g. $CH_3CH{=}CH\underset{\displaystyle O}{\underset{\|}{C}}OCH_3 + H_2\ (1\ mole)/Pd \longrightarrow$ **G**

h. **H** $+ O_3/Zn \longrightarrow \begin{matrix} CH_3 \\ CH_3 \end{matrix}\!\!>C{=}O$ (only product)

i. $CH_3CH{=}CH\underset{\displaystyle O}{\underset{\|}{C}}H$ + **I** $\longrightarrow CH_3\underset{\displaystyle OH}{\underset{|}{C}H}{-}\underset{\displaystyle Cl}{\underset{|}{C}H}\overset{\displaystyle O}{\overset{\|}{C}}H$

j. **J** $+ Br_2 \longrightarrow CH_3CHBrCHBrCO_2C_2H_5$

k. 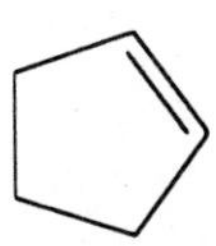 $+ \quad HOCl \longrightarrow$ **K**

l. $HO{-}\underset{\displaystyle CH_2CO_2H}{\underset{|}{\overset{\displaystyle CH_2CO_2H}{\overset{|}{C}}}}{-}CO_2H + H_2SO_4 \xrightarrow{heat}$ **L** (decolorizes Br_2/CCl_4)

m. $n\text{-}C_6H_{13}CH{=}CH_2 + H_2/\text{pressure} \xrightarrow{225°C}$ **M**

n. $n\text{-}C_6H_{13}CH{=}CH_2 + Br_2/CCl_4 \longrightarrow$ **N**

o. $n\text{-}C_6H_{13}CH{=}CH_2 + HI \longrightarrow$ **O**

p. 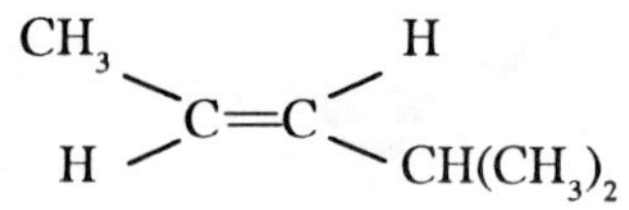 + O_3/Zn ⟶ **P** (C_2H_4O) + **Q** (C_4H_8O)

q. 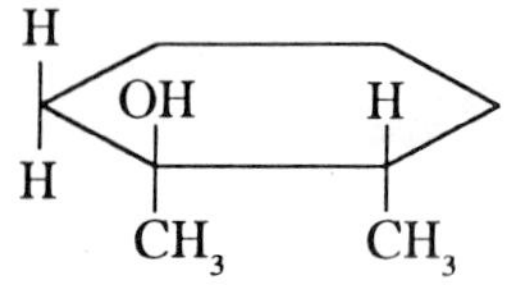 $\xrightarrow[\text{heat}]{H_2SO_4}$ **R** (major product) $\xrightarrow{H_2/Pt}$ **S**

r. 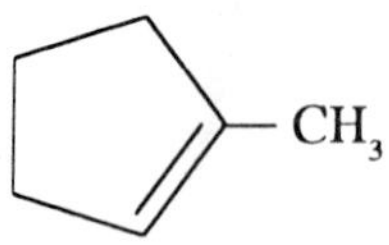 $\xrightarrow[CCl_4]{Br_2}$ **T**

s. 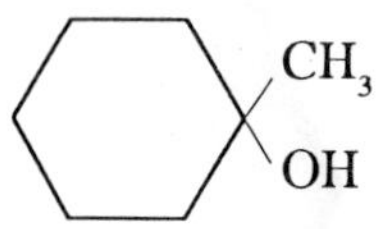 $\xrightarrow[\text{heat}]{H_2SO_4}$ **U** (major product) $\xrightarrow[\text{2. } H_2O_2/OH^-]{\text{1. } BH_3\text{—THF}}$ **V**

t. $(CH_3)_3CCH_2OH$ $\xrightarrow[\text{heat}]{H_2SO_4}$ **W** (major product)

u. 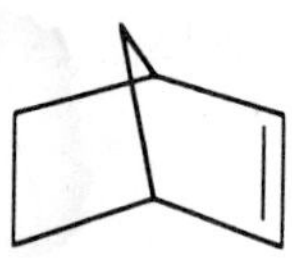 $\xrightarrow[\text{2. } H^+]{\text{1. hot } KMnO_4}$ **X**

3.7 What, if anything is wrong with the name 2,5-diethylcyclohexene?

3.8 Give the product or products expected from the reaction of 1-methylcyclohexene with each of the following:

a. H_2, Pt

b. Cl_2, CCl_4

c. $Hg(OAc)_2$ + H_2O; then $NaBH_4$

d. cold, dilute $KMnO_4$

e. $KMnO_4/OH^-$, reflux; H^+

f. O_3; then Zn /HCl

g. BH_3; then H_2O_2/OH^-

h. BH_3; then CH_3COOH/ heat

i. HBr

j. HBr /peroxides

k. HOCl

l. $C_6H_5-\overset{O}{\overset{\|}{C}}-OOH$

m. product 1 + H_3O^+

n. product (*i*) + Mg /ether

o. product *n* + H_2O

p. CH_2I_2 + Zn /Cu /ether

3.9 For each of the following pairs , which molecule is more stable and why?

a. $CH_3CH_2CH{=}CH_2$ or $CH_3-\underset{CH_3}{\underset{|}{C}}{=}CH_2$

b. 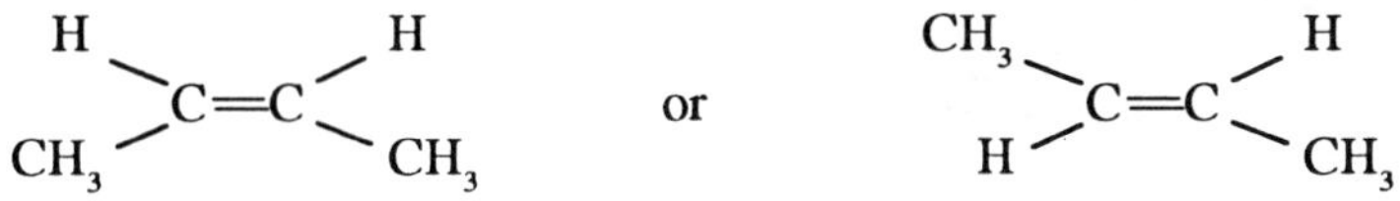

c. $CH_3CH_2CH_2CH{=}CH_2$ or $CH_3CH_2CH{=}CHCH_3$

d. $(CH_3)_3CCH{=}CH_2$ or 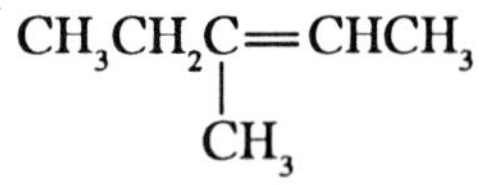

e. 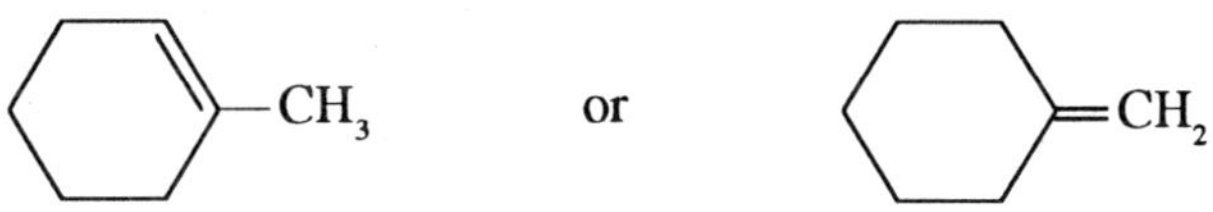

f. 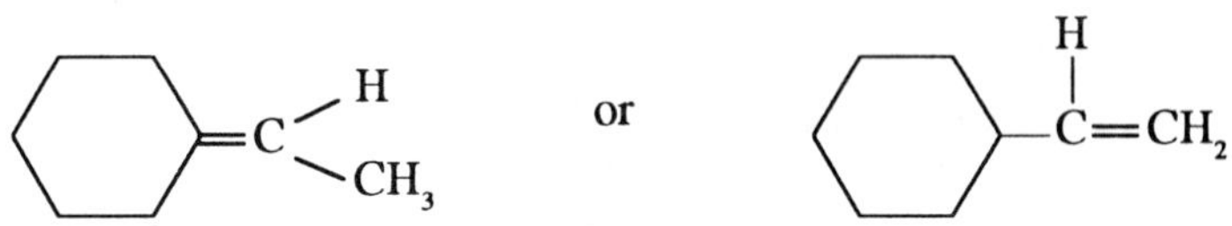

g. 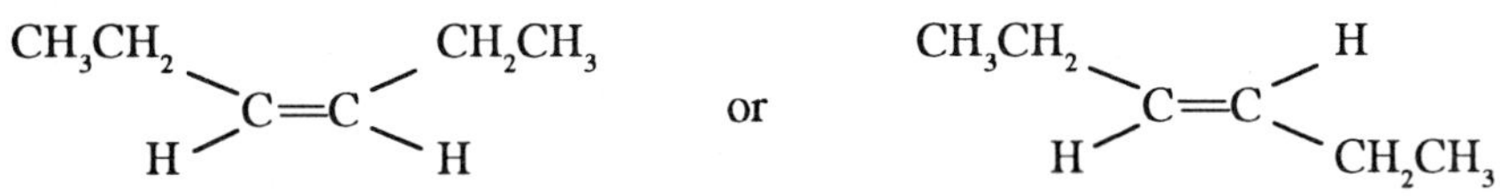

3.10 Each of the following processes involves a rearrangement in which a more stable carbocation is produced from a less stable carbocation. In each case, predict which atom or group is involved in the rearrangement.

a. $CH_3CH_2CH_2^+ \longrightarrow CH_3\overset{+}{C}HCH_3$

b. $C_6H_5CH_2CH_2CH_2OH \xrightarrow{H^+/H_2O} C_6H_5CH{=}CHCH_3$

c. 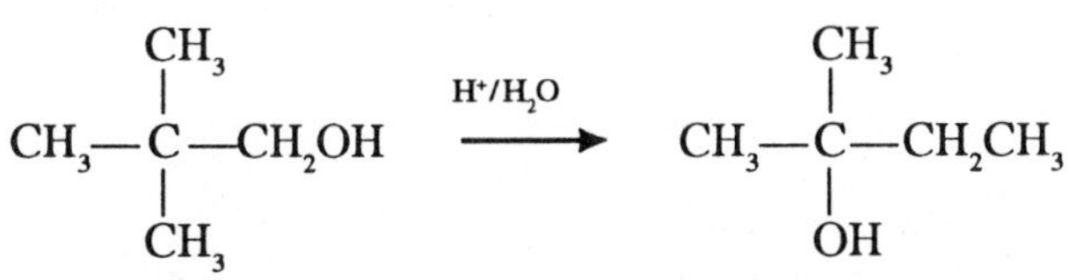

d. 

e. 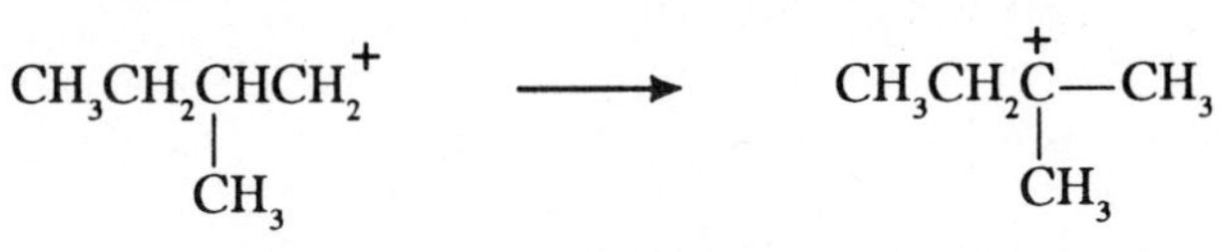

f. 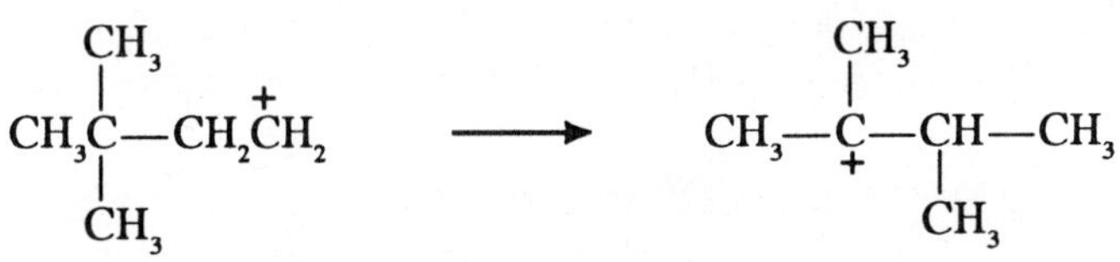

3.11 Predict the product(s) of the reaction of each of the following compounds with O_3 followed by H_2O/Zn:

a. 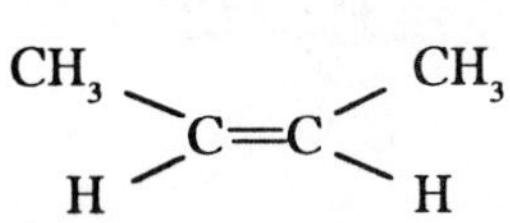

b. 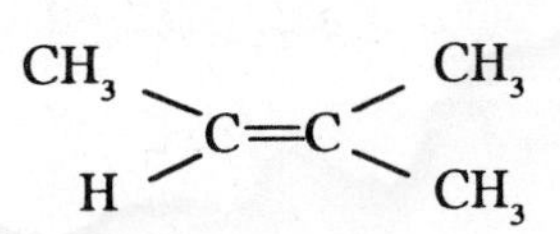

c. 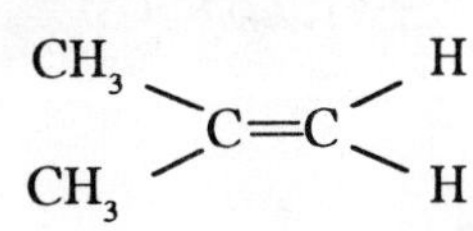

d. 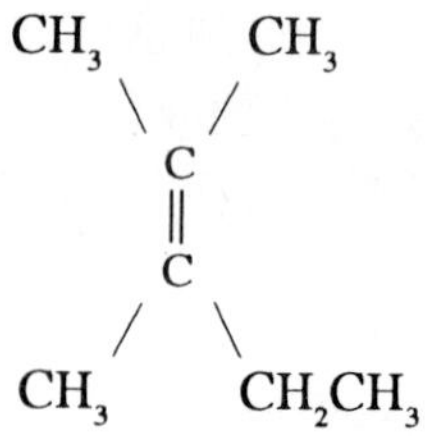

e. 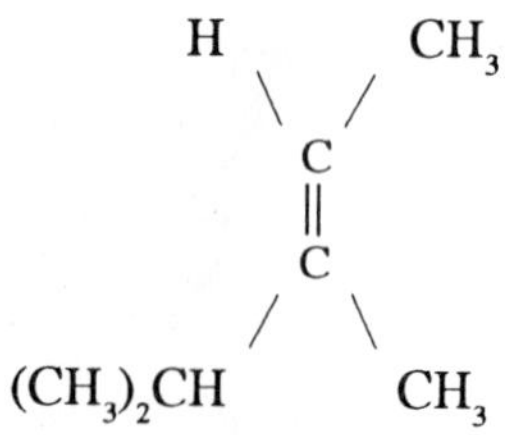

f. 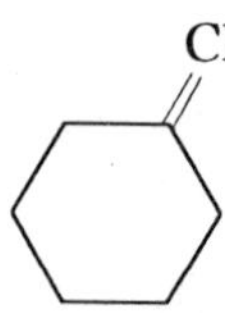

g. 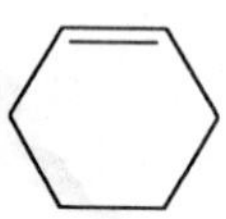

h. 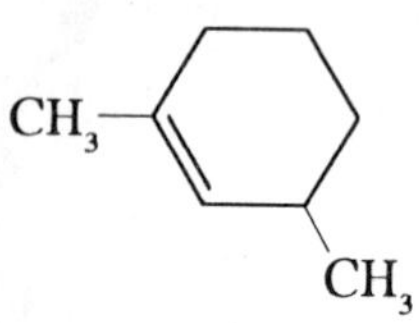

i. 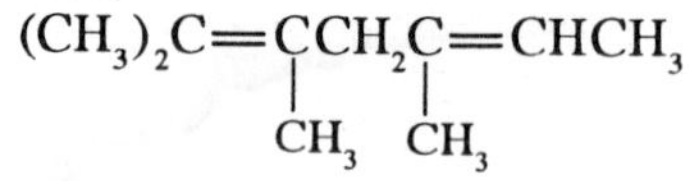

j. 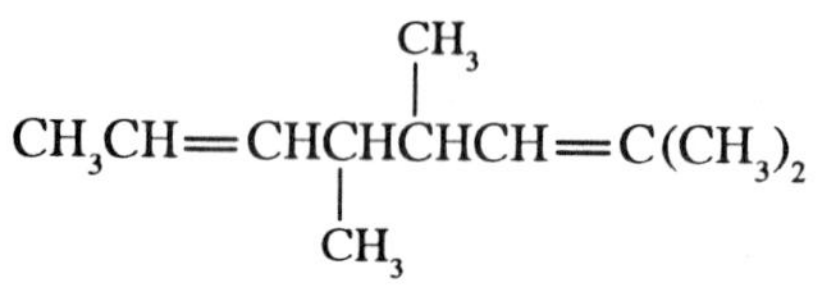

k. 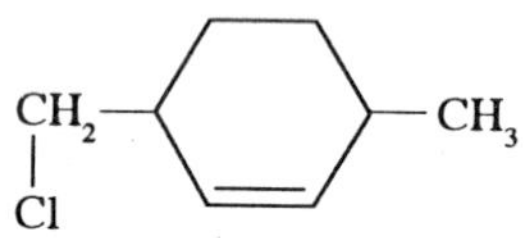

l. 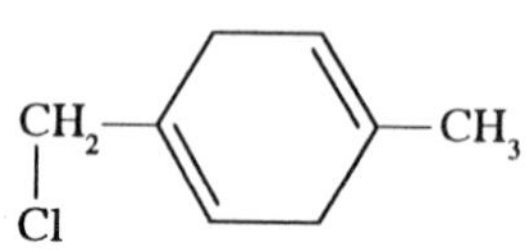

m. 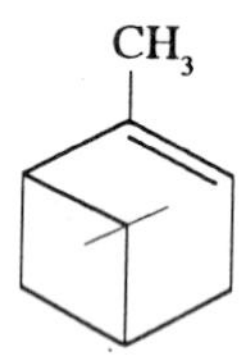

3.12 Predict the product(s) formed in the reaction of the compounds in problem 3.11 with hot $KMnO_4/OH^-$ followed by H^+/H_2O.

3.13 Write the structures of the alkenes that give the following products upon oxidative cleavage with O_3 followed by H_2O/Zn.

a. 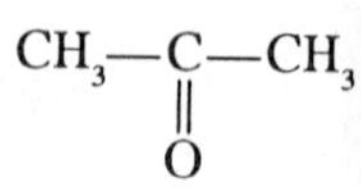 (2 moles)

b. 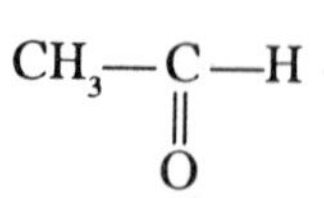 (2 moles)

c. 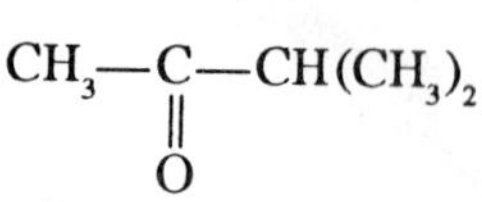 (2 moles)

d. 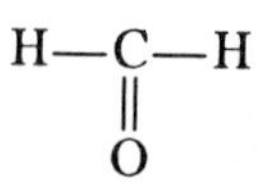 (2 moles)

e. 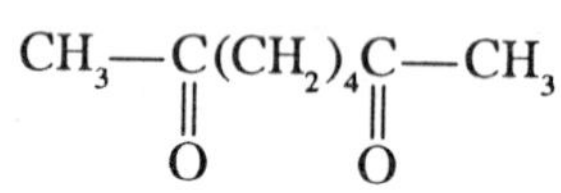

f. 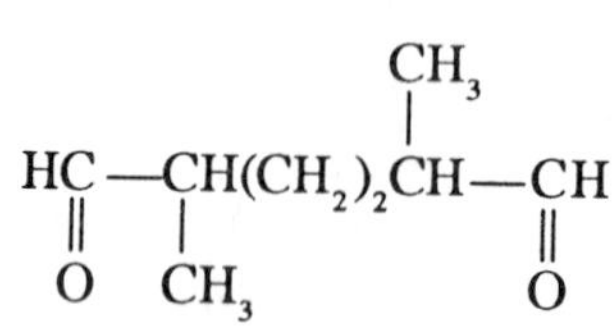

g. 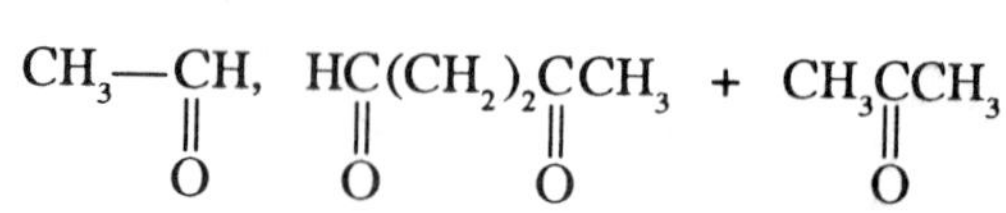

h. 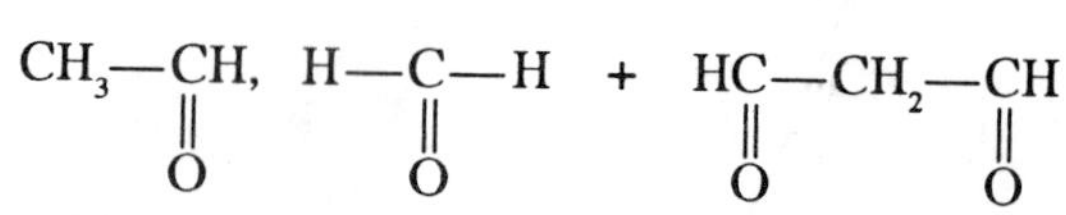

i. 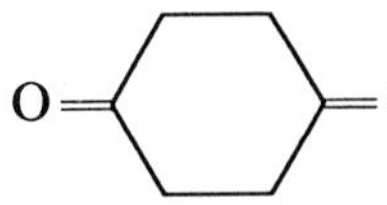 + CH_2O (2 moles)

k. 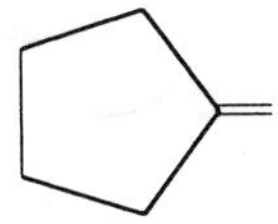 + CH_2O

j. 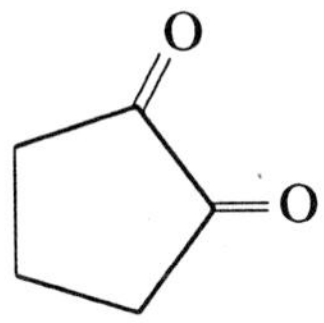 + $\underset{\displaystyle O}{\overset{\|}{HC}}CH_2\underset{\displaystyle O}{\overset{\|}{CH}}$

3.14 A 2° alcohol can be converted to a ketone by oxidation with acidic $Na_2Cr_2O_7$.

$$\underset{\text{2° alcohol}}{\underset{\displaystyle OH}{CH_3\overset{|}{C}HCH_3}} + Na_2Cr_2O_7/H^+ \longrightarrow \underset{\text{ketone}}{\underset{\displaystyle O}{CH_3\overset{\|}{C}CH_3}} + Cr^{+3}$$

In the synthesis of 5-cholesten-3-one from cholesterol, the following sequence is recommended. Explain why.

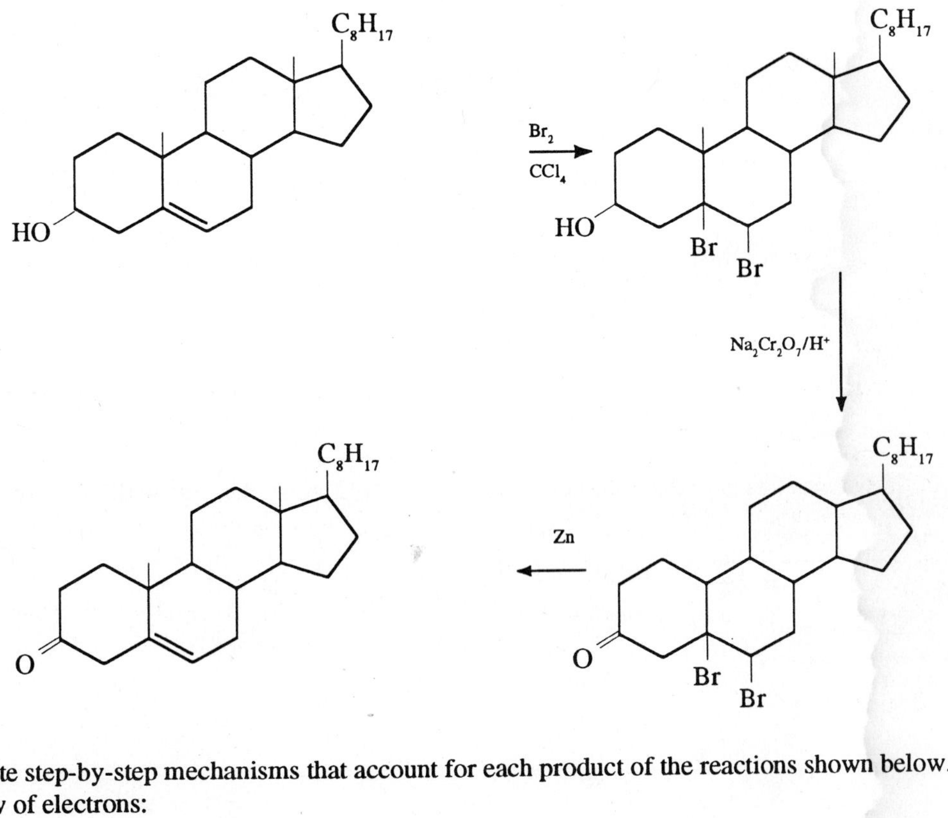

3.15 Write step-by-step mechanisms that account for each product of the reactions shown below. Use arrows to show the flow of electrons:

a. $CH_3-\overset{CH_3}{\underset{CH_3}{\overset{|}{\underset{|}{C}}}}-CH_2OH \xrightarrow{H^+/\text{heat}} (CH_3)_2C{=}CHCH_3 + H_2C{=}\underset{CH_3}{\underset{|}{C}}-CH_2CH_3$

b. $CH_3-\overset{CH_3}{\underset{H}{\overset{|}{\underset{|}{C}}}}-CH{=}CH_2 \xrightarrow{H^+/H_2O} CH_3-\overset{CH_3}{\underset{OH}{\overset{|}{\underset{|}{C}}}}-CH_2CH_3 + CH_3-\overset{CH_3}{\underset{H}{\overset{|}{\underset{|}{C}}}}-\underset{OH}{\underset{|}{C}}HCH_3$

c. 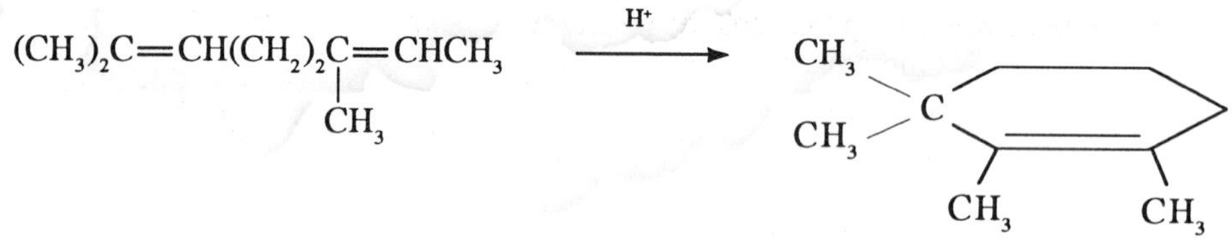

d. 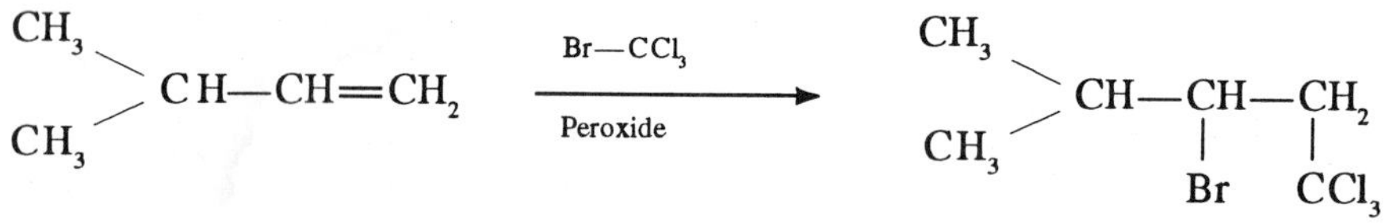

e. 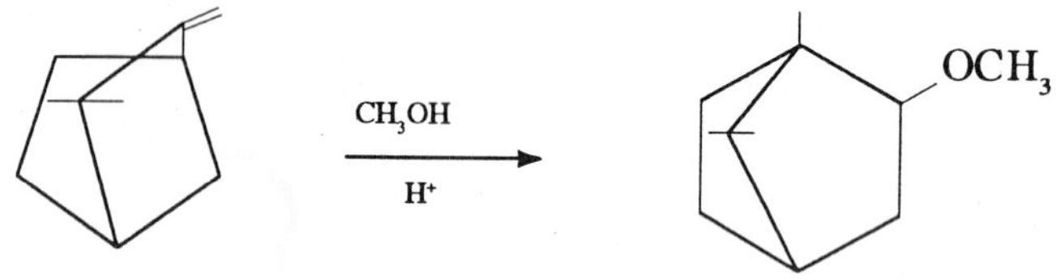

f. 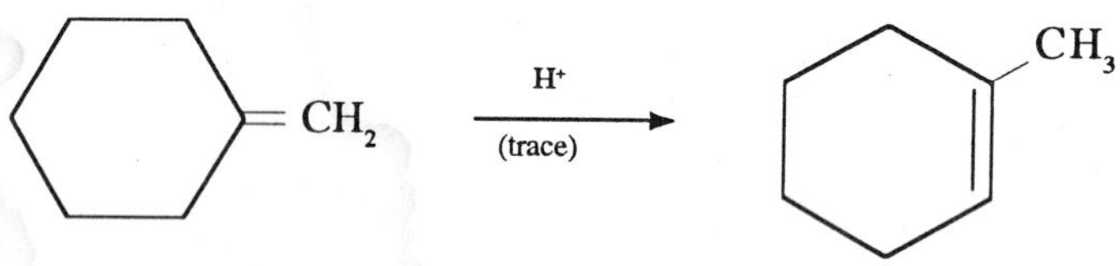

g. 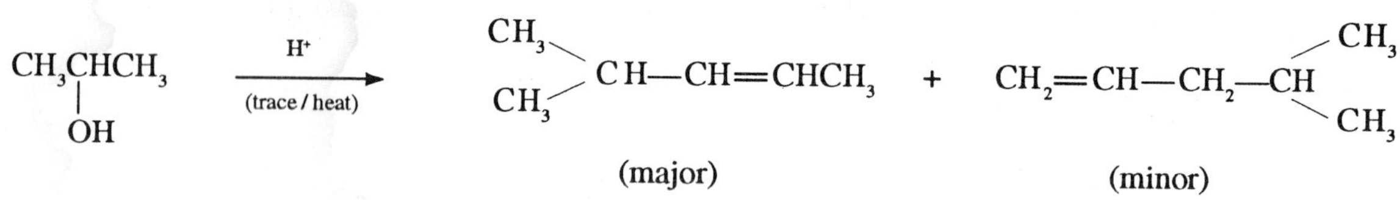

h. 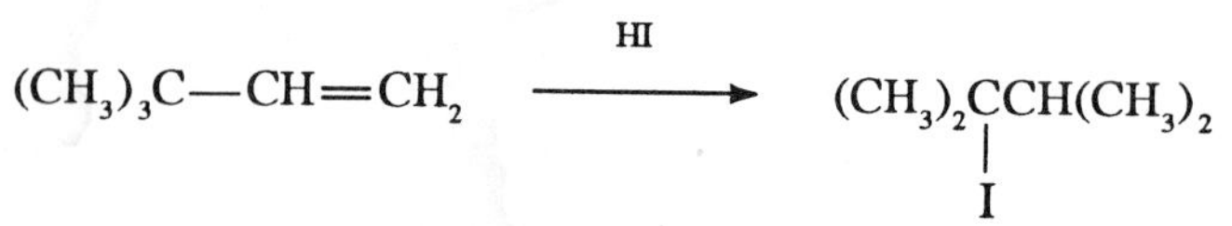

i. 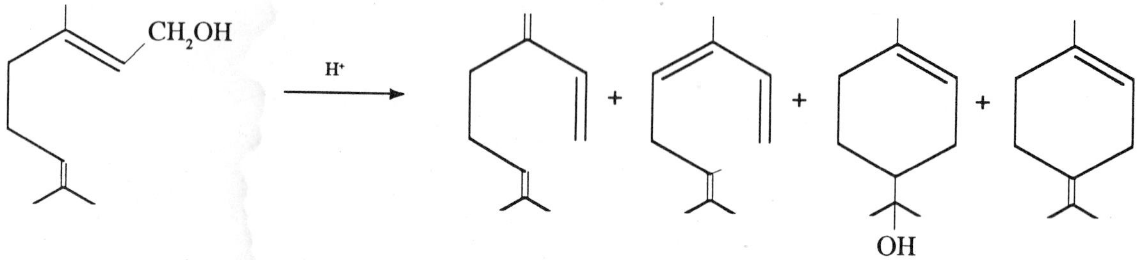

j. 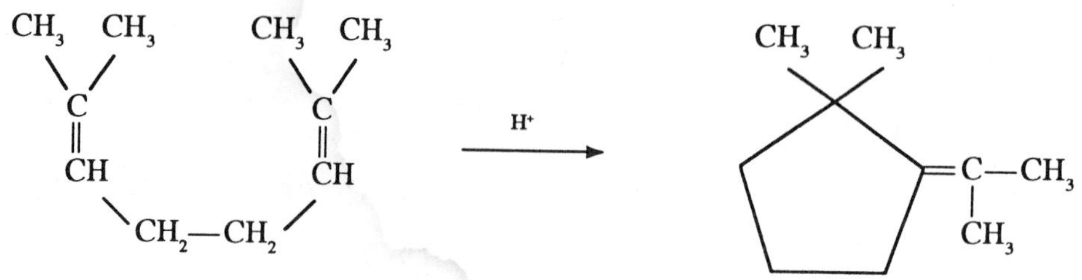

3.16 A mixture of two isomeric alkyl chlorides are formed when HCl undergoes an addition reaction with 3-methyl-1-butene. What are they? How are they formed?

3.17 Suggest structures for the compounds missing in the sequence of reactions shown below:

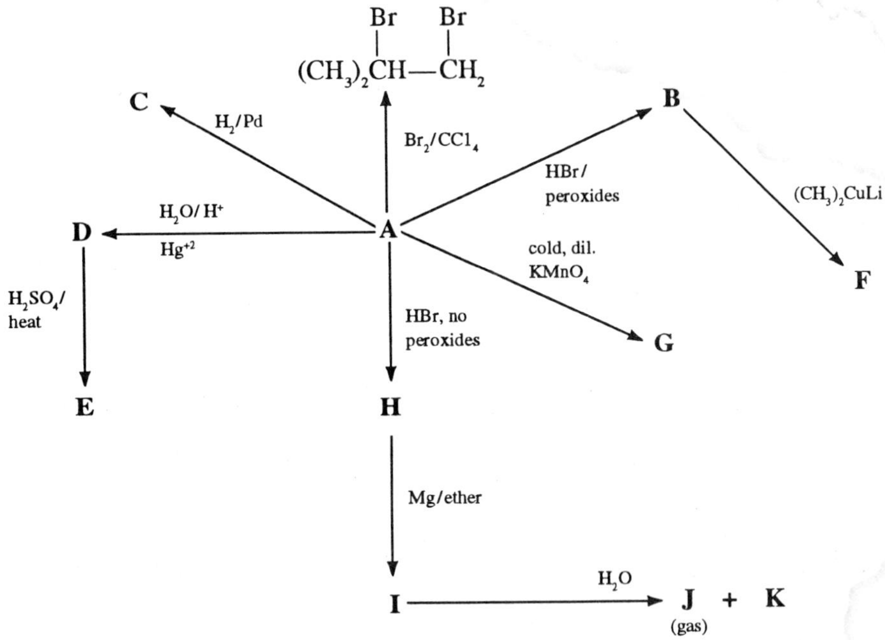

3.18 Write plausible mechanisms for the following conversions:

a.

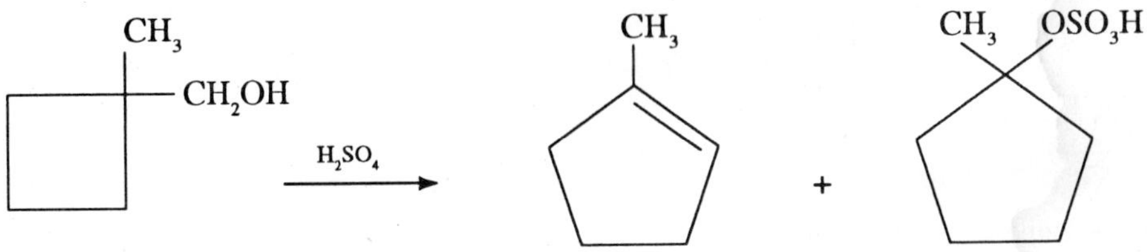

b.

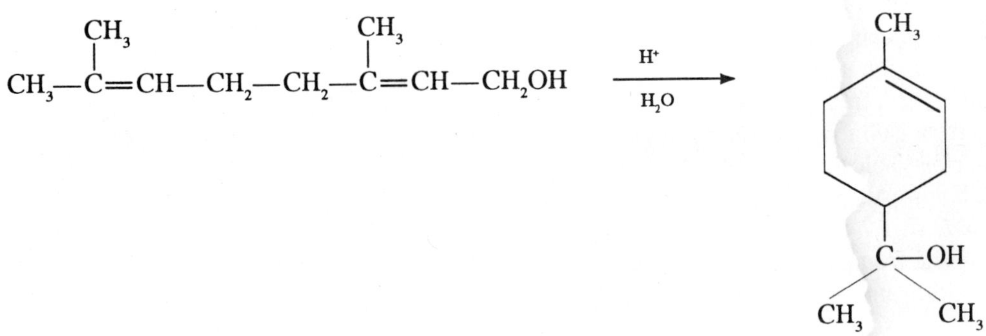

c.

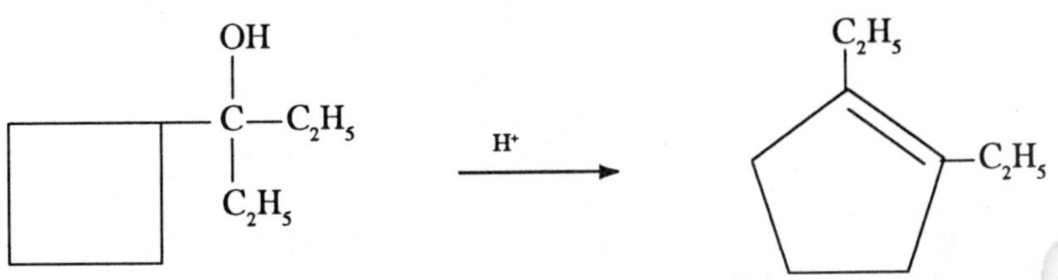

3.19 When methylene cyclobutane is treated with HCl, the addition is anti-Markovnikow; however, methylene cyclohexane reacts with HCl to give the expected Markovnikov product. Explain.

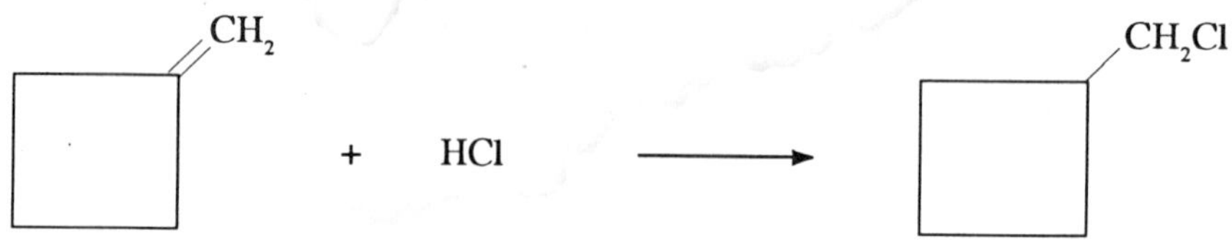

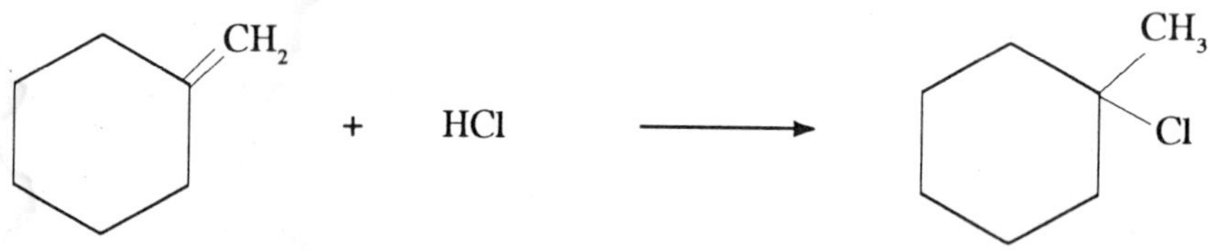

3.20 a. Write a mechanism for the generation of dichlorocarbene from chloroform and potassium *t*-butoxide.
b. Deuterated chloroform, $CDCl_3$, is recovered when the reaction of chloroform with KOH is D_2O is quenched. Explain the formation of $CDCl_3$.
c. Describe the bonding and geometry of singlet and triplet carbene.
d. The reaction of singlet carbene with an alkene is stereospecific. What product(s) would result from the reaction of (*E*)-3-chloro-4-methyl-3-hexene with singlet carbene? With triplet carbene?

3.21 Give the product, including stereochemistry, from the reaction of
a. *trans*-2-butene + $:CH_2$ (singlet)
b. *cis*-2-butene + $:CH_2$ (triplet)

3.22 What is the best way to synthesize:
a. propane from propene
b. propene from propane
c. 1-bromopropane from propane
d. 2-bromopropane from propane
e. acetone from isobutane
f. acetic acid from *n*-butane
g. *t*-butyl alcohol from isobutane
h. propylene chlorohydrin from propane
i. *t*-butyl alcohol from isobutyl alcohol
j. isobutane from *t*-butyl alcohol
k. 1-chloro-2-methylpropane from *t*-butyl alcohol
l. isobutyl alcohol from 1,2-dibromo-2-methylpropane
m. 1,3-dibromopropane from propane
n. *n*-hexane from *n*-propane

3.23 How could you separate cyclohexene from a mixture of cyclohexene and cyclohexane, which have about the same boiling point (Note: Fractional distillation is definitely ineffective.)

3.24 A 4.9 mg sample of an unsaturated hydrocarbon (known to have only one double bond) absorbed 1.12 mL of gaseous H_2 measured at 0°C and 760 mm Hg upon catalytic hydrogenation. Suggest a plausible molecular formula.

3.25 A 0.235 g sample of a hydrocarbon (Molecular Weight = 94) absorbed 168 mL of H_2 at STP. How many double bonds does this compound contain? What is its molecular formula?

3.26 a. 0.138 g of pure, unsaturated hydrocarbon absorbed 100.7 mL of H_2 at STP upon catalytic hydrogenation. The molecular weight of this compound is known to be 92. How many double bonds does it have? What is its molecular formula?

b. Ozonolysis of this hydrocarbon yielded two products:

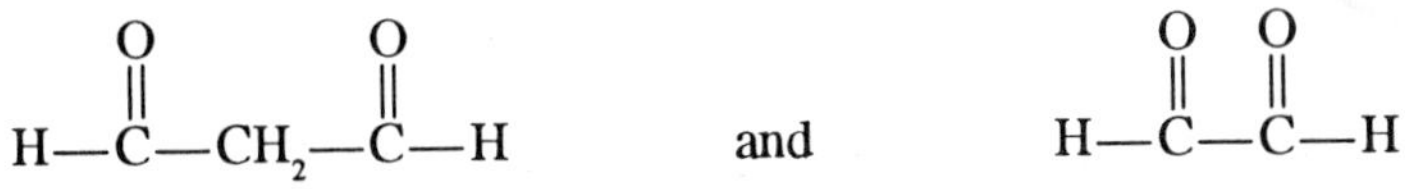

Propose a structure for the hydrocarbon.

3.27 How may grams of Br_2 will react with 50 g of 2-pentene? How much 2,3-dibromopentane will be formed in a 100% yield reaction?

3.28 Give the mechanisms to explain the formation of the indicated product(s) from the starting reagents

a. $CH_2{=}CHBr + HI \longrightarrow CH_3CHBrI$

b.

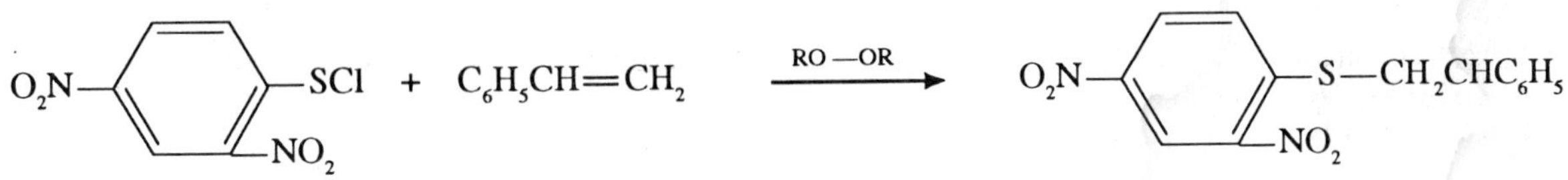

c. $(CH_3)_3\overset{+}{N}{-}CH{=}CH_2 + HI \longrightarrow (CH_3)_3\overset{+}{N}{-}CH_2CH_2I$

d. $CF_3CH{=}CH_2 + HI \longrightarrow CF_3CH_2CH_2I$

Multiple Choice Test

3.1 Which is the only name that is correct?
a. 2-ethyl-1-butene
b. 1-methyl-2-cyclohexene
c. 2-ethyl-2-butene
d. 3-ethyl-1-butene

3.2 Which of the following does not react with an alkene?
a. Lowry-Bronsted acids
b. Lewis bases
c. radicals
d. oxidizing agents (such as $KMnO_4$)

3.3 What is the order of stability of the following carbocations?

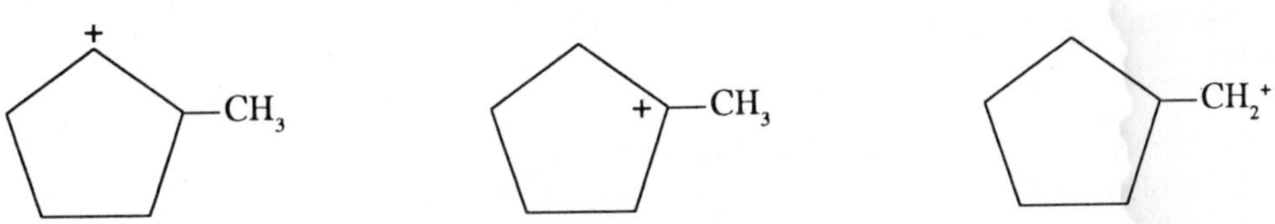

A B C

a. A > B > C
b. A > C > B
c. B > A > C
d. C > B > A

3.4 What is the order of stability of the following alkenes?

A	B	C	D
$(CH_3)_2CHCH{=}CH_2$	$(CH_3)_2C{=}CHCH_3$	(Z)—$CH_3CH{=}CHCH_3$	(E)—$CH_3CH{=}CHCH_3$

a. A > B > D > C
b. B > D > C > A
c. B > A > D > C
d. A > B > C > D

3.5 Which is the reaction that alkenes do not undergo?
a. elimination
b. addition
c. oxidation
d. polymerization

3.6 The first step in the acid catalyzed dehydration of an alcohol to yield an alkene is:
a. loss of OH^- to form a carbocation
b. loss of water to form a carbocation
c. protonation of the —OH group to form a protonated alcohol
d. loss of a proton from the alcohol to form a carbocation.

3.7 Which alcohol does not yield 1,2-dimethylcyclohexene as one of the products when dehydrated?
a. 1,2-dimethylcyclohexanol
b. 1-hydroxymethyl-2-methylcyclohexane
c. 2,2-dimethylcyclohexanol
d. 1-isopropylcyclopentanol

3.8 Which alkene is not formed when 3-methyl-1-pentanol is dehydrated?
a. 3-methyl-1-pentene
b. 3-methyl-2-pentene
c. 2-ethyl-1-butene
d. 2-methyl-1-pentene

3.9 Which alkene liberates the most heat when hydrogenated?
a. (*E*)-4,4-dimethyl-2-hexene
b. (*Z*)-4,4-dimethyl-2-hexene
c. 3,3-dimethyl-1-hexene
d. (*E*)-3,4-dimethyl-3-hexene

3.10 Hydrogenation of 1,2-dimethylcyclohexene with H_2/Pt yields
a. *cis*-1,2-dimethylcyclohexane
b. *trans*-1,2-dimethylcyclohexane
c. a 1:1 mixture of (a) and (b)
d. 1,1-dimethylcyclohexane

3.11 Which compound yields two aldehydes upon treatment with O_3/Zn?
a. 2,3-dimethyl-1-pentene
b. 3,3-dimethyl-1-pentene
c. 2,3-dimethyl-2-pentene
d. 2,4-dimethyl-2-pentene

3.12 Acid-catalyzed hydration of 2-methyl-2-butene yields
a. 2-methyl-1-butanol
b. 2-methyl-2-butanol
c. 3-methyl-2-butanol
d. 3-methyl-1-butanol

3.13 Which of the following reactions does not yield 2-bromobutane?
a. 1-butene + HBr (no peroxides)
b. 1-butene + HBr (peroxides)
c. 2-butene + HBr (no peroxides)
d. 2-butene + HBr (peroxides)

3.14 Which reagent does not yield 1,2-cyclohexanediol from cyclohexene?
a. OsO_4
b. $KMnO_4$ (cold, dilute)
c. H_2O/H^+
d. HCO_3H/H^+

3.15 Hot $KMnO_4$ oxidation of 1-methylcylopentene yields
a. $OHC(CH_2)_2CH(CH_3)CHO$
b. $OHC(CH_2)_3COCH_3$
c. $HOOC(CH_2)_3COCH_3$
d. $HOOC(CH_2)_2CH(COOH)COOH$

3.16 The intermediate in the free radical addition of HBr to isobutylene is

a. $(CH_3)_2CHCH_2\cdot$
b. $(CH_3)_3C\cdot$
c. $(CH_3)_2CBrCH_2^{\bullet}$
d. $(CH_3)_2\dot{C}CH_2Br$

3.17 Which reagent does not undergo an addition reaction with an alkene?

a. $(CH_3)_3C^+$
b. NBS
c. H_2O (in acid)
d. OsO_4

3.18 Reaction of B_2D_6 / OH^-/ H_2O_2 with isobutylene yields

a. $(CH_3)_2CDCH_2OH$
b. $(CH_3)_2CHCH_2OD$
c. $(CH_3)_2CDCH_2OD$
d. $(CH_3)_3COH$

3.19 What is the major product (**D**) of the following reaction sequence?

$$\text{isobutane} \xrightarrow{Br_2/UV} \mathbf{A} \xrightarrow{\text{alc. KOH}} \mathbf{B} \xrightarrow{\text{NBS}} \mathbf{C} \xrightarrow{D_2/Pt} \mathbf{D}$$

a. $BrCH_2\underset{\displaystyle CH_3}{\underset{|}{C}}D—CH_2D$
b. $(CH_2D)_2CBrCH_3$
c. $(CH_2D)_2CHCH_2Br$
d. $(CD_3)_3CBr$

3.20 Which reagent cannot be used to distinguish cyclohexane from cyclohexene?

a. Br_2/CCl_4
b. $KMnO_4$
c. conc. H_2SO_4
d. NaOH

Answers for Questions

3.1 a. Alkene: A hydrocarbon with one (or more) carbon-carbon double bonds.

b. π bond: When two *p* orbitals of adjacent atoms overlap sidewise (rather than end-on), their electrons form π-molecular orbitals. A π bond is much weaker than a σ bond.

c. Unsaturated hydrocarbon: A hydrocarbon that contains one or more double or triple bond(s).

d. Geometric Isomerism: Steroisomers of alkenes and cyclic compounds that occur because of hindered rotation. They are classified as either the (*Z*) or the (*E*) isomer.

e. Catalytic hydrogenation: The process whereby H_2 is added to an unsaturated compound (double or triple bonds). For instance,

$$CH_2{=}CH_2 + H_2 \xrightarrow{\text{catalyst}} CH_3—CH_3$$

Catalysts include platinum, palladium, and nickel.

f. Markovnikov's rule: In an addition reaction of a polar moiety to an alkene, a carbocation must first be formed. The positively-charged portion of the polar compound adds at the side of the double bond that produces the more stable carbocation (3° > 2° > 1°). Thus, in the reaction of a hydrogen halide HX (eg., HCl), H^+ adds to the carbon that holds the greater number of hydrogens, and X^- (eg., Cl^-) adds to the other carbon. For instance,

$$CH_3—CH{=}CH_2 + HX \longrightarrow \underset{\text{formed}}{CH_3—\underset{\displaystyle X}{\underset{|}{C}}H—CH_3} + \underset{\text{not formed}}{CH_3—CH_2—CH_2—X}$$

3.2 a. Four alkenes of molecular formula C_4H_8:

$CH_3CH_2CH{=}CH_2$

1-butene

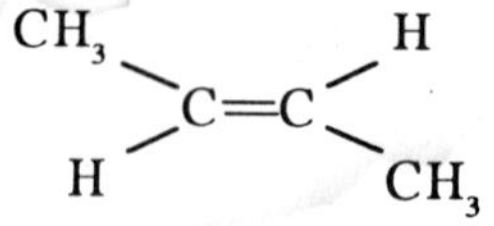

(*E*)-2-butene

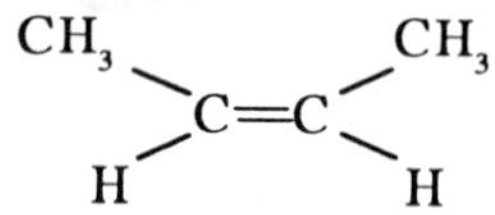

(*Z*)-2-butene

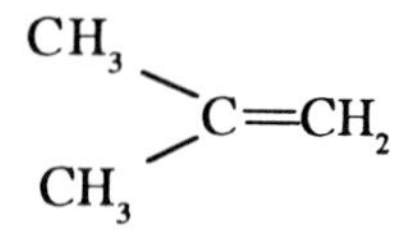

2-methylpropene

b. Six alkenes of molecular formula C_5H_8:

$CH_3CH_2CH_2CH{=}CH_2$

1-pentene

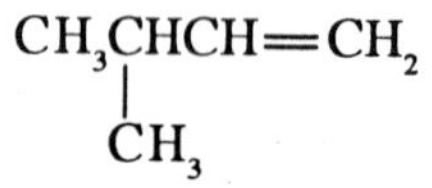

3-methyl-1-butene

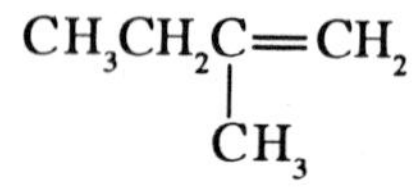

2-methyl-1-butene

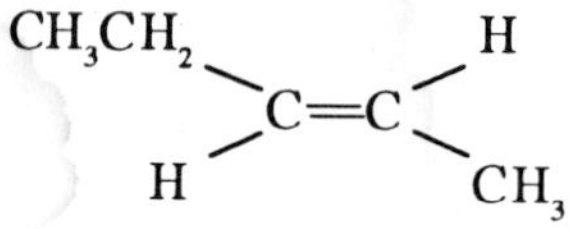

(*E*)-2-pentene

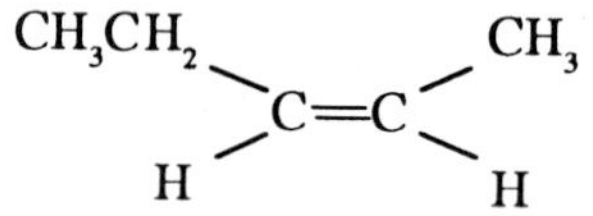

(*Z*)-2-pentene

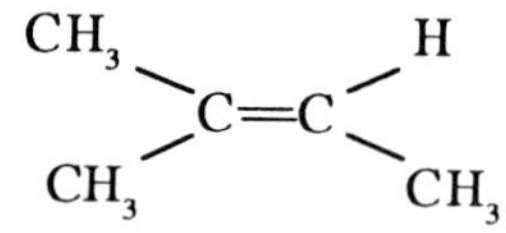

2-methyl-2-butene

c. Four chloro-2-methyl-2-butenes:

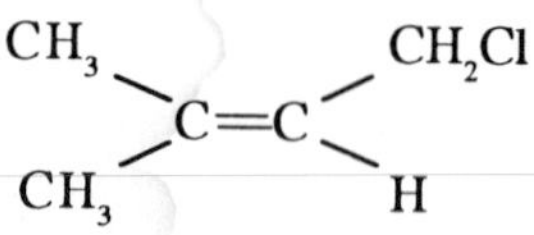

1-chloro-3-methyl-2-butene

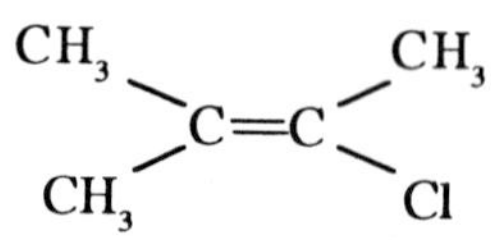

2-chloro-3-methyl-2-butene

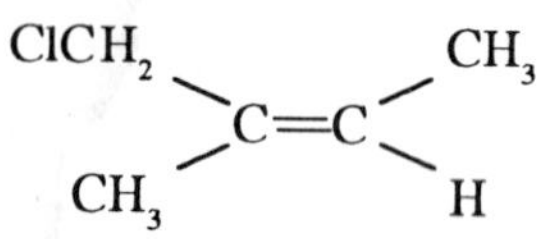

(*Z*) 1-chloro-2-methyl-2-butene

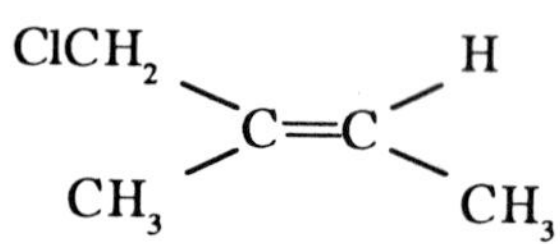

(*E*) 1-chloro-2-methyl-2-butene

d. $(CH_3)_2C{=}C(CH_3)_2$

3.3
a. 3-chloro-1-pentene
b. 3-methyl-1-butene
c. 2-methyl-1-propene (or isobutylene)
d. 3,3-dimethyl-1-butene
e. 2,3,4-trichloro-1-heptene
f. chloroethene (or chloroethylene, or vinyl chloride)
g. tetracyanoethylene (or tetracyanoethene)
h. 2,3-dimethyl-2-butene
i. 1-octene
j. (*E*)-1-chloropropene (not 3-chloro-2-propene)
k. 3-ethyl-1-methylcyclopentene
l. (*E*)-2-pentene (not 3-pentene)
m. (*E*)-4-chloro-4-methyl-2-pentene (not 2-chloro-2-methyl-3-pentene)
n. (*E*)-1-bromo-2-chloroethene
o. (*Z*)-2-octene
p. 3-propyl-1-hexene
q. (*E*)-1-fluoro-1,3-diiodo-2-methylpropene
r. 2-ethyl-1-butene
s. 1,1-dichloroethene (or vinylidine chloride)
t. 5-(4-methyl-1-cyclohexenyl)-2,5-dimethyl-2-hexene
u. (*E*)-3-methyl-3-heptene
v. (*E*)-1-bromo-3,4-diethyl-3-heptene
w. 4-nitro-1-butene

3.4

a. $H_2C{=}CHCH_2CH_3$

b. $CF_2{=}CF_2$

c. $CH_2{=}CHCH_2I$

d. 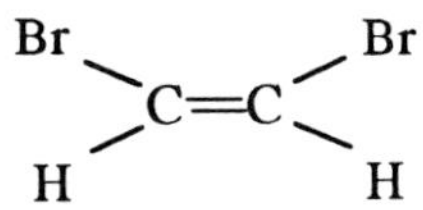

e. 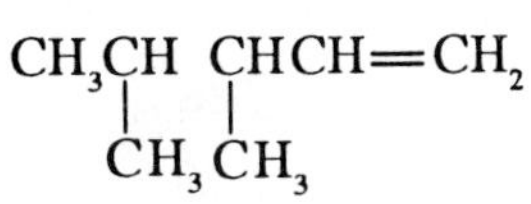

f. $CH_2{=}CHCl$

g. 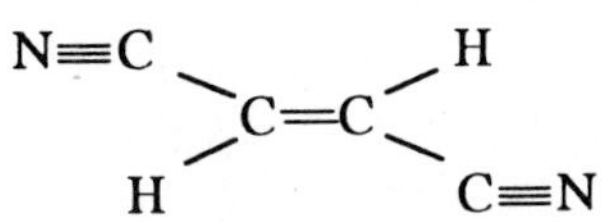

h. 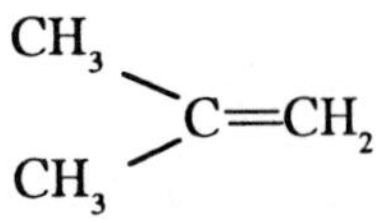

i. 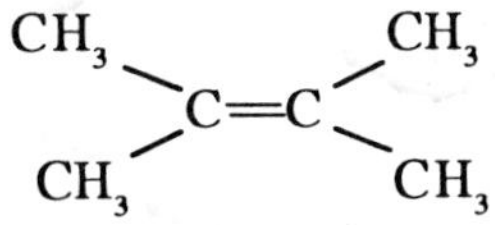

j. 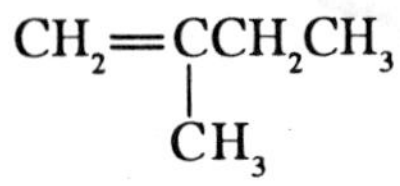

k. 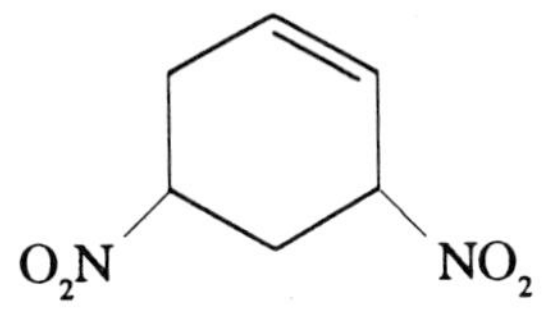

l. 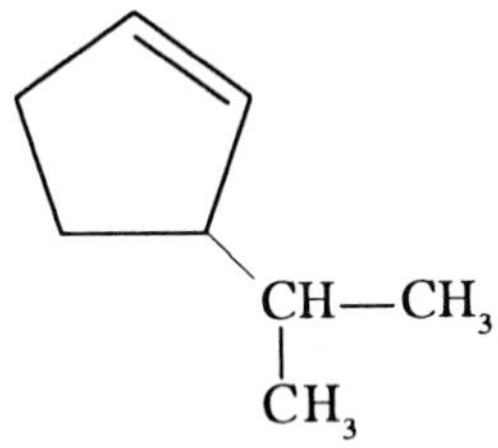

m. 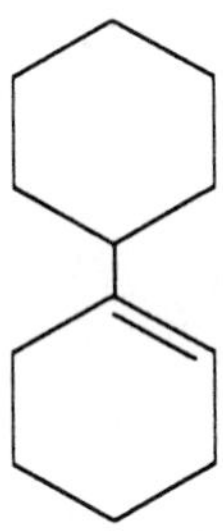

n. $CH_2{=}CHCH_2Cl$

o. 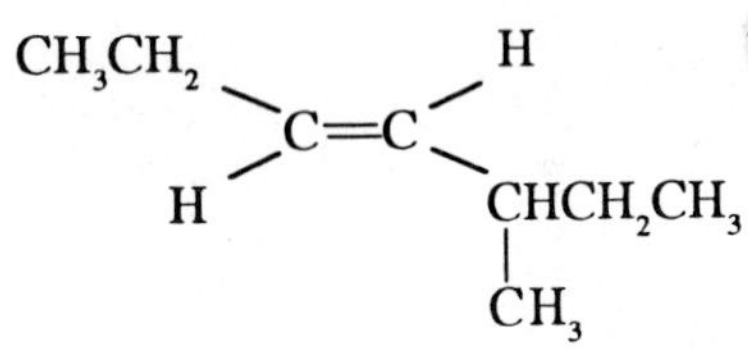

p. 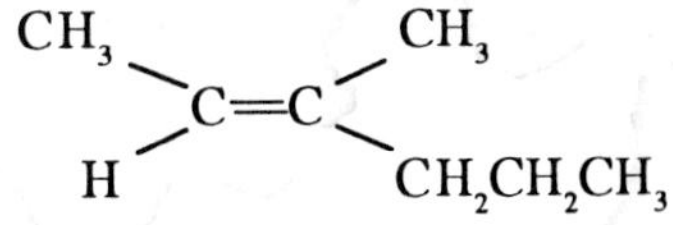

q. 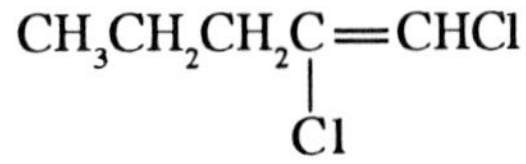

r. $(CH_2{=}CH)_4C$

s. 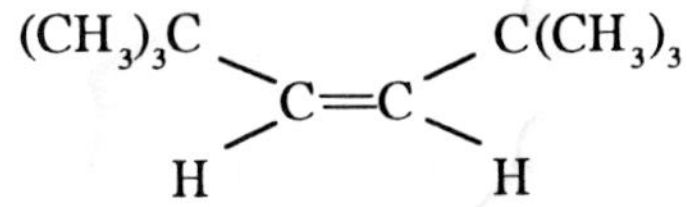

t. 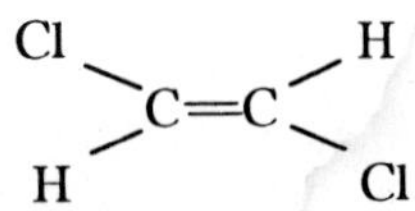

u. 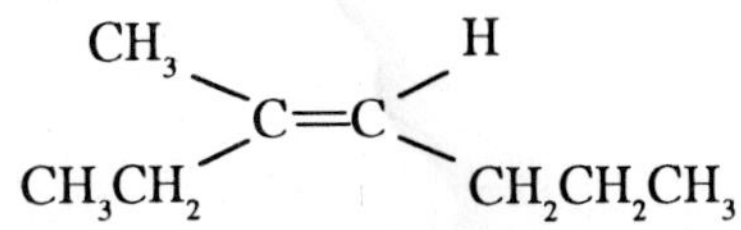

v. 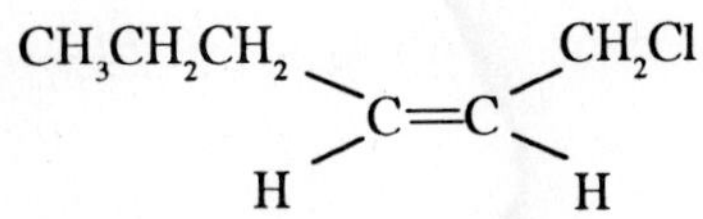

3.5 Structures of the compounds:

a. $CH_3CH{=}CHCH_3$

(yes)

b. 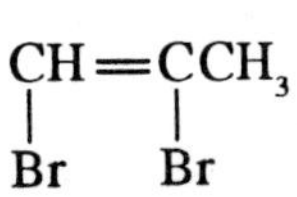

(yes)

c. $CH_2{=}CHCH_2Cl$

(no)
(also called 3-chloropropene)

d. 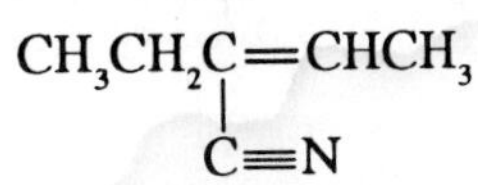

(yes)

e. 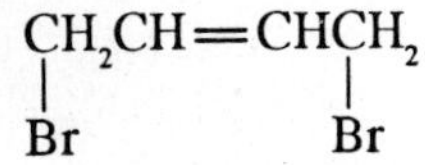

(yes)

f. 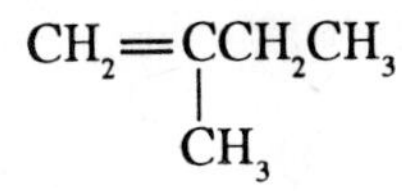

(no)

g. $CH_3CH_2CH_2CH{=}CBr_2$
(no)

h. 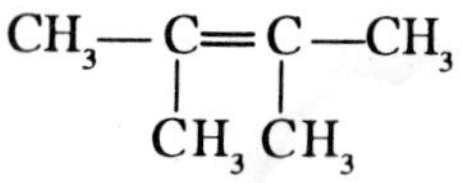
(no)

i. 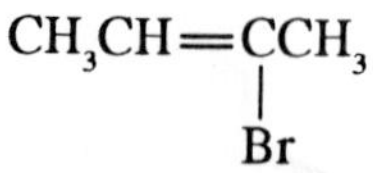
(yes)

j. 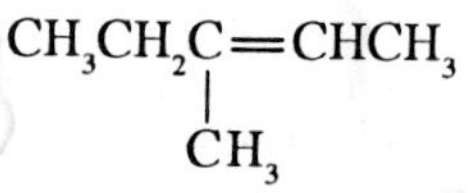
(yes)

k. 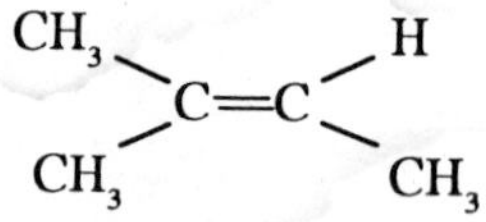
(no)
(also named 2-methyl-2-butene)

l. $CF_2{=}CF_2$
(no)
(also named tetrafluoroethene)

m. $(CH_3)_3CCH{=}CH_2$
(no)

n. 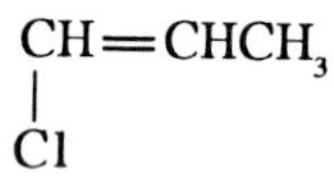
(yes)

o. 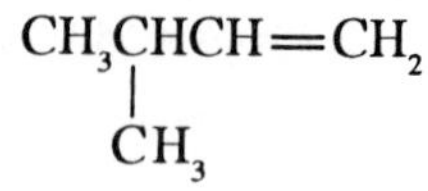
(no)

p. $(CH_3)_2C{=}CHCH_3$
(no)

q. 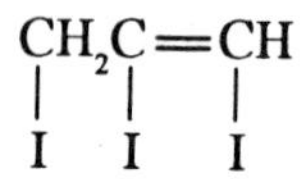
(yes)

r. 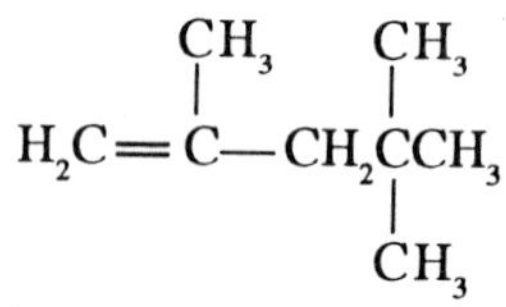
(no)

3.6 **A** = **B** = 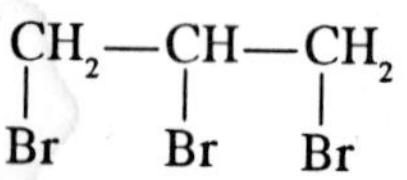

C = 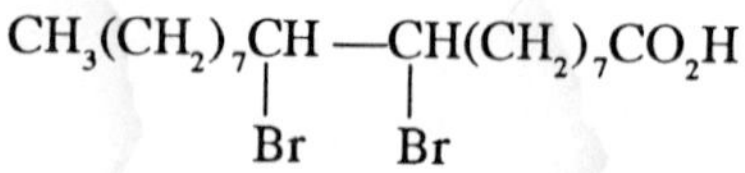

D = 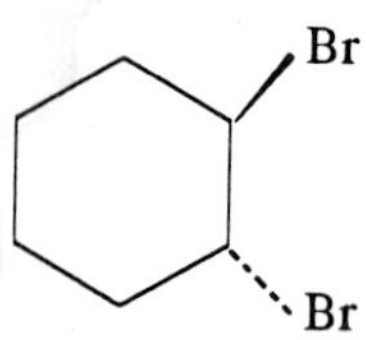

E = $HO_2C{-}CH{=}CH{-}CO_2H$

F = 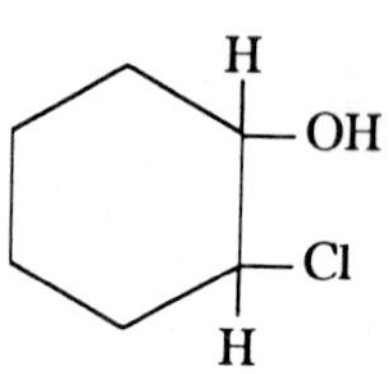
(*trans*)

G = 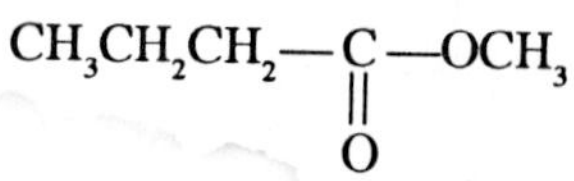

H = $(CH_3)_2C{=}C(CH_3)_2$

I = Cl_2/H_2O

J = $CH_3CH{=}CHCO_2C_2H_5$

K = 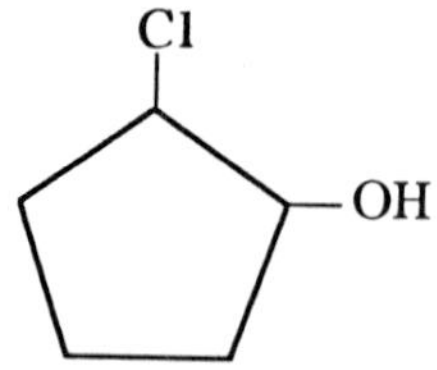
(*trans*)

L = 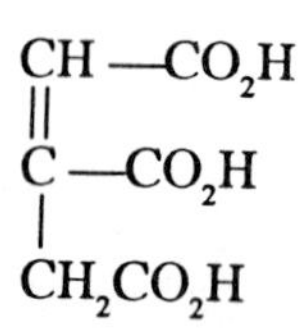

M = $n\text{-}C_6H_{13}CH_2CH_3$

N = $n\text{-}C_6H_{13}CHCH_2Br$ (with Br on the CH)

S =

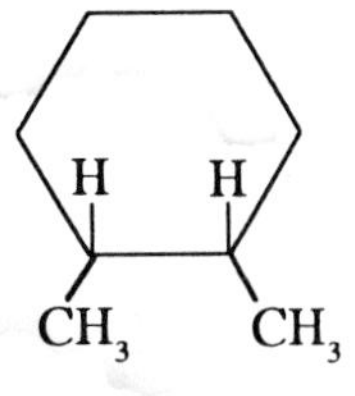

(*cis*)

O = $n\text{-}C_6H_{13}CH—CH_3$ (with I on the CH)

T =

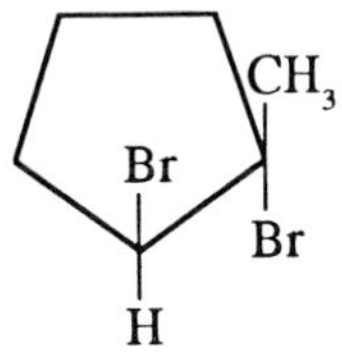

P = $CH_3—CH{=}O$

U =

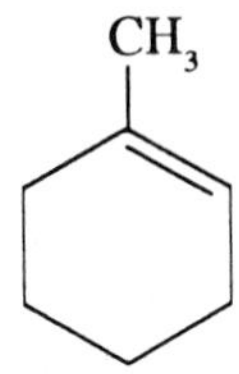

(trisubstituted alkene)

Note: **U** is not:

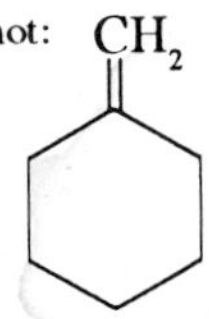

because this is a less thermodynamically stable product.

(disubstituted alkene)

Q = $(CH_3)_2CH—CH{=}O$

V =

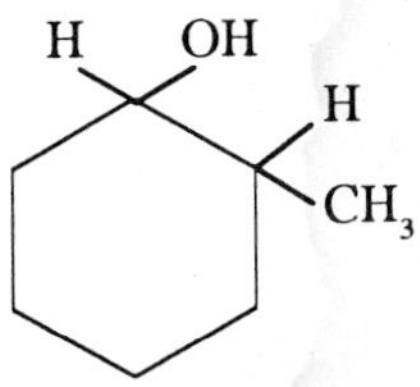

R =

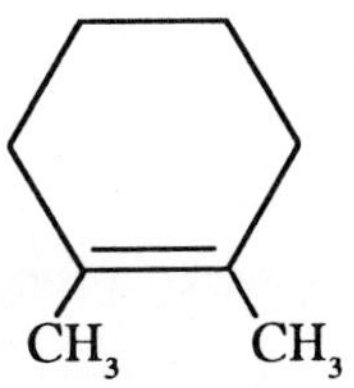

(tetrasubstituted alkene)

Note: **R** is not

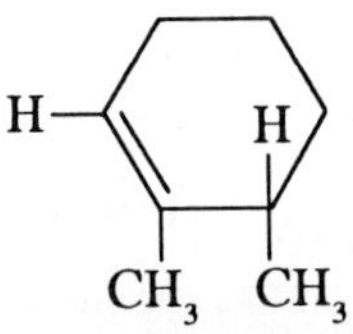

because this is a less thermodynamically stable product. (trisubstituted alkene)

W = $(CH_3)_2C{=}CH—CH_3$

X =

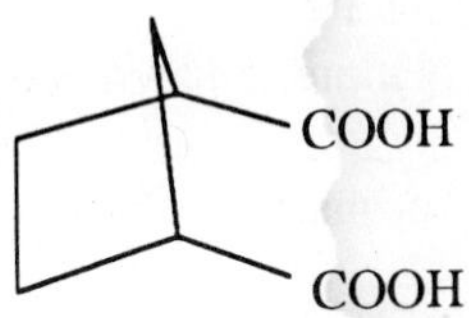

equal to

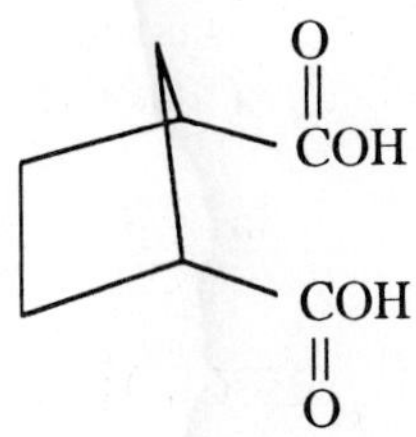

3.7 The structure is

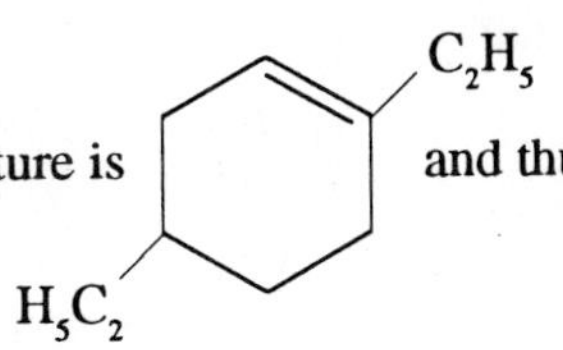

and thus should be called 1,4-diethylcyclohexene, because 1,4 is a lower set of numbers than 2,5.

3.8 a.

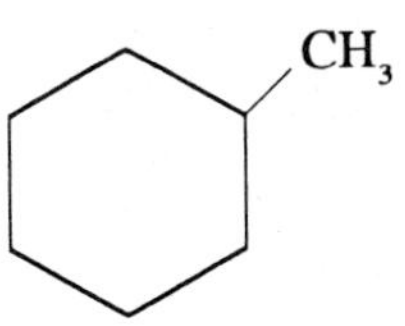

b.

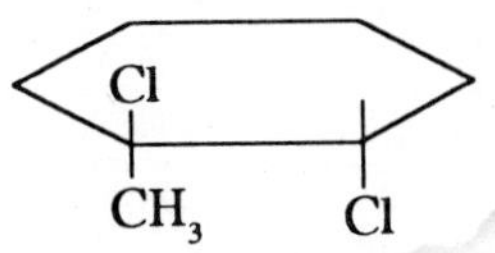

c.

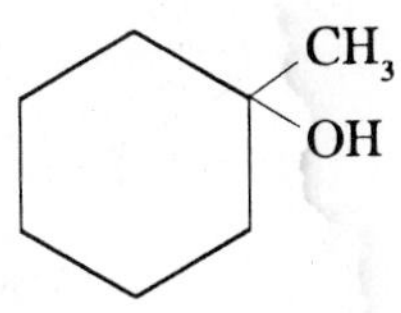

d. 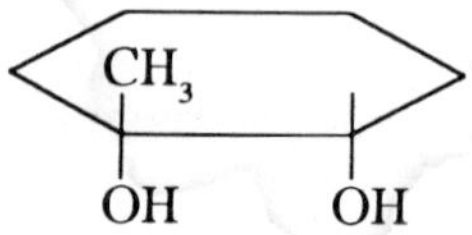

i. 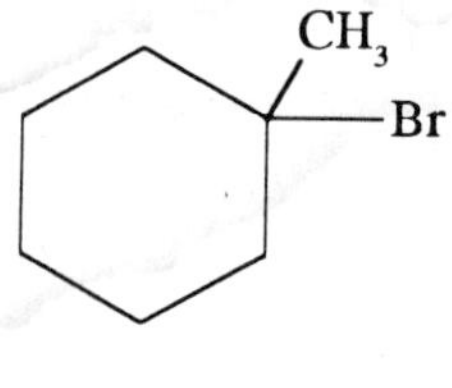

m. 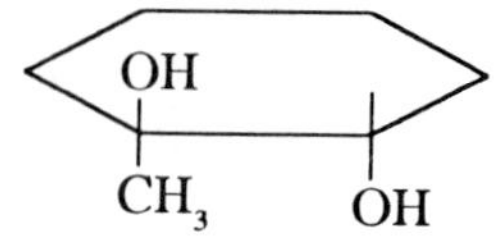

e. 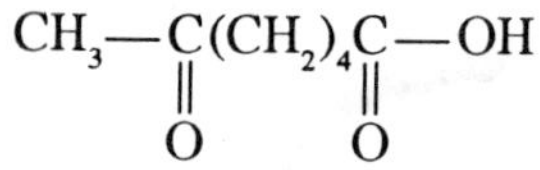

j. 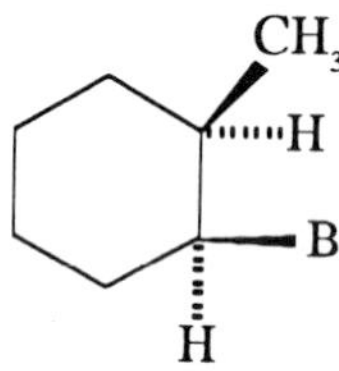 and 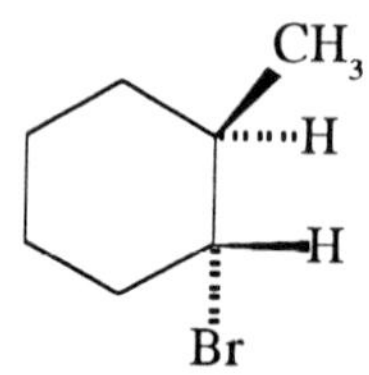

f. 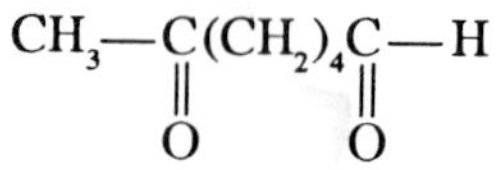

k. 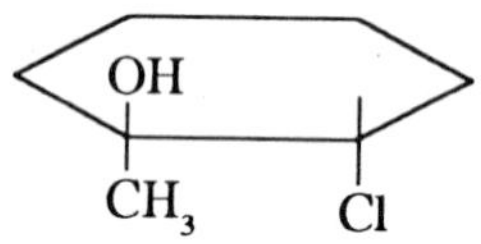

n. 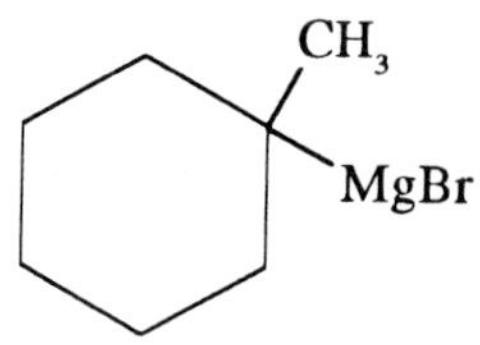

g. 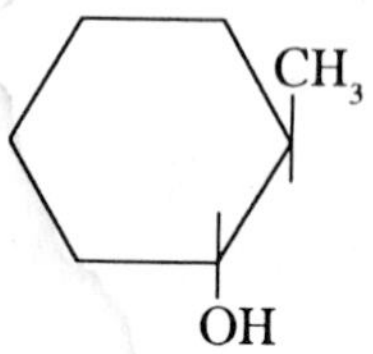

l. 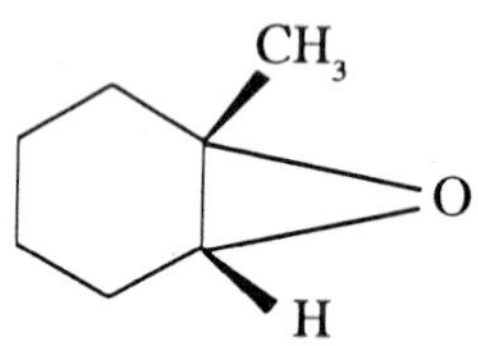

o. 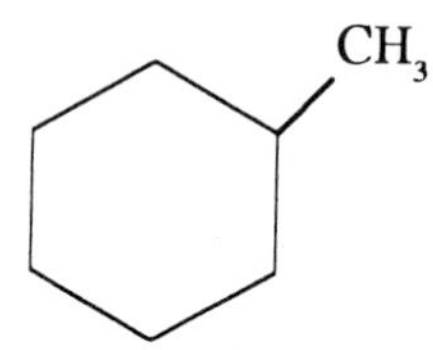

h. 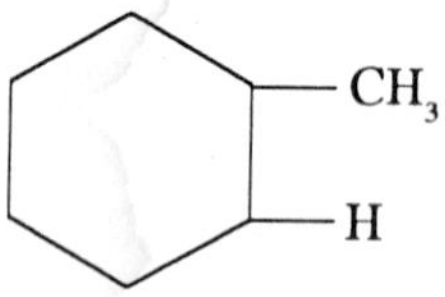

p. 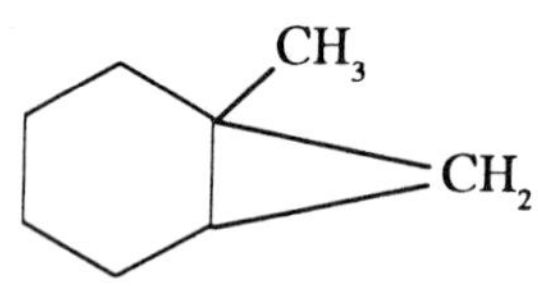

3.9 The more alkyl groups attached to the carbon of the double bond, the greater the thermodynamic stability of the compound:
$R_2C{=}CR_2 > RCH{=}CR_2 > RCH{=}CHR, R_2C{=}CH_2 > RCH{=}CH_2 > CH_2{=}CH_2$
This may depend on the number of C_{sp^2}—C_{sp^3} bonds, which is greater in the more highly substituted alkenes. *Trans* isomers are usually more stable than *cis* isomers because they are less "crowded."

a. 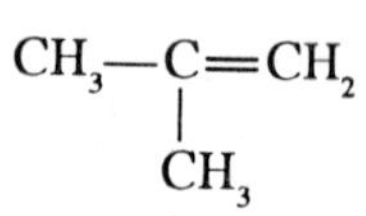 because it is a disubstituted alkene, while $CH_3CH_2CH{=}CH_2$ is a monosubstituted alkene.

b. 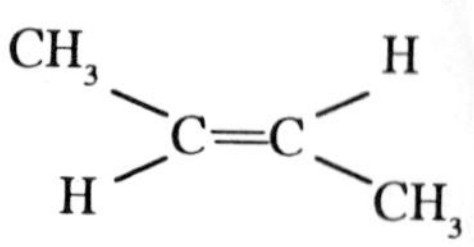 because it is a *trans* compound, which is more stable than the corresponding *cis* compound.

c. 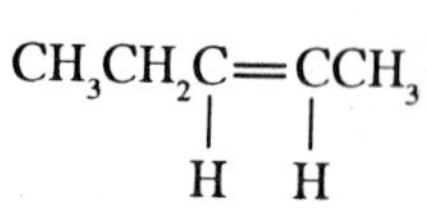 because it is a disubstituted alkene, while $CH_3CH_2CH_2CH{=}CH_2$ is a monosubstituted alkene.

d. 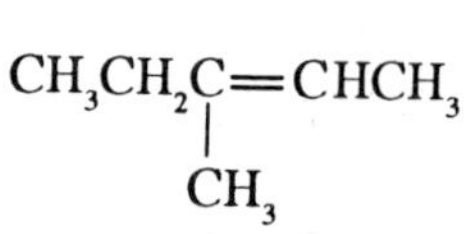 because it is a trisubstituted alkene, while $(CH_3)_3C{-}CH{=}CH_2$ is a monosubstituted alkene.

e. 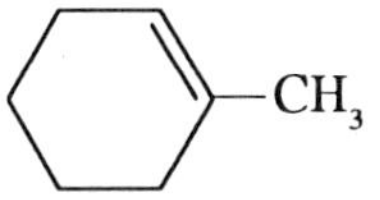 because it is a trisubstituted alkene, while 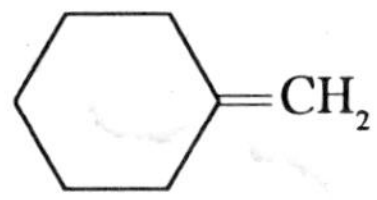 is a disubstituted alkene.

f. 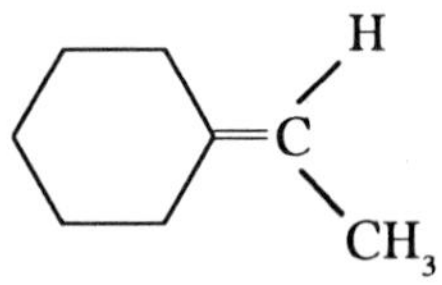 because it is a trisubstituted alkene, while 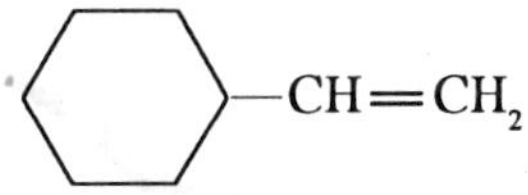 is a mono-substituted alkene.

g. 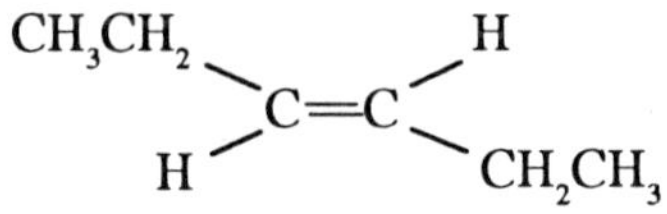 because it is a *trans* compound which is more stable than the corresponding *cis* compound.

3.10 a. H: (hydride) shift

b. H: shift

c. CH_3: (methyl) shift

d. RCH_2: shift where R= part of the ring; this results in ring enlargement.

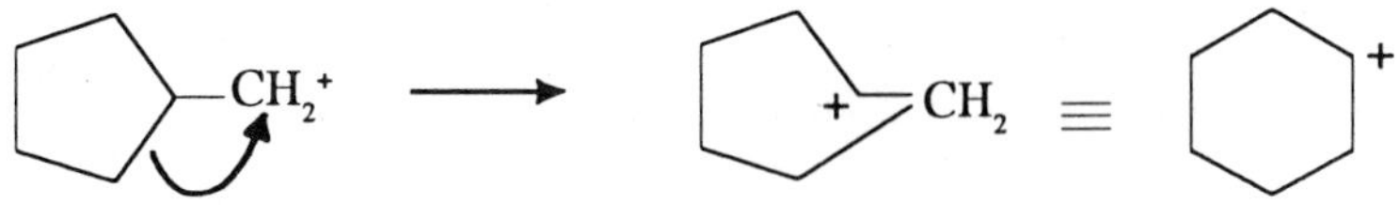

e. H: shift

f. double shift, first H: and then CH_3:

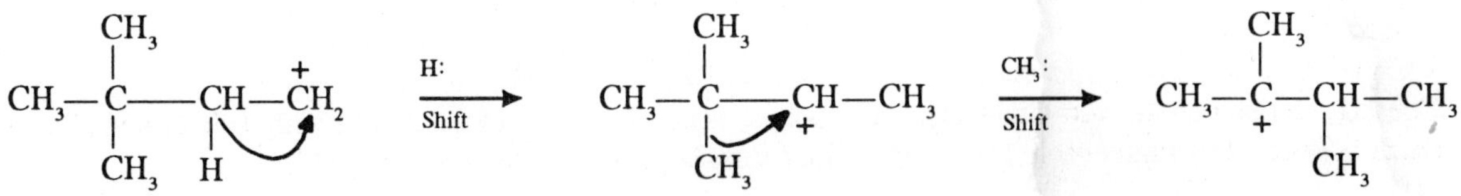

3.11 a. 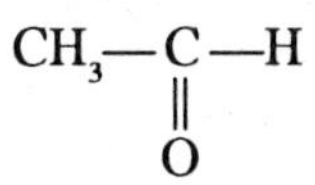 (2 moles)

b. 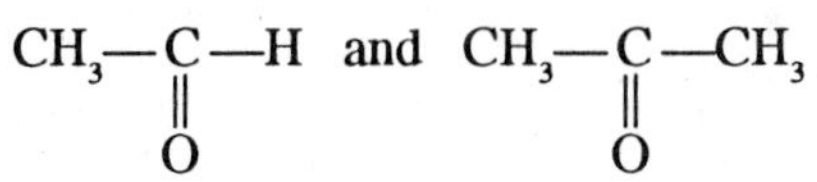

c. 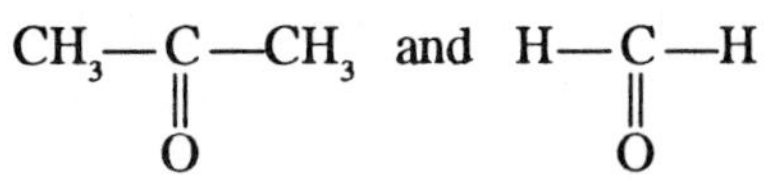

d. 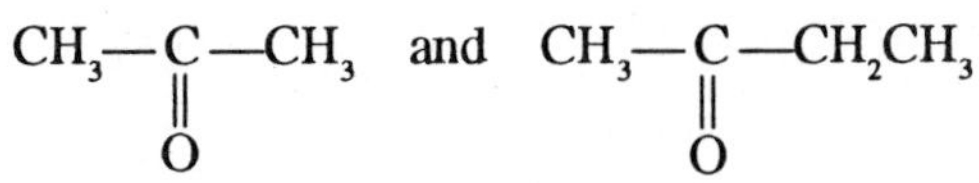

e. 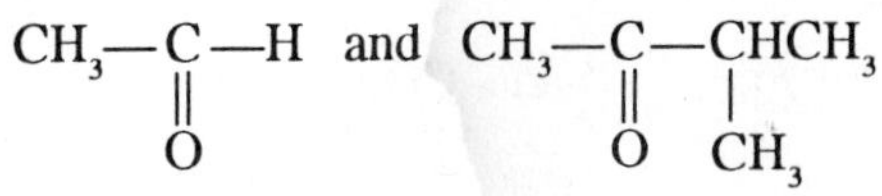

f. 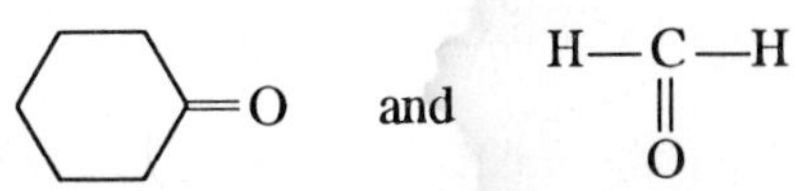

g. 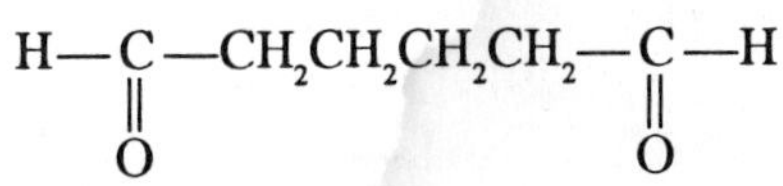

h. 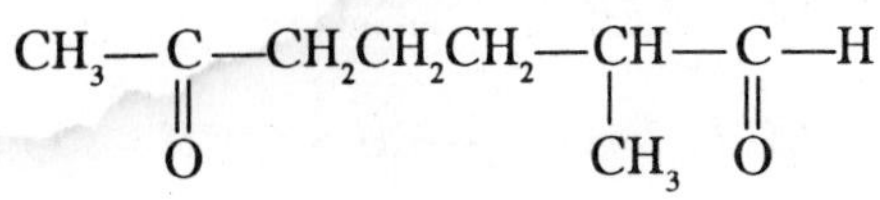

i. $CH_3-C(=O)-CH_3$, $CH_3-C(=O)CH_2C(=O)-CH_3$ and $CH_3-C(=O)-H$

j. $CH_3-C(=O)-H$, $H-C(=O)CH(CH_3)CH(CH_3)C(=O)-H$ and $CH_3-C(=O)-CH_3$

k. 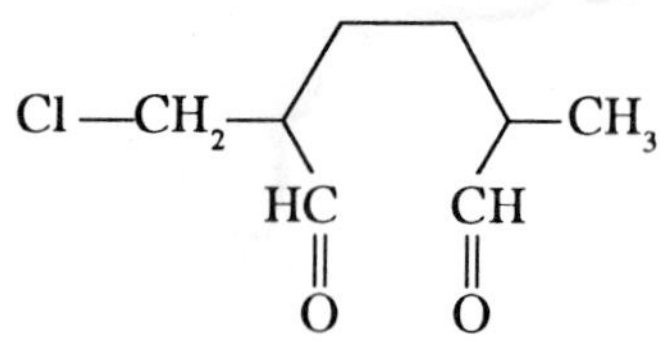

l. 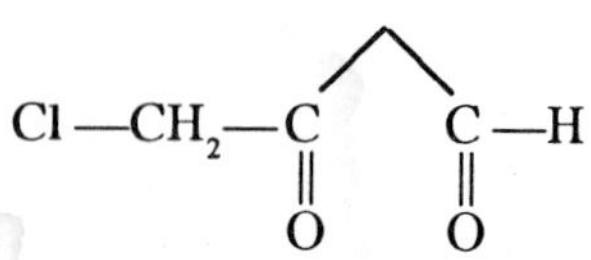 and $H-C(=O)-CH_2-C(=O)-CH_3$

m. 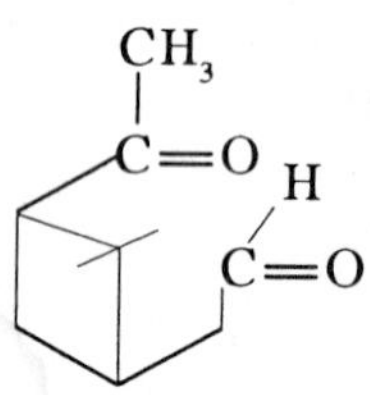

3.12 a. CH_3CO_2H (2 moles)

b. CH_3CO_2H and $(CH_3)_2C{=}O$

c. $(CH_3)_2C{=}O$ and CO_2

d. same as 3.11 (d).

e. CH_3CO_2H and $CH_3-C(=O)-CH(CH_3)_2$

f. 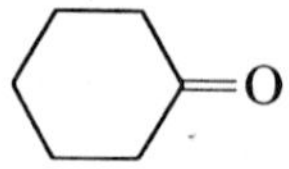 and CO_2

g. $HO_2C-CH_2CH_2CH_2CH_2-CO_2H$

h. $CH_3-C(=O)-CH_2CH_2CH_2CH(CH_3)-CO_2H$

i. $CH_3-C(=O)-CH_3$, $CH_3-C(=O)-CH_2-C(=O)-CH_3$ and CH_3CO_2H

j. CH_3CO_2H , $HO_2C-CH(CH_3)-CH(CH_3)-CO_2H$ and $CH_3-C(=O)-CH_3$

k. 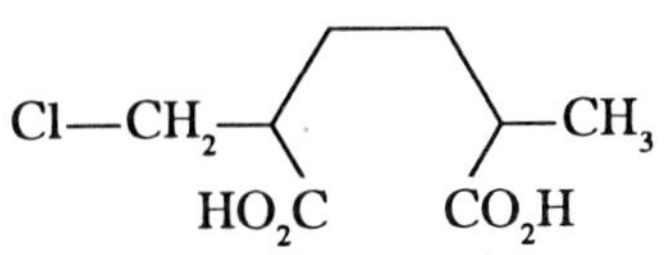

l. 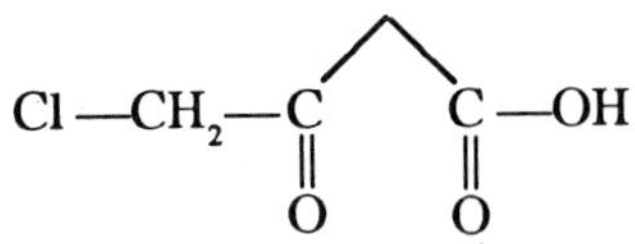 and $HO_2C-CH_2-\underset{O}{\underset{\|}{C}}-CH_3$

m. 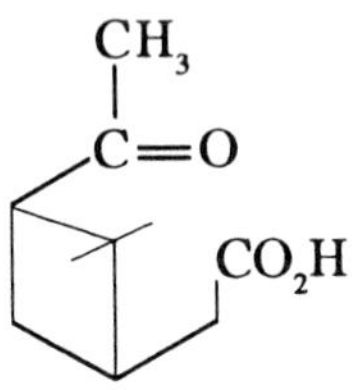

3.13 a. $(CH_3)_2C=C(CH_3)_2$

b. $CH_3CH=CHCH_3$ (*cis* or *trans*)

c. $(CH_3)_2CH-\underset{CH_3}{\underset{|}{C}}=\underset{CH_3}{\underset{|}{C}}-CH(CH_3)_2$

(*cis* or *trans*)

d. $H_2C=CH_2$

e. 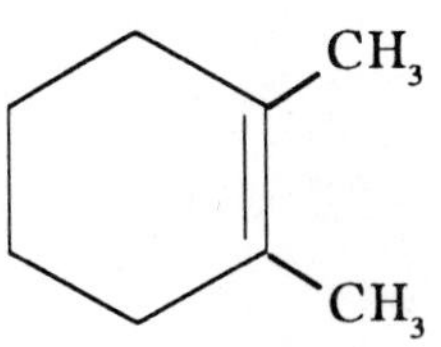

f. 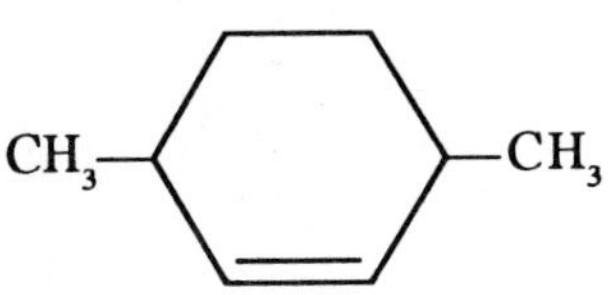

g. $CH_3-CH=CH-CH_2-CH_2-\underset{CH_3}{\underset{|}{C}}=\underset{CH_3}{\underset{|}{C}}-CH_3$

(*cis* or *trans*)

or

$(CH_3)_2C=CH-CH_2-CH_2-\underset{CH_3}{\underset{|}{C}}=CHCH_3$

(*cis* or *trans*)

h. $CH_3-CH=CH-CH_2-CH=CH_2$

(*cis* or *trans*)

i. 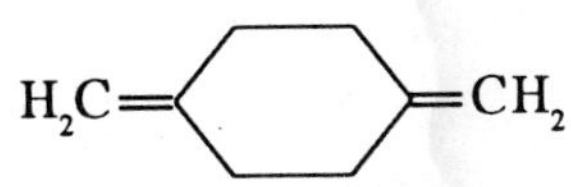

j. 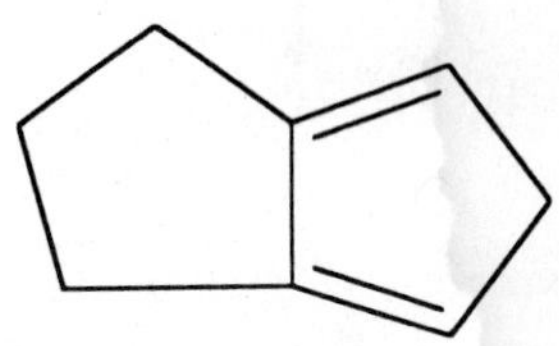

k. 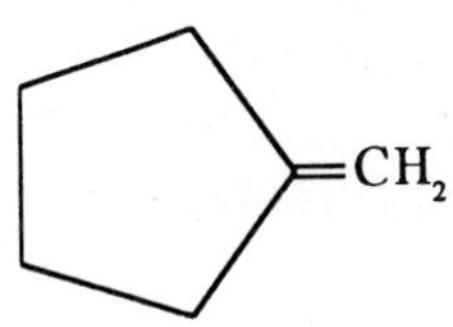

3.14 Sodium dichromate, $Na_2Cr_2O_7$, is an oxidizing agent that can break down a carbon-carbon double bond. Direct treatment of cholesterol with $Na_2Cr_2O_7/H^+$, oxidizes the double bond faster than the alcohol (–OH) functional group.

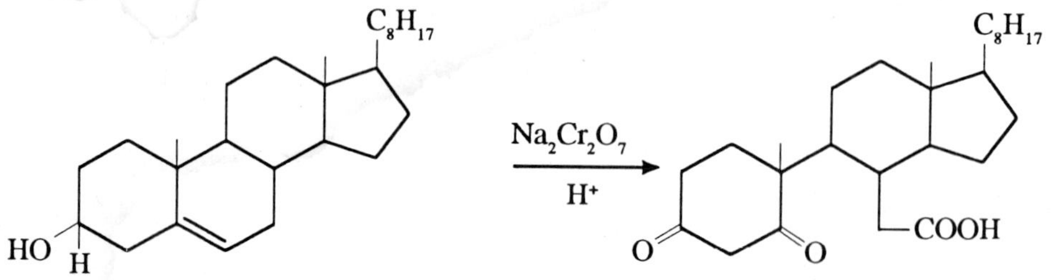

Thus, protection of the double bond with Br_2/CCl_4 is necessary. Sodium dichromate then oxidizes only the alcohol group. The double bond can later be regenerated with Zn.

3.15 a.

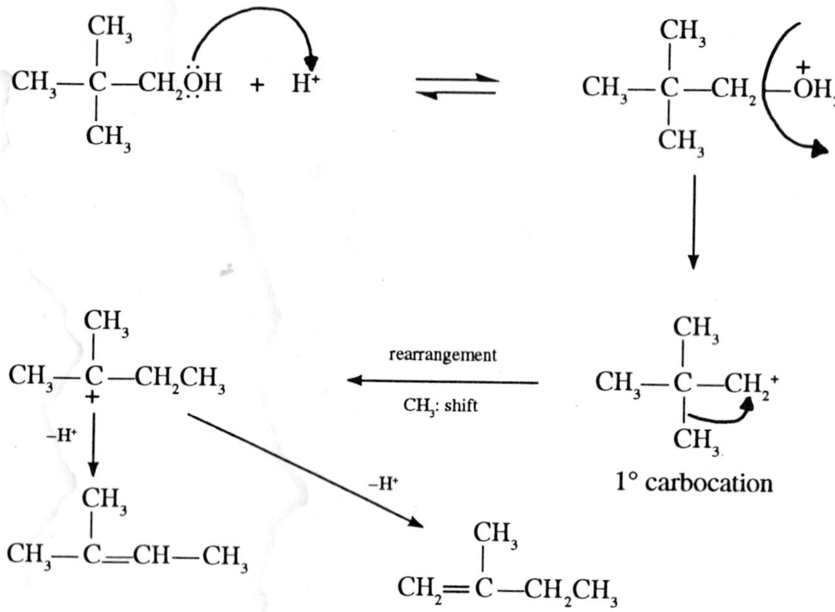

b.

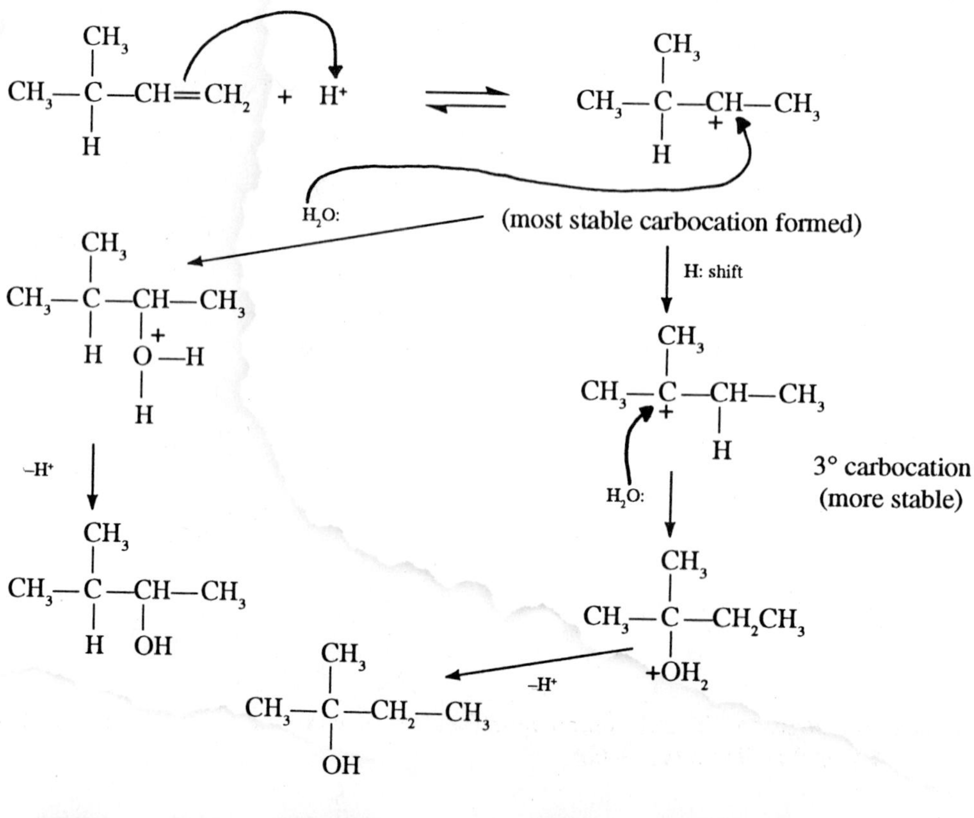

c.

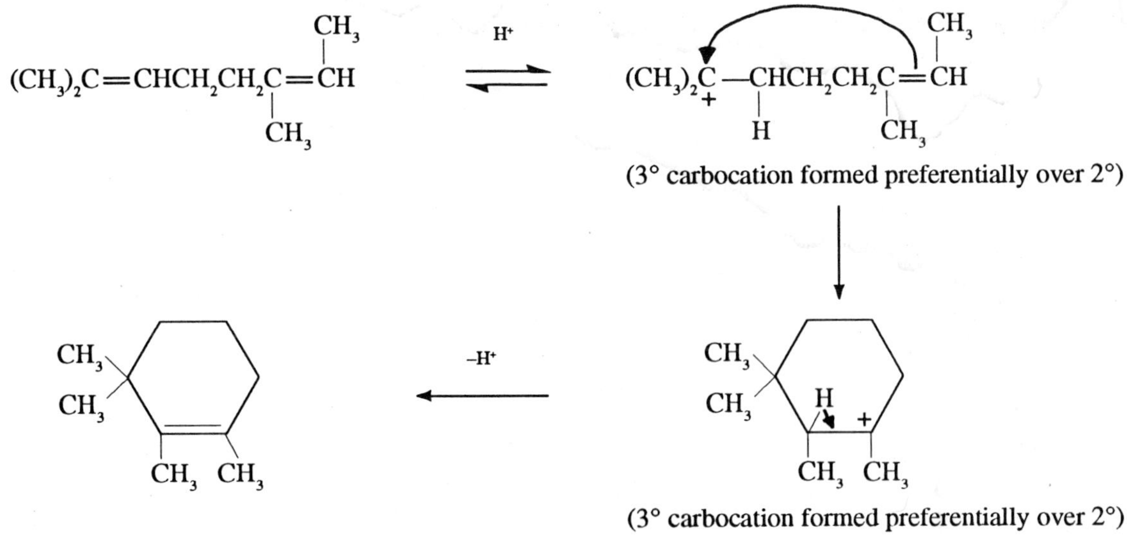

d. *Initiation steps:*

$$RO\text{—}OR \longrightarrow 2RO\cdot$$

$$RO\cdot + Br\text{—}CCl_3 \longrightarrow RO\text{—}Br + \cdot CCl_3$$

Propagation steps:

$$(CH_3)_2CH\text{—}CH\text{=}CH_2 + \cdot CCl_3 \longrightarrow (CH_3)_2CH\text{—}\dot{C}H\text{—}CH_2\text{—}CCl_3$$

$$(CH_3)_2CH\text{—}\dot{C}H\text{—}CH_2\text{—}CCl_3 \xrightarrow{BrCCl_3} (CH_3)_2CH\text{—}CH(Br)\text{—}CH_2\text{—}CCl_3 + \cdot CCl_3$$

Termination steps:

$$(CH_3)_2CH\text{—}\dot{C}H\text{—}CH_2CCl_3 + \cdot CCl_3 \longrightarrow (CH_3)_2CH\text{—}CH(CCl_3)\text{—}CH_2CCl_3$$

$$2\,(CH_3)_2CH\text{—}\dot{C}H\text{—}CH_2\text{—}CCl_3 \xrightarrow{\text{dimerization}} (CH_3)_2CH\text{—}CH(CH_2CCl_3)\text{—}CH(CH_2CCl_3)\text{—}CH(CH_3)_2$$

e.

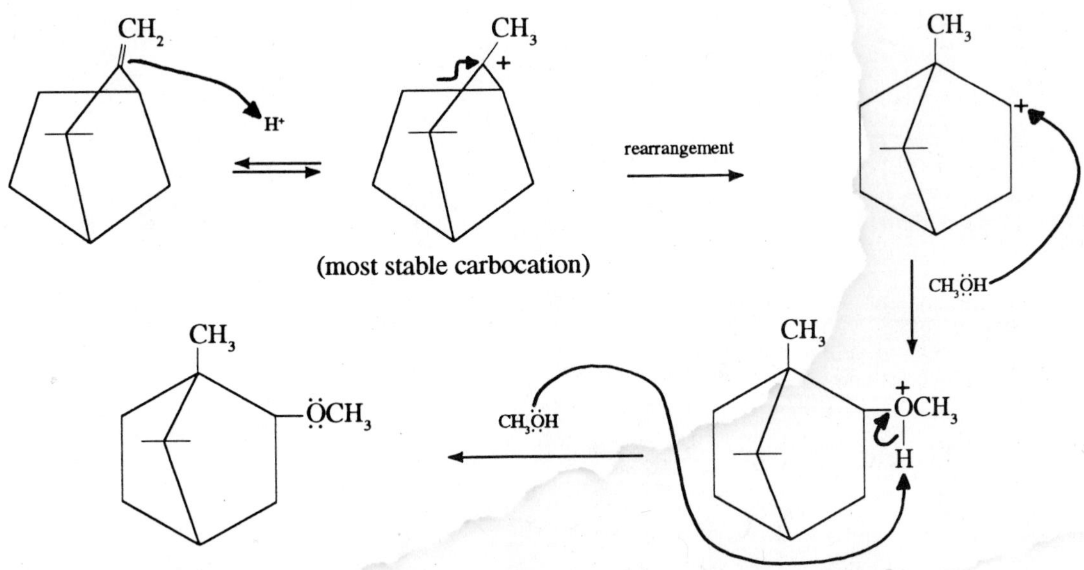

The 3° carbocation rearranges to a 2° carbocation; however, the 1,2-alkyl shift results in a ring system that has less ring strain and that is therefore more stable.

f.

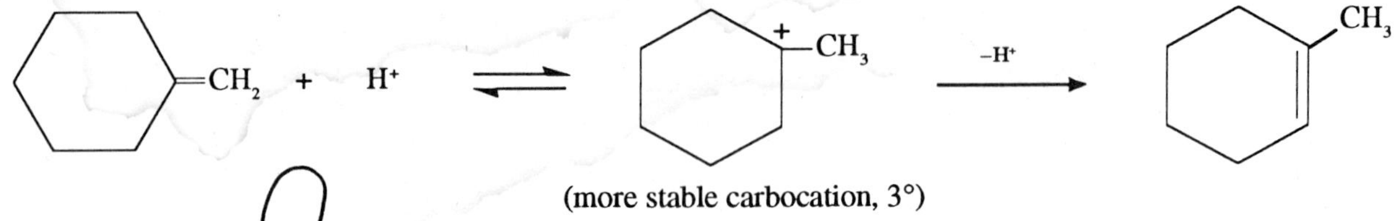

=CH2 + H+
CH3
–H+
CH3
(more stable carbocation, 3°)

g.

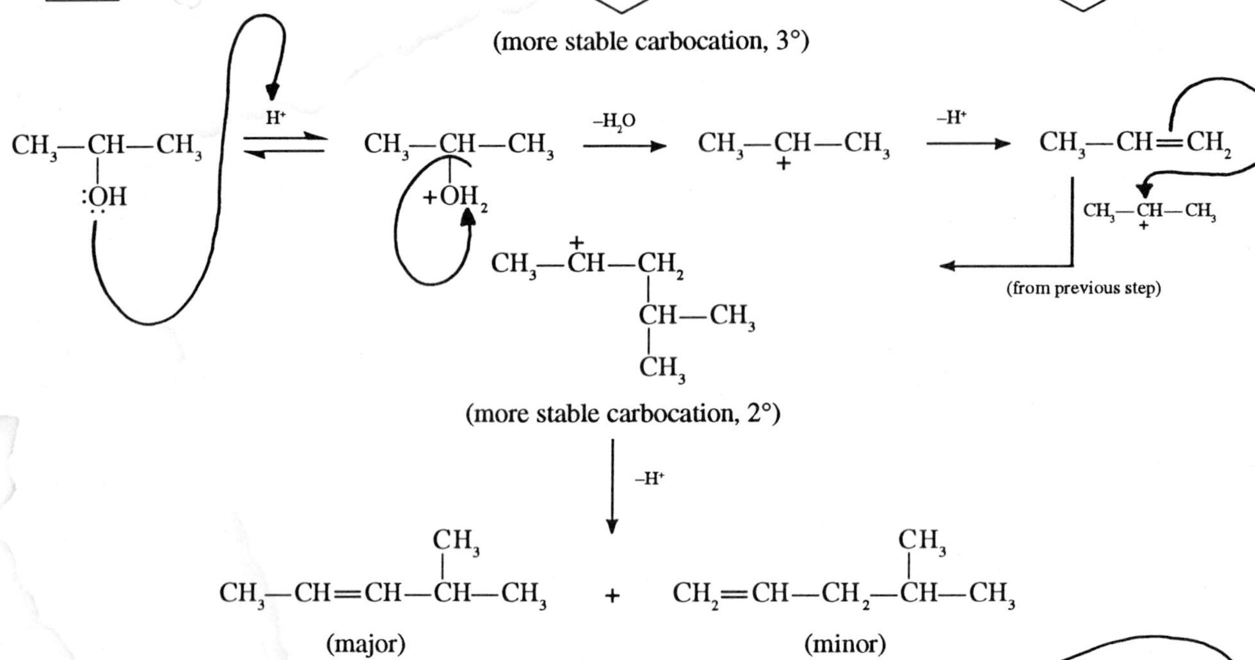

CH3—CH—CH3
:OH
H+
CH3—CH—CH3
+OH2
–H2O
CH3—CH—CH3
–H+
CH3—CH=CH2
CH3—CH—CH3
(from previous step)
CH3—CH—CH2
CH—CH3
CH3
(more stable carbocation, 2°)
–H+
CH3—CH=CH—CH—CH3
CH3
+
CH2=CH—CH2—CH—CH3
(major)
(minor)

h.

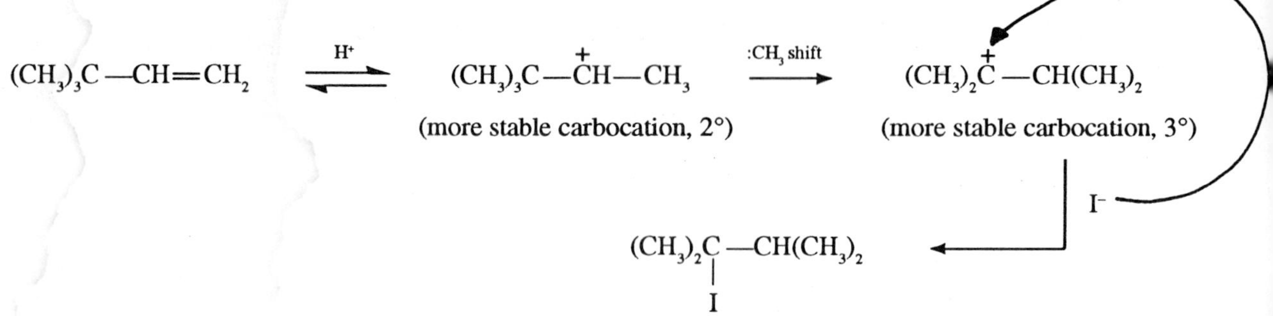

(CH3)3C—CH=CH2
H+
(CH3)3C—CH—CH3
(more stable carbocation, 2°)
:CH3 shift
(CH3)2C—CH(CH3)2
(more stable carbocation, 3°)
I–
(CH3)2C—CH(CH3)2
I

i.

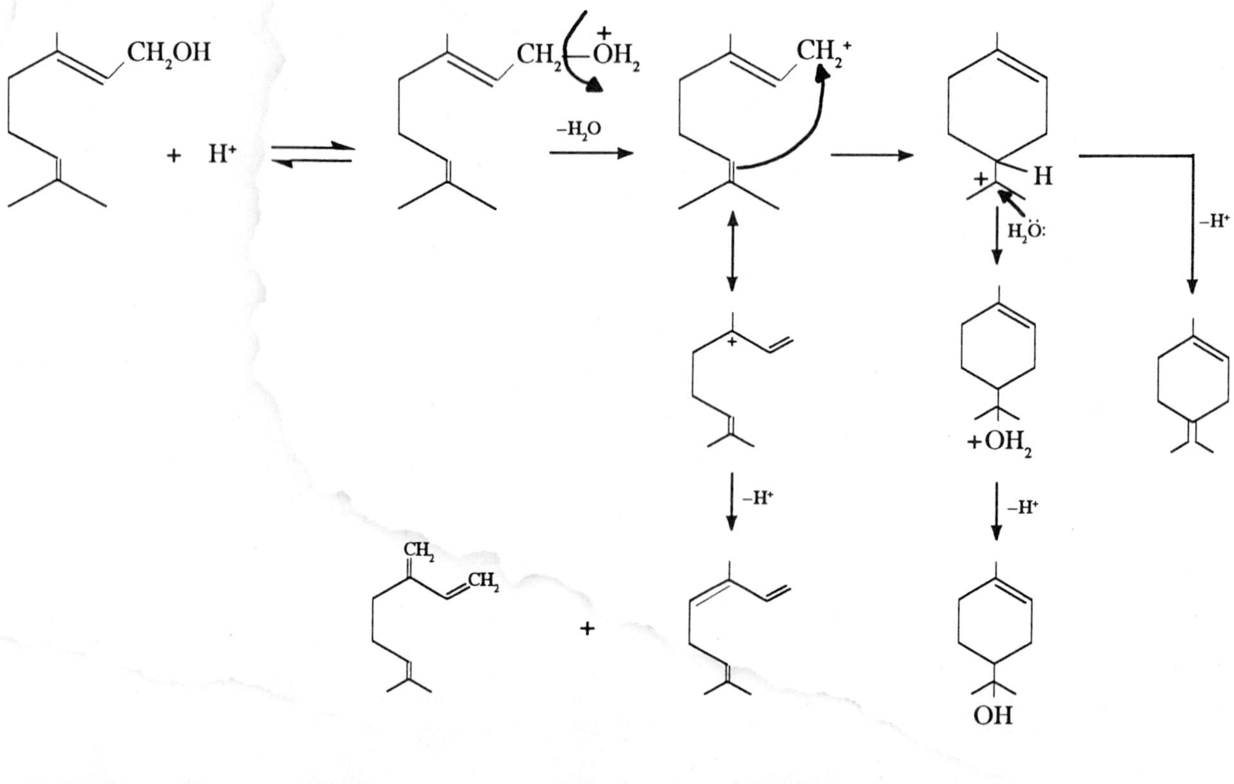

CH2OH
+ H+
CH2—OH2
–H2O
CH2+
H
H2Ö:
–H+
+OH2
–H+
–H+
CH2
CH2
+
OH

j.

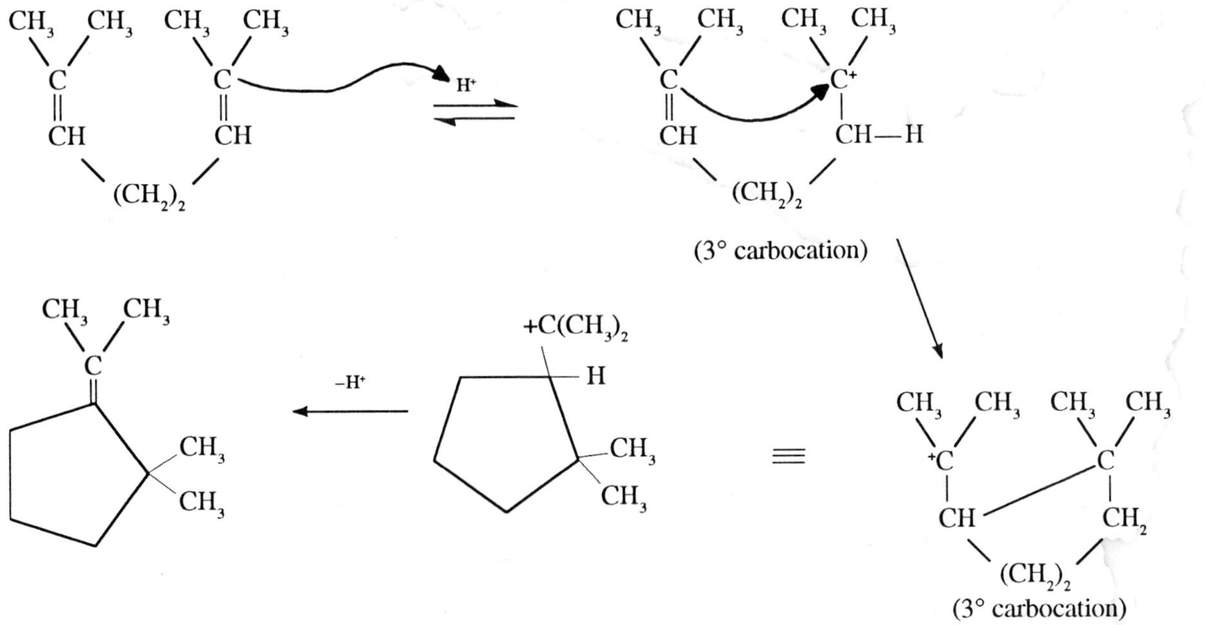

3.16 $(CH_3)_2CH—CH=CH_2$ + HCl $\longrightarrow$ $(CH_3)_2C(Cl)CH_2CH_3$ + $(CH_3)_2CHCH(Cl)CH_3$

Mechanism:

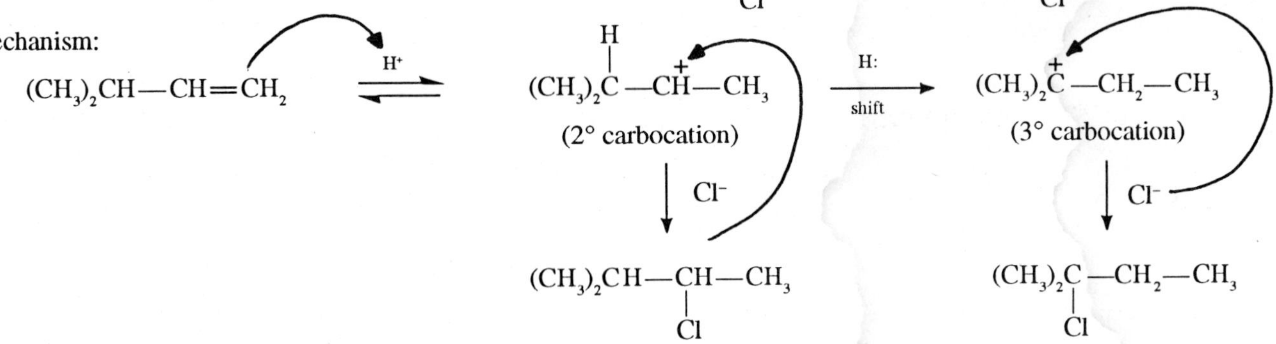

3.17 **A** = $(CH_3)_2C{=}CH_2$

B = $(CH_3)_2CHCH_2Br$

C = $(CH_3)_2CHCH_3$

D = $(CH_3)_2C(OH)CH_3$

E = $(CH_3)_2C{=}CH_2$

F = $(CH_3)_2CHCH_2CH_3$

G = $(CH_3)_2C(OH)—CH_2OH$

H = $(CH_3)_2C(Br)—CH_3$

I = $(CH_3)_3CMgBr$

J = $(CH_3)_3CH$

K = MgCl(OH)

3.18 a.

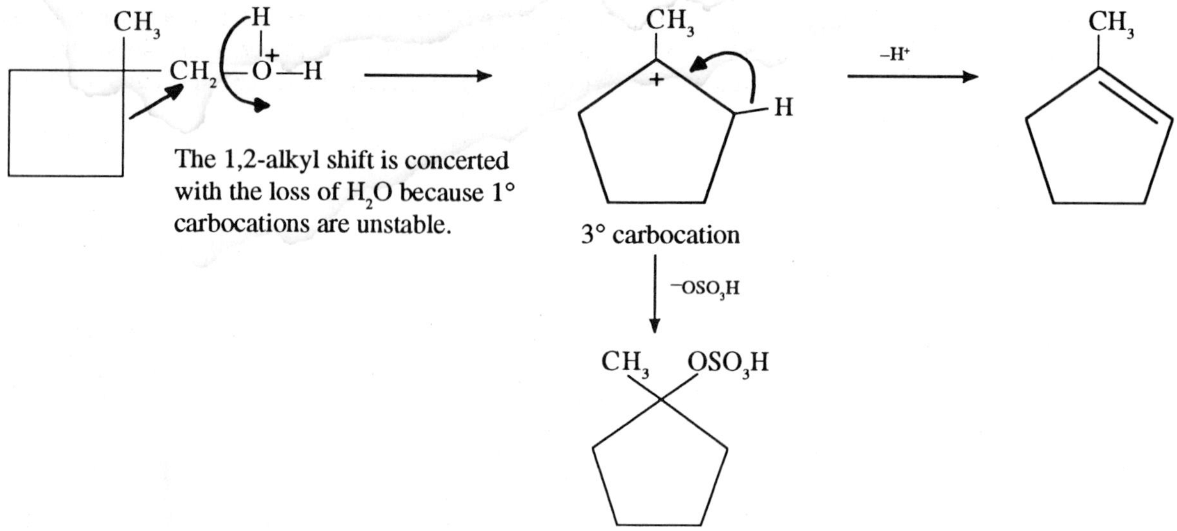
CH3
CH2
H
O—H
The 1,2-alkyl shift is concerted with the loss of H2O because 1° carbocations are unstable.
3° carbocation
–H+
–OSO3H
CH3 OSO3H

b.

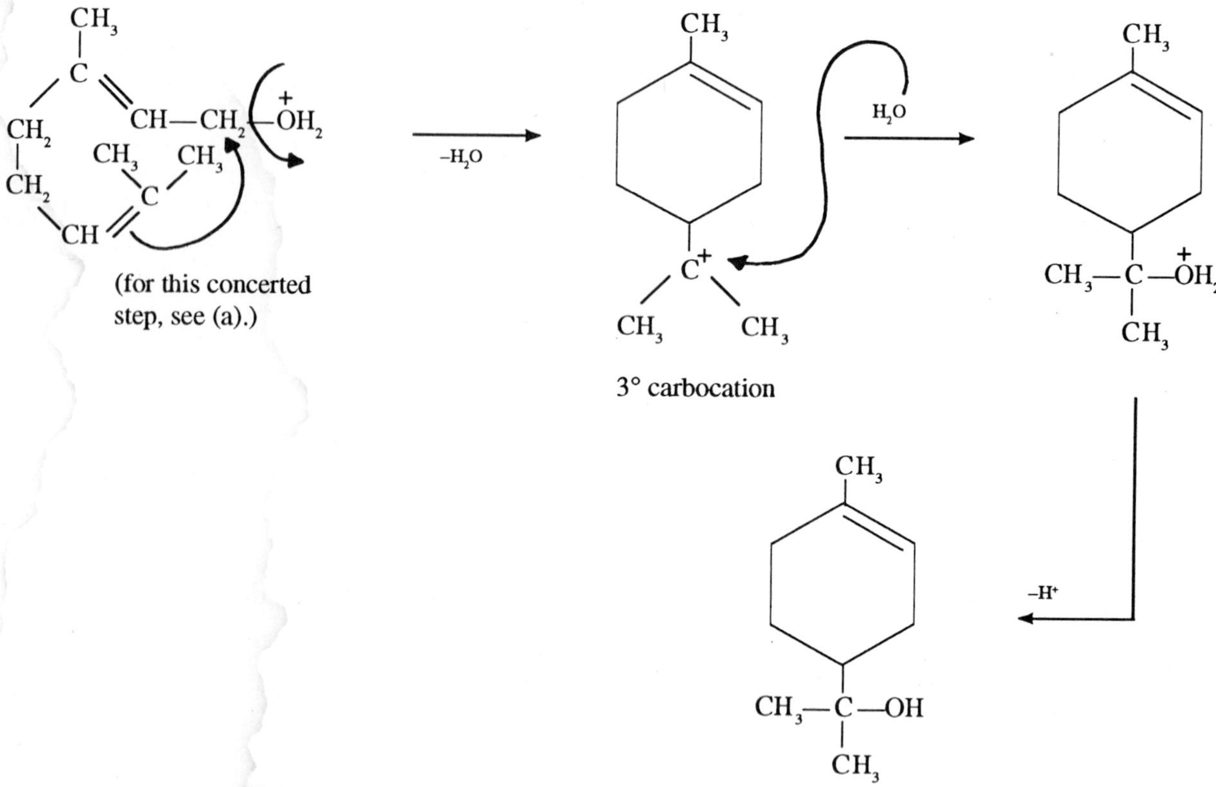
(for this concerted step, see (a).)
–H2O
H2O
3° carbocation
CH3—C—OH2
–H+
CH3—C—OH

c.

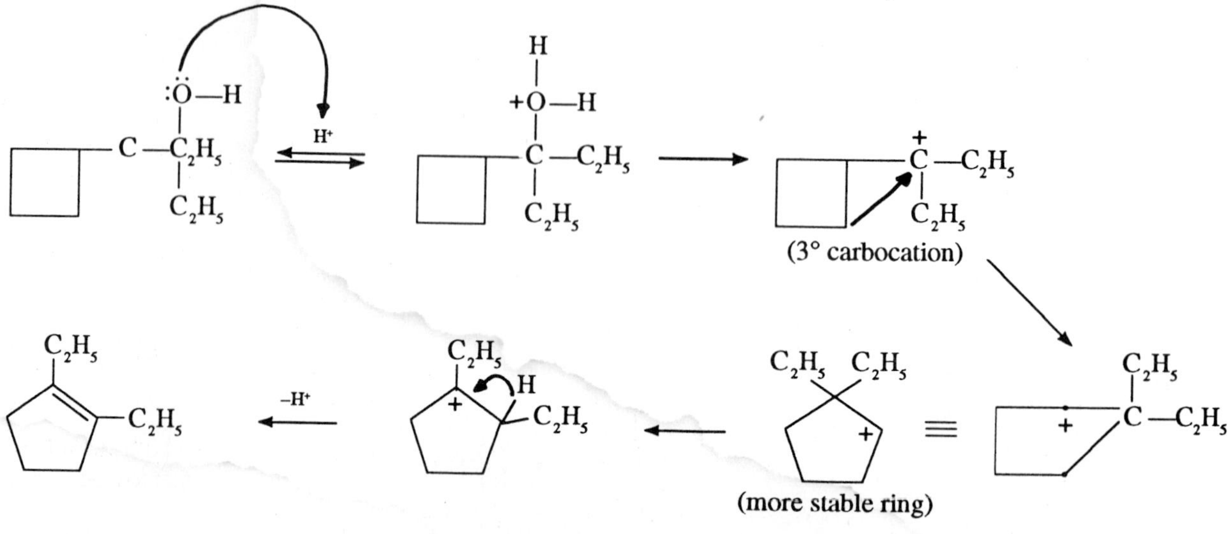
:Ö—H
C—C2H5
H+
+O—H
(3° carbocation)
(more stable ring)
–H+
C2H5

3.19 Markovnikov addition to methylene cyclobutane would result in the unstable cyclobutyl carbocation. Since bonding in carbocations is sp^2 hybridized with bond angles of 120°, carbocation formation in 3- and 4-membered rings, which have constricted bond angles of 60° and 90° respectively, is highly unfavored. Such angle strain in six membered rings is less significant, i.e.,

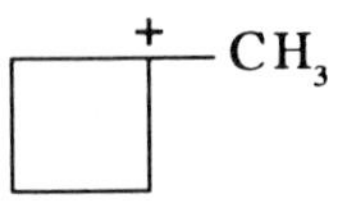

does not tend to form.

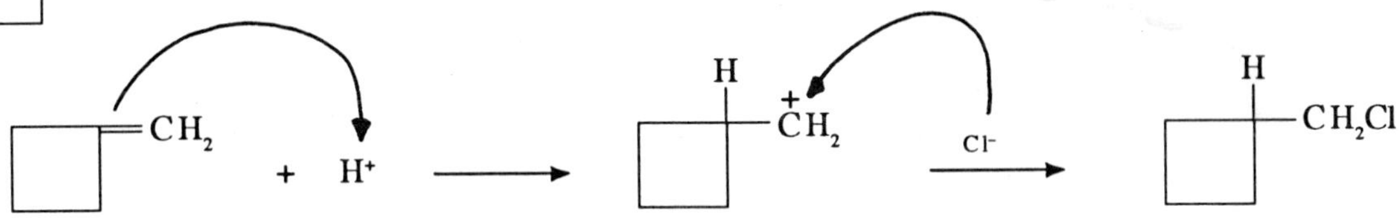

3.20 a. $HCCl_3 + (CH_3)_3CO^- \rightleftharpoons :\bar{C}Cl_3 + HOC(CH_3)_3$

$^-:CCl_3 \rightleftharpoons :CCl_2 + Cl^-$ (α-elimination)

b. The loss of a proton in the first step is reversible. In solvent (D_2O) the $:\bar{C}Cl_3$ anion can abstract D^+ to form $DCCl_3$:

$$CHCl_3 + OH^- \rightleftharpoons {}^-:CCl_3 + H_2O \overset{D_2O}{\rightleftharpoons} CDCl_3 + OD^-$$

Carbanion
(stabilized by the
electron withdrawing
chlorine atom)

c. <u>Singlet carbene</u>: sp^2 hybridized, stereospecific insertion reactions

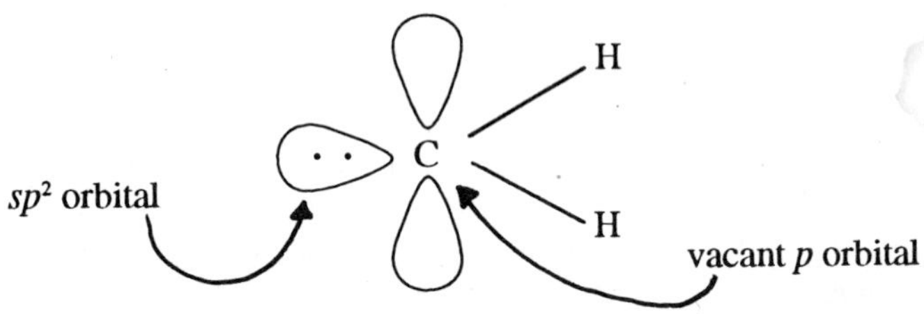

<u>Triplet carbene</u>: sp hybridized, radical character and non-stereospecific in reactions

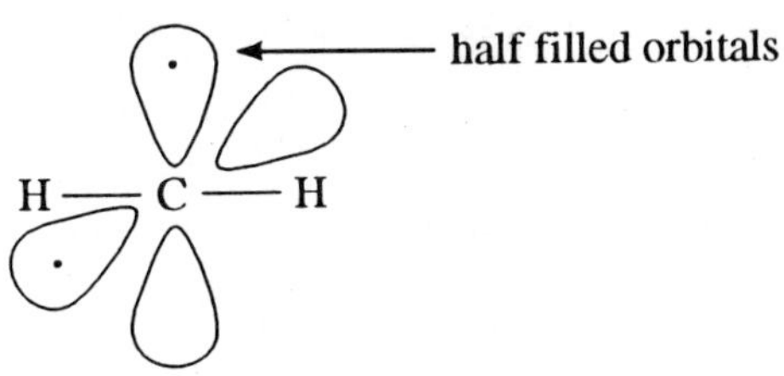

d.

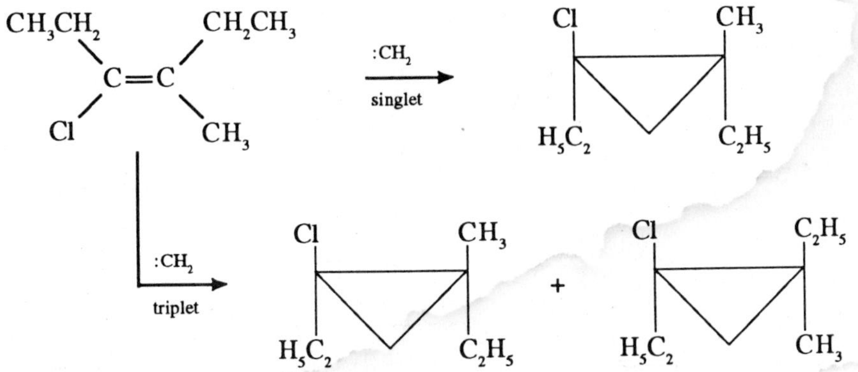

Triplet carbene proceeds via

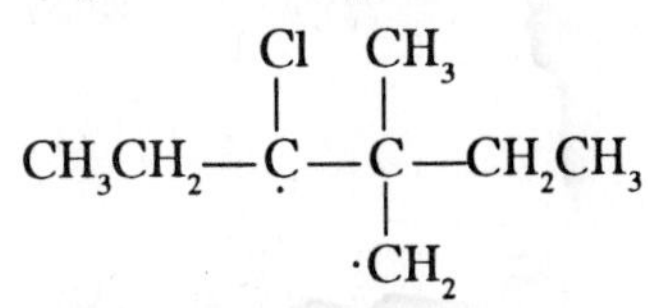

which can rotate and lose stereospecificity.

3.21 a.

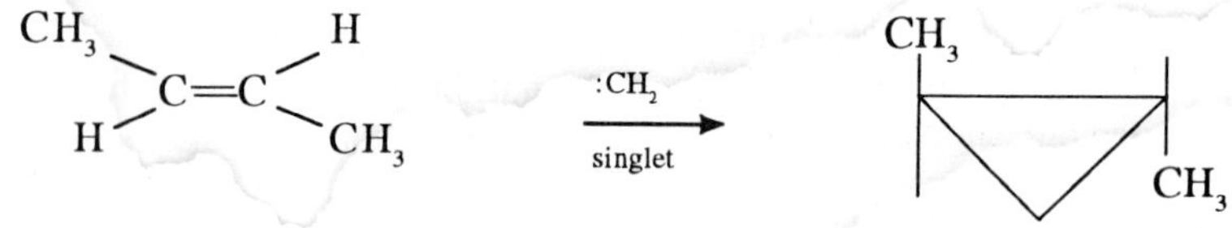

b.

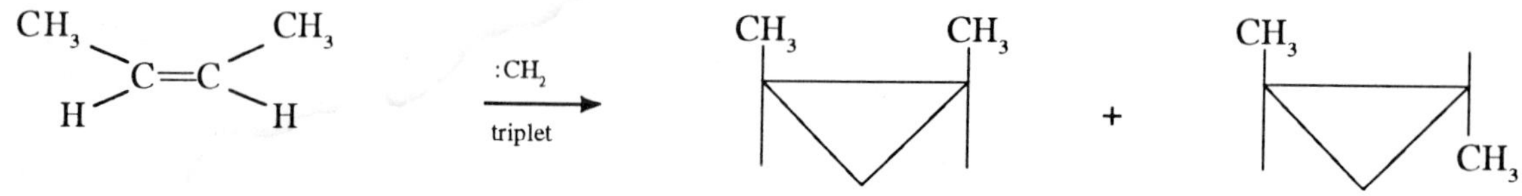

Remember, triplet carbene is not stereospecific.

3.22 a. $CH_3CH{=}CH_2$ + H_2/Pd (or Pt) ⟶ $CH_3CH_2CH_3$

b. $CH_3CH_2CH_3$ + Br_2/UV ⟶ $CH_3CH(Br)CH_3$ (major) + $CH_3CH_2CH_2Br$ (minor)

Both products, alc. KOH ⟶ $CH_3CH{=}CH_2$

c. $CH_3CH_2CH_3$ + Br_2/UV ⟶ $CH_3CHBrCH_3$ (more than 99%)

alc. KOH ↓

$CH_3CH{=}CH_2$ —HBr/peroxides⟶ $CH_3CH_2CH_2Br$

d. $CH_3CH_2CH_3$ ⟶ $CH_3CH{=}CH_2$ (as prepared in *b*) —HBr/no peroxides⟶ $CH_3CH(Br)CH_3$

Note: Direct bromination via Br_2/UV will also yield the desired product in high yield; this may be considered an acceptable synthetic path.

e. $(CH_3)_2CHCH_3$ + Br_2/UV ⟶ $(CH_3)_3C{-}Br$ —alc. KOH⟶ $(CH_3)_2C{=}CH_2$ —1. O_3; 2. Zn, H_2O (or hot $KMnO_4$)⟶ $(CH_3)_2C{=}O$

f. $CH_3CH_2CH_2CH_3$ + Br_2/UV ⟶ $CH_3CH_2CH(Br)CH_3$ (major; more than 99%)

alc. KOH ↓

$CH_3CH{=}CHCH_3$ (mixture of *cis* and *trans*) —1. $KMnO_4$, OH^- 2. H^+/H_2O⟶ CH_3CO_2H (2 moles)

g. $(CH_3)_3C{-}H$ + Br_2/UV ⟶ $(CH_3)_2CBrCH_3$ —alc. KOH⟶ $(CH_3)_2C{=}CH_2$ —H_2O/H^+⟶ $(CH_3)_2C(OH){-}CH_3$

h. $CH_3CH_2CH_3 + Br_2/UV \longrightarrow CH_3CHBrCH_3$ (major, more than 99%) $\xrightarrow{\text{alc. KOH}} CH_3CH{=}CH_2 \xrightarrow{H_2O/Cl_2} CH_3CH(OH)CH_2Cl$

i. $(CH_3)_2CH{-}CH_2OH + H^+/\text{heat} \longrightarrow (CH_3)_2C{=}CH_2 \xrightarrow[H_2O/H^+]{} (CH_3)_2C(OH){-}CH_3$

j. $(CH_3)_2C(OH){-}CH_3 + H^+/\text{heat} \longrightarrow (CH_3)_2C{=}CH_2 \xrightarrow{H_2/Pt} (CH_3)_3C{-}H$

k. $(CH_3)_2C(OH){-}CH_3 \xrightarrow[\text{heat}]{H^+} (CH_3)_2C{=}CH_2 \xrightarrow[2.\ H_2O_2/OH^-]{1.\ BH_3/THF} (CH_3)_2CH{-}CH_2{-}OH \xrightarrow{PCl_3} (CH_3)_2CH{-}CH_2Cl$

l. $(CH_3)_2C(Br)CH_2Br + \text{Zn dust} \xrightarrow{\Delta} (CH_3)_2C{=}CH_2 \xrightarrow[2.\ H_2O_2/NaOH]{1.\ BH_3/THF} (CH_3)_2CH{-}CH_2OH$

m. $CH_3CH_2CH_3 + Br_2/UV \longrightarrow (CH_3)_2CHBr \xrightarrow{\text{alc. KOH}} CH_3CH{=}CH_2 \xrightarrow{NBS} BrCH_2CH{=}CH_2 \xrightarrow{\text{HBr/peroxides}} BrCH_2CH_2CH_2Br$

(NBS= N-bromosuccinimide)

n. $CH_3CH_2CH_3 \longrightarrow CH_3CH_2CH_2Br$ (as prepared in (c)) $\xrightarrow[2.\ CuI]{1.\ Li} (CH_3{-}CH_2{-}CH_2)_2CuLi \xrightarrow{CH_3CH_2CH_2Br} CH_3CH_2CH_2{-}CH_2CH_2CH_3$

Note: The product may also be produced by the Wurtz reaction:

$$2CH_3CH_2CH_2{-}Br + 2Na \longrightarrow CH_3CH_2CH_2{-}CH_2CH_2CH_3 + 2NaBr$$

3.23 Adding Br_2 to the mixture converts cyclohexene to 1,2-dibromocyclohexane, which has a much higher boiling point than cyclohexane. The cyclohexane can be separated by distillation, and cyclohexene can be regenerated by treating 1,2-dibromocyclohexane with Zn dust.

3.24 Assuming an ideal gas behavior

PV= nRT = (mass/M.W.)RT, where

P = pressure = 1 atm	V = volume = 1.12×10^{-3} L
R = gas constant = 0.082 atm.L/K.mole	Mass = 4.9×10^{-3} g
T = temperature = 273K	M.W. = molecular weight

$$\text{Thus, M.W.} = \frac{\text{mass} \times R \times T}{P \times V} = \frac{4.9 \times 10^{-3} \times 0.082 \times 273}{1 \times 1.12 \times 10^{-3}} = 98$$

Possible formulas: C_7H_{14} (maximum one double bond) and C_8H_2 (maximum 8 double bonds). Most likely formula: C_7H_{14}.

3.25 Number of moles of hydrocarbon: 0.235/94 = 0.0025

Number of moles of H_2 = PV/RT (from the ideal gas equation)

at STP = P = 1 atm   V = 0.168 L

T = 273 K   R = 0.082 atm.L/K.mole

Using the equation, we find that the number of moles of H_2 is 0.0075. Thus, for one mole of hydrocarbon, three moles of H_2 are consumed. This is derived from the ratio 0.0075/0.0025 = 3. Therefore, 3 π bonds are present. Since the compound is a hydrocarbon, there are only carbon and hydrogen atoms present. The only formula that matches both the molecular weight and the number of double bonds is C_7H_{10}.

3.26 a. $$n_{H_2} = \frac{0.1007}{0.082 \times 273} = 4.5 \times 10^{-3} \text{ moles}$$

The number of moles of hydrocarbon is $n = \dfrac{0.138}{92} = 1.5 \times 10^{-3}$ moles

There must be 3 double bonds, because the number of moles of H_2 absorbed is three times the number of moles of hydrocarbon.

To have a molecular weight of 92, the hydrocarbon must have 7 carbon atoms and 8 hydrogen atoms, C_7H_8 (*Note:* 6 carbons would show an excess of hydrogens). There are four degrees of unsaturation: three double bonds and a ring.

b. C_7H_8 indicates a hydrogen deficiency of 8. Three double bonds account for 6 hydrogens and there is, therefore, one ring. Ozonolysis products account for 5 carbons. To account for 7 carbons, there must be 2 moles of H—C(=O)—C(=O)—H for 1 mole of H—C(=O)—CH_2—C(=O)H. Putting it together to form one ring with three double bonds.

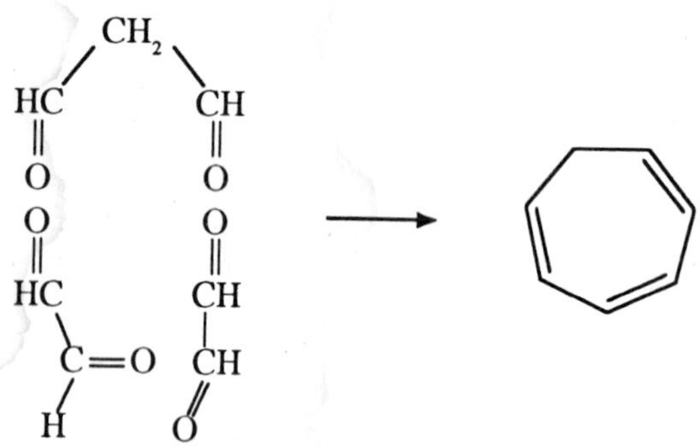

1,3,5-cycloheptatriene

3.27 $CH_3CH{=}CHCH_2CH_3$ + Br_2 ⟶ $CH_3CH(Br)$—$CH(Br)CH_2CH_3$

2-pentene

Molar ratio:	1	1	1
Mol. weight:	70	160	230
Stoichiometric molar ratio:	0.71	0.71	0.71 (expected)

Since the number of moles is mass/M.W., the mass of Br_2 = 0.71 × 160 = 114 g, and the mass of product = 0.71 × 230 = 163 g

3.28 a.

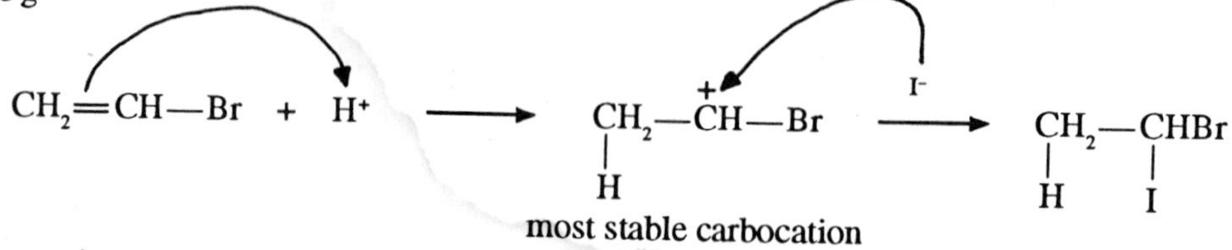

$CH_3\overset{+}{C}HBr$ is more stable than $^+CH_2CH_2Br$ because of resonance stabilization:

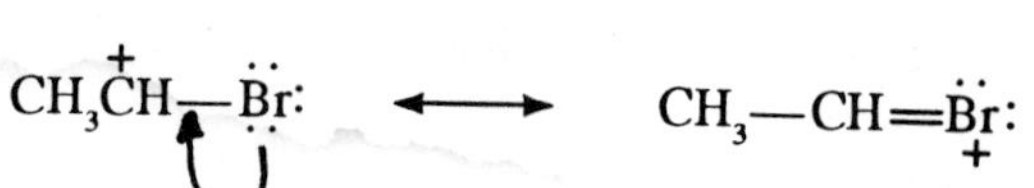

b. 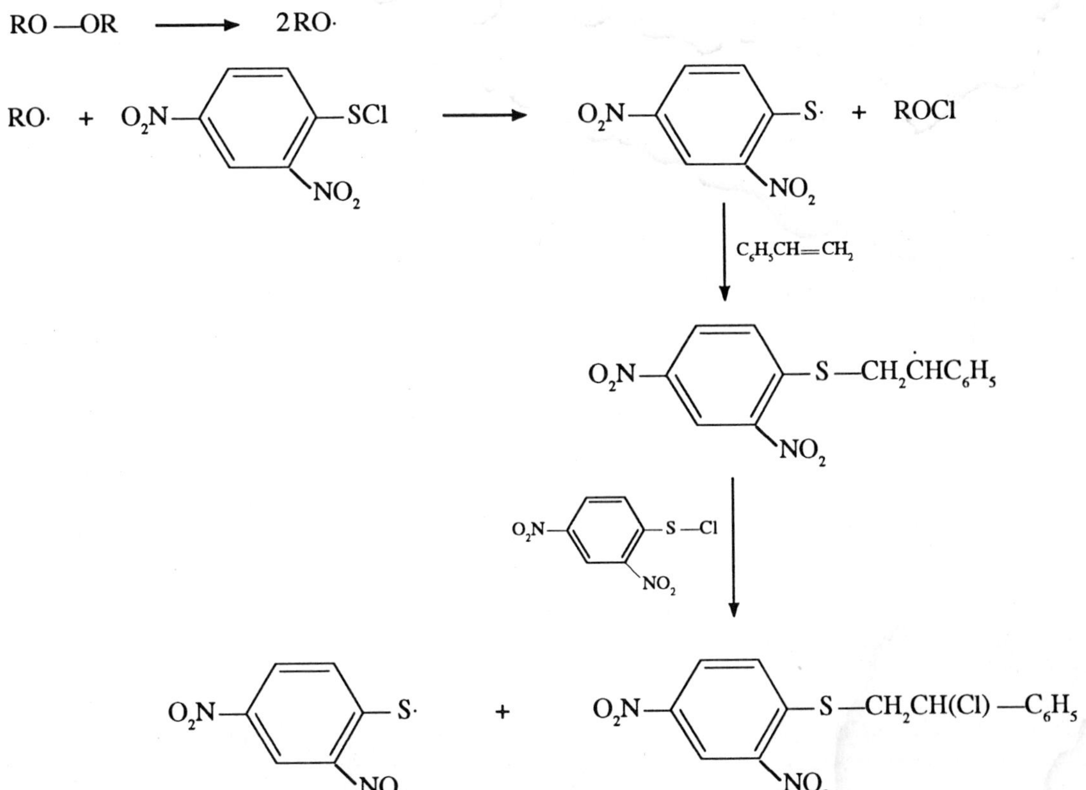

c. 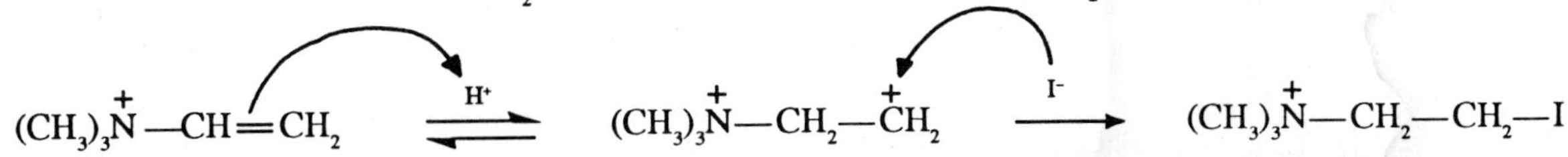

Markovnikov addition of HI does not occur because the intermediate carbocation $(CH_3)_3\overset{+}{N}—\overset{+}{C}H—CH_3$ (which has adjacent positive charges) is much less stable than the carbocation which leads to anti-Markovnikov addition.

d. 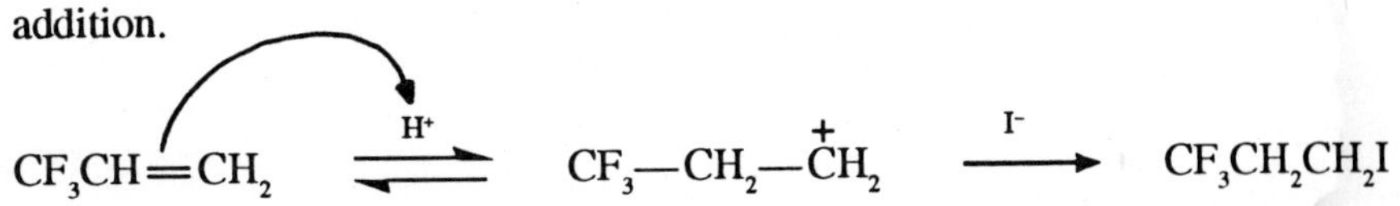

Anti-Markovnikov addition occurs, the Markovnikov intermediate, $CF_3\overset{+}{C}H—CH_3$, is less stable because the positive charge is adjacent to the very electronegative CF_3 group.

Answers for Multiple Choice Test

3.1 a	**3.6** c	**3.11** b	**3.16** d
3.2 b	**3.7** d	**3.12** b	**3.17** b
3.3 c	**3.8** d	**3.13** b	**3.18** a
3.4 b	**3.9** c	**3.14** c	**3.19** a
3.5 a	**3.10** a	**3.15** c	**3.20** d

4 Alkenes and Dienes

Questions

4.1 Identify each of the following unsaturated compounds as alkenes, alkynes, conjugated dienes, nonconjugated dienes or cumulated dienes.

a. $CH_2{=}CH{-}CH{=}CH_2$

b. $CH_2{=}C{=}CH{-}CH_3$

c. $CH_2{=}CH{-}CH_2{-}CH{=}CH_2$

d. $CH_3{-}CH{=}CH{-}CH_2CH_2CH_3$

e.

f.

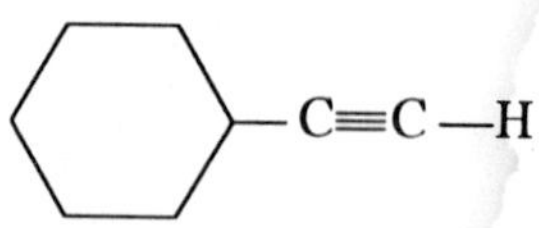

g.

h.

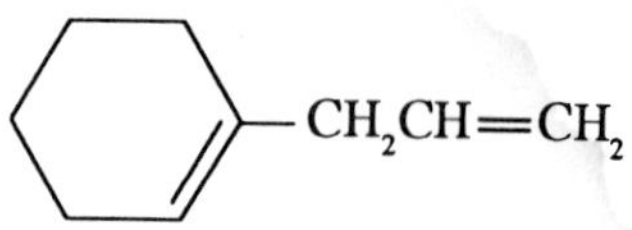

i.

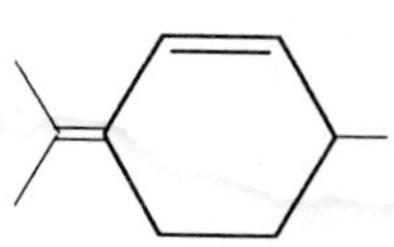

j. $CH_3C{\equiv}CCH_2CH_3$

k. $CH_2{=}C(CH_3){-}CH{=}CH{-}CH_3$

l. $CH_2{=}C{=}CH_2$

m. $CH_3{-}CH{=}C{=}CH{-}CH_3$

n.

o.

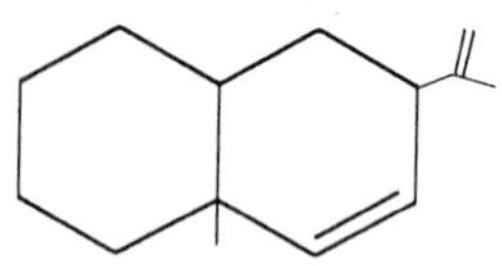

p.

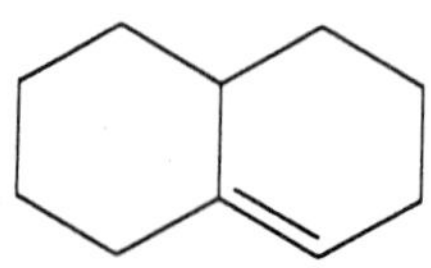

q.

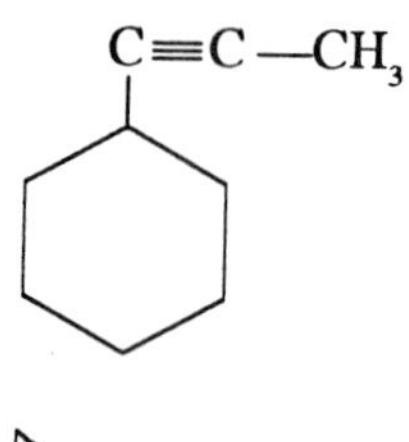

r.

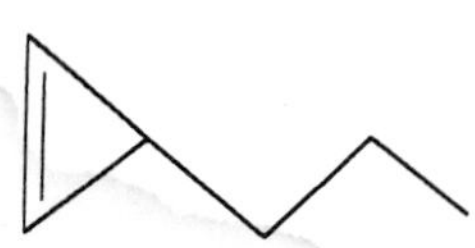

4.2 Write structures and give names for
a. all alkynes of molecular formula C_5H_8.
b. all alkynes of molecular formula C_6H_{10}.
c. all dienes of molecular formula C_5H_8 (identify them as conjugated, nonconjugated or cumulated).
d. all conjugated dienes of molecular formula C_6H_{10}.

4.3 Name the following compounds:

a. $CH_3—C{\equiv}C—CH_3$

b. 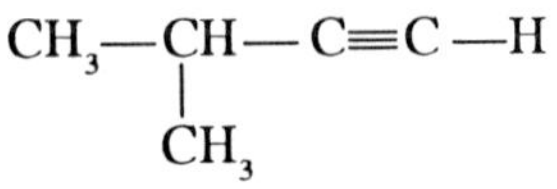

c. $CH_2{=}C{=}CH_2$

d. $HC{\equiv}CCH_2C{\equiv}CH$

e. $HC{\equiv}C^{-}Ag^{+}$

f. $ClCH_2—C{\equiv}C—CH_2Cl$

g. 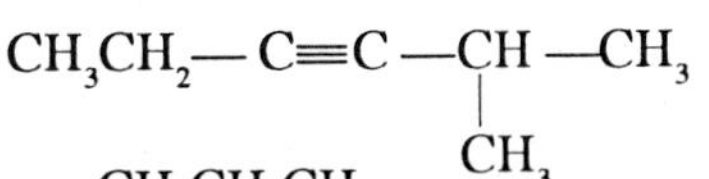

h. 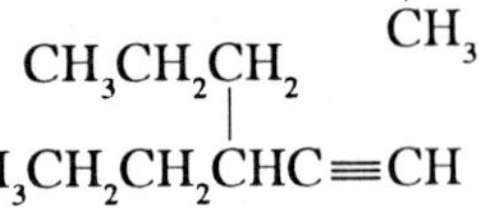

i. $CH_3C{\equiv}C^{-}Cu^{+}$

j. 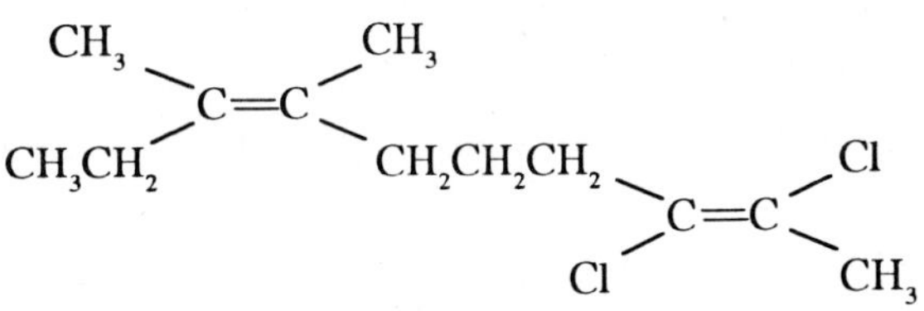

k. 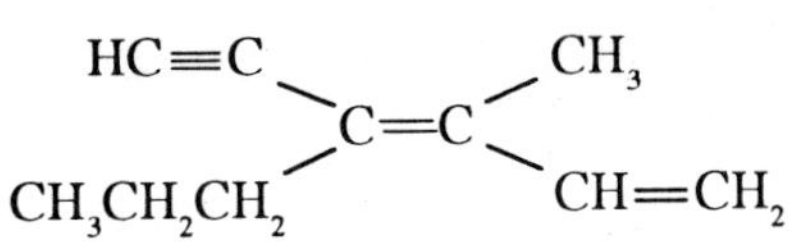

l. 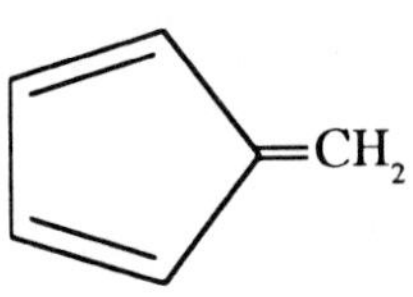

m. $ClCH_2CH_2CH_2—C{\equiv}C—CH_2CH_2CH_2Cl$

n. $(CH_3)_3C—C{\equiv}C—C(CH_3)_3$

o. 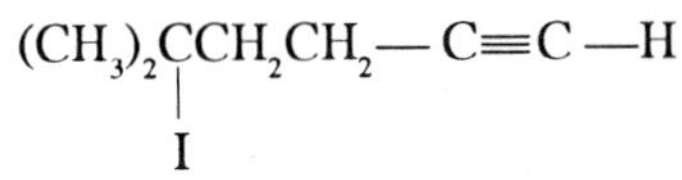

p. $HC{\equiv}C—CH_2CH{=}CH_2$

q. 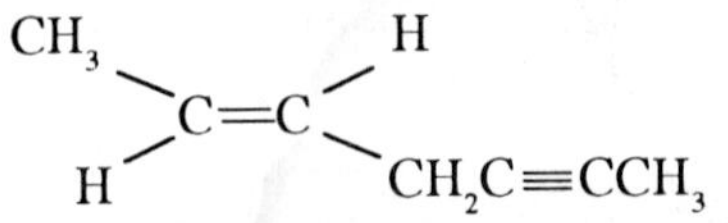

r. $CH_2{=}CH—C{\equiv}C—H$

s. 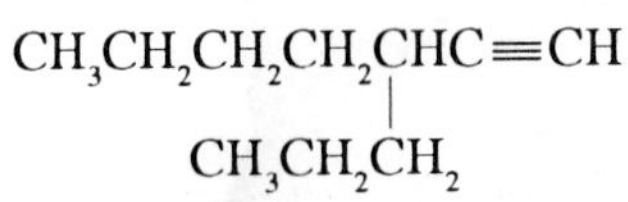

t. 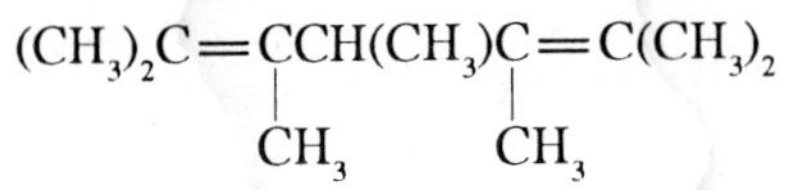

u. 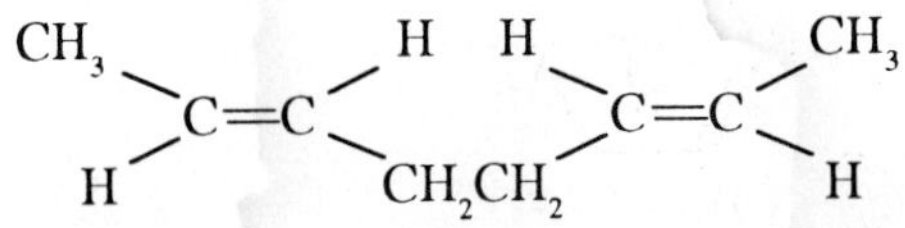

v. 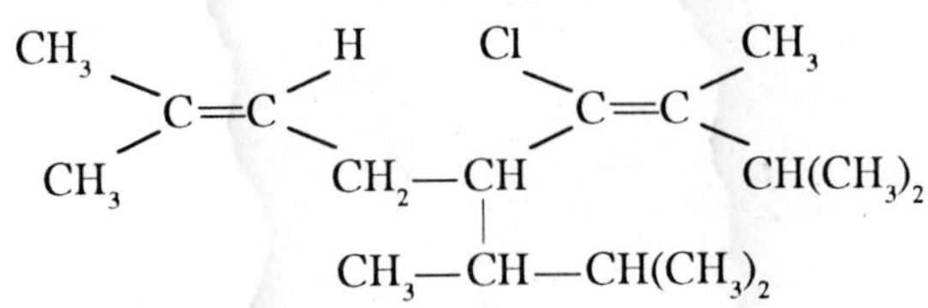

w. 

4.4 Give structures for

a. propyne
b. 2-pentyne
c. vinylacetylene
d. diethylacetylene
e. 2-methyl-1-penten-3-yne
f. 1,3-hexadiyne
g. sodium pentynide
h. 1,8-nonadiyne
i. 1-penten-4-yne
j. 1-chloro-2-butyne
k. 3,4-dimethyl-1-pentyne
l. silver butynide
m. calcium propynide
n. 2-cyano-1-hexen-3-yne
o. diallyl acetylene
p. 2-chloro-1-buten-3-yne
q. (*Z*) -3-methyl-3-penten-1-yne
r. 5,5-diiodo-1-penten-3-yne
s. 4-methyl-2-pentyne
t. 3-iodo-1-propyne
u. 1-penten-4-yne
v. (*Z*) -1,3-hexadien-5-yne
w. 4-nitro-1-butyne

4.5 Give an example of

a. a cumulene
b. a terminal alkyne
c. a vinylic chloride
d. an allylic chloride
e. a geminal dibromide
f. a vicinal difluoride
g. an enol
h. the keto form of *g*
i. a disubstituted acetylene
j. an isoprene molecule
k. a polymer of vinyl chloride

4.6 Identify the following pairs of structures as tautomers, resonance forms or non-tautomeric isomers:

a. 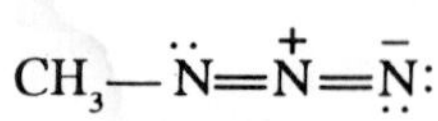 and 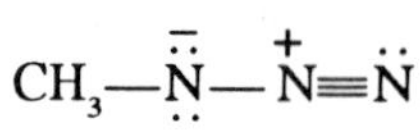

b. 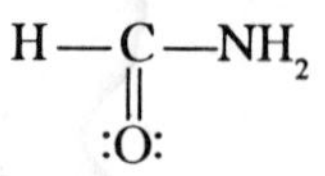 and 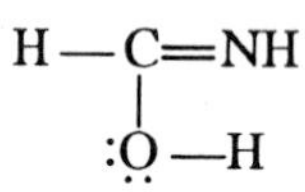

c. 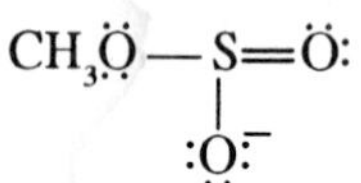 and 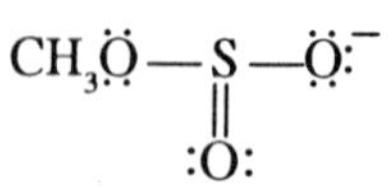

d. 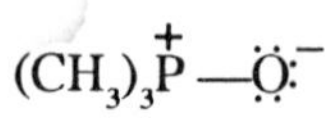 and 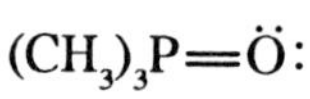

e. 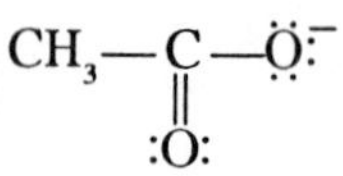 and 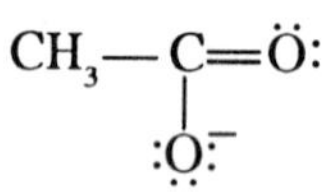

f. 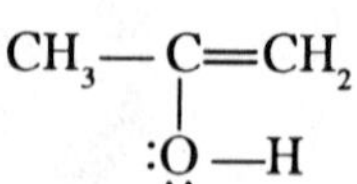 and 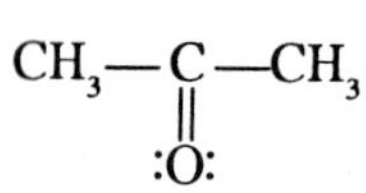

g. 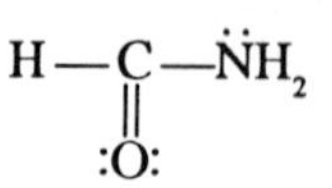 and 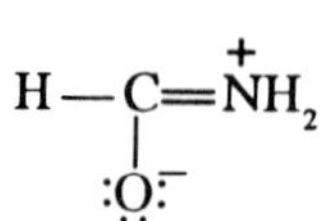

h.  and 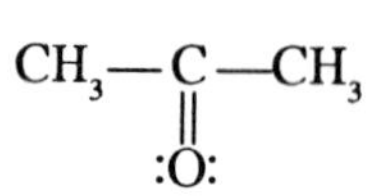

i. 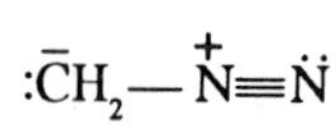 and 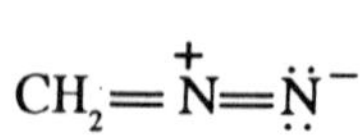

j. 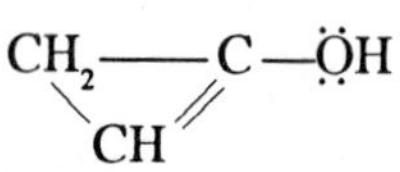 and 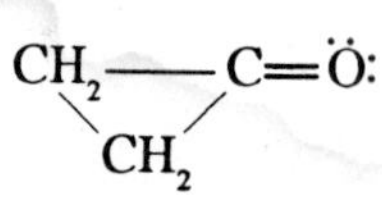

k. $CH_3—C(=\ddot{O}:)—\ddot{O}H$ and $CH_2(H\ddot{O})—C(=\ddot{O}:)—H$

l. $CH_3CH_2\ddot{O}H$ and $CH_3—\ddot{O}—CH_3$

4.7 Give structures of the products from the reaction of 1-pentyne with:

a. H_2 (1 mole/Lindlar's Catalyst)
b. H_2/Pt (2 moles)
c. Br_2/CCl_4 (1 mole)
d. Br_2/CCl_4 (2 moles)
e. HCl (gas, 1 mole)
f. HCl (gas, 2 moles)
g. Na metal
h. $AgNO_3$/ethanol
i. NaOH (aqueous)
j. $NaNH_2$
k. $H_2O/H_2SO_4/Hg^{+2}$
l. Na/liquid NH_3 (1 mole)
m.

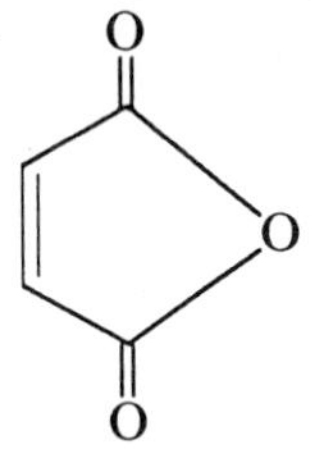

n. O_3
o. $KMnO_4$/heat

4.8 What will be the products in the reactions of the previous problem if the reactant is 3-hexyne?

4.9 Give the structure of the major product from the reactions of 1,3-butadiene with the following reageants:

a. H_2/Pt (2 moles)
b. Br_2/CCl_4 (2 moles)
c. HCl (gas, 1 mole, low T)
d. HCl (gas, 1 mole, high T)
e. H_2O/H_2SO_4
f. NaOH (aqueous)
g. Na metal
h. O_3/Zn
i. $NaNH_2$
j. $AgNO_3$
k. $KMnO_4$/heat
l. CuCl
m. H_2/Pt (1 mole, high T)
n. Br_2/CCl_4 (1 mole, high T)

4.10 Fill in the reagents and/or products missing in the reactions below:

a. $n\text{-}C_5H_{11}—C{\equiv}C—H$ + $(CH_3)_2CHMgCl$ $\xrightarrow{\text{(ether)}}$ A + B (gas)

A $\xrightarrow{\text{1. } n\text{-}C_4H_9I;\ \text{2. } H^+, H_2O}$ C (organic compound) + D

b. $HC{\equiv}CH$ $\xrightarrow{\text{1. Na};\ \text{2. } n\text{-}C_4H_9Br}$ E (organic compound)

c. $CH_3CH_2CH_2CH(Br)CH_2Br$ + $NaNH_2$ ⟶ F + G (2 moles) + $2NH_3$

d. $CF_3—C(Cl){=}C(Cl)—CF_3$ + Zn/ethanol ⟶ H + I (ppt)

e. C_2H_5Br (2 moles) $\xrightarrow{\text{1. Mg/ether (2 moles)};\ \text{2. } BrCH_2C{\equiv}CCH_2Br}$ J + K (2 moles)

f. 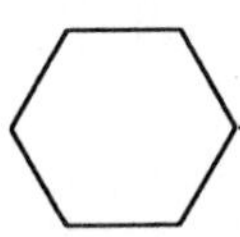—Br + Mg/ether ⟶ L $\xrightarrow{BrCH_2—C(Br){=}CH_2}$ M $\xrightarrow{NaNH_2}$ N

g.

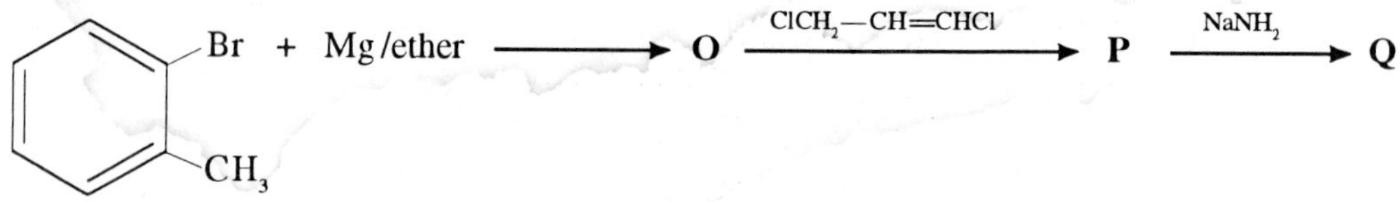

h. $CH_3(CH_2)_7CH(Br)—CH(Br)(CH_2)_7C(=O)OH + 2NaNH_2 \longrightarrow$ **R**

i. $CH_2{=}CH(CH_2)_8COOH + Br_2 \longrightarrow$ **S** $\xrightarrow[\text{2 moles}]{NaNH_2}$ **T**

j. $n\text{-}C_4H_9C{\equiv}CH + C_2H_5MgBr \longrightarrow$ **U** + **V** (gas)

k.

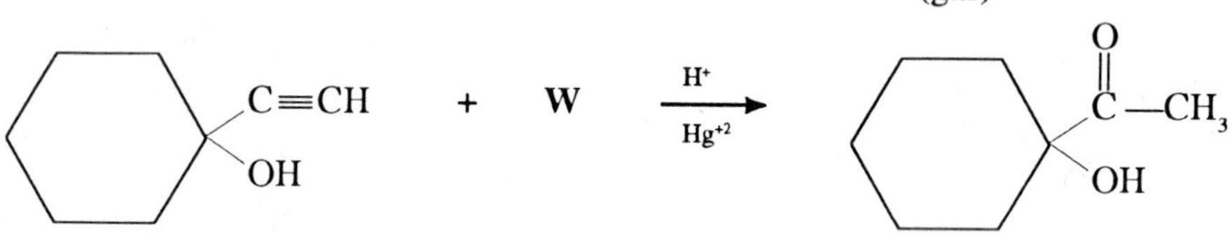

l. $C_5H_{11}C{\equiv}CCH_3 + (CH_3)_2CHMgCl \longrightarrow$ **X** $\xrightarrow{H_2O}$ **Y** (gas) + **Z**

m. $C_6H_{13}CH{=}C(Cl)CH_3 + NaNH_2 \longrightarrow$ **AA**

n. $C_6H_5C{\equiv}C^-Li^+ + (CH_3)_2CHBr \longrightarrow$ **BB**

o. $C_2H_5Br \xrightarrow[\text{2. } BrCH_2C\equiv CCH_3]{\text{1. Li/ether}}$ **CC**

p.

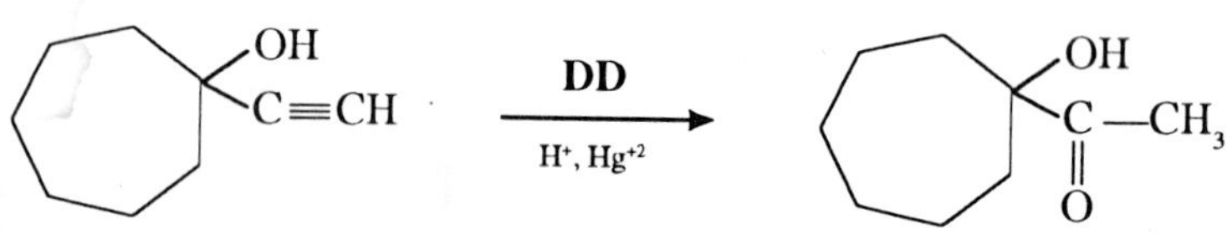

q.

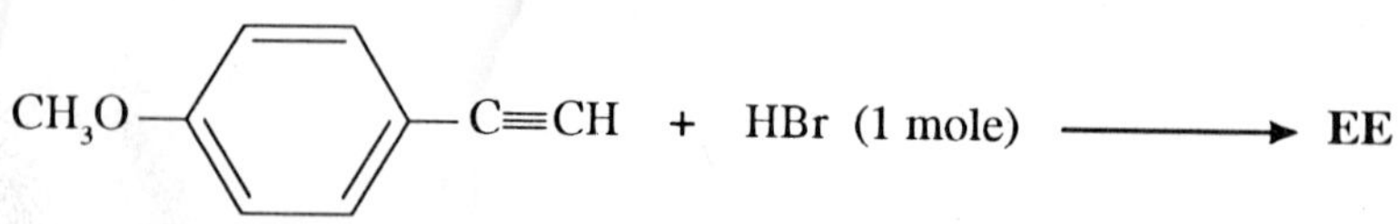

r. $C_8H_{17}CH(Br)CH_2Br$ + Zn dust $\longrightarrow$ **FF**

s. $C_6H_5C{\equiv}C^-Li^+ + H_2O \longrightarrow$ **GG** + **HH** (basic to litmus)

t. $C_4H_9C{\equiv}CC_4H_9 + \xrightarrow[\text{2. } H_2O]{\text{1. } O_3}$ **II**

u. $CH_3C{\equiv}CH \xrightarrow[\text{2. } H_2/Pt]{\text{1. HBr (1 mole)}}$ **JJ**

v.

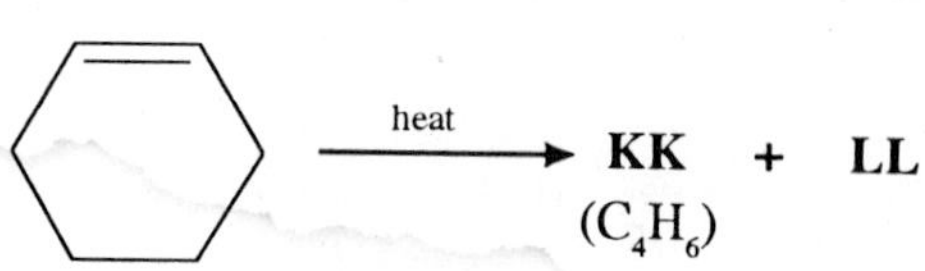

w. 2-methyl-1,3-butadiene + HCl (1 mole) $\xrightarrow[\text{temperature}]{\text{low}}$ **MM** (major) + **NN** (minor)

x. 2-methyl-1,3-butadiene + $BrCCl_3$ (1 mole) $\xrightarrow{\text{peroxides}}$ **OO** + **PP**

y. 1,3-pentadiene + HCl (1 mole) $\longrightarrow$ **QQ**

4.11 Give the major product or products expected from the reaction of 1-methyl-1,3-cyclohexadiene with each of the following:
a. $2H_2$, Pt b. HBr (1 mole) c. HBr (2 moles) d. $HCCl_3$, peroxide

4.12 Show each step in the laboratory preparation of the following compounds from cyclopentanol:
a. cyclopentene
b. 3-bromocyclopentene
c. 1,3-cyclopentadiene
d. cyclopentane
e. bromocyclopentane
f. *cis*-cyclopentanediol
g. *trans*-cyclopentanediol
h. 6,6-dichlorobicyclo[3.1.0]hexane
i. monodeuterocyclopentane (C_5H_9D)
j.

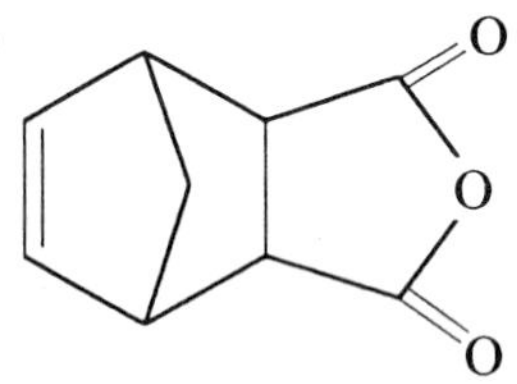

k. glutaric acid ($HO—\overset{O}{\overset{\|}{C}}(CH_2)_3\overset{O}{\overset{\|}{C}}—OH$)

4.13 Suggest plausible schemes for the following syntheses:
a. $CH_3CH(Br)CH_2D$ from acetylene
b. $(CH_3)_2C{=}O$ from acetylene
c. $CH_3CH(OH)CH_2OH$ from acetylene
d. $CH_3CH_2CH(Br)CH_3$ from acetylene
e. CH_3CHO from CaC_2
f. $CH_3CH_2C{\equiv}CCH_3$ from acetylene
g. $CH_3CH_2CH(OH)CH_3$ from acetylene
h. $CH_3C{\equiv}CH$ from propane
i. $CH_3\overset{O}{\overset{\|}{C}}CH_3$ from propane
j. $DC{\equiv}CH$ from acetylene
k. $BrCH_2CH{=}CH_2$ from acetylene
l. propyne from propylene
m. isopropyl bromide from acetylene
n. isopropyl alcohol from acetylene
o. tetrachloroethylene from acetylene

4.14 How many rings are present in each of the following?
a. $C_{22}H_{36}$ — this compound does not absorb H_2
b. $C_{22}H_{36}$ — contains one double and two triple bonds
c. $C_{22}H_{36}$ — contains one triple bond
d. $C_{14}H_{20}$ — absorbs four moles of Br_2
e. C_8H_{16} — absorbs one mole of H_2
f. C_6H_{10} — contains one triple bond
g. C_9H_{10} — contains one triple bond and one double bond
h. $C_{10}H_{14}$ — contains three double bonds

4.15 Draw the structure of 6-methyl-2-heptene and label each set of hydrogen atoms to show their relative reactivity toward bromine atoms.

4.16 Explain why the terminal hydrogen of an alkyne can be abstracted by a strong base (e.g., NH_2^- or a Grignard reagent), while that of an alkene or alkane cannot be abstracted; that is,

$$HC{\equiv}CH + NH_2^- \longrightarrow HC{\equiv}C^- + NH_3$$

$$CH_2{=}CH_2 + NH_2^- \longrightarrow \text{No reaction}$$

$$CH_3{—}CH_3 + NH_2^- \longrightarrow \text{No reaction}$$

4.17 Explain why

a. hydrogen cyanide (HC≡N) is a stronger acid than acetylene.
b. 1,2-butadiene is less stable than 1,3-butadiene.
c. the C_2—C_3 single bond of 1,3-butadiene is shorter (1.48Å) than the normal C—C single bond (1.54Å) found in alkanes.
d. when cyclopentadiene is reacted with one mole of HBr, only one product (3-bromocyclopentene) is formed; however, when cyclopentadiene is reacted with one mole of Br_2/CCl_4, two products (3,4-dibromocyclopentene and 3,5-dibromocyclopentene) are formed.
e. 1,2-addition of HCl to 1,3-butadiene predominates at low temperatures (–40°C), but 1,4-addition is favored at higher temperatures (+40°C).

4.18 Identify the simplest chemical test that would help you distinguish the compounds in each of the following pairs. Tell what you would do and see.

a. 1-hexyne and 1-hexene
b. 1-hexyne and 2-hexyne
c. 1-hexyne and 1,3-hexadiene
d. 3-hexyne and 2-chlorohexane
e. hexane and 2-hexyne
f. 1,3-hexadiene and 1,4-hexadiene
g. 1-hexyne and 1-chloro-3-hexyne

4.19 Give products in the following Diels-Alder reactions:

a.
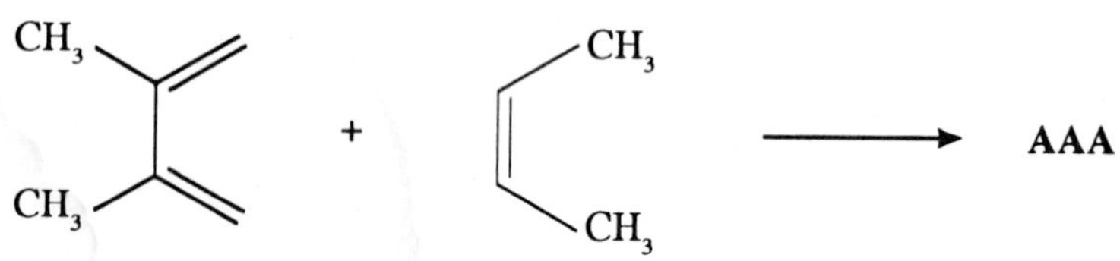

b.
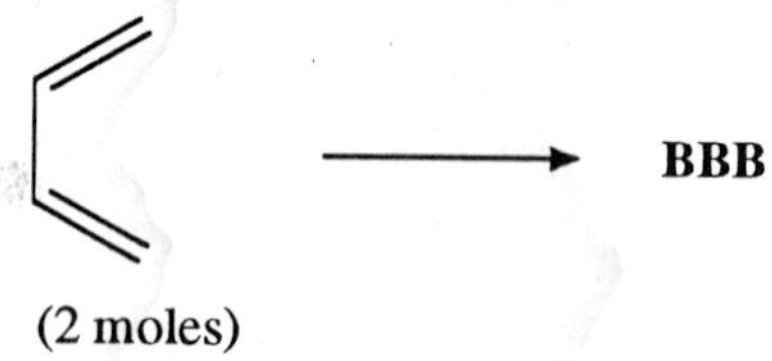

c.
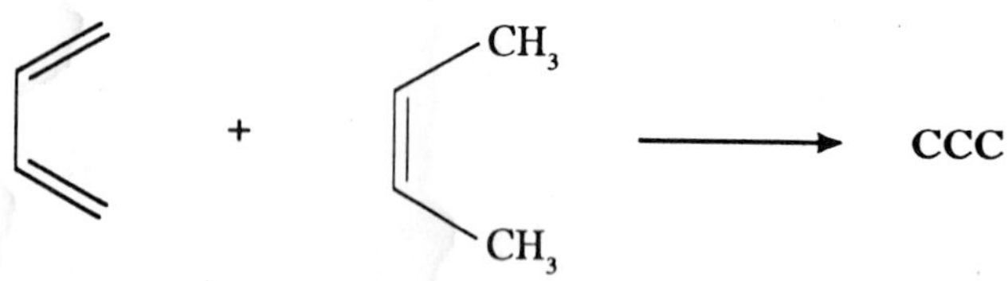

d.
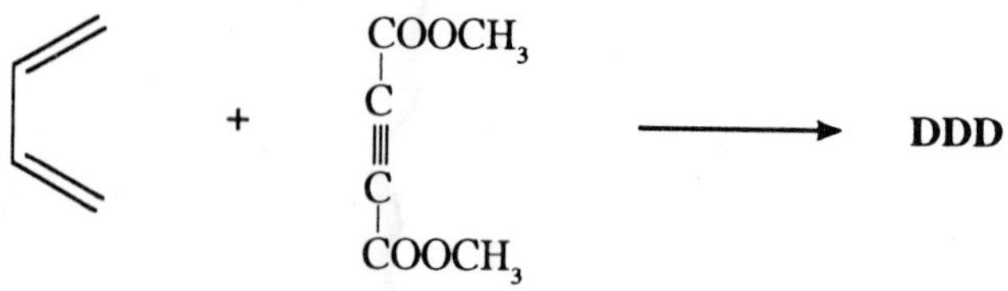

e.
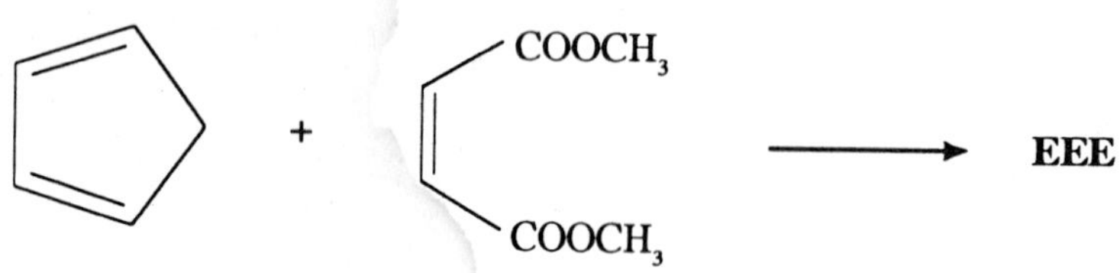

f.
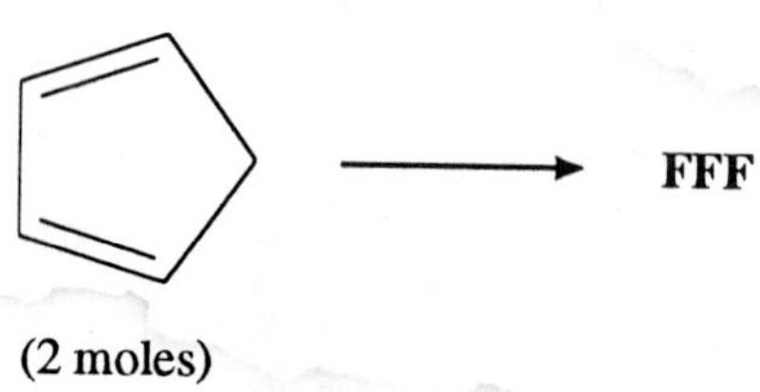

g. 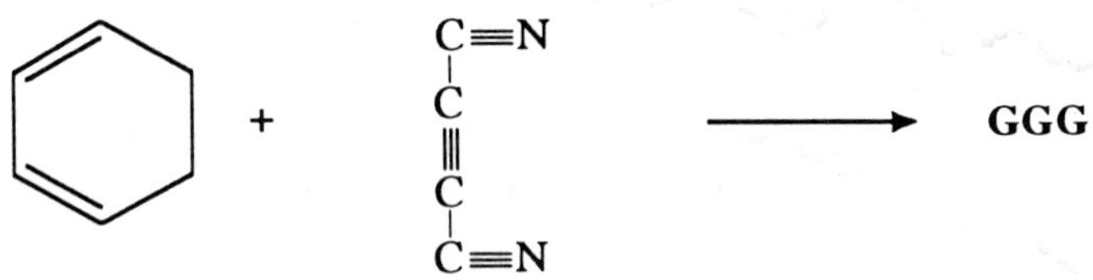

4.20 Which reactants, if any, would yield the following Diels-Alder adducts?

a. 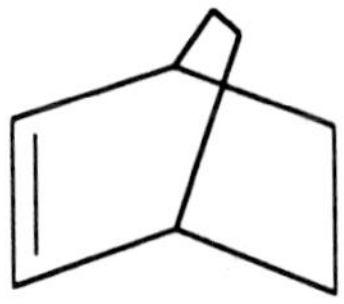

b. 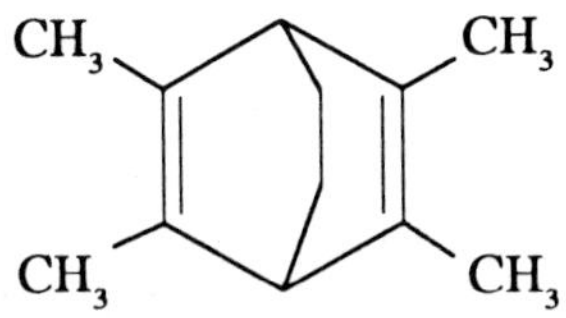

c. 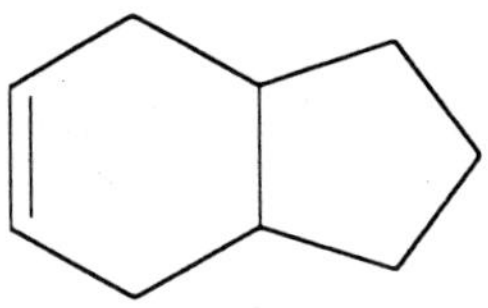

d. 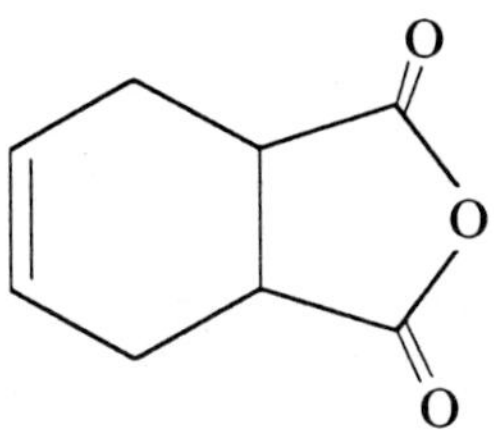

e. 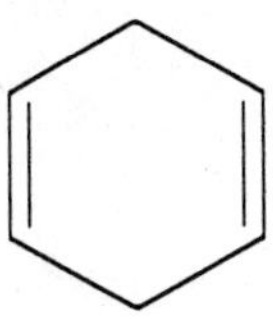

f. 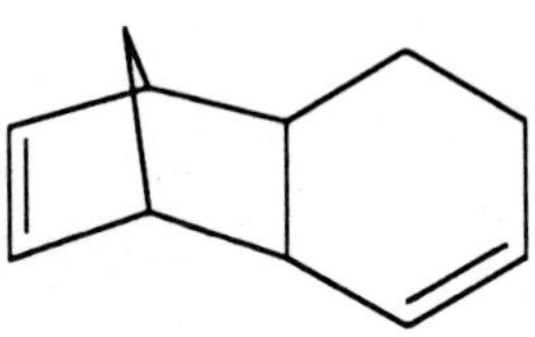

g. 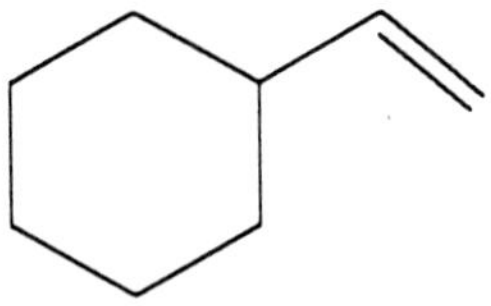

h. 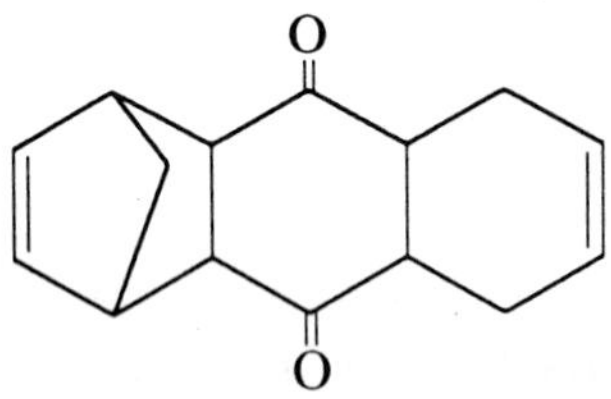

i. 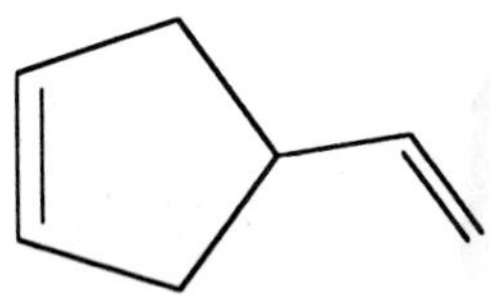

j. 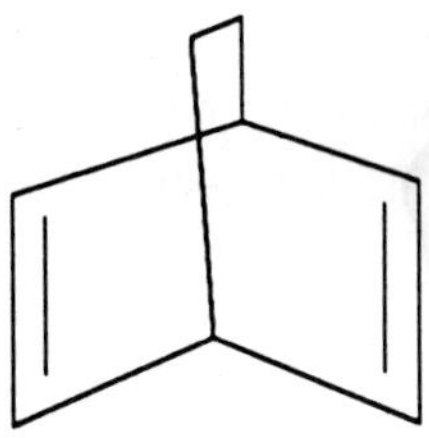

k. 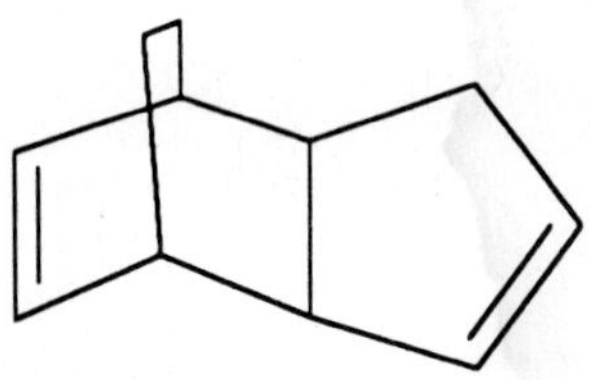

4.21 β-Selinene ($C_{15}H_{24}$) is a natural product isolated from the oil of celery. It absorbs two moles of hydrogen to give a saturated hydrocarbon. Treatment of β-selinene with hot $KMnO_4$ yields

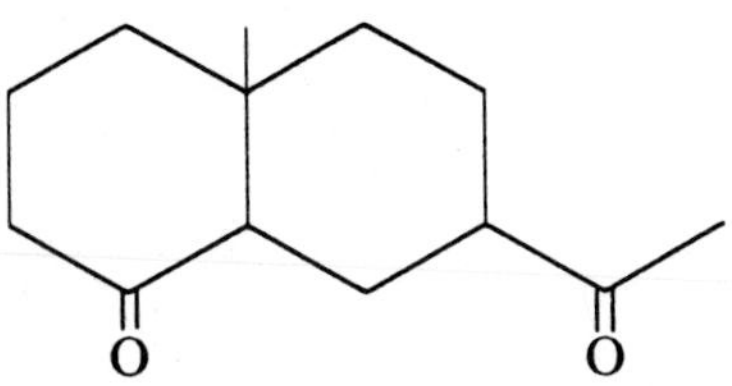

a. How many double bonds are there in β-selinene?
b. How many rings are there in β-selinene?
c. What would be the products formed when O_3/Zn is used instead of hot $KMnO_4$?
d. What is the structure of β-selinene?
e. Does β-selinene obey the isoprene rule?

4.22 Myrcene ($C_{10}H_{16}$) is a natural product that absorbs three moles of hydrogen. Ozonolysis followed by hydrolysis in the presence of zinc yields the following compounds:

$$CH_3\underset{\underset{O}{\|}}{C}CH_3 \qquad \underset{\underset{O}{\|}}{HC}CH_2CH_2\underset{\underset{O}{\|}}{C}\,\underset{\underset{O}{\|}}{CH} \qquad H\underset{\underset{O}{\|}}{CH}$$

(2 moles)

a. How many double bonds are there in myrcene?
b. How many rings are there in myrcene?
c. What are the possible structures for myrcene?
d. Bearing in mind that myrcene obeys the isoprene rule, which is the correct structure for myrcene?

4.23 γ-Terpinene ($C_{10}H_{16}$) is a natural product that absorbs two moles of hydrogen. Ozonolysis followed by hydrolysis in the presence of zinc yields the following compounds:

$$CH_3{-}\underset{\underset{O}{\|}}{C}{-}CH_2{-}\underset{\underset{O}{\|}}{CH} \qquad \underset{\underset{O}{\|}}{HC}{-}CH_2{-}\underset{\underset{O}{\|}}{C}{-}CH(CH_3)_2$$

a. How many double bonds are there in γ-terpinene?
b. How many rings are there in γ-terpinene?
c. What are the possible structures for γ-terpinene?
d. Bearing in mind that γ-terpinene obeys the isoprene rule, which is the correct structure for γ-terpinene?

4.24 Suggest a plausible mechanism that would explain the formation of the indicated products. Use arrow to show the flow of electrons:

a.

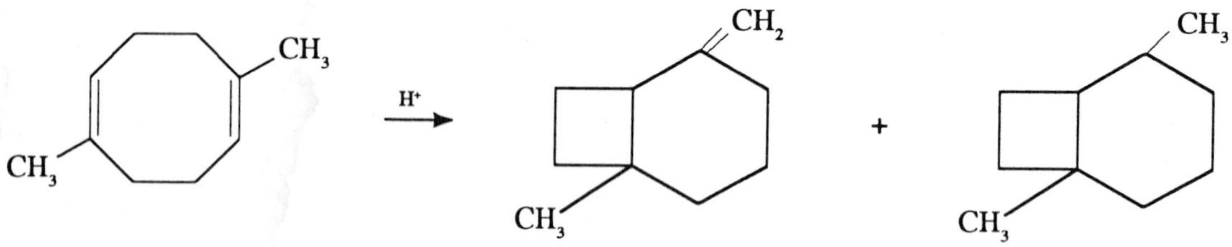

b.

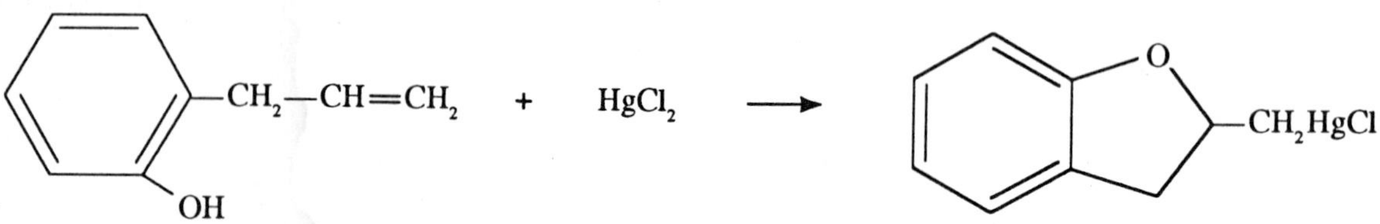

c. $CH_3CH_2CH_2C{\equiv}C{-}H \;+\; HCl$ (2 moles) $\longrightarrow CH_3CCl_2CH_2CH_2CH_3$ (formed)

$\longrightarrow ClCH_2{-}CH(Cl){-}CH_2CH_2CH_3$ (not formed)

4.25 What factor is responsible for the fact that alkyne formation is favored over diene formation when a vicinal dihalide is treated with a base?

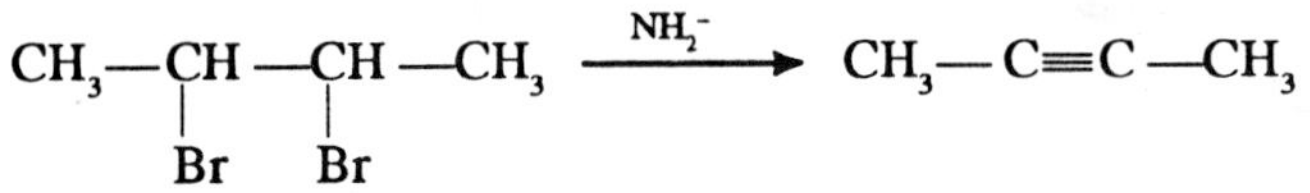

4.26 Arrange the following unsaturated compounds in order of their reactivity to addition of HBr, from most to least reactive: 2-butene, propyne, isoprene, 1,3-butadiene.

4.27 Match the heat of hydrogenation with the proper hydrocarbon:

$\Delta H°$ (kcal/mole) CH	Hydrocarbon CH
–27	(*E*)-1,3-pentadiene
–28	1,4-pentadiene
–30	2-methyl-1-butene
–57	2-methyl-2-butene
–60	3-methyl-1-butene

4.28 Which of each pair undergoes the addition reaction faster?

a.

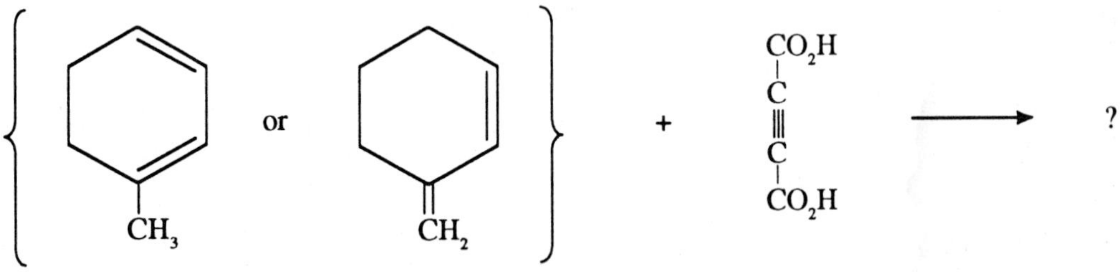

b.

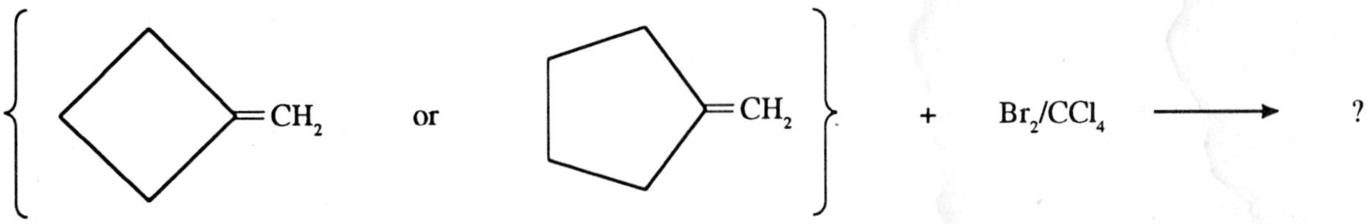

c.

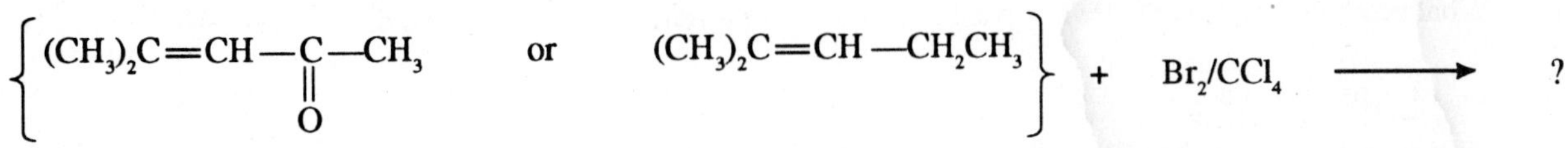

4.29 Electrophilic additions to conjugated systems are 1,2- and 1,4-additions. Generally the 1,2-product is less thermodynamically stable and is formed in smaller amounts. However, Br_2 addition to 1,4-diphenyl-1,3-butadiene yields only the 1,2-product. Why?

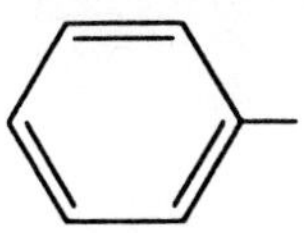 = phenyl

Similarly the addition of Br_2 to 1,3,5-hexatriene yields only the 1,2- and 1,6-adducts but no 1,4-adduct. Explain.

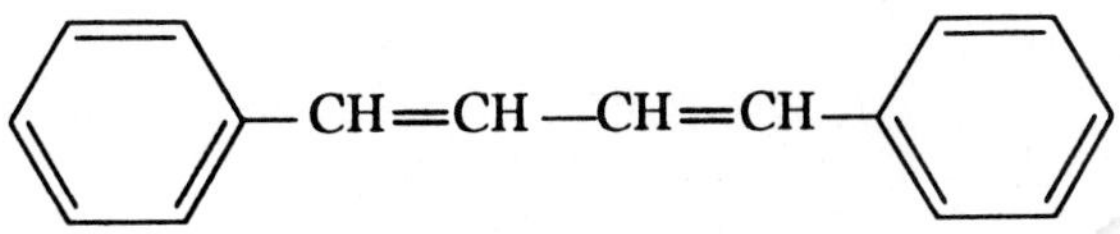

1,4-diphenyl-1,3-butadiene

4.30 Identify the isoprene units in the natural products shown below:

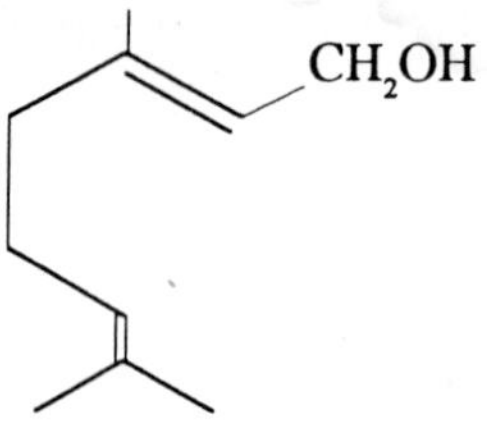

Geraniol

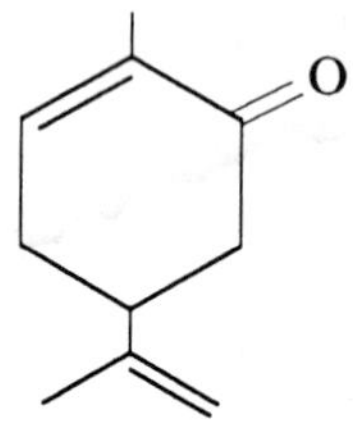

Carvone

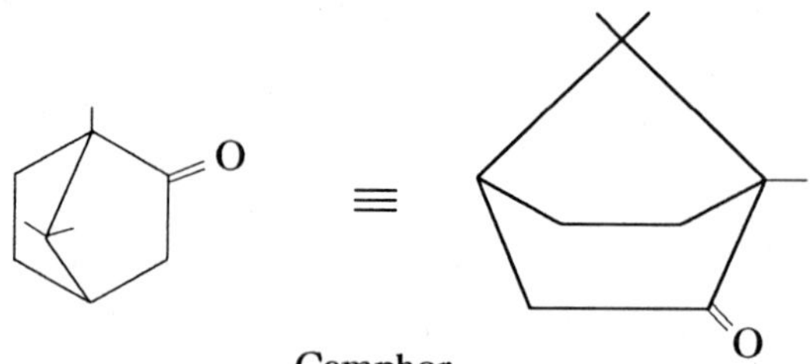

Camphor

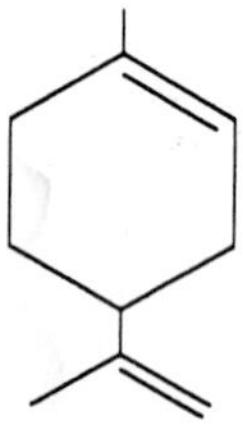

Limonene

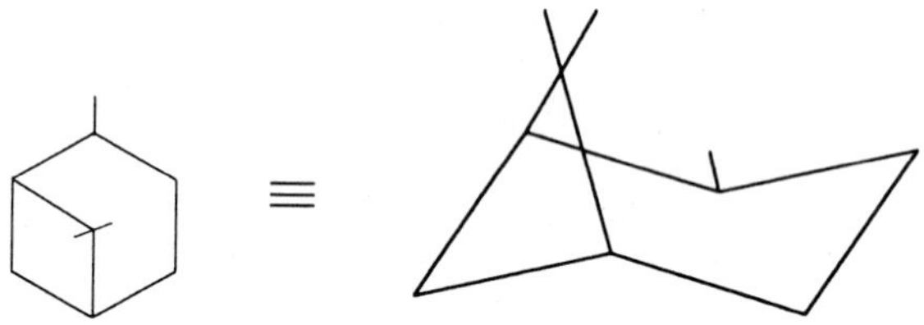

Pinane

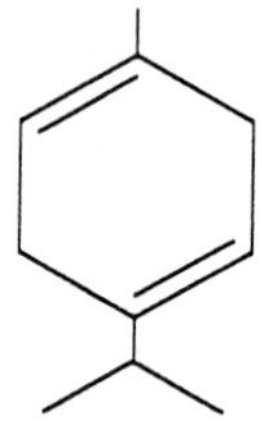

γ-Terpinene

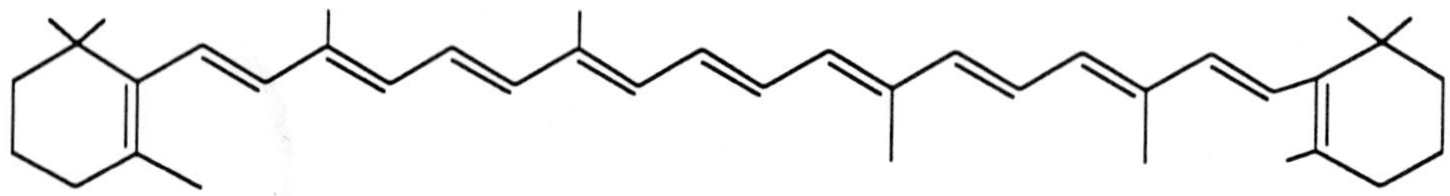

β–Carotene

Myrcene

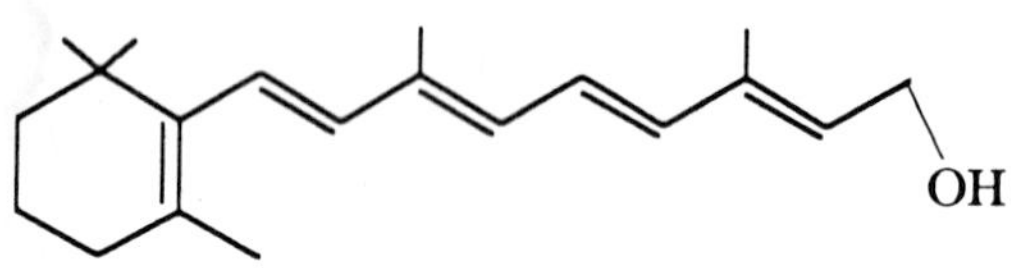

Vitamin A

4.31 What products are expected when each of the following reacts with O_3 followed by Zn/H^+?

a. 1,4-pentadiene

b. 1,3-cyclopentadiene

c. 1,3-cyclohexadiene

d. 2-methyl-1,3-butadiene

e. $HOCH_2—CH{=}C(CH_3)CH_2CH_2CH{=}C(CH_3)_2$

f. $(CH_3)_2C{=}CH—CH_2CH_2C({=}CH_2)—CH{=}CH_2$

g.

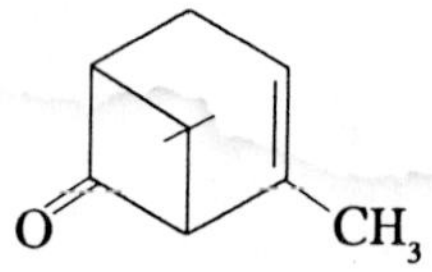

Multiple Choice Test

4.1 What is the hybridization of the carbons labelled **A, B** and **C**?

$$\underset{\textbf{A}}{(CH_3)_2C}{=}\underset{\textbf{B}}{CHCH_2}\underset{\textbf{C}}{C}{\equiv}CCH_3$$

a. *sp*, sp^3, sp^2
b. *sp*, sp^2, sp^3
c. sp^2, sp^3, *sp*
d. sp^2, sp^2, *sp*

4.2 How many carbons lie in a straight line in 2-methyl-3-hexyne?
a. 2
b. 3
c. 4
d. 6

4.3 What is the only correct statement about cyclohexyne?
a. It is acidic.
b. It has four 109° C—C—C bond angles.
c. It is planar.
d. It is too strained to be synthesized.

4.4 The name of the compound below is

$CH_3CH(CH_2CH_3)CH_2C{\equiv}CC(CH_3)_3$

a. 6-ethyl-2,2-dimethyl-3-heptyne
b. 2,2,6-trimethyl-3-octyne
c. *t*-butylisopropylacetylene
d. 2,6-dimethyl-3-heptyne

4.5 The sequence of steps needed for the conversion of 1-butene to butyne is
a. HBr/peroxides; then alc. KOH
b. HCl/no peroxides; then $NaNH_2$/liq. NH_3
c. Br_2/CCl_4; then $NaNH_2$/liq. NH_3
d. $KMnO_4$/cold; then H_2SO_4/heat

4.6 Acid-catalyzed hydration of propyne yields
a. $(CH_3)_2C{=}O$
b. $CH_2{=}C(OH)CH_3$
c. $(CH_3)_2C(OH)_2$
d. CH_3CH_2CHO

4.7 Hot $KMnO_4$ treatment of a compound yields $(CH_3)_2C{=}O$, CH_3COOH and $HOOCCH_2COOH$. What is its structure?
a. $CH_3CH{=}CHCH_2CH{=}C(CH_3)_2$
b. $(CH_3)_2C{=}CHCH_2C{\equiv}CCH_3$
c. $(CH_3)_2C{=}CHCH_2C{\equiv}CCH_3$
d. either (a) or (b)

4.8 The most stable carbocation is
a. $CH_2{=}CH\overset{+}{C}H_2$
b. $CH_3CH{=}\overset{+}{C}H_2$
c. $CH_3\overset{+}{C}{=}CH_2$
d. $CH_3CH_2\overset{+}{C}H_2$

4.9 The name of the compound below is

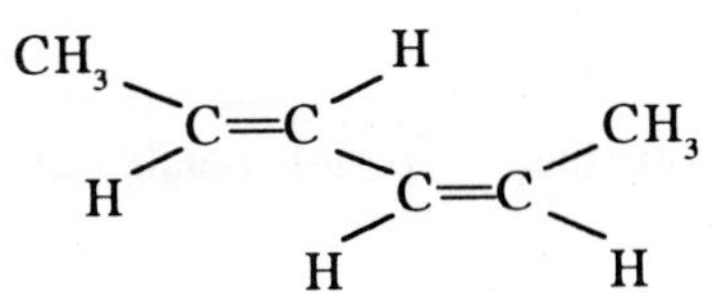

a. (*Z*), (*Z*)-2,4-hexadiene
b. (*E*), (*E*)-2,4-hexadiene
c. (*E*), (*Z*)-2,4-hexadiene

4.10 Which is the correct order of decreasing basicity in the anions

HO–	NH_2^-	HC≡C–	$CH_3CH_2^-$	$CH_2{=}CH^-$
A	**B**	**C**	**D**	**E**

a. A > B > C > D > E
b. D > E > B > C > A
c. C > B > D > A > E
d. D > E > C > B > A

4.11 The correct order of decreasing acidity in the compounds below is

H_2O	CH_3CH_3	NH_3	$CH_2{=}CH_2$	$HC{\equiv}CH$
A	**B**	**C**	**D**	**E**

a. A > E > C > D > B
b. A > C > E > D > B
c. E > A > C > B > D
d. A > E > D > B > C

4.12 Which reagent can distinguish between 1-hexyne and 1-hexene?
a. $KMnO_4$
b. H_2SO_4
c. Br_2/CCl_4
d. $AgNO_3$

4.13 What is the correct order of decreasing heats of hydrogenation of the following dienes:

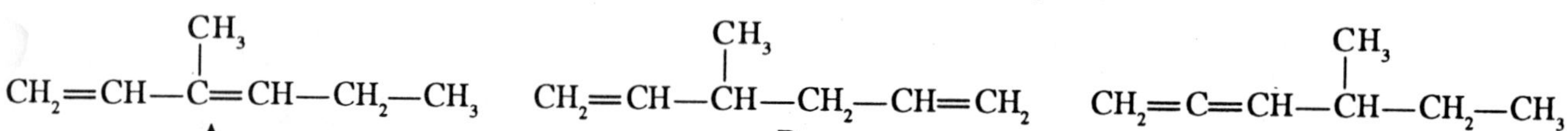

A **B** **C**

a. A > B > C
b. A > C > B
c. B > A > C
d. C > B > A

4.14 Addition of HCl (1 mole, high temperature) to 1,3-butadiene yields ________ as the major product.
a. 4-chloro-1-butene
b. 3-chloro-1-butene
c. 1-chloro-2-butene
d. 3-chloro-2-butene

4.15 Addition of one mole of HCl to 1,3-butadiene, followed by one mole of bromine yields
a. 1,2-dibromo-3-chlorobutane
b. 2,3-dibromo-1-chlorobutane
c. 1,2-dibromo-4-chlorobutane
d. both (a) and (b)

4.16 The reagents needed to prepare 4-cyanocyclohexene are

ethylene	1,3-butadiene	1-butyne	cyanoethene	1-cyano-1,3-butadiene	2-cyano-1,3-butadiene
A	**B**	**C**	**D**	**E**	**F**

a. A + E
b. B + D
c. C + D
d. A + F

4.17 The sequence of reagents needed for the conversion of 1-bromopropane to propyne is
a. alc. KOH/heat; then Br_2/CCl_4; then $NaNH_2/NH_3$
b. $(CH_3)_3CO^-K^+$/heat; then HBr; then $NaNH_2/NH_3$
c. alc. KOH/heat; then Br_2/CCl_4; then H_2SO_4
d. alc. KOH; then H_2/Pt; then $NaNH_2/NH_3$

4.18 The reagent needed to convert 2-butyne to *trans*-2-butene is
a. H_2/Pt
b. Li/NH_3
c. H_2/Pd-Lindlar
d. HBr, then Zn dust

4.19 Compound **A** $C_{12}H_{16}OCl_2$ reacts with Br_2/CCl_4 to yield $C_{12}H_{16}OCl_2Br_2$. **A**, therefore, has ____ double bonds and ____ rings.
a. 1, 3
b. 2, 3
c. 2, 2
d. 1, 2

4.20 What is the result of reacting 2-hexyne with Na metal followed by CH_3Br (excess)
a. $CH_3CH_2CH_2CH{=}C(CH_3)_2$
b. $CH_3CH_2CH_2C(CH_3){=}CHCH_3$ (*Z* & *E*)
c. $CH_3CH_2CH_2CH_2C(CH_3)_3$
d. no reaction

Answers for Questions

4.1 Alkenes: d, p, r
Conjugated dienes: a, e, i, k, n
Non-conjugated dienes: c, g, h, o
Cumulated dienes: b, l, m
Alkynes: f, j, q

4.2 a. Three alkynes of molecular formula C_5H_8:

$CH_3CH_2CH_2C{\equiv}CH$

1-pentyne

$CH_3CH_2C{\equiv}CCH_3$

2-pentyne

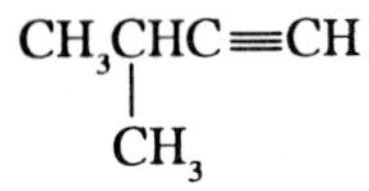

3-methyl-1-butyne

b. Seven alkynes of molecular formula C_6H_{10}:

$CH_3CH_2CH_2CH_2C{\equiv}CH$

1-hexyne

$CH_3CH_2CH_2C{\equiv}CCH_3$

2-hexyne

$CH_3CH_2C{\equiv}CCH_2CH_3$

3-hexyne

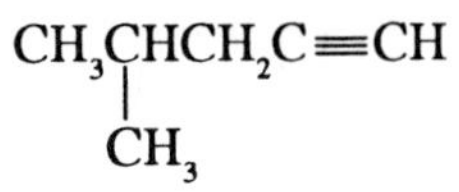

4-methyl-1-pentyne

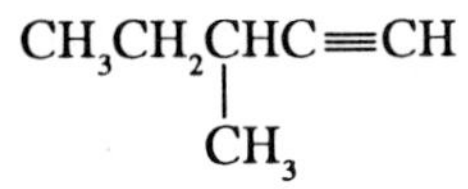

3-methyl-1-pentyne

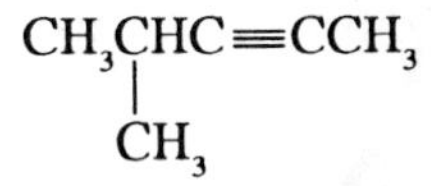

4-methyl-2-pentyne

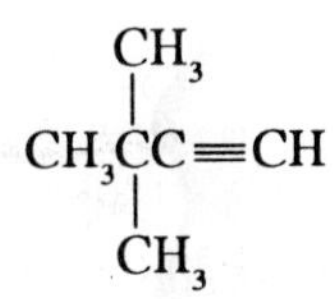

3,3-dimethyl-1-butyne

c. Seven dienes of molecular formula C_5H_8:

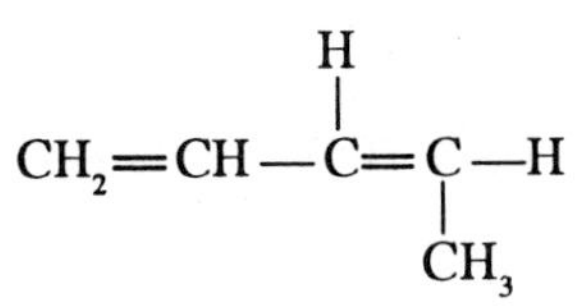

(*Z*)-1,3-pentadiene
(conjugated)

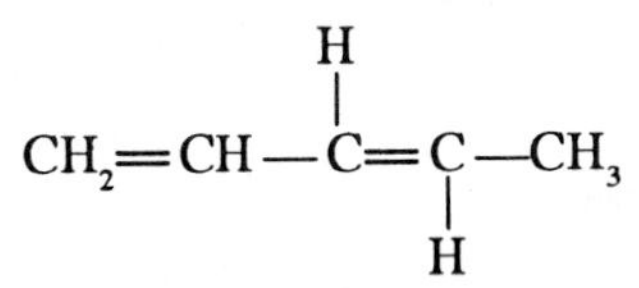

(*E*)-1,3-pentadiene
(conjugated)

$CH_2{=}CH{-}CH_2{-}CH{=}CH_2$

1,4-pentadiene
(non-conjugated)

$CH_2{=}C{=}CH{-}CH_2CH_3$

1,2-pentadiene
(cumulated)

$CH_2{=}C(CH_3)CH{=}CH_2$

2-methyl-1,3-butadiene
(conjugated)

$CH_3{-}CH{=}C{=}CH{-}CH_3$

2,3-pentadiene
(cumulated)

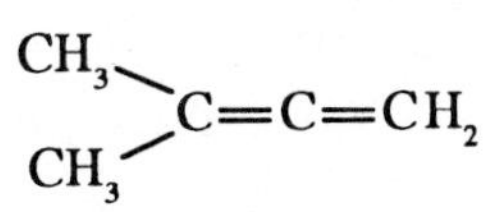

3-methyl-1,2-butadiene
(cumulated)

d. Twelve conjugated dienes of molecular formula C_6H_{10}:

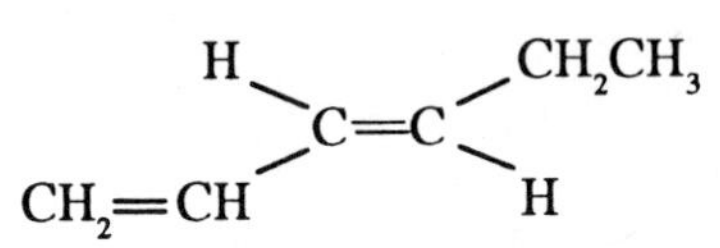

(*E*)-1,3-hexadiene

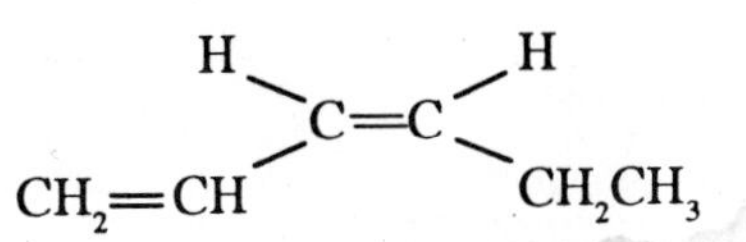

(*Z*)-1,3-hexadiene

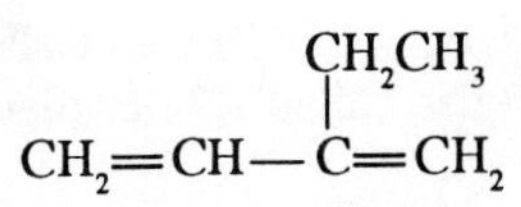

2-ethyl-1,3-butadiene

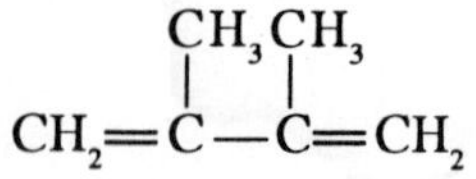

2,3-dimethyl-1,3-butadiene

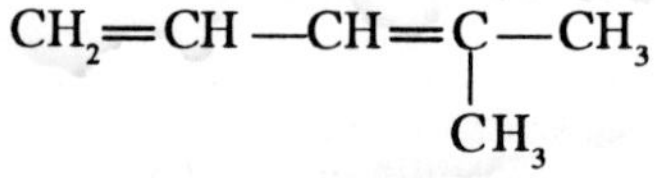

4-methyl-1,3-pentadiene

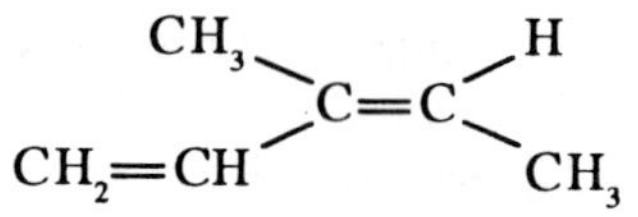

(*Z*)-3-methyl-1,3-pentadiene

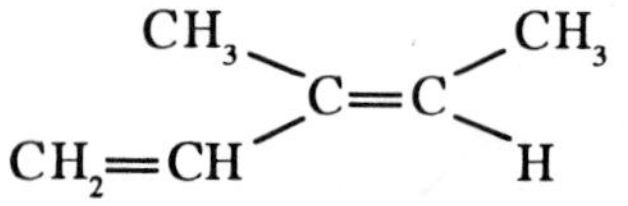

(*E*)-3-methyl-1,3-pentadiene

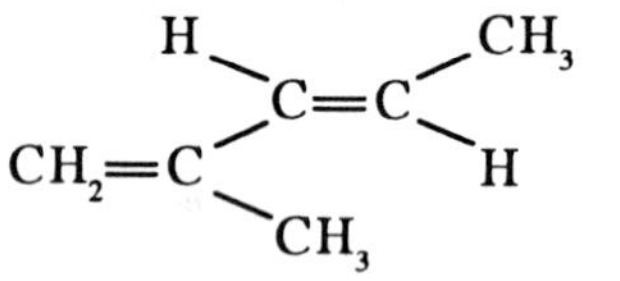

(*E*)-2-methyl-1,3-pentadiene

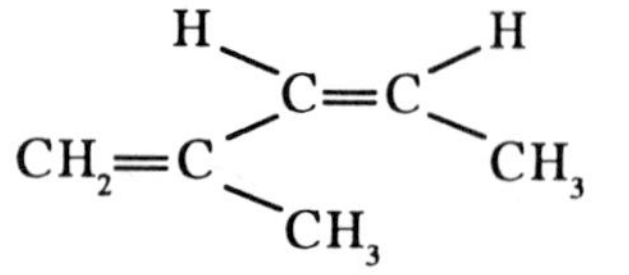

(*Z*)-2-methyl-1,3-pentadiene

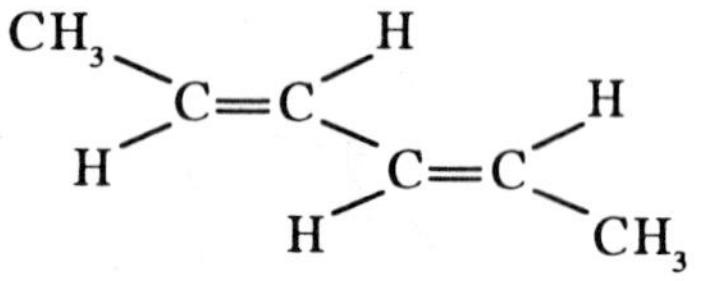

(*E*), (*E*)-2,4-hexadiene

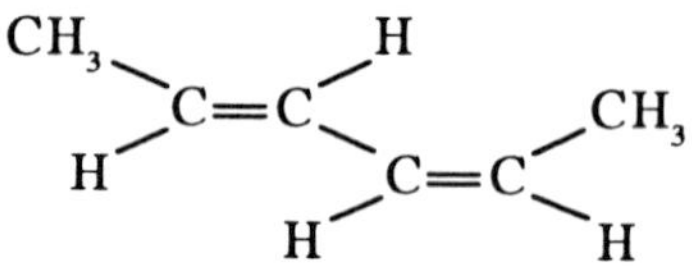

(*E*), (*Z*)-2,4-hexadiene

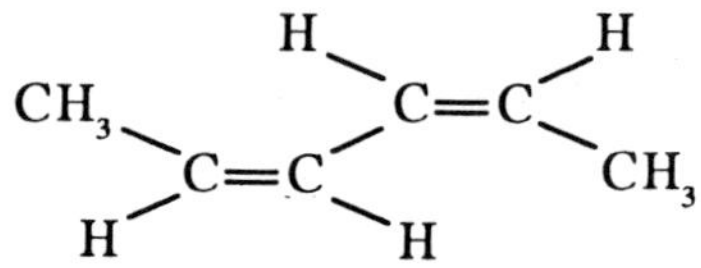

(*Z*), (*Z*)-2,4-hexadiene

4.3 a. 2-butyne
b. 3-methyl-1-butyne
c. propadiene (or allene)
d. 1,4-pentadiyne
e. silver acetylide (or silver ethynide)
f. 1,4-dichloro-2-butyne
g. 2-methyl-3-hexyne
h. 3-propyl-1-hexyne
i. copper (I) propynide (or cuprous propynide)
j. (2*E*, 7*Z*)-2,3-dichloro-7,8-dimethyl-2,7-decadiene
k. (*E*)-3-methyl-4-propyl-1,3-hexadien-5-yne
l. 5-methylene-1,3-cyclopentadiene
m. 1,8-dichloro-4-octyne
n. 2,2,5,5-tetramethyl-3-hexyne (or di-*t*-butyl acetylene)
o. 5-iodo-5-methyl-1-hexyne
p. 1-penten-4-yne
q. (*E*)-2-hepten-5-yne
r. 1-buten-3-yne
s. 3-propyl-1-heptyne
t. 2,3,4,5,6-pentamethyl-2,5-heptadiene
u. (2*E*,6*E*)-octadiene
v. (6*E*)-6-chloro-2,7,8-trimethyl-5-(1,2-dimethylpropyl)-2,6-nonadiene
w. 1,6-cyclooctadien-3-yne

4.4 a. $CH_3C{\equiv}CH$
b. $CH_3C{\equiv}CCH_2CH_3$
c. $CH_2{=}CHC{\equiv}CH$ (also called 1-buten-3-yne)
d. $CH_3CH_2C{\equiv}CCH_2CH_3$ (also called 3-hexyne)
e. $CH_2{=}C(CH_3)C{\equiv}CCH_3$
f. $HC{\equiv}CC{\equiv}CCH_2CH_3$
g. $CH_3CH_2CH_2C{\equiv}C^-Na^+$
h. $HC{\equiv}C(CH_2)_5C{\equiv}CH$
i. $CH_2{=}CHCH_2C{\equiv}CH$
j. $ClCH_2C{\equiv}CCH_3$
k. $HC{\equiv}CCH(CH_3)CH(CH_3)_2$
l. $CH_3CH_2C{\equiv}C^-Ag^+$
m. $(CH_3C{\equiv}C^-)_2Ca^{+2}$
n. 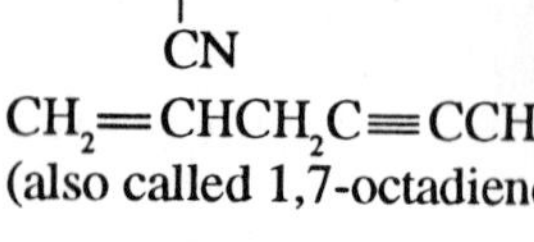
o. $CH_2{=}CHCH_2C{\equiv}CCH_2CH{=}CH_2$ (also called 1,7-octadiene-4-yne)
p. 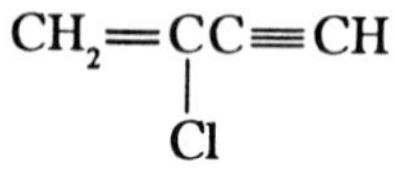
q. 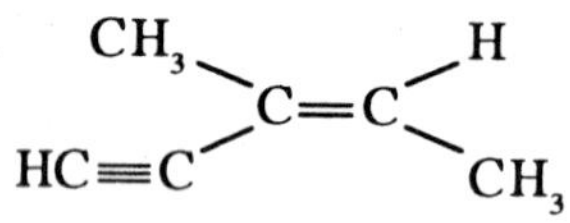
r. $CH_2{=}CHC{\equiv}CCHI_2$
s. $CH_3C{\equiv}CCH(CH_3)_2$
t. $HC{\equiv}CCH_2I$
u. $CH_2{=}CHCH_2C{\equiv}CH$
v. 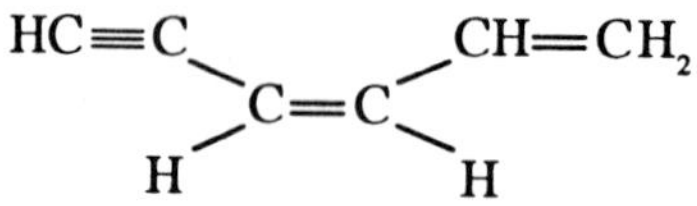
w. $HC{\equiv}CCH_2CH_2NO_2$

4.5 a. $CH_2{=}C{=}CH_2$
b. $CH_3C{\equiv}CH$
c. $CH_2{=}CHCl$
d. $CH_2{=}CHCH_2Cl$
e. $CH_3CBr_2CH_3$
f. $CH_3CH(F)CH_2F$
g. $CH_3{-}C(OH){=}CH_2$
h. CH_3COCH_3
i. $CH_3{-}C{\equiv}C{-}CH_3$
j. 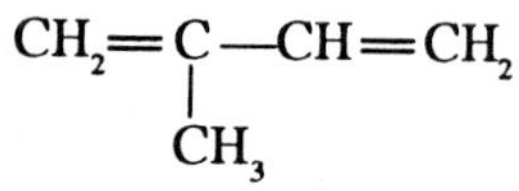
k. 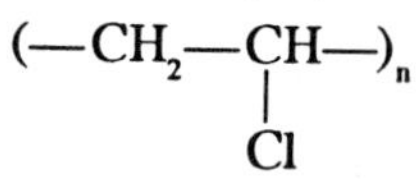

4.6 Tautomers: b, f, j
Resonance forms: a, c, d, e, g, i
Non-tautomeric isomers: h, k, l

4.7
a. $CH_3CH_2CH_2CH{=}CH_2$
b. $CH_3CH_2CH_2CH_2CH_3$
c. $CH_3CH_2CH_2CBr{=}CHBr$ (trans)
d. $CH_3CH_2CH_2CBr_2CHBr_2$
e. $CH_3CH_2CH_2C(Cl){=}CH_2$
f. $CH_3CH_2CH_2CCl_2CH_3$
g. $CH_3CH_2CH_2C{\equiv}C^-Na^+ + \frac{1}{2}H_2$
h. $CH_3CH_2CH_2C{\equiv}C^-Ag^+ + HNO_3$
i. no reaction
j. $CH_3CH_2CH_2C{\equiv}C^-Na^+ + NH_3$
k. $CH_3CH_2CH_2COCH_3$
l. $CH_3CH_2CH_2CH{=}CH_2$
m. no reaction
n. $CH_3CH_2CH_2COOH + CO_2 + H_2O$
o. $CH_3CH_2CH_2COOH + CO_2 + H_2O$

4.8
a. 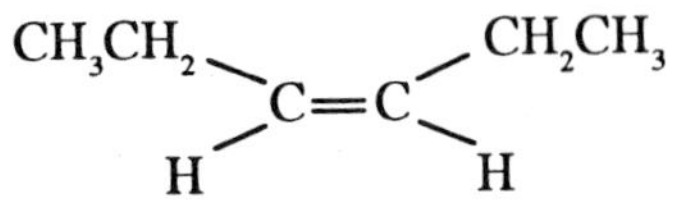
b. $CH_3CH_2CH_2CH_2CH_2CH_3$
c. 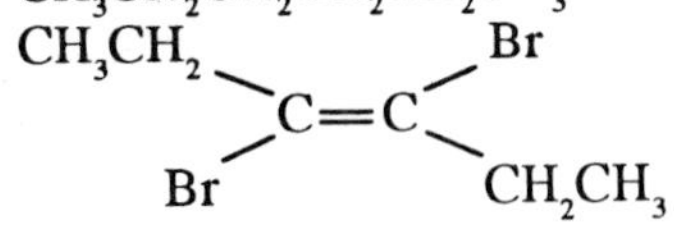
d. $CH_3CH_2CBr_2CBr_2CH_2CH_3$
e. 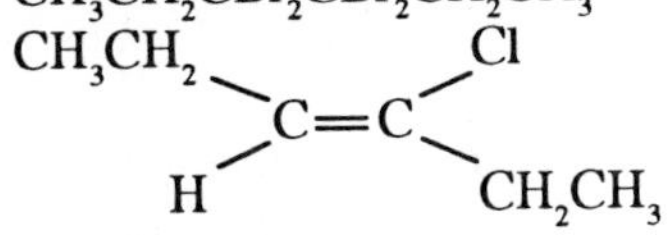 and 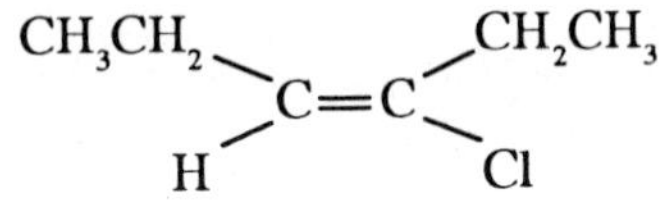
f. $CH_3CH_2CCl_2CH_2CH_2CH_3$
g. no reaction
h. no reaction
i. no reaction
j. no reaction
k. 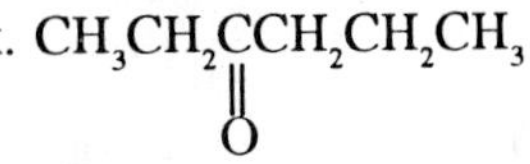
l. 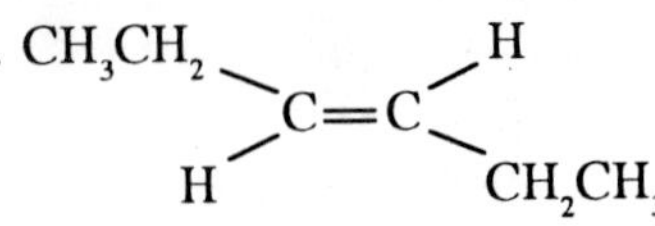
m. no reaction
n. CH_3CH_2COOH (2 moles)
o. CH_3CH_2COOH (2 moles)

4.9
a. $CH_3CH_2CH_2CH_3$
b. $BrCH_2CH(Br)CH(Br)CH_2Br$
c. $CH_3CH(Cl)CH{=}CH_2$ (major product)
d. 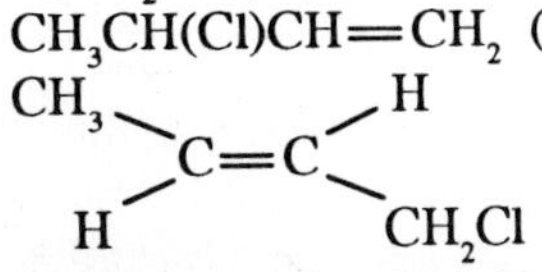 (major product)
e. $CH_3CH(OH)CH(OH)CH_3$
f. no reaction
g. no reaction
h. $H_2C{=}O$ (2 moles) + 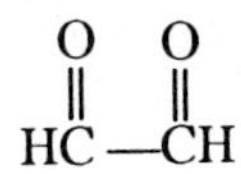
i. no reaction
j. no reaction
k. $2CO_2 + HOOC—COOH$
l. no reaction
m. $CH_3—CH{=}CH—CH_3$
n. $BrCH_2—CH{=}CH—CH_2Br$

4.10
A = $(n\text{-}C_5H_{11})C{\equiv}CMgCl$
B = $(CH_3)_2CH_2$
C = $(n\text{-}C_5H_{11})—C{\equiv}C—(n\text{-}C_4H_9)$
D = MgCl I
E = $HC{\equiv}C—(n\text{-}C_4H_9)$
F = $CH_3CH_2CH_2C{\equiv}CH$
G = NaBr
H = $CF_3—C{\equiv}C—CF_3$
I = $ZnCl_2$
J = $C_2H_5CH_2C{\equiv}CCH_2C_2H_5$
K = $MgBr_2$
L = 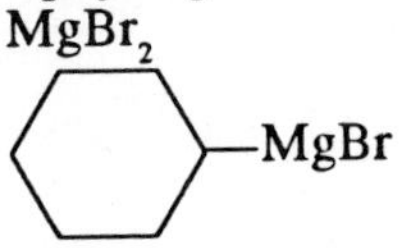
M = 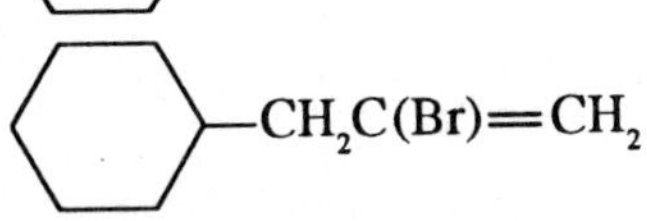
N = 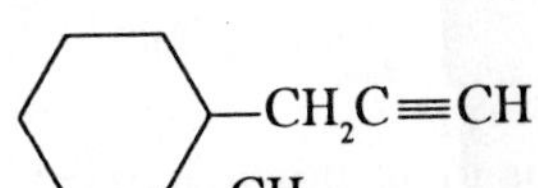
O = 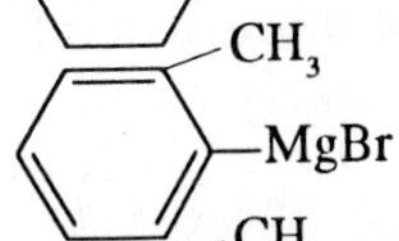
P = 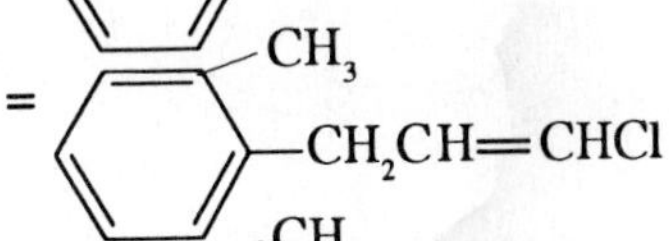
Q = 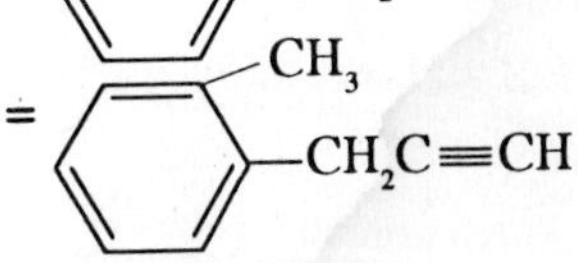
R = $CH_3(CH_2)_7C{\equiv}C(CH_2)_7COOH$
S = $BrCH_2CH(Br)(CH_2)_8COOH$
T = $HC{\equiv}C(CH_2)_8COOH$
U = $n\text{-}C_4H_9C{\equiv}CMgBr$
V = C_2H_6

W = H_2O
X = no reaction
Y = $CH_3CH_2CH_3$
Z = MgCl(OH)
AA = $C_6H_{13}C{\equiv}CCH_3$
BB = $(CH_3)_2CHC{\equiv}CC_6H_5$
CC = $C_2H_5CH_2C{\equiv}CCH_3$
DD = H_2O
EE =

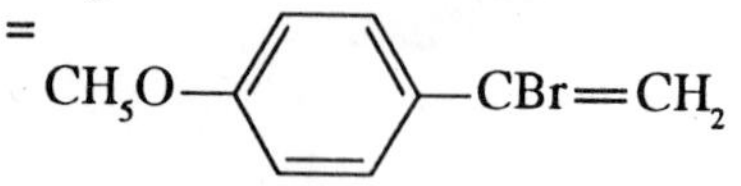

FF = $C_8H_{17}CH{=}CH_2$
GG = $C_6H_5C{\equiv}CH$
HH = LiOH
II = C_4H_9COOH
JJ = $CH_3CHBrCH_3$

KK =
LL =

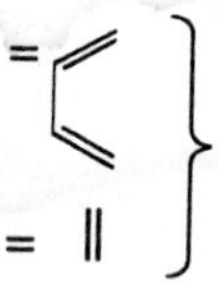

reverse Diels-Alder reaction

MM = $CH_3{-}CCl(CH_3){-}CH{=}CH_2$

NN = $CH_3{-}C(CH_3){=}CH{-}CH_2Cl$

OO = $Cl_3CCH_2{-}CBr(CH_3){-}CH{=}CH_2$

PP = $Cl_3CCH_2{-}C(CH_3){=}CH{-}CH_2Br$

QQ = $CH_3CHClCH{=}CHCH_3$

4.11 a.

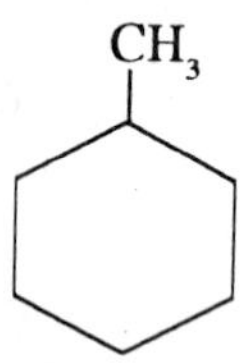

b.

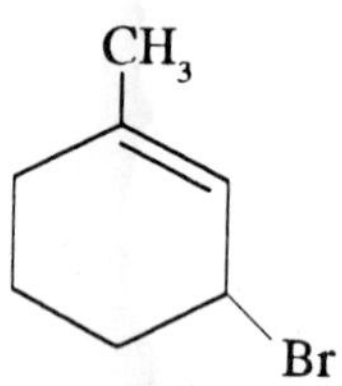

(1,2-addition)

+

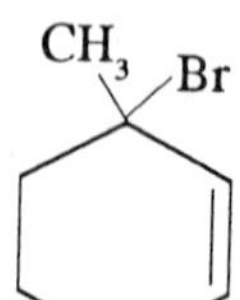

(1,4-addition)

The reaction proceeds via the *most stable* carbocation intermediate:

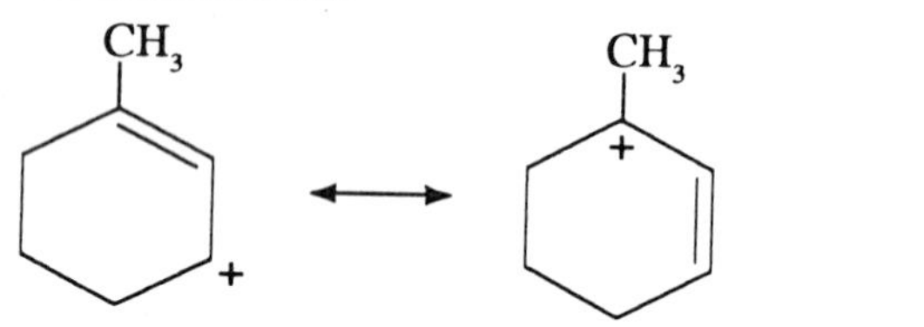

(especially stable, 3°, allylic)

c.

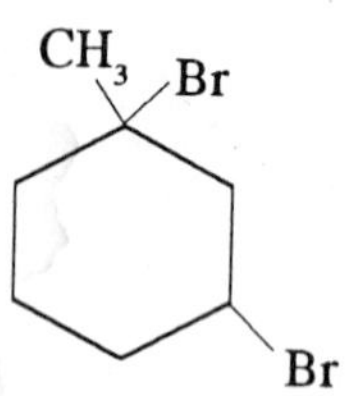

+

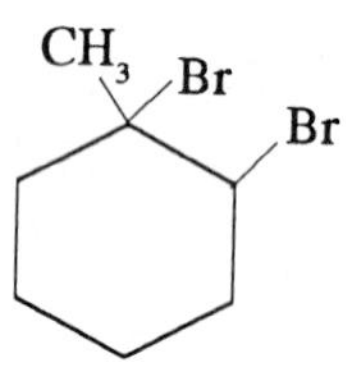

Markovnikov addition to the products of (b)

d.

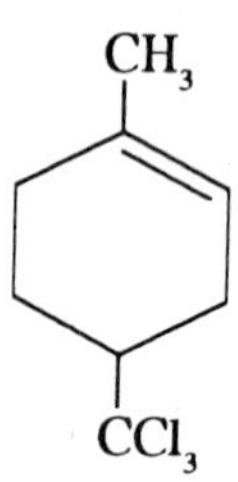

+

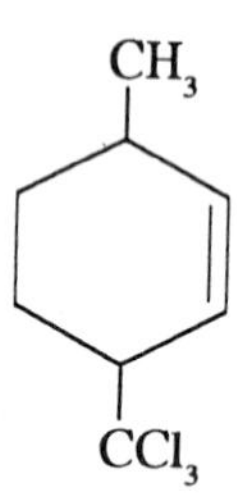

The reaction proceeds via the *most stable* radical intermediate, with the following mechanism:

$$R{-}O{-}O{-}R \longrightarrow 2RO^{\cdot}$$

$$RO^{\cdot} + CHCl_3 \longrightarrow R{-}OH + {}^{\cdot}CCl_3$$

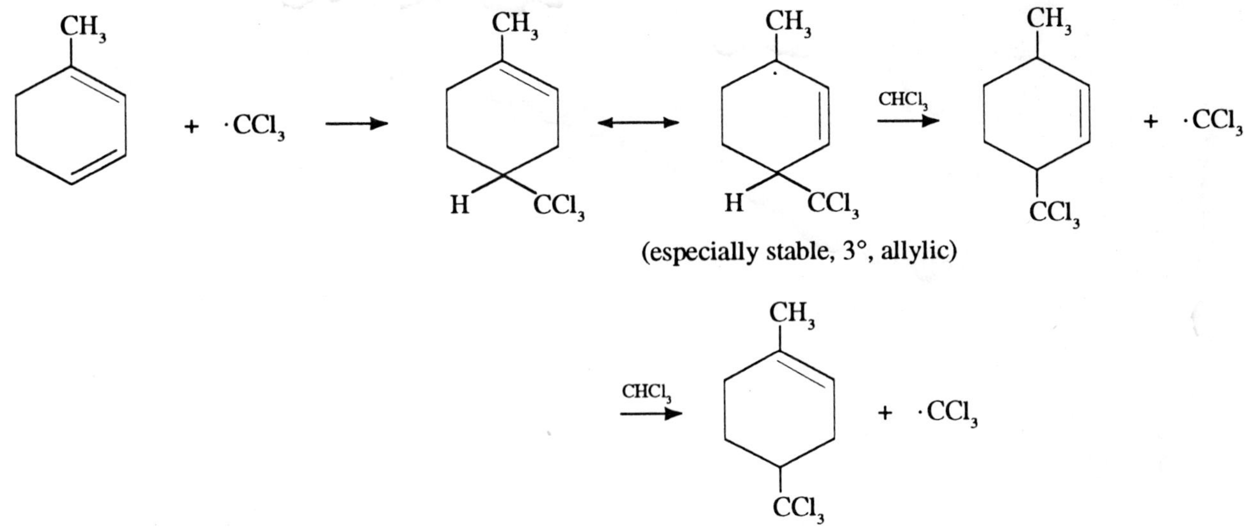
CH3
+ ·CCl3
H
CCl3
(especially stable, 3°, allylic)
CHCl3
+ ·CCl3

4.12 a.

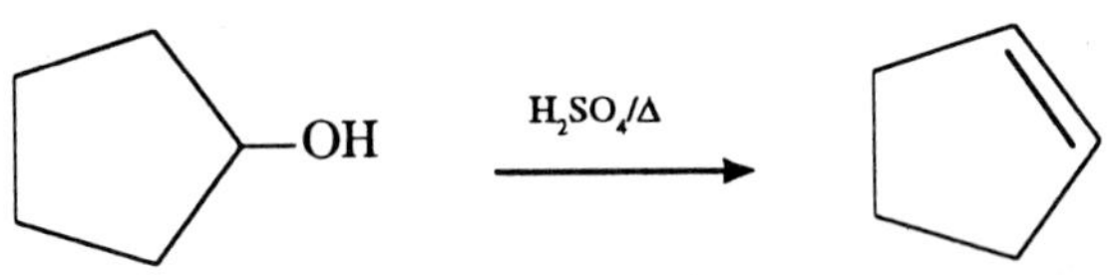
OH
H2SO4/Δ

b.

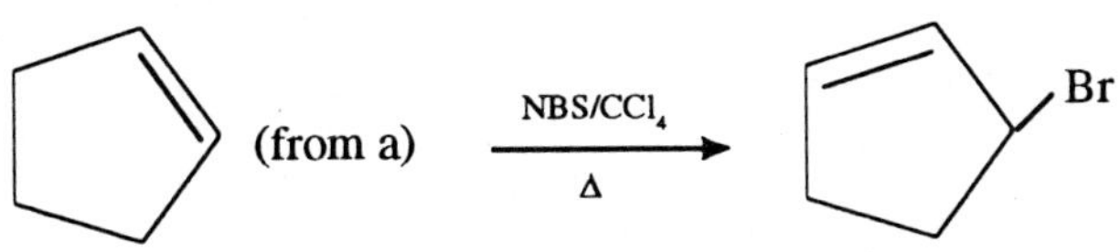
(from a)
NBS/CCl4
Δ
Br

c.

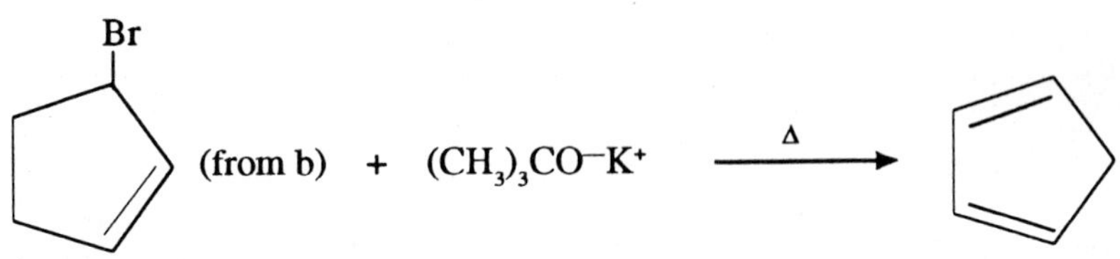
Br
(from b) + (CH3)3CO−K+
Δ

d.

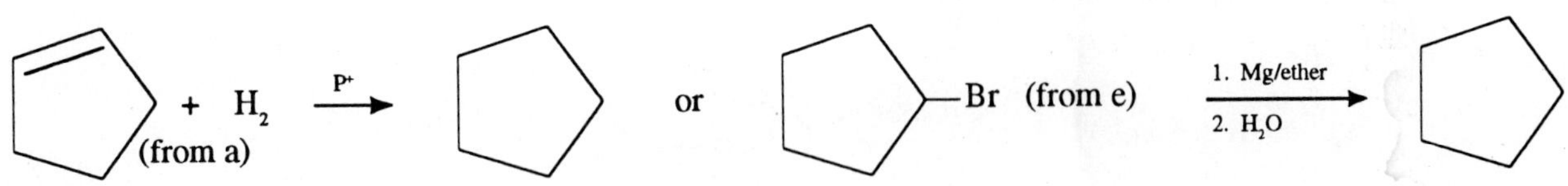
+ H2
(from a)
P+
or
Br (from e)
1. Mg/ether
2. H2O

e.

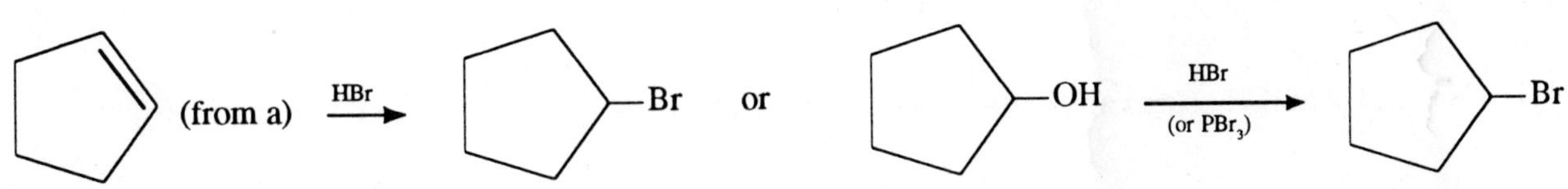
(from a)
HBr
Br
or
OH
HBr
(or PBr3)
Br

f.

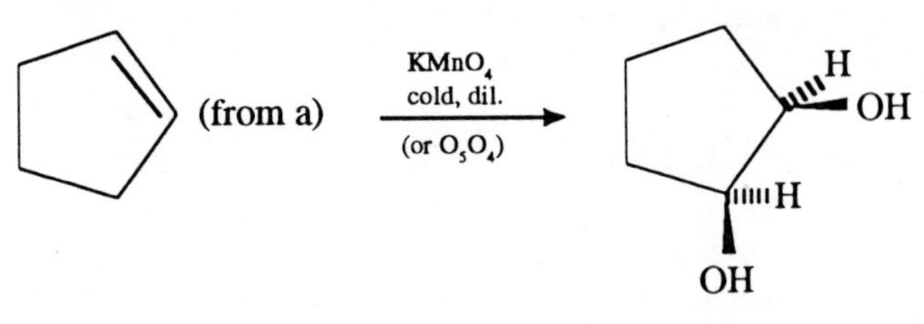
(from a)
KMnO4
cold, dil.
(or O5O4)
H
OH
H
OH

g.

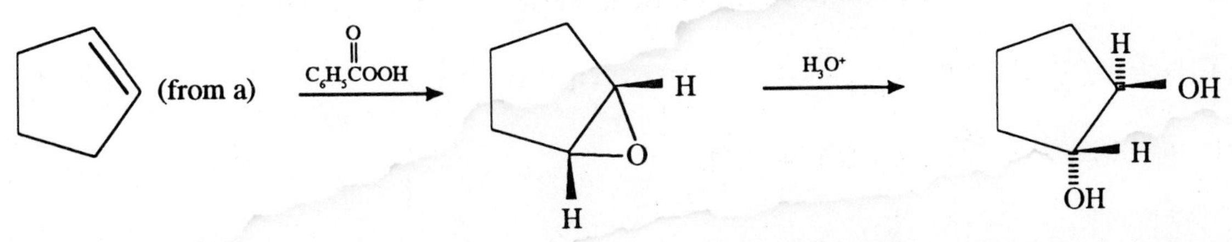
(from a)
O
C6H5COOH
H
O
H
H3O+
H
OH
H
OH

h. 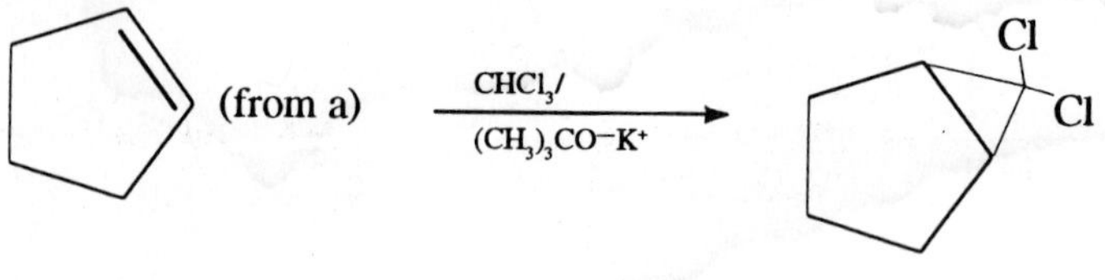

i. 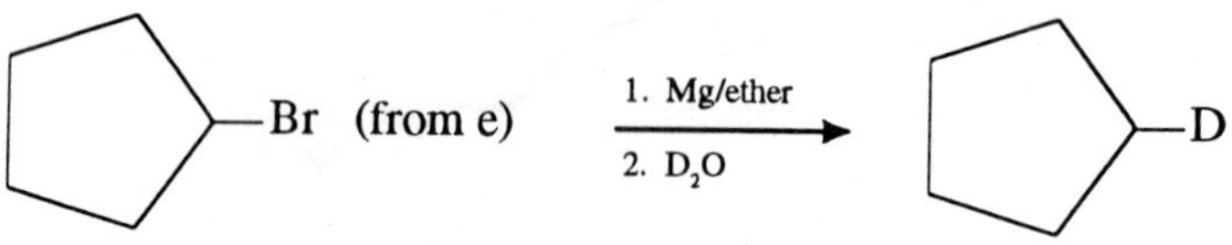

j. 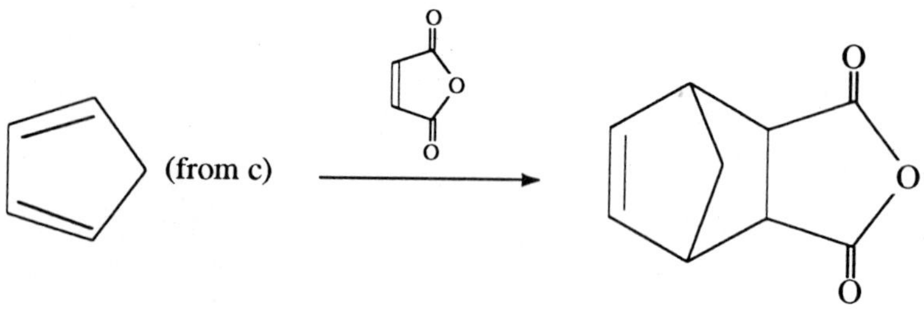

k. 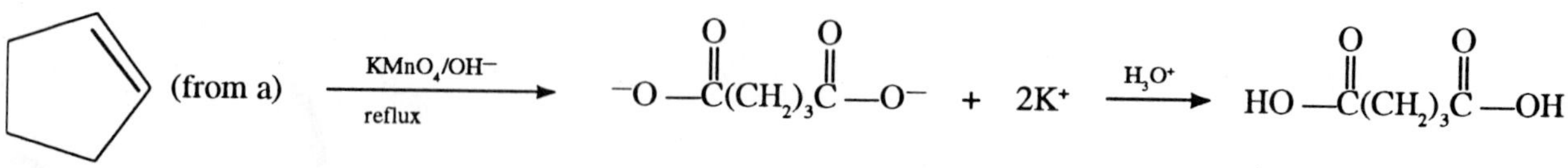

4.13 a. $HC{\equiv}CH + NaNH_2 \longrightarrow HC{\equiv}C^-Na^+ \xrightarrow{CH_3I} HC{\equiv}C{-}CH_3 \xrightarrow{H_2/\text{Lindlar's catalyst}} H_2C{=}CH{-}CH_3$

$H_2C{=}CH{-}CH_3 \xrightarrow{\text{DBr/no peroxides}} DH_2C{-}CH(Br){-}CH_3$

b. $CH_3C{\equiv}CH$ (from a) $\xrightarrow[Hg^{+2}]{H_2O/H^+} CH_3{-}\overset{O}{\overset{\|}{C}}{-}CH_3$

c. $CH_3C{\equiv}CH$ (from a) $\xrightarrow[\text{Lindlar's catalyst}]{H_2\ (1\ \text{mole})} CH_3CH{=}CH_2 \xrightarrow[\text{cold, dilute}]{KMnO_4/OH^-} CH_3\underset{OH}{\underset{|}{C}H}CH_2OH$

d. $CH_3C{\equiv}CH$ (from a) $\xrightarrow{NaNH_2} CH_3{-}C{\equiv}C^-Na^+ \xrightarrow{CH_3I} CH_3{-}C{\equiv}C{-}CH_3$

$CH_3{-}C{\equiv}C{-}CH_3 \xrightarrow[\text{catalyst}]{H_2\ (1\ \text{mole})\ \text{Lindlar's}} CH_3{-}CH{=}CH{-}CH_3 \xrightarrow{HBr} CH_3CH_2CHBrCH_3$

e. $CaC_2 + H_2O \longrightarrow HC{\equiv}CH \xrightarrow[Hg^{2+}]{H_2O/H^+} CH_3CHO$

f. $CH_3C{\equiv}CH$ (from a) $\xrightarrow[2.\ CH_3CH_2I]{1.\ NaNH_2} CH_3C{\equiv}CCH_2CH_3$

g. $CH_3CH{=}CHCH_3$ (from d) $\xrightarrow{H_2O/H^+} CH_3CH(OH)CH_2CH_3$

h. $CH_3CH_2CH_3 \xrightarrow[UV]{Br_2} CH_3\underset{}{\overset{Br}{C}}HCH_3 \xrightarrow[EtOH]{KOH} CH_3CH{=}CH_2$ *Note:* "EtOH = ethyl alcohol"

$CH_3CH{=}CH_2 \xrightarrow{Br_2/CCl_4} CH_3\underset{Br}{C}HCH_2Br \xrightarrow[2.\ H^+]{1.\ 3NaNH_2} CH_3C{\equiv}CH$

i. $CH_3C{\equiv}CH$ (from h) $\xrightarrow[Hg^{2+}]{H_2O/H^+} CH_3\overset{O}{\overset{\|}{C}}CH_3$

j. $HC{\equiv}CH \xrightarrow{NaNH_2} HC{\equiv}C^-Na^+ \xrightarrow{DCl} HC{\equiv}CD + NaCl$

k. $CH_3C{\equiv}CH$ (from a) $\xrightarrow[\text{Lindlar's catalyst}]{H_2\ (1\ mole)} CH_3CH{=}CH_2 \xrightarrow{NBS} Br{-}CH_2CH{=}CH_2$

l. $CH_3CH{=}CH_2 \xrightarrow{Br_2/CCl_4} CH_3\underset{Br}{C}HCH_2Br \xrightarrow[2.\ H^+]{1.\ 3NaNH_2} CH_3C{\equiv}CH$

m. $CH_3CH{=}CH_2$ (from k) $\xrightarrow{HBr} CH_3\underset{Br}{C}HCH_3$

n. $CH_3CH{=}CH_2$ (from k) $\xrightarrow{H_2O/H^+} CH_3\underset{OH}{C}HCH_3$

o. $HC{\equiv}CH \xrightarrow[CCl_4]{2Cl_2} Cl_2CH{-}CHCl_2 \xrightarrow[EtOH]{KOH} Cl_2C{=}CHCl \xrightarrow{Cl_2/CCl_4} Cl_3C{-}CHCl_2 \xrightarrow[EtOH]{KOH} Cl_2C{=}CCl_2$

4.14 If the compounds in a, b, and c were completely saturated, the formula would be $C_{22}H_{46}$. Each compound has a deficiency of 10 hydrogens. A double bond accounts for 2, a triple bond accounts for 4, and a ring accounts for 2 hydrogens.

a. 5 rings account for the hydrogen deficiency.
b. No rings present.
c. 3 rings present.
d. Absorption of four moles of bromime indicates the presence of four double bonds or two triple bonds, or of two double bonds and one triple bond. Therefore, there must be one ring since there are 10 hydrogens missing in the original compound (compared to the open chain alkane $C_nH_{2n+2} = C_{14}H_{30}$).
e. No rings
f. No rings
g. Two rings
h. One ring

4.15

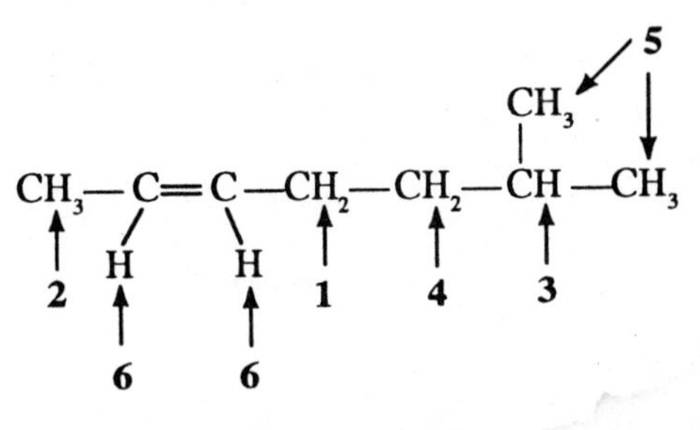

6-methyl-2-heptene

1. are 2° and allylic
2. are 1° and allylic
3. is 3°
4. are 2°
5. are 1°
6. are vinylic

4.16 An *s* orbital is of lower energy than a *p* orbital. Thus the more *s* character a hybrid orbital has, the less energy it contains. There is a difference in the amounts of *s* character of the three families below

Family	Hybridization	Percentage *s* character
Alkanes	sp^3	25.0
Alkenes	sp^2	33.3
Alkynes	sp	50.0

Consequently, a carbanion at an *sp* carbon is more stable than a carbanion at an sp^2 or sp^3 carbon. This means that $HC{\equiv}C^-$ (greatest *s* character) is a weaker base, and that its conjugate acid ($HC{\equiv}CH$) is a stronger acid.

4.17 a. Hydrogen cyanide is a stronger acid than acetylene because of the electronegative nature of the nitrogen atom. This makes its conjugate base ($^-C{\equiv}N$) more stable and weaker. Consequently the corresponding acid ($HC{\equiv}N$) is stronger.

b. The two double bond systems of 1,2-butadiene (and other cumulenes) are at right angles to each other. Consequently, it is impossible for it to have the overlapping of *p* orbitals seen in conjugated systems such as 1,3-butadiene. Overlapping of *p* orbitals in conjugated systems permits delocalization of the π electrons over more than two nuclei.

c. In 1,3-butadiene (and in other conjugated systems) extended overlapping of *p* orbitals results in shortening of the C_2—C_3 single bond which then exhibits some double-bond character. The shorter bond distance results because the bond is a product of sp^2-sp^2 hybridization rather than the sp^3-sp^3 hybridization in alkanes. The partial double-bond character can be rationalized by showing contributing resonance structures:

$$\overset{+}{C}H_2{-}CH{=}CH{-}\overset{-}{C}H_2 \longleftrightarrow CH_2{=}CH{-}CH{=}CH_2 \longleftrightarrow \overset{-}{C}H_2{-}CH{=}CH{-}\overset{+}{C}H_2$$

d.

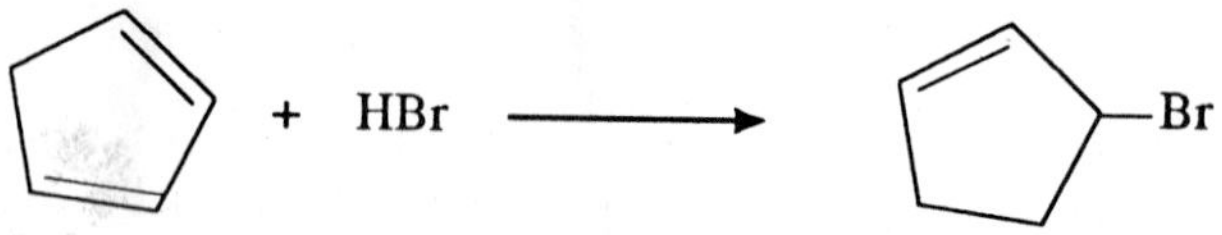

3-bromocyclopentene

Mechanism:

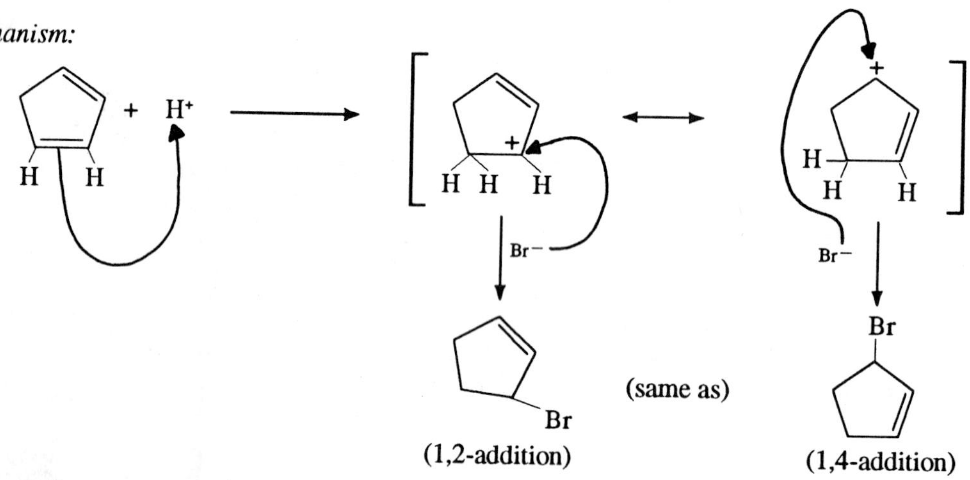

1,2- and 1,4-addition result in the formation of the same product, 3-bromocyclopentene. Whereas,

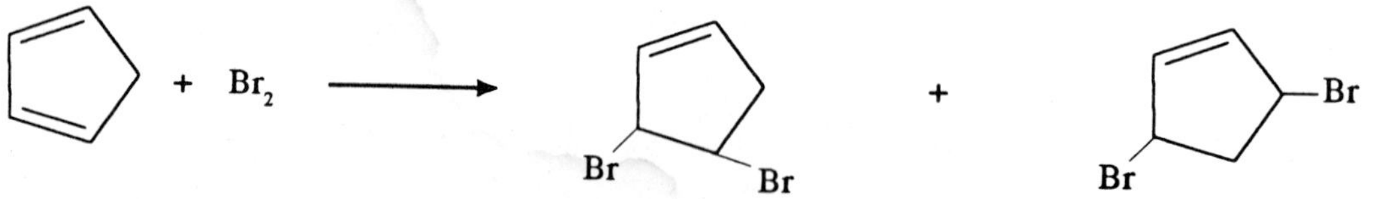

3,4-dibromocyclopentene 3,5-dibromocyclopentene

Mechanism:

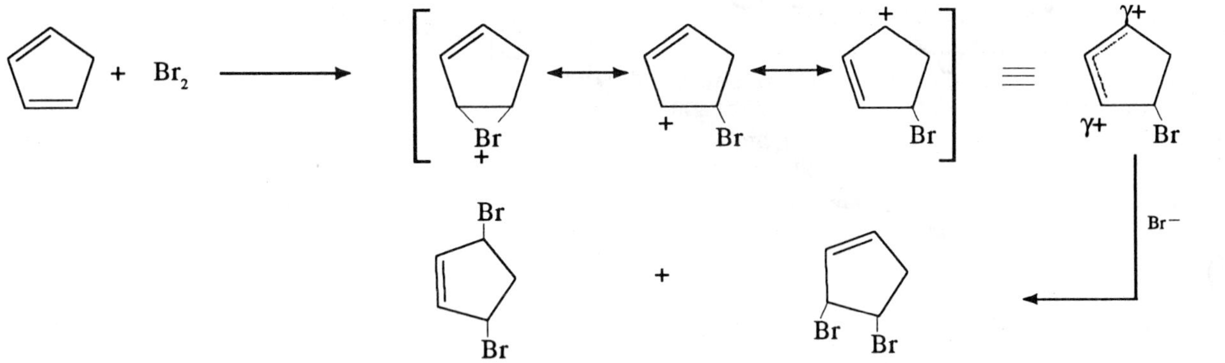

3,5-dibromocyclopentene (1,4-addition) 3,4-dibromocyclopentene (1,2-addition)

1,2- and 1,4-addition result in the formation of two different products.

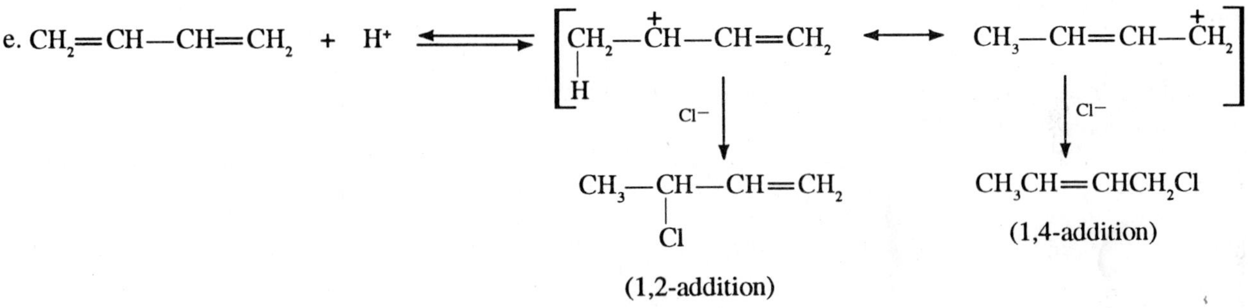

The 1,2-product is favored at low temperatures because its formation requires a lower energy of activation. At higher temperatures, the thermodynamically more stable 1,4-product predominates, and the product distribution is equilibrium (thermodynamically) controlled.

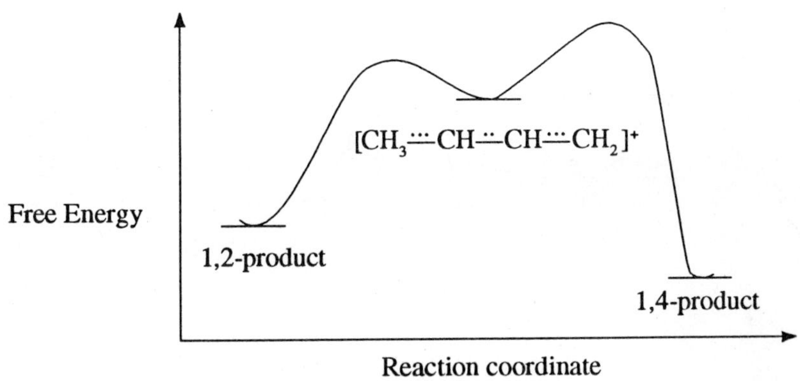

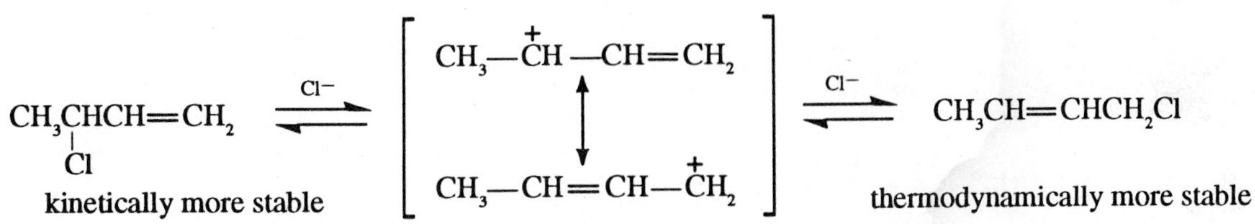

4.18 a. Addition of $AgNO_3$/ethanol to hexyne will yield a white precipitate according to the equation

$$CH_3(CH_2)_3C{\equiv}CH \;+\; Ag^+ \longrightarrow \underset{\text{white ppt}}{CH_3(CH_2)_3C{\equiv}C^-Ag^+} \;+\; H^+$$

No reaction will occur with 1-hexene

b. Only terminal alkynes (RC≡CH) react with $AgNO_3$; therefore, only 1-hexyne will give a white precipitate (see (a)).
c. 1-Hexyne will react with $AgNO_3$, while 1,3-hexadiene will not (see(a)).
d. 3-Hexyne will decolorize Br_2/CCl_4 (orange solution), while 2-chlorohexane will not. Also, 3-hexyne will react with the purple $KMnO_4$ solution to yield a black MnO_2 precipitate. 2-Chlorohexane does not react with $KMnO_4$.
e. 2-hexyne reacts with both Br_2/CCl_4 and $KMnO_4$, while hexane does not (see (d)).
f. Only conjugated dienes react with alkenes in Diels-Alder reactions. Therefore, only 1,3-hexadiene will react with a dienophile (e.g., maleic anhydride) to yield a Diels-Alder adduct (ppt)

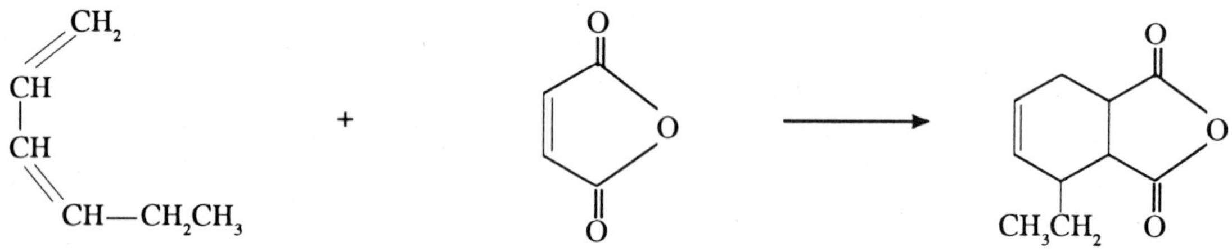

1,3-hexadiene   maleic anhydride   Diels-Alder adduct

g. 1-Hexyne will not show the presence of chlorine in the sodium fusion test, while 1-chloro-2-hexyne will. *Note*: Both compounds give a precipitate when reacted with $AgNO_3$. The Beilstein flame test for halogen can be used for confirmation of the presence or absence of chlorine.

4.19 **AAA** =

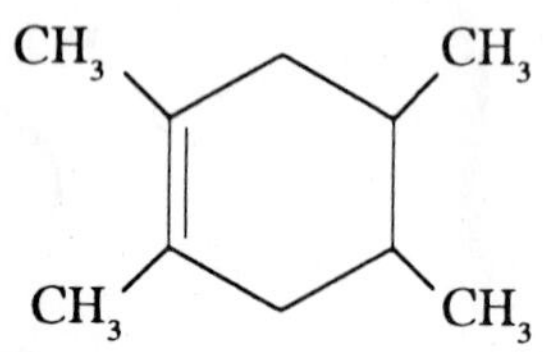

EEE =

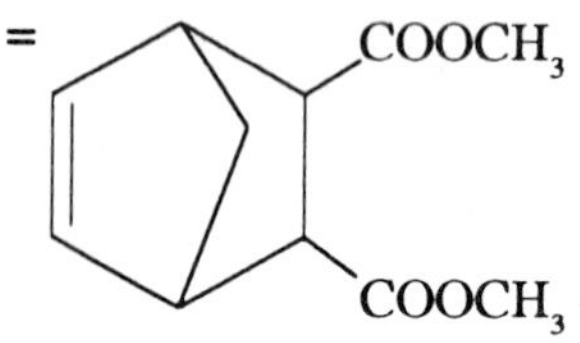

BBB =

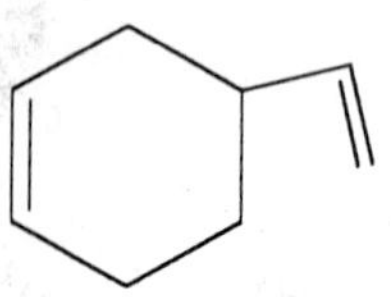

FFF =

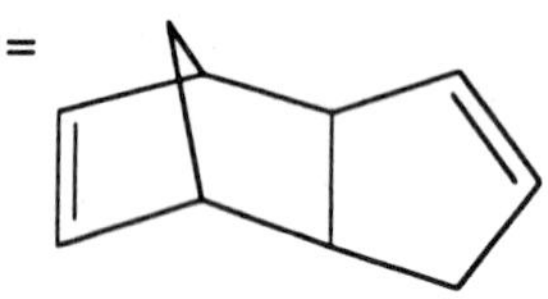

CCC =

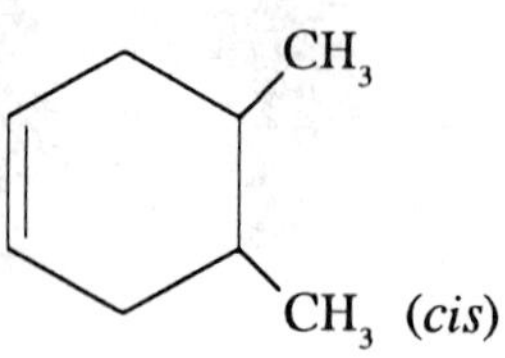

GGG =

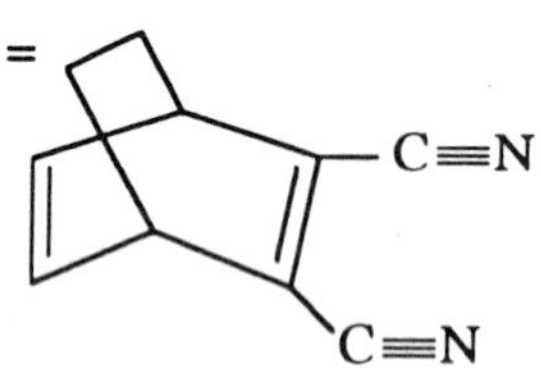

DDD =

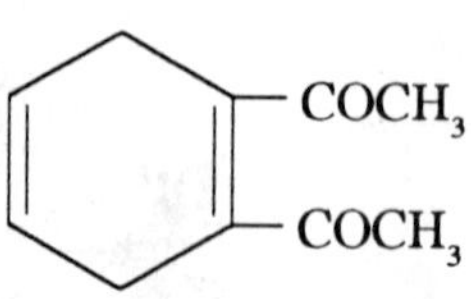

4.20 a.

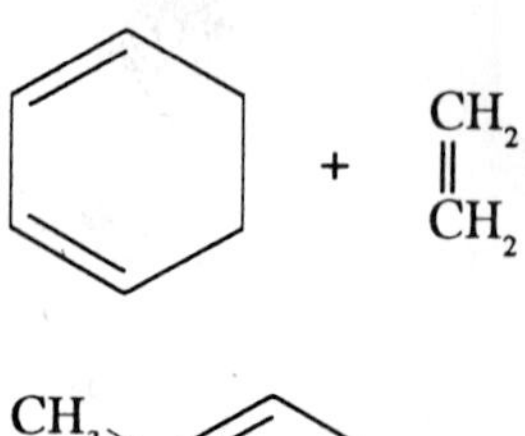

c.

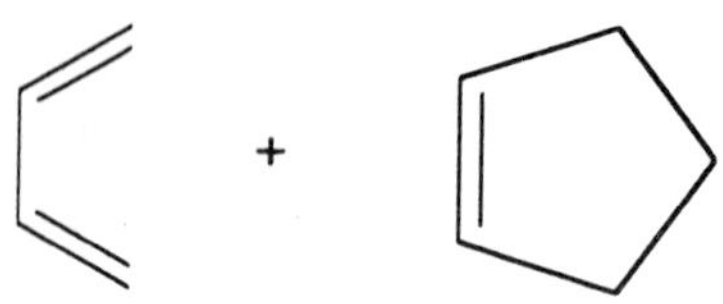

b.

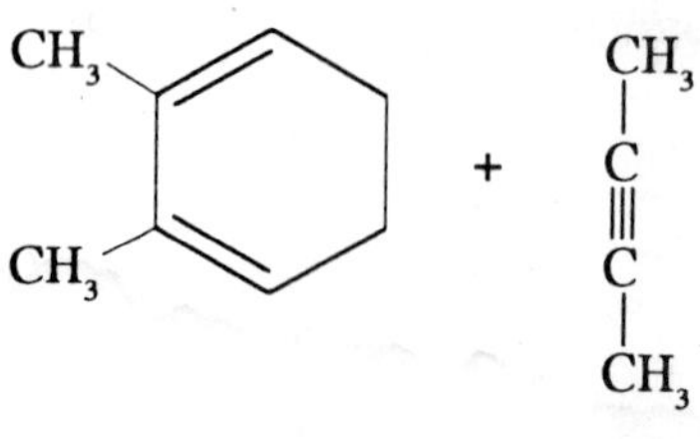

d.

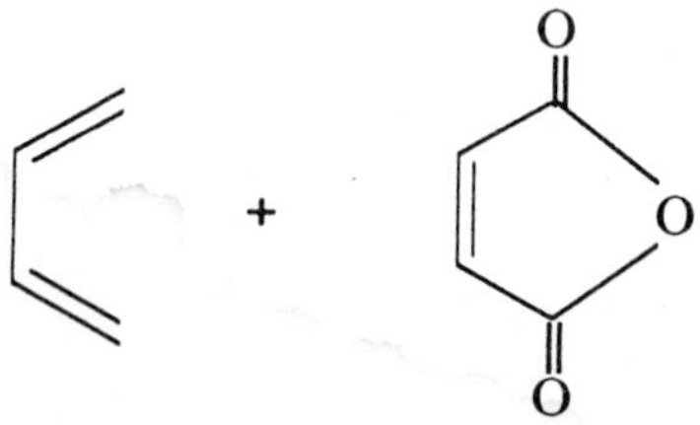

e.

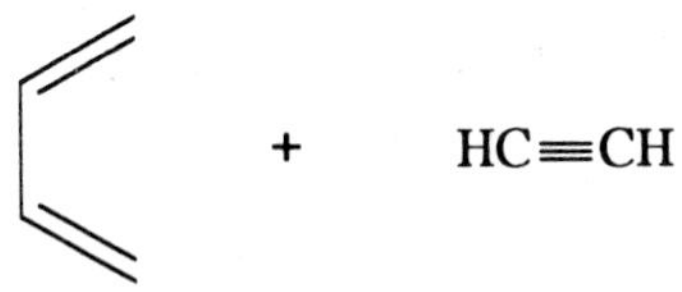

f.

g. This compound cannot be synthesized directly via a Diels-Alder reaction because the six-membered ring does not have a carbon-carbon double bond.

h.

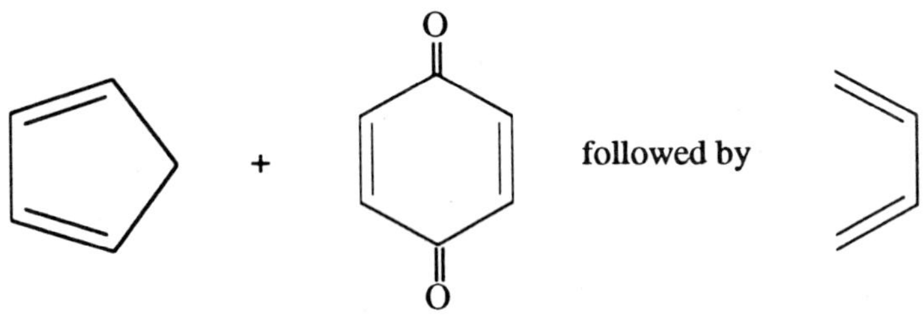

i. This compound can not be made directly via a Diels-Alder reaction, since there is no six-membered ring in the molecule.

j.

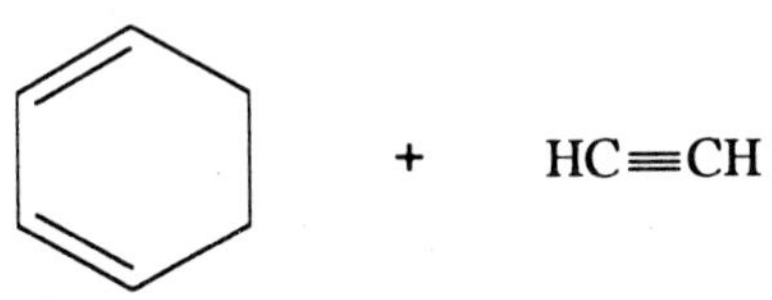

k.

4.21 a. two

b. two

c. Since two carbons are missing from the isolated product from reaction with hot $KMnO_4$, one is led to the conclusion that there must be two terminal alkene (methylene) groups. These should yield formaldehyde (CH_2O) upon oxidation, which is readily converted into H_2CO_3 (CO_2 & H_2O) by the harsh $KMnO_4$ conditions. Consequently, the products of O_3/Zn oxidation of β-selinene are

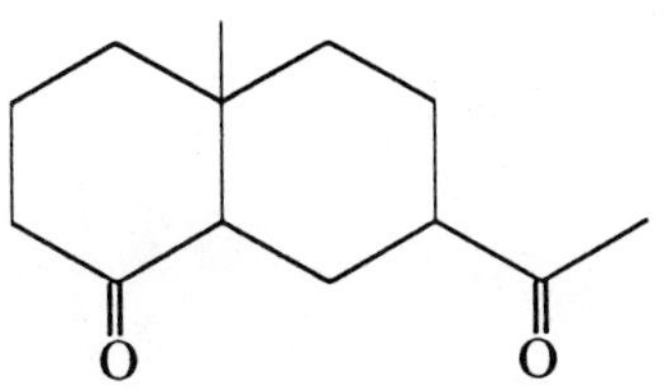

and HCHO

(two moles)

d. The structure of β-selinene is

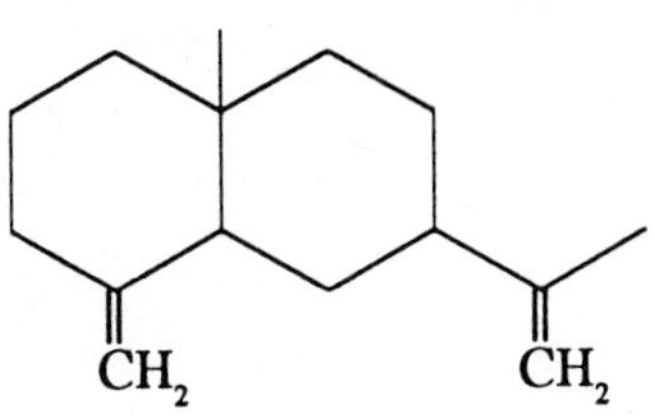

e. Yes, it does, as shown by the dotted lines which delineate the isoprene units.

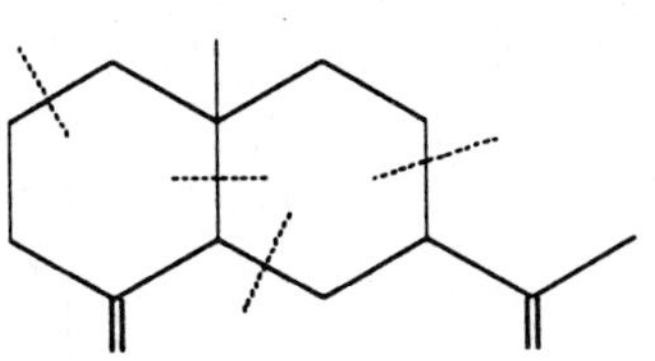

4.22 a. three
b. zero

c. 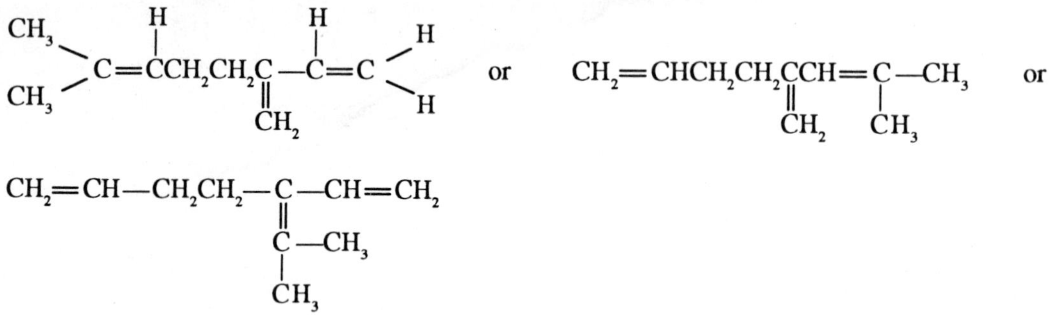

d. 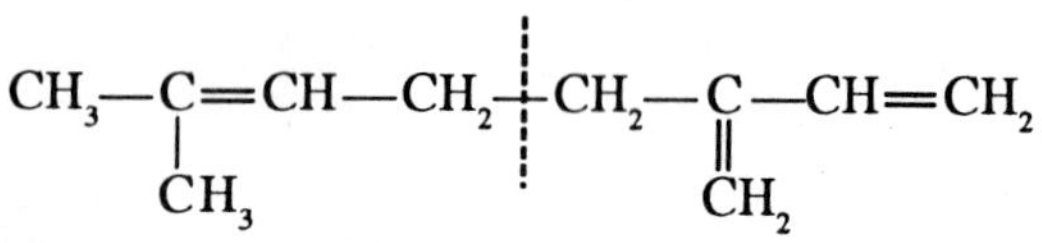

4.23 a. two
b. one

c. 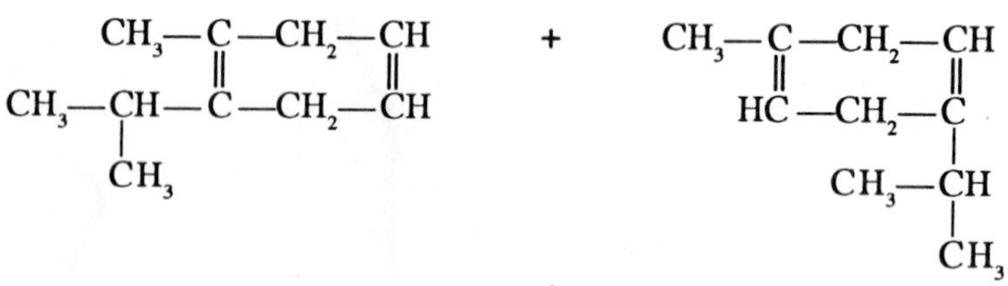

d. 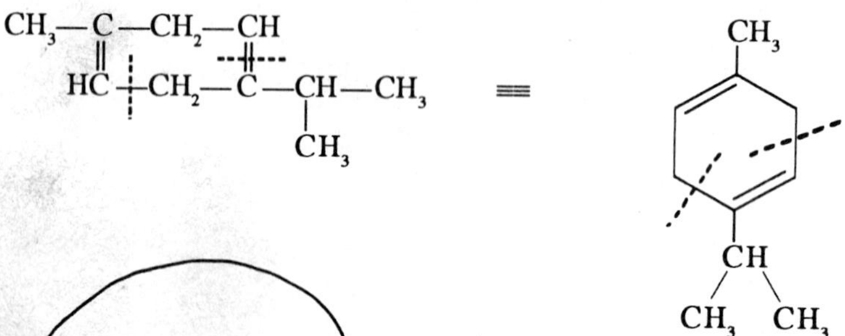

4.24 a. 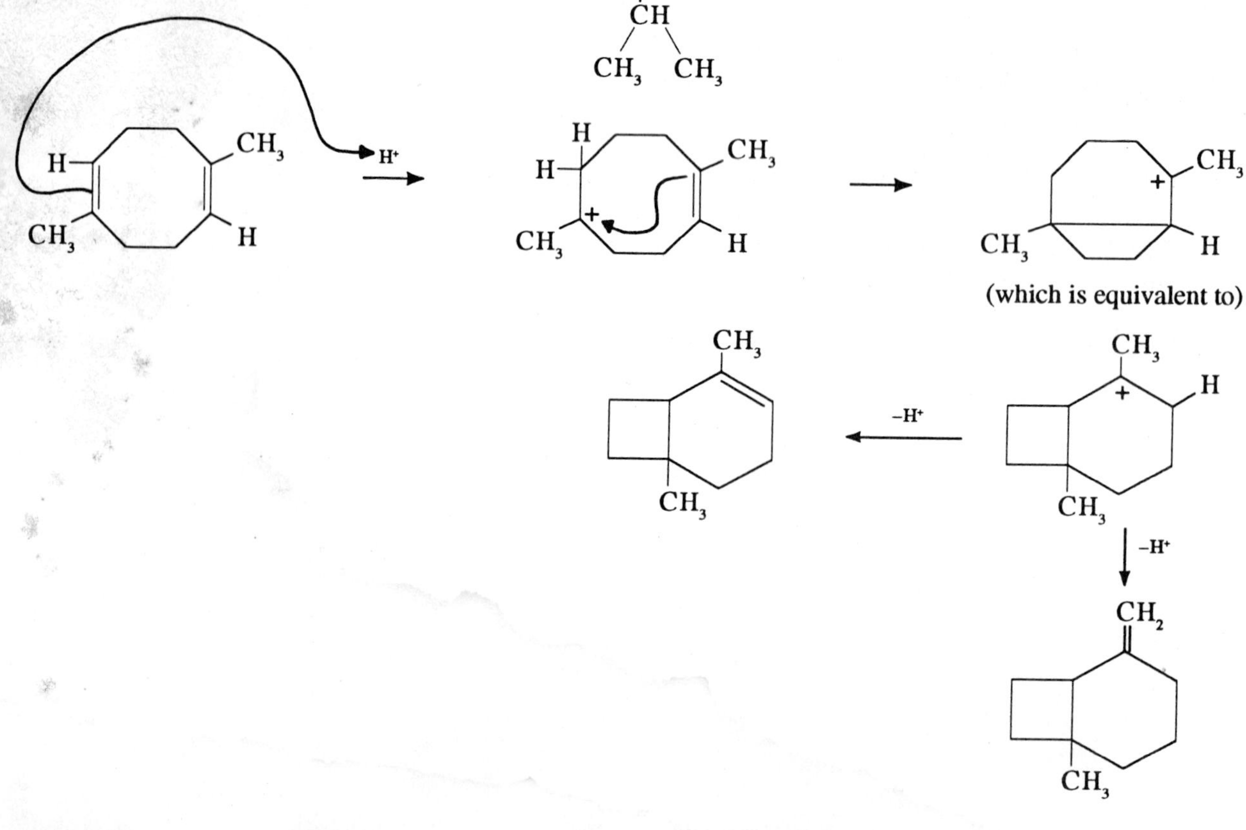

b.

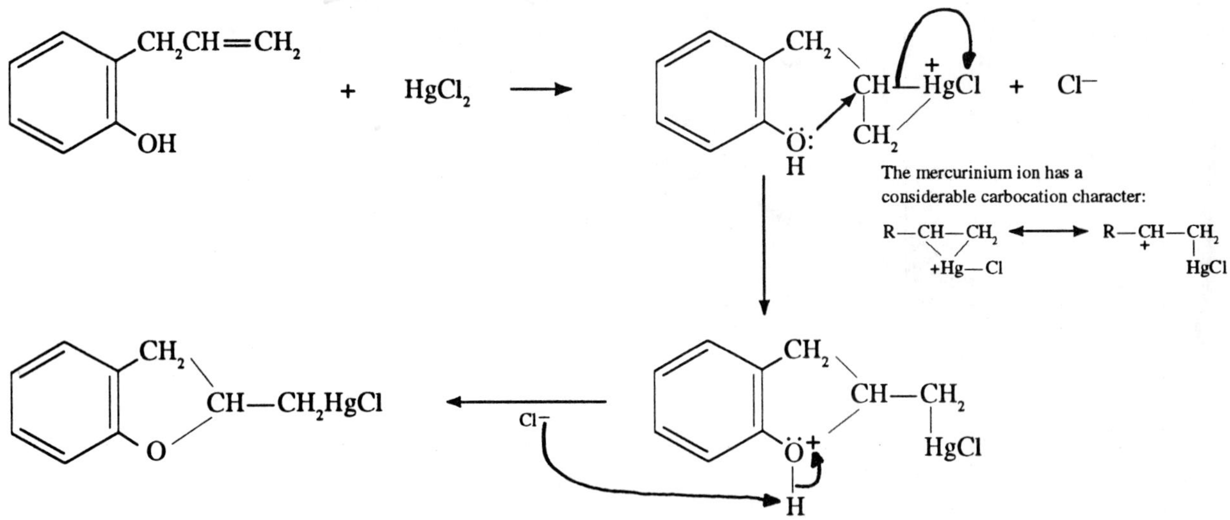

c. Markovnikov addition results in the more stable carbocation intermediates

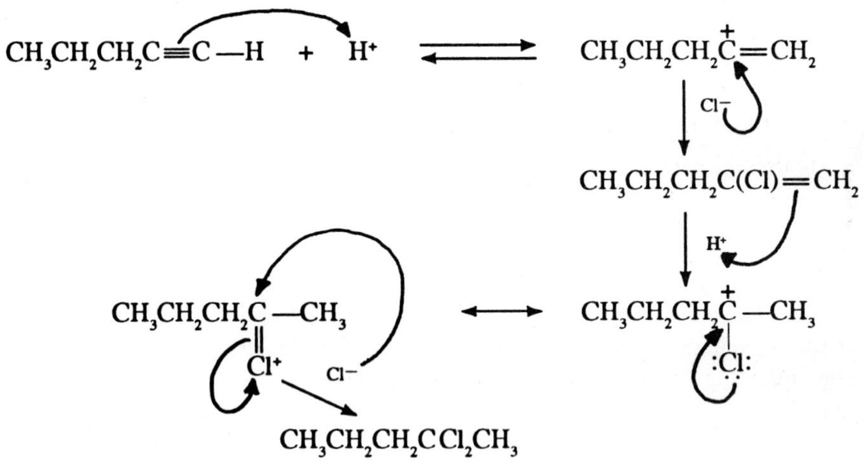

4.25 The electronegativity of the bromine atom increases the acidity of the starred hydrogens in the structure below, thus making them more susceptible to removal by a base. For diene formation, the base must abstract the less acidic neighboring hydrogens that are on the methyl groups.

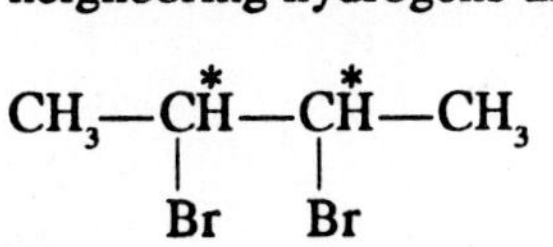

Thus, the formation of alkyne over diene depends on the relative rates at which the hydrogens are abstracted by the base, and the reaction is rate-controlled (i.e., kinetically controlled).

4.26 The relative rates are in order of most to least stable carbocation

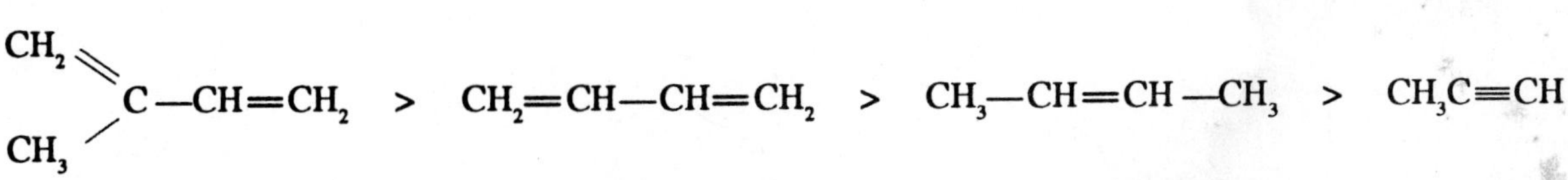

isoprene butadiene 2-butene propyne

4.27

Hydrocarbon	Structure	ΔH° (kcal/mole)
2-methyl-2-butene	$CH_3-\underset{\underset{CH_3}{\mid}}{C}=CHCH_3$ (trisubstituted alkene)	–27
2-methyl-1-butene	$CH_2=\underset{\underset{CH_3}{\mid}}{C}-CH_2CH_3$ (disubstituted alkene)	–28
3-methyl-1-butene	$CH_2=CH-\underset{\underset{CH_3}{\mid}}{C}HCH_3$ (monosubstituted alkene)	–30
E-1,3-pentadiene	$CH_2=CH-CH=CH-CH_3$ (conjugated diene)	–54
1,4-pentadiene	$CH_2=CH-CH_2-CH=CH_2$ (non-conjugated diene)	–60

Heats of hydrogenation can be used to measure the relative stabilities of alkenes when the alkenes can be reduced to the same alkane. In these cases, the smaller the value of ΔH°, the more thermodynamically stable the alkene. The first three alkenes listed above yield isopentane on catalytic hydrogenation and thus can be compared. The most highly substituted alkene, 2-methyl-2-butene, has the smallest value of ΔH° and is the most thermodynamically stable of the three alkenes. Conjugated dienes are more stable than non-conjugated dienes as shown by the ΔH° values of (*E*)-1,3-pentadiene and 1,4-pentadiene.

4.28 a. This is a Diels-Alder reaction. The diene must be in the S-*cis* conformation

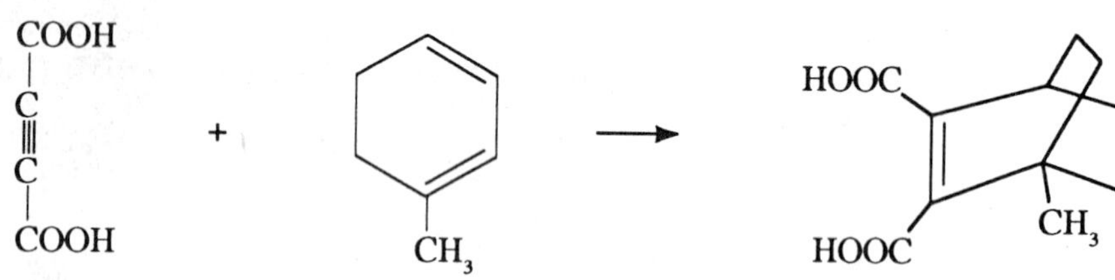

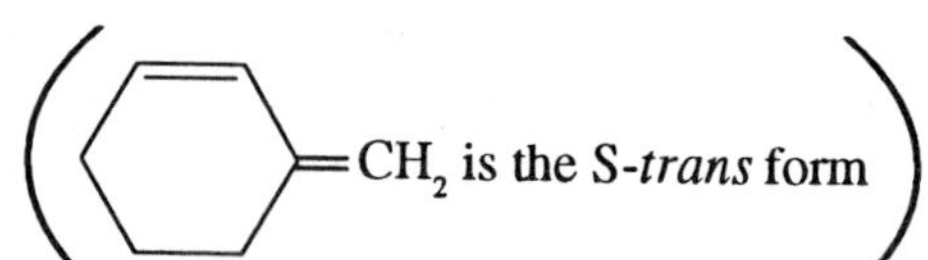

b.

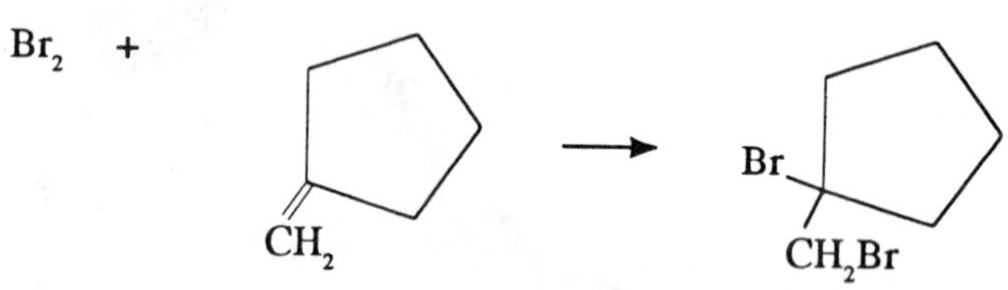

There is less angle strain in the intermediate bromonium ion.

c. $(CH_3)_2C{=}CH{-}CH_2CH_3 + Br_2 \longrightarrow (CH_3)_2C(Br){-}CH(Br){-}CH_2CH_3$

There are two ways to look at the difference in reactivity.

1. Br_2 addition to $(CH_3)_2C{=}CH{-}C(=O){-}CH_3$ is slowed by the electron-withdrawing effect of the carbonyl group as expressed by the resonance structures:

$$(CH_3)_2C{=}CH{-}C(=O){-}CH_3 \longleftrightarrow (CH_3)_2\overset{+}{C}{-}CH{=}C(O^-){-}CH_3$$

2. The electron withdrawing effect of the carbonyl group destabilizes the intermediate bromonium ion and the transition state leading to it:

$$(CH_3)_2{-}C{-}C{-}C(=O){-}CH_3 \ (\text{bridged by } Br^+) \quad \text{is less stable than} \quad (CH_3)_2{-}C{-}C{-}CH_2{-}CH_3 \ (\text{bridged by } Br^+)$$

4.29 In both instances the observed products are more highly conjugated and, therefore, more stable.

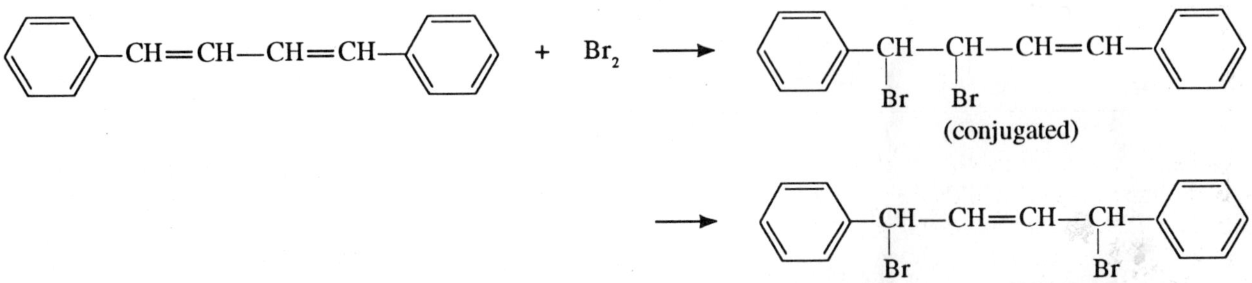

Similarly,

$$CH_2{=}CH{-}CH{=}CH{-}CH{=}CH_2 + Br_2$$

$$\longrightarrow CH_2Br{-}CHBr{-}CH{=}CH{-}CH{=}CH_2 + CH_2Br{-}CH{=}CH{-}CH{=}CH{-}CH_2Br$$

while $\longrightarrow CH_2Br{-}CH{=}CH{-}CHBr{-}CH{=}CH_2$

(non-conjugated)

4.30

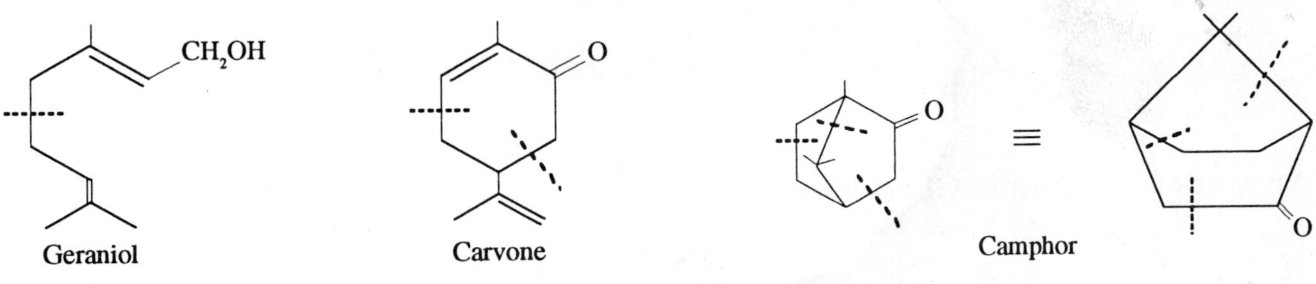

Geraniol Carvone Camphor

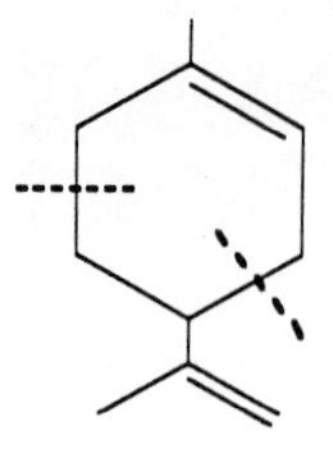

Limonene

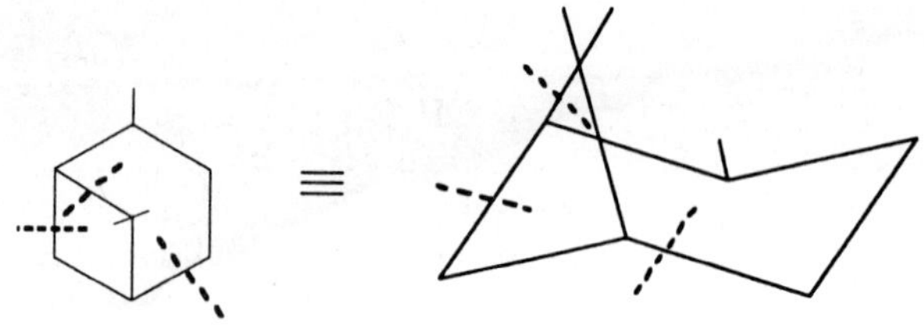

Pinane

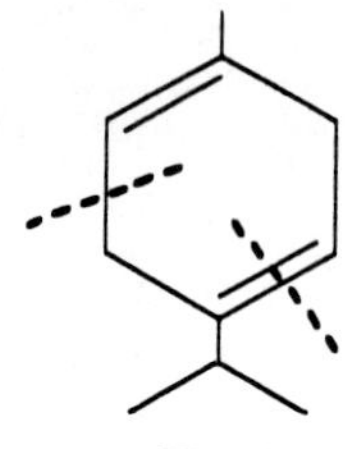

γ-Terpinene

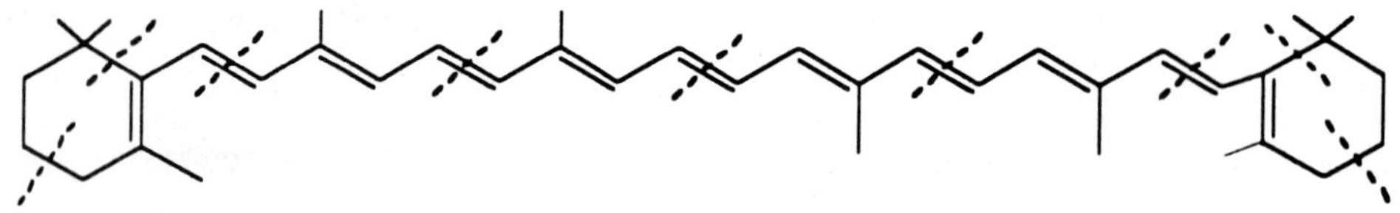

β–Carotene

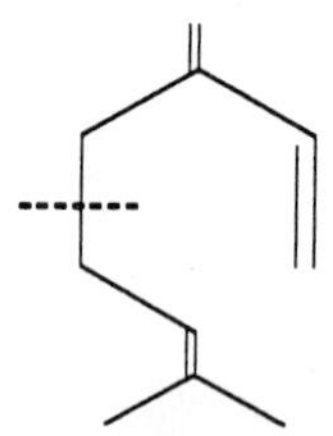

Myrcene

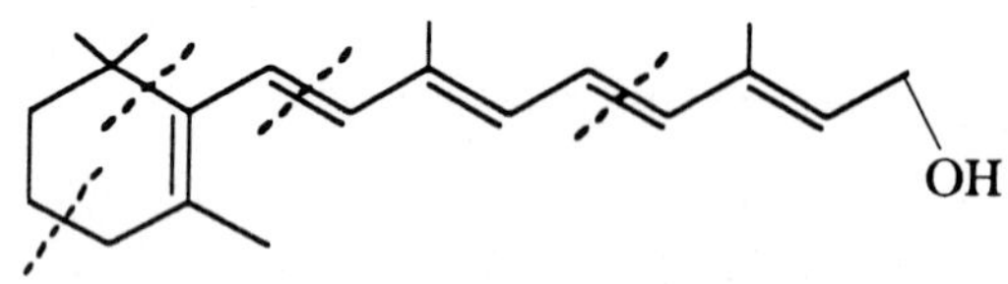

Vitamin A

4.31 a. HCHO (2 moles) + OHC—CH_2—CHO

b. OHC—CHO + OHC—CH_2—CHO

c. OHC$(CH_2)_2$CHO + OHC—CHO

d. HCHO (2 moles) + $CH_3\underset{\underset{O}{\|}}{C}\underset{\underset{O}{\|}}{C}H$

e. $HOCH_2$—CHO + $CH_3\underset{\underset{O}{\|}}{C}(CH_2)_2\underset{\underset{O}{\|}}{C}H$ + O=C$(CH_3)_2$

f. $(CH_3)_2C{=}O$ + $H\underset{\underset{O}{\|}}{C}CH_2CH_2\underset{\underset{O}{\|}}{C}\underset{\underset{O}{\|}}{C}H$ + HCHO (2 moles)

g.

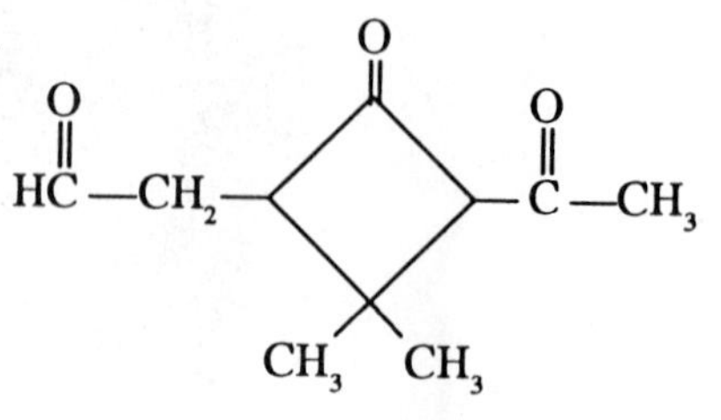

Answers for Multiple Choice Test

4.1 c	**4.6** a	**4.11** a	**4.16** b
4.2 c	**4.7** d	**4.12** d	**4.17** a
4.3 d	**4.8** a	**4.13** d	**4.18** b
4.4 b	**4.9** c	**4.14** c	**4.19** a
4.5 c	**4.10** b	**4.15** d	**4.20** d

5 Stereochemistry

Questions

5.1 Which of the following compounds have chiral carbon(s)?

a. 1-chloropropane
b. 2-chloropropane
c. 1,2-dichloropropane
d. 1-chlorobutane
e. 2-chloropentane
f. 3-chloropentane
g. 2-butanol
h. 1,2-butanediol
i. 3-chloro-2-butanol
j. 2,2-dichloropropane
k. 2,2-dichlorobutane
l. 3-hexanol
m. chlorocyclopropane
n. cyclobutanol
o. 1,1-dichlorocyclopentane
p. 1,3-cyclobutanediol
q. 2,3-dimethylpentane
r. 3,3-dimethylhexane

5.2 How many stereoisomers are possible for each of the following:

a. 2-butanol
b. 2,3-butanediol
c. 3-chloro-2-butanol
d. $H_2N—CH_2—COOH$
e. 1,2-dibromocyclopentane
f. 1,2-diiodocyclopropane

5.3 Draw the formula for an *unsaturated* alkyl chloride (C_5H_9Cl) that shows

a. neither geometric isomerism nor optical activity.
b. both geometric isomerism and optical activity.
c. geometric isomerism but not optical activity.
d. optical activity but not geometric isomerism.

5.4 Draw structures for compounds **A** through **J**.

a. **A** ($C_5H_8O_2$) $\xrightarrow{H_2/Pt}$ **B** ($C_5H_{10}O_2$)

A: optically active carboxylic acid; B: optically active

b. **C** ($C_6H_8O_2$) $\xrightarrow{2\ H_2/Pt}$ **D** ($C_6H_{12}O_2$)

C: optically active carboxylic acid carrying a triple bond; D: optically inactive carboxylic acid

c. **E** the only compound of molecular formula $C_4H_{10}O$ that has a chiral carbon atom.

d. **F** ($C_5H_{10}O$) $\xrightarrow{H_2/Pt}$ **G** ($C_5H_{12}O$)

optically active alcohol → optically inactive alcohol

e. **H** ($C_6H_{11}Cl$) $\xrightarrow[2.\ Zn/H^+]{1.\ O_3}$ **I** (C_3H_6O) + **J** (C_3H_5ClO)

optically active → optically inactive + optically active

5.5 Draw stereochemical formulas for all possible stereoisomers of the following compounds. Label pairs of enantiomers and *meso* compounds. Which of the compounds are optically active?

a. 1,2-dibromoethane
b. 1,2-propanediol
c. 2,3-butanediol
d. 1,2,3,4-tetrabromobutane

5.6 Arrange the following atoms or groups in each set according to the sequencing rules proposed by Cahn, Ingold, and Prelog.

a. —Br —Cl —I —F

b. —H $—CH_3$ $—CH_2CH_3$ $—C(CH_3)_3$ $—CH(CH_3)_2$

c. $—CH_2CH_2CH_2CH_3$ $—CH(CH_3)CH_2CH_3$ $—CH_2CH(CH_3)CH_3$ $—C(CH_3)_3$

d. $—CH(CH_3)CH_2CH_2CH_3$ $—CH_2CH(CH_3)CH_2CH_3$ $—CH_2C(CH_3)_2CH_3$

e. $—CH{=}CH—CH_3$ $—CH_2CH_2CH_3$ $—CH(CH_3)_2$

f. $—CH{=}CH_2$ $—C({=}O)—H$ $—C({=}N—H)—H$ $—C({=}S)—H$

g. $—C{\equiv}CH$ $—CH{=}CH_2$ $—CH_2CH_3$

h. $—CH(Cl)CH_2CH_3$ $—CH_2CH_2CH_2Cl$ $—CH_2CH(Cl)CH_3$

i. $—NH_2$ —OH —SH $—CH_3$

j. $(CH_3)_2CH$ $—C{\equiv}CH$ $—C({=}O)—OH$ —OH $—CH{=}CH_2$ 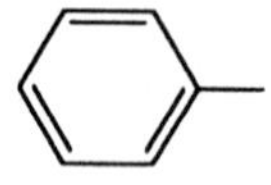 $(CH_3)_3C—$

5.7 Name the following compounds (include the R/S designations):

a.

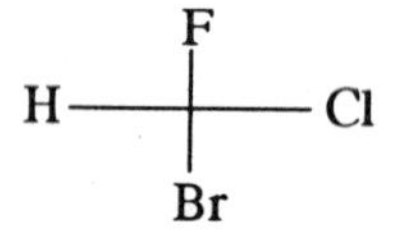

b.

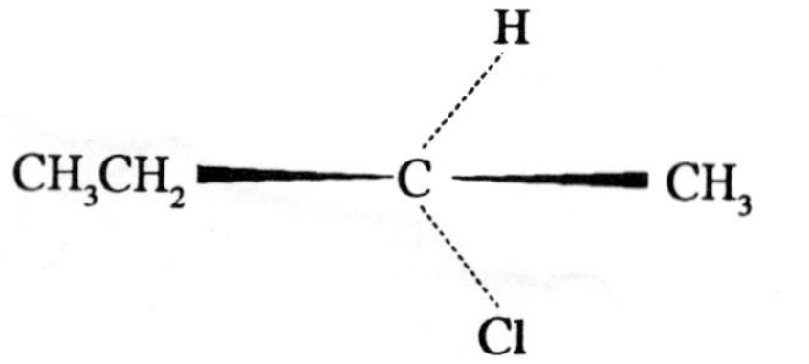

c.

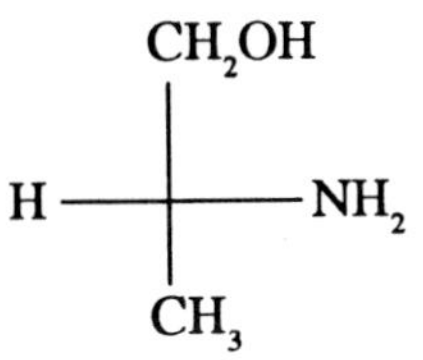

d.

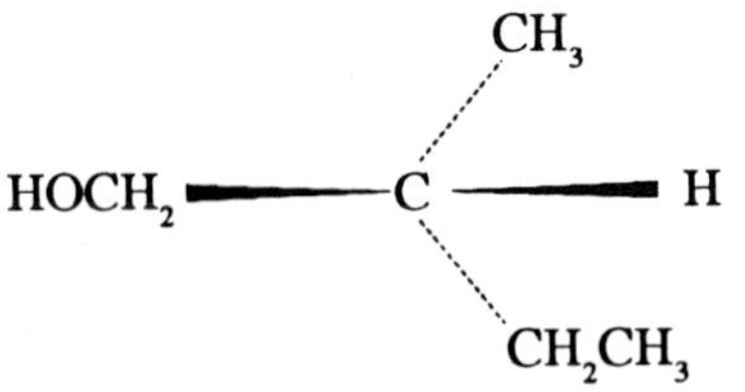

e.

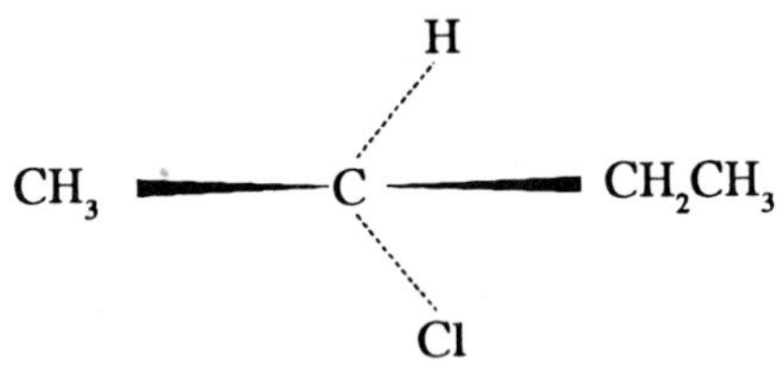

f.

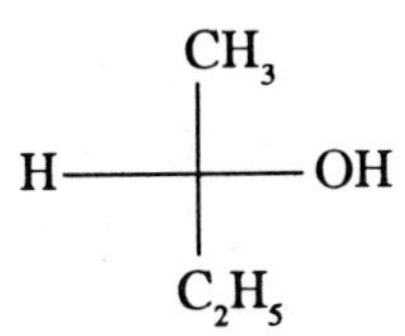

g.

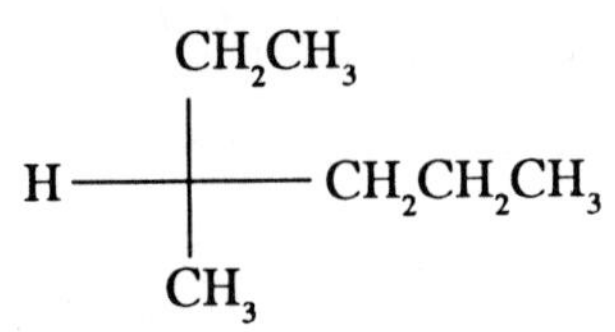

h.

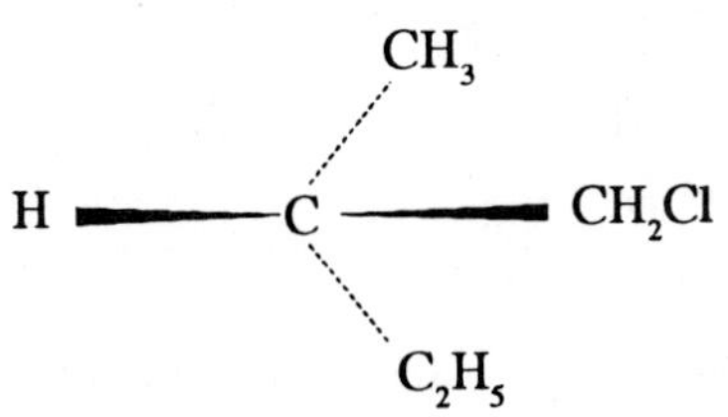

i.

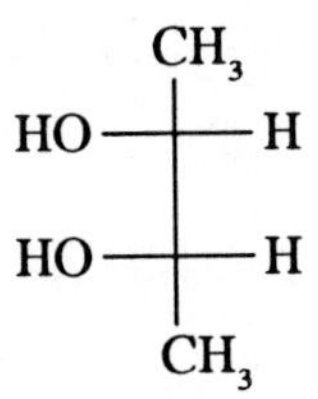

j.

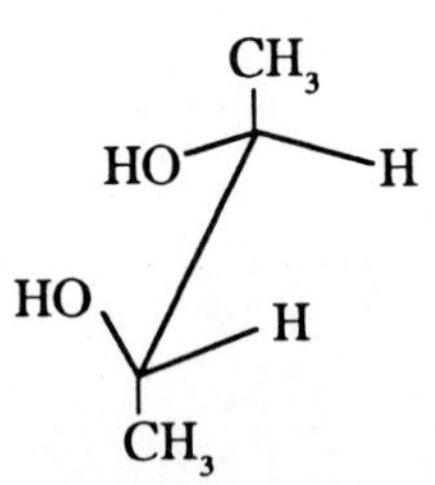

k.

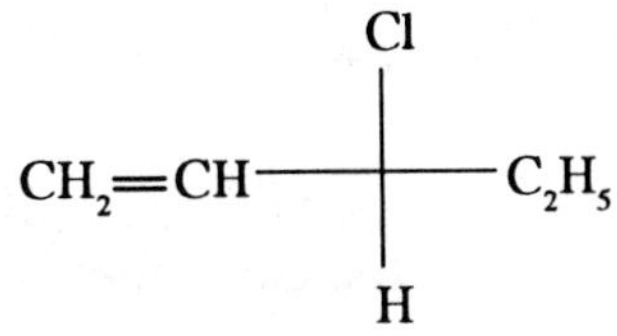

l.

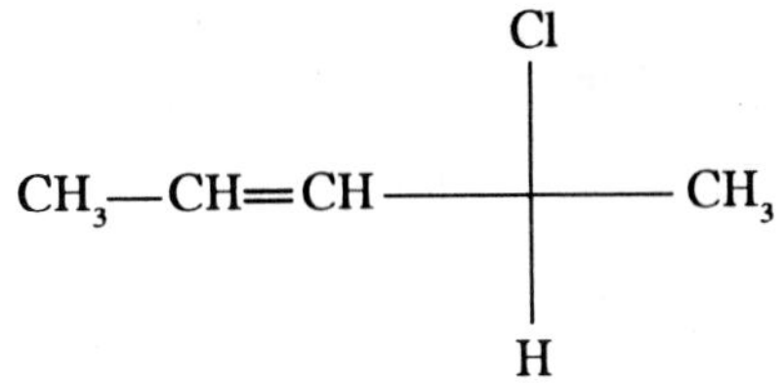

m.

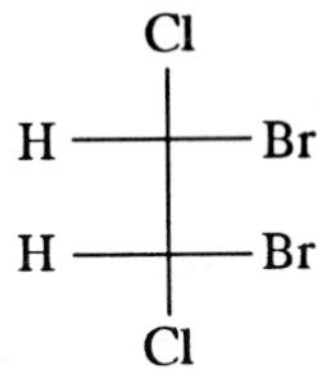

n.

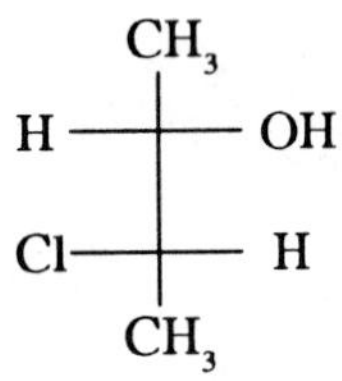

o.

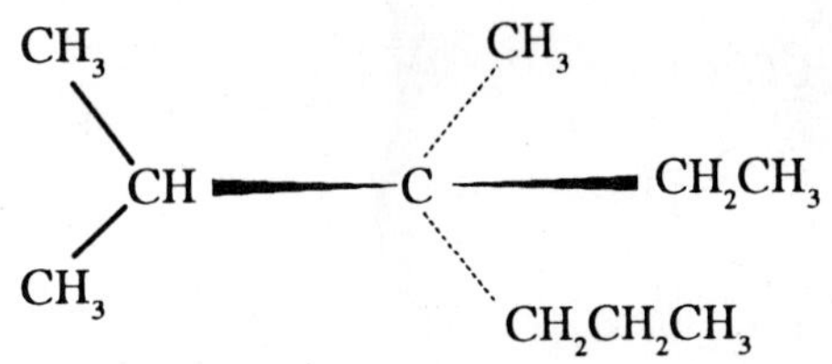

p.

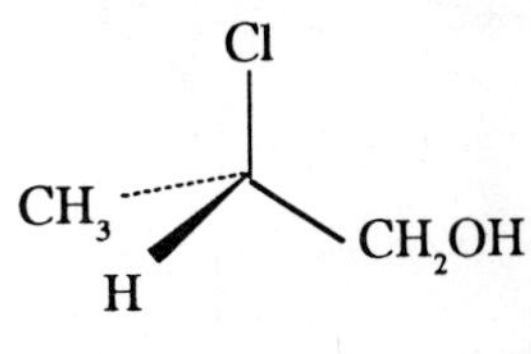

q.

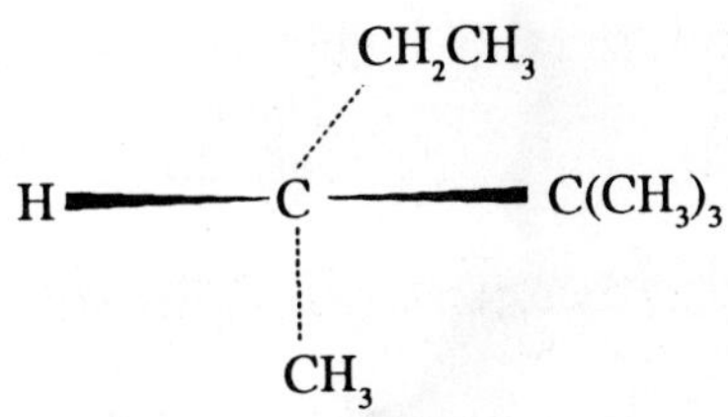

5.8 Write *all* structures that correspond to each molecular formula given below, marking chiral carbons (if any) with an asterisk and indicating enantiomeric pairs.

a. pentanes, C_5H_{12}
b. hexanes, C_6H_{14}
c. heptanes, C_7H_{16}
d. alcohols, C_4H_9OH
e. dibromoderivatives of *n*-butane
f. bromochloropropanes

5.9 Identify the following pairs as enantiomers, diastereomers, structures of the same compound, or constitutional isomers:

a. 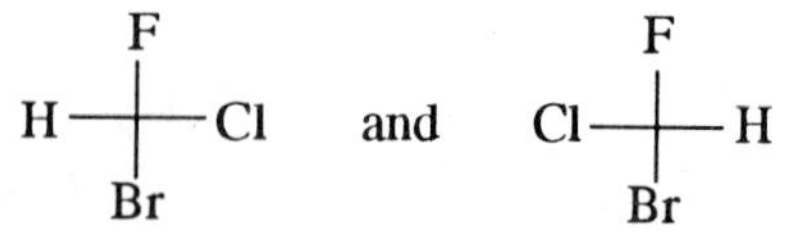

b. 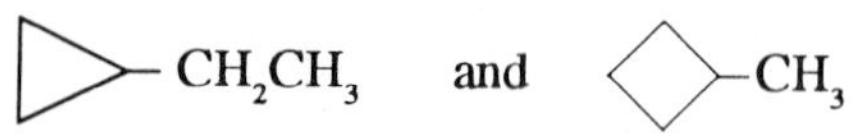

c. 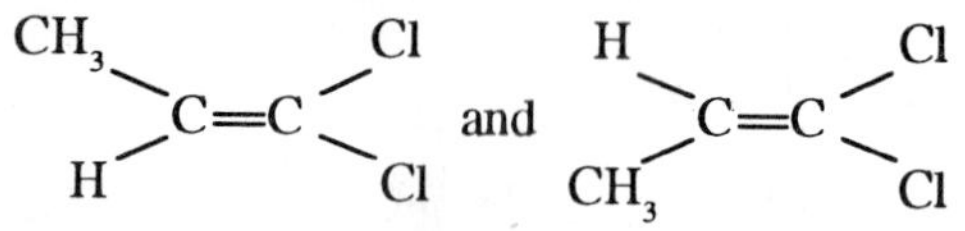

d. 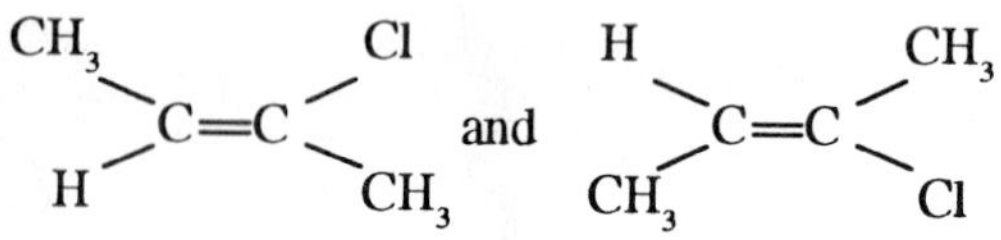

e. 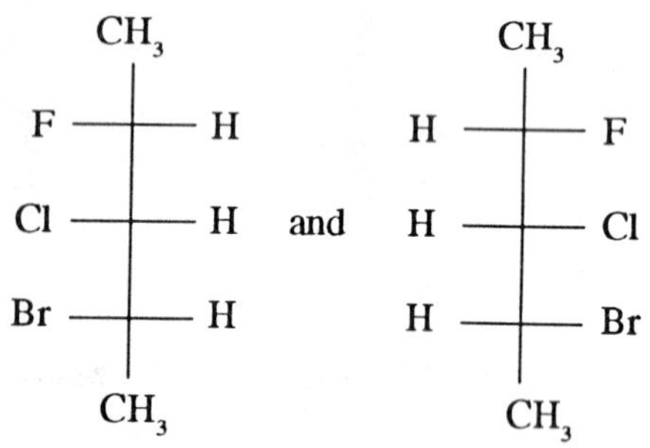

f. 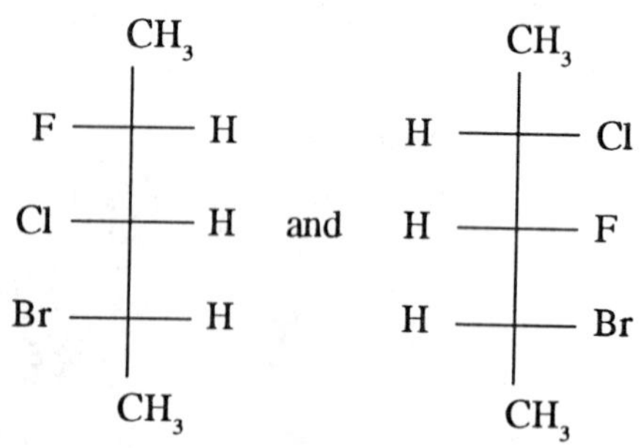

g. 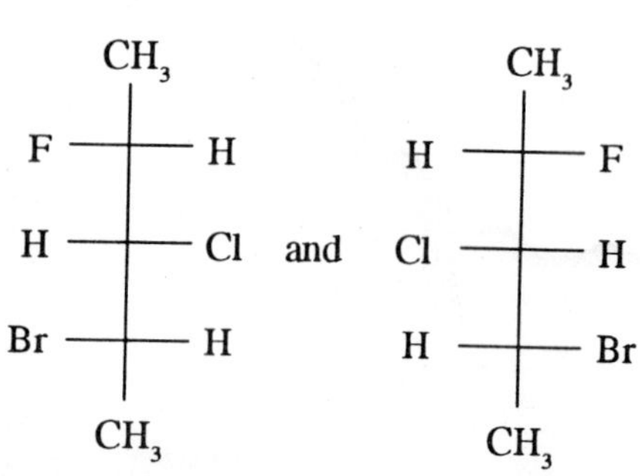

h. 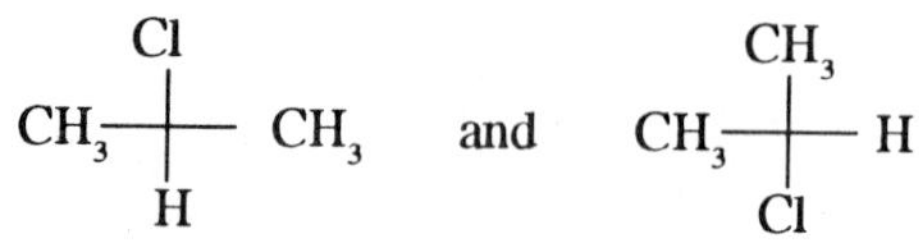

i. CH_3CHCl_2 and $ClCH_2CH_2Cl$

j. 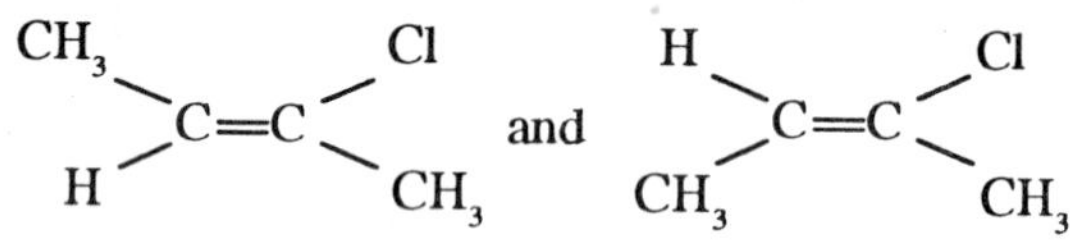

k. 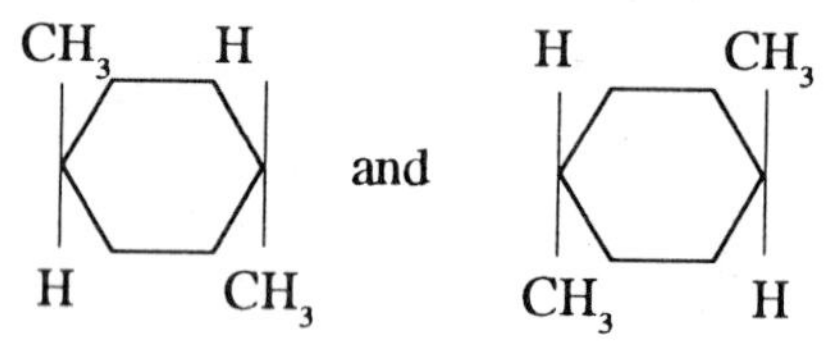

l. 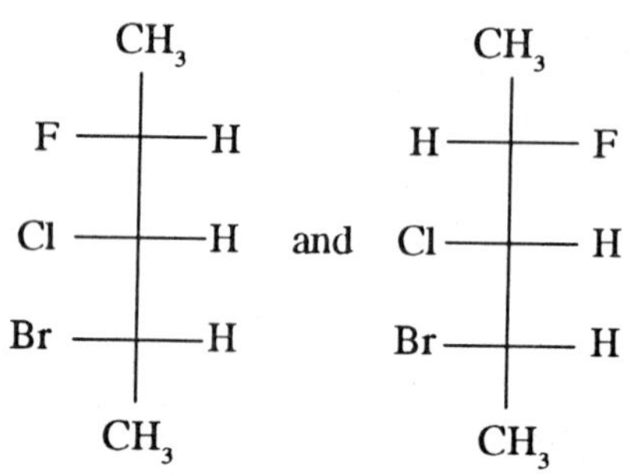

m. 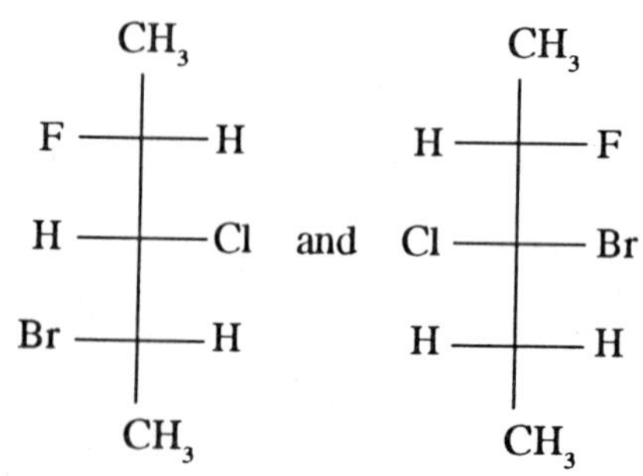

n. 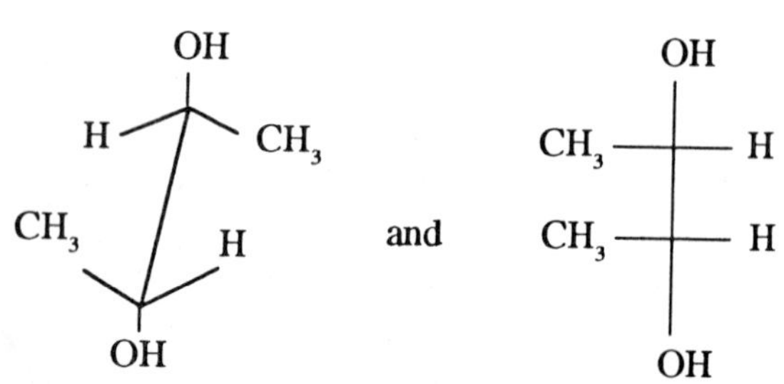

o.

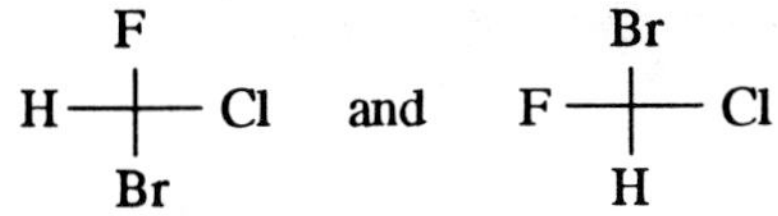

p.

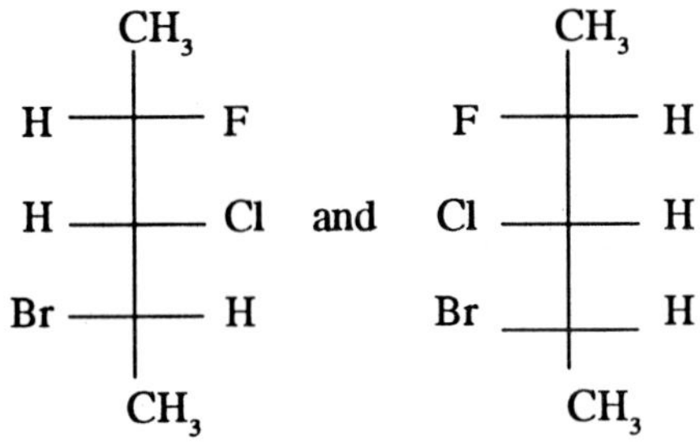

q.

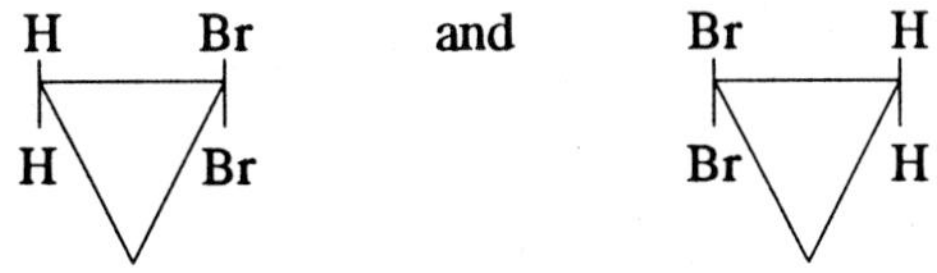

r.

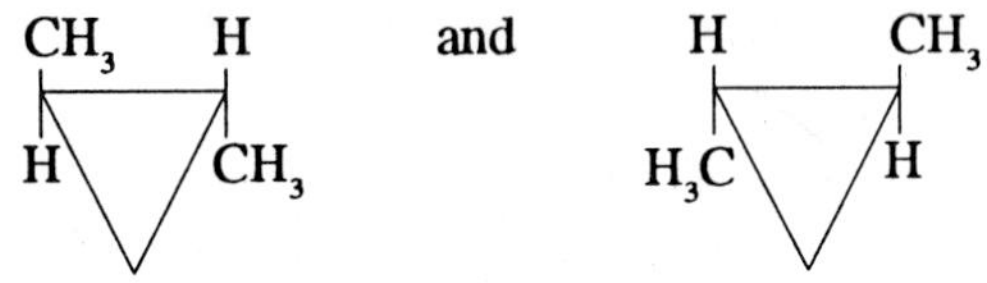

s.

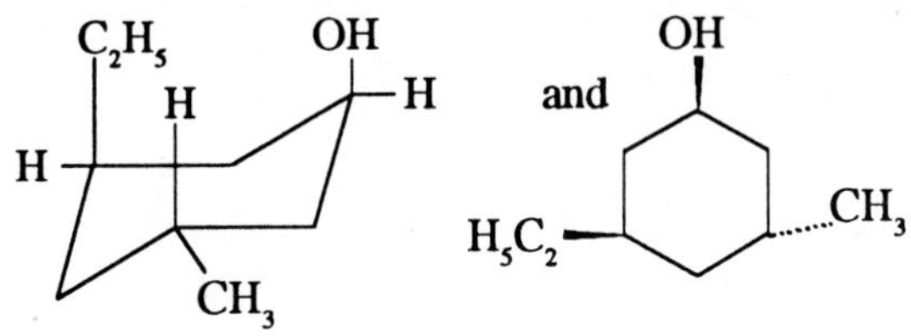

t.

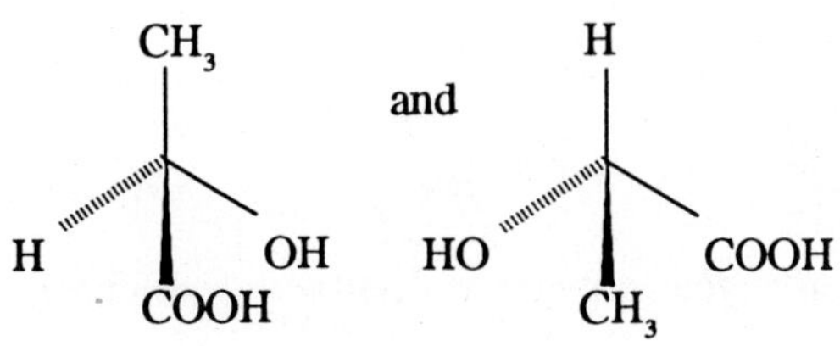

u.

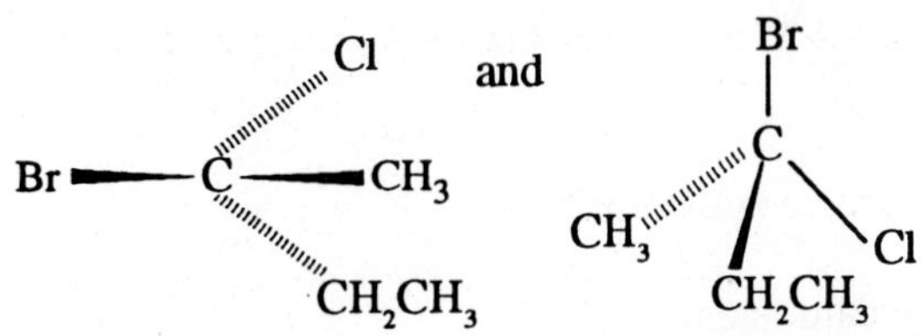

v.

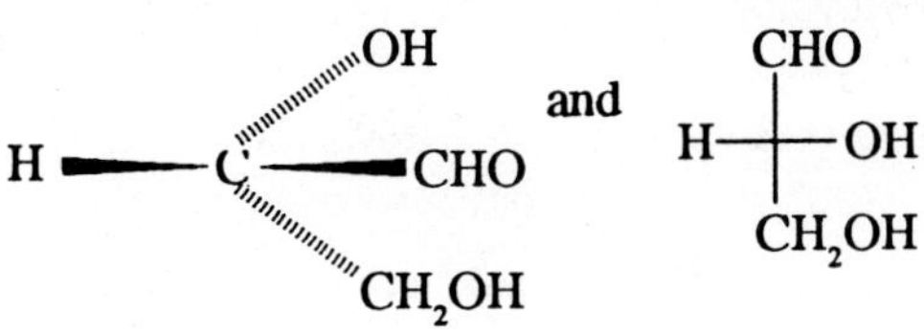

w.

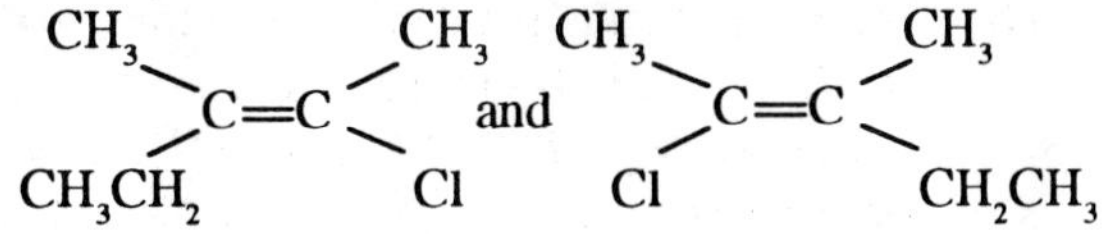

x.

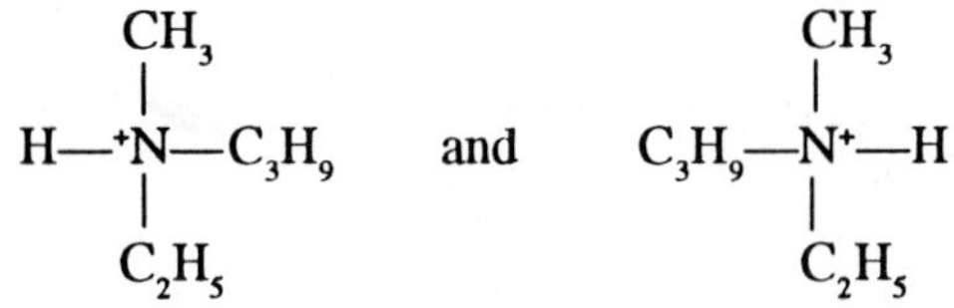

y.

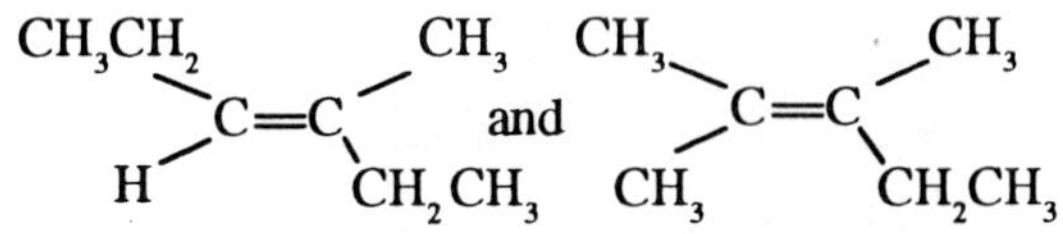

z.

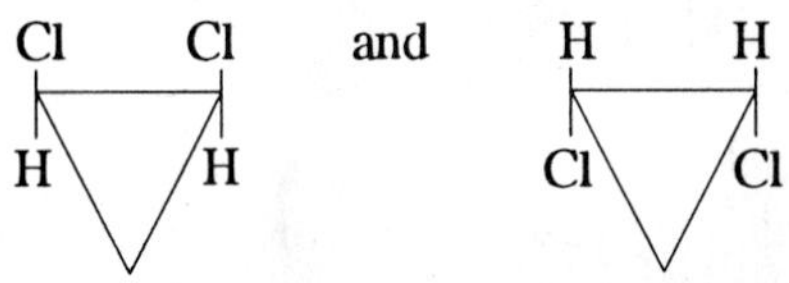

aa.

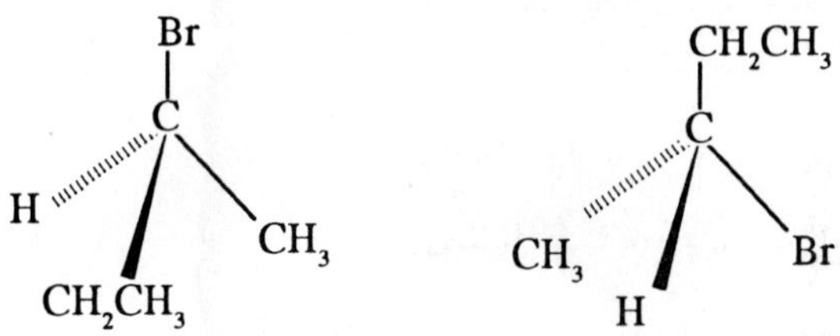

bb.

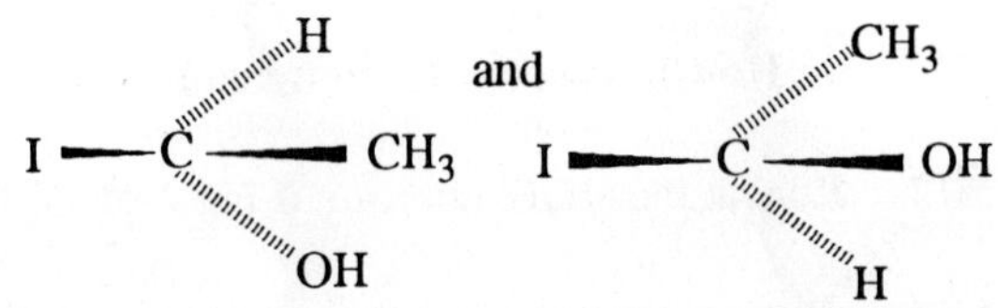

cc.

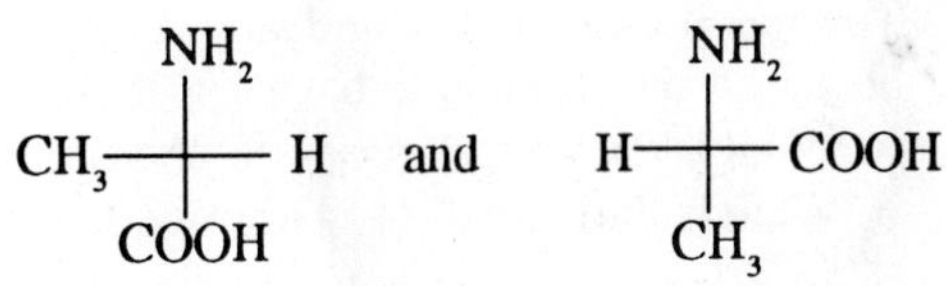

dd.

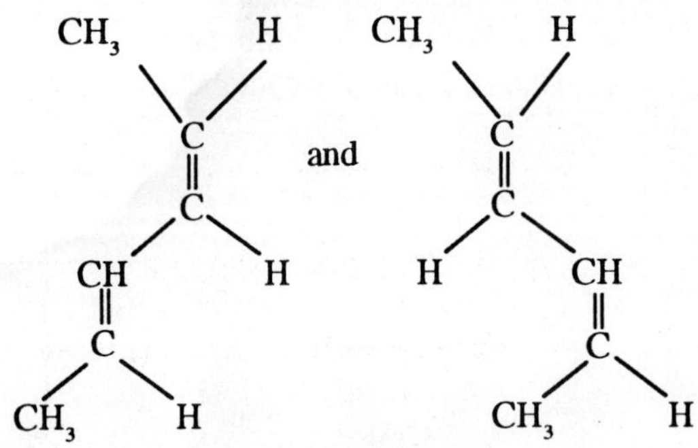

ee. 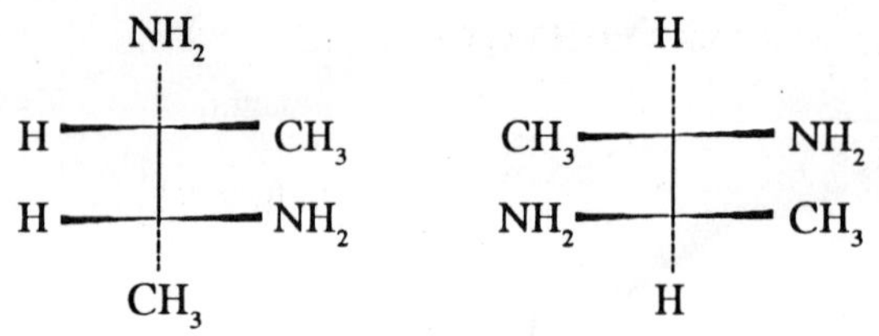

hh. 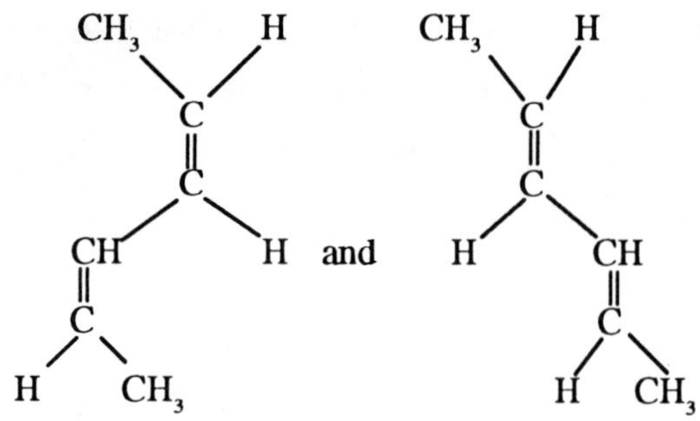

ff. 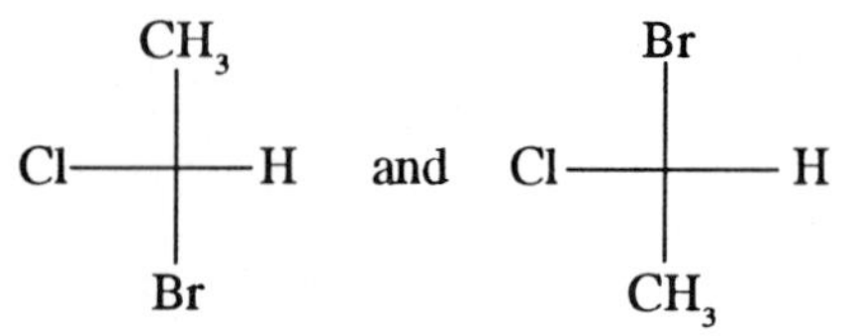

ii. 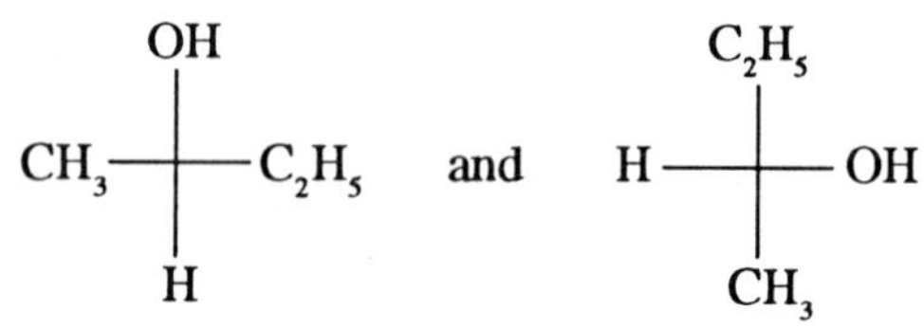

gg. 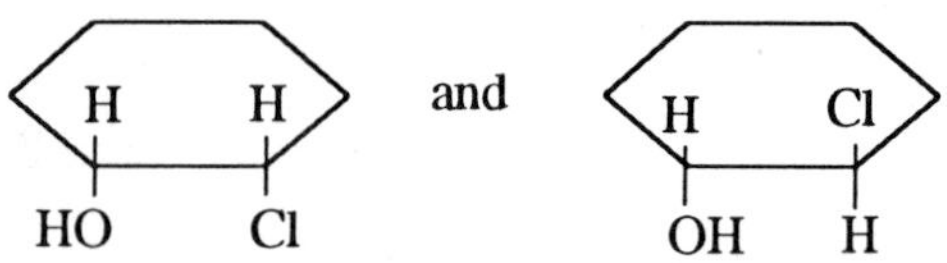

5.10 There are four steroeoisomers that correspond to the formula 1,2,3,4,5-pentachlorocyclopentane. Indicate which are optically active and which are optically inactive.

5.11 (*R*)-2-chloropentane is monochlorinated using UV light and the reaction products are separated by fractional distillation.

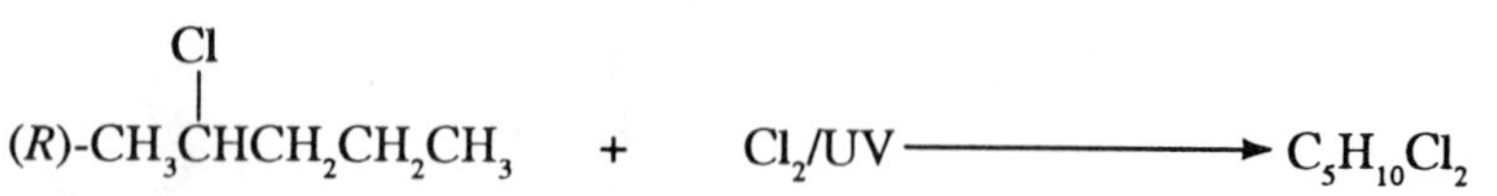

a. How many stereochemical isomers are formed? Give structures, names and assign *R*/*S* labels to each. Which are optically active?
b. How many distillation fractions will be collected?

5.12 Repeat the above question using 2-methylbutane.

5.13 The smallest isolable cyclic alkyne is cyclooctyne. Why haven't cyclic alkynes of smaller ring size been synthesized?

5.14 The specific rotation of optically active (+) -2-bromooctane is +21.6°. A mixture of (+) -2-bromooctane and (–) -2-bromooctane is found to have a specific rotation of +10°.
a. What is the optical purity of the sample?
b. What is the ratio of (+) -2-bromooctane to (–) -2-bromooctane in the mixture?

5.15 Which dimethylcyclobutane can be resolved into a pair of enantiomers?
a. 1,1-dimethylcyclobutane
b. *cis*-1,2-dimethylcyclobutane
c. *trans*-1,2-dimethylcyclobutane
d. *cis*-1,3-dimethylcyclobutane
e. *trans*-1,3-dimethylcyclobutane

5.16 Complete the following reactions indicating the proper stereochemistry of the products:

a. 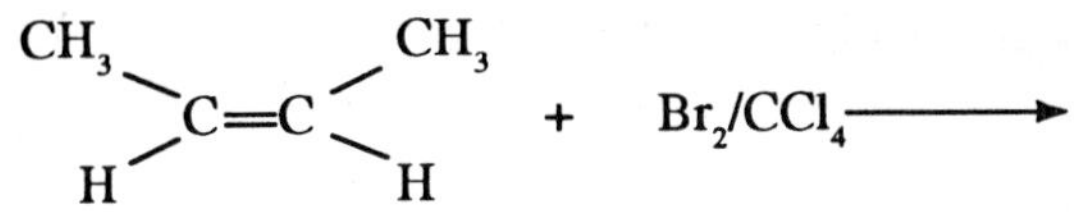

b. 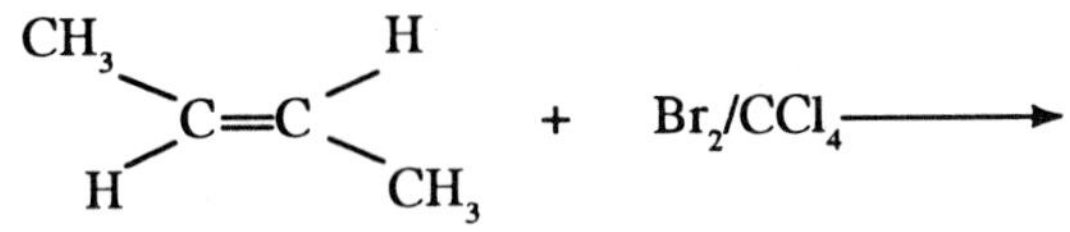

c. 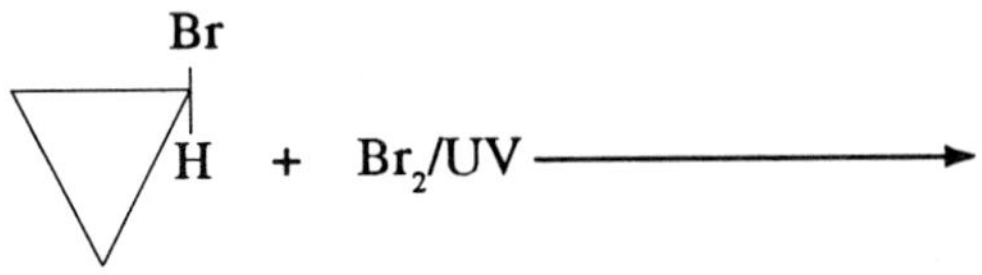

d. trans-1,2-dichlorocyclopropane + Cl_2/UV ⟶
(racemic mixture)

e. 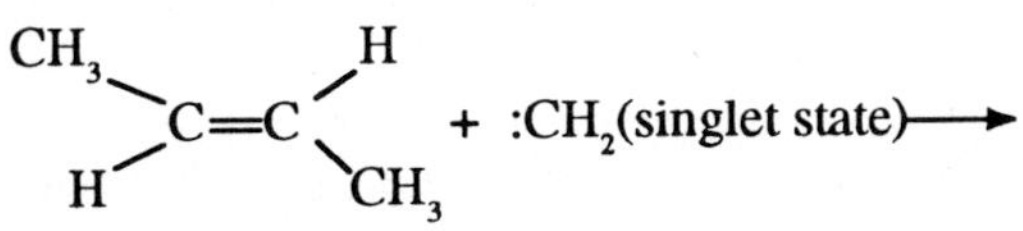

f. 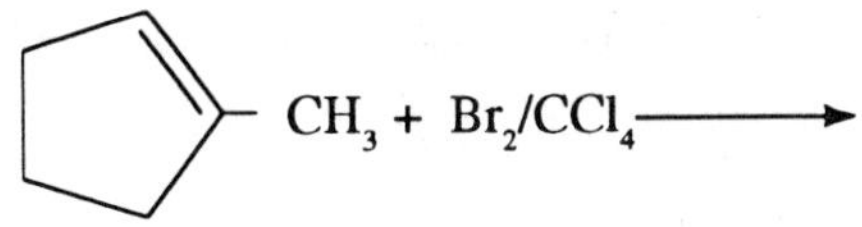

g. 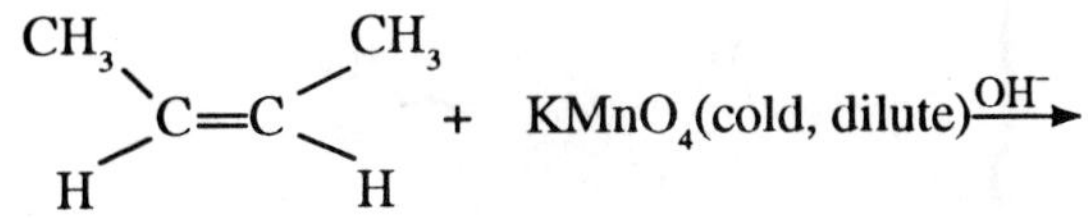

h. 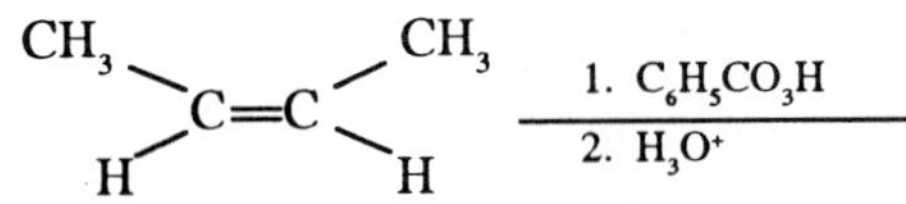

i. 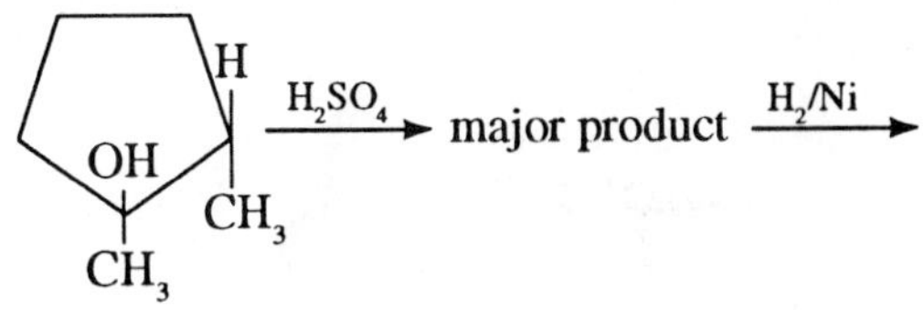

5.17 How many pairs of enantiomers and/or *meso* compounds are possible for the following:

a. $CH_3CH(OH)CH(Cl)CH(OH)CH_3$

b. $CH_3(CHOH)_4CH_3$

c. 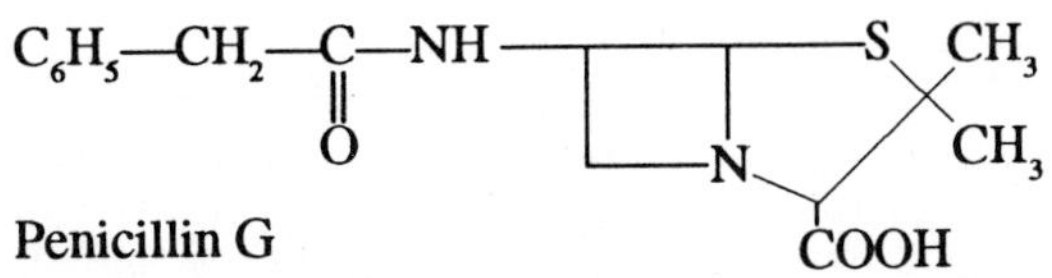

Penicillin G

d. 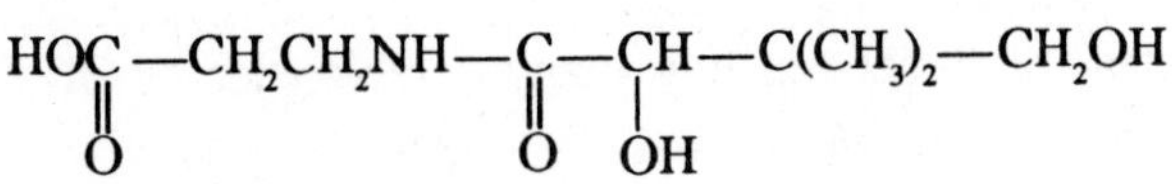

Pantothenic acid

e. 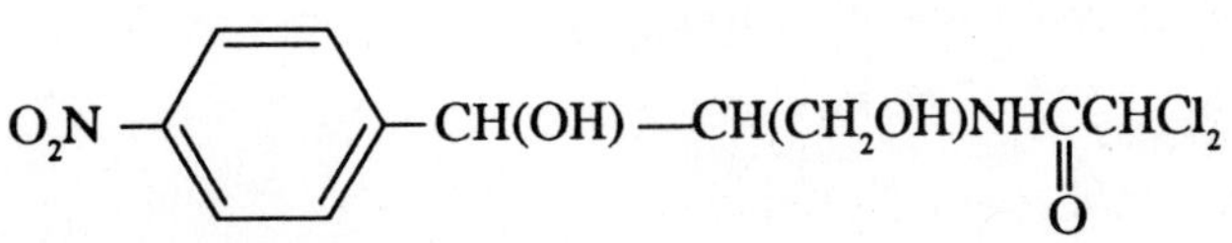

Chloramphenicol

5.18 Give structures of all the unknown reactants and products in the following reactions below. Give the *R*/*S* designation for **B** and **C**.

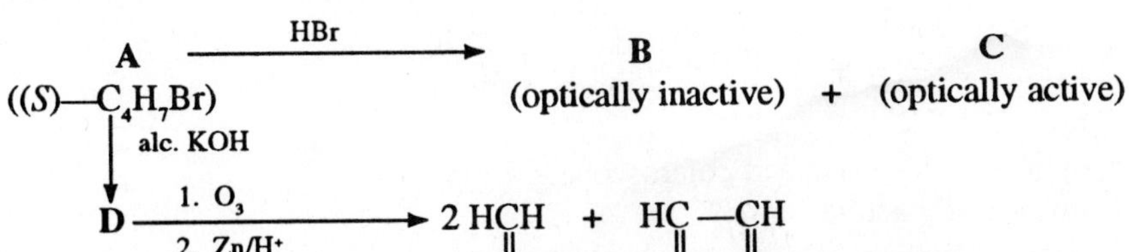

5.19 (*R*)-4-bromo-3-methyl-1-butene is levorotary.
a. Draw its stereochemical structure.
b. When this compound is treated with HBr/peroxides, 1,4-dibromo-2-methylbutane is obtained. Write the stereochemical structure of the product and assign *R*/*S* labels.

5.20 Complete the following reactions showing the stereochemical results:

a. 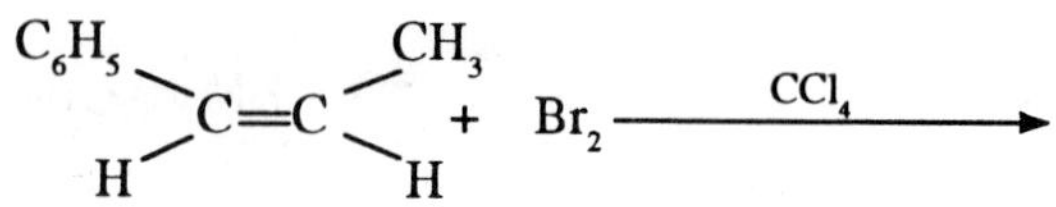

c. 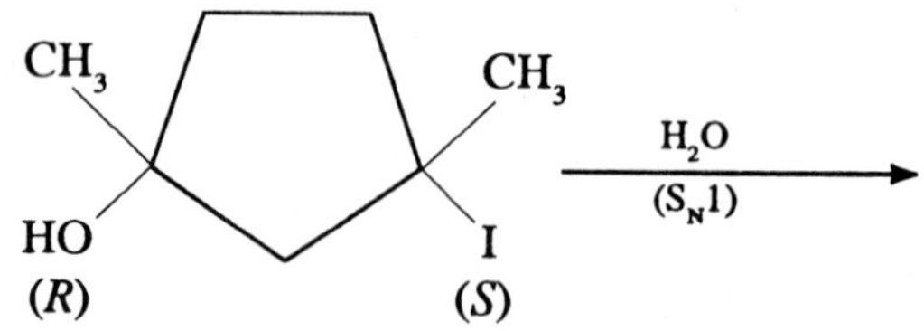

b. 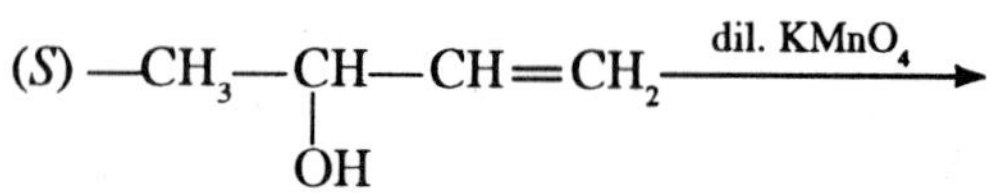

d. 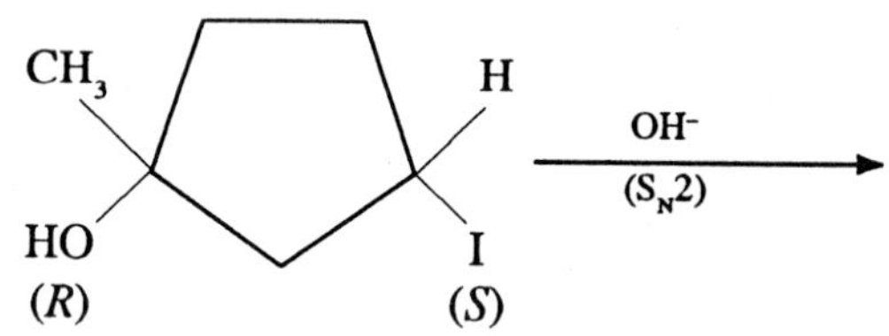

Multiple Choice Test

5.1 How many optical isomers of 1,2-dichlorocyclohexane are there?
a. one c. three
b. two d. four

5.2 Which of the following is *not* true for enantiomers?
A. They have the same boiling point.
B. They have the same melting point.
C. They have the same chemical reactivity with archiral reagents.
D. They have the same reactivity with chiral reagents.
E. They have the same density.
F. They have the same specific rotation.
a. C and F c. F
b. D and F d. all but F

5.3 If carbon compounds had a square planar structure instead of tetrahedral, how many stereoisomers of CH_2Cl_2 would be possible?
a. One d. four
b. two e. five
c. three

5.4 Which of the following compounds have chiral carbons?
A. 1-bromobutane
B. 3-bromohexane
C. *t*-butyl chloride
D. *sec*-butyl chloride
E. isobutane
a. B c. B and D
b. C d. all but C

5.5 Which of the following statements is correct?
a. The instrument used for measuring optical activity is called polaroscope.
b. All compounds with chiral centers are optically active.
c. Both (a) and (b)
d. Neither (a) nor (b)

5.6 The reaction of 2-methylbutane with chlorine in the presence of UV light will yield_____compounds of molecular formula $C_5H_{11}Cl$, of which_____are optically active. The total number of fractions isolated by distillation is_____.
a. 4, 2, 4
b. 6, 4, 4
c. 6, 4, 2
d. 4, 4, 4

5.7 A 50:50 mixture of a pair of enantiomers
a. is optically inactive.
b. may or may not be optically active.
c. has the same boiling point as that of the individual enantiomers.
d. is called a racemic mixture.
e. all but b.

5.8 Which is/are optically inactive?
a. a racemic mixture
b. a *meso* compound
c. every achiral compound
d. a 50-50 mixture of *R* and *S* enantiomers
e. all of the above

5.9 Resolution of racemic lactic acid, $CH_3CH(OH)COOH$, can be accomplished by
a. fractional vacuum distillation of the free acid.
b. fractional crystallization of the free acid.
c. fractional crystallization of the sodium salt of the acid, followed by regeneration of the acid.
d. fractional crystallization of the reaction product with the optically active strychnine salt of the acid, followed by regeneration of the acid.
e. Lactic acid cannot be resolved since it is a *meso* compound.

5.10 A sample of optically pure (+)-2-bromooctane weighing 5 g was dissolved in 20 mL of a CCl_4 solution. Its observed optical rotation was + 5.4° measured in a 10 cm polarimeter tube. What is the specific rotation of (+)-2-bromooctane?
a. +15.6°
b. +21.6°
c. +27.0°
d. +54.0°
e. +108°

5.11 A second sample of (+)-2-bromooctane which was not optically pure was found to have a specific rotation of +7.2°. In this sample, what is the ratio of (+) to (–) isomers?
a. 3/1
b. 2/1
c. 1/1
d. 1/2
e. 1/3

5.12 What would the addition of bromine to cyclopentene yield?
a. one *dl* pair
b. two *dl* pairs
c. one *meso* compound
d. (a) and (c)
e. (b) and (c)

5.13 The correct stereochemical designation for this compound is_____-3,4-hexanenediol

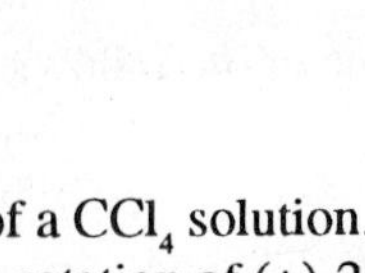

a. (3*R*, 4*R*)
b. (3*S*, 4*S*)
c. (3*R*, 4*S*)
d. (3*R*)
e. (4*S*)

5.14 The Pt-catalyzed addition of H_2 to 1,2-dimethylcyclopropene will result in the formation of
a. one *R* isomer.
b. one *S* isomer.
c. a *dl* pair.
d. a *meso* compound.
e. both (c) and (d).

5.15 Arrange the following three groups in the order of highest to lowest priority.

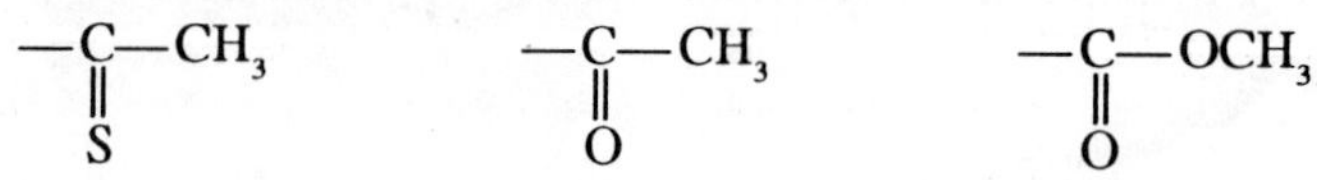

A **B** **C**

a. A > B > C
b. C > B > A
c. A > C > B
d. B > C > A

5.16 How many configurational isomers are possible for a compound with the structure $CH_3CH{=}C(CH_3)CH(CH_3)CH(CH_3)$? (with CH_3CHBr attached to the last CH)

a. 2^3
b. 2^4
c. 2^5
d. 2^6
e. 2^7

5.17 How many configurational isomers are possible for 3-chloro-2,4-pentanediol?

a. three
b. four
c. five
d. six
e. seven

5.18 In the previous question, the number of *meso* compounds is _____ and the number of *dl* pairs is _____.

a. 0, 2
b. 1, 1
c. 2, 1
d. 2, 2
e. 1, 3

5.19 Which one of the following alkanes can be prepared by the hydrogenation of only three different alkenes (in separate reactions)?

a. isopentane
b. 2-methylpentane
c. methylcyclohexane
d. isobutane

5.20 The relationship between these compounds is that they are

a. superimposable without bond rotation.
b. diastereomers.
c. enantiomers.
d. conformational isomers.

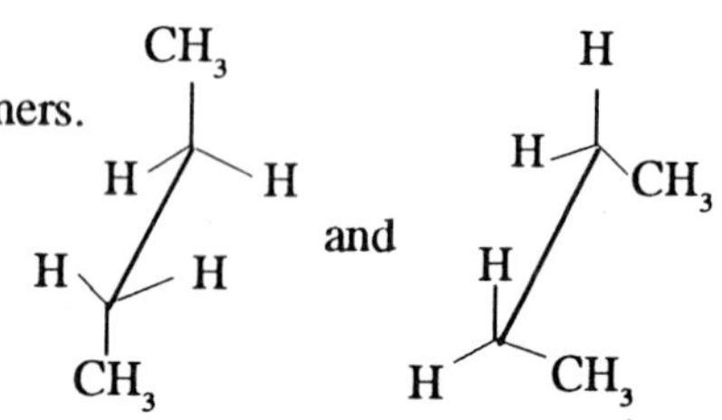

Answers for Questions

5.1 Compounds with chiral carbon(s): c, e, g, h, i, l, q

c. 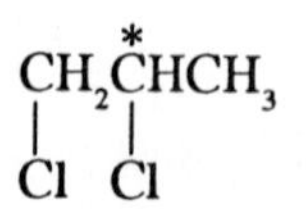

i. 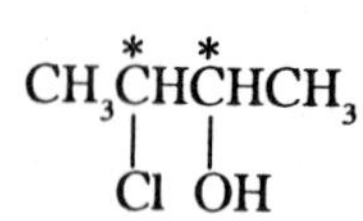

e. 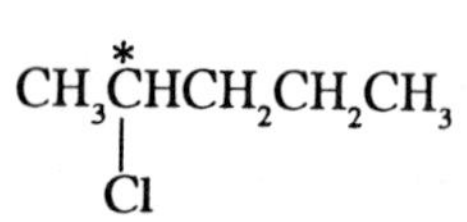

l. 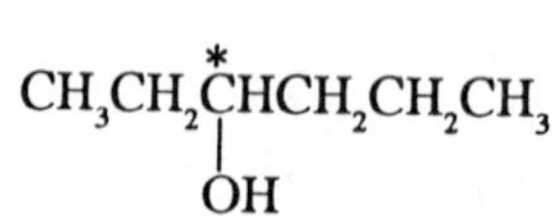

g. 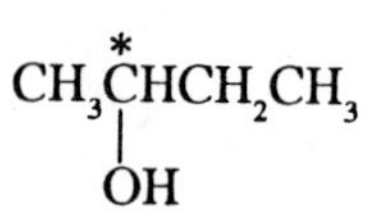

q. 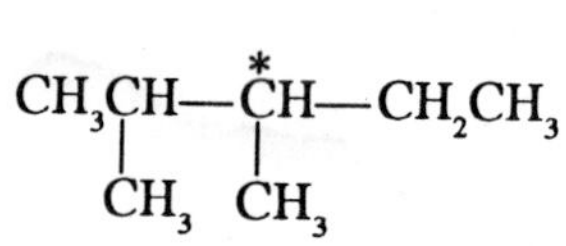

h. 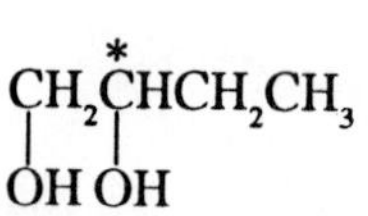

5.2 a. *two*: (*R*)- and (*S*)-2-butanol.
b. *three*: (2*R*, 3*R*), (2*S*, 3*S*) and *meso* (2*R*, 3*S* or 2*S*, 3*R*)-butanediol.
c. *four*: (2*R*, 3*R*), (2*S*, 3*S*), (2*R*, 3*S*) and (2*S*, 3*R*)-3-chloro-2-butanol.
d. *one*: achiral
e. *three*: one *cis*-(*R*,*S* or *S*,*R*) and two *trans*-(*R*,*R* and *S*,*S*)-1,2-dibromocyclopentane
f. *three*: same as (e)

5.3 The answers below are *not* the only structures that fulfill the requirements of the question.

a. $(CH_3)_2C{=}C(CH_3)Cl$

b. $CH_3CH{=}CH{-}\overset{*}{C}H(Cl){-}CH_3$ (Z and E)

c. $CH_3CH{=}CH{-}CH_2{-}CH_2Cl$ (Z and E)

d. $H_2C{=}C(CH_3){-}\overset{*}{C}H(Cl){-}CH_3$

5.4 Optically active centers are designated with an asterisk.

A $CH_2{=}CH{-}\overset{*}{C}H(CH_3){-}COOH$

B $CH_3CH_2{-}\overset{*}{C}H(CH_3){-}COOH$

C $CH{\equiv}C{-}\overset{*}{C}H(CH_2CH_3){-}COOH$

D $CH_3CH_2{-}CH(CH_2{-}CH_3){-}COOH$

E $CH_3{-}\overset{*}{C}H(OH){-}CH_2CH_3$

F $CH_2{=}CH{-}\overset{*}{C}H(OH){-}CH_2CH_3$

G $CH_3CH_2{-}CH(OH){-}CH_2CH_3$

H $(CH_3)_2C{=}CH{-}\overset{*}{C}H(Cl){-}CH_3$

or

$CH_3CH_2CH{=}CH{-}\overset{*}{C}H(Cl){-}CH_3$

I $CH_3{-}C({=}O){-}CH_3$ or $CH_3CH_2{-}CH({=}O)$

J $CH_3{-}\overset{*}{C}H(Cl){-}CH({=}O)$

5.5 a. $CH_2(Br){-}CH_2(Br)$ No optical isomerism is possible; the compound exists as only one compound.

b. $CH_2(OH){-}\overset{*}{C}H(OH){-}CH_3$ Only one chiral carbon; the compound can exist as two stereoisomers, which are enantiomers; both are optically active.

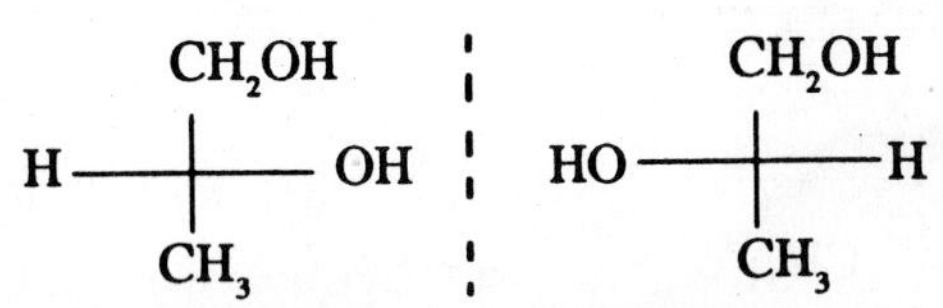

c. $\overset{*}{C}H_3CH(OH)-\overset{*}{C}H(OH)CH_3$ — more exactly, $CH_3\overset{*}{C}H(OH)-\overset{*}{C}H(OH)CH_3$

Two chiral carbons; the compound can exist as three stereoisomers: one pair of enantiomers and one *meso* compound.

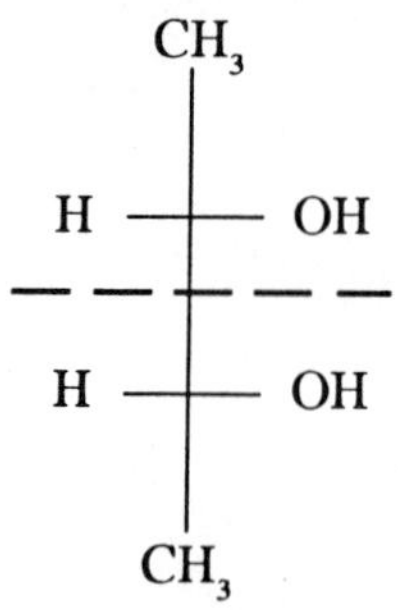

meso
(optically inactive)

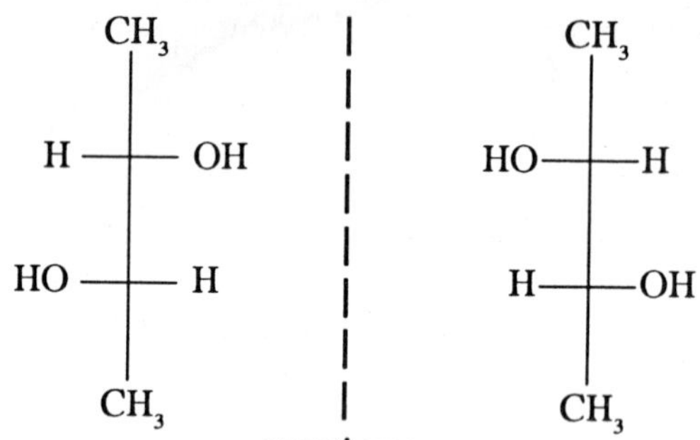

enantiomers
(optically active)

d. $CH_2(Br)-\overset{*}{C}H(Br)-\overset{*}{C}H(Br)-CH_2(Br)$

(same situation as in *c*; two chiral carbons, and thus three stereoisomers: one pair of enantiomers and one *meso* compound)

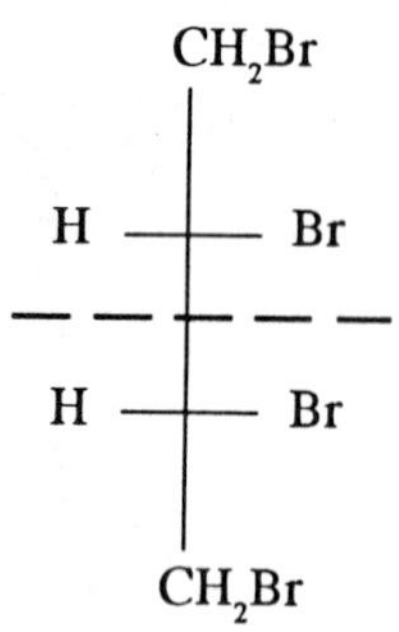

meso
(optically inactive)

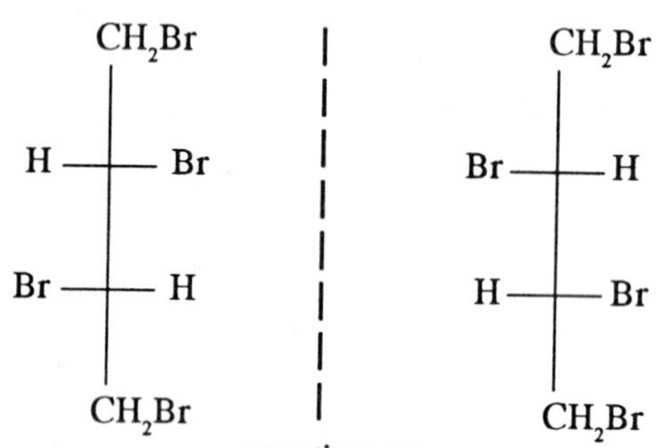

enantiomers
(optically active)

5.6 a. $-I$ > $-Br$ > $-Cl$ > $-F$

b. $-C(CH_3)_3$ > $-CH(CH_3)_2$ > $-C_2H_5$ > $-CH_3$ > $-H$

c. $-C(CH_3)_3$ > $-CH(CH_3)-CH_2CH_3$ > $-CH_2-CH(CH_3)-CH_3$ > $-(CH_2)_3CH_3$

d. $-CH(CH_3)CH_2CH_2CH_3$ > $-CH_2C(CH_3)(CH_3)CH_2CH_3$ > $-CH_2CH(CH_3)CH_2CH_3$

e. $-CH=CH-CH_3$ > $-CH(CH_3)_2$ > $-CH_2CH_2CH_3$

f. $-C(=S)-H$ > $-C(=O)-H$ > $-C(=N-H)-H$ > $-CH=CH_2$

g. $-C\equiv CH$ > $-CH=CH_2$ > $-CH_2CH_3$

h. $-CH(Cl)CH_2CH_3$ > $-CH_2CH(Cl)CH_3$ > $-CH_2CH_2CH_2Cl$

i. $-SH$ > $-OH$ > $-NH_2$ > $-CH_3$

j. —OH > $-\overset{O}{\overset{\|}{C}}-OH$ > 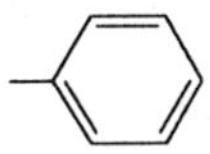 > —C≡CH > $(CH_3)_3C-$ > $-CH{=}CH_2$ > $(CH_3)_2CH$

5.7
a. (*R*)-bromochlorofluoromethane
b. (*R*)-2-chlorobutane
c. (*R*)-2-amino-l-propanol
d. (*R*)-2-methyl-l-butanol
e. (*S*)-2-chlorobutane
f. (*S*)-2-butanol
g. (*R*)-3-methylhexane
h. (*S*)-1-chloro-2-methylbutane
i. (*R*,*S*)-2,3-butanediol
j. (*S*,*S*)-2,3-butanediol
k. (*S*)-3-chloro-l-pentene
l. (*S*)-4-chloro-2-pentene
m. (*R*,*S*) or *meso*-1,2-dibromo-1,2-dichloroethane
n. (*S*,*S*)-3-chloro-2-butanol
o. (*S*)-3-ethyl-2,3,-dimethylhexane
p. (*S*)-2-chloro-1-propanol
q. (*R*)-2,2,3-trimethylpentane

5.8 a. Pentanes, C_5H_{12}: No chiral carbons are present in any of the compounds.

$CH_3CH_2CH_2CH_2CH_3$

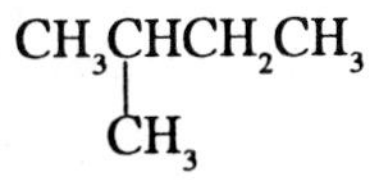

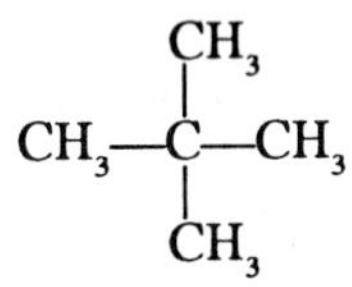

b. Hexanes, C_6H_{14}: No chiral carbons are present in any of the compounds.

$CH_3CH_2CH_2CH_2CH_2CH_3$

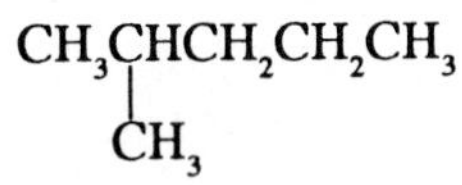

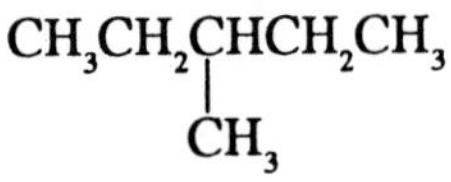

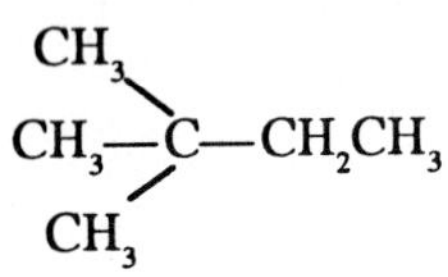

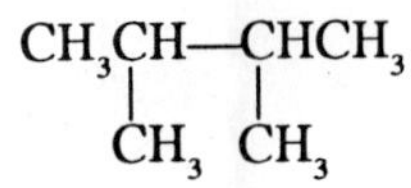

c. Heptanes, C_7H_{16}: Eleven compounds; seven achiral and two pairs of enantiomers.

$CH_3CH_2CH_2CH_2CH_2CH_2CH_3$ $(CH_3)_2CHCH_2CH_2CH_2CH_3$

$(CH_3)_3CCH_2CH_2CH_3$ $CH_3CH_2C(CH_3)_2CH_2CH_3$

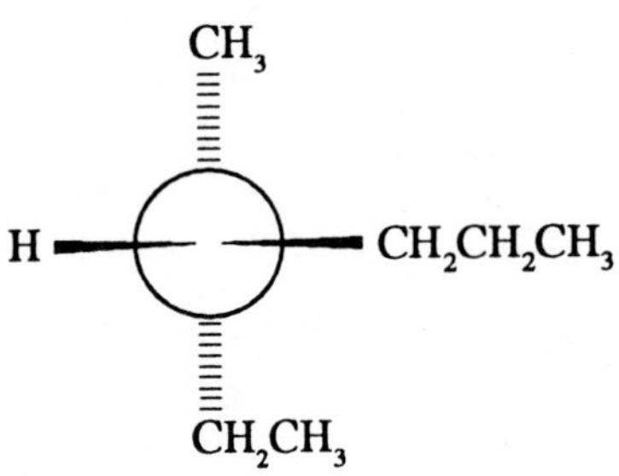

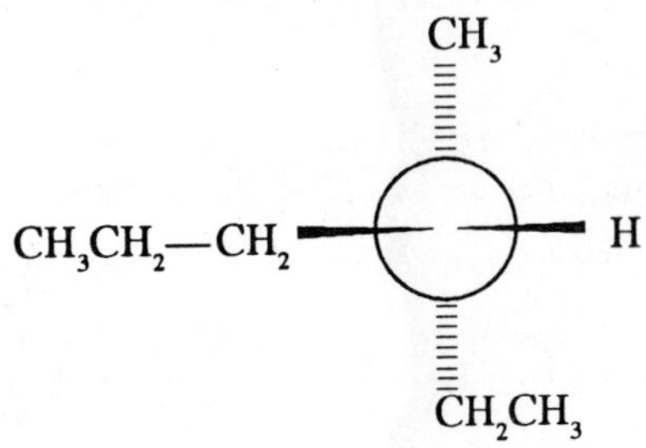

enantiomers of

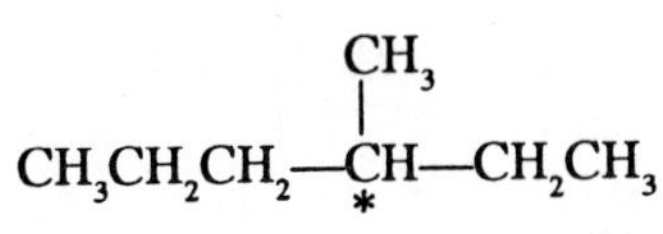

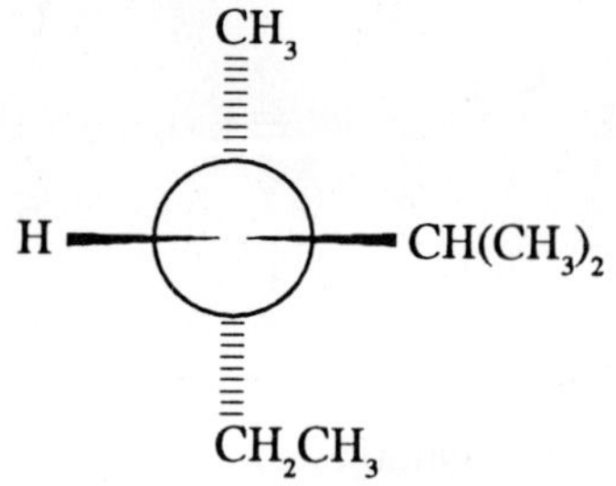

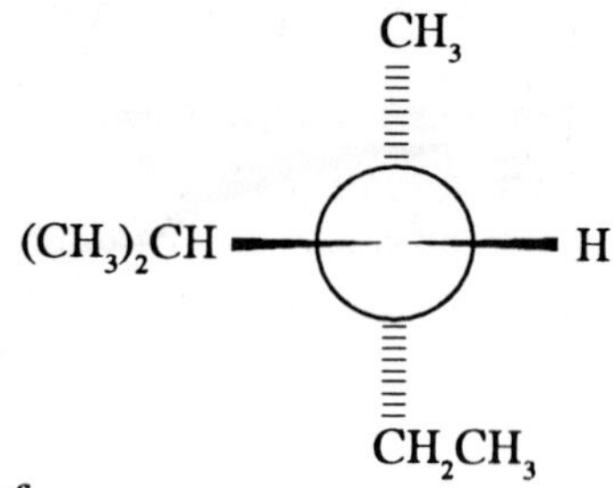

enantiomers of

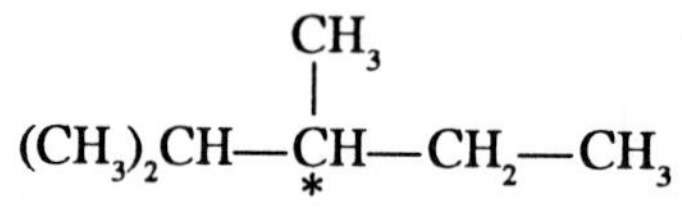

$(CH_3)_2CH—CH_2—CH(CH_3)_2$ $(CH_3)_3C—CH(CH_3)_2$ $(CH_3CH_2)_3CH$

d. Alcohols, C_4H_9OH: Five compounds; three optically inactive and one pair of enantiomers.

$CH_3CH_2CH_2CH_2OH$ $(CH_3)_3C—OH$ $(CH_3)_2CHCH_2OH$

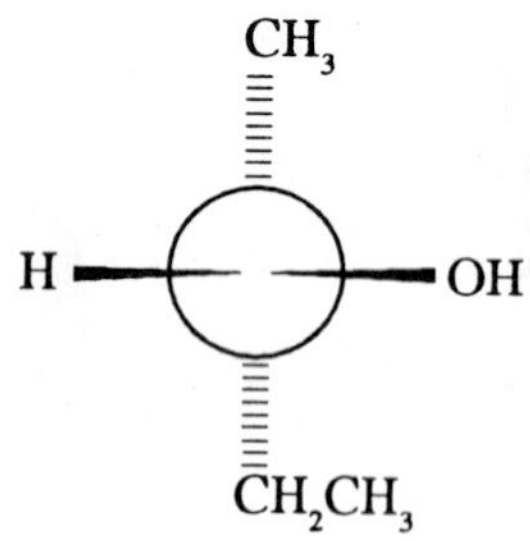

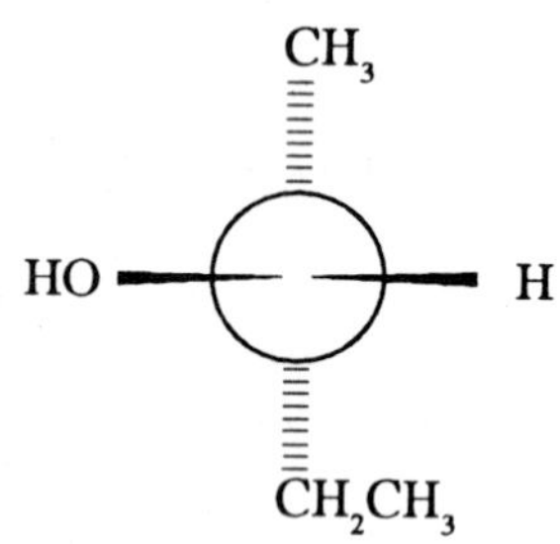

enantiomers of

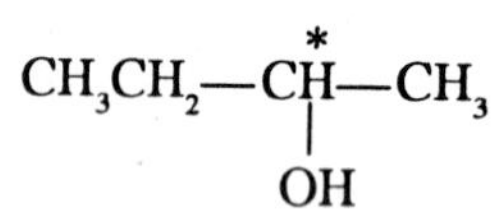

e. Dibromoderivatives of *n*-butane: Ten compounds: Three achiral, three pairs of enantiomers and one *meso* form.

$CH_3CH_2CH_2CHBr_2$ $CH_3CBr_2CH_2CH_3$ $BrCH_2CH_2CH_2CH_2Br$

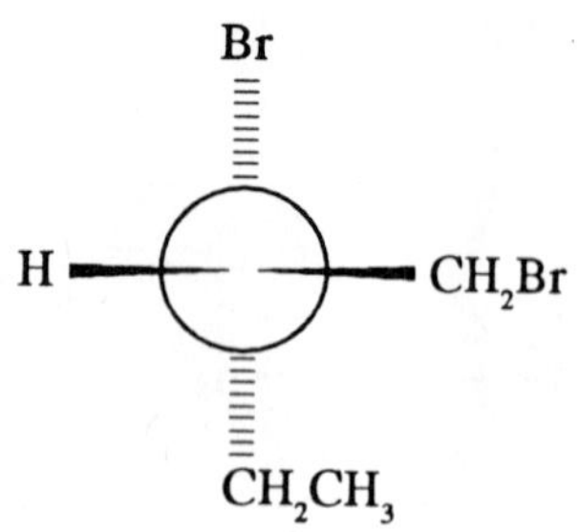

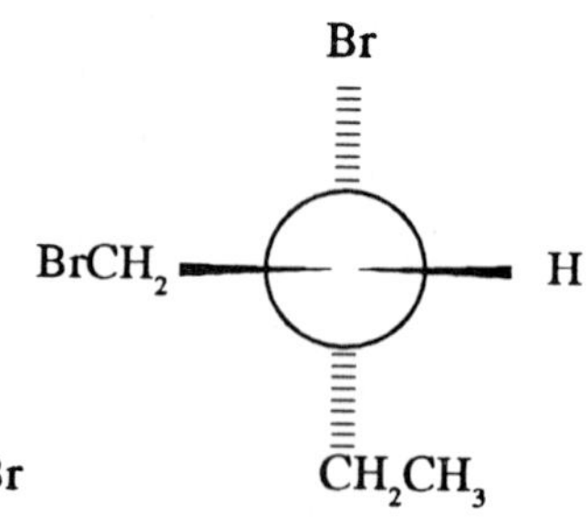

enantiomers of

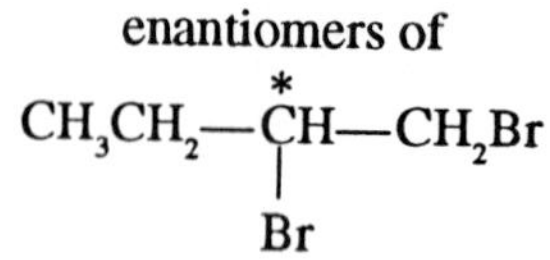

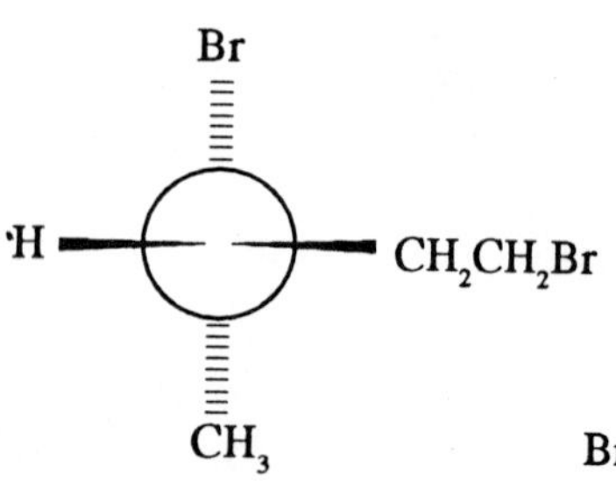

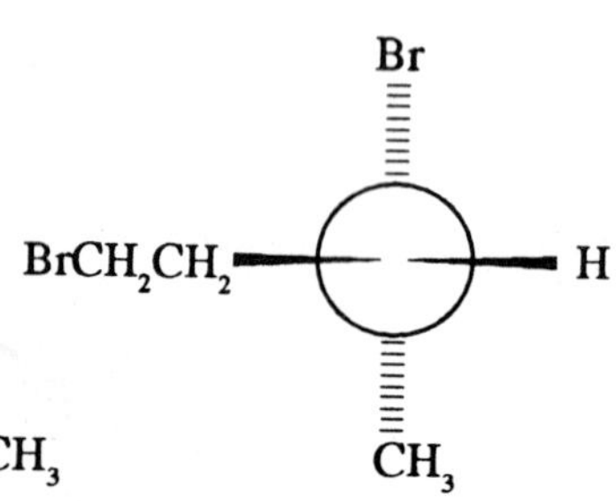

enantiomers of

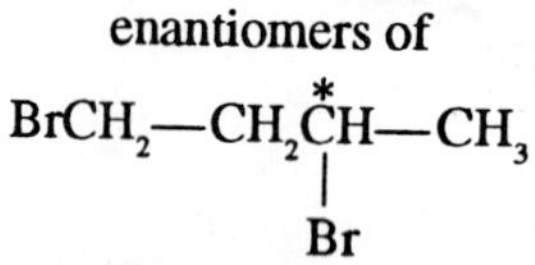

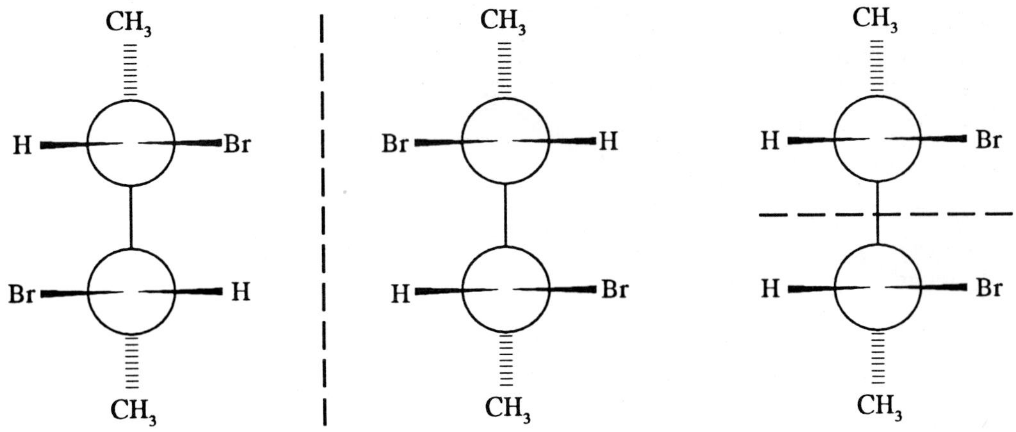

enantiomers and *meso* compound of CH_3—$\overset{*}{C}H$—$\overset{*}{C}H$—CH_3
Br Br

f. Bromochloropropanes: Eight compounds; two achiral and three pairs of enantiomers.

$BrCH_2CH_2CH_2Cl$ CH_3CBrCH_3
Cl

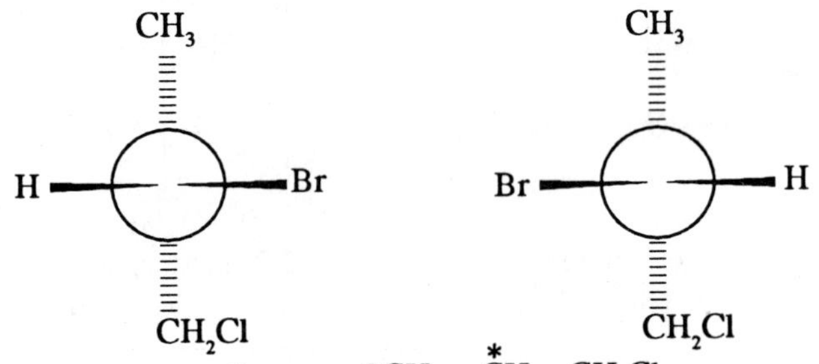

enantiomers of CH_3—$\overset{*}{C}H$—CH_2Cl
Br

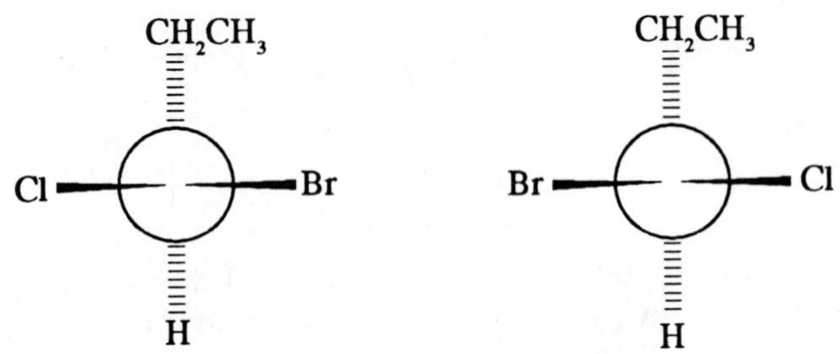

enantiomers of $CH_3CH_2\overset{*}{C}HClBr$

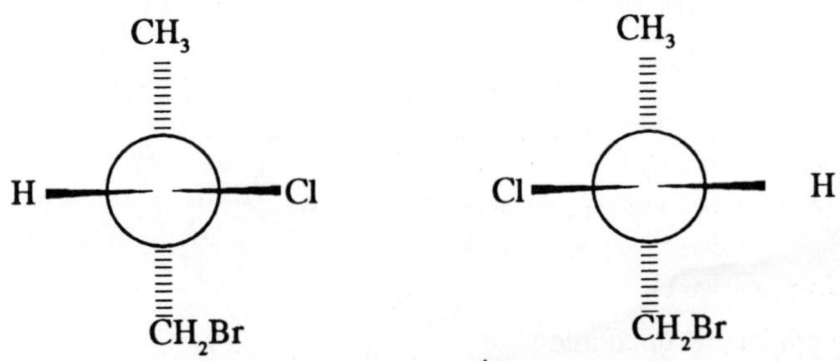

enantiomers of CH_3—$\overset{*}{C}H$—CH_2Br
Cl

5.9 Enantiomers: a, e, g, r, t, u, v, x, aa, ff
Diastereomers: j, l, p, gg, hh
Same Compound: c, d, h, k, n, o, q, s, w, z, bb, cc, dd, ee, ii
Constitutional isomers: b, i, f, m, y

5.10

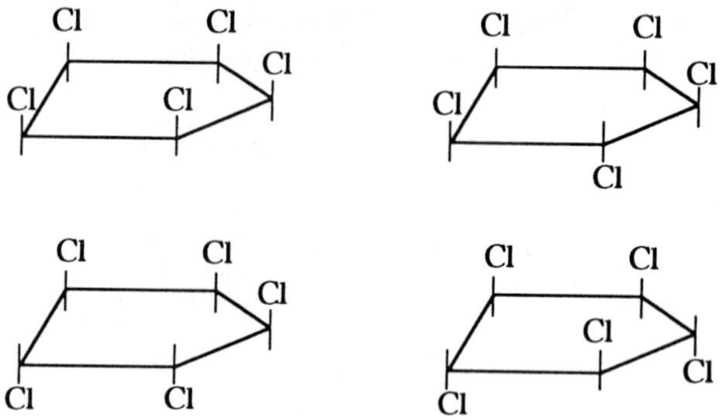

All are optically inactive. Each contains a plane of symmetry which makes it a *meso* compound. (Draw each structure's mirror image to clarify the fact that they *are* superimposable.)

5.11 a.

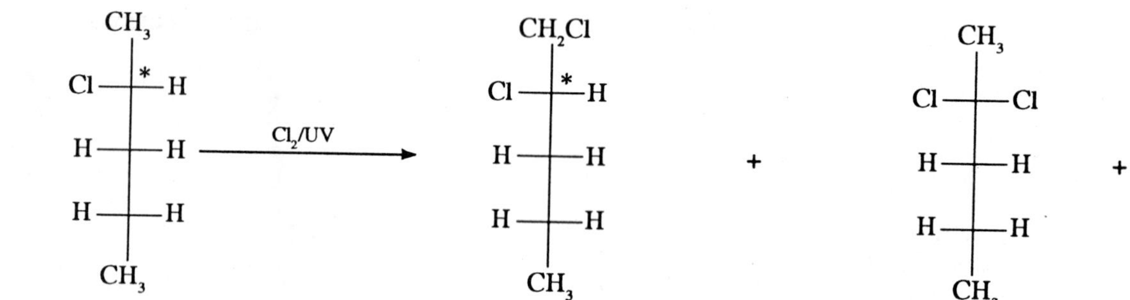

(*R*)-2-chloropentane (optically active) — (*S*)-1,2-dichloropentane (optically active) — 2,2-dichloropentane (achiral; optically inactive)

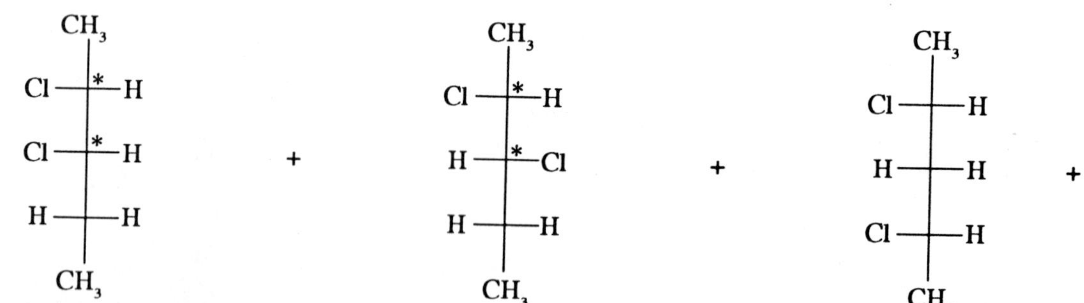

(2*R*,3*S*)-dichloropentane (optically active) — (2*R*,3*R*)-dichloropentane (optically active) — (2*R*,4*S*)-dichloropentane (meso-optically inactive)

(2*R*,4*R*)-dichloropentane (optically active) — (*R*)-1,4-dichloropentane (optically active)

b. There is a total of seven isomers, all of them diastereomers; therefore, fractional distillation yields seven fractions.

5.12 a.

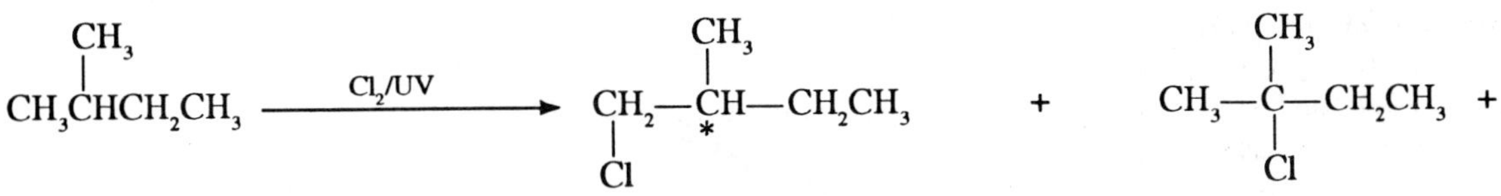

2-methylbutane optically inactive | 1-chloro-2-methylbutane (racemic mixture) | 2-chloro-2-methylbutane (achiral)

2-chloro-3-methylbutane (racemic mixture) | 1-chloro-3-methylbutane (achiral)

b. There is a total of six isomers. Careful fractional distillation will yield four fractions: two achiral and two racemic mixtures (*R* and *S*). None of these *fractions* is optically active, but there are four *individual* optically active compounds (*R* or *S*) and two optically inactive compounds (the achiral ones).

5.13 Too much angle strain; *sp* hybridization requires 180° angles, which small rings cannot accomodate.

5.14 a. Optical purity = (+10/+21.6) × 100 = 46%

b. The mixture is 46% optically pure and 54% racemic. Thus, the composition is
46% + (54% ÷ 2) = 73% (+)-2-bromooctane
100% – 73% = 27% (–)-2-bromooctane

5.15 Only c.

a.

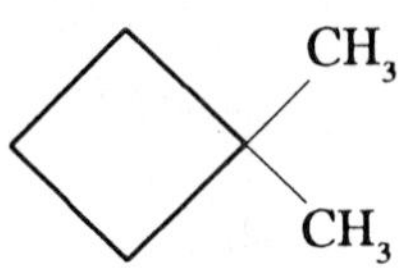

(achiral)

d.

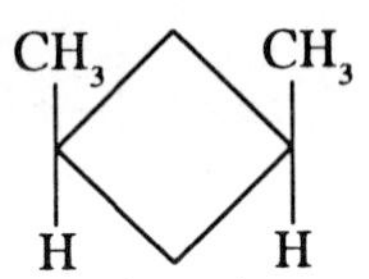

(*meso*)

b.

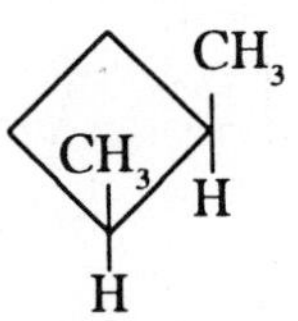

(*meso*)

e.

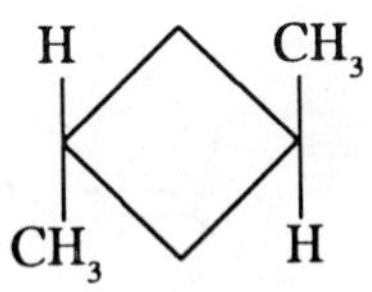

(*meso*)

c.

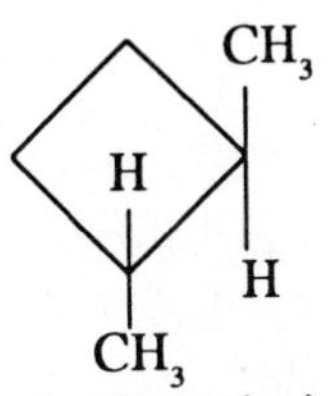

(optically active)

5.16 a.

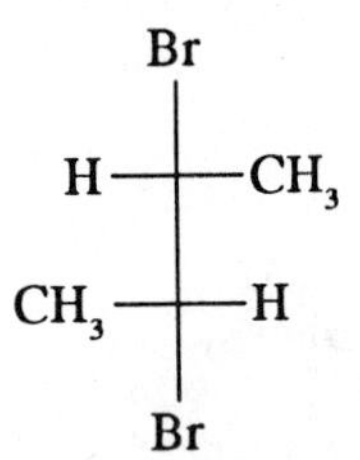

(*R*,*R*)

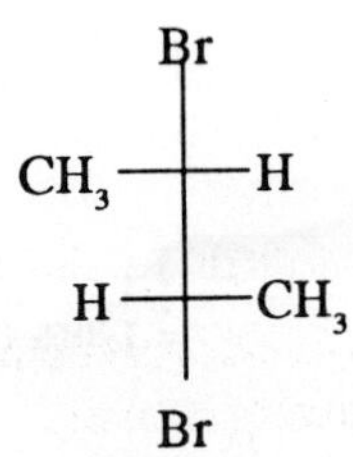

(*S*,*S*)

(racemic mixture)

b.

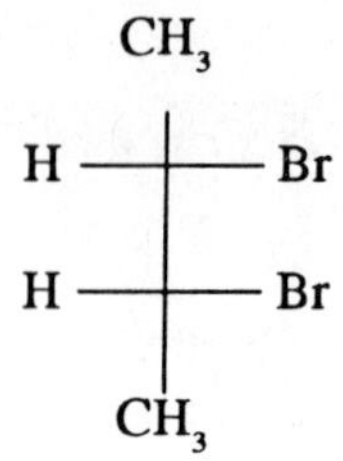

(*R,S*)
(*meso*)

c.

1,1-dibromocyclopropane (achiral) + *cis*-1,2-dibromocyclopropane (*meso*) + *trans*-1,2-dibromocyclopropane (racemic pair)

d.

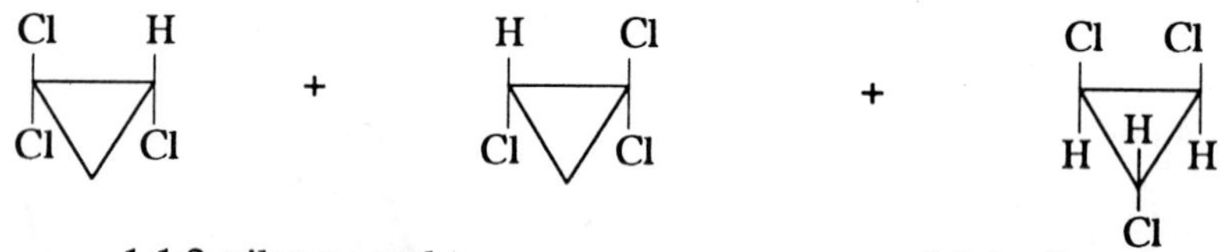

1,1,2-tribromocyclopropane (racemic) + 1,2,3-tribromocyclopropane (achiral)

e.

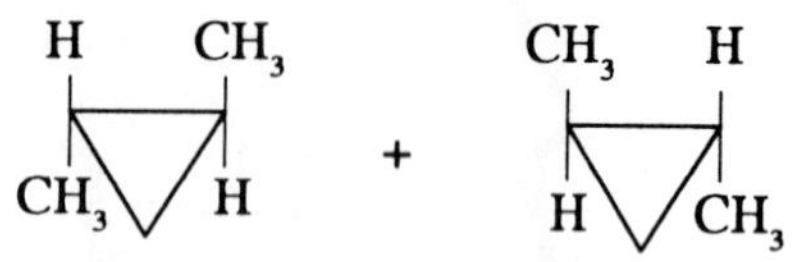

trans-1,2-dimethylcyclopropane
(racemic pair)

f.

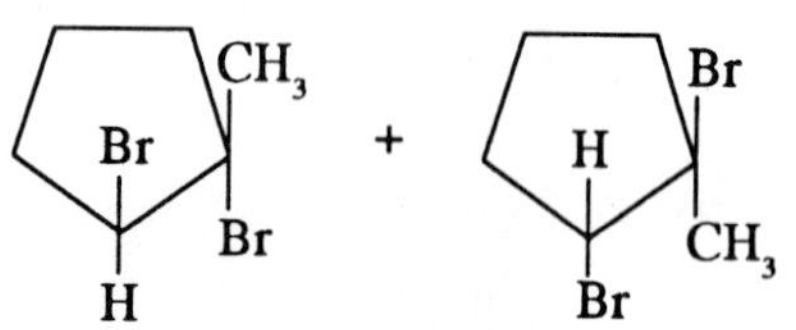

(racemic pair)

g.

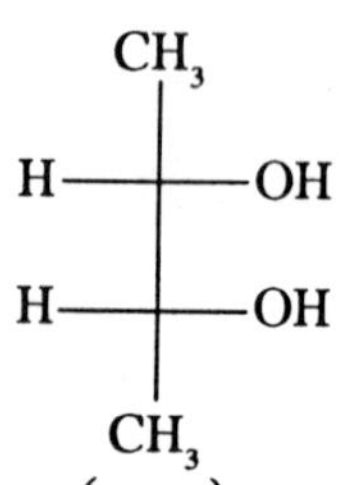

(*meso*)

h.

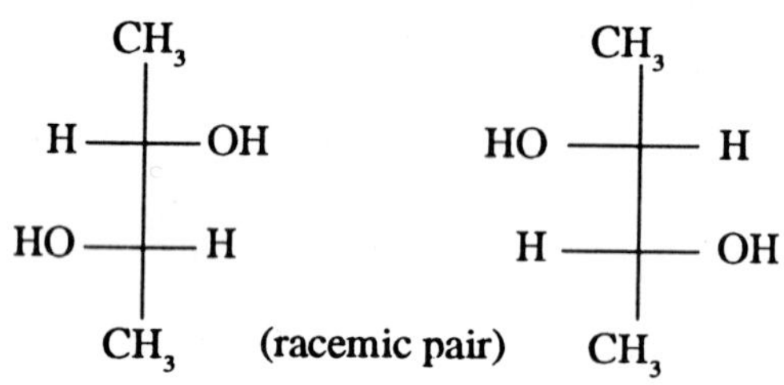

(racemic pair)

i.

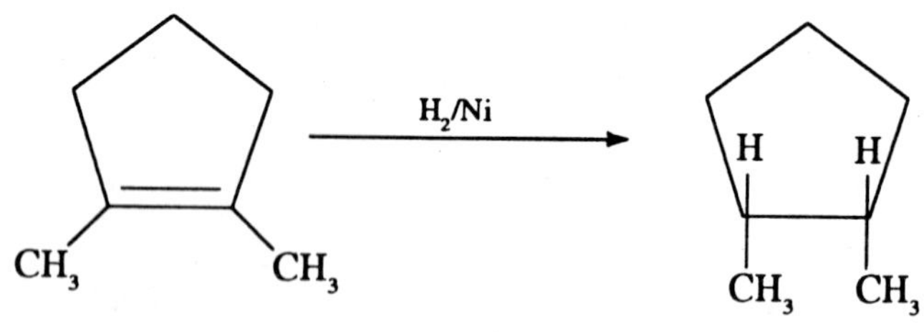

(major product)
(tetrasubstituted alkene)

(*cis*)
(*meso*)

5.17 a. 1 pair of enantiomers and 2 *meso*.
b. 4 pairs of enantiomers and 2 *meso*.
c. 4 pairs of enantiomers (3 chiral carbons)
d. 1 pair of enantiomers (1 chiral carbon)
e. 2 pairs of enantiomers (2 chiral carbons)

5.18 **A** (C_4H_7Br) has either one double bond or ring (the corresponding open chain alkane is C_4H_{10}). Since HBr addition to **A** results in the formation of two configurational isomers **B** and **C**, **A** has a double bond. From the ozonolysis products, the structure of **D** is

$H_2C{=}CH{-}CH{=}CH_2$
butadiene

Since **A** is optically active its structure is

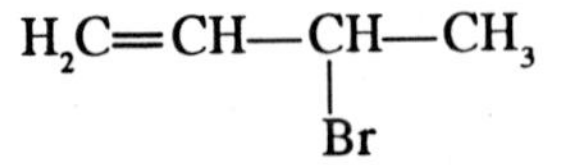

Thus, the sequence of reactions is

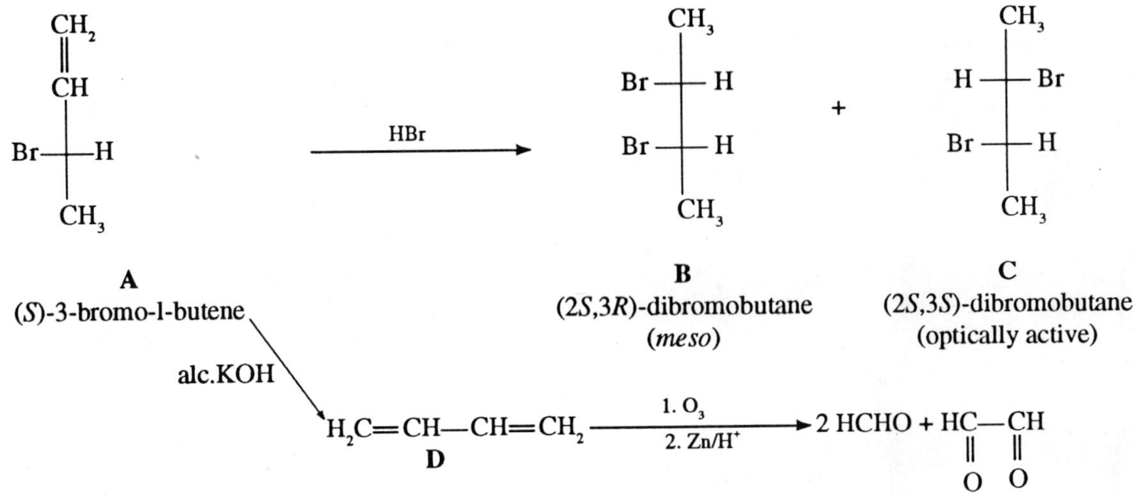

A (*S*)-3-bromo-l-butene; **B** (2*S*,3*R*)-dibromobutane (*meso*); **C** (2*S*,3*S*)-dibromobutane (optically active); **D**

5.19 a.

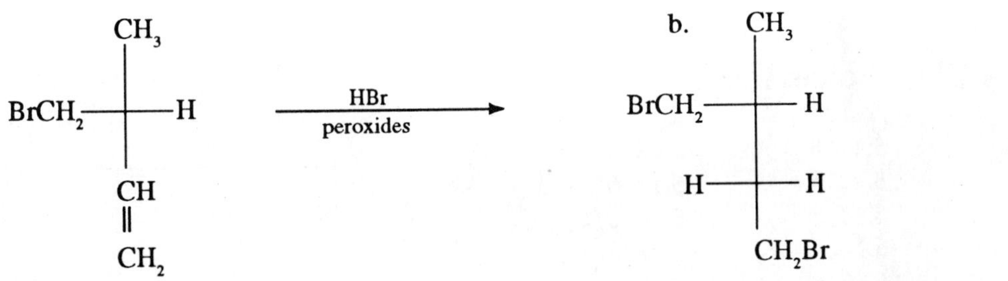

(*R*)-4-bromo-3-methyl-1-butane

b. (*R*)-1,4-dibromo-2-methylbutane

5.20 a. *Anti* addition of the Br_2 yields an enantiomeric pair

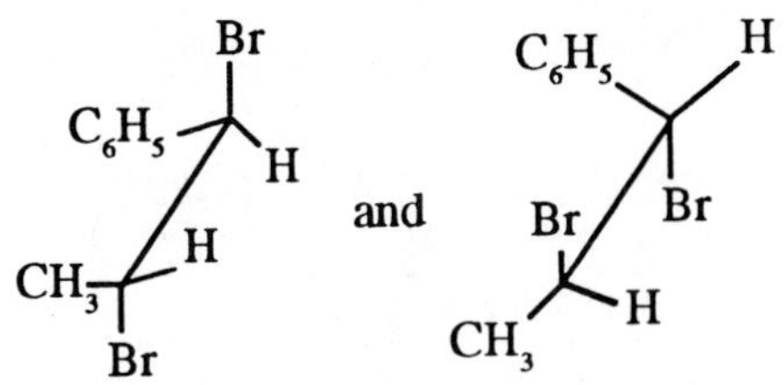

(R, R) and (S, S)

b.

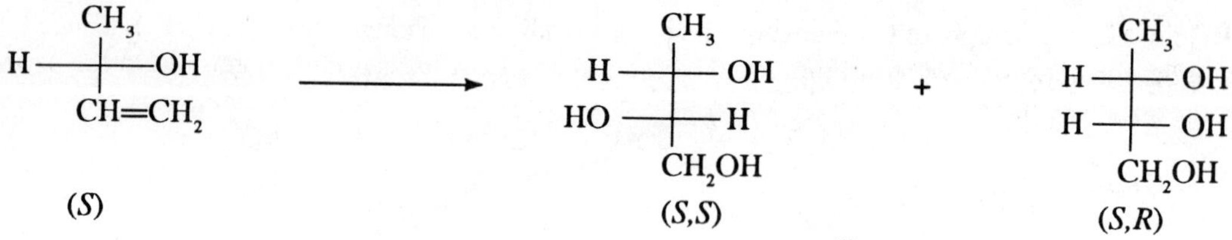

diastereomers

c.

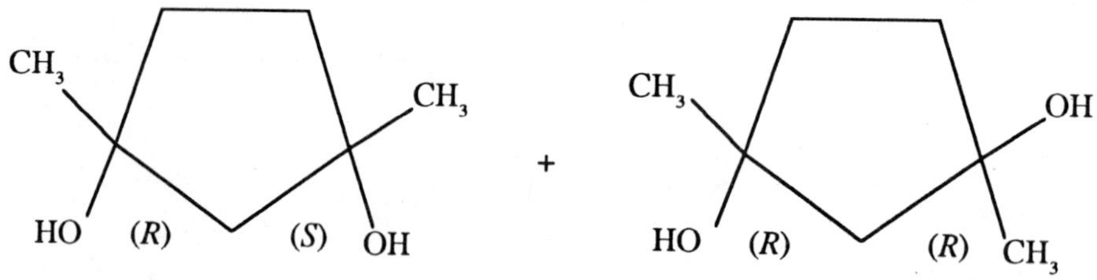

diastereomers formed via the carbocation

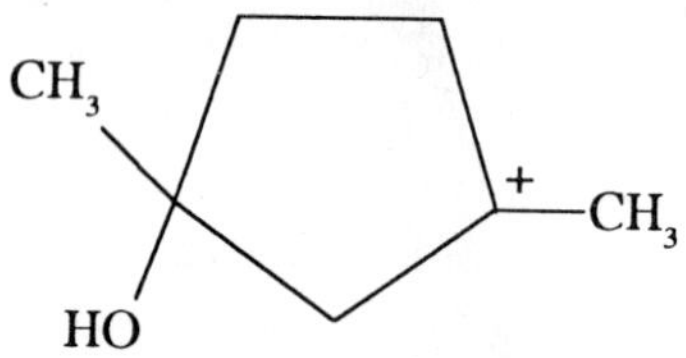

d.

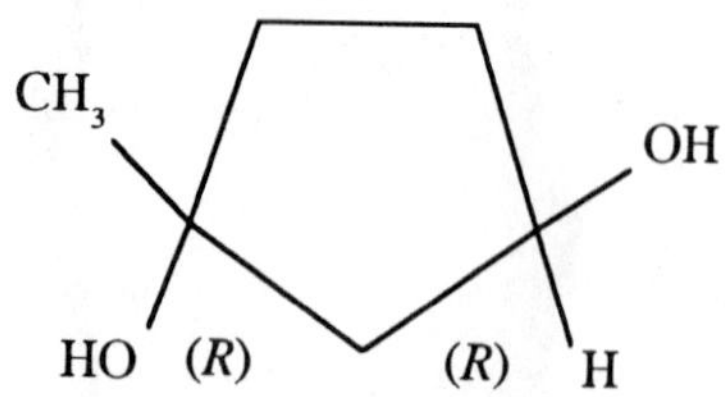

via S_N2 inversion

Answers for Multiple Choice Test

5.1 c	**5.6** b	**5.11** b	**5.16** a
5.2 b	**5.7** e	**5.12** a	**5.17** b
5.3 b	**5.8** e	**5.13** c	**5.18** c
5.4 c	**5.9** d	**5.14** d	**5.19** a
5.5 d	**5.10** b	**5.15** c	**5.20** d

6 Alkyl Halides

Questions

6.1 Define the following terms:

a. nucleophilic substitution
b. concerted process
c. reaction rate
d. aprotic solvent
e. unimolecular reaction
f. S_N1 reaction
g. S_N2 reaction

6.2 Identify the products or reagents missing in the reactions below:

a. $n\text{-}C_{12}H_{25}Br + NaN_3 \longrightarrow$ **A** + NaBr
b. $(C_2H_5)_3C\text{—}OH + HCl \longrightarrow$ **B** (gives a ppt. when reacted with $AgNO_3$)
c. $Na_2S + 2CH_3CH_2Br \longrightarrow$ **C** + **D** (inorganic salt)
d. $(CH_3)_2CHBr + NaSCN \longrightarrow$ **E** + **F** (inorganic salt)
e. $3CH_3OH + PI_3 \longrightarrow$ **G** + **H** (acidic to litmus)

f. $CH_3\text{—}C(Cl)\text{=}CHCH_2Cl + Na_2CO_3 \xrightarrow{H_2O}$ **I** (C_4H_7ClO)

g. 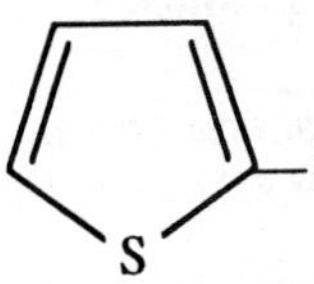

–SMgI + $CH_3I \longrightarrow$ **J** + **K** (inorganic salt)

h. $ClCH_2CH_2CH_2CH_2Cl$ + **L** $\longrightarrow$ 

+ 2NaCl

i. $CHCl_3 + 3\,C_2H_5O^-Na^+ \longrightarrow$ **M** ($C_7H_{16}O_3$) + **N** (inorganic salt)
j. $BrCH_2CH_2CH_2Br + C_6H_5O^-Na^+$ (1 mole) $\longrightarrow$ **O** + **P** (inorganic salt)
k. $(CH_3)_3COH + HBr \longrightarrow$ **Q** (C_4H_9Br) + **R**
l. $3C_{16}H_{33}OH$ + **S** $\longrightarrow 3C_{16}H_{33}I + H_3PO_3$

m. $2CH_3CH_2OH \xrightarrow{\mathbf{T}} CH_3CH_2\text{—}O\text{—}CH_2CH_3 + H_2O$
n. $n\text{-}C_{12}H_{25}Br + KCN \longrightarrow$ **U** + **V** (inorganic salt)

o. 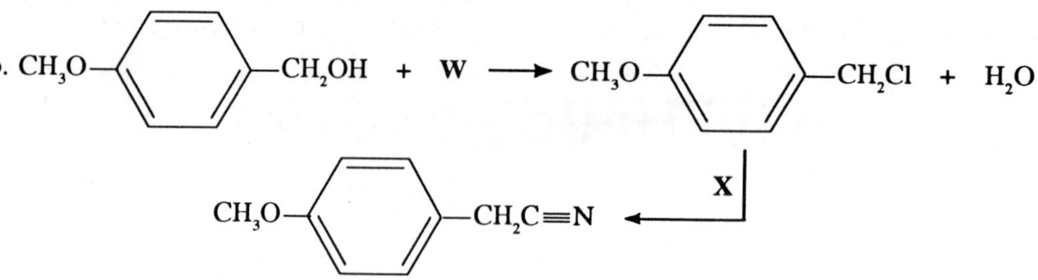

p. $HC{\equiv}C^-Na^+$ + **Y** ⟶ $HC{\equiv}CCH_2CH_2CH_2CH_3$ + NaBr
q. $(n\text{-}C_4H_9)_2CuLi$ + **Z** ⟶ $n\text{-}C_6H_{14}$ + CuBr + LiBr
r. $C_6H_5CH_2Cl$ + **AA** ⟶ $C_6H_5CH_2NHC_6H_5$ + HCl
s. $C_6H_5CH_2Cl$ + **BB** ⟶ $C_6H_5CH_2CN$ + NaCl
t. $CH_3(CH_2)_2CH_2OH$ + PCl_3 ⟶ **CC** + H_3PO_3
u. $ClCH_2CH_2CH_2Br$ + KCN (1 mole) ⟶ **DD** (C_4H_6NCl) + KBr

v. 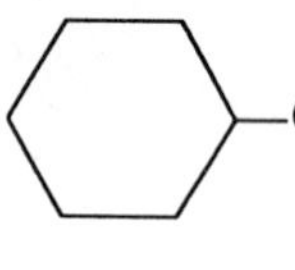 —OH + H_2SO_4 $\xrightarrow{\text{heat}}$ **EE** (decolorizes Br_2/CCl_4)

w. CH_3CH_2Br + $CH_3\overset{\displaystyle O}{\overset{\|}{C}}O^-Ag^+$ ⟶ **FF** + **GG** ($C_4H_8O_2$)

x. $CH_3CH_2\underset{\displaystyle OH}{\underset{|}{C}}H\underset{\displaystyle CH_3}{\underset{|}{C}}HCH_2\ddot{N}(CH_3)_2$ + HBr ⟶ **HH** ($C_7H_{16}NBr$)

y. $n\text{-}C_8H_{17}$—I + **II** ⟶ $n\text{-}C_8H_{17}$—CN
z. $n\text{-}C_8H_{17}$—Cl + $Na^{36}Cl$ ⟶ **JJ** (yields a ppt. when reacted with $AgNO_3$)

6.3 Consider the equation:

$$R{-}L + Nu^- \longrightarrow R{-}Nu + L^-$$

Compare the S_N1 and S_N2 reactions in terms of the following:
a. overall mechanism (how many steps?)
b. reaction mechanism (individual steps)
c. order of reaction
d. effect on rate of doubling the nucleophile's concentration
e. effect on rate of doubling the substrate's concentration
f. rate equation
g. favorable medium (acidic or basic medium?)
h. stereochemistry of products
i. solvent effects
j. carbon at which reaction occurs
k. competitive reactions
l. catalysis
m. leaving group ability (L-halogen)
n. importance of nucleophile strength

6.4 Suggest structures that satisfy each of the descriptions below:
a. the simplest optically active deuterated alcohol that would undergo S_N2 reactions with hydrogen halides.
b. an optically active alcohol ($C_7H_{16}O$) that would not racemize in hot acid.
c. an alcohol more likely to rearrange than to substitute or eliminate in acidic solutions.

6.5 Explain why
a. allyl halides, $CH_2{=}CH{-}CH_2{-}X$ (where X= Cl or Br), undergo S_N1 substitution even though they are 1° halides.
b. both 3,3-dimethyl-2-butanol and 2,3-dimethyl-2-butanol yield 2-chloro-2,3-dimethylbutane when reacted with concentrated HCl.
c. $AgNO_3$ accelerates the rate of an S_N1 substitution reaction of an alkyl bromide.

d. Method A provides higher yields than Method B, even though both reactions yield the same product:

Method A

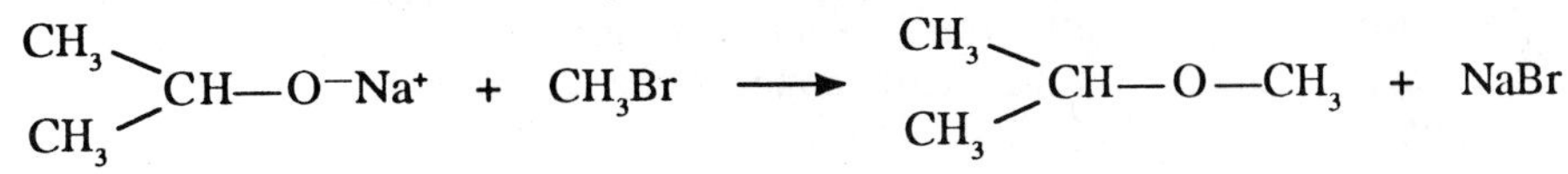

Method B

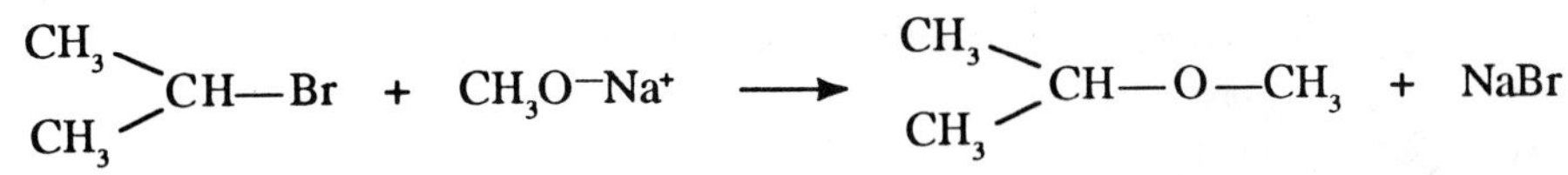

e. acetic acid behaves as a base in 100% sulfuric acid, and as an acid in 50% aqueous acid solutions.

f. (+)-2,3-dimethyl-3-pentanol racemizes when heated in formic acid:

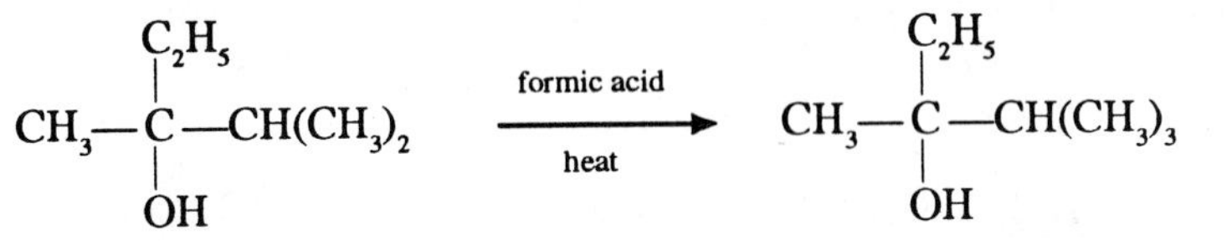

(optically active) (optically inactive; racemic)

6.6 Suggest mechanisms that would explain the formation of the products in the following reactions:

a.

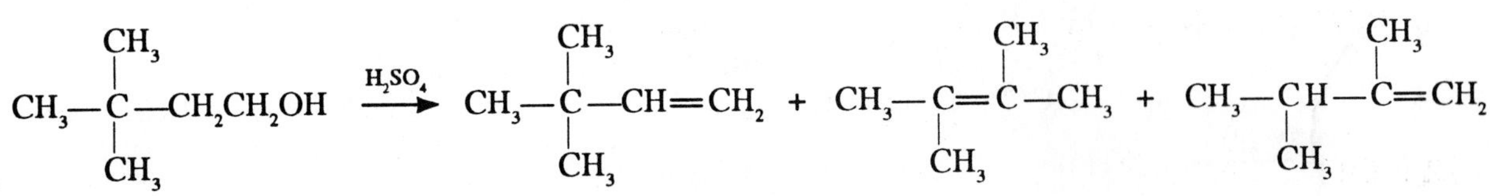

b.

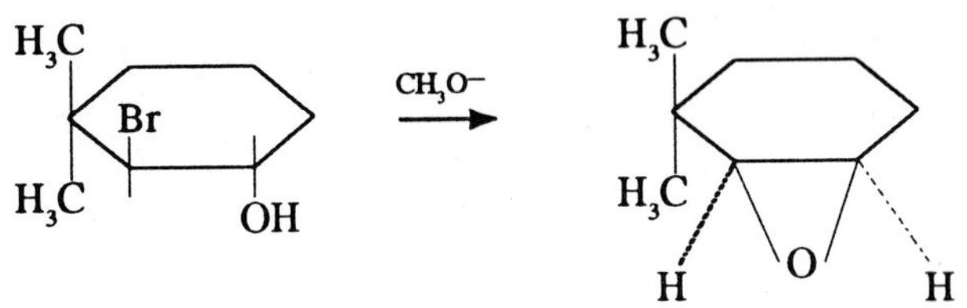

c.

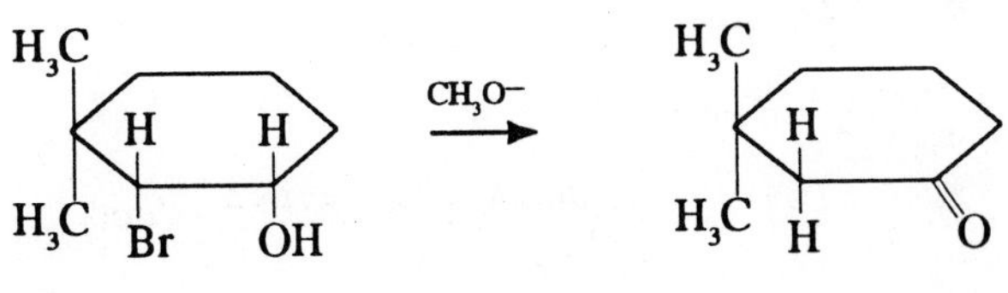

d.

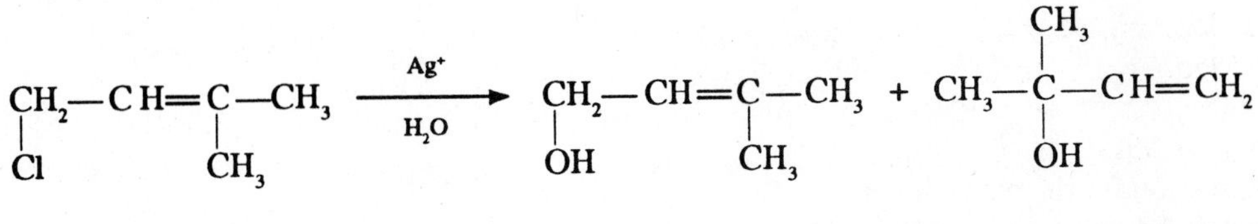

e.

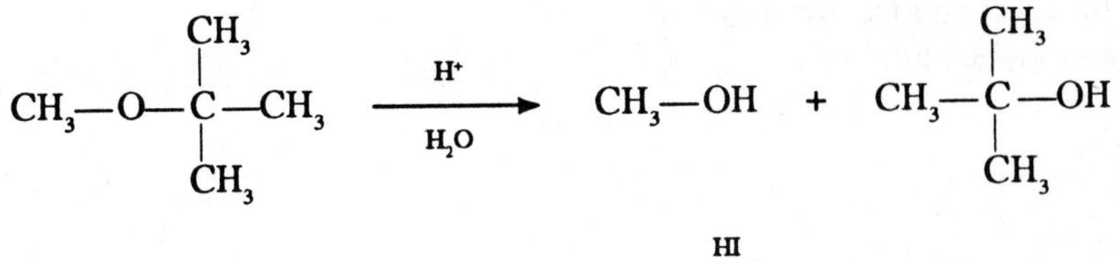

f.

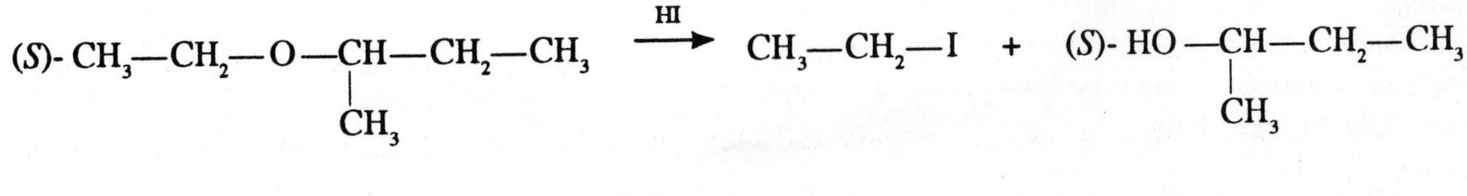

6.7 How would you prepare
a. isobutyl bromide from isobutyl alcohol (two methods).
b. 2-methylpropanenitrile (i.e., isopropyl cyanide) from propane.
c. 2-methoxy-2-methylpropane from isobutane (two methods).
d. (*E*)-2-chloro-2-butene

6.8 For reactions involving alkyl halides, what reaction conditions
a. favor S_N2 over E^2 reaction products?
b. favor S_N1 over S_N2 reaction products?
c. favor E^2 over S_N2 reaction products?

6.9 Arrange the compounds in each of the following sets in order of decreasing reactivity towards the indicated reagents:
a. $CH_3CH(Br)CH_2CH_3$, $(CH_3)_3CBr$, $CH_3CH_2CH_2CH_2Br$ towards $AgNO_3$/ethanol.

b. $CH_3CH(Br)CH_2CH_3$, $(CH_3)_3CBr$, $CH_3CH_2CH_2CH_2Br$ towards NaI/acetone.

c. $(CH_3)_3CCl$, $(CH_3)_3CBr$, $(CH_3)_3CI$ towards methanol to yield $(CH_3)_3C—OCH_3$.
d. H_2O , C_2H_5OH , n-C_4H_9OH towards $(CH_3)_3CBr$ in a solvolysis reaction.
e. 0.5M CH_3O^- , 1M CH_3O^- , 2M CH_3O^- towards CH_3CH_2Br to yield $CH_3CH_2—O—CH_3$.
f. CH_3O^- , $(CH_3)_3CO^-$, OH^- towards CH_3I to yield $CH_3—O—R$ (where, R = $CH_3—$, $(CH_3)_3C—$, H— respectively).

6.10 Compare the relative rates of S_N1 solvolysis for

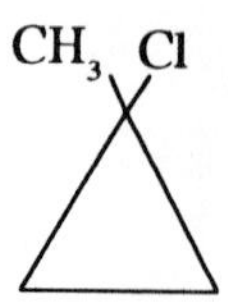

A

, 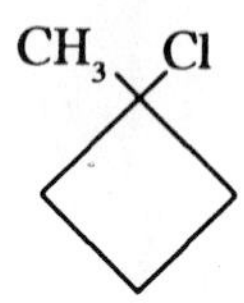

B

and $(CH_3)_3C—Cl$

C

6.11 Consider the following reaction mechanism:

$$\text{I} \quad R—Cl \underset{k_2}{\overset{k_1}{\rightleftharpoons}} R^+ + Cl^-$$

$$\text{II} \quad R^+ + Nu^- \xrightarrow{k_3} R—Nu$$

a. Derive a rate expression for the mechanism.
b. What condition must be true for the reaction to show first-order kinetics?
c. What effect will the following have on the rate?
(i) : addition of Cl^-
(ii) : addition of Br^-
(iii) : using a more polar solvent
(iv) : using a stronger nucleophile.

6.12 Primary alkyl halides of the type $R—\ddot{O}—CH_2X$ undergo S_N1 reactions, while other primary alkyl halides react via S_N2 mechanisms. Explain.

6.13 *Trans*-3-bromocyclohexane carboxylate **A** undergoes an intramolecular nucleophilic substitution to give lactone **B**. The *cis*-isomer **C** does not react in the same way. Explain.

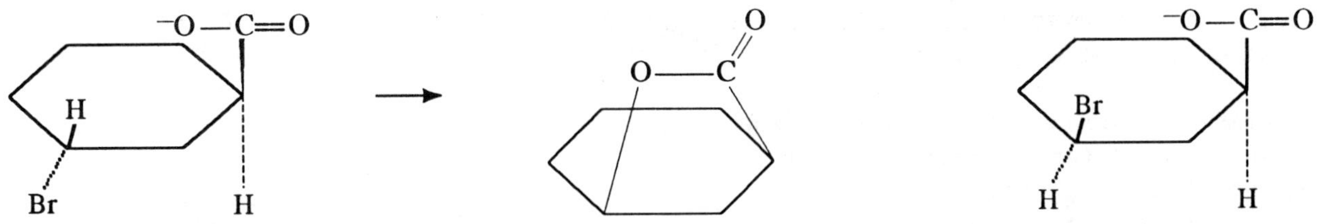

A **B** **C**

6.14 Treatment of anisole with HI yields methyl iodide and phenol, and not iodobenzene and methanol. Explain.

anisole phenol methyl iodide

6.15 Nitrogen mustards hydrolyze by the following mechanism:

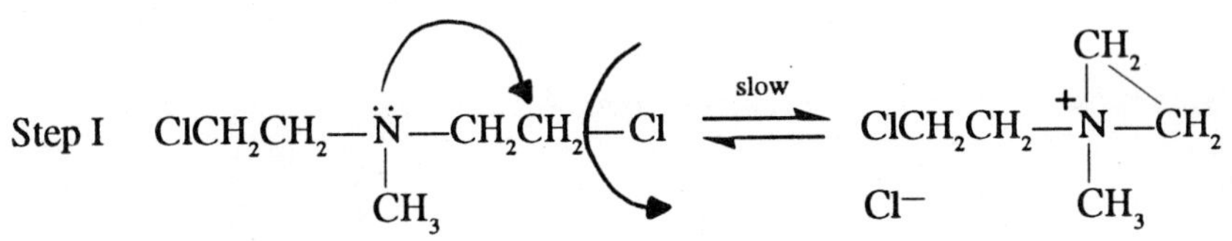

nitrogen mustard

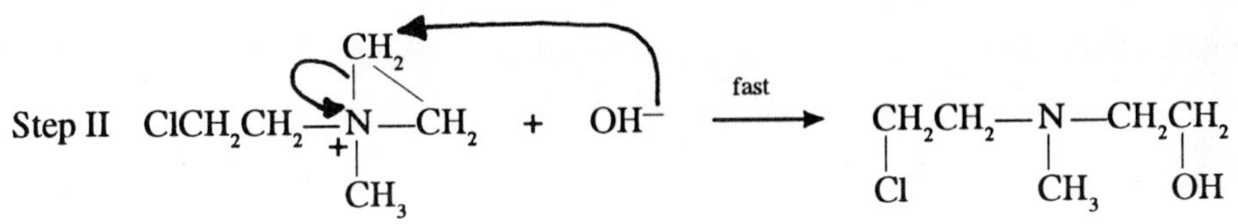

a. Is the reaction first or second order?

b. Predict the effect on the rate of the following:
   (i) : an increase in the concentration of OH^-.
   (ii) : the use of a less polar solvent.
   (iii) : an increase in the Cl^- concentration.
   (iv) : the use of I^- instead of OH^- in Step II.
   (v) : an increase in reaction temperature.

c. This hydrolysis is many times faster than the hydrolysis of other 1° halides. Suggest a reason.

6.16 To study the E^2 mechanism, a solution of 2-bromobutane, ethanol, D_2O, and KOH was refluxed and then quenched in ice. The recovered 2-bromobutane contained no deuterium. How does this experiment descredit the following two-step mechanism?

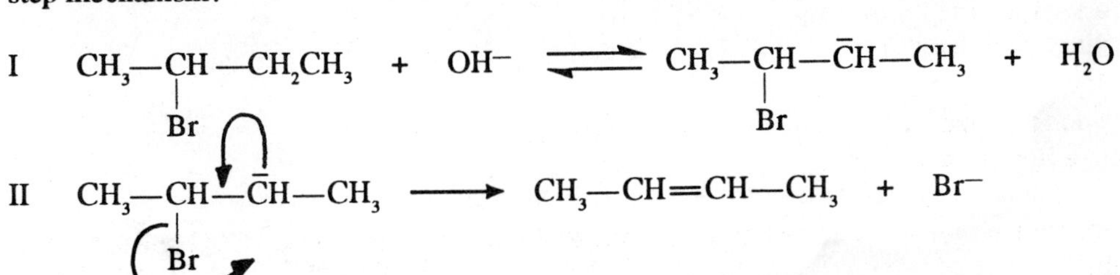

6.17 The E^2 mechanism is characterized by a deuterium isotope effect. What does this suggest about the mechanism?

6.18 Explain why the reaction of Br_2 with cyclohexene in the presence of I^- yields *trans*-1,2-dibromocyclohexane and
a. *trans*-1,2-diiodocyclohexane,
b. *cis*-1,2-diiodocyclohexane,
c. *cis*-1-bromo-2-iodocyclohexane,
d. *trans*-1-bromo-2-iodocyclohexane.

6.19 S_N1 reactions of cyclohexyl halides proceed faster when the position of the leaving group is *axial* than when it is *equatorial*. Consider the nature of the transition state and the relative stabilities of starting materials to explain why *cis*-4-*t*-butylcyclohexyl bromide is more reactive than the *trans*-isomer toward S_N1 reactions.

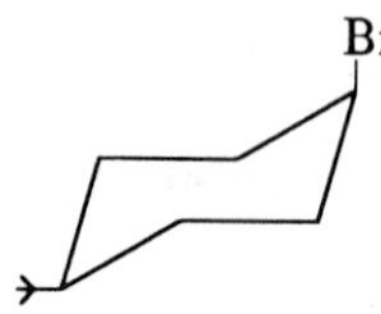

undergoes S_N1 reactions faster than

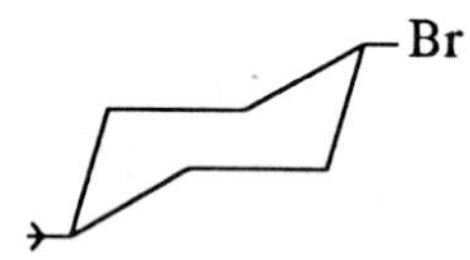

6.20 Arrange the following alkyl halides from most to least reactive in an S_N2 reaction.

$CH_3—CH(CH_3)—CH_2—CH_2Br$ $\quad$ $CH_3—CH_2—CH(CH_3)—CH_2Br$ $\quad$ $CH_3—CH_2—CH_2—CH(CH_3)—Br$ $\quad$ $CH_3—CH_2—C(CH_3)_2—Br$

6.21 Arrange the following alkyl halides from most reactive to least reactive in an S_N1 reaction.

$(CH_3)_3C—I$ $\quad$ $(CH_3)_3C—Br$ $\quad$ $CH_3CH_2—CH(CH_3)—I$ $\quad$ $CH_3CH_2—CH(CH_3)—Br$

6.22 S_N2 reactions proceed faster in the solvent dimethylsulfoxide (CH_3SOCH_3) than in ethanol or ethanol-water solvents. Explain.

6.23 Consider the following two S_N2 reactions:

(2*S*, 3*S*)-2-bromo-3-phenylbutane $\xrightarrow{I^-}$

(2*R*, 3*R*)-2-bromo-3-phenyl butane $\xrightarrow{I^-}$

a. What is the stereochemical realtionship of
  i. the starting materials
  ii. the products
b. Do the reactions proceed at the same or at different rates?
c. Can the products be separated by distillation techniques?

Multiple Choice Test

6.1 Which compound is a tertiary alkyl halide?

a. $(CH_3)_3CH_2Br$

c. 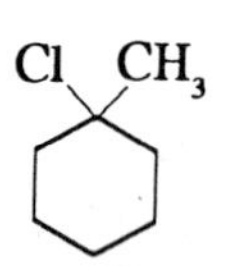

b. $(CH_3CH_2)_2C(CH_3)CH_2Br$

d. 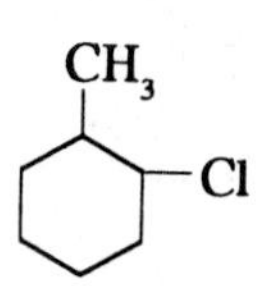

6.2 What is the configuration of the product in the reaction below?

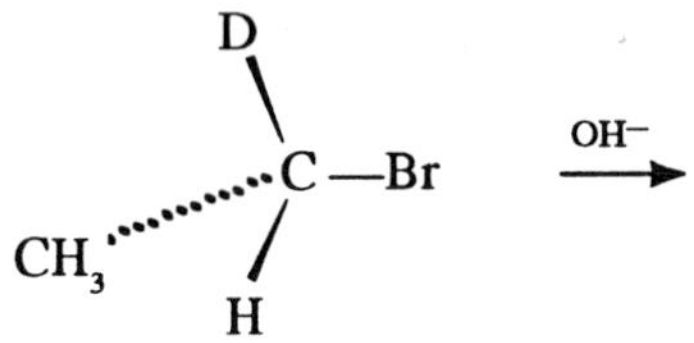

a. *R*
b. *S*
c. racemic
d. *E*

6.3 Which one is not true for the S_N2 reaction: **A** + **B** ⟶ products?
a. Rate = k[A][B].
b. Reactivity order is RI > RBr > RCl.
c. The reaction proceeds with racemization.
d. The process is concerted.

6.4 Which of the following compounds would undergo S_N2 most rapidly?
a. 1-chloropentane
b. 2-chloropentane
c. 2-chloro-2-methylbutane
d. neopentylchloride

6.5 Which compound would undergo E^1 reaction the fastest?
a. 1-chlorobutane
b. 2-chlorobutane
c. 2-chloro-2-methylbutane
d. neopentylchloride

6.6 Which one of the following would not react with $(CH_3)_2\overset{+}{C}H$?
a. H_2O
b. NH_3
c. Br^-
d. $AlCl_3$

6.7 The solvolysis of (*R*)-2-bromoheptane leads to
a. inversion.
b. retention.
c. racemization.
d. (a) and (c)

6.8 Which compound would react the fastest with concentrated HCl to yield an alkyl halide?
a. 2-pentanol
b. 2,2-dimethyl-1-butanol
c. 3-methylcyclopentanol
d. 2-methyl-2-pentanol

6.9 Which species of each pair would be the better leaving group?

A HO^- or **B** H_2O **C** CH_3COO^- or **D** $CH_3CH_2O^-$

a. A and C
b. A and D
c. B and C
d. B and D

6.10 The leaving group in the synthesis of 1-bromopropane from 1-propanol and HBr is
a. H_2O
b. OH^-
c. H_3O^+
d. HBr

6.11 The rate law in the solvolysis of *t*-butyl chloride, R—Cl, is
a. Rate = k[R—C][H_2O]
b. Rate = k[R—Cl]
c. Rate = $k[R—Cl]^2$
d. Rate = $k[R—Cl][H_2O]^{-1}$ (where R = $(CH_3)_3C$—)

6.12 Which compound will not yield 1-chlorobutane from 1-butanol?
a. $SOCl_2$
b. $AlCl_3$
c. PCl_3
d. (a) and (b)

6.13 Which of the following reactions would not lead to the formation of the indicated product?
a. $CH_3CH_2O^- + CH_3OH \longrightarrow CH_3CH_2OCH_3 + OH^-$
b. $CH_3S^-Na^+ + CH_3CH_2I \longrightarrow CH_3CH_2SCH_3 + NaI$
c. $NaN_3 + CH_3CH_2CH_2Br \longrightarrow CH_3CH_2CH_2N_3 + NaBr$
d. $(CH_3)_2NH + CH_3Br \longrightarrow (CH_3)_3NH^+Br^-$

6.14 Which compound will produce only one alkene from E^2 elimination?

a. 1-bromo-3-methylbutane
b. 2-bromopentane
c. 3-bromopentane
d. 2-bromo-2-methylbutane

6.15 What is the first step in the reaction of cyclohexanol with HBr to produce 1-bromocyclohexane?

a. nucleophilic attack of Br^- to displace OH^-
b. loss of OH^- to form the carbocation
c. protonation of the –OH group to form $-OH_2^+$
d. concerted mechanism: as the C—OH bond breaks, the C—Br bond is formed

6.16 Which alcohol would react the fastest with HBr to form an alkyl halide?

a. 3-methyl-1-butanol
b. 2-methyl-2-butanol
c. 3-methyl-2-butanol
d. 2-methyl-1-butanol

6.17 The name of the compound below is ____________ cyclohexane

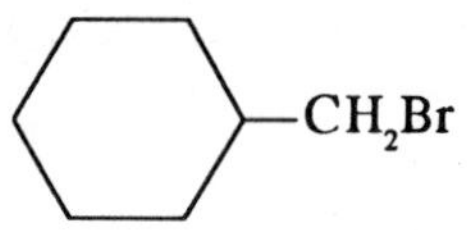

a. 1-bromo-1-methyl
b. 1-bromo-2-methyl
c. bromomethyl
d. methylbromo

6.18 In the substitution reaction of CH_3Br with OH^-, doubling the concentrations of both the substrate and the nucleophile leads to

a. no change
b. doubling of the rate
c. tripling of the rate
d. quadrupling of the rate

6.19 What is the leaving group in the reaction below

A B C D

a. A
b. B
c. C
d. D

6.20 An optically pure 2° alcohol with a specific rotation of +24° was converted to the tosylate with *p*-toluenesulfonyl chloride. Subsequent treatment of the tosylate with OH^- ion yielded a 2° alcohol. What is the specific rotation of that alcohol?

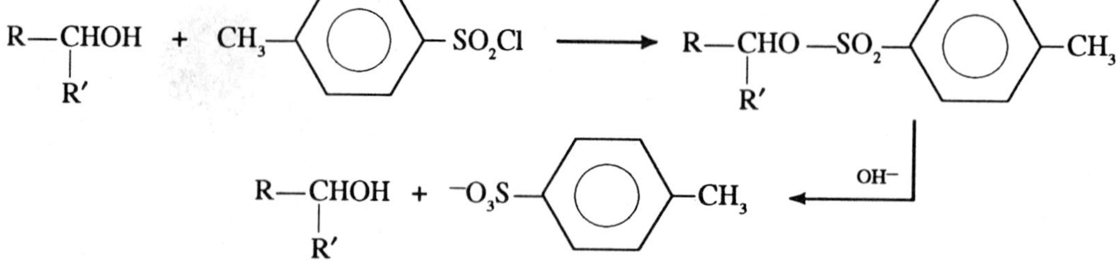

a. +24°
b. –24°
c. 0°
d. +12°

Answers for Questions

6.1 a. *Nucleophilic substitution*: A reaction in which a *nucleophile* (a Lewis base) displaces a *leaving group* (usually a halide ion or other fragment) from a molecule. The general equation is:

$$Nu:^- \;+\; R—L \longrightarrow R—Nu \;+\; :L^-$$

b. *Concerted reaction*: A single-step reaction in which bond making and bond breaking occur simultaneously, as in S_N2 reactions.

c. *Reaction rate*: The amount of reactant consumed per unit time *or* the amount of product formed per unit time.

d. *Aprotic solvent*: A solvent whose molecules lack hydrogens capable of hydrogen-bond formation. Aprotic solvents include acetone, $(CH_3)_2C{=}O$; dimethylsulfoxide (DMSO), $(CH_3)_2S{=}O$; dimethylformamide (DMF), $HC({=}O)N(CH_3)_2$; and benzene, C_6H_6.

e. *Unimolecular reaction*: A reaction in which only one species is involved in the bond making and/or breaking in the activated complex. For instance, the ionization step of the S_N1 reaction:

$$(CH_3)_3C—Cl \longrightarrow (CH_3)_3C^+ \;+\; Cl^-$$

f. S_N1 *reaction*: A unimolecular first-order nucleophilic substitution reaction. The S_N1 mechanism is a nonconcerted, two-step process in which the rate-determining step involves the formation of a carbocation.

g. S_N2 *reaction*: A bimolecular second-order nucleophilic substitution reaction. The S_N2 mechanism is a concerted, one-step process characterized by "backside" nucleophile attack and inversion of configuration, if the reactant is an optically active compound.

6.2
A $n\text{-}C_{12}H_{25}N_3$
B $(C_2H_5)_3C—Cl$
C $(CH_3CH_2)_2S$
D 2NaBr
E $(CH_3)_2CHSCN$
F NaBr
G $3CH_3I$
H H_3PO_3
I $CH_3C(Cl){=}CHCH_2OH$
J 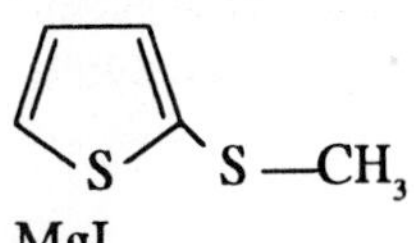
K MgI_2
L Na_2S
M $HC(OC_2H_5)_3$
N 3NaCl
O $C_6H_5O—CH_2CH_2CH_2Br$
P NaBr
Q $(CH_3)_3C—Br$
R H_2O
S PI_3
T H_2SO_4
U $n\text{-}C_{12}H_{25}C{\equiv}N$

V KBr
W HCl or $SOCl_2$ or PCl_3
X NaCN or KCN
Y $n\text{-}C_4H_9Br$
Z $2CH_3CH_2Br$
AA $C_6H_5NH_2$
BB NaCN
CC $CH_3(CH_2)_2CH_2Cl$
DD $Cl—CH_2CH_2CH_2—CN$
EE 
FF AgBr
GG $CH_3CH_2OC({=}O)CH_3$
HH 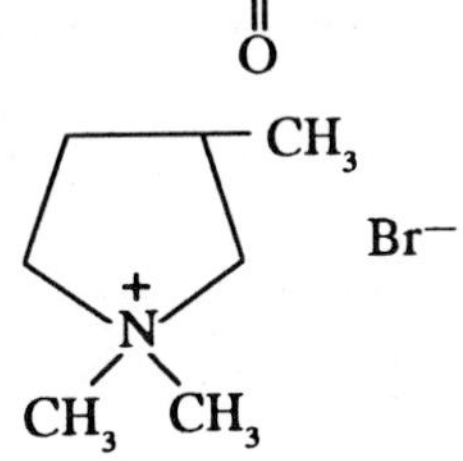

via 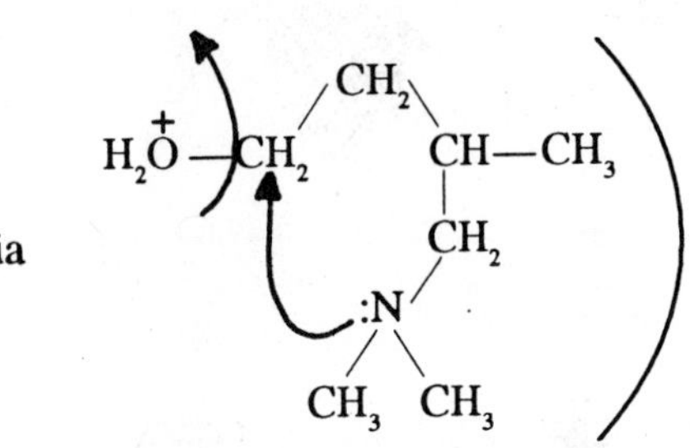

II NaCN or KCN
JJ $n\text{-}C_8H_{17}—^{36}Cl$

6.3

	S_N1	S_N2
a. Mechanism	Two steps (via carbocation)	One step, concerted
b. Mechanism steps	$R—L \rightleftharpoons R^+ + L^-$ $R^+ + Nu^- \longrightarrow R—Nu$	$Nu^- + R—L \longrightarrow Nu—R + L^-$
c. Order of reaction	First	Second
d. Doubling the nucleophile's concentration	No effect on rate	Doubles the rate
e. Doubling the substrate's concentration	Doubles the rate	Doubles the rate
f. Rate equation	Rate = k[R—L]	Rate = k[R—L] [Nu⁻]
g. Favorable medium	Acid (generally)	Base (generally)
h. Stereochemistry of of products	Racemization, often with a considerable amount of inversion	Inversion
i. Solvent effects	Favored by ionizing (polar) solvents	Polarity of solvent (generally) insignificant; hydrogen-bonding solvents decrease the rate, and aprotic solvents increase the rate
j. Carbon at which reaction occurs	Stable carbocation favors reaction: 3° > 2° > 1°	Steric hindrance unfavorable to reaction: 1° > 2° > 3°
k. Competitive reactions	Elimination and rearrangement	Elimination; no rearrangement
l. Catalysis	H^+ (alcohols and ethers); Ag^+ (alkyl halides)	None usually; can sometimes be catalyzed by H^+
m. Leaving group ability	Favored by good leaving groups: for halides $I^- > Br^- > Cl^- > F^-$	Favored by good leaving groups: $I^- > Br^- > Cl^- > F^-$
n. Importance of nucleophile strength	Generally not important, may be weak; solvent is often the nucleophile.	Favored by strong nucleophiles

6.4 a. 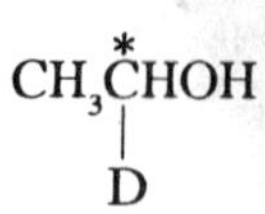

b. 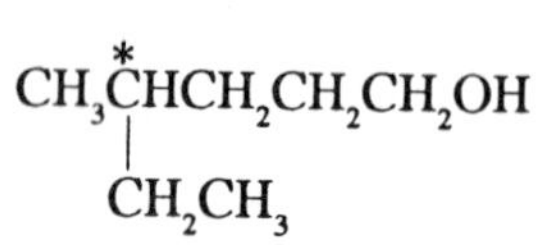

c. 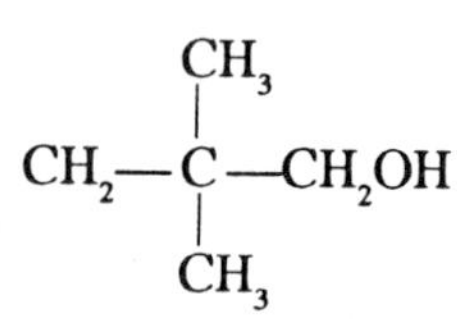

6.5 a. Allyl halides form a stable allylic carbocation (with two contributing resonance structures) and can further react with the nucleophile (Nu⁻)

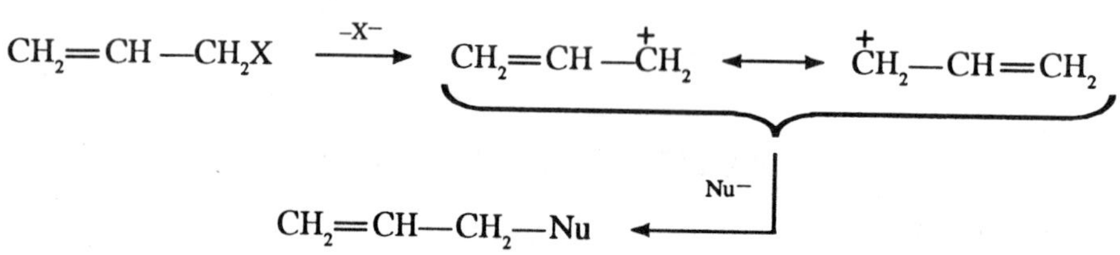

b. In acidic solutions, 2° alcohols rearrange, but 3° alcohols usually do not.

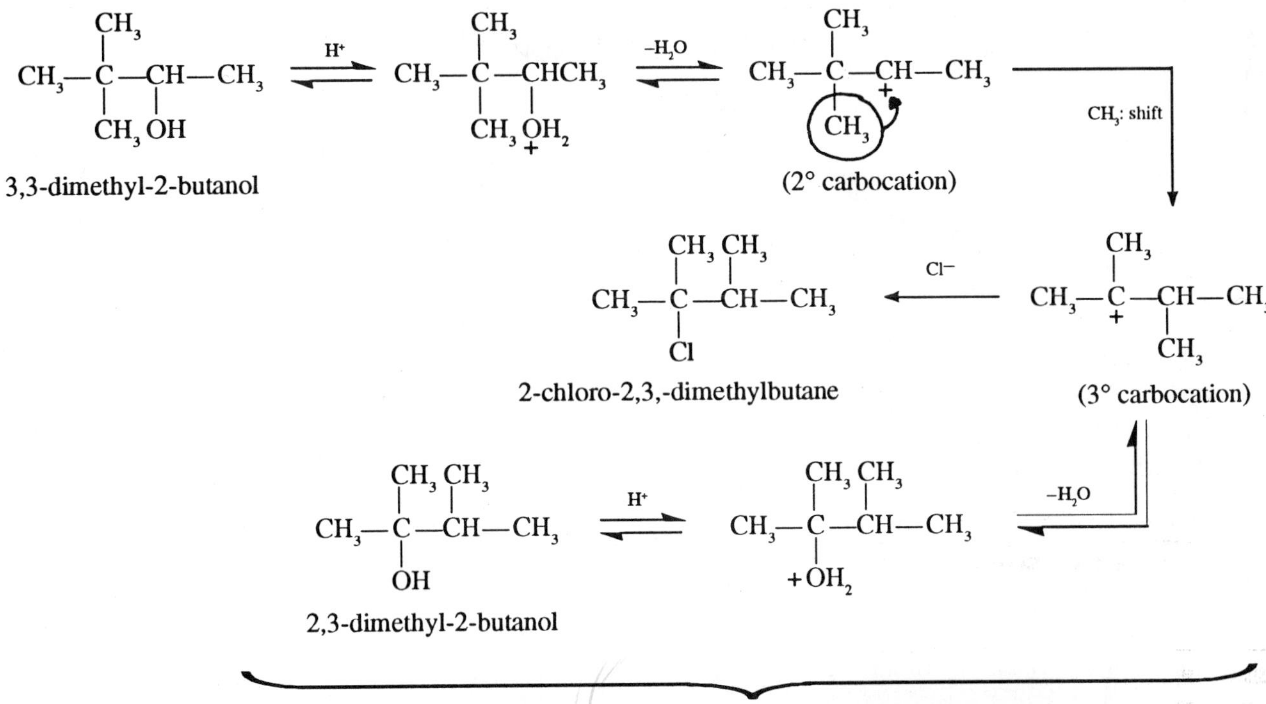

Overall mechanism, illustrates how 2,3-dimethyl-2-butanol and 3,3-dimethyl-2-butanol react with HCl to form the product 2-chloro-2,3-dimethylbutane.

c. Ag^+ assists in the removal of the halide to form the carbocation and the precipitate AgBr. S_N1 reactions involve the formation of the carbocation.

$$R{-}Br + Ag^+ \rightleftharpoons \overset{\delta+}{R}\text{---}Br\text{---}\overset{\delta+}{Ag} \longrightarrow R^+ + AgBr$$

d. Method A involves the attack of a strong base on a 1° alkyl halide in a typical S_N2 reaction. Method B involves the attack of a strong base on a 2° alkyl halide, which results in E^2 elimination as the major path of the reaction.

$$(CH_3)_2CHBr + CH_3ONa \xrightarrow{E^2} \underset{\text{(major)}}{CH_2{=}CH{-}CH_3} + CH_3OH + NaBr$$

2° alkyl halide

$$\xrightarrow{S_N2} \underset{\text{(minor)}}{(CH_3)_2CH{-}OCH_3} + NaBr$$

e. H_2SO_4 is a stronger acid than acetic acid, which in turn is a stronger acid than water. Consequently, in a solvent consisting of 100% H_2SO_4, the proton donor will be H_2SO_4 and acetic acid will serve as the base. In a 50–50 acetic acid and water solution, acetic acid will be the proton donor and water will act as the proton acceptor.

100% H_2SO_4 solution: $\underset{\text{stronger acid}}{H_2SO_4} + CH_3CO_2H \rightleftharpoons CH_3C\overset{+}{O}_2H_2 + HSO_4^-$

50% aqueous acetic acid: $\underset{\text{stronger acid}}{CH_3CO_2H} + H_2O \rightleftharpoons CH_3CO_2^- + H_3O^+$

f. Protonation of the alcohol leads to a planar, sp^2-hybridized carbocation that H_2O can attack from either side (racemization).

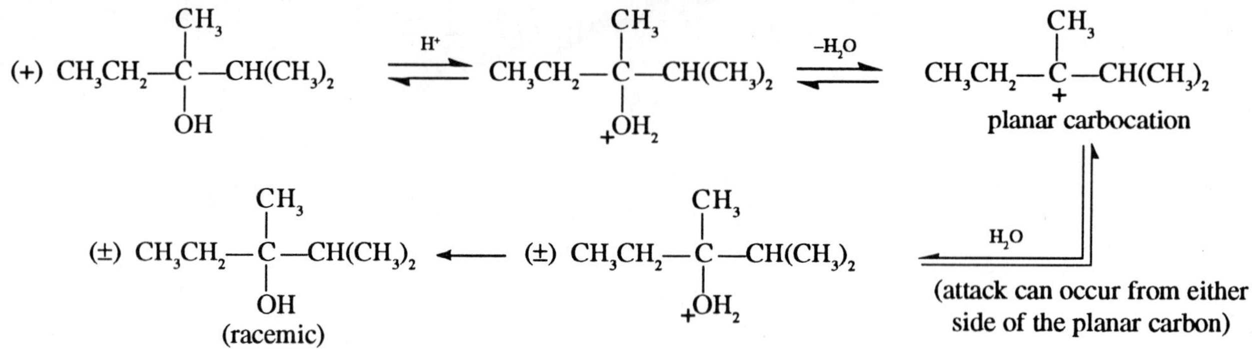

6.6 a.

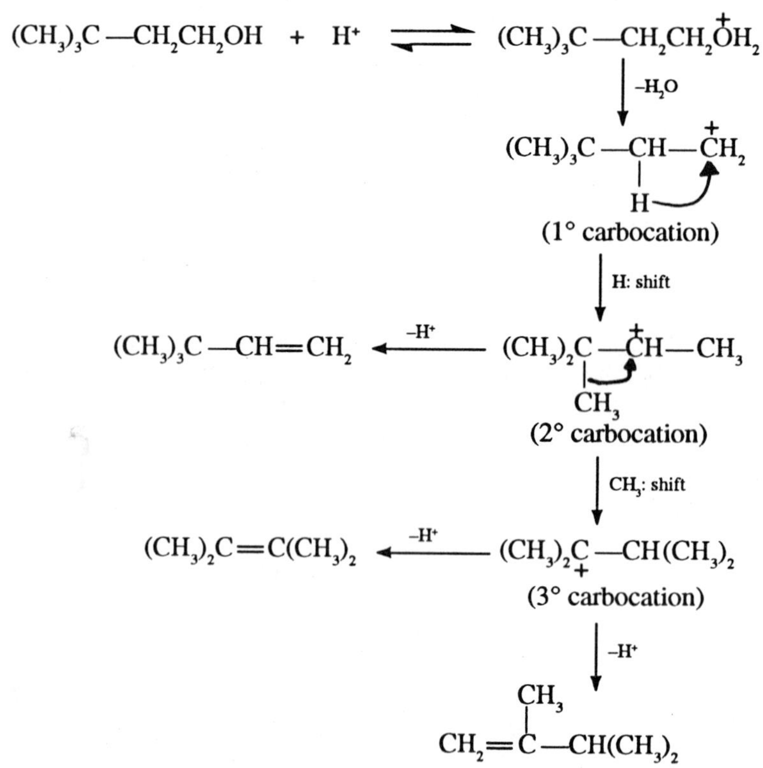

b.

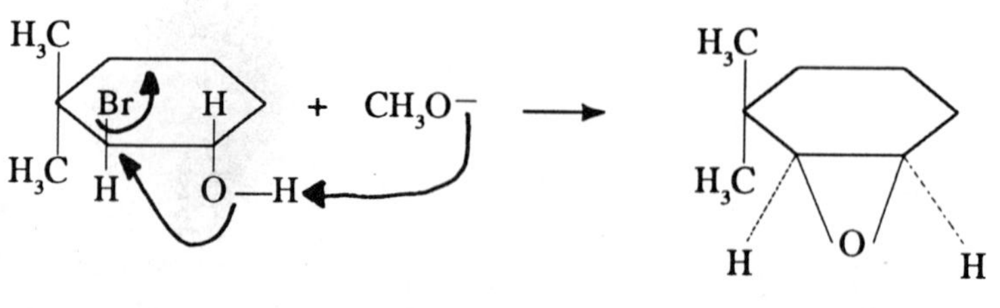

NOTE: For elimination to occur, the carbon atom carrying the bromine atom has to be attacked from the backside as in an S_N2 reaction.

c.

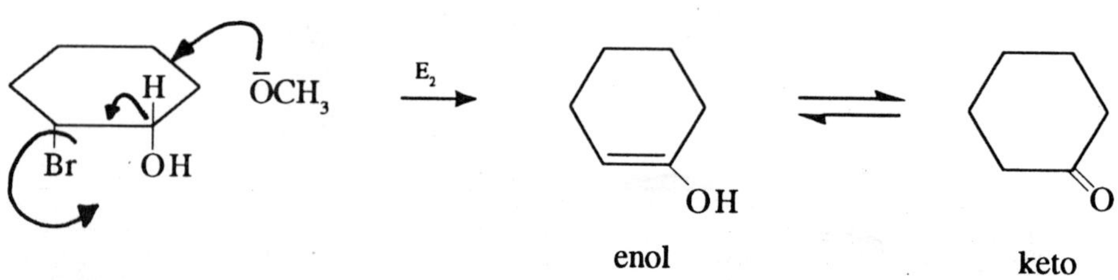

NOTE: In this case, the –OH and –Br are *cis* to each other, and thus elemination as in (b) cannot occur

d. 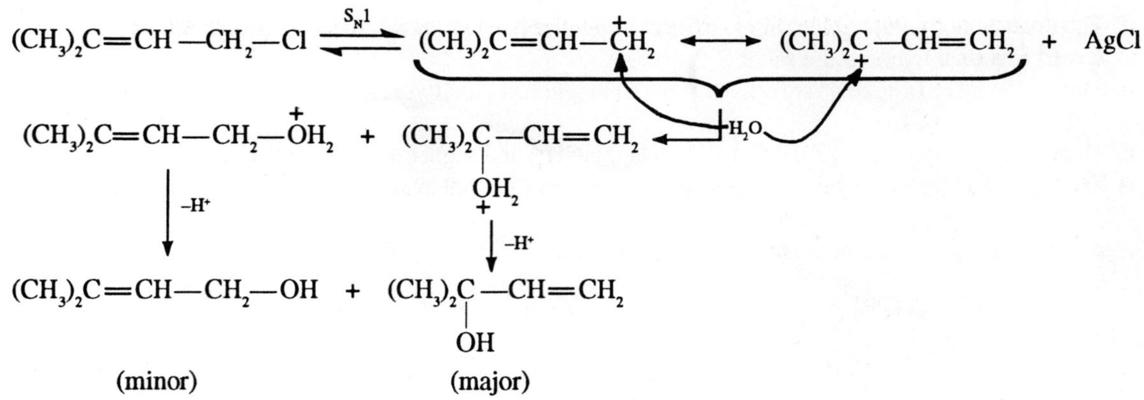

Formation of the allylic carbocation is assisted by Ag^+

e. 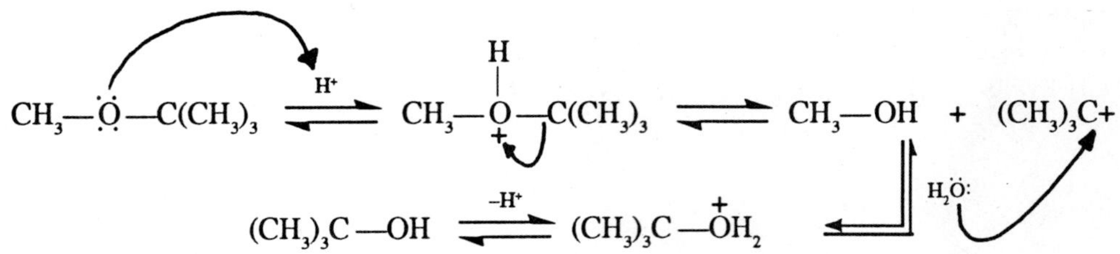

f. 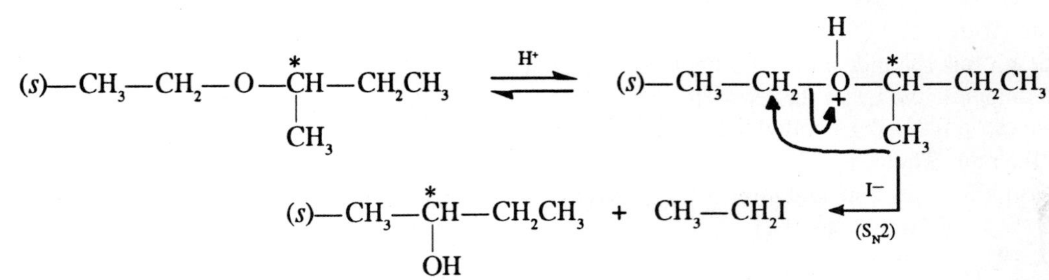

(retention of configuration)

The reaction proceeds with retention of configuration since none of the bonds of the chiral carbon (marked with a *) are broken.

6.7 a. $(CH_3)_2CHCH_2OH \xrightarrow{PBr_3} (CH_3)_2CH{-}CH_2Br$

or $(CH_3)_2CH{-}CH_2{-}OH \xrightarrow{H_2SO_4} (CH_3)_2C{=}CH_2 \xrightarrow[\text{Peroxide}]{HBr} (CH_3)_2CH{-}CH_2Br$

b. 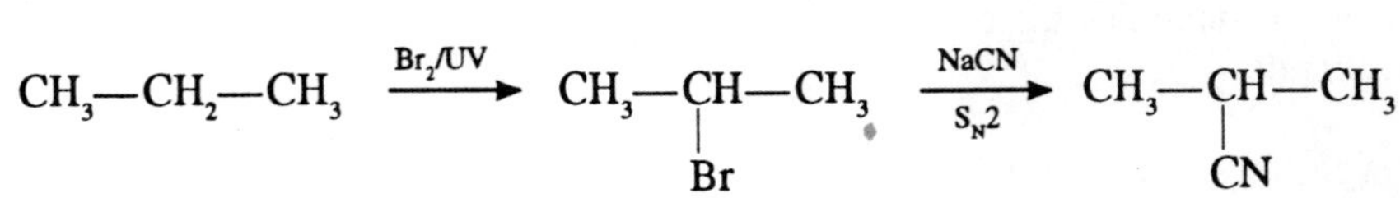

c. $(CH_3)_3CH \xrightarrow{Br_2/UV} (CH_3)_3CBr \xrightarrow[CH_3CH_2OH]{CH_3CH_2O^-Na^+} (CH_3)_2C{=}CH_2 \xrightarrow[\text{2. } NaBH_4]{\text{1. } Hg\,(O{-}\overset{O}{\overset{\|}{C}}{-}CF_3)_2/CH_3OH/THF} (CH_3)_3{-}C{-}OCH_3$

or $(CH_3)_2C{=}CH_2 \xrightarrow[H^+]{H_2O} (CH_3)_3C{-}OH \xrightarrow{NaH} (CH_3)_3CO^-Na^+ \xrightarrow{CH_3I} (CH_3)_3C{-}OCH_3$

d. 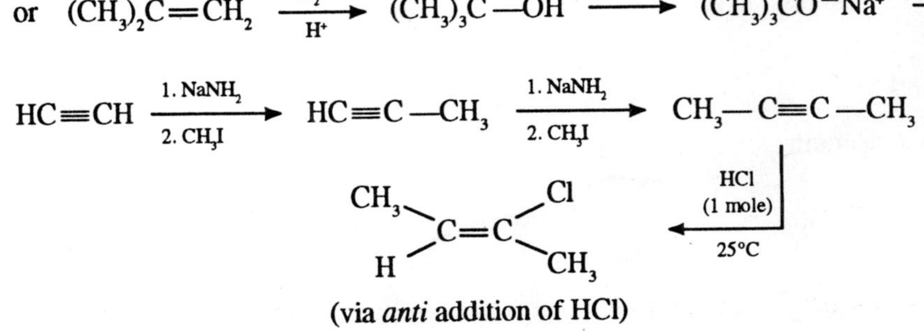

(via *anti* addition of HCl)

6.8 a. The presence of good nucleophiles, which are relatively weak bases (for example, I^- rather than OH^-) and the lowest practical temperature favor S_N2 over E^2.

b. The use of polar (ionizing) solvents, weak nucleophiles (usually the solvent), and 3° substrates favor S_N1 over S_N2.

c. The use of strong bases (KOH, CH_3CH_2ONa, $NaNH_2$) and higher temperatures favor E^2 over S_N2. Sterically hindered bases such as potassium *t*-butoxide in *t*-butyl alcohol are especially useful.

6.9 a. $AgNO_3$/ethanol assists the formation of a carbocation. Therefore, the S_N1 pathway is the most likely. Thus,

$(CH_3)_3C{-}Br$ (3°) > $CH_3CHCH_2CH_3$ with Br on C-2 (2°) > $CH_3CH_2CH_2CH_2Br$ (1°)

b. NaI/acetone favors S_N2 substitution. Thus,

$CH_3CH_2CH_2CH_2Br$ (1°) > $CH_3CHCH_2CH_3$ with Br on C-2 (2°) > $(CH_3)_3C{-}Br$ (3°)

c. This is an S_N1 reaction in which the rate determining step is the loss of the halide ion. Since as leaving groups, $I^- > Br^- > Cl^-$ then $(CH_3)_3C{-}I > (CH_3)_3C{-}Br > (CH_3)_3C{-}Cl$

d. The use of a 3° alkyl halide and a weak nucleophile indicates an S_N1 process. The greater the polarity of the solvent/nucleophile the faster the reaction. Thus, $H_2O > C_2H_5OH > n\text{-}C_4H_9OH$

e. A strong nucleophile and a 1° alkyl halide indicates an S_N2 reaction. Increasing the nucleophile's concentration increases the rate. Thus 2M CH_3O^- > 1M CH_3O^- > 0.5M CH_3O^-

f. This is another S_N2 reaction, and increasing the strength of the nucleophile (the reactant) increases the rate. Thus, $(CH_3)_3CO^- > CH_3O^- > OH^-$

6.10 S_N1 solvolysis proceeds through a carbocation intermediate, and the relative rates reflect the relative ease of carbocation formation. Carbocations formed from **A** and **B** have bond angles of about 60° and 90° respectively, instead of the normal, more stable 120°. Thus, based on the stabilities of the carbocations, the relative rates of S_N1 solvolysis is in the order **C** > **B** > **A**.

6.11 a. Rate = $d[R{-}Nu]/dt$
$= k_3[R^+][Nu^-]$

To write a rate expression using measurable concentrations, we make a *steady state assumption* when substituting in the concentration of R^+: We assume that the concentration of R^+ is *constant* (i.e., the rate of its formation is equal to the rate of its consumption). Thus,

$$\frac{d[R^+]}{dt} = 0 = k_1[RCl] - k_2[R^+][Cl^-] - k_3[R^+][Nu^-]$$

From the above equation,

$$[R^+] = \frac{k_1[RCl]}{k_2[Cl^-] + k_3[Nu^-]}$$

Substitution in the original rate expression gives

$$\text{Rate} = \frac{k_1 k_3[RCl][Nu^-]}{k_2[Cl^-] + k_3[Nu^-]}$$

Dividing both numerator and denominator by $k_3[Nu^-]$

$$\text{Rate} = \frac{k_1[RCl]}{k_2[Cl^-]/k_3[Nu^-] + 1}$$

b. If $k_3 >>> k_2$, the rate simplifies to:
Rate = k_1[RCl], since k_2/k_3 approaches 0.
This requires that step II be much faster than step I, and that essentially as soon as the carbocation R^+ is formed, it reacts to form the product, R—Nu.

c. (i) Cl^- will slow the reaction via the common ion effect. (ii) Br^- is not a common ion and will not behave like Cl^- (in fact, it will increase the rate by increasing the polarity of the medium). (iii) A more polar solvent assists ionization and, therefore, increases the rate. (iv) The strength of the nucleophile has no effect on S_N1 rates because the carbocation intermediate is so reactive (in fact, strong nucleophiles are usually strong bases and are involved more in E^2 reactions).

6.12 Primary alkyl halides form stable carbocation in which the electron pair of oxygen is involved:

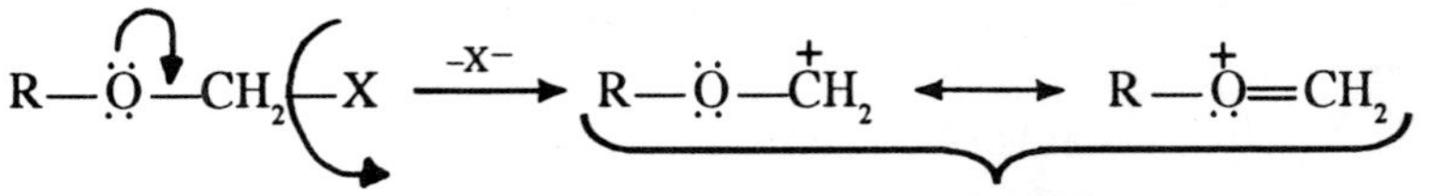

Resonance-stabilized carbocation

6.13 The S_N2 mechanism requires a backside nucleophilic attack. Only in the *trans* configuration is the carboxylate anion *anti* to the bromide.

6.14 The reaction of anisole and HI is an S_N2 type:

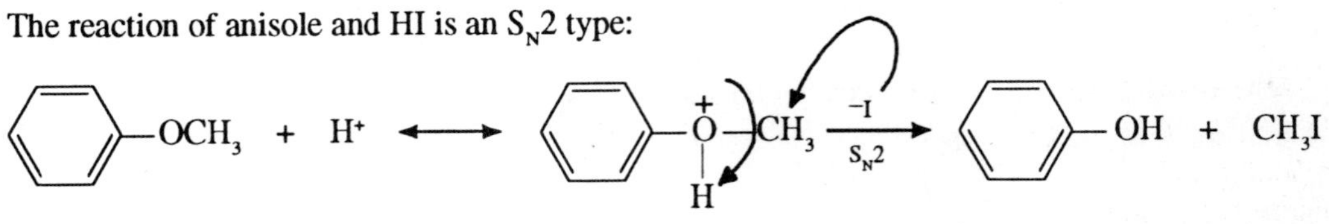

Backside attack of I^- cannot occur at the benzene ring but does take place at the methyl group.

6.15 a. Although the reaction is S_N2-like, it is only a first order reaction because the slow step involves only the substrate and not the nucleophile.

b. (i) Higher concentration of hydroxide ion would have no effect. (ii) A less polar solvent would slow the reaction by retarding ionization. (iii) Cl^- would slow the reaction by the common ion effect. (iv) I^- would have no effect because the rate depends on the first step. (v) The rate would increase because the kinetic energy increases with temperature.

c. The reaction is assisted by nucleophilic attack of the properly positioned electron pair on a neighboring nitrogen atom. The effect is sometimes called *anchimeric assistance* or *neighboring group effect*.

6.16 The carbanion postulated in the reversible first step would react with D_2O to incorporate deuterium in 2-bromobutane, ie,

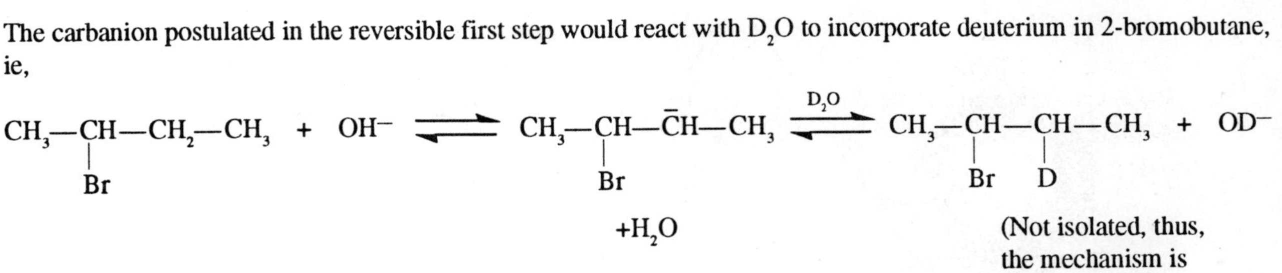

(Not isolated, thus, the mechanism is discredited.)

6.17 Breakage of the C—H bond is part of the slow step. This fact, along with the information (from 6.16) that there is no hydrogen-deuterium exchange, suggests a one-step concerted mechanism

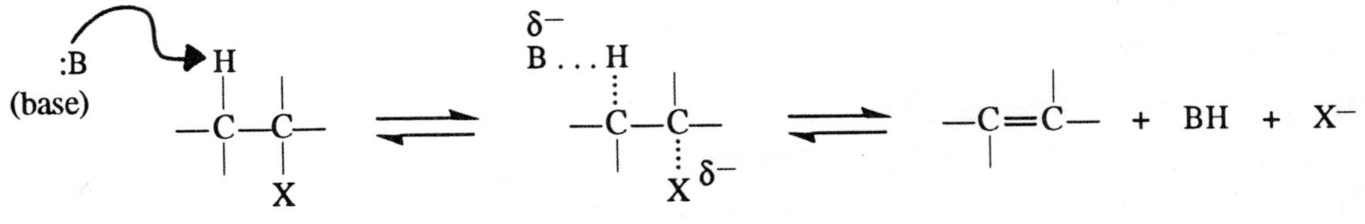

6.18

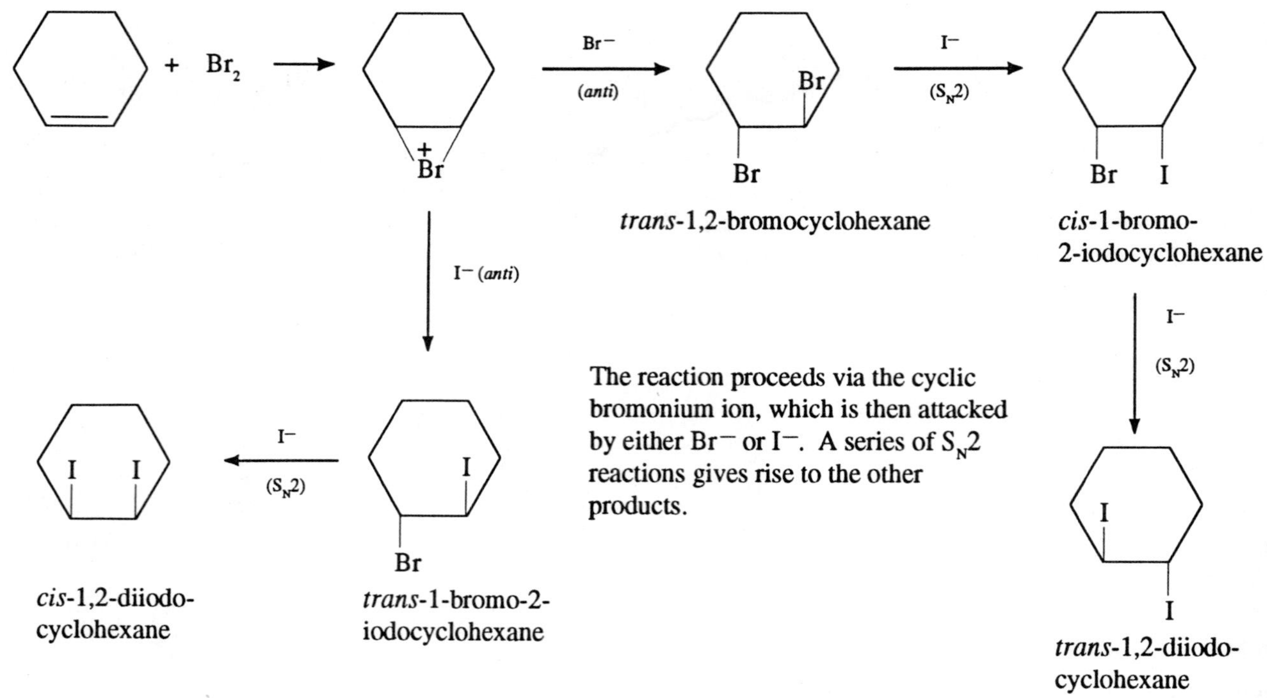

The reaction proceeds via the cyclic bromonium ion, which is then attacked by either Br^- or I^-. A series of S_N2 reactions gives rise to the other products.

6.19 In S_N1 reactions, ionization of the substrate to the carbocation results in the same transition state for both axial and equatorial leaving groups.

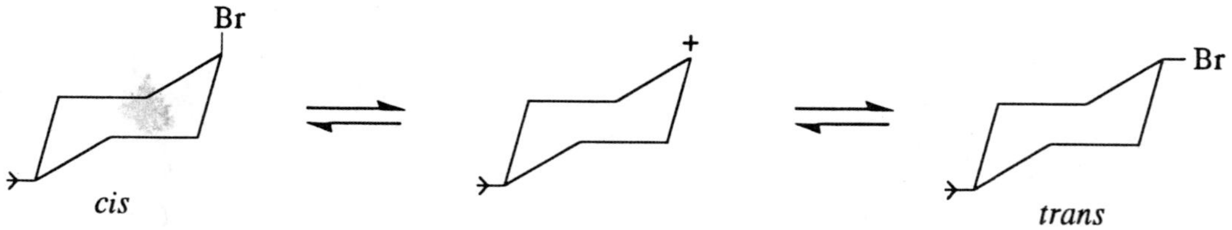

Since the transition states are the same, to compare axial and equatorial *leaving groups*, we must examine the relative energies of the starting molecules. The energy of the equatorial conformer is lower than that of the axial. This results in a larger energy of activation for equatorial isomers, and thus in slower reactions.

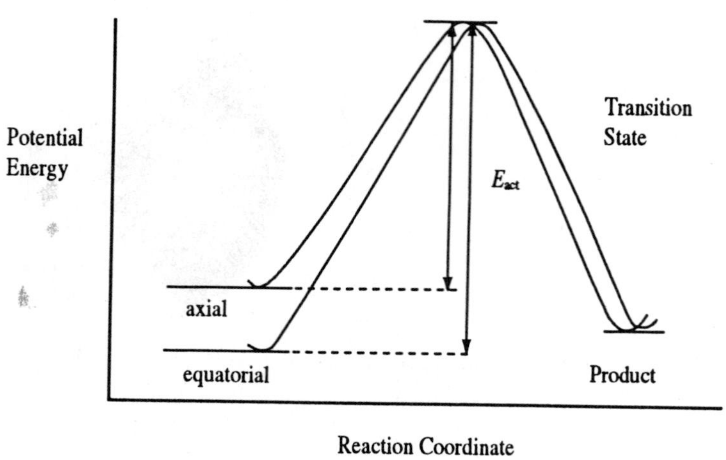

6.20

$$CH_3{-}CH(CH_3){-}CH_2{-}CH_2{-}Br > CH_3{-}CH_2{-}CH(CH_3){-}CH_2{-}Br > CH_3{-}CH_2{-}CH_2{-}CH(CH_3){-}Br > CH_3{-}CH_2{-}C(CH_3)_2{-}Br$$

The least hindered alkyl halide reacts fastest.

6.21

$$(CH_3)_3C{-}I > (CH_3)_3C{-}Br > CH_3{-}CH_2{-}CH(CH_3){-}I > CH_3{-}CH_2{-}CH(CH_3){-}Br$$

I^- is a better leaving group than Br^- and 3° alkyl halides form more stable carbocations than do 2° alkyl halides.

6.22 DMSO is an *aprotic* solvent, which does not hydrogen bond to the nucleophile. Ethanol and water hydrogen bond to the nucleophile and deactivate it. S_N2 reactions are many times faster in aprotic solvents than in alcohol solvents.

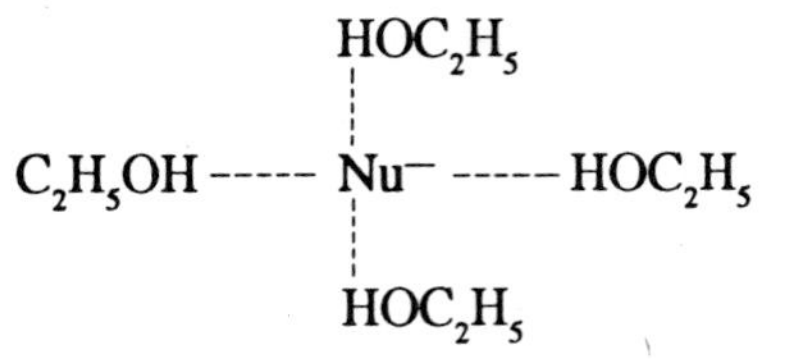

6.23 a. The starting materials are enantiomers; the products are (2*R*,3*S*)-2-iodo-3-phenylbutane and (2*S*,3*R*)-2-iodo-3-phenylbutane. They are also enantiomers.
b. Enantiomers react at the same rate with achiral reagents.
c. Enantiomers cannot be separated by distillation.

Answers for Multiple Choice Test

6.1 c	**6.6** d	**6.11** b	**6.16** b
6.2 b	**6.7** d	**6.12** b	**6.17** c
6.3 c	**6.8** d	**6.13** a	**6.18** d
6.4 a	**6.9** c	**6.14** a	**6.19** c
6.5 c	**6.10** a	**6.15** c	**6.20** b

7 Benzene and Aromaticity

Questions

7.1 Give structures for the following compounds:

a. *m*-nitrobenzenesulfonic acid
b. benzyl chloride
c. *p*-aminobenzoic acid
d. allylbenzene
e. (*E*)-3-(4-chlorophenyl)4-hexen-1-yne
f. vinylbenzene
g. 3,5-dinitroacetophenone
h. phenylacetylene
i. *m*-nitrosobenzaldehyde
j. *o*-chloroanisole
k. 2,4,6-trinitrophenol (or picric acid)
l. *m*-xylene
m. *p*-aminobenzaldehyde
n. benzene sulfonamide

7.2 Give structures for the following polynuclear aromatics and aromatic heterocyclic compounds:

a. 2-chloronaphthalene
b. 9-bromophenanthrene
c. 3,5-dinitropyridine
d. 3-methoxyfuran
e. 9,10-dichloroanthracene
f. 3-nitrosopyrrole
g. 2,5-diethylthiophene
h. dibenzofuran
i. 3,6-di(chloromethyl)-1-nitronaphthalene
j. dibenzothiophene
k. N-methylpyrrole
l. pyridine N-oxide
m. 2-acetylfuran

7.3 Name the following compounds:

a.
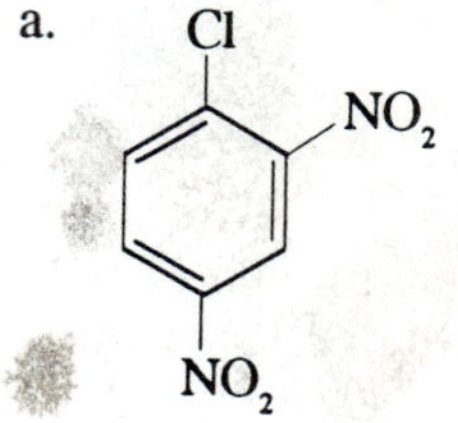

c.
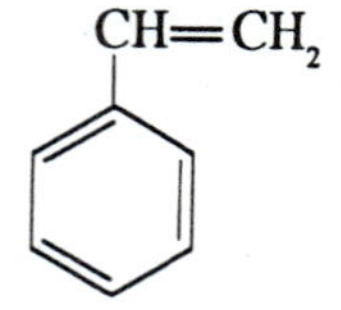

e.
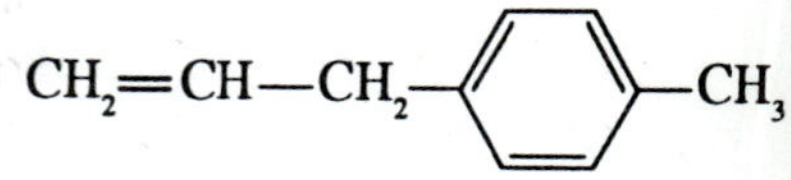

b.
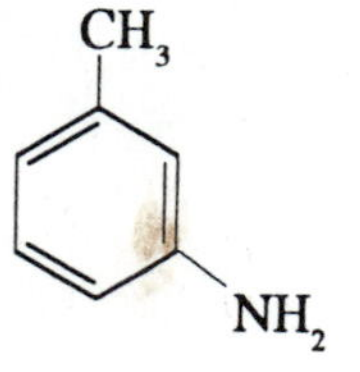

d.
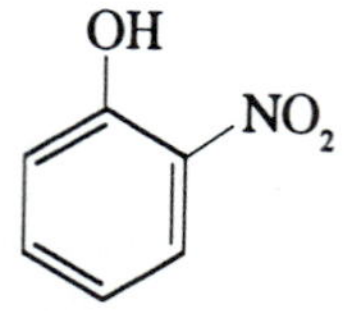

f.
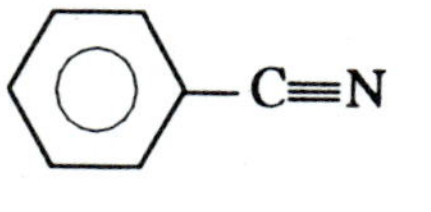

g. 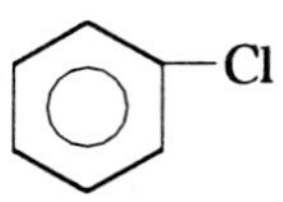

l. 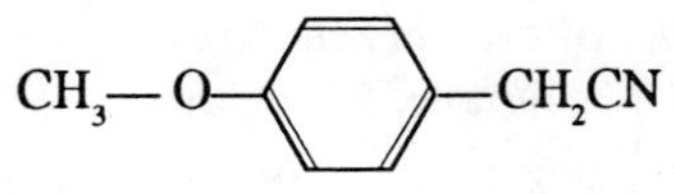

q. 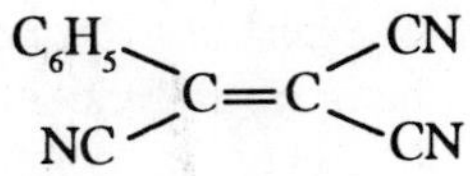

h. 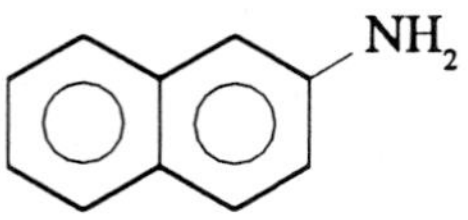

m. 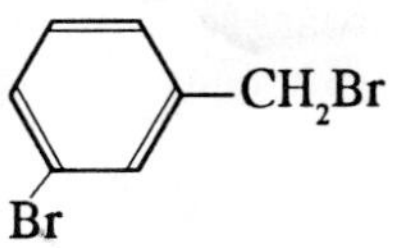

r. 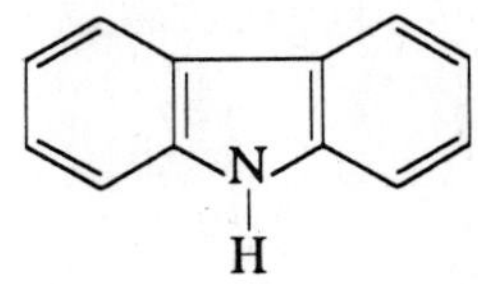

i. 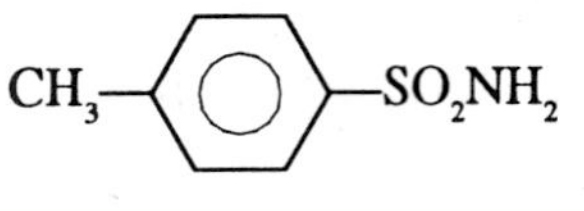

n. 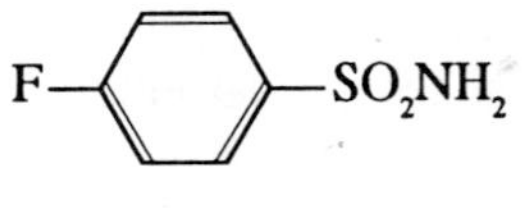

s. 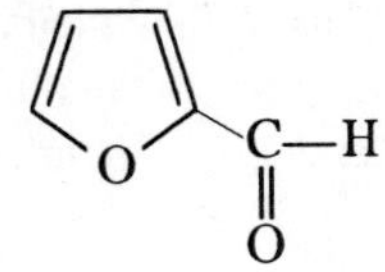

j. 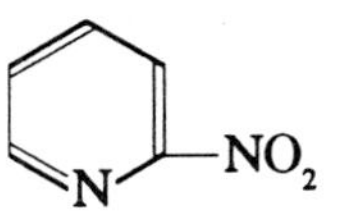

o. 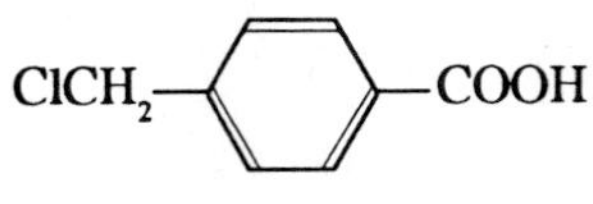

t. 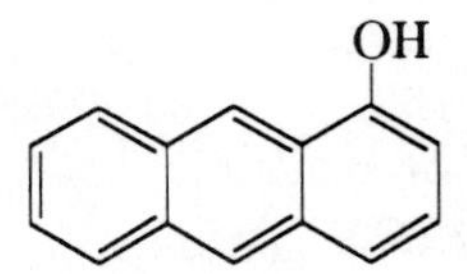

k. 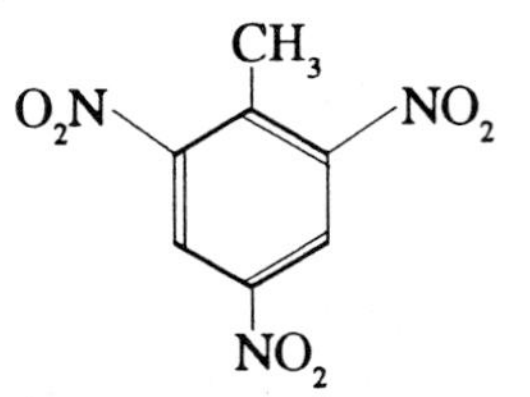

p. 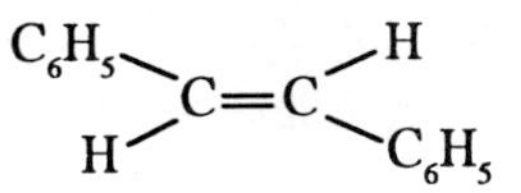

u. 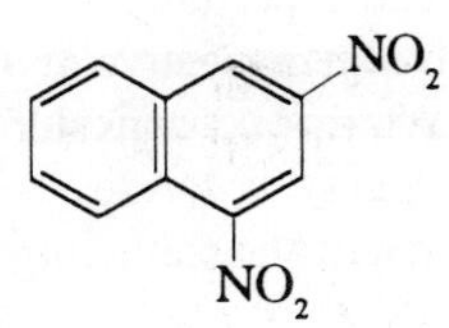

Note: The term mononitro derivative that appears in the next two questions refers to all benzene-nitrated compounds (isomers) possible, not just compounds that can be synthesized by the nitration (HNO_3/H^+) reaction.

7.4 How many mononitroderivatives are possible for
a. benzyl bromide
b. 1,2-dibromobenzene
c. *m*-xylene
d. *p*-bromotoluene
e. 2-chlorobromobenzene
f. cumene (2-phenylpropane)

7.5 Using the molecular formulas, give structures for the following benzene derivatives:
a. (C_8H_{10}) that can have only one mononitro derivative
b. (C_8H_{10}) that can have two possible mononitro derivatives
c. (C_8H_{10}) that can have three possible mononitro derivatives
d. (C_9H_{12}) that can have only one mononitro derivative
e. (C_9H_{12}) that can have two possible mononitro derivatives
f. (C_9H_{12}) that can have three possible mononitro derivatives
g. (C_9H_{12}) that can have four possible mononitro derivatives

7.6 Three tribromobenzenes are known with melting points of 44°C, 87°C and 119°C. When nitrated separately, they yield three, two, and one mononitro derivatives respectively. Give the structures of the three tribromobenzenes and assign the melting point of each.

7.7 Fill in the blanks in the paragraph below.

The ________ π-electrons present in benzene combine to form ________ molecular orbitals, of which ________ are bonding and ________ antibonding. The lowest energy molecular orbital has ________ nodal planes while the highest energy antibonding molecular orbital has ________. The *p* electrons completely fill the bonding orbitals in accordance with the ________ principle and satisfy Hückel's ________ rule.

7.8 List the criteria for aromaticity.

7.9 a. The heat of hydrogenation ($\Delta H°$) of cyclohexene to cyclohexane is –28.4 kcal/mole. The corresponding value for benzene to cyclohexane is –49.8 kcal/mole. Using these values, calculate an empirical resonance energy for benzene.

b. $\Delta H°$ for cyclooctene to cyclooctane is –23.3 kcal/mole, while that for 1,3,5,7-cyclooctatetraene to cyclooctane is –100.9 kcal/mole. Calculate an empirical resonance energy for 1,3,5,7-cyclooctatetraene.

1,2,5,7-cyclooctatetraene

c. Theoretical heats of combustion can be calculated by adding up the contributions of the different types of bonds that are present in the molecule. Using the values below, calculate a heat of combustion for both Kekulé benzene and 1,3,5,7-cyclooctatetrene.

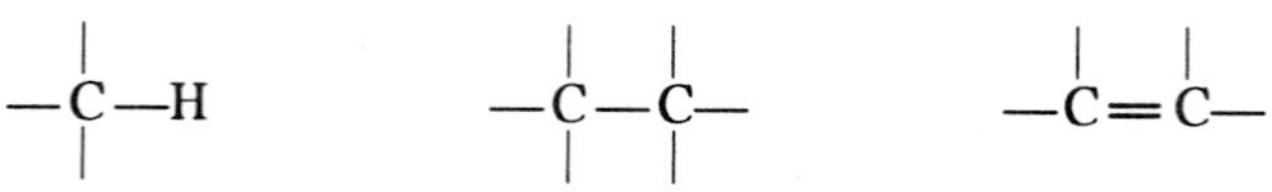

54 kcal/mole   49.3 kcal/mole   117.4 kcal/mole

d. The experimental heats of combustion for benzene and cyclooctatetraene are –789.1 kcal/mole and –1099 kcal/mole, respectively. Compare these values with those calculated above.

e. Would you consider cyclooctatetraene to show aromatic behavior?

7.10 Which of the following compounds are aromatic? (Consider possible resonance forms that fit the $4n+2$ rule.)

a.
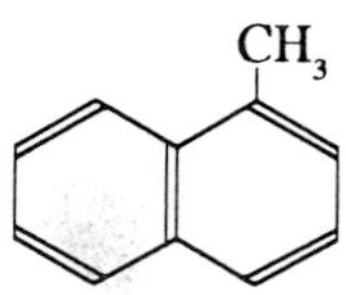

b.
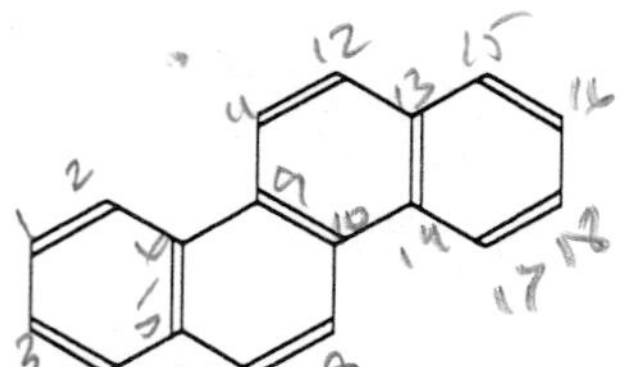

c.
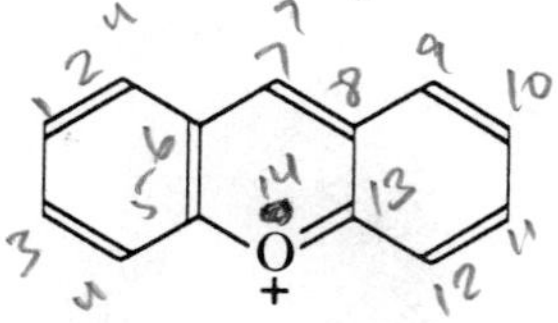

d.

e.
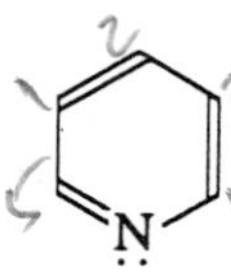

f.
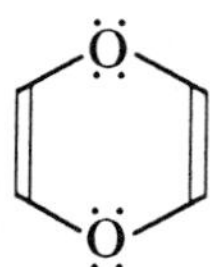

g.
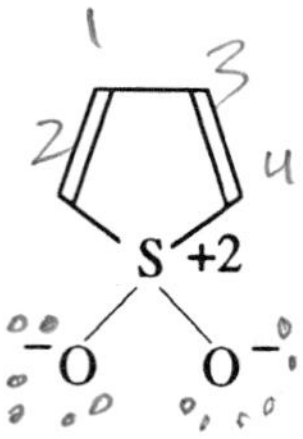

h.
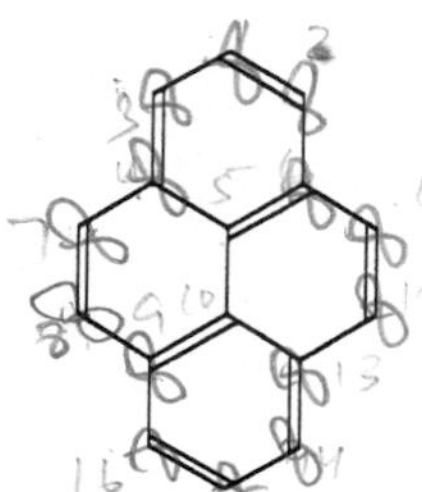

i.
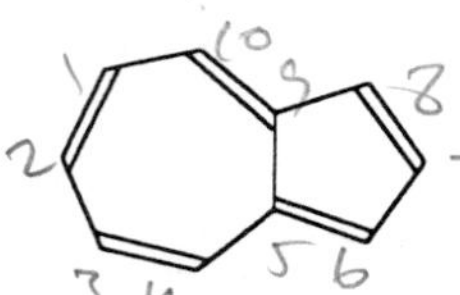

j.
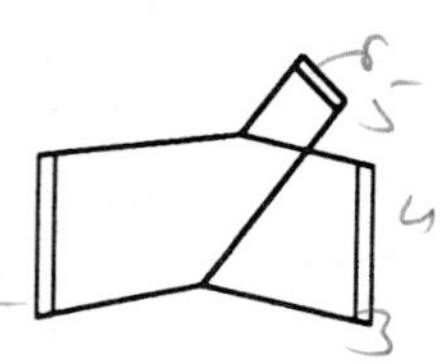

k.

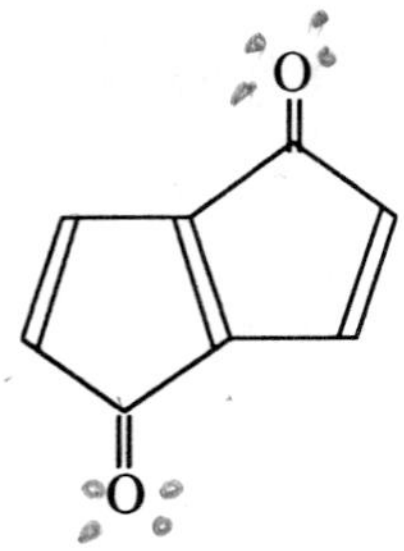

p.

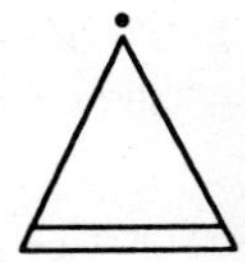

l.

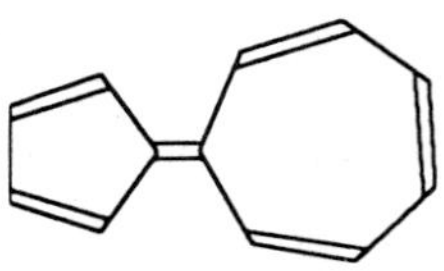

q.

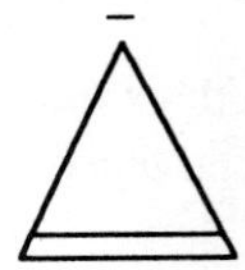

m.

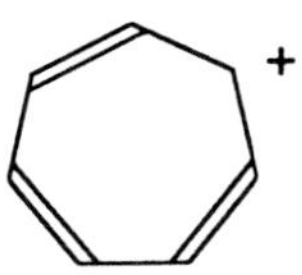

r.

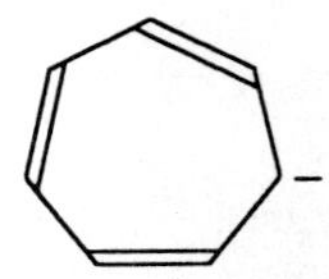

n.

s.

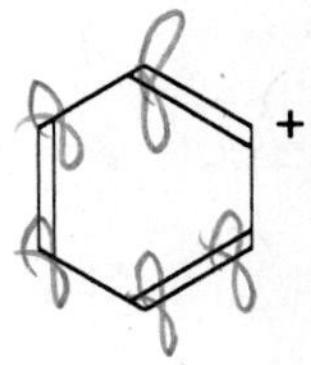

o.

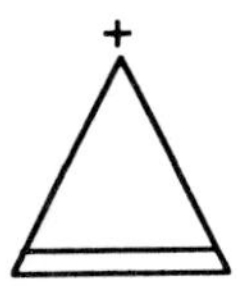

t.

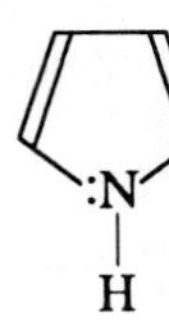

7.11 Why is pyrrole a much weaker base than pyrrolidine.

pyrrole

pyrrolidine

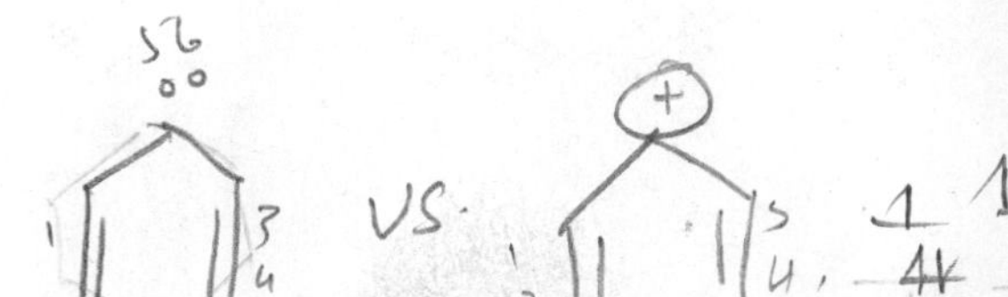

7.12 Compare the relative stabilities of the cyclopentadienyl anion and the cyclopentadienyl cation.

7.13 [10]Annulene, $C_{10}H_{10}$, satisfies the $4n+2$ rule, yet it does not show aromatic properties. Explain.

7.14 Diphenylcyclopropenone has a greater dipole moment and is more basic than benzophenone. Explain.

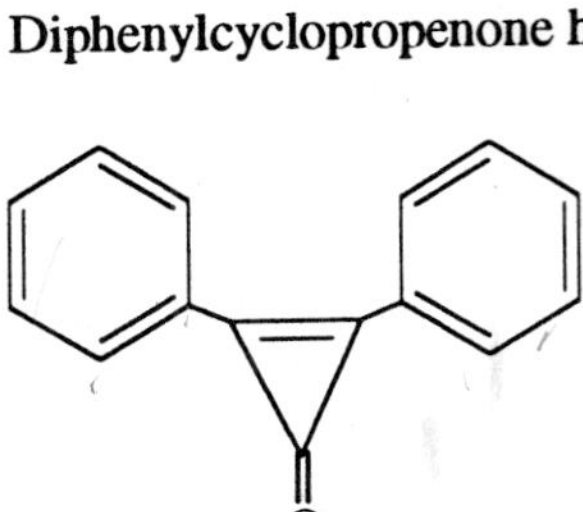

diphenyleyclopropenone

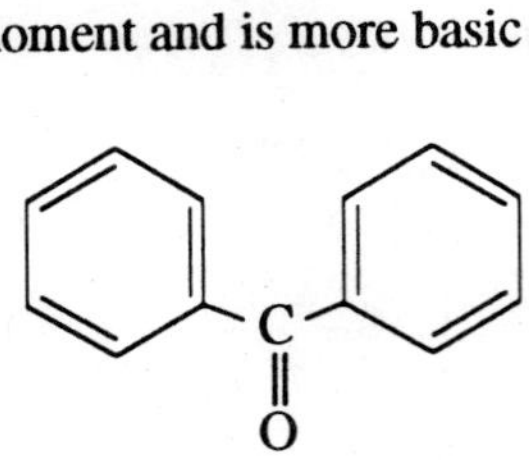

benzophenone

7.15 The blue compound azulene shown below has a dipole moment. In acidic medium, only one ring gets protonated. Which one? Account for the existence of a dipole moment.

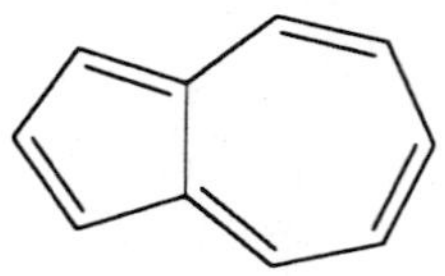

azulene

7.16 Give all possible Kekulé structures for
a. naphthalene b. anthracene c. phenanthrene

7.17 a. Propose a mechanism and calculate ΔH° for the radical chlorination of benzene:

$$C_6H_6 + Cl_2 \xrightarrow{h\nu} C_6H_5Cl + HCl$$

Bond Dissociation Energies:
C_6H_5—H = 112 kcal/mole C_6H_5—Cl = 86 kcal/mole
Cl_2 = 58 kcal/mole HCl = 103 kcal/mole

b. Radical chlorination is not a good method for the preparation of chlorobenzene. Why?

7.18 The dipole moment of compound **A** is 6.870 D.
a. Predict the dipole moment of compound **B**.
b. Would you expect the dipole moment of **C** to be larger, smaller, or the same as that of **A**?
c. Actually, the dipole moment of **C** is only 4.110 D. Suggest a reason to explain why it is smaller than that of **A**.

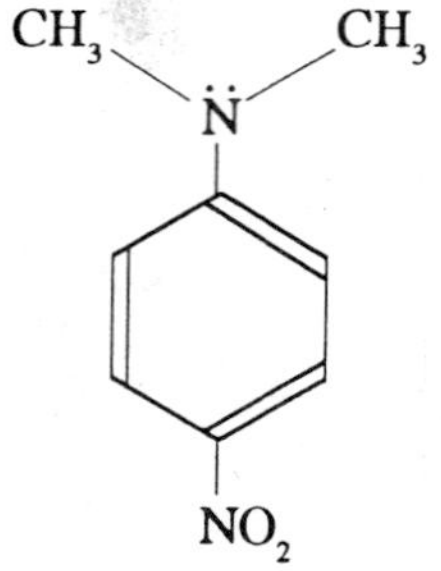

A

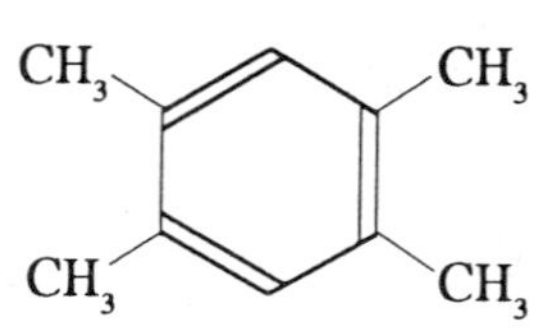

B

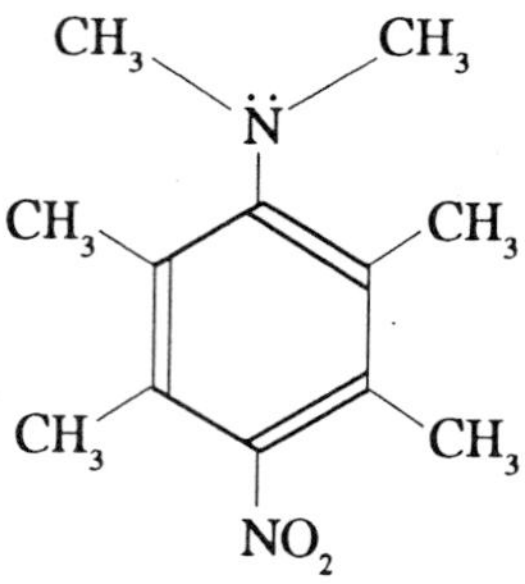

C

7.19 Explain why
a. *p*-Methoxybenzyl bromide is hydrolyzed much easier than *p*-nitrobenzyl bromide under S_N1 conditions.
b. Friedel-Crafts *alkylation* of benzene with 2-bromopropane requires excessive amounts of benzene. On the other hand, Friedel-Crafts *acylation* of the same compound with propanoyl chloride, CH_3CH_2COCl, does not need excess benzene. Both routes, however need a catalytic amount of aluminum or ferric chloride.
c. *n*-butylbenzene can be synthesized using either of the following methods:

Method 1 $$C_6H_6 + CH_3(CH_2)_2CH_2Cl \xrightarrow{AlCl_3} C_6H_5{-}CH_2(CH_2)_2CH_3$$

Method 2 $$C_6H_6 + CH_3(CH_2)_2C({=}O){-}Cl \xrightarrow{AlCl_3} C_6H_5{-}C({=}O)(CH_2)_2CH_3 \xrightarrow{H_2/\text{Catalyst}} C_6H_5{-}CH_2(CH_2)_2CH_3$$

Method 2 however, leads to much higher yields even though it is a two-step synthesis.

d. Nitrobenzene can occasionally be used as a solvent in Friedel-Crafts alkylations.
e. Direct nitration of aniline yields, among other products, *m*-nitroaniline.
f. 2,4-Dinitrochlorobenzene undergoes nucleophilic aromatic substitution with aniline. Aniline, however, does not react with 3,5-dinitrochlorobenzene.
g. Nitration of naphthalene produces 1-nitro (and not 2-nitro) naphthalene as the major product.
h. Benzene undergoes bromination only when a Lewis catalyst (such as ferric chloride) is present, and yields bromobenzene. On the other hand, phenanthrene adds Br_2 and yields 9,10-dibromophenanthrene.
i. Electrophilic aromatic substitution of chlorobenzene leads to *o*- and *p*-substituted products.

7.20 Suggest plausible mechanisms that would explain the formation of the indicated products. Use arrows to show the movement of electrons.

a.
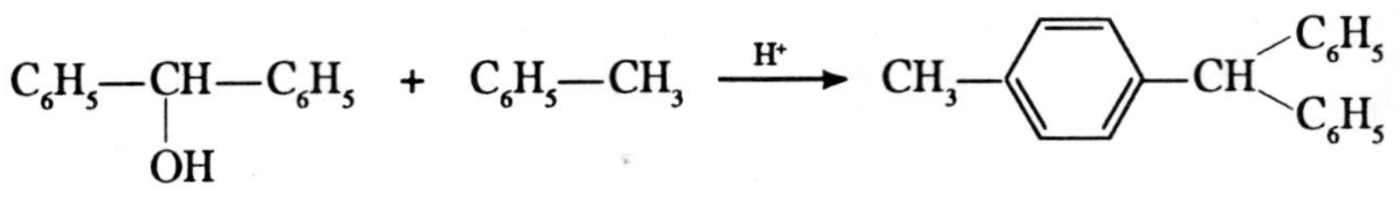

b.
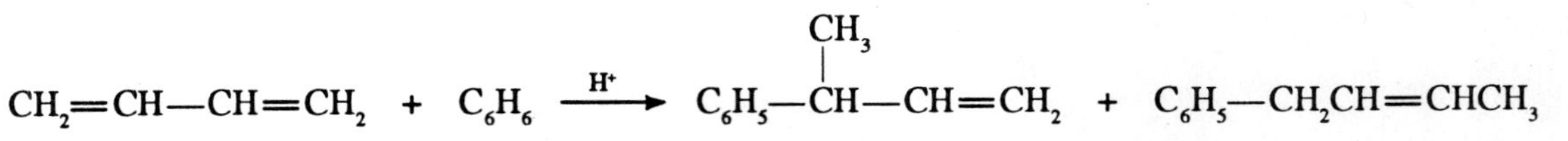

c.
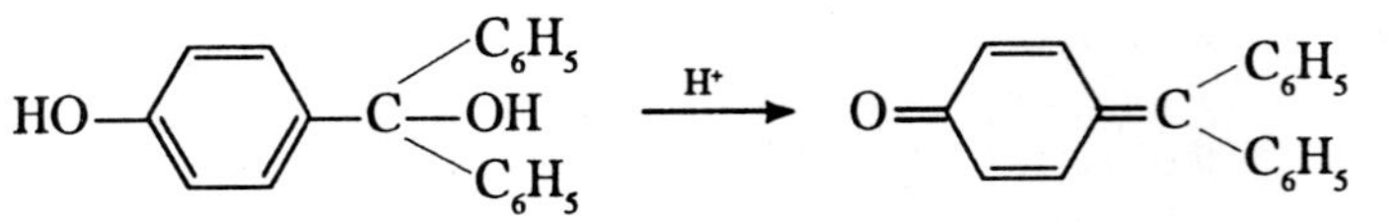

d.
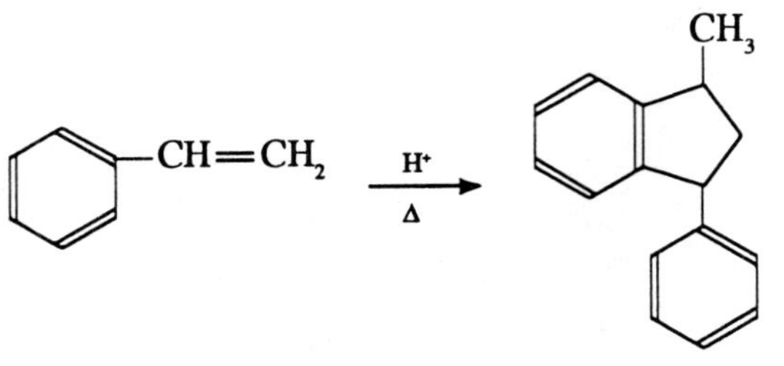

e.
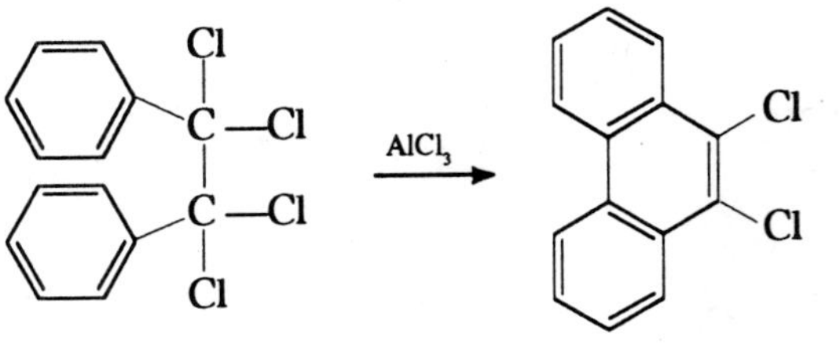

f.
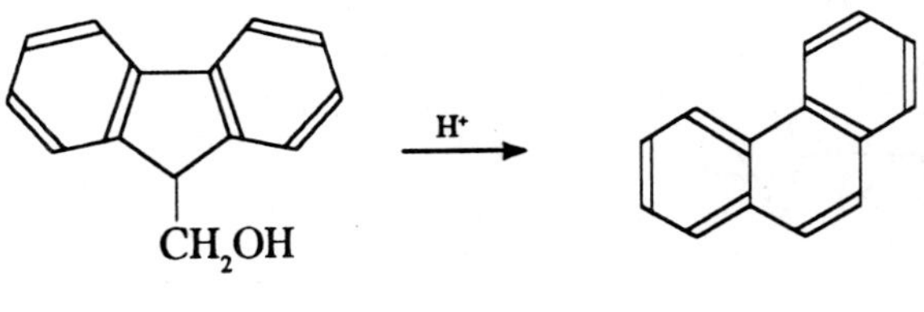

g.
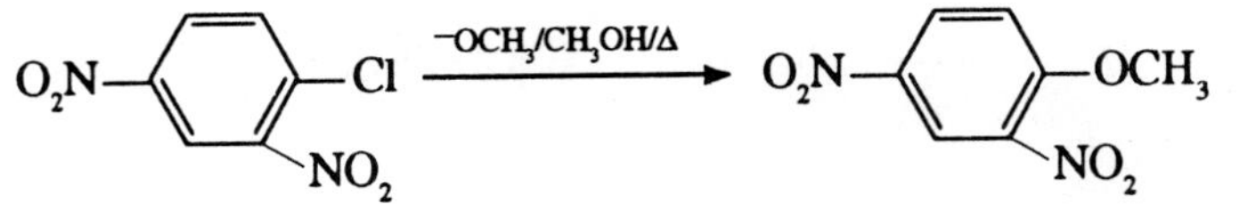

7.21 Indicate clearly with an arrow the *favored* position of substitution in the nitration of the following compounds

a.
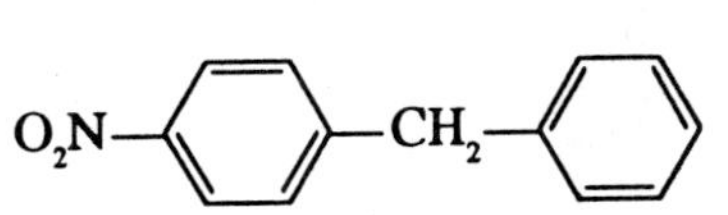

b.
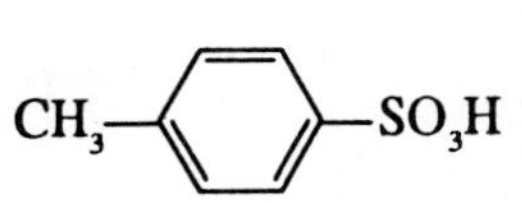

c.

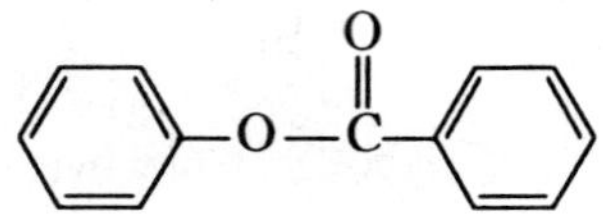

d.

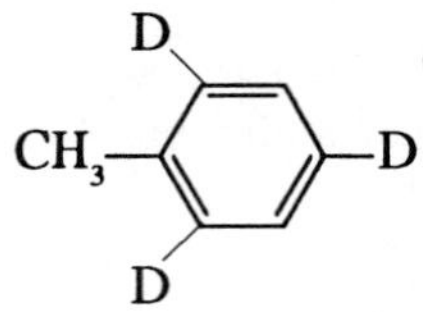

(D=deuterium)

e.

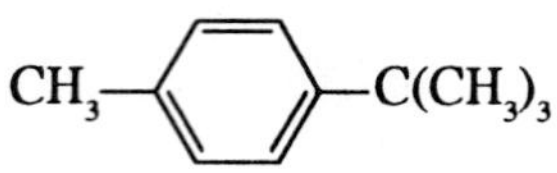

f.

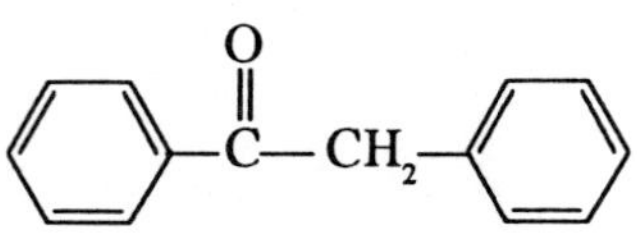

g.

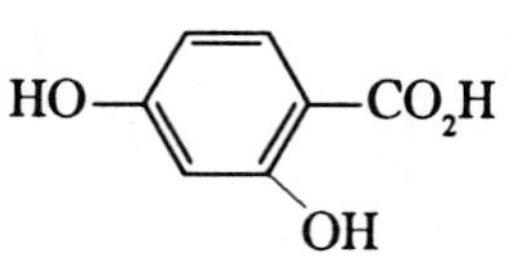

h.

i.

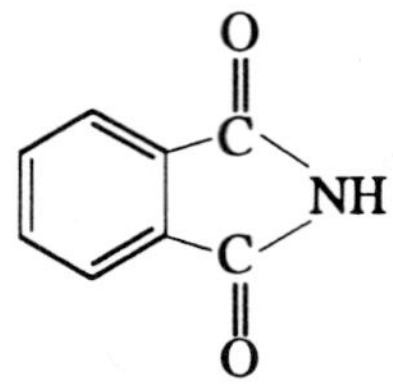

j.

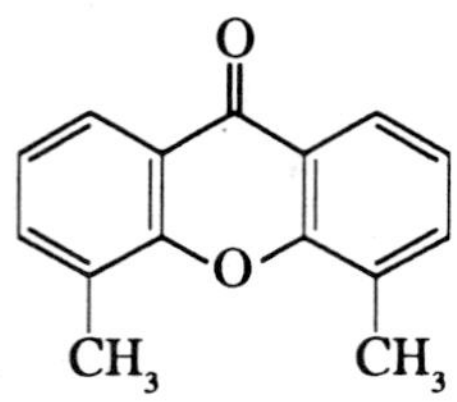

k.

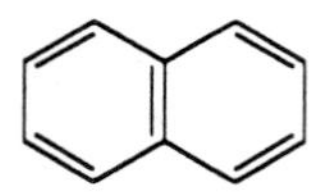

l.

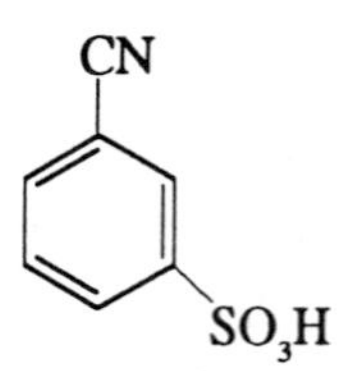

7.22 Identify compounds **A** through **U** in the sequence of reactions below.

a.

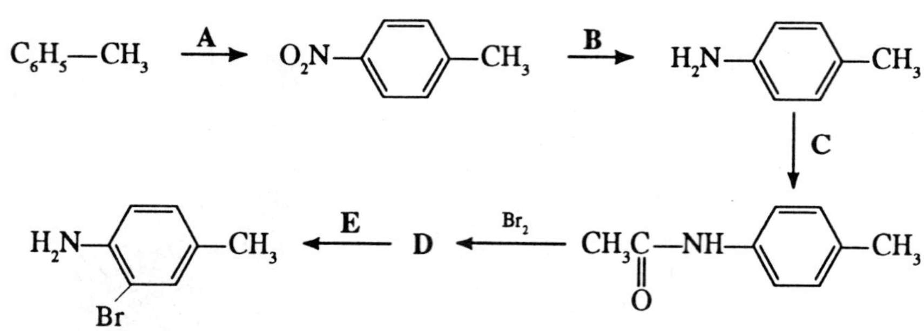

b.

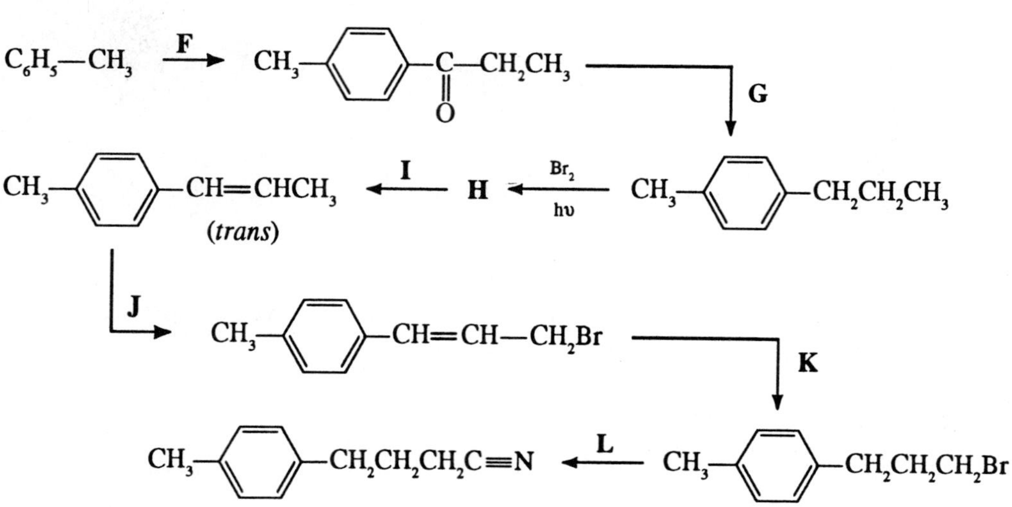

c.

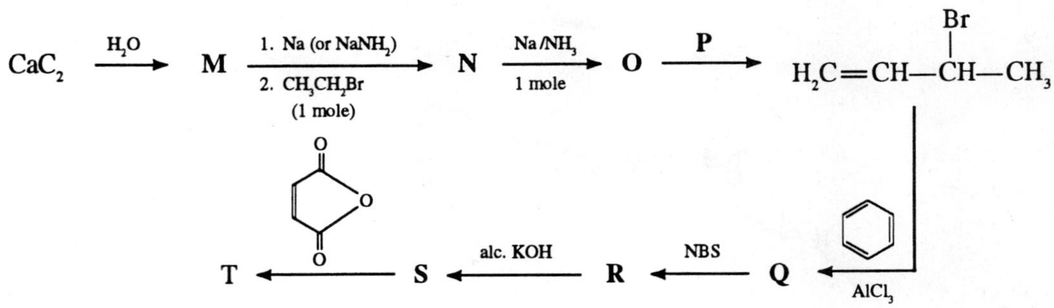

d.

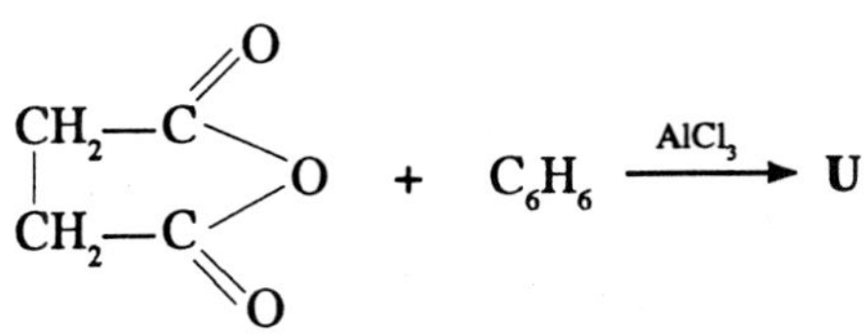

7.23 Give the reactants and/or products in the reactions below

a.

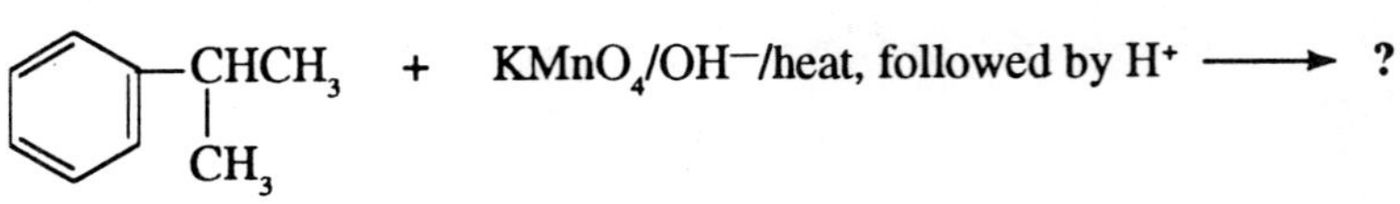

b.

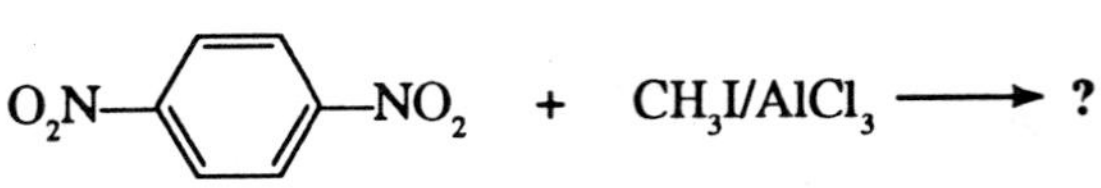

c.

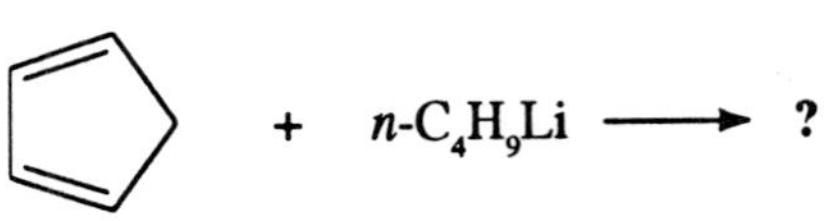

d.

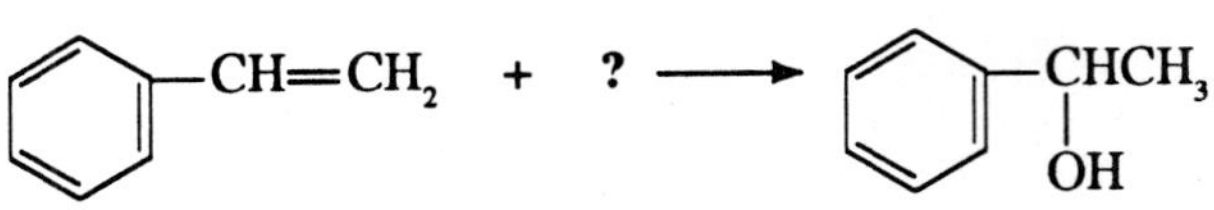

e.

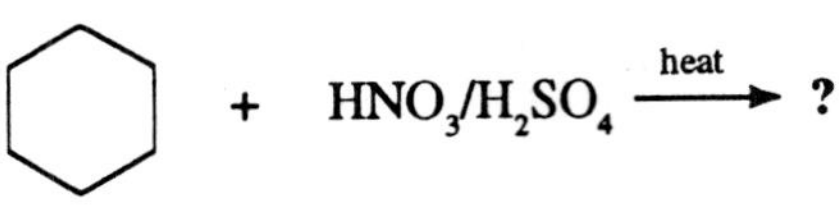

f.

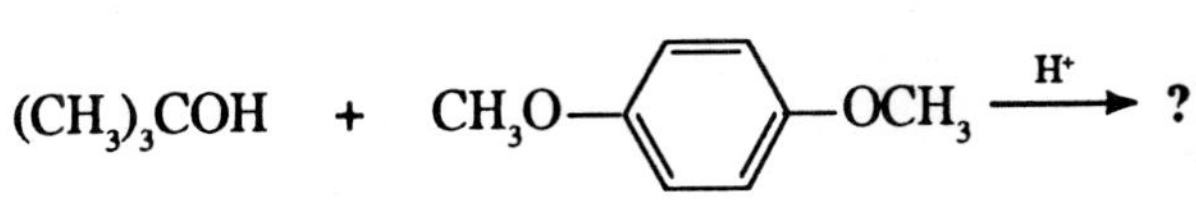

g.

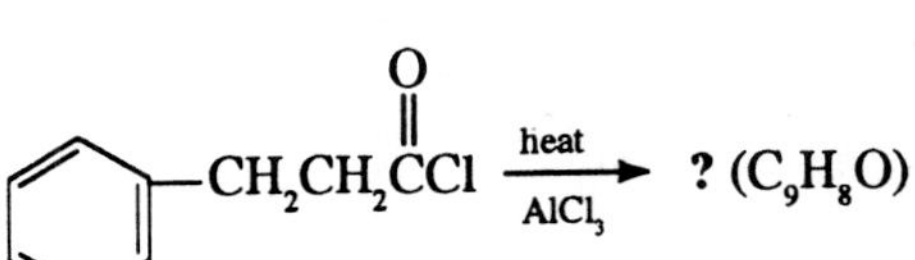

h.

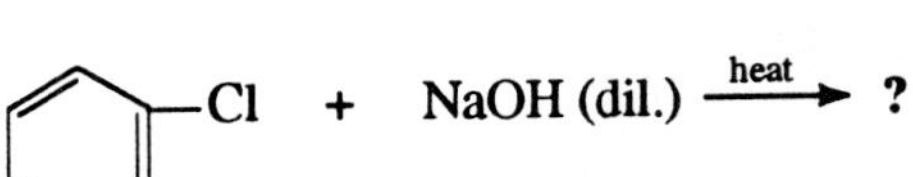

i.

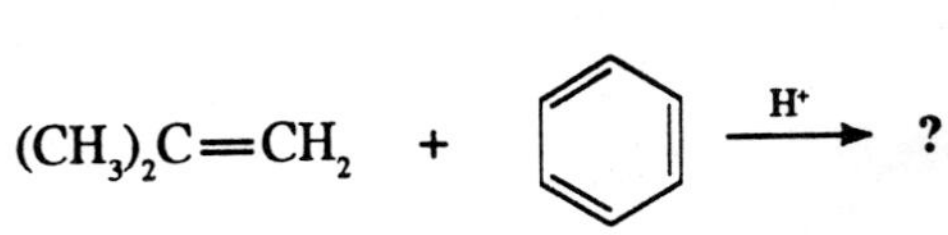

j.

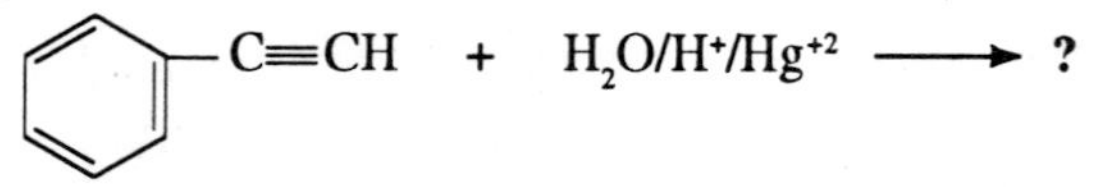

k.

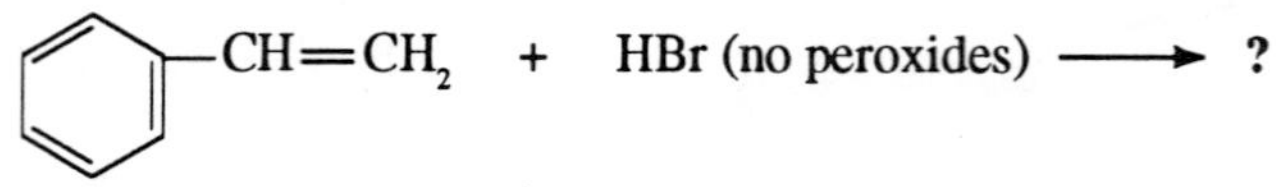

l.

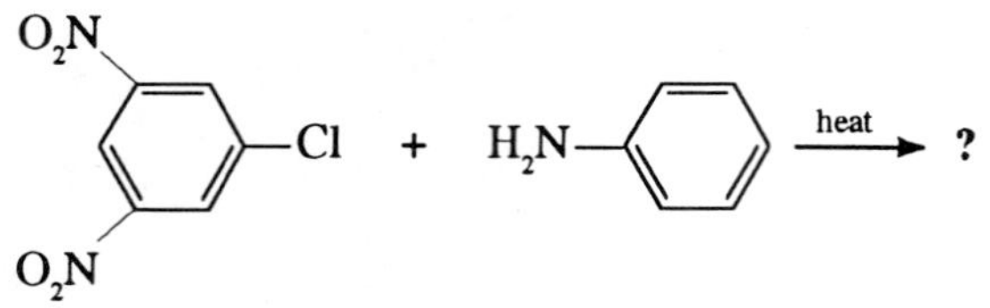

m.

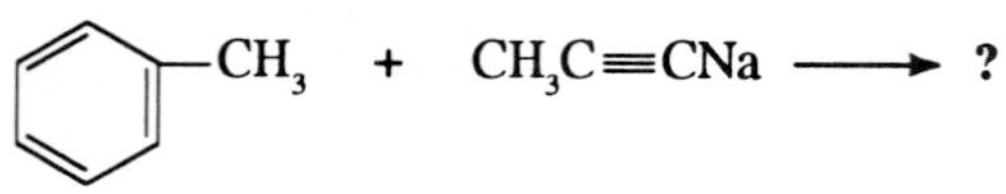

n.

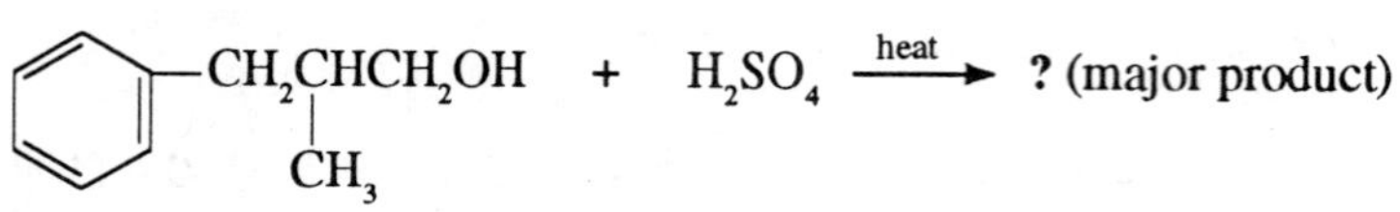

o.

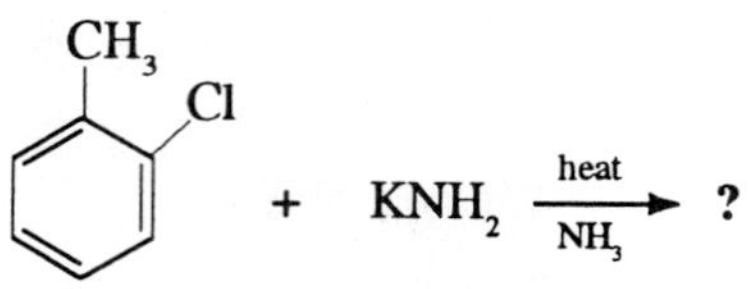

p.

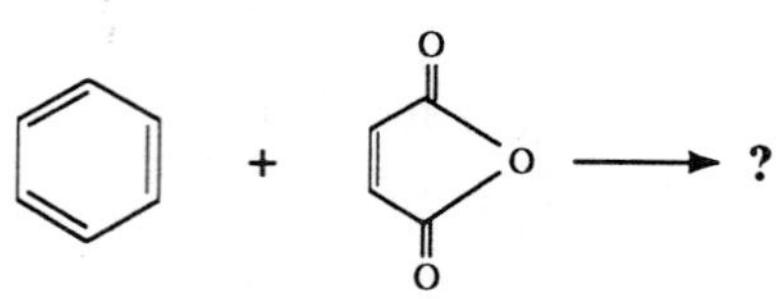

q.

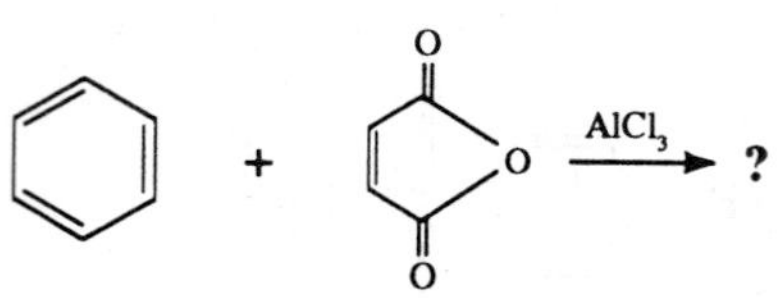

r.

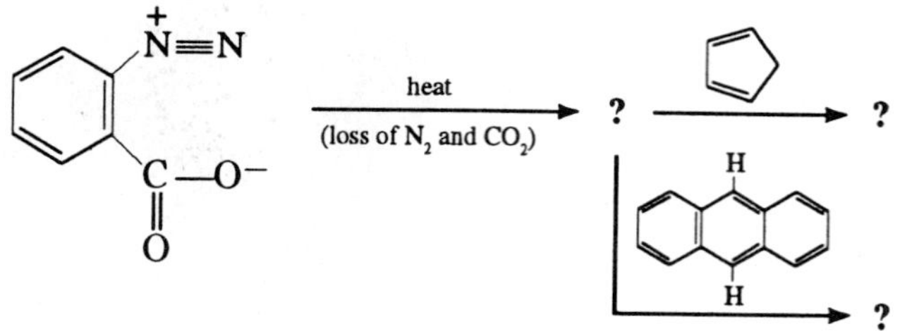

7.24 Suggest reasonable routes that would lead to the synthesis of the following compounds from benzene.

a. *p*-nitrobenzoic acid
b. *m*-nitrobenzoic acid
c. *o*-nitrobenzoic acid
d. *p*-deuterated toluene
e. *o*-aminotoluene
f. 1-bromo-2-phenylethane
g. 2-phenyl-2-propanol
h. *n*-butylbenzene
i. *p*-benzylchlorobenzene
j. *p*-nitroaniline
k. 1-bromo-1-phenylbutane
l. 2-amino-5-nitrotoluene
m. *m*-deuterated ethylbenzene

Multiple Choice Test

7.1 The electrophile in the nitration reaction of benzene is _______.

a. HNO_3 c. NO_2
b. HNO_3/H_2SO_4 d. NO_2^+

7.2 The electrophile in the sulfonation reaction of benzene is _______.

a. SO_3^+ c. H_3O^+
b. SO_3/H_2SO_4 d. $[SO_3H]^+$

7.3 Which of the following can serve as catalyst in Friedel-Crafts reactions?

a. $FeCl_3$ c. $SnCl_4$
b. $AlCl_3$ d. all of the above

7.4 The function of $FeBr_3$ in Friedel-Crafts alkylations is

a. to form a complex with benzene, thus increasing its reactivity.
b. to complex with the carbocation, thus stabilizing it.
c. to abstract the halide from the alkyl halide, thus forming a carbocation.
d. to abstract the proton from the σ complex and regenerate the aromatic ring.

7.5 The reaction of isobutyl bromide with benzene in the presence of $AlCl_3$, yields _______ as the major product.

a. 2-phenylbutane c. isobutylbenzene
b. *n*-butylbromide d. *t*-butylbenzene

7.6 Which of the following reactants would form an arene when reacted with benzene

a. $(CH_3)_2C{=}CH_2/H^+$ d. only (a) and (b)
b. $(CH_3)_3COH/H^+$ e. all of the above
c. $(CH_3)_2CH_2Cl/AlCl_3$

7.7 The reagent(s) in the synthesis of α-bromomethylbenzene from toluene is/are

a. Br_2/UV c. Br_2/Fe
b. HBr d. Br_2/CCl_4

7.8 Which of the following is *not* a resonance form in the nucleophilic aromatic substitution reaction below?

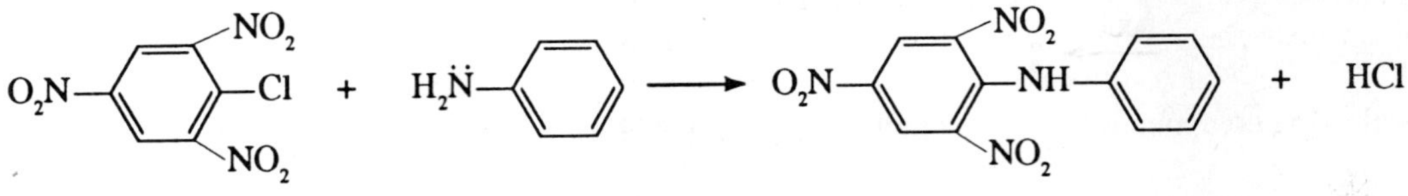

a.

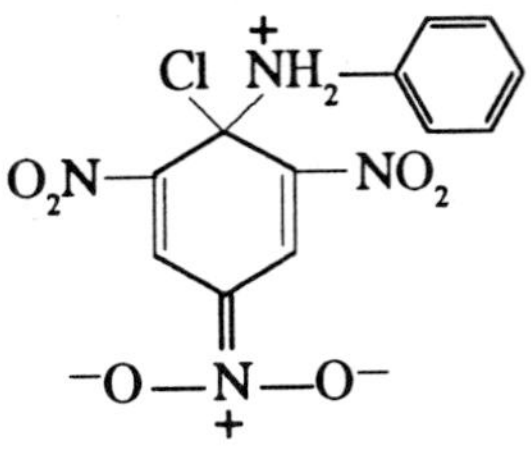

c.

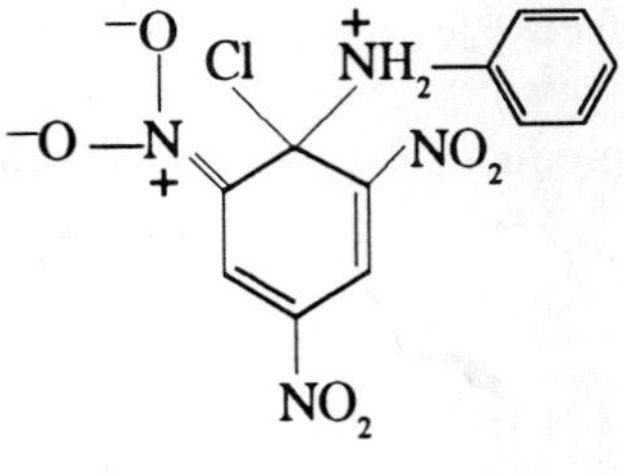

b.

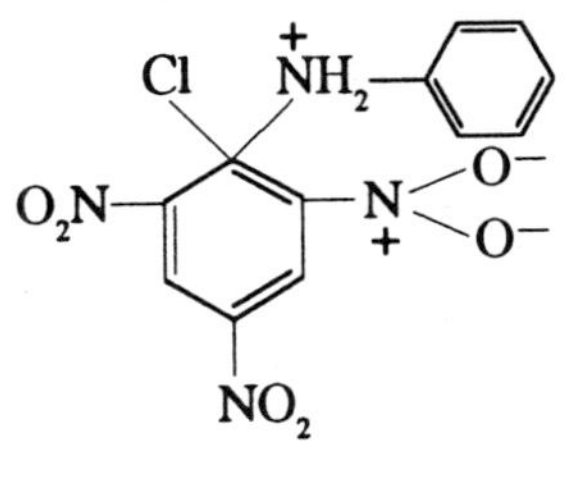

d.

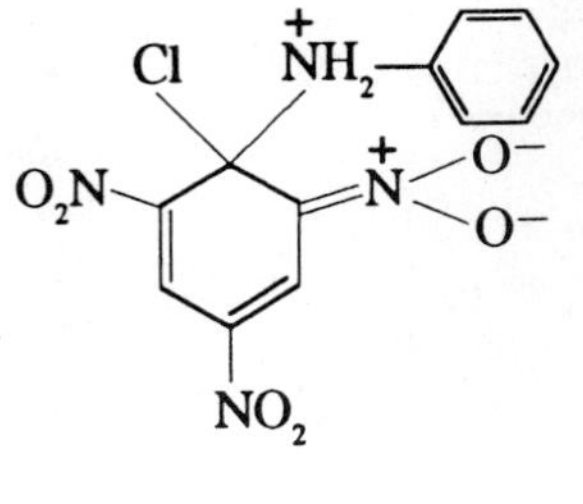

7.9 The major mononitration product of the following compounds is

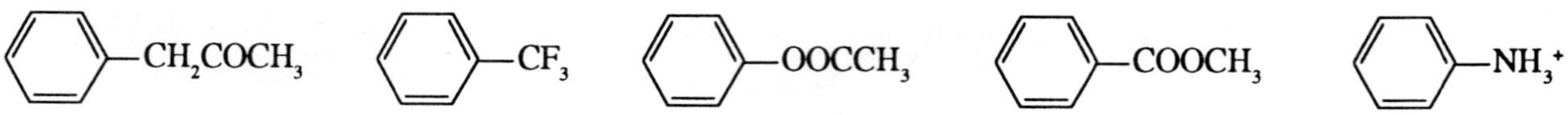

a.	*o–/p–*	*m–*	*o–/p–*	*m–*	*m–*
b.	*m–*	*o–/p–*	*o–/p–*	*m–*	*m–*
c.	*o–/p–*	*o–/p–*	*m–*	*m–*	*m–*
d.	*m–*	*m–*	*o–/p–*	*o–/p–*	*o–/p–*

7.10 The common name for 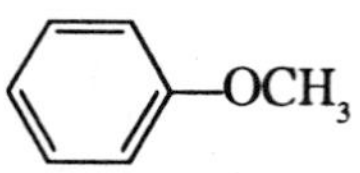 is

a. aniline
b. toluene
c. anisole
d. acetophenone

7.11 The presence of chlorine on the benzene ring ________ the ring and directs the electrophile ________.
a. activates; *o–/p–*
b. deactivates; *o–/p–*
c. activates; *m–*
d. deactivates; *m–*

7.12 Which of the following is not a *meta*-directing substituent in electrophilic aromatic substitutions?
a. $—N(CH_3)_3^+$
b. $—NO_2$
c. —C≡N
d. $—NHCOCH_3$

7.13 The most reactive and the least reactive compound towards nitration are

benzene	acetophenone	toluene	chloromethyl benzene	chlorobenzene	phenol
A	**B**	**C**	**D**	**E**	**F**

a. F/B
b. A/B
c. C/E
d. F/D

7.14 Which of the following compounds will undergo a nucleophilic aromatic reaction the fastest?
a. 3-nitrochlorobenzene
b. 2,4-dinitrochlorobenzene
c. 2,4-dimethylchlorobenzene
d. 3,5-dinitrochlorobenzene

7.15 Which of the following compounds would react with $NaNH_2$/liq. NH_3 to yield two products through the benzyne intermediate?
a. 2-chloro-1,3-dimethylbenzene
b. 2-chloro-1,4-dimethylbenzene
c. 1-chloro-3,4,5-trimethylbenzene
d. *o*-chlorotoluene

7.16 What is/are the reaction product(s) of *p*-chlorotoluene with $NaNH_2/NH_3$ (liq.)?
a. *p*-methylaniline
b. *m*-aminotoluene
c. α-aminomethylbenzene
d. both (a) and (b)

7.17 The sequence of reactions below yields

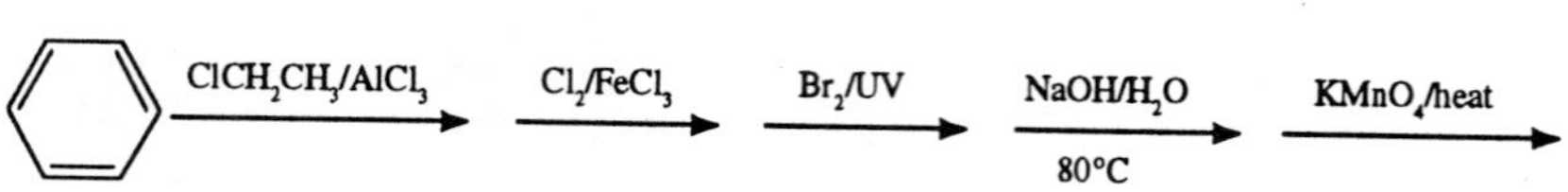

a.
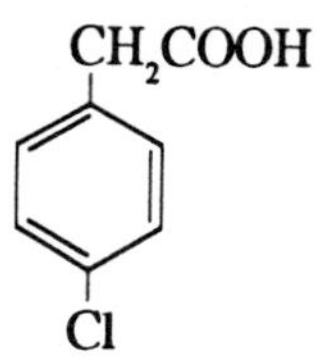

b.
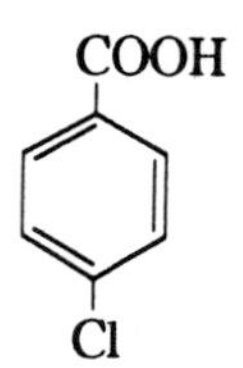

c.
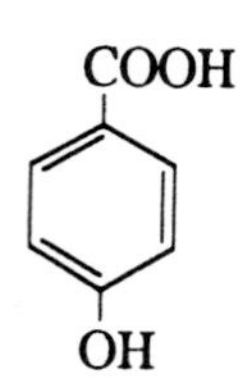

d.
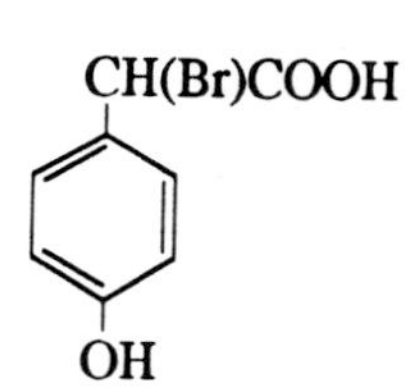

7.18 Which of the following reagents will react with bromobenzene?

KI/acetone	Mg/ether	$NaCN/H_2O$	$NaNH_2/NH_3$	NaOH in ethanol, 80°C.
A	**B**	**C**	**D**	**E**

a. B, C, and D
b. B and E
c. B and D
d. only B
e. all of the above

7.19 Benzyne can be described as a(n) ________.
a. electrophile
b. nucleophile
c. stable species
d. radical

7.20 Which is the incorrect resonance formula in the nitration of anisole?

a. 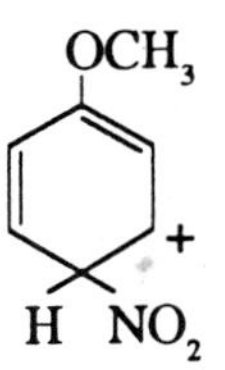
b. 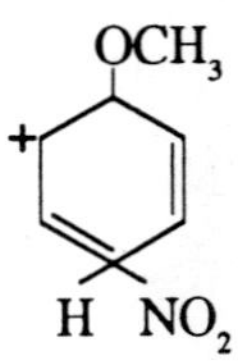
c. 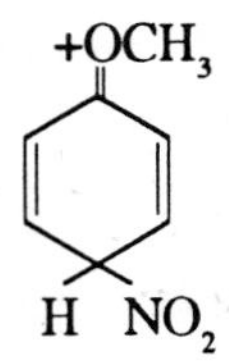
d. 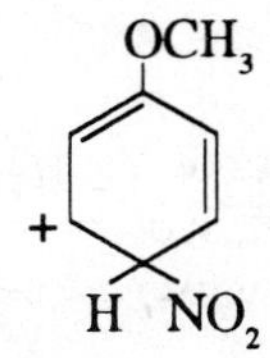

Answers for Questions

7.1 a. 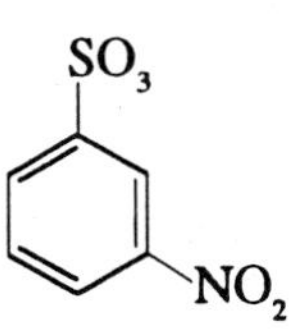

b. 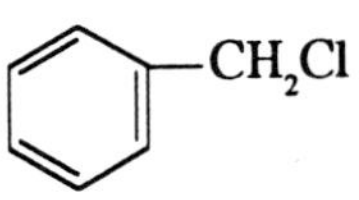

c. 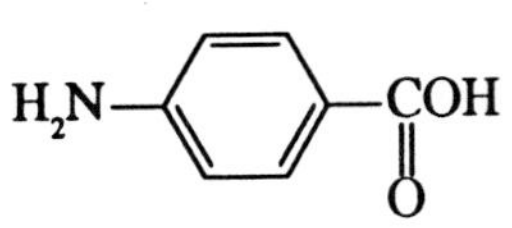

d. 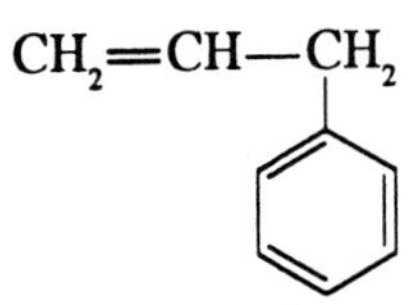

e. 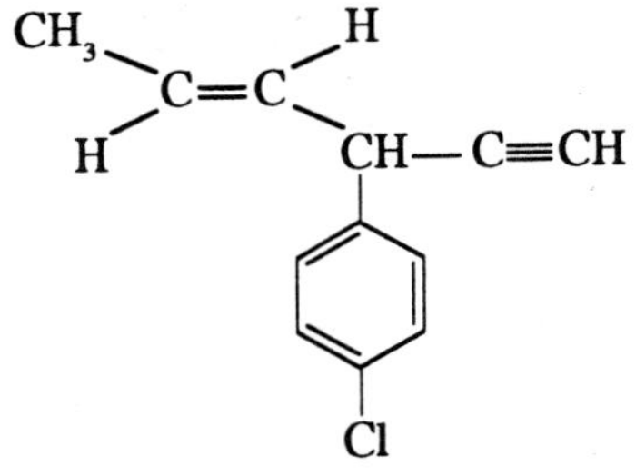

f. 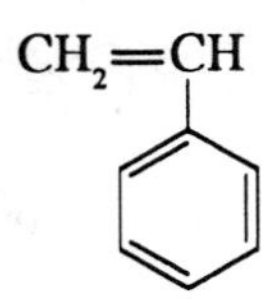

g. 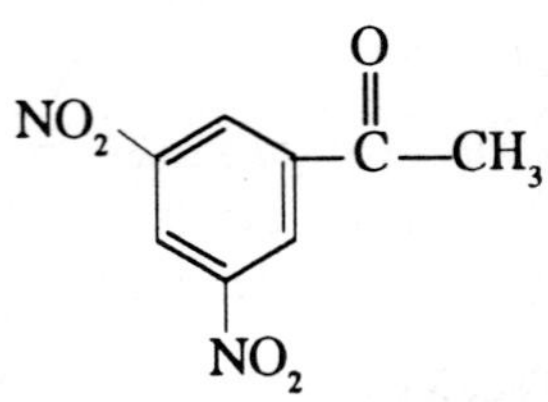

h. 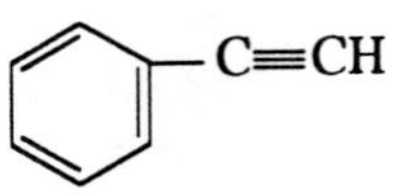

i. 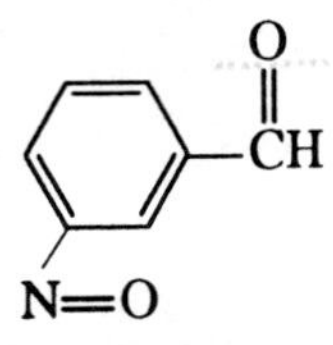

j. 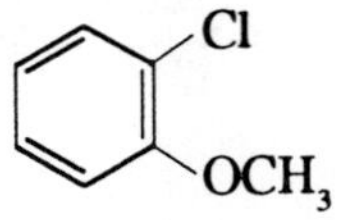

k. 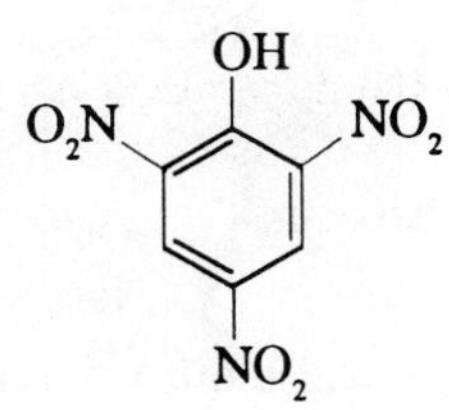

l. 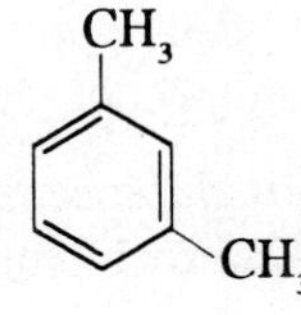

m. 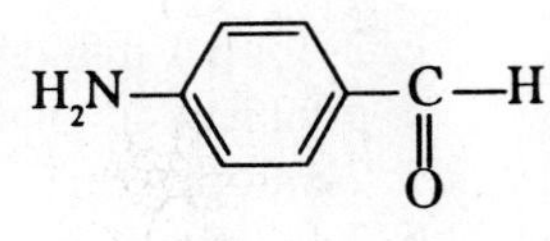

n. 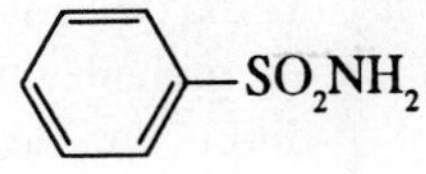

7.2 a. 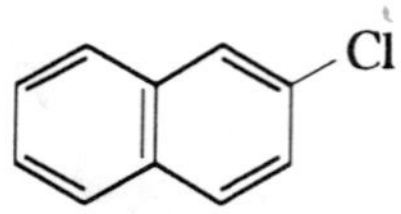

f. 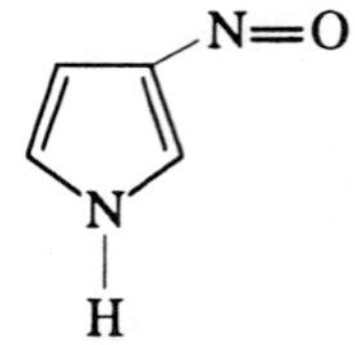

k. 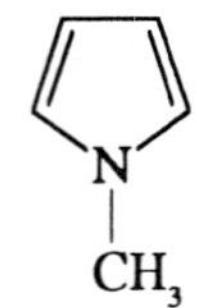

b. 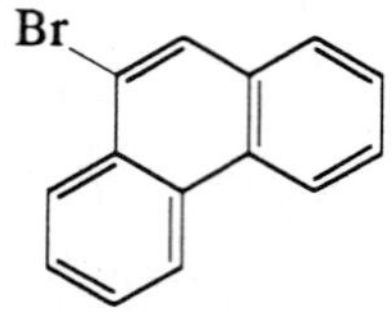

g. 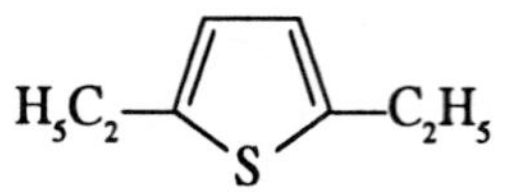

l. 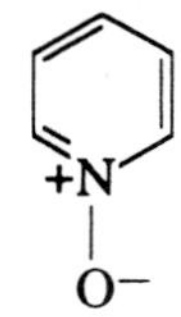

c. 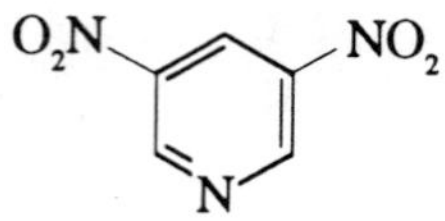

h. 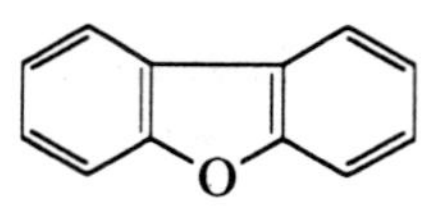

m. 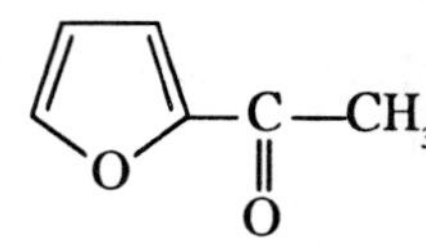

d. 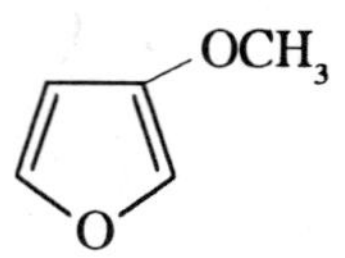

i. 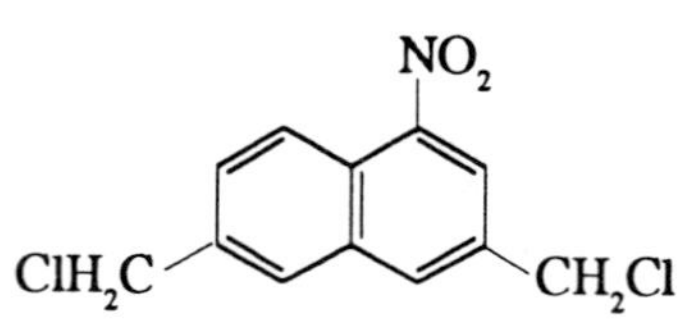

e. 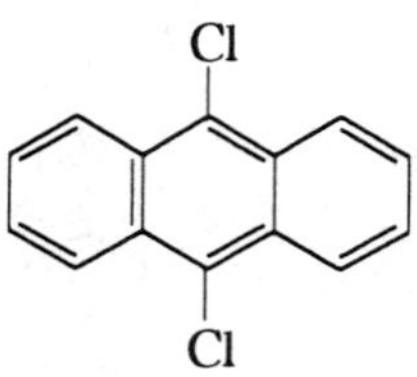

j. 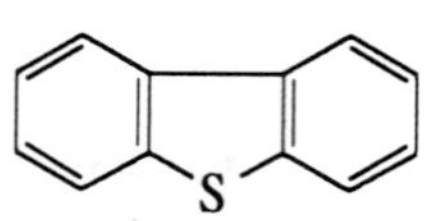

7.3 a. 2,4-dinitrochlorobenze
b. *m*-aminotoluene (or *m*-methylaniline or *m*-toluidine)
c. phenylethene (or styrene or vinylbenzene)
d. *o*-nitrophenol (or *o*-hydroxynitrobenzene)
e. *p*-allyltoluene
f. cyanobenzene (or benzonitrile)

g. chlorobenzene (and not benzyl chloride, which has the formula 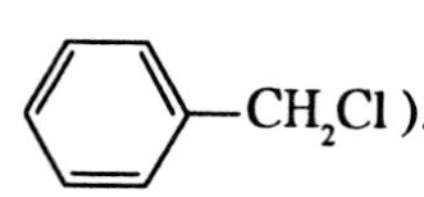).

h. 2- (or β-) aminonaphthalene
i. *p*-toluenesulfonamide
j. 2-nitropyridine
k. 2,4,6-trinitrotoluene
l. *p*-methoxyphenylcyanomethane (or *p*-methoxyphenylacetonitrile)
m. *m*-bromobenzyl bromide
n. *p*-fluorobenzenesulfonamide
o. *p*-(chloromethyl)benzoic acid
p. *trans*-1,2-diphenylethene
q. 1,1,2-tricyanovinylbenzene (or 1,1,2-tricyanophenylethene)
r. dibenzopyrrole
s. 2-formylfuran
t. 1-hydroxyanthracene
u. 2,4-dinitronaphthalene

7.4 a. 3 d. 2
b. 2 e. 4
c. 3 f. 3

7.5 a. 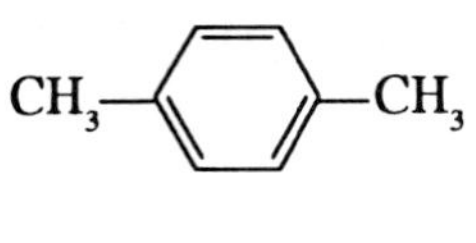

b. 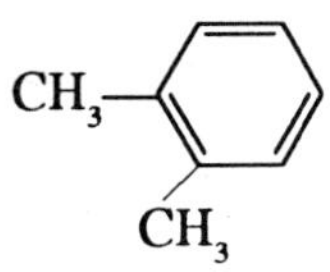

c. 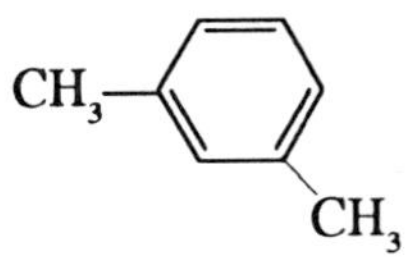 or 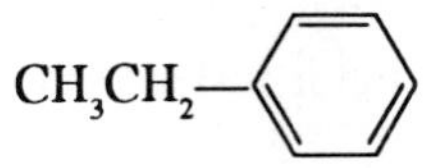

d. 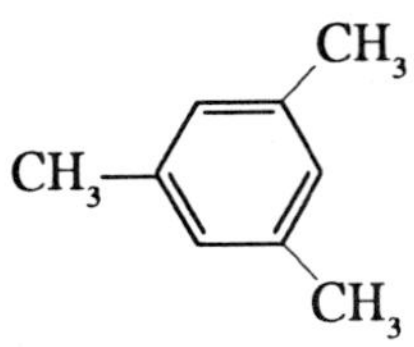

e. 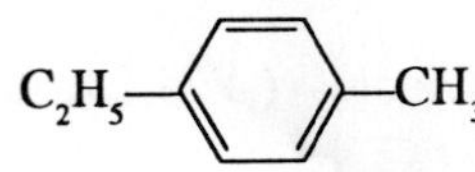 or 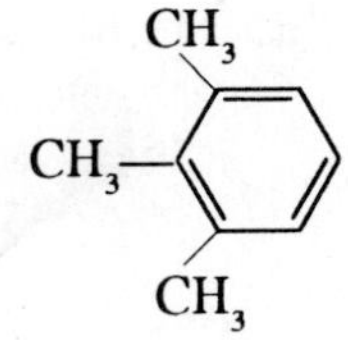

f. 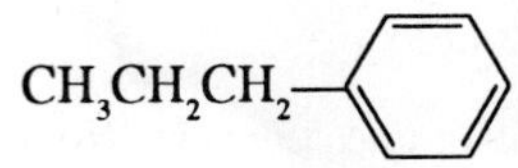 or 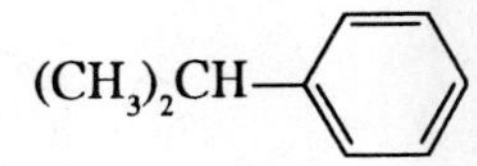 or 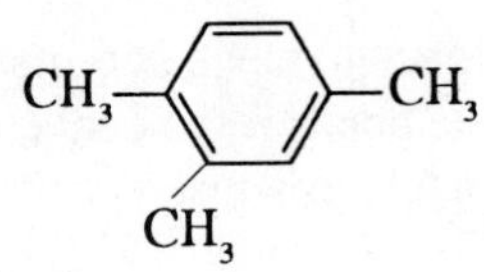

g. 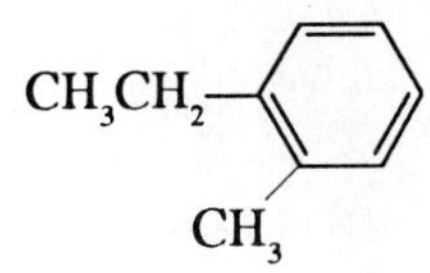 or 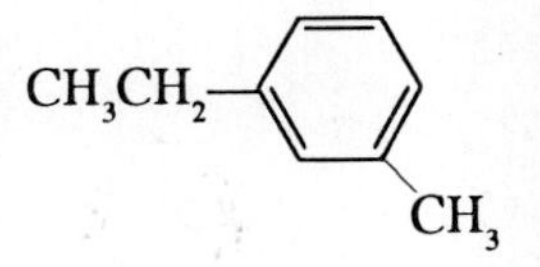

7.6 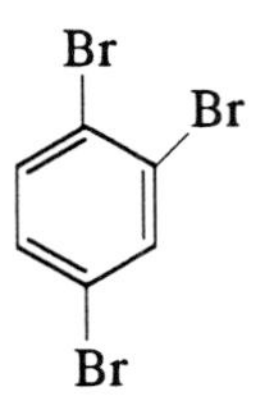

mp 44°C

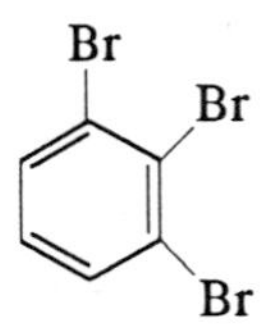

mp 87°C

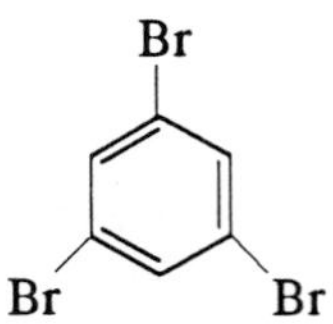

mp 119°C

7.7 Six, six, three, three, no, three, exclusion, $4n+2$

7.8 Aromatic compounds are cyclic, planar molecules with $4n+2$ π electrons (n = 0, 1, 2...) contained in a continuous system of overlaping *p* orbitals. Aromatics possess large resonance energies, undergo substitution reactions rather than addition reactions, and exhibit a ring current (i.e., a downfield shift of aromatic protons in their nmr spectra).

7.9 a. 3 × –28.4 kcal/mole = –85.2 kcal/mole expected
Resonance energy = –49.8 kcal/mole – (–85.2) = 35.4 kcal/mole

b. 4 × –23.3 kcal/mole = –93.2 kcal/mole expected
Resonance energy = –100.9 kcal/mole – (–93.2 kcal/mole) = –7.7 kcal/mole

The negative value indicates that the double bonds *resist* conjugation. Compounds of this type have $4n$ instead of $4n+2$ π electrons and are referred to as antiaromatic compounds.

c. Benzene

6 × –54.0 kcal/mole	= –324.0 kcal/mole
3 × –49.3 kcal/mole	= –147.9 kcal/mole
3 × –117.4 kcal/mole	= –352.2 kcal/mole
Expected value	= –824.1 kcal/mole

Cyclooctatetraene

8 × –54.0 kcal/mole	= –432.0 kcal/mole
4 × –49.3 kcal/mole	= –197.2 kcal/mole
4 × –117.4 kcal/mole	= –469.6 kcal/mole
Expected value	= –1098.8 kcal/mole

d. The difference between benzene's actual heat of combustion and the expected valve calculated in (c) is the resonance energy:
R.E. = –789.1 kcal/mole – (–824.1 kcal/mole)
= +35 kcal/mole

And for cyclooctatetraene:
R.E. = –1099 kcal/mole – (–1098.8 kcal/mole)
= –0.2 kcal/mole

The values for resonance energies from both methods are in reasonable agreement.

e. Cyclooctatetraene does not show aromatic properties. The molecule is a $4n$ rather than a $4n+2$ system. The four double bonds resist conjugation and act independently (i.e., they undergo addition reactions). Cyclooctatetraene is not planar as a result of the distortion of the double bonds, but is instead tub-shaped.

7.10 Aromatic compounds: a, b, c, e, h, i, k, l, m, n, o, t
Note: Compound (h) has fourteen π electrons in its periphery. Compounds (k) and (l) are better written as

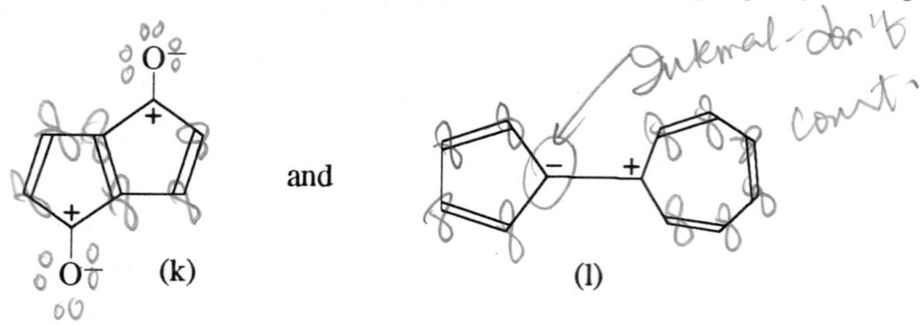

As "internal" ions, they possess $4n+2$ π electrons with continuous overlap. If models of the heterocyclic molecules are built, it is easier to see whether the nonbonded electron pairs are involved in ring aromaticity.

7.11 Pyrrole is aromatic. Nitrogen's lone pair of electrons is part of the aromatic sextet. As a result these electrons are not available to acids.

7.12 The cyclopentadienyl anion contains six π electrons and possesses aromatic stability. On the other hand, the cation has only four π electrons, which is not enough to fill the bonding molecular orbitals.

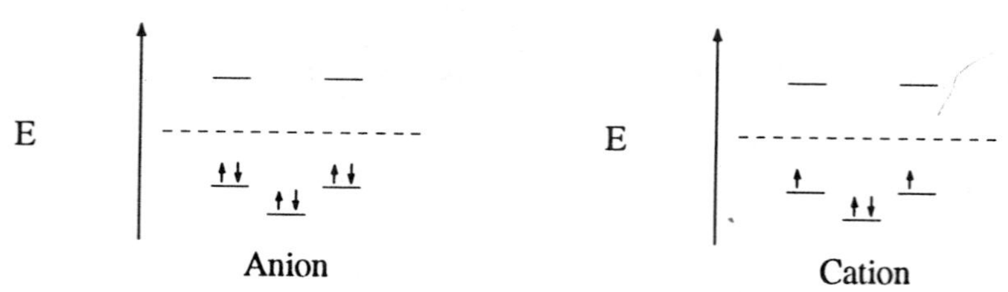

7.13 Although [10]annulene satisfies the $4n+2$ rule, it is not planar; thus it lacks an uninterupted system of overlapping p orbitals. It is not planar because of steric interaction between the two inner hydrogen atoms:

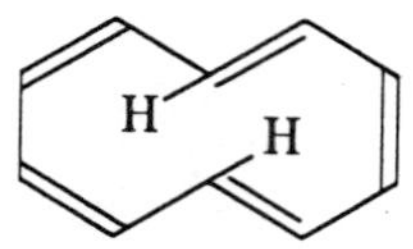

7.14 The dipole moment of diphenylcyclopropenone is due to resonance structure **B**, a major contributor because of its stability which is due to a $4n+2$ (2)aromatic ring system.

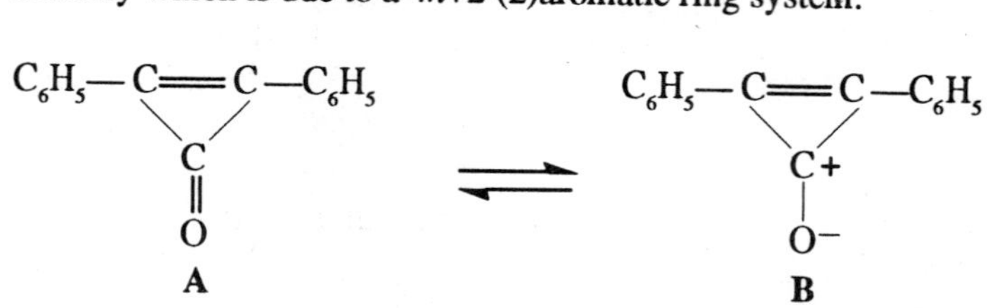

The basicity of this molecule is caused by the same phenomenon. Benzophenone has no such stabilizing factor.

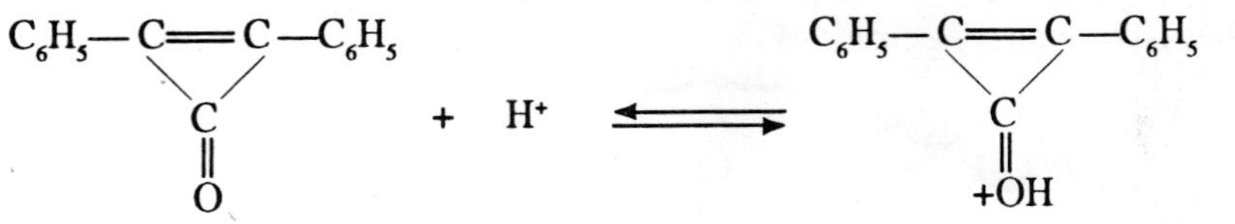

7.15 Azulene is an aromatic compound consisting of fused cyclopentadienyl anion and tropyliun cation rings.

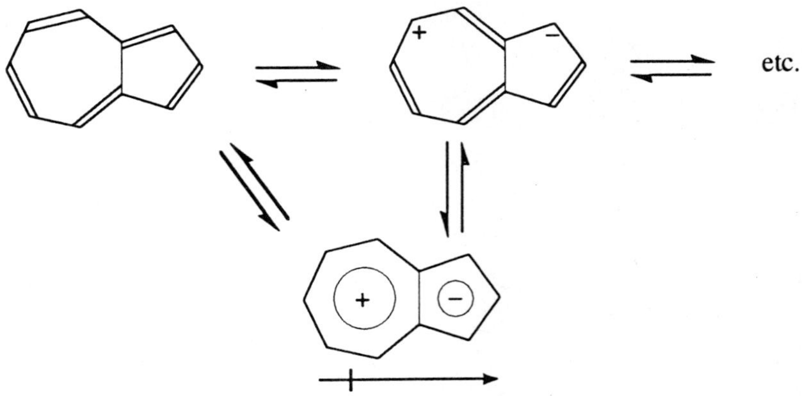

The direction of the dipole is from the 7-membered ring to the 5-membered ring. Thus, protonation (as well as aromatic substitution) occurs in the 5-membered ring.

7.16 a. Three Kekulé structures for naphthalene:

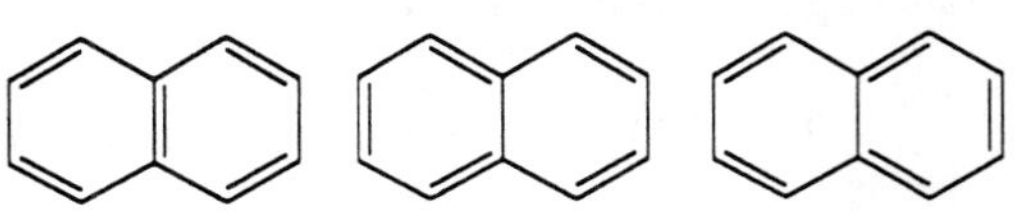

b. Four Kekulé structures for anthracene:

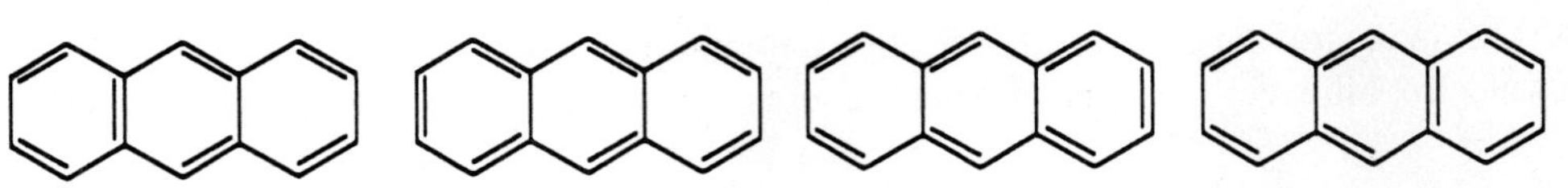

c. Five Kekulé structures for phenanthrene:

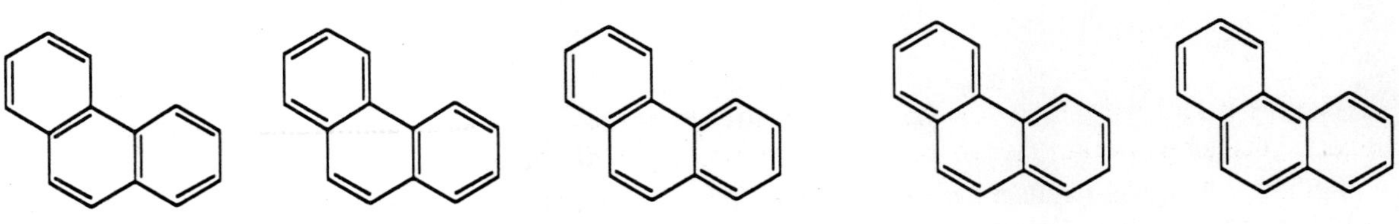

7.17 a. $Cl_2 \xrightarrow{h\nu} 2Cl\cdot$

Reaction	$\Delta H°$
$Cl^{\cdot} + C_6H_6 \longrightarrow C_6H_5^{\cdot} + HCl$	+ 9 kcal/mole
$C_6H_5^{\cdot} + Cl_2 \longrightarrow C_6H_5Cl + Cl\cdot$	−28 kcal/mole
	$\Delta H°$ = −19 kcal/mole

b. In practice, radical chlorination is a poor method for the preparation of chlorobenzene. The first propagation step requires a large activation energy.

7.18 a. Compound **B** should have a dipole moment of zero.

b. Since compound **C** looks like an "addition" of compounds **A** and **B**, it might be predicted that its dipole is the addition of dipoles **A** and **B**, or 6.870 + 0.000 = 6.870 D.

c. Interactions between the two CH_3– groups attached to nitrogen with the two CH_3– groups at the *ortho* positions, force the dimethylamino group out of the plane of the benzene ring. The result is that there is less overlap between the lone pair of electrons on nitrogen with the π electrons of the ring.

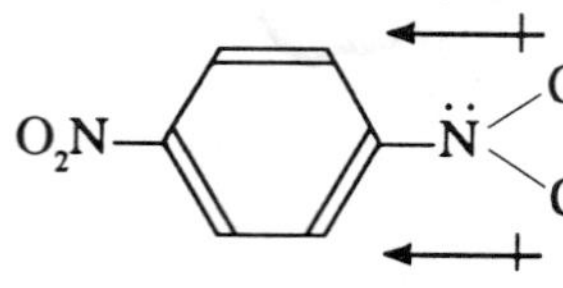

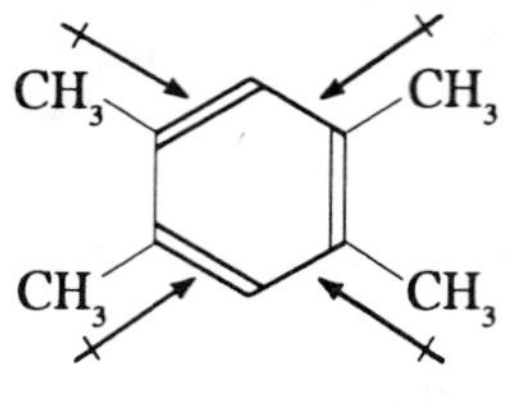

dipole moment = 0

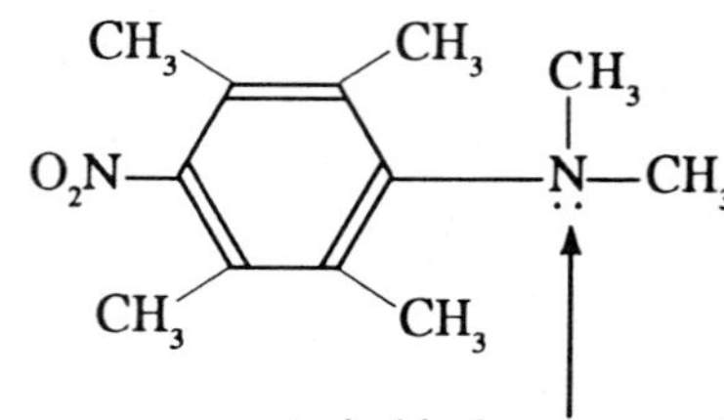

steric hindrance

7.19 a. Under S_N1 conditions, *p*-methoxybenzyl bromide forms a carbocation that is stabilized by resonance:

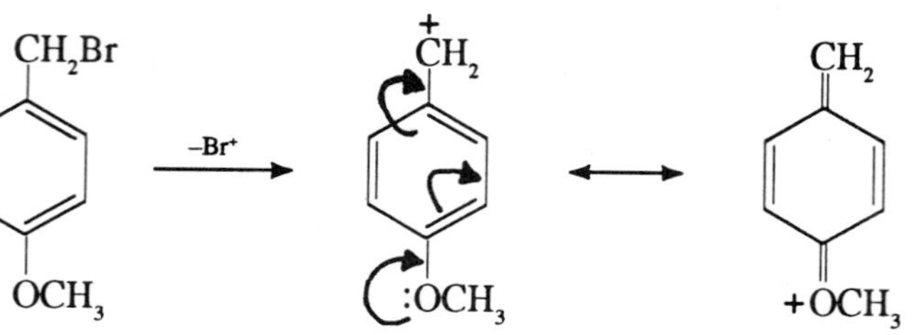

p-Nitrobenzyl bromide forms a carbocation that is not stabilized by resonance because of the electron withdrawing effect of the $-NO_2$ group.

b. The Friedel-Crafts alkylation leads to the formation of a compound whose benzene ring is more activated towards electrophilic reagents than benzene because of the electron-donating isopropyl group, $(CH_3)_2CH-$. Thus, polysubstitution is very likely to occur:

$$C_6H_6 + (CH_3)_2CHBr \xrightarrow[\text{acid}]{\text{Lewis}} C_6H_5\text{—}CH(CH_3)_2 + HBr$$

more activating benzene ring

The Friedel-Crafts acylation, on the other hand, leads to the formation of a compound whose benzene ring is less activated towards electrophilic reagents than benzene because of the electron-withdrawing propanoyl group, $CH_3CH_2\overset{|}{C}{=}O$. The ketone formed has a less activated ring than benzene itself.

$$C_6H_6 + CH_3CH_2\underset{\underset{O}{\|}}{C}\text{—}Cl \xrightarrow[\text{acid}]{\text{Lewis}} C_6H_5\text{—}\underset{\underset{O}{\|}}{C}\text{—}CH_2CH_3 + HCl$$

less activated benzene ring

c. Method 1, a Friedel-Crafts alkylation, involves the formation of $CH_3(CH_2)_2\overset{+}{C}H_2$ (a 1° carbocation) that rearranges to $CH_3CH_2\underset{+}{C}HCH_3$ (a 2° carbocation), thus reducing the yield of the desired product. The carbocation formed in Method 2 does not rearrange because it is stabilized by resonance:

$$CH_3(CH_2)_2\underset{+}{C}{=}\ddot{O}: \longleftrightarrow CH_3(CH_2)_2C{\equiv}\underset{+}{O}:$$

d. Nitrobenzene possesses a highly deactivated benzene ring because of the electron-withdrawing $-NO_2$ group.
e. The amino group $:NH_2$ of aniline is basic, and it activates the ring by promoting resonance; thus it acts as an *ortho-/para*-director. In acidic medium, the amino group is protonated:

$$C_6H_5\text{—}NH_2 + H^+ \longrightarrow C_6H_5\text{—}NH_3^+$$

The NH_3^+ group does not have the electron pair available and deactivates the ring by its inductive effect; thus it acts as a *meta*-director.

f. In nucleophilic aromatic substitution, the aromatic ring must be significantly deactivated by the presence of groups such as the nitro group, $-NO_2$. However, the reaction will be possible only if the intermediate is stabilized by resonance.

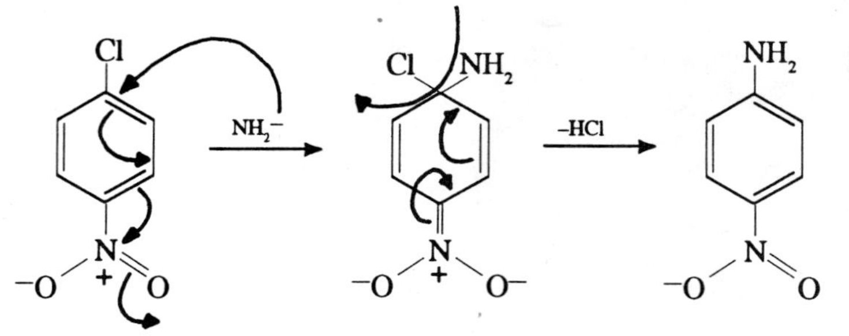

g.

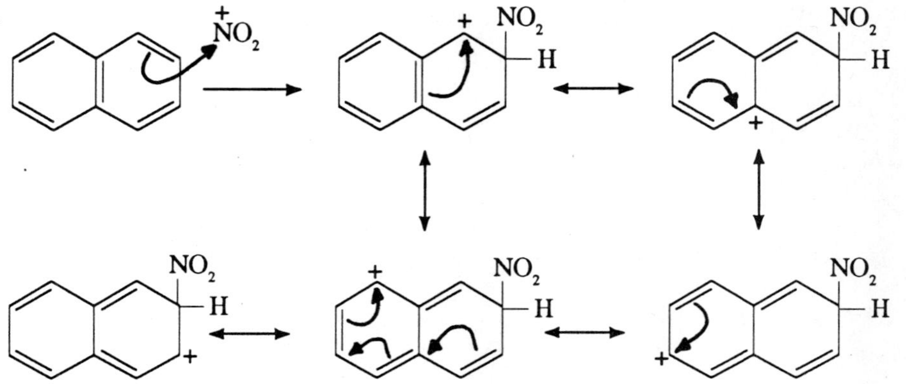

h. The intermediate in the bromination of benzene is not aromatic, and substitution (addition of Br^+, loss of H^+) should occur to preserve the aromatic nature of the ring.

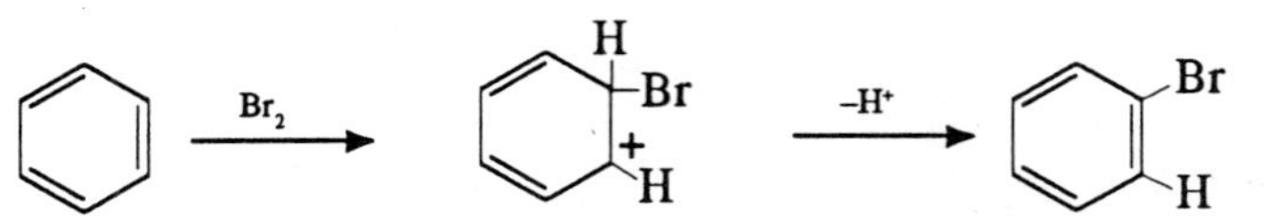

The intermediate in the bromination of phenanthrene is aromatic. The product has two benzene rings with a higher resonance energy per ring than either of phenanthrene's rings:

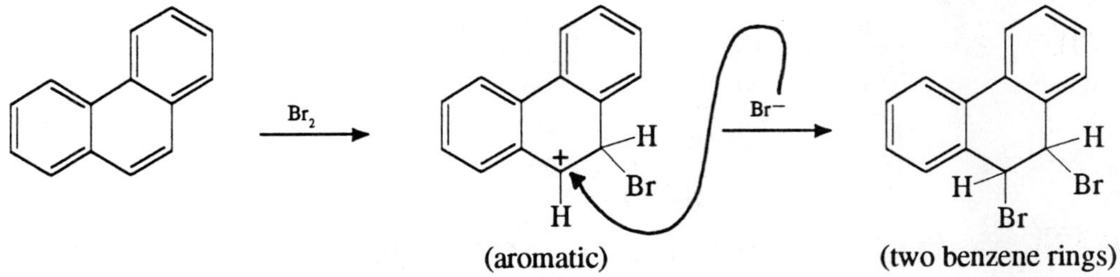

i. The chlorine atom is electron-withdrawing by its inductive effect (which would make it a *m*-director), but is also electron-donating because it promotes resonance (which would make it an *o*-/*p*-director). Since the resonance factor is much more significant, an electrophilic aromatic substitution would produce *o*-/*p*-products.

7.20 a.

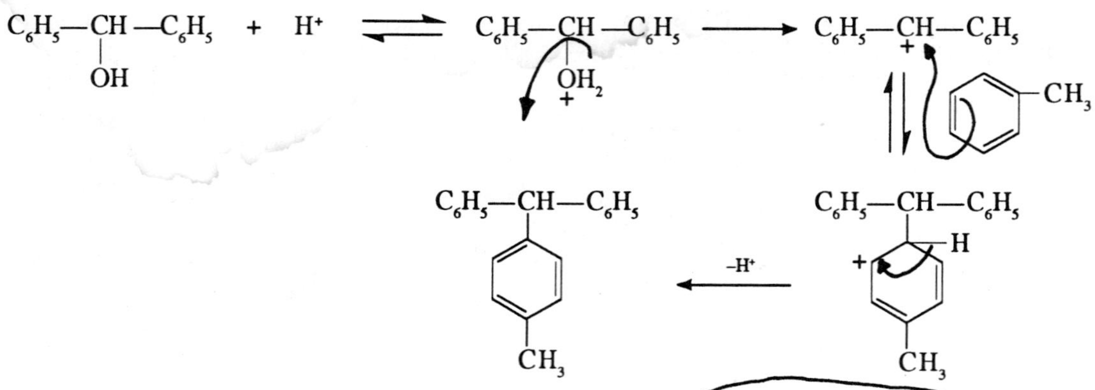
C6H5—CH—C6H5 + H+
OH
C6H5—CH—C6H5
OH2
C6H5—CH—C6H5
CH3
C6H5—CH—C6H5
H
–H+
CH3
C6H5—CH—C6H5
CH3

b.

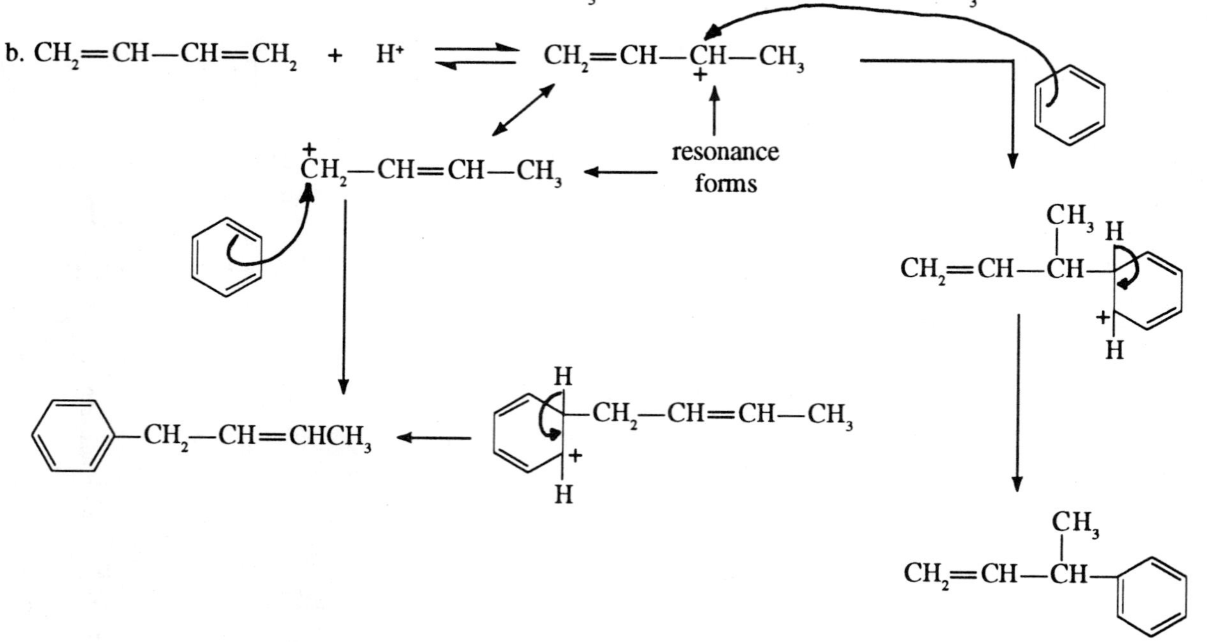
CH2=CH—CH=CH2 + H+
CH2=CH—CH—CH3
resonance
forms
CH2—CH=CH—CH3
CH3
H
CH2=CH—CH—
H
H
CH2—CH=CH—CH3
H
CH2—CH=CHCH3
CH3
CH2=CH—CH—

c.

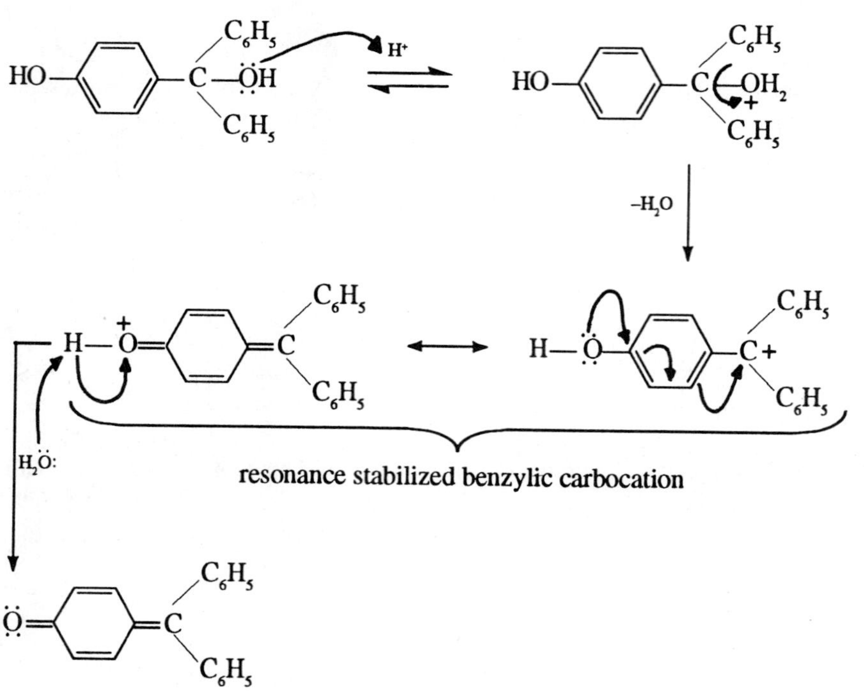
HO
C6H5
C—OH
C6H5
H+
HO
C6H5
C
OH2
C6H5
–H2O
C6H5
H—O=
=C
C6H5
H—O
C6H5
C+
C6H5
H2O:
resonance stabilized benzylic carbocation
O=
C6H5
=C
C6H5

d.

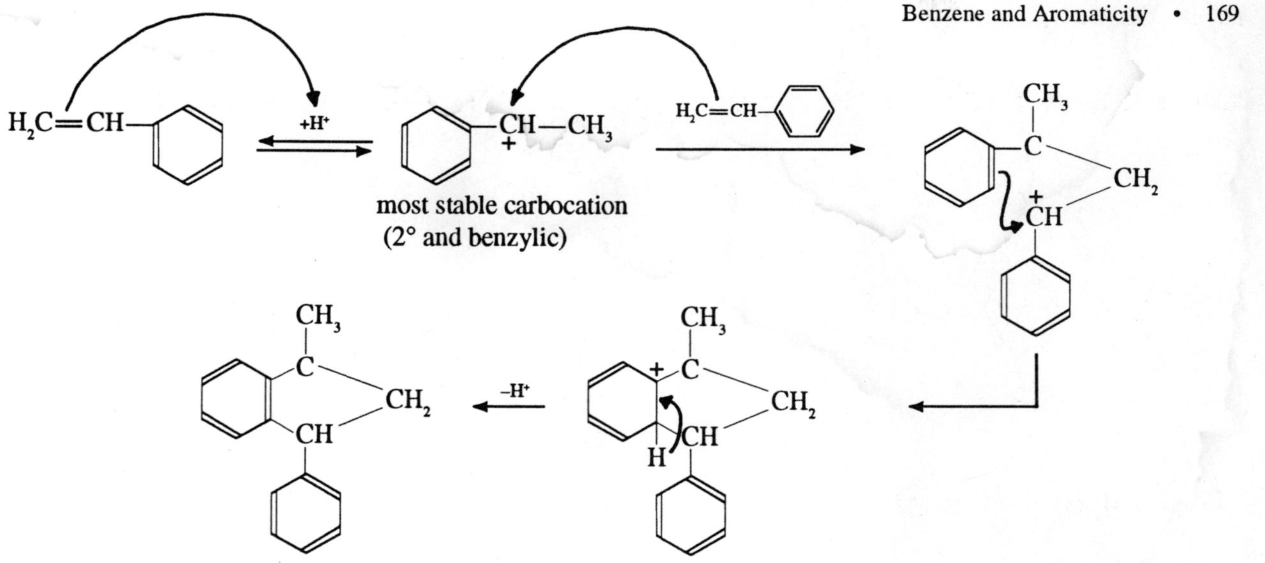
H2C=CH–
+H+
CH—CH3
+
most stable carbocation
(2° and benzylic)
H2C=CH–
CH3
C
CH2
CH
–H+
H

e.

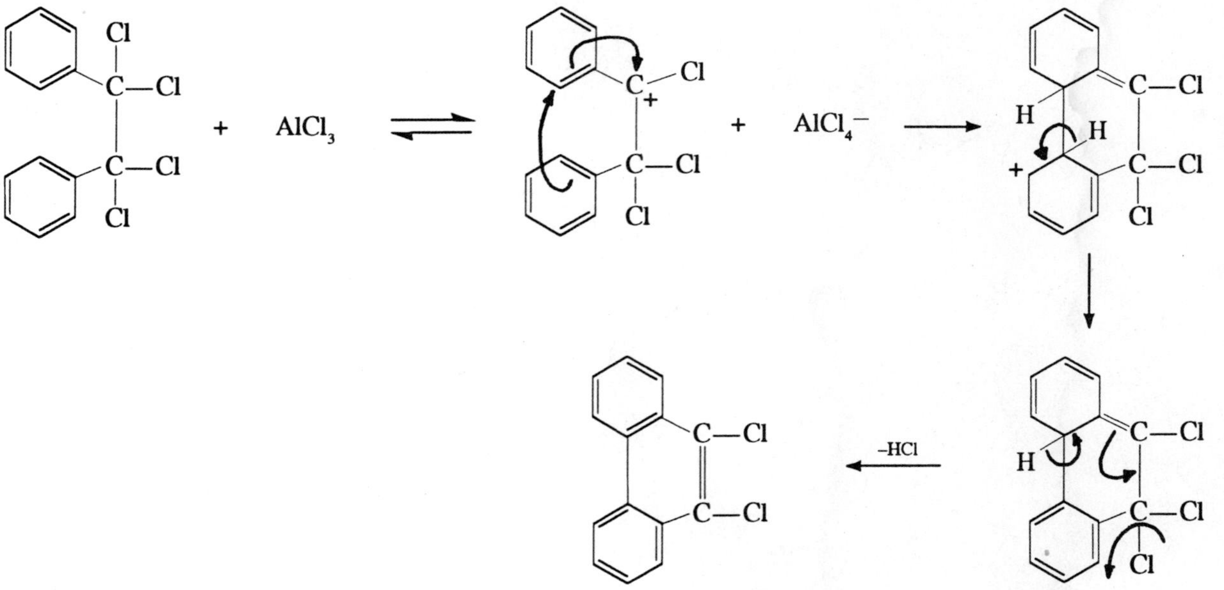
Cl
C—Cl
C—Cl
Cl
+
AlCl3
AlCl4−
H
–HCl

f.

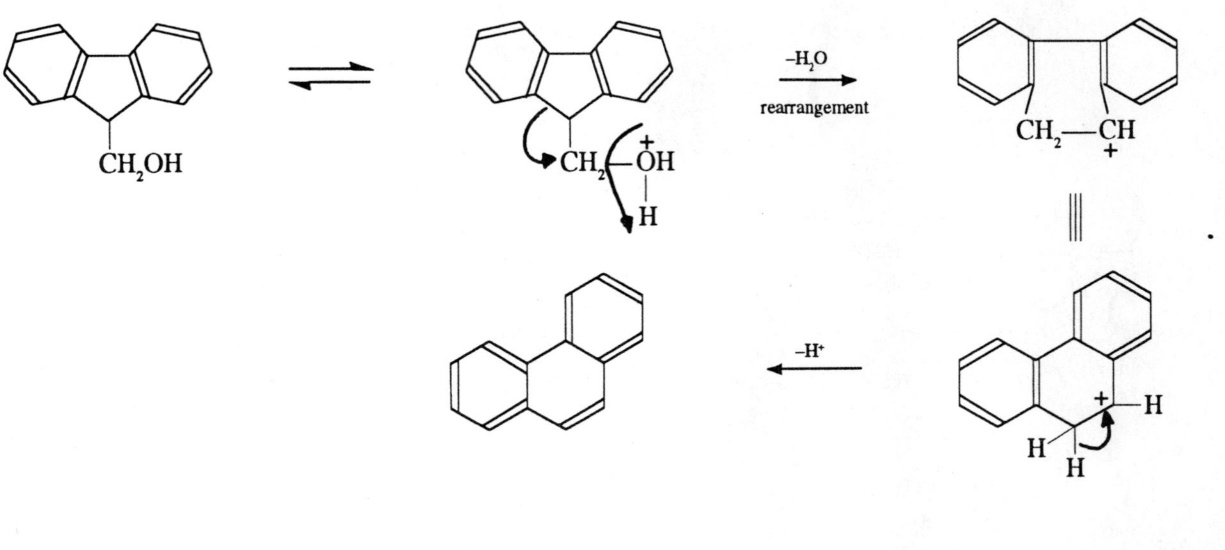
CH2OH
CH2
OH
H
–H2O
rearrangement
CH2—CH
+
–H+
H

g. 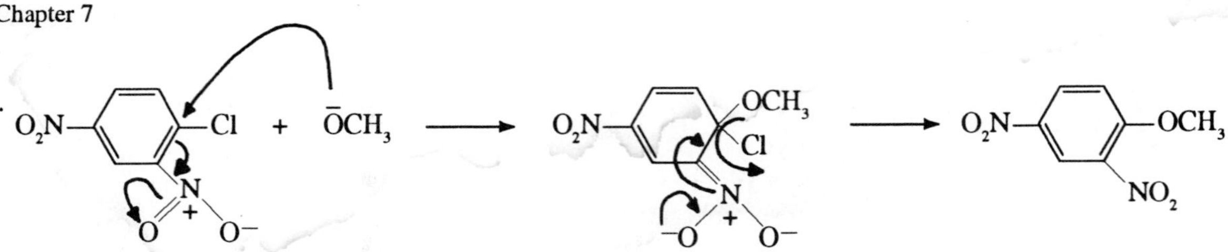

7.21 a. 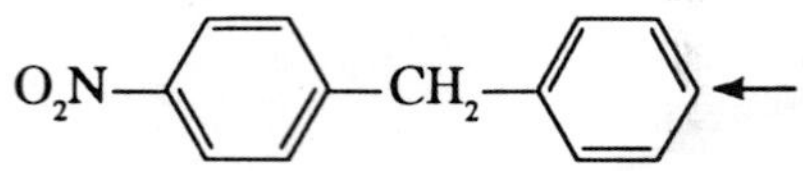

b. 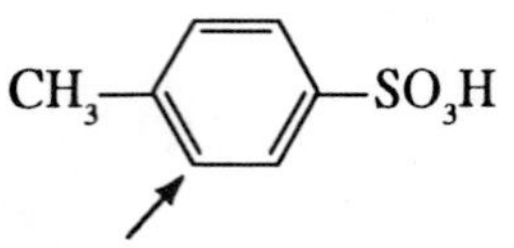

c. 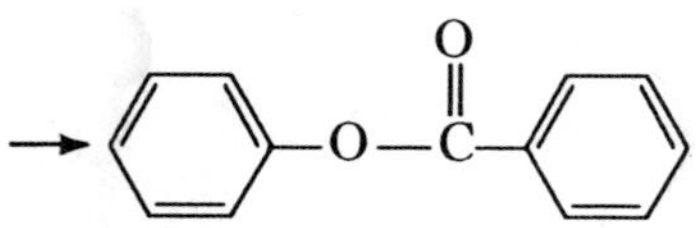

d. 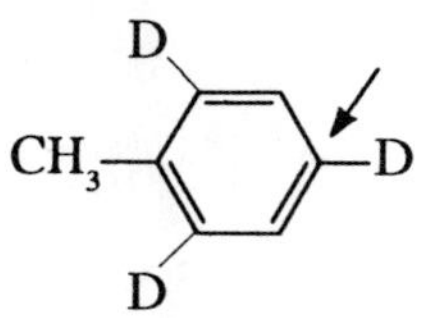

e. 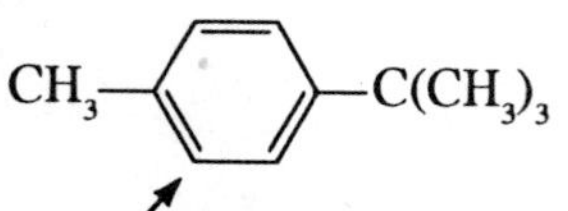
(less sterically hindered)

f. 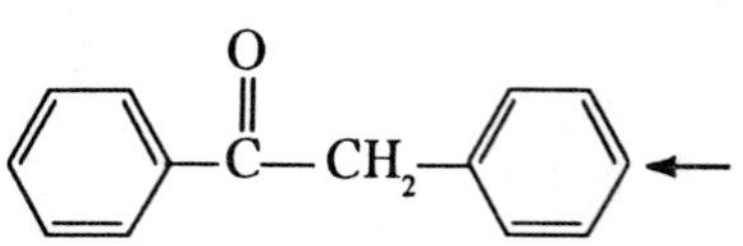

g. 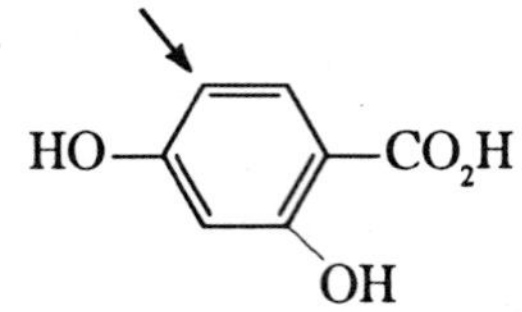

h. 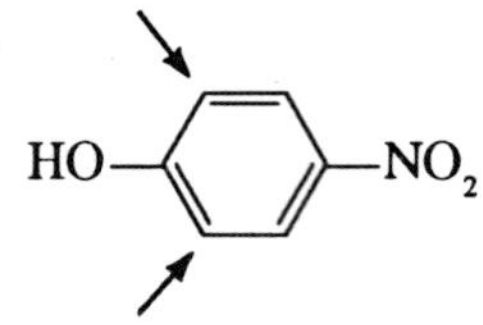

i. 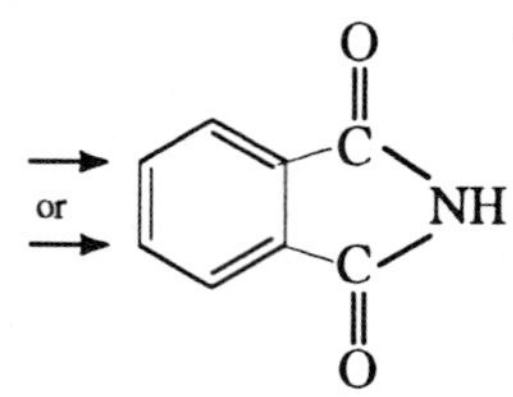

j. 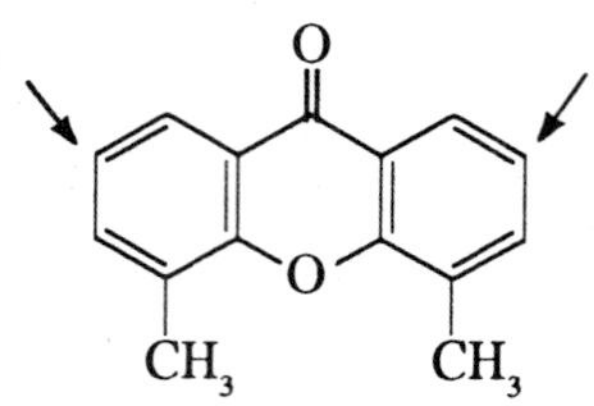

k. 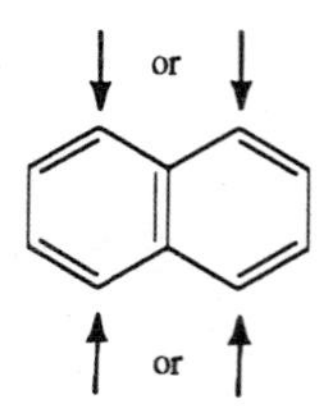

l. 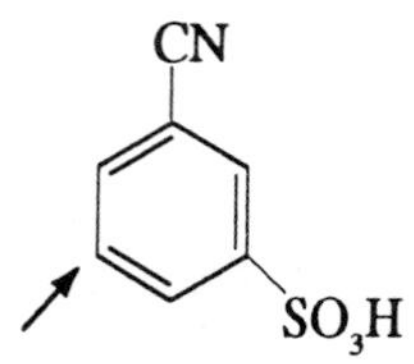

7.22 **A** = HNO_3/H_2SO_4

B = Sn/HCl

C = CH_3CCl (C=O) or $(CH_3C{-})_2O/H^+$ (C=O)

D = 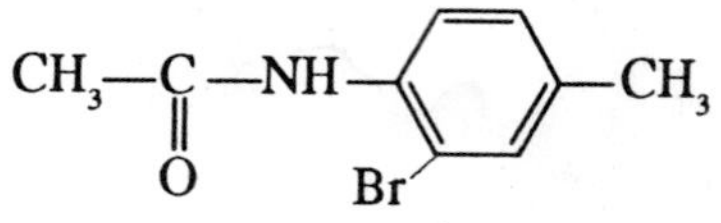

E = H_3O^+/heat

F = $CH_3CH_2CCl/AlCl_3$ (C=O)

G = Zn(Hg)/HCl

H = 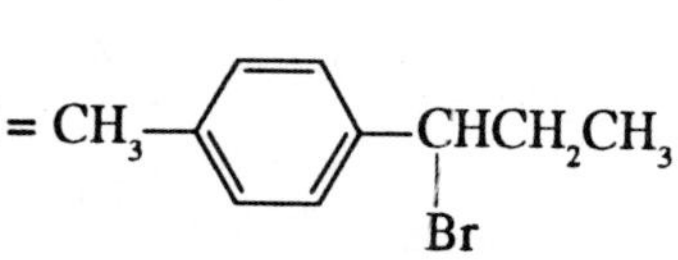

I = alc. KOH

J = NBS

K = H_2/Pt

L = $NaCN/H_2O$

M = $HC\equiv CH$

N = $CH_3CH_2C\equiv CH$

O = $CH_3CH_2CH{=}CH_2$

P = NBS

Q = 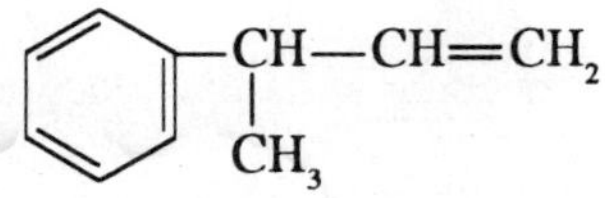

R = 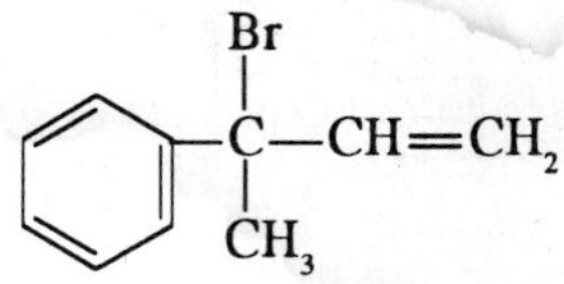

S = 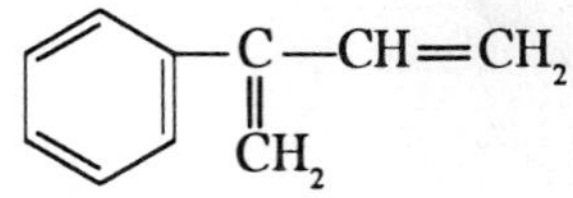

T = 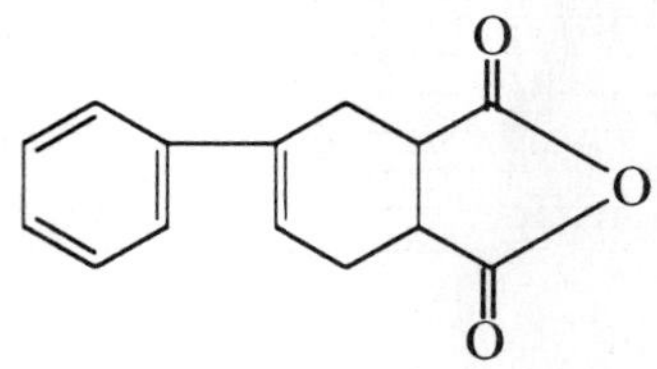

U = 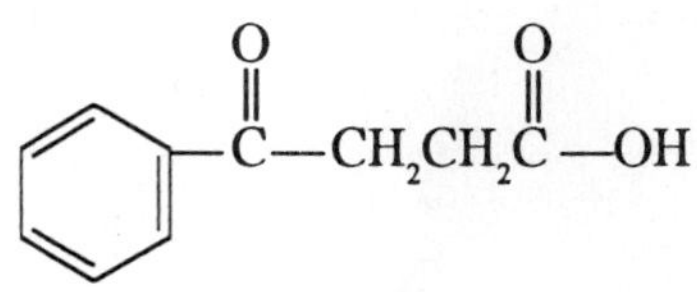

7.23 a. 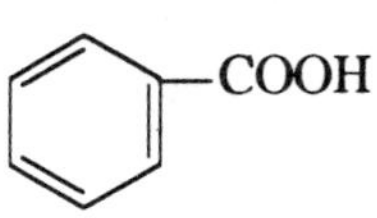 + MnO_2 + CO_2 + H_2O

b. No reaction (ring too deactivated).

c. 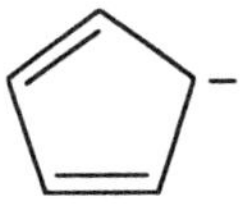 $-$ Li^+ (stable, aromatic system) + n-C_4H_{10}

d. H_2O/H^+

e. No reaction (ring not aromatic)

f. 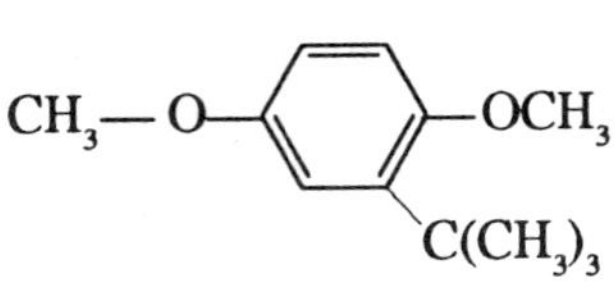 + H_2O

g. 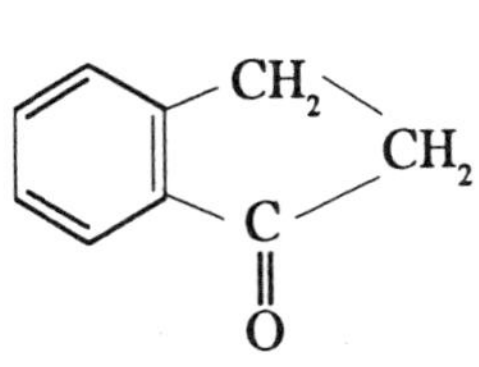 + HCl

h. No reaction (compound is an aryl halide)

i. 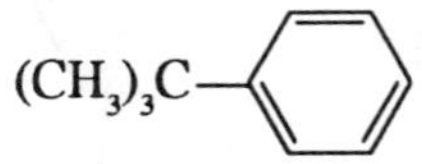

(Friedel-Crafts-like reaction)

j. 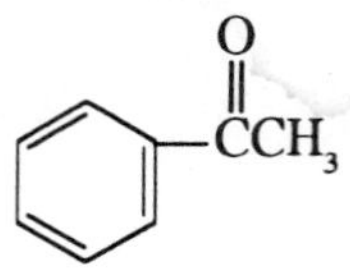

(keto form more favorable than enol form: 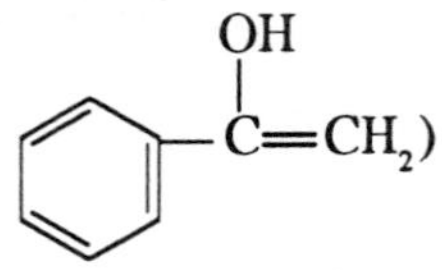

)

k. 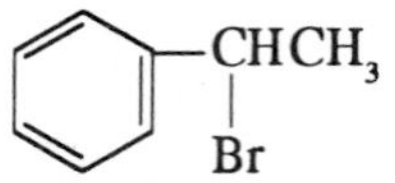

l. No reaction (nucleophilic aromatic substitution not possible because of nonstabilized intermediate)

m. No reaction

n. 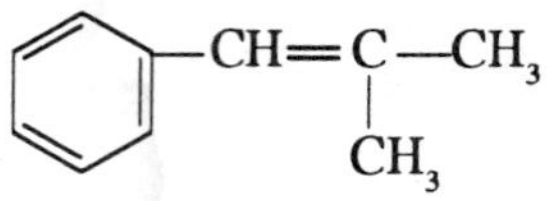

o. 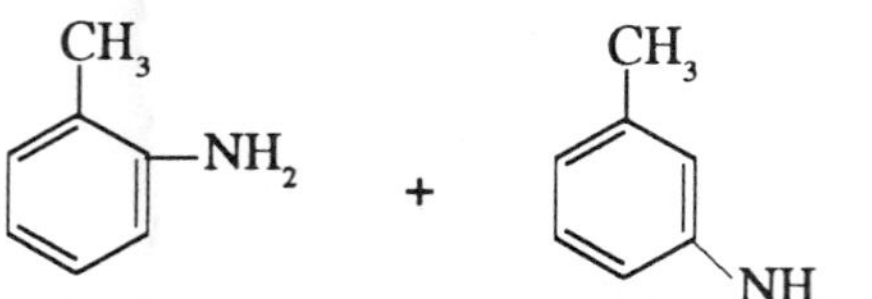

(via benzyne)

p. No reaction (No Diels-Alder reaction because benzene is aromatic)

q. 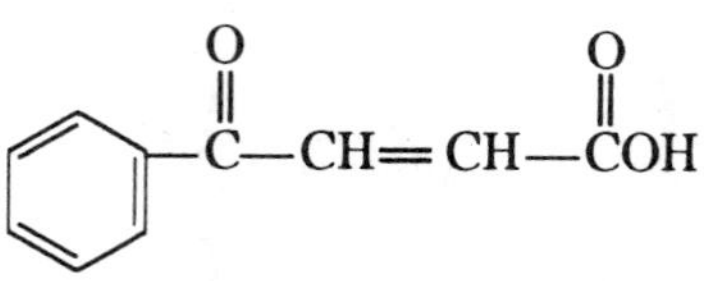

(Friedel-Crafts acylation)

r. 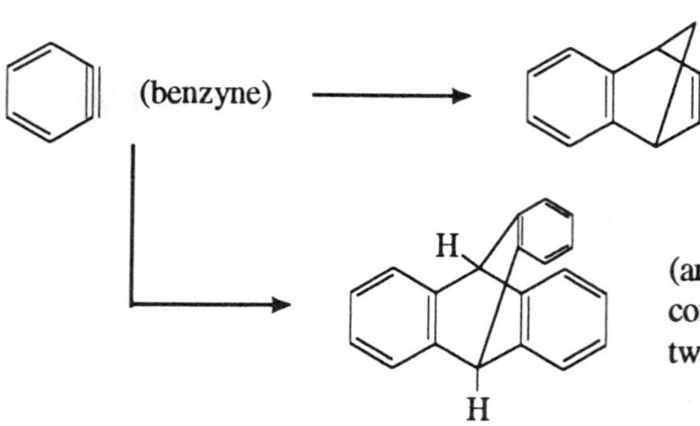

(benzyne acts as the dienophile)

(anthracene reacts at positions 9 and 10, thus converting benzyne to benzene and producing two benzene rings from the anthracene moiety)

7.24 a. 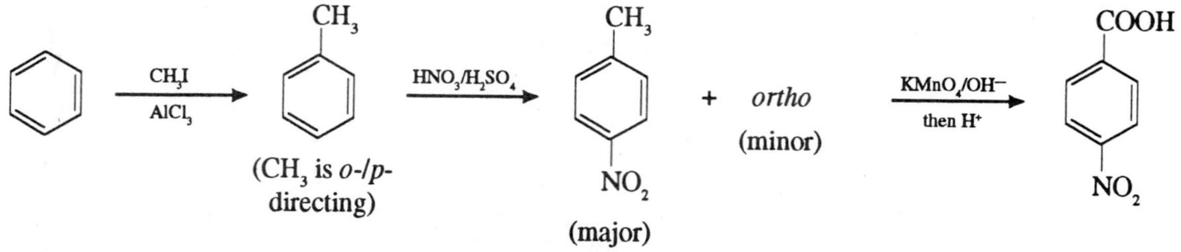

b.

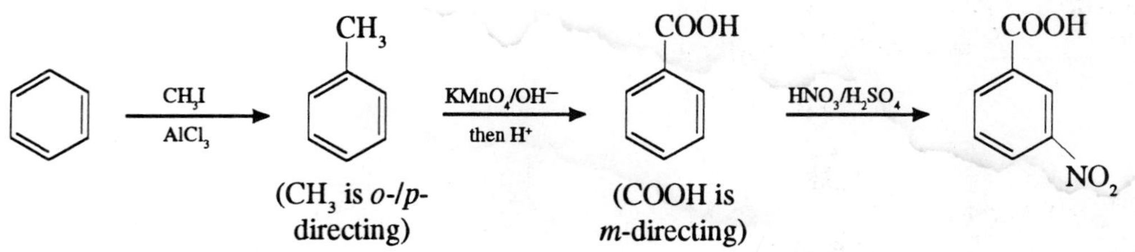

c. Since direct nitration of toluene will yield predominantly *p*-nitrotoluene, the *para* position must be blocked first until the *o*-position is nitrated.

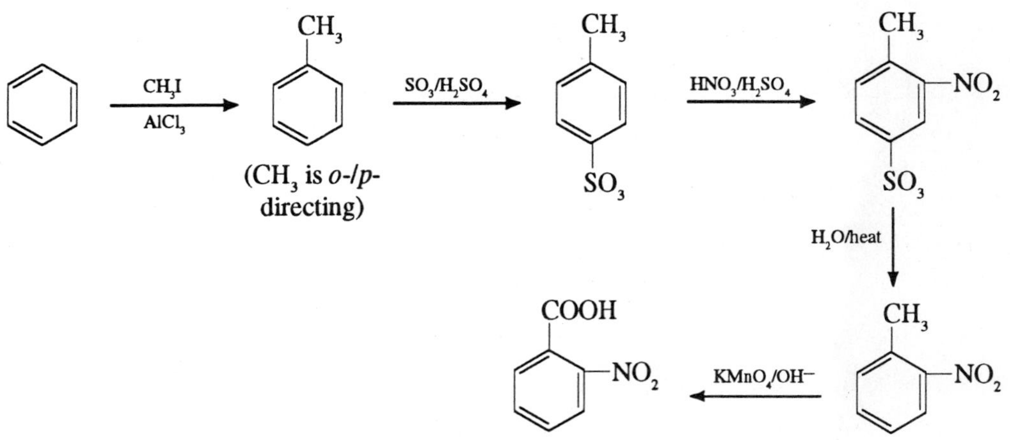

d.

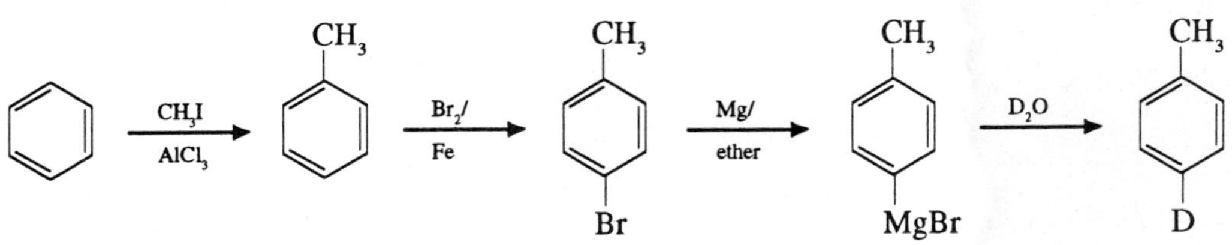

e. Prepare 2-nitrotoluene as in (c), then treat it with Sn/HCl.

f.

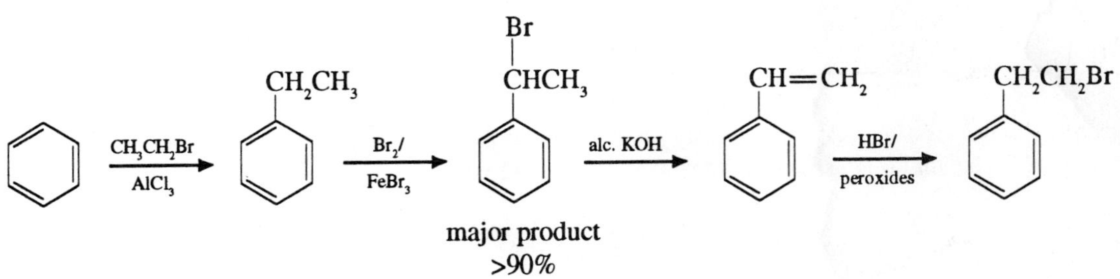

g.

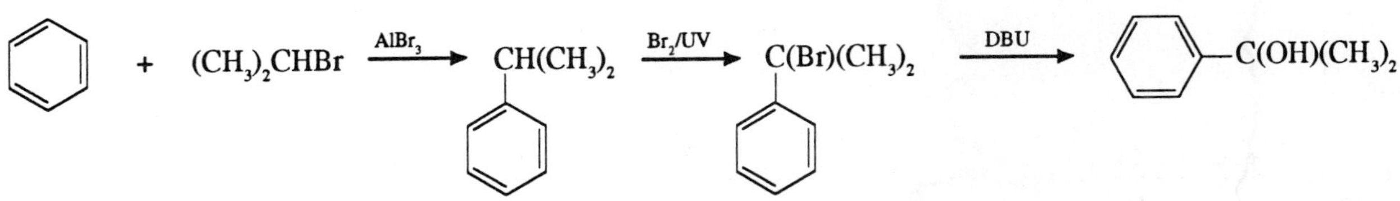

h.

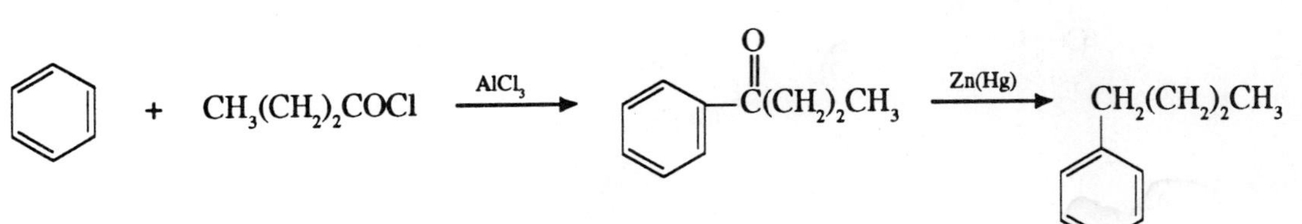

i.

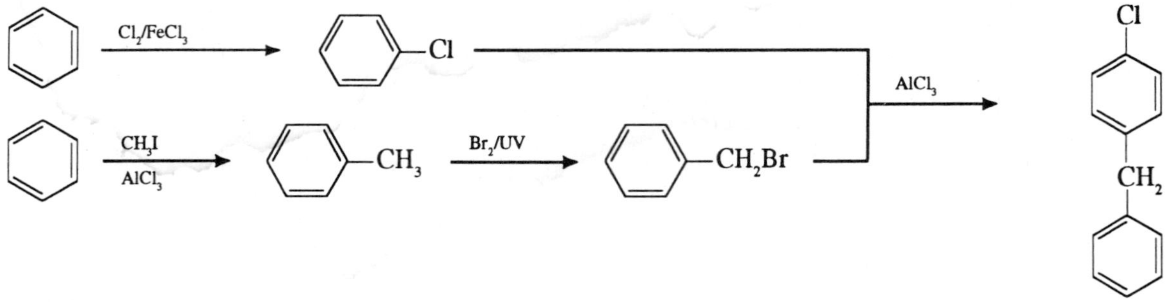

j.

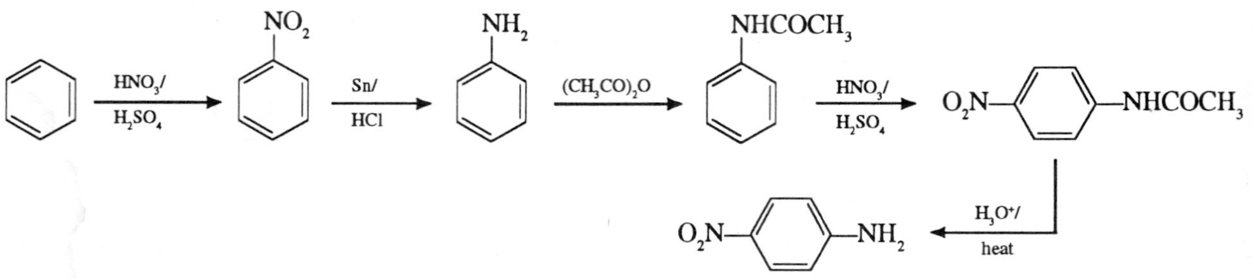

k. Prepare 1-phenylbutane as in (h), then brominate using Br_2/UV.

l. Prepare *o*-aminotoluene as in (e); then

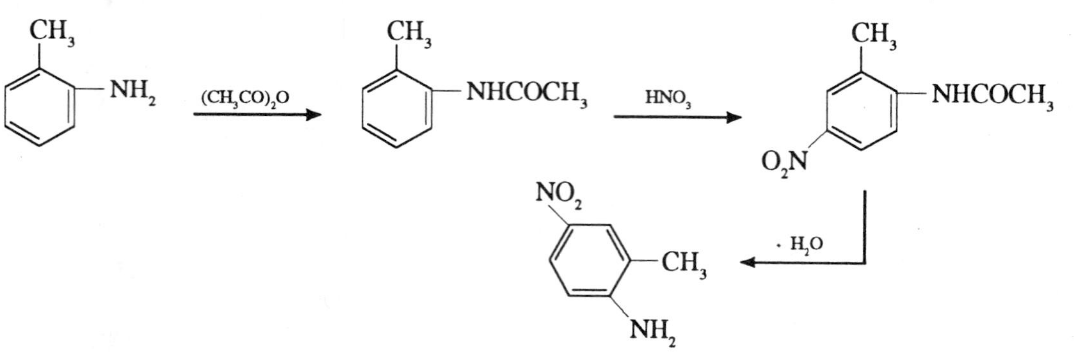

The acetylation process (first step) is necessary before the nitration (second step) to avoid multiple nitration and ring oxidation. The amino group can be regenerated by acid hydrolysis (third step).

m.

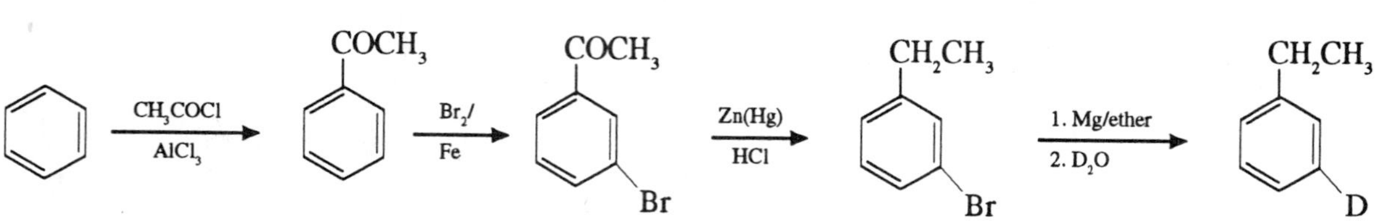

Answers for Multiple Choice Test

7.1 d	**7.6** e	**7.11** b	**7.16** d
7.2 d	**7.7** a	**7.12** d	**7.17** b
7.3 d	**7.8** b	**7.13** a	**7.18** c
7.4 c	**7.9** a	**7.14** b	**7.19** a
7.5 d	**7.10** c	**7.15** d	**7.20** b

8 Spectroscopy

Questions

Note: A concise listing of the significant ^{1}H NMR chemical shifts, IR wavenumber values, and isotope mass and relative abundances are given in Appendix I.

8.1 Suggest a reasonable structure for each of the following fragment cations frequently observed in mass spectroscopy:

a. $m/e = 29$ from a compound containing only C and H
b. $m/e = 43$ from a compound containing only C and H
c. $m/e = 91$ from a compound containing only C and H
d. $m/e = 77$ from a compound containing only C and H
e. $m/e = 43$ from a compound containing C, H and O
f. $m/e = 59$ from a compound containing C, H and O
g. $m/e = 105$ from an aromatic oxygen-containing compound

8.2 Propose structures for the following neutral molecules commonly lost in mass spectral fragmentation:

a. $m/e = 18$, from an oxygen-containing compound
b. $m/e = 28$, from an oxygen-containing, organic compound
c. $m/e = 80$, from a compound whose $P^{\dot{+}}/P^{\dot{+}}{+}2$ abundance ratio is almost equal to 1
d. $m/e = 36$, from a compound whose $P^{\dot{+}}/P^{\dot{+}}{+}2$ abundance ratio is approximately equal to 3
e. $m/e = 28$, from a hydrocarbon

8.3 Give a formula consistent with the given $P^{\dot{+}}$ isotope clusters

a. $P^{\dot{+}}$ = 182 (100%)
$P^{\dot{+}} + 1$ = 183 (9.2%)
$P^{\dot{+}} + 2$ = 184 (9%)
$P^{\dot{+}} + 3$ = 185 (1%)

b. $P^{\dot{+}}$ = 176 (100%)
$P^{\dot{+}} + 1$ = 177 (6.2%)
$P^{\dot{+}} + 2$ = 178 (96%)
$P^{\dot{+}} + 3$ = 179 (6.1%)

c. $P^{\dot{+}}$ = 138 (100%)
$P^{\dot{+}} + 1$ = 139 (5.2%)
$P^{\dot{+}} + 2$ = 140 (35%)
$P^{\dot{+}} + 3$ = 141 (2.2%)

d. $P^{\dot{+}}$ = 228 (100%)
$P^{\dot{+}} + 1$ = 229 (5.0%)
$P^{\dot{+}} + 2$ = 230 (198%)
$P^{\dot{+}} + 3$ = 231 (100%)
$P^{\dot{+}} + 4$ = 232 (96%)
$P^{\dot{+}} + 5$ = 233 (5%)

e. $P^{\dot{+}}$ = 149 (50%)
$P^{\dot{+}} + 1$ = 150 (3.1%)
$P^{\dot{+}} + 2$ = 151 (34%)
$P^{\dot{+}} + 3$ = 152 (2.1%)
$P^{\dot{+}} + 4$ = 153 (3.2%)

f. $P^{\dot{+}}$ = 173 (100%)
$P^{\dot{+}} + 1$ = 174 (6.2%)
$P^{\dot{+}} + 2$ = 175 (98%)
$P^{\dot{+}} + 3$ = 176 (6%)

8.4 Match the IR peak that would help distinguish between the following pairs of compounds:

	Compounds		Peak (cm^{-1})
a.	$CH_3CH_2CH_2OH$ and $(CH_3)_2CHOH$	**A**	2200
b.	$CH_3CH_2CH_2OH$ and $CH_3OCH_2CH_3$	**B**	1380 (doublet)
c.	$CH_3CH_2COCH_3$ and $CH_2{=}CHCOCH_3$	**C**	3300
d.	$CH_3CH_2C{\equiv}CH$ and $CH_2{=}CHCH_3$	**D**	1695
e.	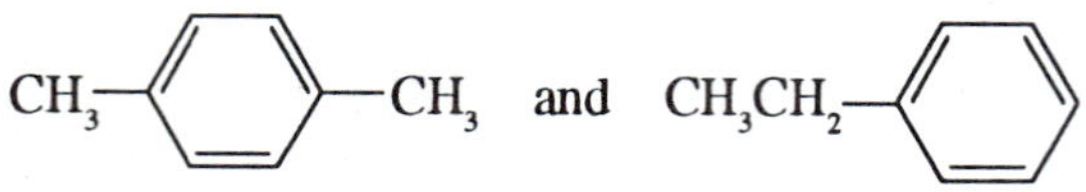	**E**	1100
f.	(*E*)-2-butene and (*Z*)-2-butene	**F**	810
g.	$CH_3CH(OH)(CH_2)_2CH_3$ and $CH_3(CH_2)_3CH_2OH$	**G**	980
h.	$CH_3CH_2CH_2CH_2OH$ and $CH_3CH_2C{\equiv}CH$	**H**	1710
i.	$CH_3CH_2COCH_2CH_3$ and $CH_3CH_2OCH_2CH_3$		
j.	$CH_3CH_2CH_2C{\equiv}CH$ and $CH_3CH_2C{\equiv}CCH_3$		

8.5 Identify compounds **A** through **D** in the sequence of reactions below using the given IR spectra. Identify the peaks that are characteristic for each functional group.

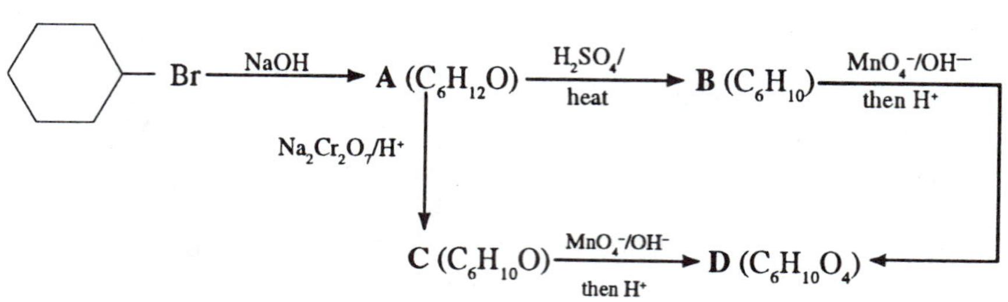

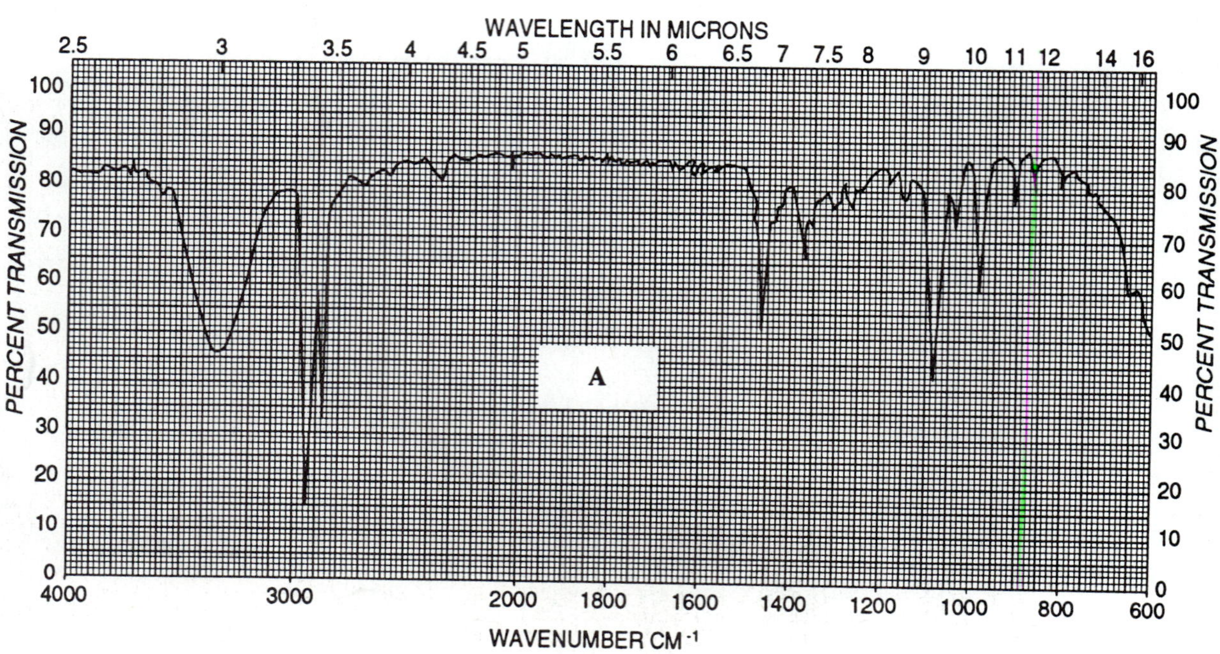

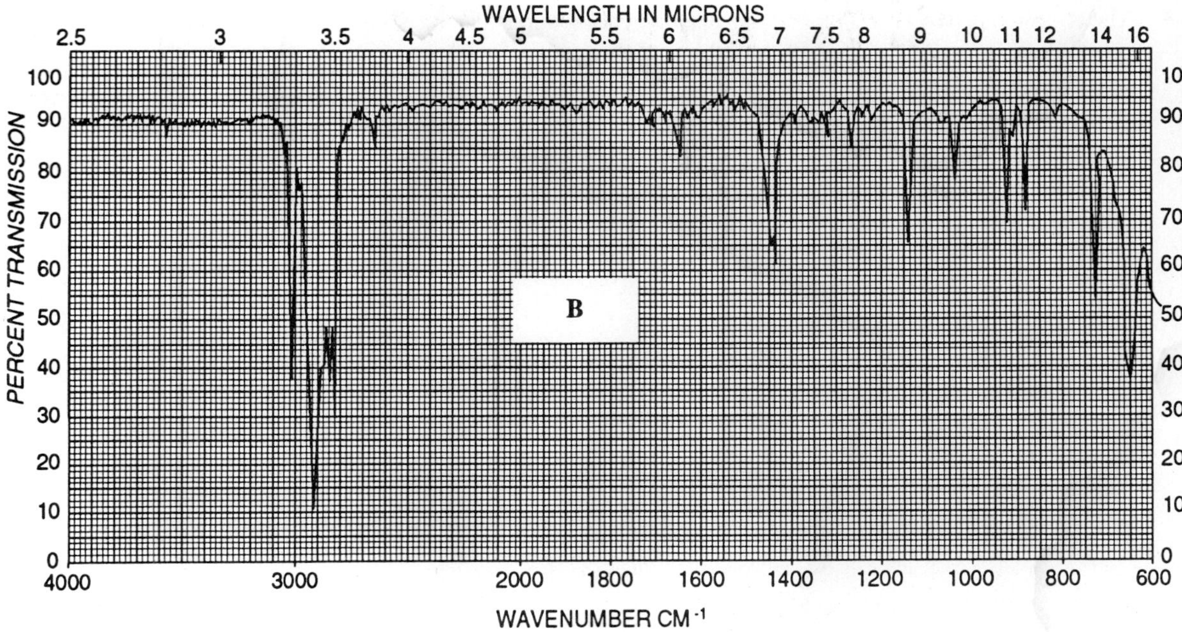
WAVELENGTH IN MICRONS
PERCENT TRANSMISSION
B
WAVENUMBER CM⁻¹

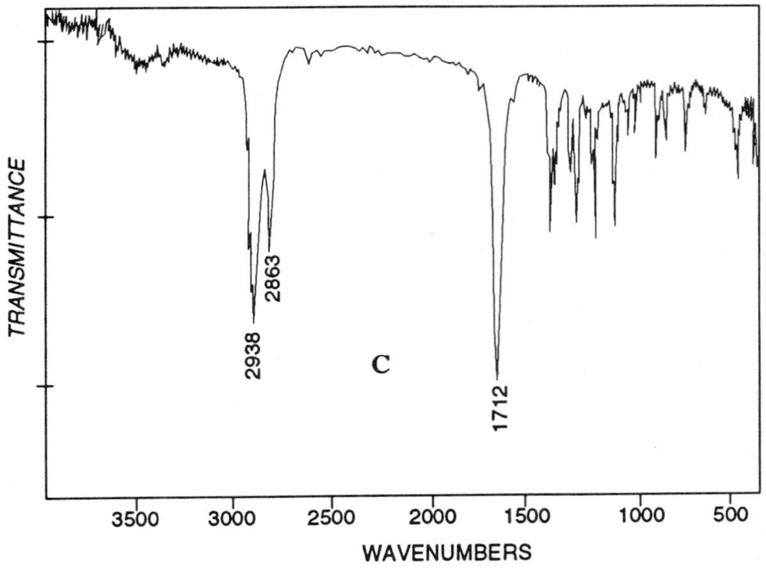
TRANSMITTANCE
2938
2863
1712
C
WAVENUMBERS

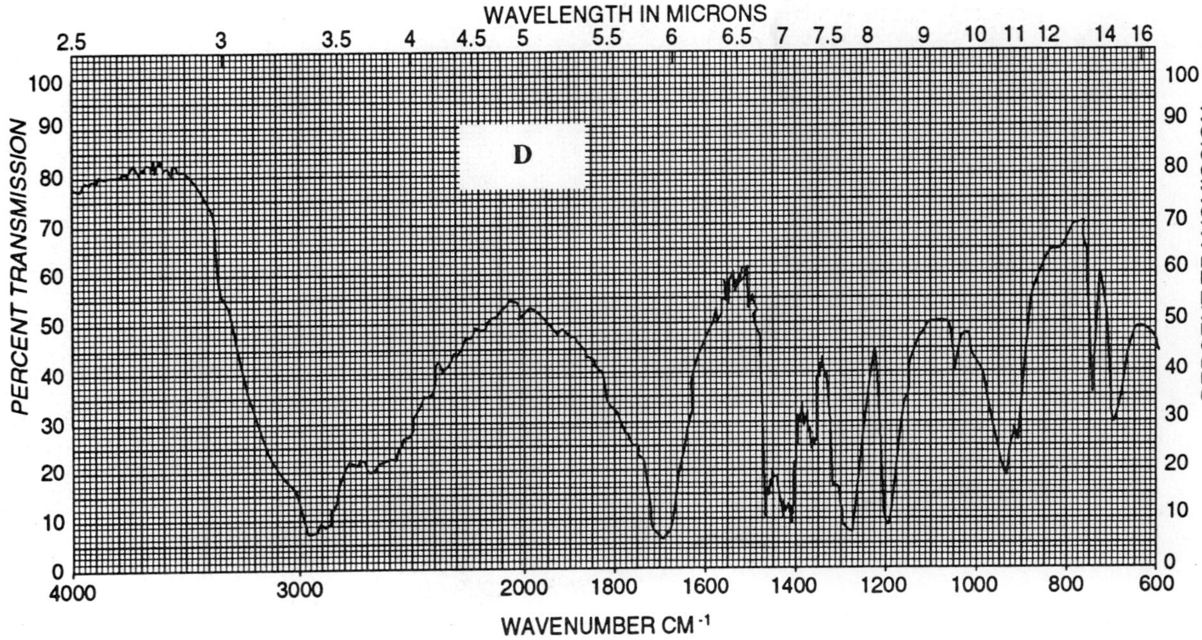
WAVELENGTH IN MICRONS
PERCENT TRANSMISSION
D
WAVENUMBER CM⁻¹

8.6 The following spectra correspond to four of the compounds listed. Show your reasoning in assigning structures for these spectra.

A isobutyraldehyde, $(CH_3)_2CHCHO$

B ethylbenzene,  C_2H_5

C butyric acid, $CH_3(CH_2)_2COOH$

D benzylmethylketone,  CH_2COCH_3

E benzoic acid, 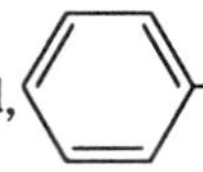 COOH

F 4-methyl-3-penten-2-one, $(CH_3)_2C{=}CHCOCH_3$ (mesityl oxide)

G ethylphenylketone, C_2H_5—CO— 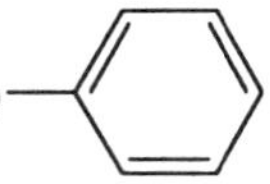

a.

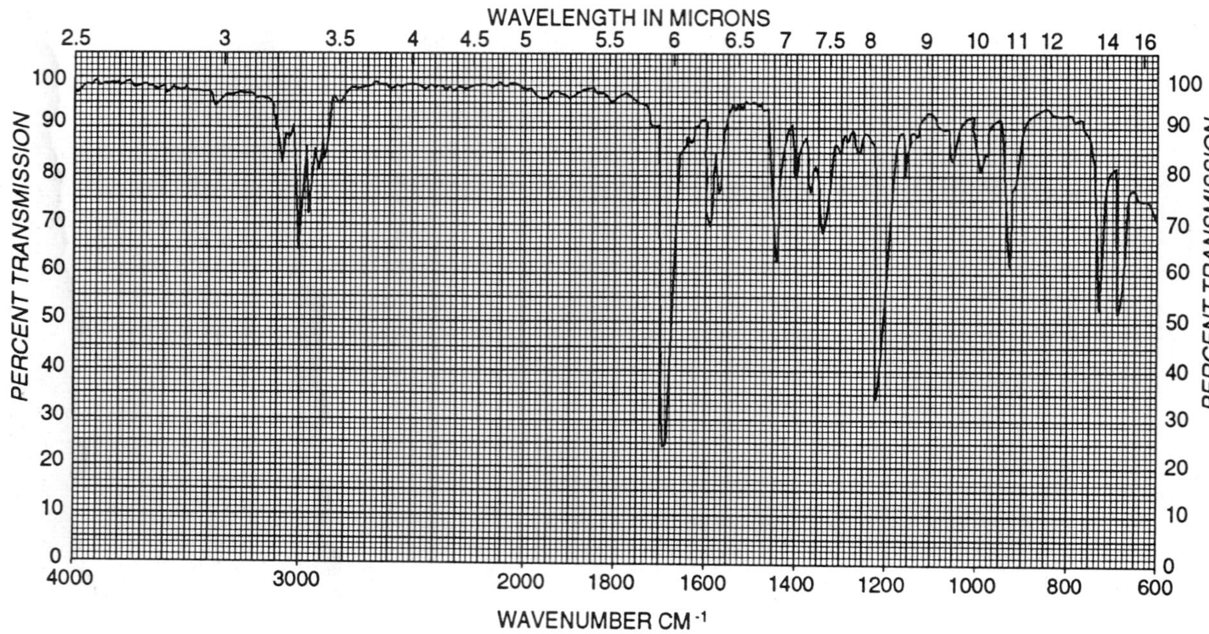

b.

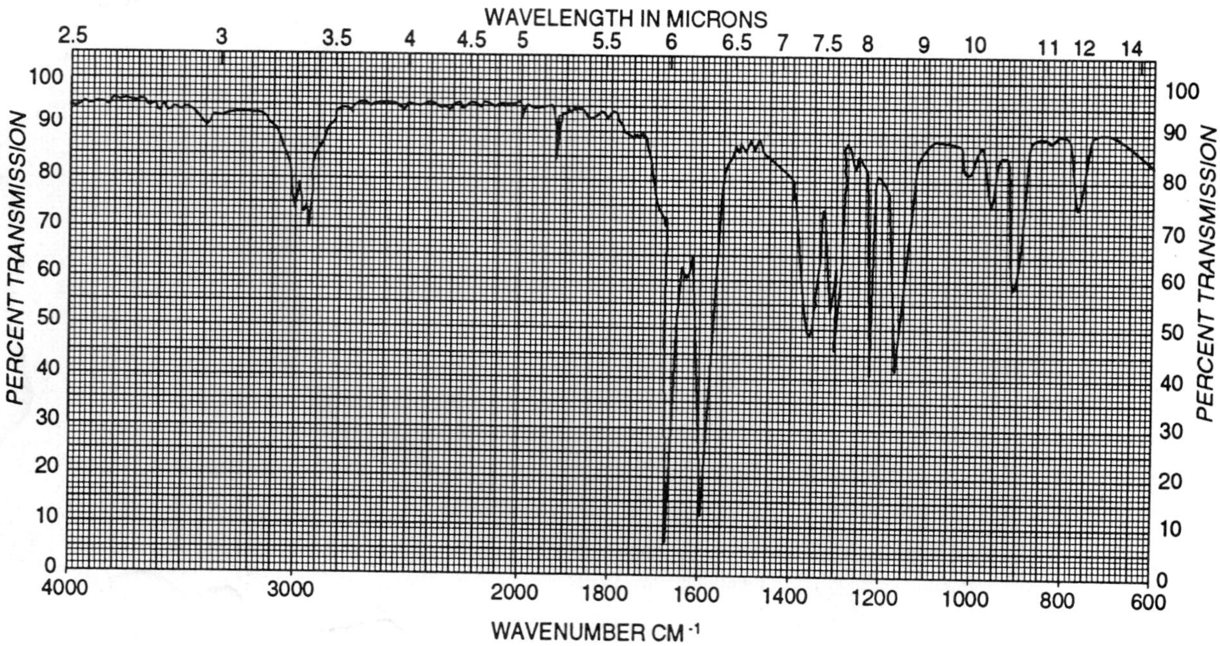

c.

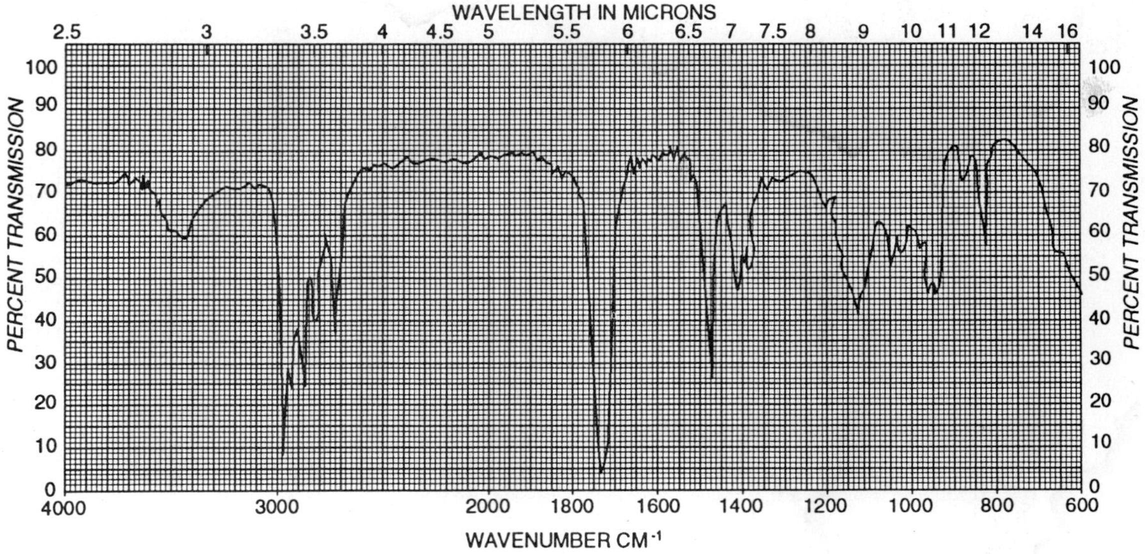

d.

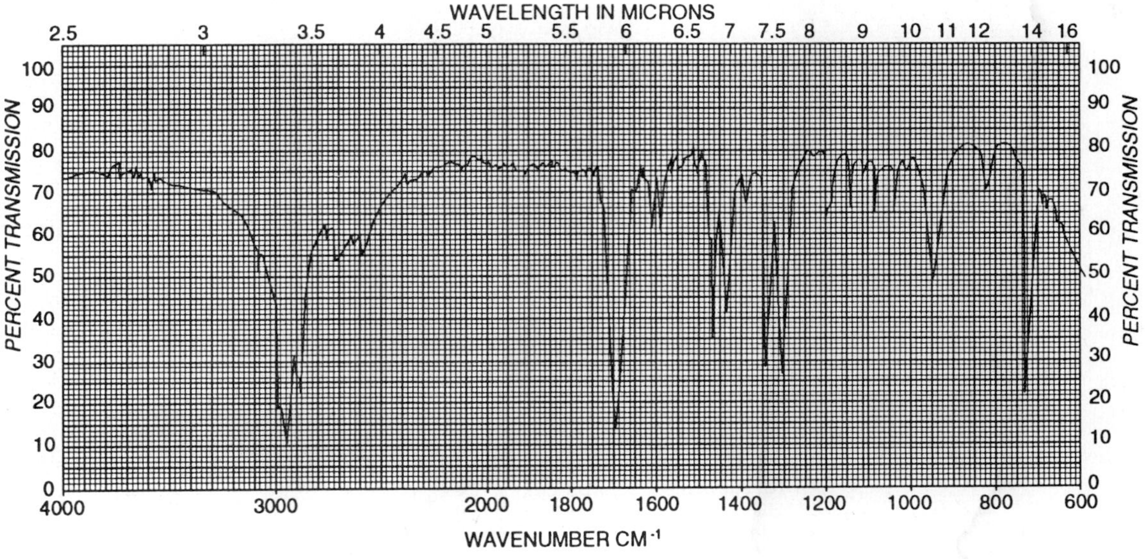

8.7 Identify the compounds whose IR spectra are seen below.

a.

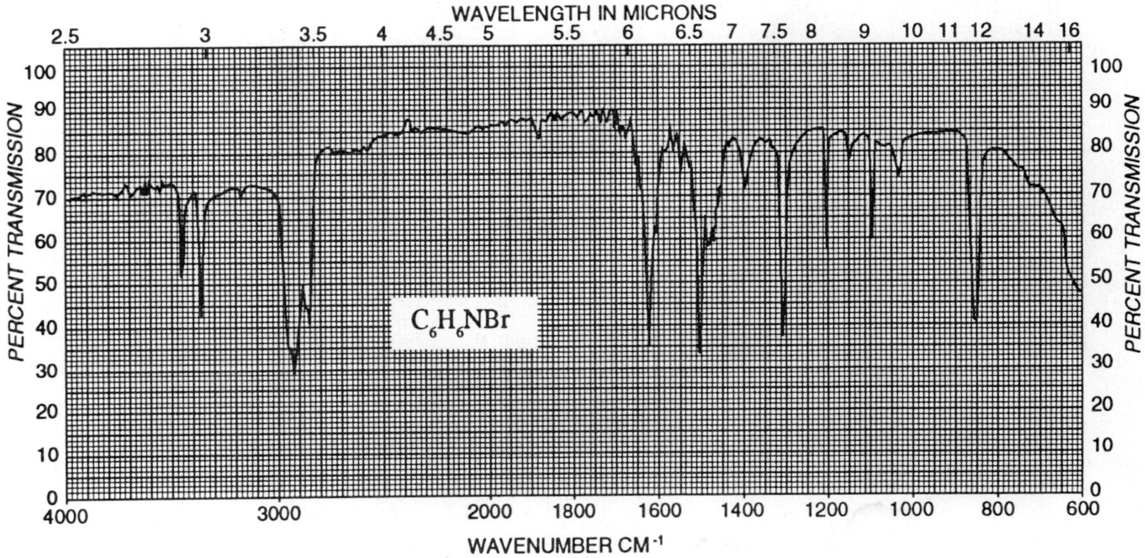

b.

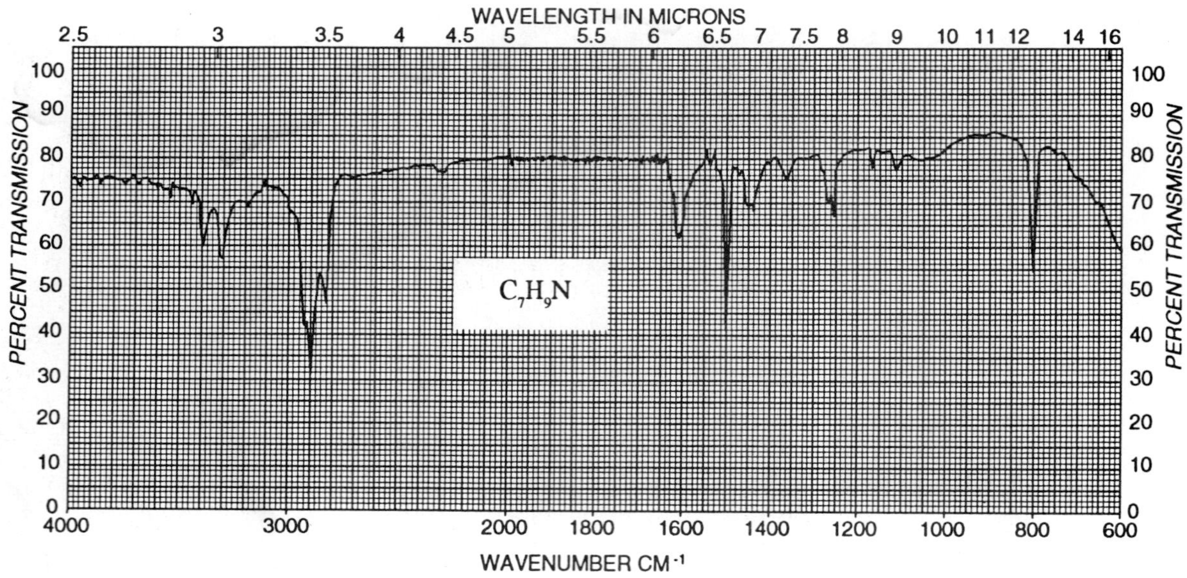
WAVELENGTH IN MICRONS
2.5 3 3.5 4 4.5 5 5.5 6 6.5 7 7.5 8 9 10 11 12 14 16
PERCENT TRANSMISSION
100 90 80 70 60 50 40 30 20 10 0
C_7H_9N
4000 3000 2000 1800 1600 1400 1200 1000 800 600
WAVENUMBER CM-1

c.

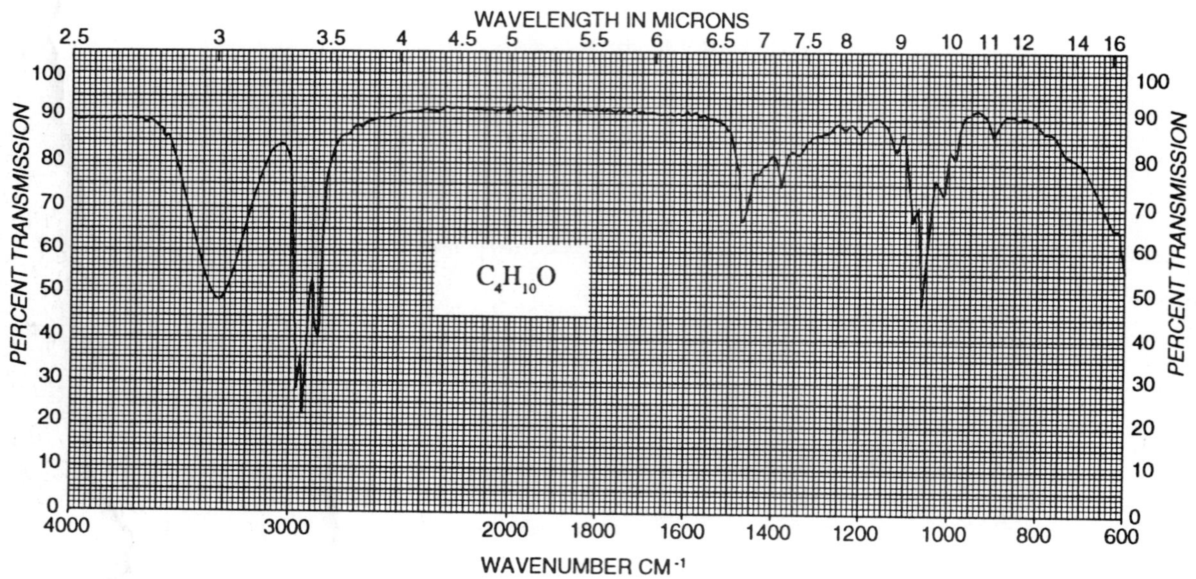
WAVELENGTH IN MICRONS
2.5 3 3.5 4 4.5 5 5.5 6 6.5 7 7.5 8 9 10 11 12 14 16
PERCENT TRANSMISSION
100 90 80 70 60 50 40 30 20 10 0
$C_4H_{10}O$
4000 3000 2000 1800 1600 1400 1200 1000 800 600
WAVENUMBER CM-1

d.

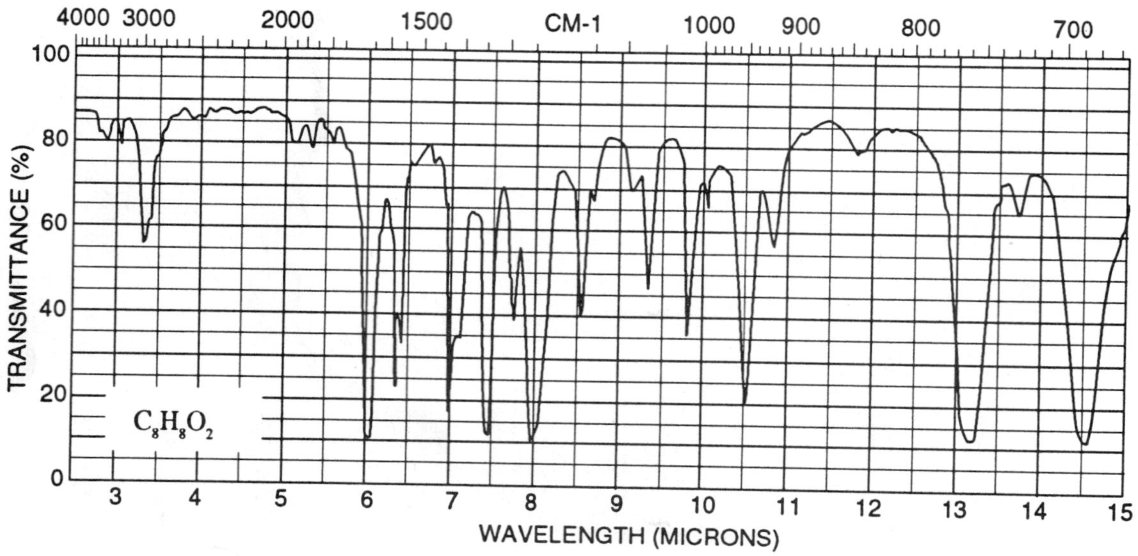
4000 3000 2000 1500 CM-1 1000 900 800 700
TRANSMITTANCE (%)
100 80 60 40 20 0
$C_8H_8O_2$
3 4 5 6 7 8 9 10 11 12 13 14 15
WAVELENGTH (MICRONS)

e.

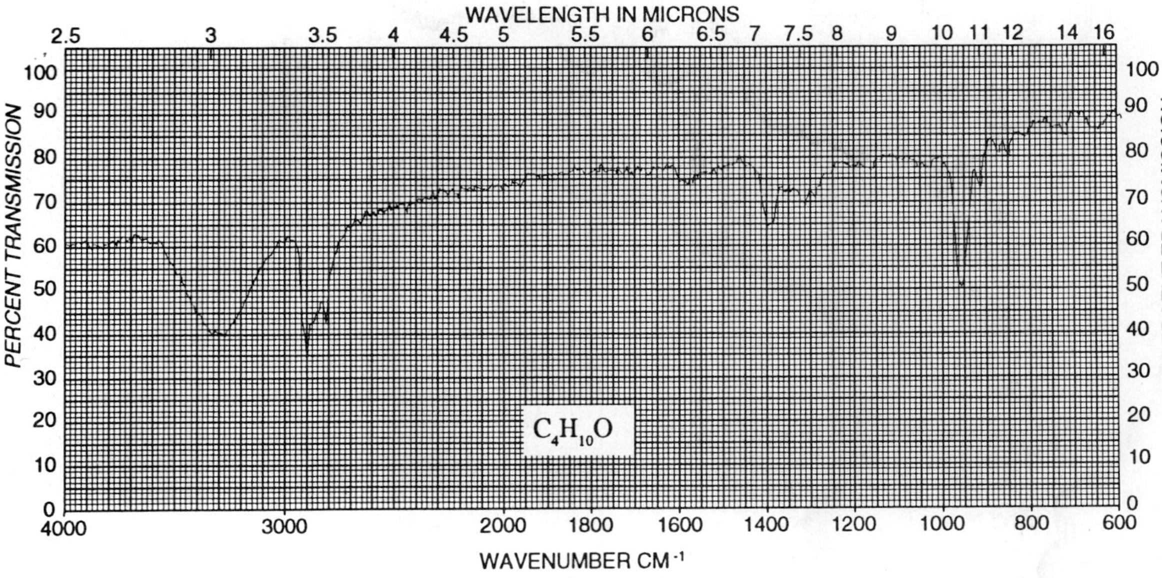
WAVELENGTH IN MICRONS
2.5 3 3.5 4 4.5 5 5.5 6 6.5 7 7.5 8 9 10 11 12 14 16
PERCENT TRANSMISSION
100 90 80 70 60 50 40 30 20 10 0
C4H10O
4000 3000 2000 1800 1600 1400 1200 1000 800 600
WAVENUMBER CM-1

f.

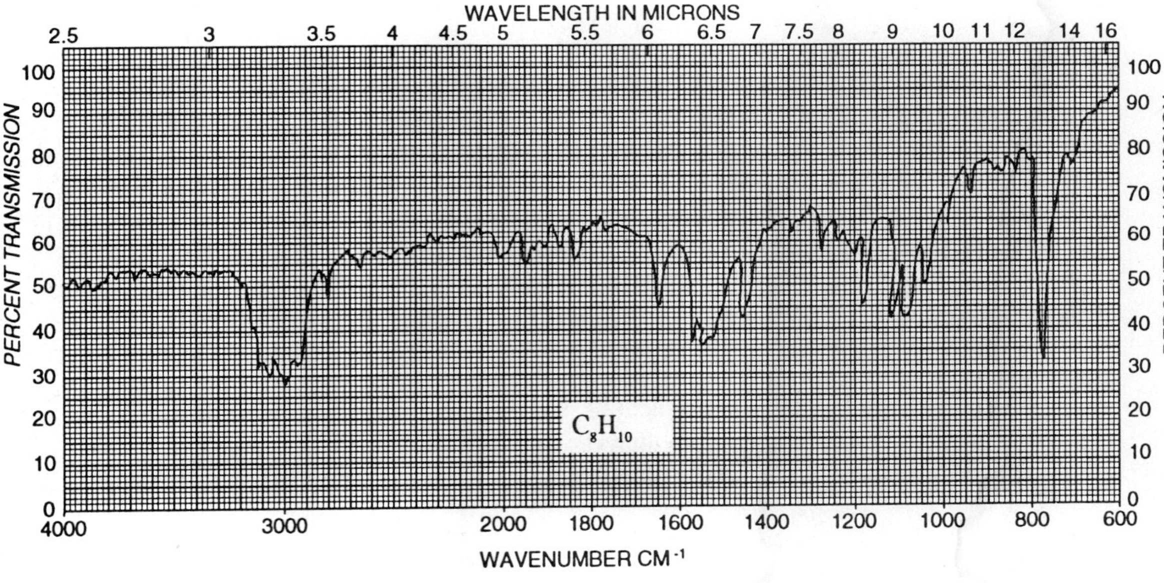
WAVELENGTH IN MICRONS
2.5 3 3.5 4 4.5 5 5.5 6 6.5 7 7.5 8 9 10 11 12 14 16
PERCENT TRANSMISSION
100 90 80 70 60 50 40 30 20 10 0
C8H10
4000 3000 2000 1800 1600 1400 1200 1000 800 600
WAVENUMBER CM-1

g.

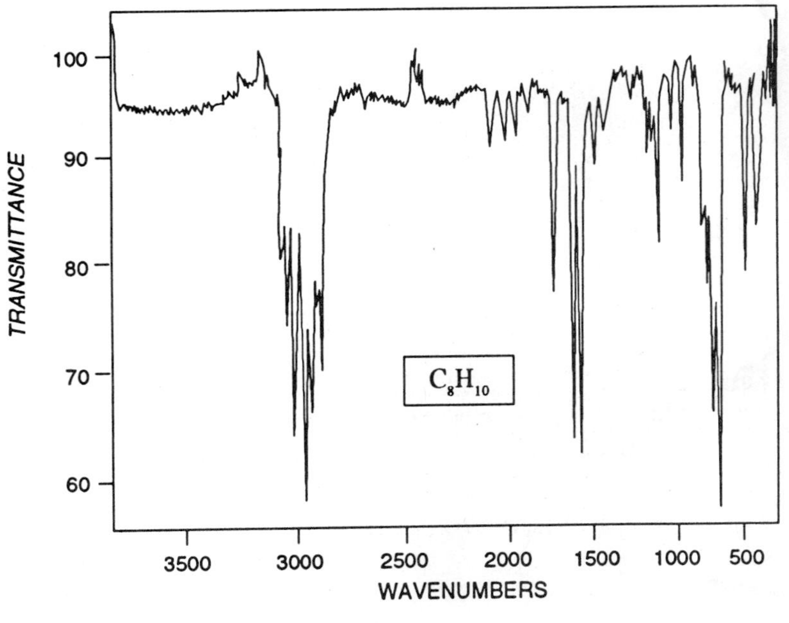
TRANSMITTANCE
100 90 80 70 60
C8H10
3500 3000 2500 2000 1500 1000 500
WAVENUMBERS

8.8 Identify the compounds of the given molecular formulas that show

a. *One ¹H NMR signal:*

C_2H_6O **A**	C_3H_4 **B**	$C_2H_4Br_2$ **C**
C_4H_6 **D**	C_5H_{12} **E**	C_3H_6 **F**

b. *Two ¹H NMR signals:*

$C_3H_6Cl_2$ **A**	$C_3H_6O_2$ **B**	$C_3H_8O_2$ **C**	C_2H_5OCl **D**

8.9 How many ¹H NMR signals will each of the compounds below show?

a. $(CH_3)_3CCH_2CH_3$
b. $CH_3COCOOCH_3$
c. $CH_3COCH_2COCH_3$
d. $CH_3COCH_2CH_2COCH_3$
e. $CH_3COCH_2CH_2COOCH_3$
f. $CH_3COCH_2CH(Cl)COCH_3$

8.10 Suggest structures consistent with the following ¹H NMR data:

a. C_4H_8O
triplet (1.0 ppm, 3H)
singlet (2.1 ppm, 3H)
quartet (2.4 ppm, 2H)

b. $C_4H_8O_2$
triplet (1.2 ppm, 3H)
singlet (2.1 ppm, 3H)
quartet (4.1 ppm, 2H)

c. C_8H_8O
singlet (2.4 ppm, 3H)
multiplet (7.5 ppm, 5H)

d. C_2H_5BrO
singlet (3.3 ppm, 3H)
singlet (4.9 ppm, 2H)

e. C_7H_8O
singlet (3.7 ppm, 3H)
multiplet (7.0 ppm, 5H)

f. $C_{10}H_{14}$
singlet (1.3 ppm, 9H)
singlet (7.3 ppm, 5H)

g. C_4H_9Br
doublet (1.1 ppm, 6H)
multiplet (1.9 ppm, 1H)
doublet (3.4 ppm, 2H)

h. $C_{10}H_{13}Cl$
singlet (1.6 ppm, 6H)
singlet (3.1 ppm, 2H)
multiplet (7.3 ppm, 5H)

i. $C_5H_8O_3$
triplet (1.2 ppm, 3H)
singlet (2.4 ppm, 3H)
quartet (4.2 ppm, 2H)

j. C_3H_7ClO
triplet (1.4 ppm, 3H)
quartet (3.8 ppm, 2H)
singlet (5.6 ppm, 2H)

k. $C_9H_{11}Br$
quintet (2.2 ppm, 2H)
triplet (2.8 ppm, 2H)
triplet (3.4 ppm, 2H)
multiplet (7.2 ppm, 5H)

l. $C_{10}H_{14}$
doublet (0.9 ppm, 6H)
multiplet (1.9 ppm, 1H)
doublet (2.4 ppm, 2H)
singlet (7.2 ppm, 5H)

m. $C_4H_8O_3$
triplet (1.3 ppm, 3H)
quartet (3.7 ppm, 2H)
singlet (4.2 ppm, 2H)
singlet (11.0 ppm, 1H)

n. $C_6H_{12}O_2$
singlet (1.2 ppm, 6H)
singlet (2.1 ppm, 3H)
singlet (2.8 ppm, 2H)
singlet (3.7 ppm, 1H)

8.11 Free radical chlorination of propane using 1 mole of C_3H_8 and 2 moles of chlorine gives a complex mixture of chlorintated products. Careful fractional distillation of the product mixture leads to the isolation of dichlorinated compounds **A**, **B**, **C** and **D**. Identify the structures using the ¹H NMR spectra data below:

A = triplet (1.2 ppm, 3H)
quintet (1.9 ppm, 2H)
triplet (5.8 ppm, 1H)

B = doublet (1.4 ppm, 3H)
doublet (3.8 ppm, 2H)
sextet (4.3 ppm, 1H)

C = singlet (2.4 ppm, 6H)

D = quintet (2.2 ppm, 2H)
triplet (3.7 ppm, 4H)

8.12 Identify each compound using the following data:

Compound	Composition	Mass Spectroscopy m/e (%)	IR peaks (cm^{-1})	^1H NMR (type, ppm, H)
a.		$P^{\dot{+}}$ 214 (100%) $P^{\dot{+}}$ + 1 215 (4.7%) $P^{\dot{+}}$ + 2 216 (198%) $P^{\dot{+}}$ + 3 217 (9.4%) $P^{\dot{+}}$ + 4 218 (96%)		singlet 3.4 (2H) singlet 1.9 (6H)
b. $C_{14}H_{22}$			1610, 1500, 1435, 810	7.1 and 1.2 ppm (relative ratio 2:9)
c. MW 69	69.6% C 10.1% H 20.3% N		2250, 1380 (doublet)	multiplet 2.7 (1H) doublet 1.4 (6H)
d. MW 74	64.9% C 13.5% H		1435, 1380 (doublet), 1100	multiplet 3.7 (1H) singlet 3.3 (3H) doublet 1.2 (6H)
e. MW 264	31.8% C 2.3% H 59.8% Br		3300, 1580, 1480, 1435, 1180	singlet 7.1 (2H) singlet 5.0 (1H) singlet 2.1 (3H)
f. MW 214	22.4% C 3.7% H 73.8% Br			multiplet 4.3 (1H) triplet 3.5 (2H) quartet 2.4 (2H) doublet 1.8 (3H)
g.		$P^{\dot{+}}$ 137 (100%) $P^{\dot{+}}$ + 1 138 (9.1%) $P^{\dot{+}}$ + 2 139 (0.3%)	3500, 3400, 1620, 1510, 1250, 810	doublet of doublets 6.8 (4H) quartet 4.0 (2H) singlet 3.0 (2H) triplet 1.3 (3H)
h.		$P^{\dot{+}}$ 150 (100%) $P^{\dot{+}}$ + 1 151 (6%) $P^{\dot{+}}$ + 2 152 (98%)		triplet 3.2 (2H) mult 1.9 (3H) doublet 1.0 (6H)

8.13 Identify the compounds whose spectra are seen below:

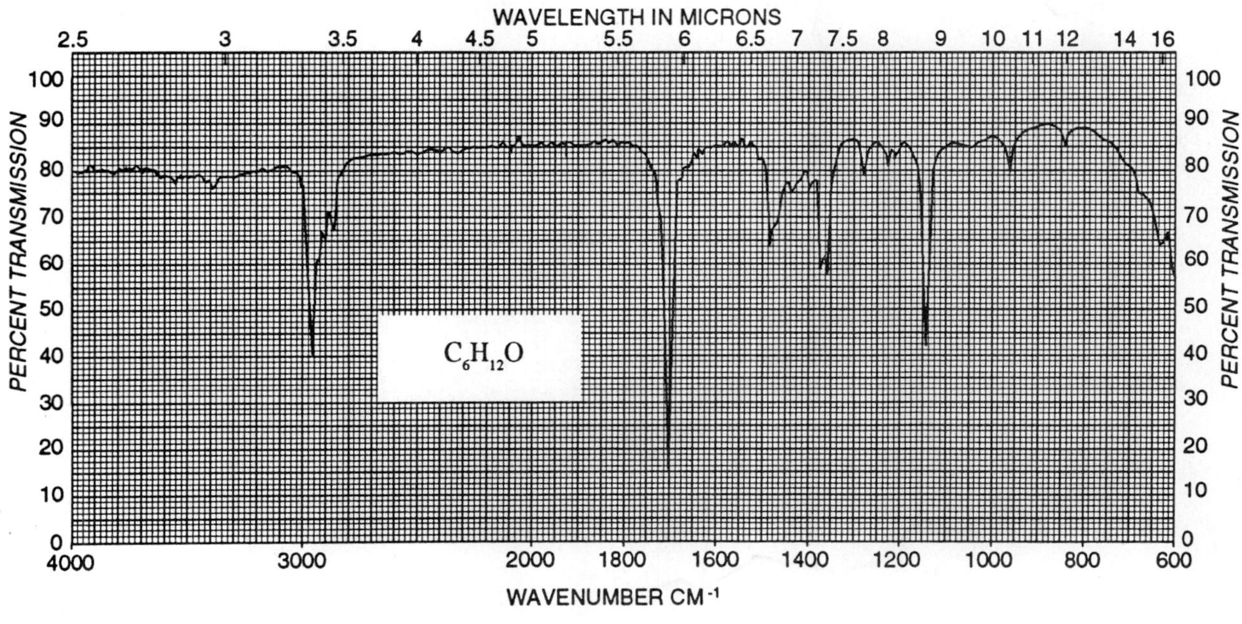

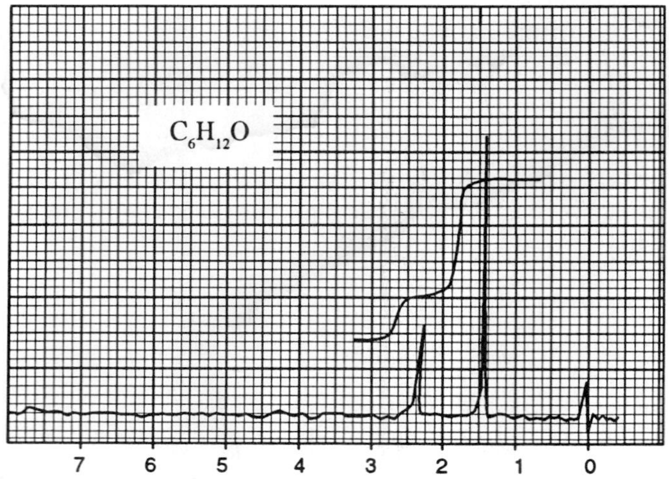

$C_6H_{12}O$
7
6
5
4
3
2
1
0

b.

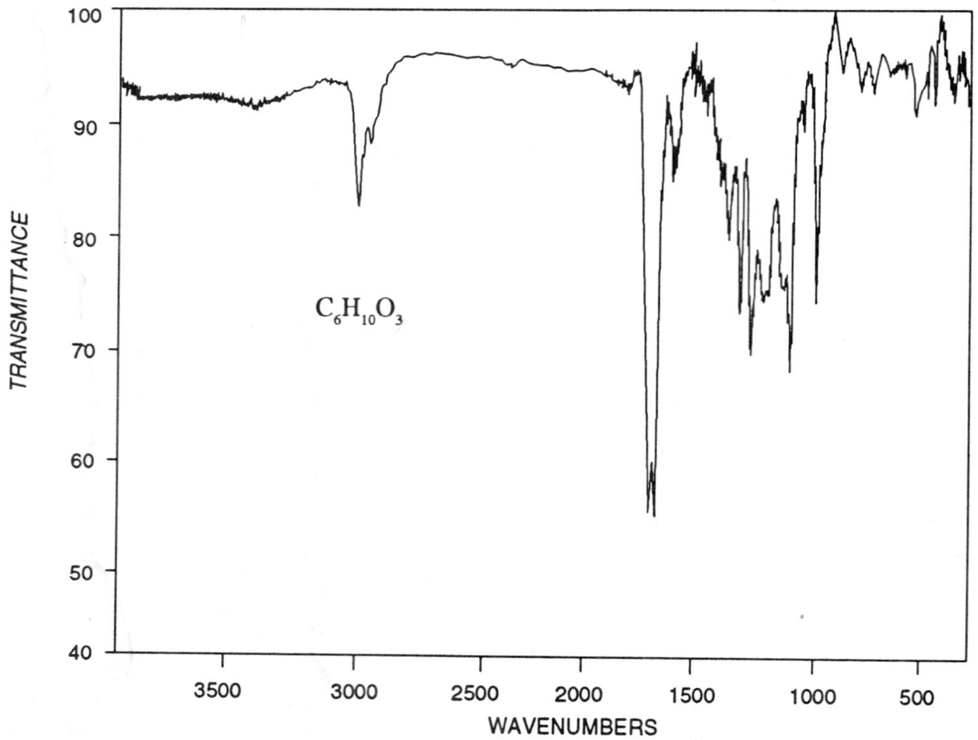

TRANSMITTANCE
100
90
80
70
60
50
40
$C_6H_{10}O_3$
3500
3000
2500
2000
1500
1000
500
WAVENUMBERS

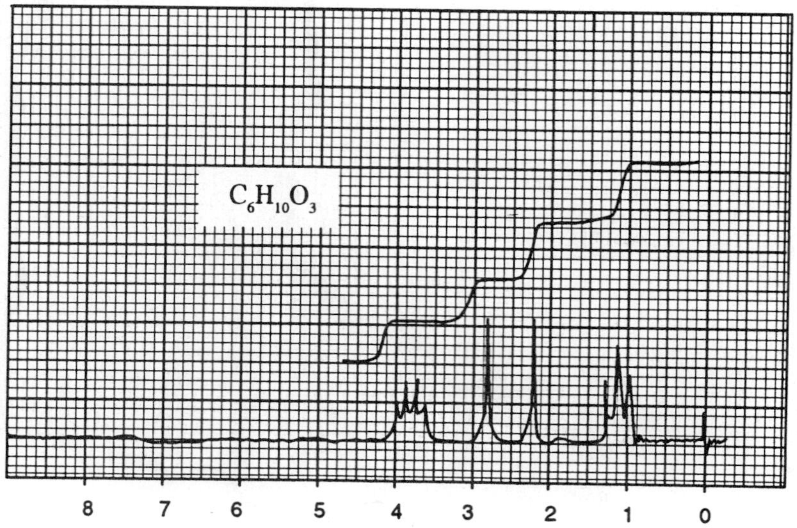

$C_6H_{10}O_3$
8
7
6
5
4
3
2
1
0

c.

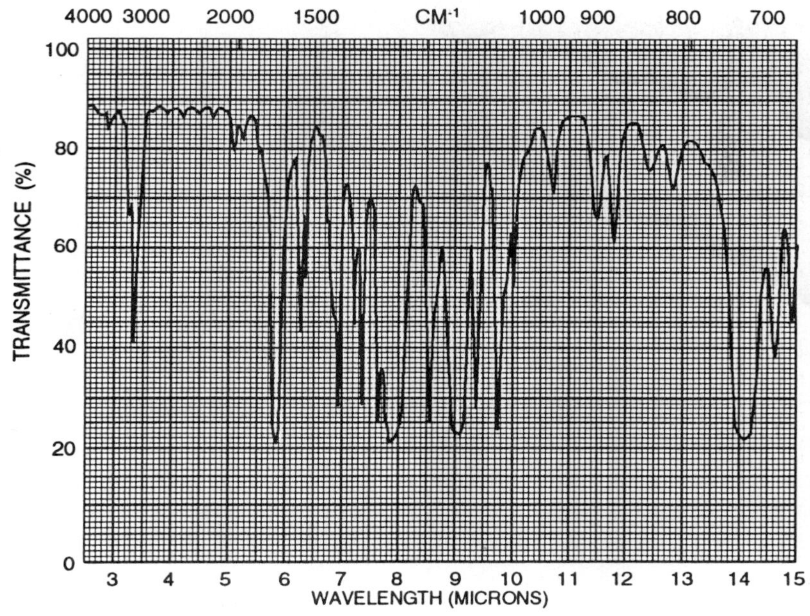

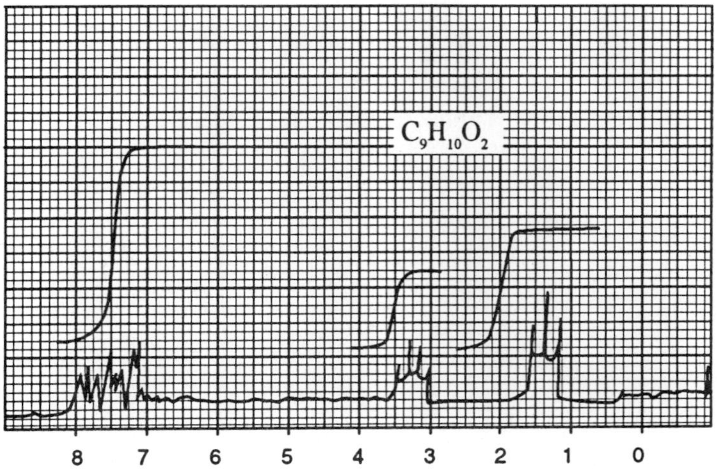

d. *Mass Spectroscopy data:*
$P^{\dot{+}}$ = 136 (100%) $P^{\dot{+}}$ + 1 = 137 (9%) $P^{\dot{+}}$ + 2 = 138 (0.7%) *also peaks at:* m/e = 77, 91, 119

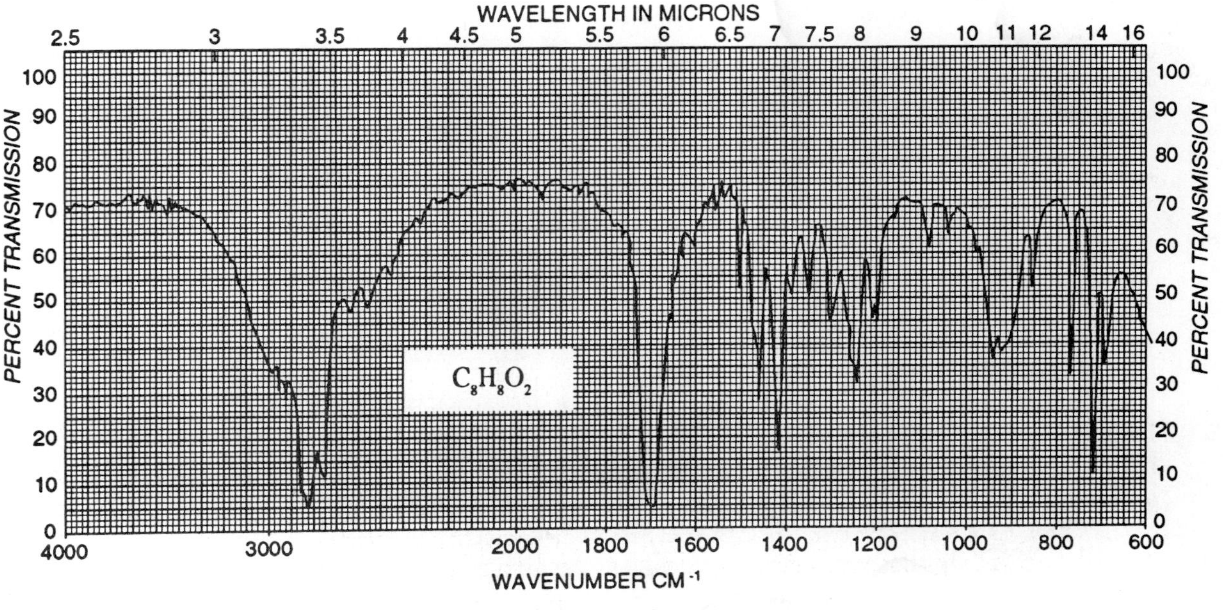

e.

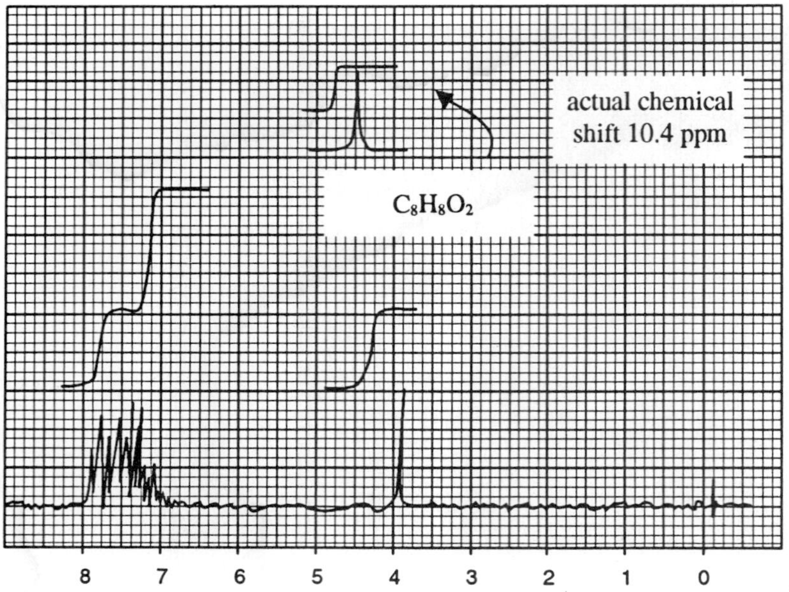

Multiple Choice Test

8.1 Which of the following methyl groups will show the most downfield ¹H NMR chemical shift?
a. CH_3I b. CH_3OH c. CH_3COCH_3 d. CH_3CH_3

8.2 How many ¹H NMR signals will 2-chloro-3-methyl-2-butene exhibit?
a. one b. two c. three d. four

8.3 The ¹H NMR spectrum of $CH_3OCH_2CH_2OCH_3$ shows
a. two singlets (integration 3 : 2)
b. one singlet and one triplet (integration 3 : 2)
c. two singlets and two triplets (integration 3 : 3: 2 : 2)
d. one singlet and two triplets (integration 6 : 2 : 2)

8.4 Which of the following compounds has/have a signal disappear when D_2O is added?

CH_3OH	CH_3COOH	C_6H_5OH	CH_3CH_2I
A	**B**	**C**	**D**

a. **A** b. **A** and **C** c. all but **D** d. all of the above

8.5 Which of the following compounds has an odd $P^{+\cdot}$?

CH_3NH_2	$BrCH_2CH_3$	CH_3N	CH_3NHNH_2
A	**B**	**C**	**D**

a. **A** b. **A** and **C** c. all but **B** d. all of the above

8.6 Which compound of molecular formula C_4H_9Cl displays the following NMR spectrum:
1.1 ppm (doublet, 6H), 1.90 ppm (multiplet, 1H), 3.35 ppm (doublet, 2H)?

a. 1-chlorobutane
b. isobutyl chloride
c. *t*-butyl chloride
d. *sec*-butyl chloride

8.7 The molecular ion cluster of a compound shows the following ($P^{+\cdot}$)/($P^{+\cdot}$ + 2)/($P^{+\cdot}$ + 4) relative ratio of intensities: 1: 2 : 1. The molecular formula of the compound contains_____atom(s).
a. one nitrogen b. one bromine c. two chlorine d. two bromine

8.8 Which alcohol of molecular formula $C_4H_{10}O$ exhibits the following IR peaks (in cm^{-1}): 3300, 1380 (doublet), 1058?
a. 1-butanol b. 2-butanol c. 2-methyl-1-propanol d. *t*-butyl alcohol

8.9 Which compound(s) would show an IR peak in the 3600–3200 cm^{-1} region?

$CH_3CH_2C{\equiv}N$	$(CH_3)_3COH$	$CH_3CONHCH_3$	$CH_3C{\equiv}CH$
A	**B**	**C**	**D**

a. **B** b. **B** and **D** c. all but **A** d. all of the above

8.10 Addition of H_2O/H^+ to styrene would produce a compound that will show a peak at ______ cm^{-1} on an IR spectrum.
a. 1740 b. 3300 c. 1695 d. 1050

8.11 UV bromination of ethylbenzene will form a compound that will show three signals (integration ______) in ^1H NMR.
a. 5 : 1 : 3 b. 5 : 2 : 2 c. 4 : 2 : 3 d. 5 : 1 : 2

8.12 Which of the following compounds would have an off-resonance decoupled ^{13}C NMR spectrum that shows two quartets, a triplet, and a doublet?
a. $CH_3CH_2OCH_2CH_2CH_3$ c. $CH_3OCH_2C(CH_3)_3$
b. $CH_3CH_2OCH(CH_3)_2$ d. $CH_3CH_2CH_2OH$

8.13 Which of the following compounds would have an off-resonance decoupled ^{13}C NMR spectrum that shows two doublets and a singlet in the 120–140 ppm region?
a. *o*-xylene b. *m*-xylene c. *p*-xylene d. ethylbenzene

8.14 Which compound ($C_6H_{12}O_2$) shows two singlets at 1.3 and 2.1 ppm with a corresponding relative integration ratio of 3 : 1 in ^1H NMR?
a. $CH_3OOCC(CH_3)_3$ c. $(CH_3O)_2C{=}C(OCH_3)_2$
b. $CH_3COOC(CH_3)_3$ d. $CH_3COCH_2CH_2COCH_3$

8.15 The loss of 28 amu in mass spectroscopy is due to the loss of
a. ethyl b. hydrogen cyanide c. carbon monoxide d. acetylene

8.16 The energy levels involved in UV spectroscopy are ______ levels.
a. vibrational b. rotational c. nuclear d. electronic

8.17 Which of the following compounds shown below would have the highest value for λ_{max} in UV.
a. 1,5-octadiene b. octene c. 1,4,7-octatriene d. 2,4-octadiene

8.18 Which of the following technique(s) can readily distinguish between $CH_3COOCH_2CH_3$ and $CH_3OOCCH_2CH_3$?

UV	IR	^1H NMR	^{13}C NMR	Mass spectroscopy
A	**B**	**C**	**D**	**E**

a. **C** and **D** b. **C**, **D**, and **E** c. all but **B** d. all but **A**

8.19 The presence of a peak at *m/e* = 105 in mass spectroscopy is expected for
a. $C_6H_5COOCH_3$ b. $C_6H_5OOCCH_3$ c. *p*-$CH_3{-}C_6H_4Cl$ d. $(CH_3)_3C{-}C_6H_5$

8.20 The IR peak that distinguishes between a 1° amine and a 1° amide is at ______ cm^{-1}.
a. 3450 and 3350 (two peaks) b. 2200 c. 1650 d. 1350

Answers

8.1 a. $C_2H_5^{+\cdot}$ e. $[CH_3C{\equiv}O]^{+\cdot}$

b. $C_3H_7^{+\cdot}$ f. $C_2H_3O_2^{+\cdot}$ or $C_3H_7O^{+\cdot}$

c. $C_6H_5CH_2^{+\cdot}$ or  g. $[C_6H_5C{\equiv}O]^{+\cdot}$

d. $C_6H_5^{+\cdot}$

8.2 a. H_2O b. CO c. HBr d. HCl e. $CH_2{=}CH_2$

8.3 a. Two sulfur atoms (intensity of $P^{+\cdot} + 2$ peak approximately equal to $2 \times 4.2\%$). Eight carbons: $(100/1.1) \times (P^{+\cdot} + 1) \div (P^{+\cdot}) = 8$. Leftover mass is $182 - [(2 \times 32) + (8 \times 12)] = 22$, which is too much for hydrogen atoms to account for. A reasonable formula would involve one oxygen and six hydrogens, $C_8H_6S_2O$. An odd number of nitrogens is not possible since $P^{+\cdot}$ is even.

b. The $(P^{+\cdot} + 2)/P^{+\cdot}$ intensity ratio suggests the presence of one bromine. There are five carbons $(100/1.1) \times (P^{+\cdot} + 1) \div (P^{+\cdot}) = 5$. Leftover mass is $176 - [79 + (5 \times 12)] = 37$, which is too much for hydrogen atoms to account for. Two oxygens or two nitrogens would give two reasonable structures: $C_5H_5BrO_2$ and $C_5H_9BrN_2$, respectively.

c. The $(P^{+\cdot} + 2)/P^{+\cdot}$ intensity ratio suggests the presence of one chlorine. There are four carbons $(100/1.1) \times (P^{+\cdot} + 1) \div (P^{+\cdot}) = 4$. Leftover mass is $138 - [35 + (4 \times 12)] = 55$, too much for hydrogen atoms to account for. Three oxygens give a reasonable structure: $C_4H_7ClO_3$. Another possibility would involve two nitrogens (even $P^{+\cdot}$, even number) and an oxygen, $C_4H_{11}N_2OCl$.

d. The $(P^{+\cdot})/(P^{+\cdot} + 2)/(P^{+\cdot} + 4)$ intensity ratio (1 : 2 : 1) suggests two bromine atoms. There are four carbons $(100/1.1) \times (P^{+\cdot} + 1) \div (P^{+\cdot}) = 4$. Leftover mass is $228 - [(2 \times 79) + (4 \times 12)] = 22$, which is too much for hydrogen atoms to account for. Once more, oxygen will help solve the problem. Formula: $C_4H_6Br_2O$.

e. Odd $P^{+\cdot}$ suggests the presence of one nitrogen. The $(P^{+\cdot})/(P^{+\cdot} + 2)/(P^{+\cdot} + 4)$ intensity ratio suggests two chlorines. There are five carbons $(100/1.1) \times (P^{+\cdot} + 1) \div (P^{+\cdot}) = 5$. Leftover mass is $149 - [14 + (2 \times 35) + (5 \times 12)] = 5$, which is accounted for by hydrogens. Formula: $C_5H_5NCl_2$.

f. Odd $P^{+\cdot}$ suggests the presence of one nitrogen. The $(P^{+\cdot})/(P^{+\cdot} + 2)$ intensity ratio suggests one bromine. There are five carbons $(100/1.1) \times (P^{+\cdot} + 1) \div (P^{+\cdot}) = 5$. Leftover mass is $173 - [14 + 79 + (5 \times 12)] = 20$, too much for hydrogens to account for. A reasonable formula would involve one oxygen and four hydrogens, C_5H_4NBrO.

8.4
a. **B** (1380, *iso* group) or **E** (1100, 2° alcohol)
b. **C** (3300, —OH)
c. **D** (1695, conjugated >C=O)
d. **A** (2200, C≡C)
e. **F** (810, *p*-substitution)
f. **G** (980, *trans* or *E*)
g. **E** (1100, 2° alcohol)
h. **A** (2200, —C≡C)
i. **H** (1700, >C=O)
j. **C** (3300, —C≡C—H)

8.5 **A** Number of double bonds and/or rings: 1.
There must be a ring (since the starting material is an alicyclic compound). The peak at 3350 cm^{-1} is due to an —OH, indicating an alcohol, which is a secondary one (peak at 1090 cm^{-1}). The natural conclusion is that **A** is

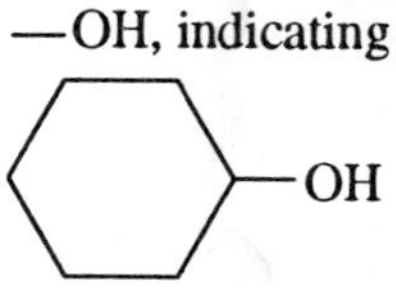

B Number of double bonds and/or rings: 2.
Comparison of **A** and **B** shows absence of the —OH peak (at 3350 cm^{-1}) and appearance of two peaks at 3030 and 1650 cm, indicating the formation of a double bond. Therefore, **B** is

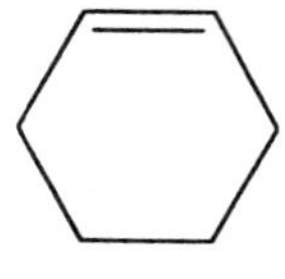

C Number of double bonds and/or rings: 2.
Comparison of **A** and **C** shows absence of —OH peak (at 3300 cm^{-1}) and appearance of a new peak at 1712 cm^{-1}, indicating the presence of a carbonyl C=O. Therefore, **C** is

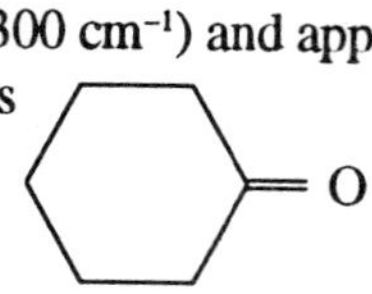

D Number of double bonds and/or rings: 2.
Comparison of **B** and **D** shows the appearance of a carboxylic acid group (broad peak in the range 3300–2500 cm^{-1} (—OH) and 1720–1680 cm^{-1} (C=O). The broadness of the carbonyl peaks recommends the presence of two —COOH groups. Therefore, **D** is $HO_2C(CH_2)_4CO_2H$

8.6 a. Absence of an —OH peak (3300 cm^{-1}) eliminates **C** and **E**. Presence of a C=O peak (1680 cm^{-1}) eliminates **B**. The peak is below 1700 cm^{-1}, which suggests conjugation of the C=O. This eliminates **A** and **D**, leaving only **F** and **G**. Mesityl oxide has a C=C, which is absent in this spectrum (no peak at 1620–1640 cm^{-1}). This compound is, therefore, $C_6H_5{-}CO{-}C_2H_5$.

b. Absence of an —OH peak (3300 cm^{-1}) eliminates **C** and **E**. Presence of a C=O peak (1680 cm^{-1}) eliminates **B**. Compounds **A** and **D** are ruled out because their C=O is not conjugated. From the remaining compounds **F** and **G** only the former one has an alkene C=C (1620 cm^{-1}), which is present in the spectrum. Therefore, this compound is $(CH_3)_2C{=}CH{-}\underset{\underset{O}{\|}}{C}{-}CH_3$.

c. Absence of an —OH peak (3300 cm^{-1}) eliminates **C** and **E**. Presence of a C=O peak (1720 cm^{-1}) rules out **B**. Structures **F** and **G** are not possible since their carbonyl is conjugated. This leaves only **A** (an aldehyde) and **D** (a ketone). The peaks at 2850 and 2950 cm^{-1} suggest an aldehyde. This, plus the absence of a peak at 1600 cm^{-1} which would indicate a benzene ring (which is present in **D**) and the presence of a doublet at 1380 cm^{-1} (*iso* group) suggest that this compound is $(CH_3)_2CH{-}CHO$.

d. The broad peak in the 3300–2600 cm^{-1} range and the peak at 1690 cm^{-1} suggest a carboxylic acid. This rules out **A, B, D, F,** and **G**. From the remaining compounds **C** and **E**, only benzoic acid has a benzene ring (peaks at 1605, 1590, and 1500 cm^{-1}) and a conjugated carbonyl (1690 cm^{-1}). Therefore, this compound is C_6H_5COOH.

8.7 a. Number of double bonds and/or rings: 4.
Peaks at 1620 and 1500 cm^{-1} (benzene ring).
Peak at 840 cm^{-1} (*p*-substitution).
Doublet at 3480–3390 cm^{-1} (—NH_2 group).
Therefore, the compound is *p*-Br—C_6H_4—NH_2

b. Number of double bonds and/or rings: 4.
Peaks at 1620 and 1520 cm^{-1} (benzene ring).
Peak at 820 cm^{-1} (*p*-substitution).
Peak at 1450 cm^{-1} (CH_3—).
Peaks at 3320 and 3400 cm^{-1} (NH_2—)
Therefore, the compound is: *p*-$CH_3C_6H_4$—NH_2

c. Number of double bonds and/or rings: 0.
Peak at 3600 cm^{-1} (—OH, alcohol).
Peak at 1060 cm^{-1} (C—O, 1° alcohol).
Absence of 1380 cm^{-1} doublet (no *iso* group).
Therefore, the compound is $CH_3CH_2CH_2CH_2OH$.

d. Number of double bonds and/or rings: 5.
Peaks at 1610 and 1495 cm^{-1} (benzene ring).
Peaks at 750 cm^{-1} and 690 cm^{-1} (monosubstitution).
Peaks at 1740 cm^{-1} (C=O ester).
Peak at 1435 cm^{-1} (CH_3).
Therefore the compound is $C_6H_5COOCH_3$

e. Number of double bonds and/or rings: 0.
Peaks at 3300 cm^{-1} (—OH) and 1050 cm^{-1} (C—O) indicate RCH_2OH.
Peak (doublet) at 1380 cm^{-1} indicate $(CH_3)_2CH$.
Therefore, the compound is: $(CH_3)_2CHCH_2OH$.

f. Number of double bonds and/or rings: 4.
Peaks at 1620 and 1510 cm^{-1} (benzene ring).
Peak at 770 cm^{-1} (*o*-substitution).
Therefore, the compound is

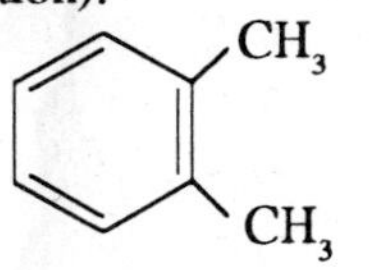

g. Number of double bonds and/or rings: 4.
Peaks at 1600–1500 cm^{-1} range (benzene ring).
Peak pattern (900-600): monosubstitution.
Therefore, the compound is

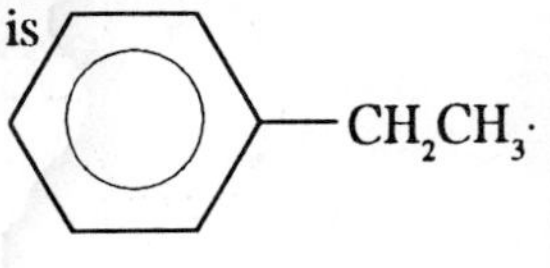

8.8 a.

A $CH_3{-}O{-}CH_3$ **C** $BrCH_2{-}CH_2Br$ **E** $(CH_3)_4C$

B $CH_2{=}C{=}CH_2$ **D** $CH_3{-}C{\equiv}C{-}CH_3$ **F**

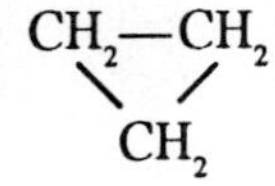

b.

A $\underset{Cl}{CH_2}{-}CH_2{-}\underset{Cl}{CH_2}$ **C** $CH_3O{-}CH_2{-}OCH_3$

B $CH_3{-}\underset{\underset{O}{\|}}{C}{-}OCH_3$ **D** $ClCH_2{-}O{-}CH_3$ or CH_3CH_2OCl

8.9 a. three b. two c. two d. two e. four f. four

8.10

a. $CH_3CH_2\underset{\underset{O}{\|}}{C}CH_3$

b. $CH_3CH_2O\underset{\underset{O}{\|}}{C}CH_3$

c. $C_6H_5\underset{\underset{O}{\|}}{C}CH_3$

d. $BrCH_2—OCH_3$

e. $C_6H_5—OCH_3$

f. $C_6H_5C(CH_3)_3$

g. $(CH_3)_2CH—CH_2Br$

h. $(CH_3)_2\underset{\underset{Cl}{|}}{C}—CH_2C_6H_5$

i. $CH_3—\overset{\overset{O}{\|}}{C}—\overset{\overset{O}{\|}}{C}—OCH_2CH_3$

j. $CH_3CH_2—O—CH_2Cl$

k. $C_6H_5CH_2CH_2CH_2Br$

l. $C_6H_5CH_2CH(CH_3)_2$

m. $CH_3CH_2—O—CH_2COOH$

n. $(CH_3)_2\underset{\underset{OH}{|}}{C}—CH_2—\underset{\underset{O}{\|}}{C}—CH_3$

8.11 A = $CH_3CH_2CHCl_2$ B = $CH_3—\underset{\underset{Cl}{|}}{CH}—CH_2Cl$ C = $(CH_3)_2CCl_2$ D = $ClCH_2—CH_2—CH_2Cl$

8.12 a. *Mass spectroscopy data:* The compound has two bromine atoms (from the $(P^{+\cdot})/(P^{+\cdot} + 2)/(P^{+\cdot} + 4)$ data) and a maximum of four carbon atoms. The molecular formula is $C_4H_8Br_2$ (no points of unsaturation) *NMR data:* The compound has the following groups: $(CH_3)_2C<$ (6H, singlet at 1.9 ppm) and $\rightarrow C—CH_2—$ (2H, singlet at 3.4 ppm). Thus, this compound is $(CH_3)_2C(Br)CH_2Br$

b. The compound has four points of unsaturation. *IR data:* Groups present: *p*-substituted (810 cm^{-1}) benzene ring (1610 and 1500 cm^{-1}) with at least one CH_3 group (1435 cm^{-1}). *NMR data:* Actual ratio of hydrogen atoms should be 4 : 18 (since the total number of hydrogen atoms is 22). The 18 hydrogen atoms suggest two $(CH_3)_3C—$ groups.

Thus, this compound is 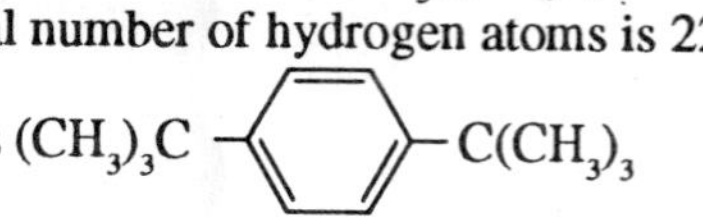

c. *Composition data:* Molecular formula is C_4H_7N. *IR data:* Presence of $(CH_3)_2CH$ (1380 cm^{-1}, doublet) and $—C{\equiv}N$ (2250 cm^{-1}). *NMR data:* Presence of $(CH_3)_2CH—$ (6H, doub. at 1.4 ppm) Thus, this compound is $(CH_3)_2CHC{\equiv}N$

d. *Composition data:* Molecular formula is $C_4H_{10}O$ (oxygen accounts for the remaining 21.6%); no points of unsaturation. *IR data:* Presence of CH_3 (1435 cm^{-1}), $(CH_3)_2CH$ (1380 cm^{-1}) and ether linkage (1100 cm^{-1}, and also because there is no peak for —OH or >C=O). *NMR data:* Presence of one $(CH_3)_2CH$ group. Thus, this compound is $(CH_3)_2CH—O—CH_3$.

e. *Composition data:* Molecular formula is $C_7H_6Br_2O$ (four points of unsaturation). *IR data:* Presence of —OH (3300 cm^{-1}), benzene ring (1580 and 1480 cm^{-1}), and CH_3 (1435 cm^{-1}) groups. *NMR data:* Presence of a tetrasubstituted ring, with two equivalent hydrogens (singlet). Also presence of $CH_3—$ attached on the benzene ring and —OH (left over from the molecular formula).

Thus, this compound is either 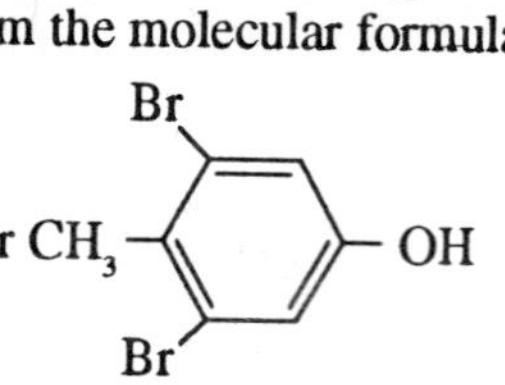

or 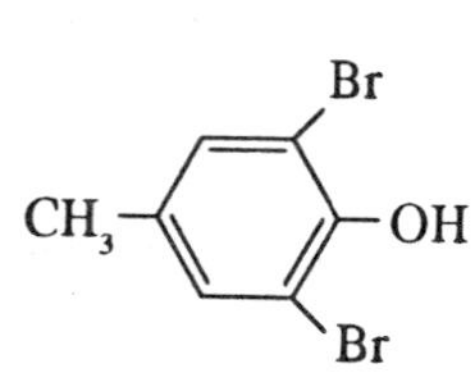

It is not as easy to distinguish between the two structures, but because of the rather downfield aromatic chemical shift (7.1 ppm), one would be inclined to choose

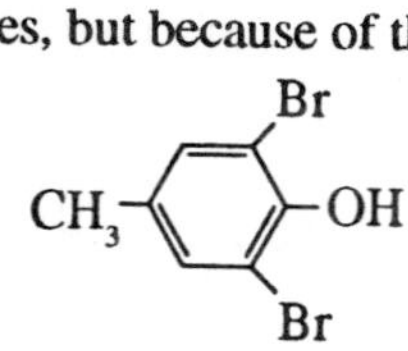

f. *Composition data:* Compound has molecualr formula $C_4H_8Br_2$; no points of unsaturation. *NMR data:* Presence of $\mathbf{CH_3}$—CH (1.8 ppm) and **CHBr** (4.3 ppm, the most downfield signal). These two groups should be attached (only one 1H signal is present). The leftover groups should be two CH_2 groups. Thus, this compound is

$$CH_3—\underset{\displaystyle Br}{\underset{|}{CH}}—CH_2—CH_2Br$$

g. *Mass Spectroscopy data:* Compound has one nitrogen (odd $P^{\dot{+}}$), a maximum of eight carbons ($P^{\dot{+}} + 1 = 9.1\%$) and one oxygen ($P^{\dot{+}} + 2 = 0.3\%$). The molecular formula is $C_8H_{11}NO$ (four points of unsaturation). *IR data:* Presence of —NH_2 (3500 and 3400 cm^{-1}), *p*-substituted (810 cm^{-1}) benzene ring (1620 and 1510 cm^{-1}), and ether linkage (1250 cm^{-1}). *NMR data:* Presence of —O—CH_2—CH_3 (—O$\mathbf{CH_2}$$CH_3$ at 4.0 ppm and —O$CH_2$$\mathbf{CH_3}$ at 1.3 ppm) and *p*-substitution (doublet of doublet).
Thus, this compound is CH_3CH_2—O 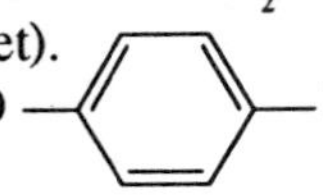 NH_2

h. *Mass Spectroscopy data:* Compound has one bromine ($P^{\dot{+}}/P^{\dot{+}} + 2$ ratio) and a maximum of five carbon atoms. The molecular formula is $C_5H_{11}Br$. *NMR data:* Presence of $(\mathbf{CH_3})_2CH$ (6H, doublet at 1.0 ppm) and —$CH_2\mathbf{CH_2}Br$ (2H triplet at 3.2, the most downfield signal). Thus, this compound is $(CH_3)_2CHCH_2CH_2Br$.

8.13 a. One point of unsaturation. *From IR:* There are no peaks due to an —OH group (3300 cm^{-1}) or a C=C (1620–1640 cm^{-1}), but there is a C=O peak at 1710 cm^{-1}. *From NMR:* The integration ratio is 3 : 9. Both sets of peaks are singlets, which suggest a methyl group and a *t*-butyl group. The compound is therefore: $CH_3—\underset{\displaystyle O}{\underset{\|}{C}}—C(CH_3)_3$

b. Two points of unsaturation. *From IR:* There are no peaks due to $\gt C=C\lt$ (1640 cm^{-1}) or —C≡C— (2200 cm^{-1}). The two peaks at 1750–1720 cm^{-1} suggests the possibliity of two carbonyls (one ester and one aldehyde or ketone). There is no peak that corresponds to the —OH group of an alcohol or acid (3300 cm^{-1}). *From NMR:* Singlet at 2.2 ppm (3H) suggests $CH_3—\underset{\displaystyle O}{\underset{\|}{C}}—$. Quartet/triplet (2H/3H) at 3.8 and 1.2 ppm suggests CH_3CH_2—O—.

Left over are —CH_2— (singlet) and ($\gt C=O$). Thus the compound is $CH_3\underset{\displaystyle O}{\underset{\|}{C}}—CH_2—\underset{\displaystyle O}{\underset{\|}{C}}—OCH_2CH_3$. The compound cannot be $CH_3\underset{\displaystyle O}{\underset{\|}{C}}—\overset{*}{C}H_2O—\underset{\displaystyle O}{\underset{\|}{C}}—CH_2CH_3$ because the —$\overset{*}{C}H_2$— would have shown a chemical shift of approximately 5.5 to 6 ppm.

c. Five points of unsaturation. *From IR:* There are no peaks corresponding to —OH (3300 cm^{-1}), $\gt C=C\lt$ (1620–1640 cm^{-1}), or —C≡C— (2200 cm^{-1}). There are peaks corresponding to $\gt C=O$ (1740 cm^{-1}, ester) and benzene ring (1600 and 1480 cm^{-1}). *From NMR:* The integration ratio is 5 : 2 : 3, which suggests a monosubstituted benzene ring (7.5 ppm, 5H) and an ethyl group (triplet, quartet). Using the IR data (ester) there are only two possible structures: $C_6H_5—\underset{\displaystyle O}{\underset{\|}{C}}—O—CH_2CH_3$ and $C_6H_5—O—\underset{\displaystyle O}{\underset{\|}{C}}—CH_2CH_3$

The first structure is more likely, because its —CH_2— chemical shift is at 3.5 ppm, while that of the latter structure is around 2.2 ppm. Therefore, the structure is $C_6H_5—\underset{\displaystyle O}{\underset{\|}{C}}—OCH_2CH_3$

d. *From mass spectroscopy:* The compound has a maximum of eight carbons and the following moiety:

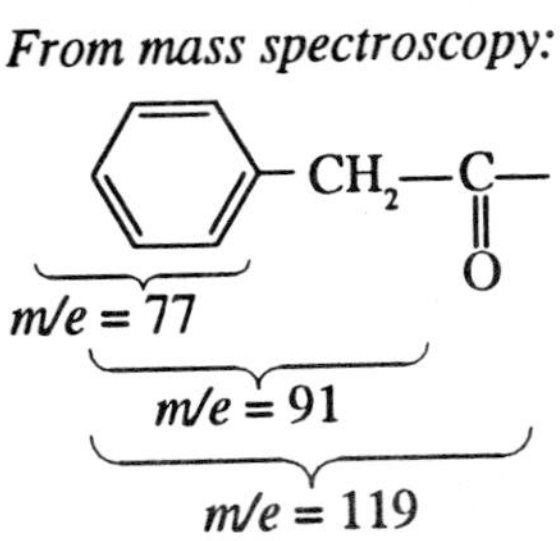

The leftover mass is due to —OH. *From IR:* The spectrum supports the presence of a benzene ring (1600 and 1500 cm^{-1}), —COOH (3300–2600 cm^{-1} and 1700 cm^{-1}) and —CH_2— (1410 cm^{-1}). The compound is therefore: $C_6H_5CH_2COOH$

Answers for Multiple Choice Test

8.1 b	**8.6** b	**8.11** a	**8.16** d
8.2 c	**8.7** d	**8.12** b	**8.17** d
8.3 a	**8.8** c	**8.13** a	**8.18** b
8.4 c	**8.9** c	**8.14** b	**8.19** a
8.5 b	**8.10** b	**8.15** c	**8.20** c

9 Alcohols, Phenols, Ethers, and Sulfur Compounds

Questions

9.1 Provide a name for each of the following compounds:

a. $(CH_3)_2C(OH)C(OH)(CH_3)_2$

b. 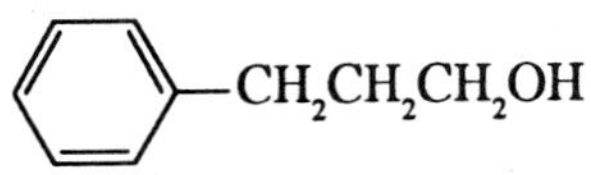

c. 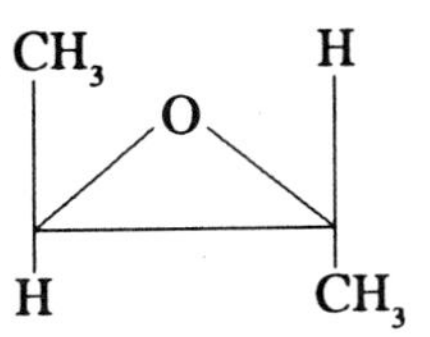

d. $CH_2{=}CHCH_2OH$

e. 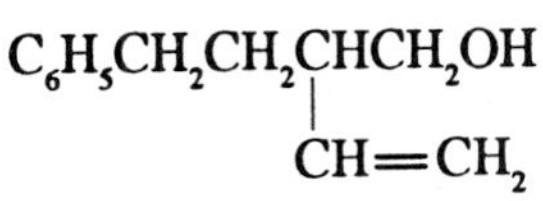

f. 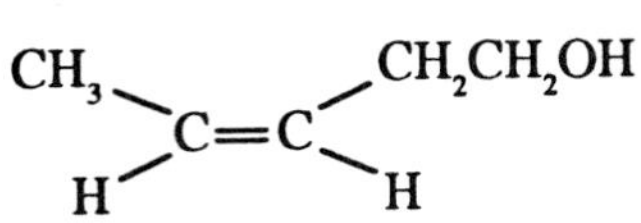

g. 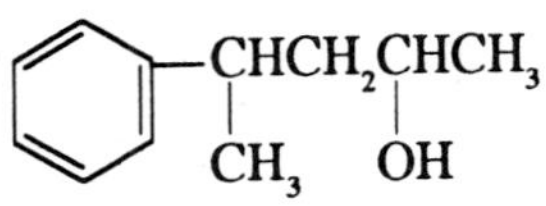

h. $(CH_3)_2CHCH_2SH$

i. 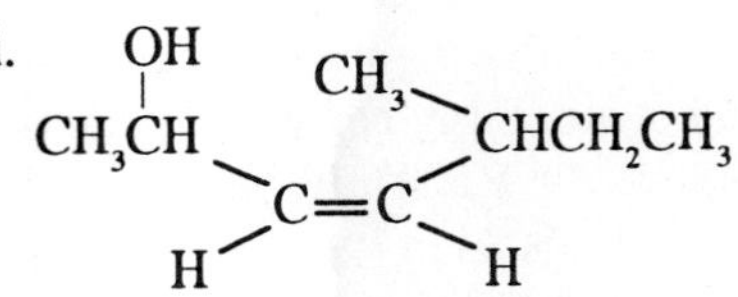

j. 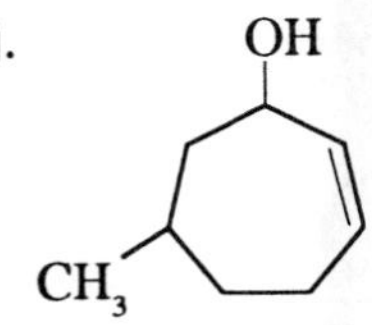

k. 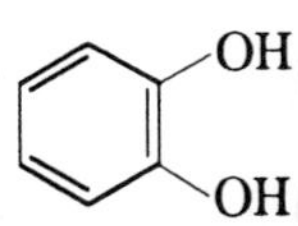

l. 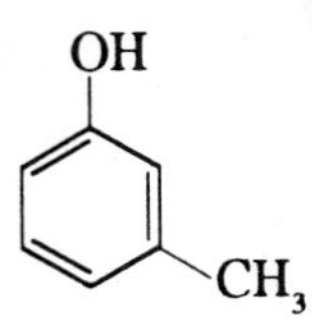

m. $(CH_3)_3CO^-K^+$

n. 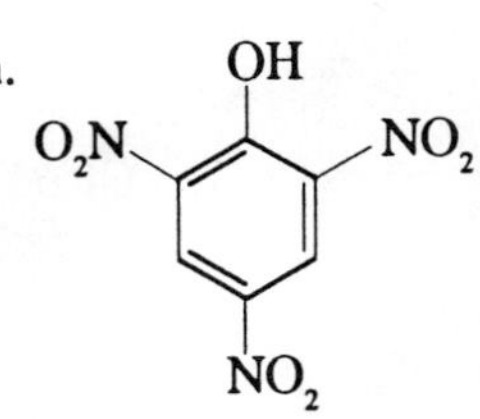

9.2 Write formulas for the following and provide alternative names:

a. *n*-propyl carbinol
b. *sec*-butyl alcohol
c. neopentyl alcohol
d. propylene oxide
e. tetrahydrofuran
f. methylhydroquinone
g. salicylic acid
h. diphenyl carbinol
i. *p*-methyl anisole
j. pinacol

9.3 Identify the family of each of the following:

a. $CH_3SOCH_2CH_3$

b. 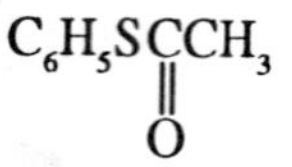

c. $(CH_3)_3CCH_2SH$

d. $(CH_3)_2SO_2$

e. 

f. 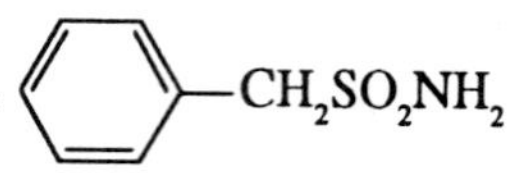

g. $(CH_3)_2CHSSCH(CH_3)_2$

h. $CH_3SCH_2CH_2CH_3$

9.4 Identify the following compounds:

a. $C_4H_{10}O$; positive iodoform test
b. the only primary alcohol to give a positive iodoform test
c. $C_9H_{12}O$; negative iodoform, immediate reaction with Lucas, positive Jones
d. $C_9H_{12}O$; negative Jones, immediate reaction with Lucas
e. $C_8H_{10}O$; positive Jones, negative iodoform, negative Lucas
f. $C_9H_{12}O$; positive iodoform, immediate reaction with Lucas
g. $C_8H_{10}O$; negative iodoform, negative Lucas, negative Jones
h. $C_8H_{10}O$; soluble in dil. NaOH, IR peak at 810 cm^{-1}
i. $C_5H_{12}O$; IR peak at 3300 cm^{-1}; three singlets in ^1H NMR
j. $C_9H_{12}O$; negative iodoform, positive Jones, optically active, negative Lucas, IR peak at 1050 cm^{-1}
k. $C_9H_{12}O$; positive iodoform, optically active, hot $KMnO_4$ oxidation yields benzoic acid

9.5 Explain why each of the following synthetic routes is inefficient:

a. 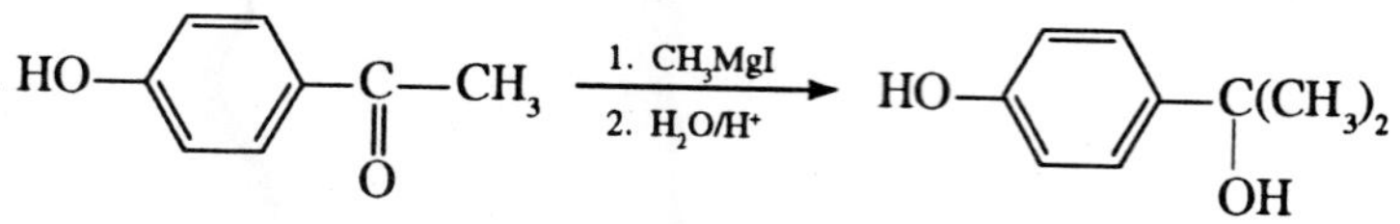

b. 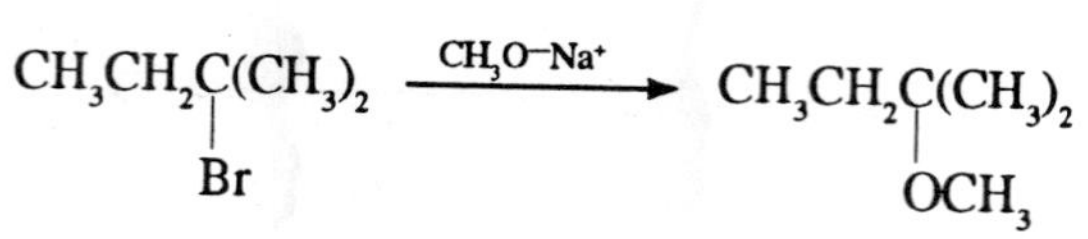

c. $CH_3CH_2CH_2OH \xrightarrow{KMnO_4/OH^-} CH_3CH_2CHO$

d. 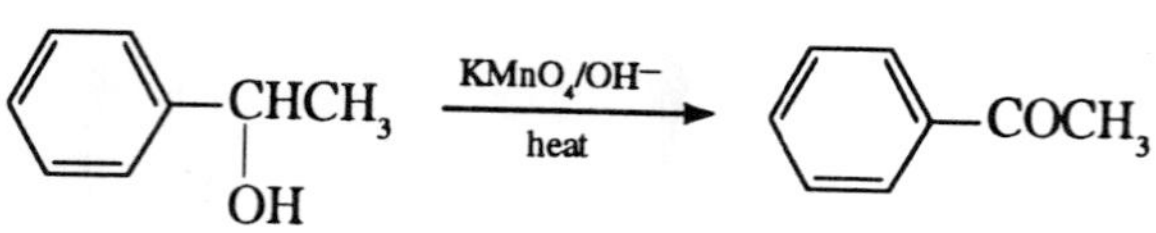

e. 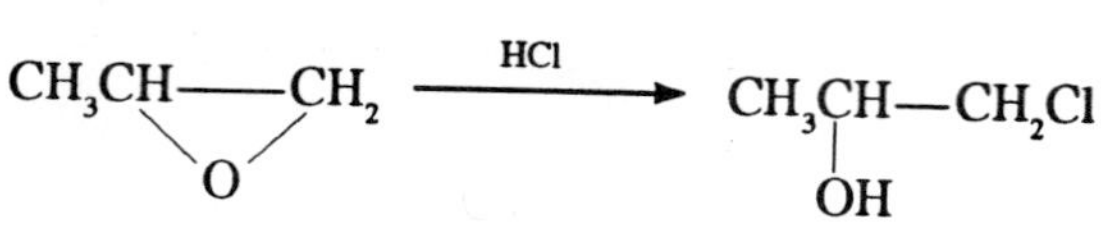

f. $(CH_3)_3CCH{=}CH_2 \xrightarrow{H^+/H_2O} (CH_3)_3CH(OH)CH_3$

9.6 Identify the structures below using the given clues:

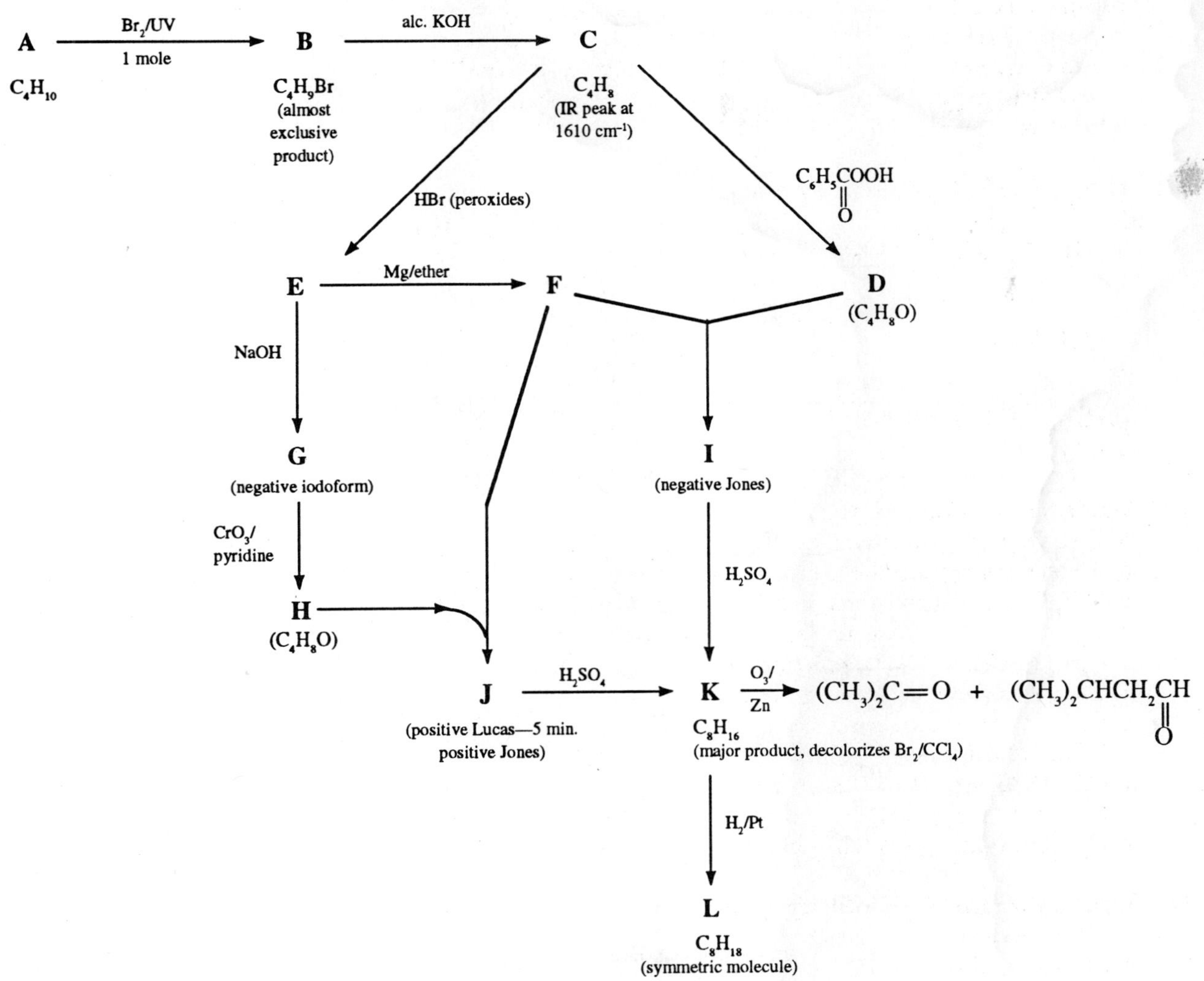

9.7 Complete the following reactions. If there is no reaction, state so.

a. 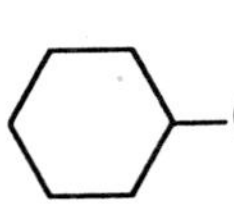 $-CH_2OH \xrightarrow[\text{pyridine}]{CrO_3}$ **A**

b. $(CH_3)_3C{-}OH \xrightarrow[\text{heat}]{KMnO_4/OH^-}$ **B**

c. $(CH_3)_2CH{-}CH_2OH \xrightarrow[\text{NaOH}]{I_2/KI}$ **C**

d. 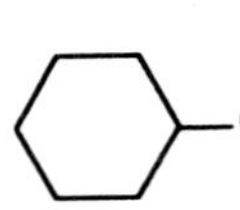 $-CH_2OH \xrightarrow{PBr_3}$ **D**

e. $n\text{-}C_6H_{13}CH_2OH \xrightarrow[\text{excess}]{Cr_2O_7^{-2}/H^+}$ **E**

f. $C_6H_5CH{=}CHCHO \xrightarrow{LiAlH_4}$ **F**

g. 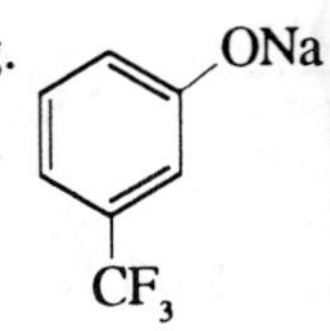

$\xrightarrow{CH_2{=}CHCH_2Br}$ **G**

h. $CH_3CH_2CH_2OH \xrightarrow{NaBH_4}$ **H**

i. $C_6H_5CH{=}CHCH_2OH \xrightarrow[\text{room temperature}]{MnO_2/CCl_4}$ **I**

j. $(CH_3)_2CH-$ 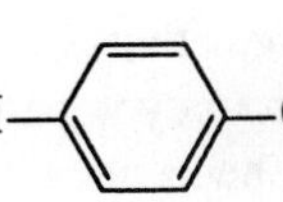 $-OH \xrightarrow[\text{pressure}]{H_2/Ni,}$ **J** ($C_9H_{18}O$)

k. 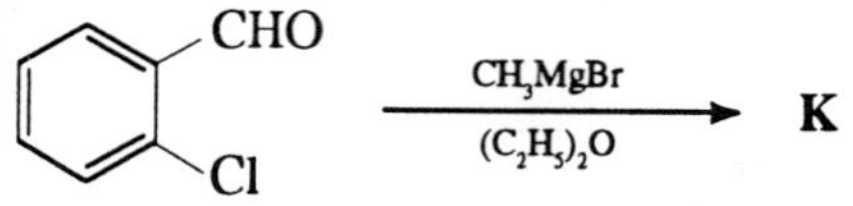

l. 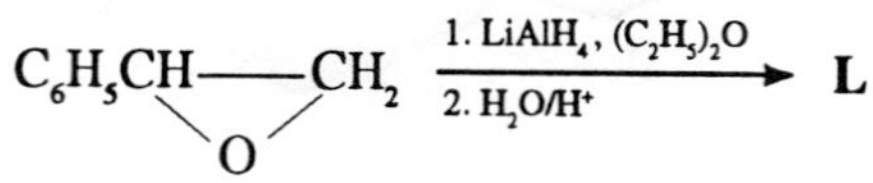

m. 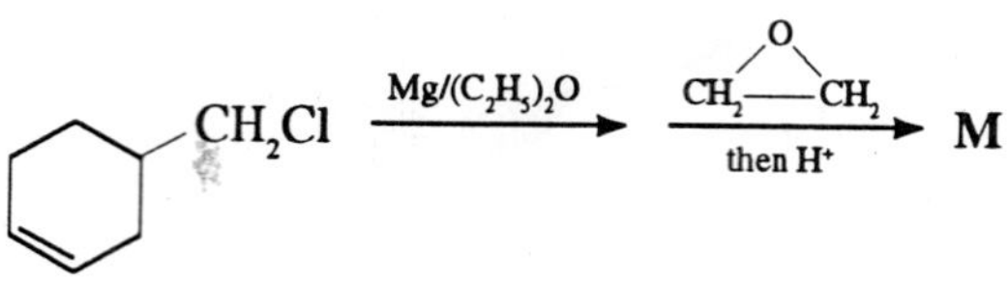

n. $C_6H_{13}MgBr \xrightarrow[\text{2. } H_2O/H^+]{\text{1. } (C_6H_5)_2C{=}O} N$

o. 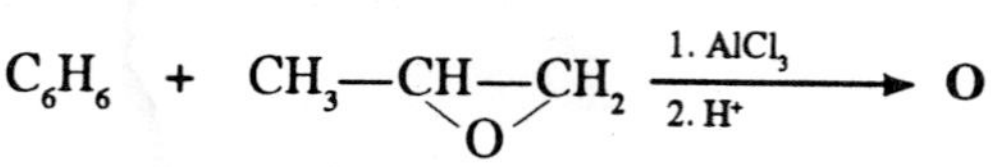

p. $(CH_3)_2CHCH_2{-}C({=}O){-}CH_3 \xrightarrow{H_2/Pt} P$

q. $(C_6H_5)_2C{=}O \xrightarrow{NaBH_4} Q$

r. $(CH_3)_2C(OH)CH_2CH_3 \xrightarrow[NaOH]{I_2/KI} R$

s. 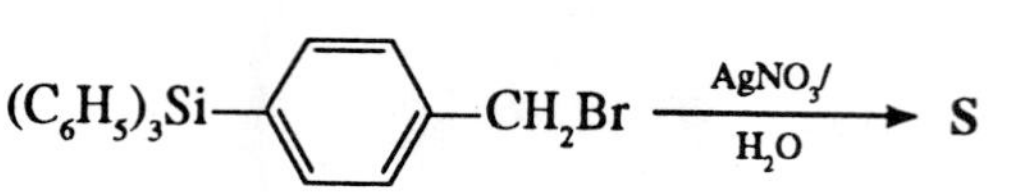

(immediate reaction with Lucas reagent)

t. $n\text{-}C_3H_7C(CH_3)(C_2H_5)OH \xrightarrow{HCl} T$

u. 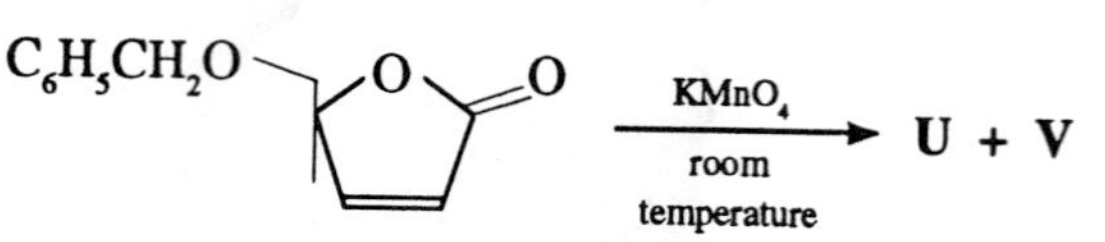

v. 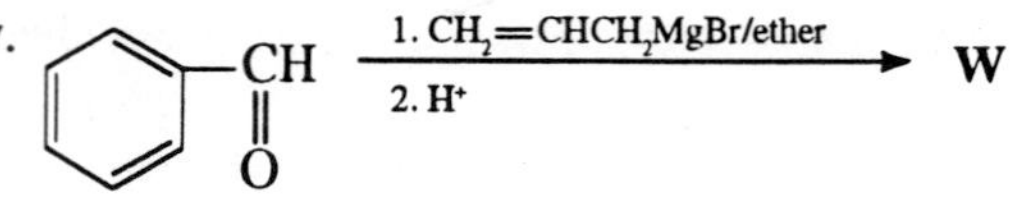

w. $n\text{-}C_6H_{13}CHO \xrightarrow[\text{2. } H^+]{\text{1. } LiAlH_4\text{, Ether}} X$

x. $n\text{-}C_6H_{13}CHO \xrightarrow{H_2/Pt} Y$

y. $C_6H_5CHO \xrightarrow[\text{then } H^+]{C_6H_5Li} Z$

z. $HO{-}CH_2{-}CN + (CH_3)_2SO_4 + NaOH \longrightarrow AA$

aa. $CCl_3(CH_2)_3CH_2OH \xrightarrow[H_2O/\text{heat}]{KMnO_4} BB$

bb. 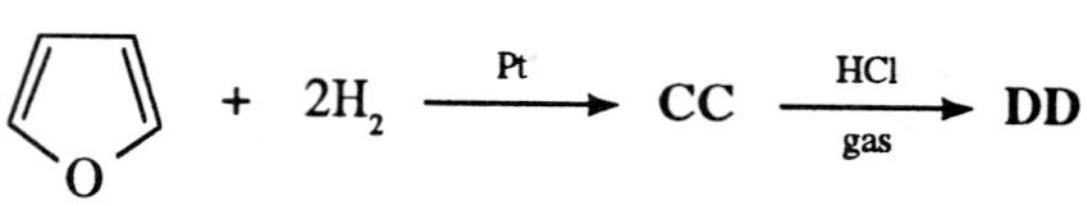

cc. 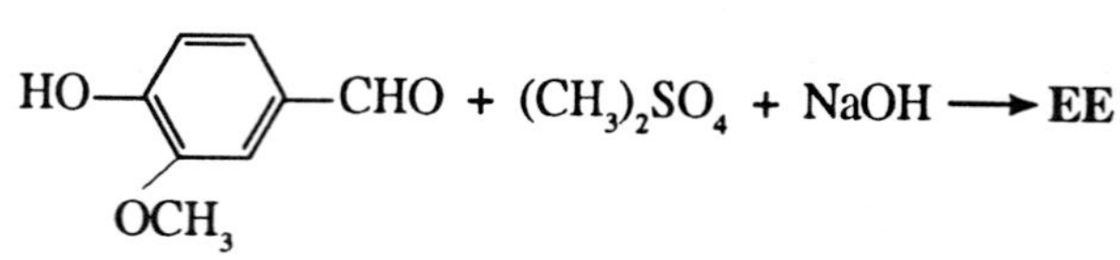

dd. 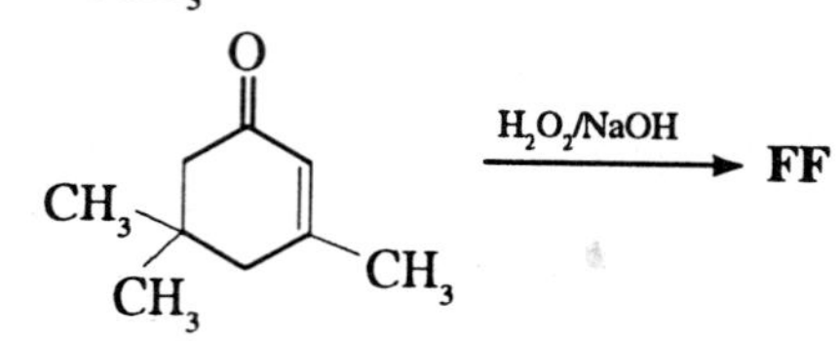

ee. $(C_6H_5)_2C(OH){-}C(OH)(C_6H_5)_2 \xrightarrow[-H_2O]{H_2SO_4} GG\ (C_{26}H_{20}O)$

ff. $(CH_3)_2CH{-}CH(OH){-}CH_3 \xrightarrow[NaOH]{I_2/KI} HH$

9.8 Suggest the simplest synthesis of the following compounds from the indicated reactant(s) as your only organic sources:

a. 3-phenyl-2-butanol from benzene and ethanol.

b. $(CH_3)_2C{=}CHCH_3$ from alcohols of four carbons or less.

c. 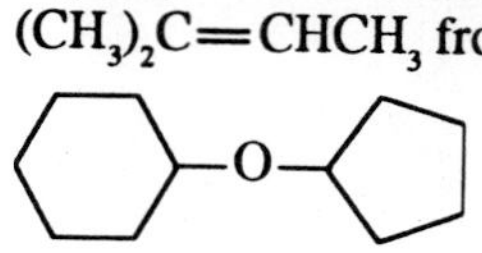 from bromobenzene and cyclopentane

d. $CH_3COCH{=}CHCH_3$ from propene.

e. 3-methyl-3-pentanol from ethanol.

f. 2-cyclopentyl ethanol from cyclopentane and ethanol.

g. ethyl phenyl ether from benzene and ethyl bromide.

9.9 Explain why structure **A** and not structure **B** is isolated in the reaction

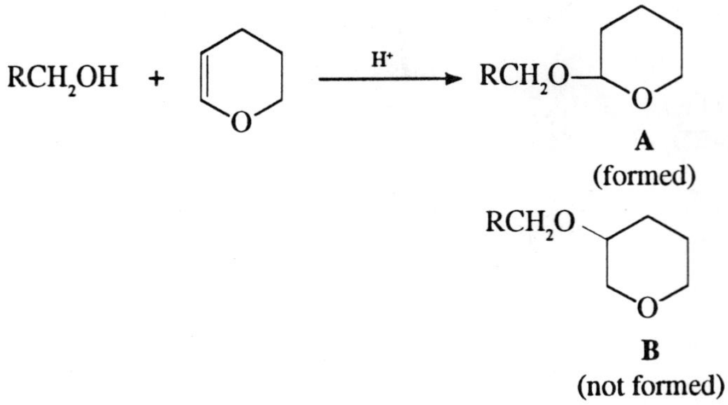

Multiple Choice Test

Questions 9.1 through 9.20 ask for choices of reactants or products in various reactions. The availability of a subsequent water workout in acid or basic medium is assumed.

9.1 What reactants would be used to prepare *t*-butyl alcohol?

a. $(CH_3)_3CCl$ + (1:1) H_2O:THF/K_2CO_3

b. CH_3CH——CH_2 + CH_3MgI (with O bridging the two carbons)

c. $(CH_3)_3CMgBr$ + CO_2

d. $(CH_3)_2C{=}CH_2$ + B_2H_6 + H_2O_2

9.2 What would be the main product in the following reaction?

$(CH_3)_2CHCH(OH)CH_3$ + HBr ⟶

a. 2-bromo-3-methylbutane

b. 2-bromopentane

c. 2-bromo-2-methylbutane

d. 1-bromo-3-methylbutane

9.3 What reagent is necessary for the conversion of 4,4-dimethyl-1-pentanol to 4,4-dimethyl-1-chloropentane?

a. NaCl

b. HCl (high temperature)

c. PCl_3

d. $SOCl_2$

e. both (c) and (d)

9.4 The addition of $H_2O/H^+/Hg(OAc)_2$, followed by $NaBH_4$, to 2,3-dimethyl-1-butene yields

a. 2,3-dimethyl-1-butanol

b. 2,3-dimethyl-2-butanol

c. 3,3-dimethyl-2-butanol

d. 3,3-dimethyl-1-butanol

9.5 Give the best route for the synthesis of this alcohol:

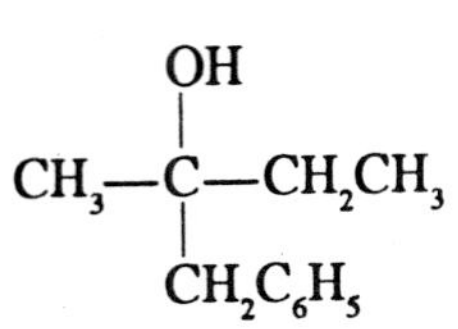

a. $C_2H_5COOCH_3$ + C_6H_5MgBr ⟶

b. $C_2H_5COCH_3$ + $C_6H_5CH_2MgBr$ ⟶

c. $CH_3COCH_2C_6H_5$ + $CH_3CH_2CH_2MgBr$ ⟶

d. $CH_3CH_2CH(CH_3)MgBr$ + C_6H_5CHO ⟶

9.6 What should be used to achieve the following conversion:

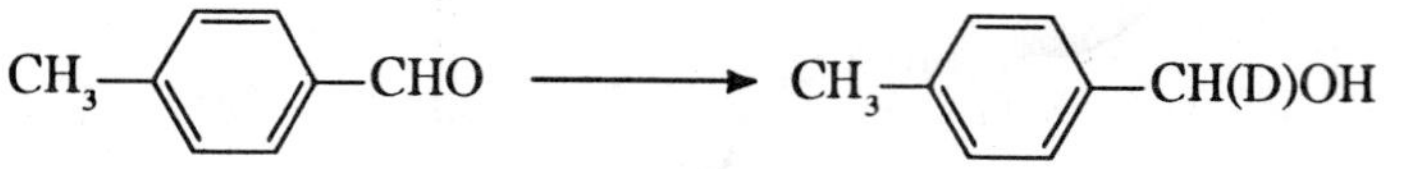

(D = deuterium)

a. $NaBH_4$, followed by D_3O^+
b. H_2/Pt, followed by D_2O
c. $KMnO_4/NaOH$ followed by $LiAlD_4$ and H_3O^+
d. $NaBD_4$ followed by H_3O^+

9.7 Which of the following processes does not yield an alcohol?
a. C_6H_5CHO + $LiAlH_4$ ⟶
b. $C_6H_5COC_2H_5$ + $NaBH_4$ ⟶
c. $C_6H_5COCH_3$ + Zn (Hg)/HCl ⟶
d. C_6H_5CHO + 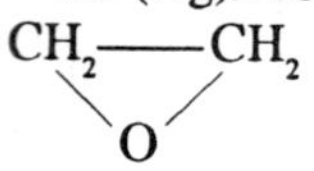 ⟶

9.8 Which of the following Grignard reactions does *not* yield the indicated product?
a. $2\ CH_3CH_2MgI$ + $CH_3COOC_2H_5$ ⟶ $(C_2H_5)_2C(OH)CH_3$
b. CH_3CH_2MgI + C_6H_5COOH ⟶ $C_6H_5COC_2H_5$
c. CH_3CH_2MgI + C_6H_5CHO ⟶ $C_6H_5CH(OH)C_2H_5$
d. CH_3CH_2MgI + $(C_6H_5)_2C{=}O$ ⟶ $(C_6H_5)_2C(OH)C_2H_5$

9.9 Which of the following compounds would yield a ketone when reacted with $Na_2Cr_2O_7/H_2SO_4$?
a. 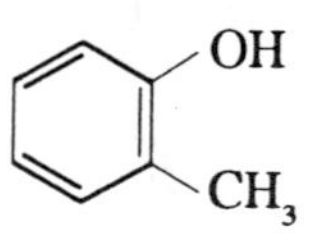

b. $(CH_3)_2C(OH)H$
c. $(CH_3)_2C(OH)CH_3$
d. $(CH_3)_3CCH_2OH$
e. $CH_3OCH_2CH_3$

9.10 Reaction of *n*-butylmagnesium bromide with compound **A** followed by H^+/H_2O workout, yields compound **B**, which reacts with the Lucas reagent after five minutes. A possible structure for **A** would be
a. acetaldehyde b. acetone c. oxirane d. formaldehyde

9.11 The product **B** in question 9.10 would then be
a. 2-methyl-2-hexanol b. 2-hexanol c. 1-hexanol d. 1-pentanol

9.12 The product(s) of the reaction of 2-methyl-1,2-pentanediol with HIO_4 is (are)
a. $CH_3(CH_2)_2COCH_3$ + HCHO
b. $CH_3(CH_2)_2COOH$ + HCOOH
c. $CH_3(CH_2)_2CH(CH_3)COOH$
d. $CH_3(CH_2)_2COCH_3$ + HCOOH

9.13 The reaction of *t*-butyl bromide and sodium methoxide yields
a. $(CH_3)_2C{=}CH_2$ + CH_3OH + NaCl
b. $(CH_3)_3COCH_3$ + NaCl
c. $(CH_3)_3CO^-Na^+$ + CH_3Cl
d. $ClCH_2C(CH_3)_2O^-Na^+$

9.14 Which combination of reagents would afford the highest yield of methyl phenyl ether

$CH_3O^-Na^+$	C_6H_5Br	CH_3I	$C_6H_5O^-Na^+$
A	**B**	**C**	**D**

a. A and C
b. B and D
c. A and B
d. C and D

9.15 Which reagent is used in the cleaving of ethers?
a. $LiAlH_4$
b. hot HI
c. hot $KMnO_4/H^+$
d. conc. NaOH

9.16 Which of the following diols would not react with HIO_4?
a. 2,3-hexanediol
b. *trans*-1,2-cyclohexanediol
c. *cis*-1,2-cyclohexanediol
d. 2,4-pentanediol

9.17 The reaction product of tetrahydrofuran with hot $KMnO_4$ is
a. $HOOC(CH_2)_2COOH$
b. $HO(CH_2)_4OH$
c. $OHC(CH_2)_2CHO$
d. no reaction, since ethers are stable toward oxidizing agents

9.18 The reaction of propylene oxide with *excess* concentrated HCl yields
a. 1-chloro-2-propanol
b. 2-chloro-1-propanol
c. 1,2-dichloropropane
d. 1-chloro-1-propane

9.19 The reaction of $C_2H_5O^-Na^+/C_2H_5OH$ with cyclopropane yields
a. ethyl *n*-propyl ether
b. ethyl isopropyl ether
c. diethyl ether
d. no reaction

9.20 Which conversion would *not* be feasible by the use of H_2O_2?
a. sufide to sulfoxide
b. thiol to disulfide
c. sulfone to sulfoxide
d. sulfide to sulfone

Answers for Questions

9.1 a. pinacol (or 2,3-dimethyl-2,3-butanediol)
b. 3-phenyl-1-propanol
c. *trans*-(or (*E*)-) 2,3-dimethyl oxirane
d. allyl alcohol (or 2-propen-1-ol or vinyl carbinol)
e. 5-phenyl-3-hydroxymethyl-1-pentene
f. (*Z*)-3-penten-1-ol
g. 4-phenyl-2-pentanol
h. isobutyl mercaptan (or 2-methyl-1-propanethiol)
i. (Z)-5-methyl-3-hepten-2-ol
j. 6-methyl-2-cyclohepten-1-ol
k. catechol (or 1,2-dihydroxybenzene)
l. *m*-cresol (or 3-methylphenol)
m. potassium *t*-butoxide
n. 2,4,6-trinitrophenol (or picric acid)

9.2 a. $CH_3CH_2CH_2—CH_2OH$ (1-butanol; *n*-butyl alcohol)
b. $CH_3CH(OH)CH_2CH_3$ (2-butanol; ethyl methyl carbinol)
c. $(CH_3)_3CCH_2OH$ (2,2-dimethyl-1-propanol; *t*-butyl carbinol)
d. $CH_3—CH——CH_2$ (methyl oxirane)
O

e. 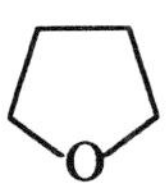 (tetramethylene oxide; oxacyclopentane)

f. 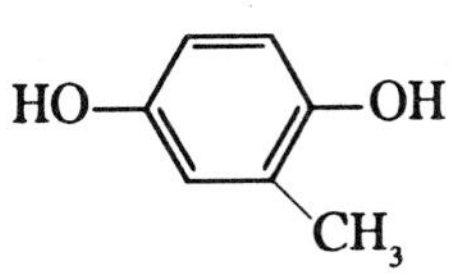

(2,5-dihydroxytoluene)

g. 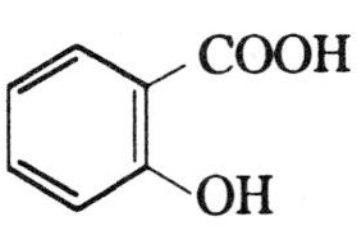

(*o*-hydroxybenzoic acid)

h. 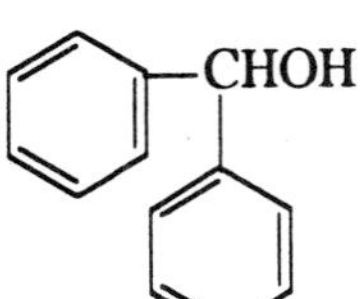

(diphenyl methanol)

i. 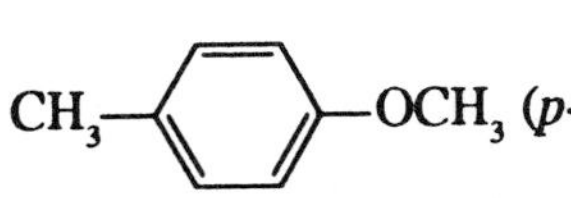

(*p*-methoxytoluene)

j. $(CH_3)_2C(OH)C(OH)CH_3$ (2,3-dimethyl-2,3-butanediol)

9.3 a. sulfoxide
b. thioester
c. mercaptan (thiol)
d. sulfone
e. sulfonic acid
f. sulfonamide
g. disulfide
h. sulfide (thioether)

9.4 a. $CH_3CH_2CH(OH)CH_3$

b. CH_3CH_2OH

c. C_6H_5—$CH(OH)CH_2CH_3$
(The alcohol is benzylic. The easily formed stable benzyl carbocation will give an immediate Lucas test)

d. $C_6H_5C(OH)(CH_3)_2$

e. $C_6H_5CH_2CH_2OH$

f. 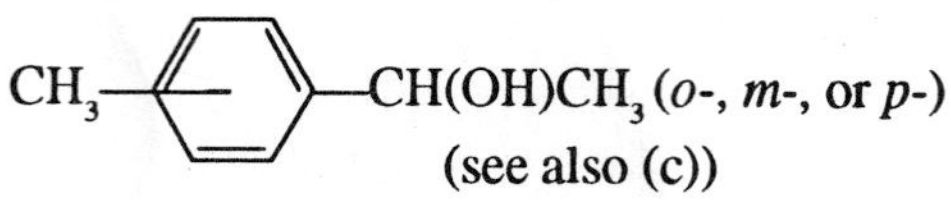
(see also (c))

g. 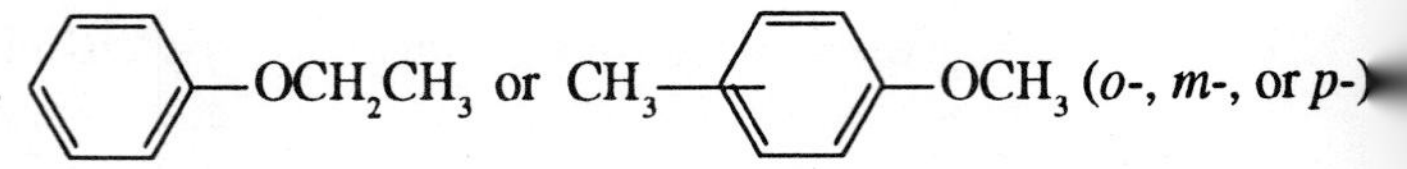

h. 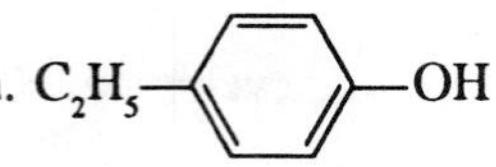

i. $(CH_3)_3CCH_2OH$

j. 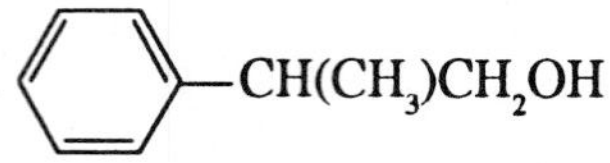

k. 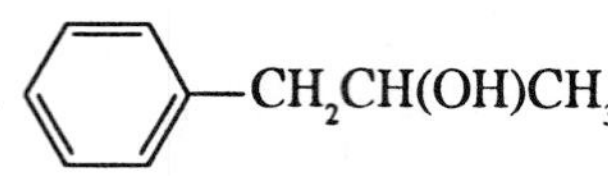

9.5 a. The Grignard reagent will react with the acidic –OH group:

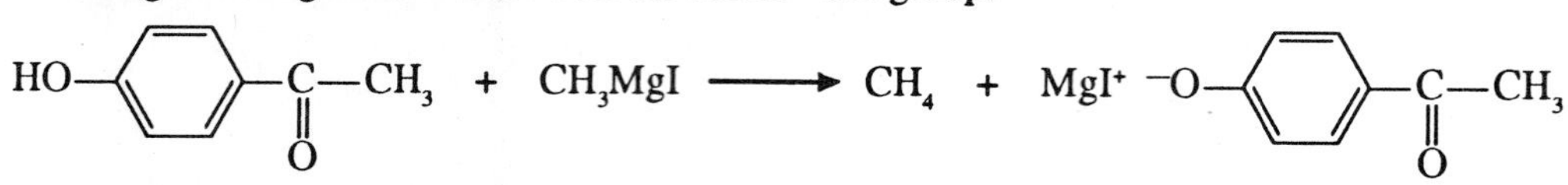

b. The reactant is a 3° alkyl halide that will react with the strong base ($CH_3O^-Na^+$) to give an elimination product:

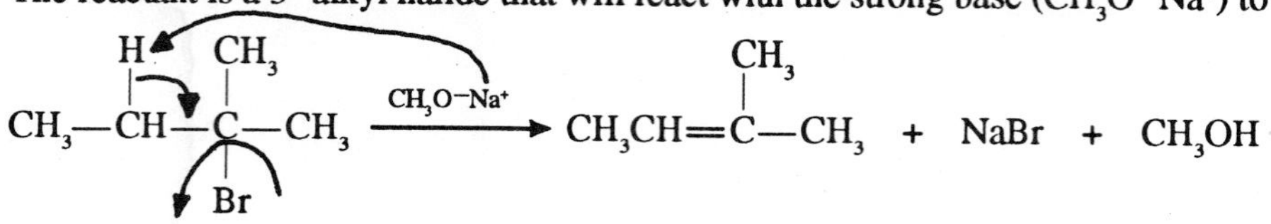

c. $KMnO_4$ oxidizes alcohols directly to carboxylic acids. The intermediate aldehyde cannot be isolated:

$$CH_3CH_2CH_2OH + KMnO_4/OH^- \longrightarrow CH_3CH_2COO^- + MnO_2$$

d. The reactant is a benzylic alcohol which is easily oxidized to benzoic acid by $KMnO_4$, a strong oxidizing agent.
e. The medium is acidic and the reaction leads to a product via the most stable carbocation. The expected product is

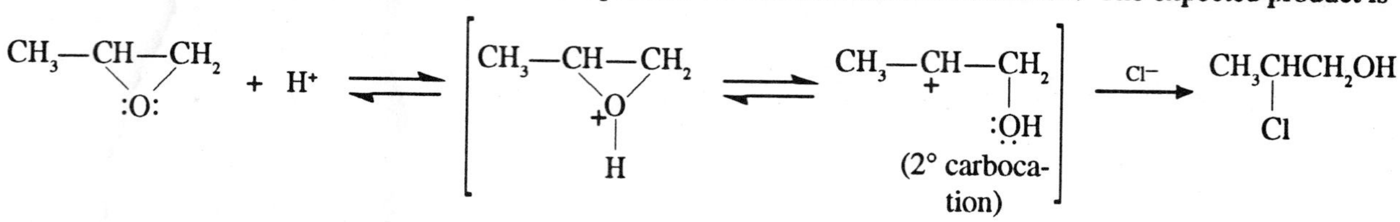

f. The major product is the one formed via the most stable carbocation

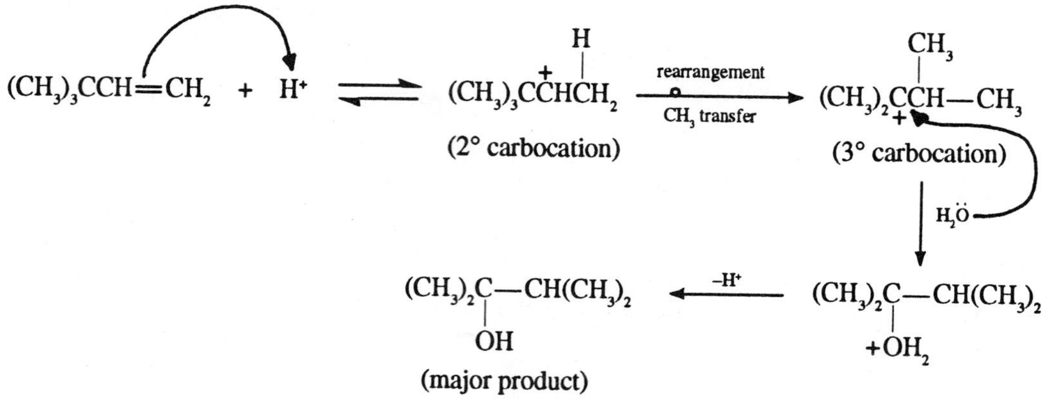

9.6 **A** $(CH_3)_3C{-}H$

B $(CH_3)_3C{-}Br$

C $(CH_3)_2C{=}CH_2$

D 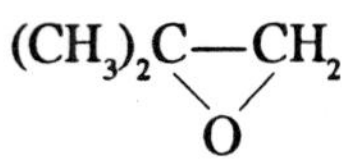

E $(CH_3)_2CH{-}CH_2Br$

F $(CH_3)_2CH{-}CH_2MgBr$

G $(CH_3)_2CH{-}CH_2OH$

H $(CH_3)_2CH{-}CHO$

I $(CH_3)_2CHCH_2CH_2C(CH_3)_2$
OH

J $(CH_3)_2CHCH_2{-}CH{-}CH(CH_3)_2$
OH

K $(CH_3)_2CHCH_2CH{=}C(CH_3)_2$

L $(CH_3)_2CHCH_2CH_2CH(CH_3)_2$

9.7 **A** 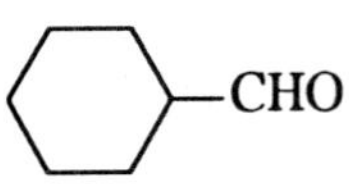

B No reaction (3° alcohol)

C No reaction

D 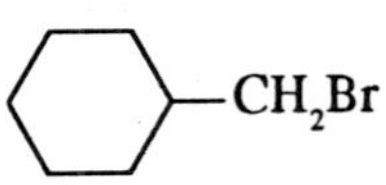

E n-$C_6H_{13}COOH$

F $C_6H_5CH_2CH_2CH_2OH$

G 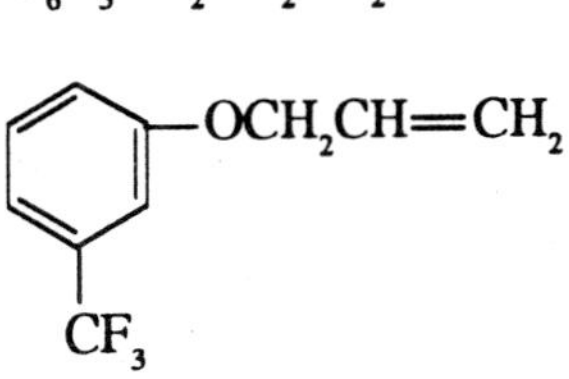

H No reaction ($NaBH_4$ reacts only with ketones and aldehydes)

I $C_6H_5CH{=}CHCHO$

J 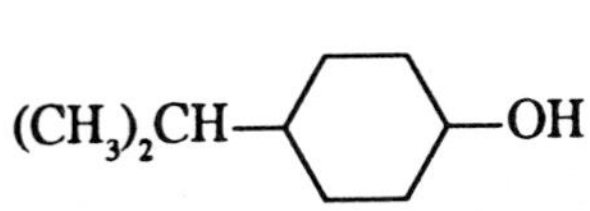

K 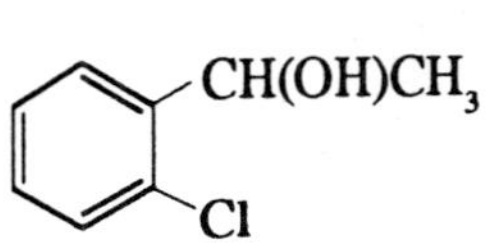

L $C_6H_5CH{-}CH_3$
OH

M 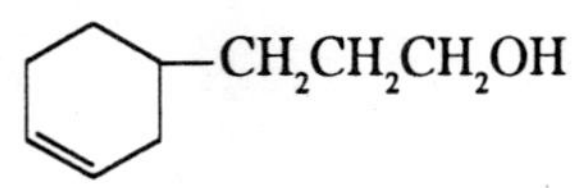

N $C_6H_{13}C(C_6H_5)_2$
OH

O $C_6H_5{-}CH{-}CH_2OH$
CH_3

P $(CH_3)_2CHCH_2CHCH_3$
OH

Q $(C_6H_5)_2CHOH$

R No reaction

S 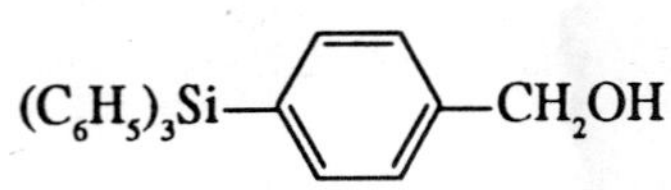
(benzyl alcohol)

T n-$C_3H_7C(CH_3)Cl$
C_2H_5

U 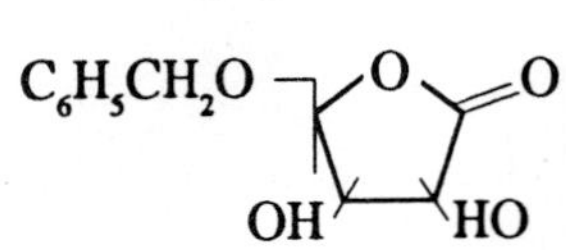

V 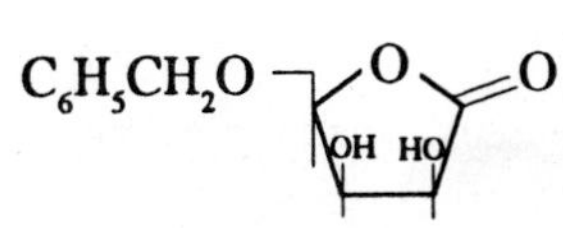

(syn addition)

W 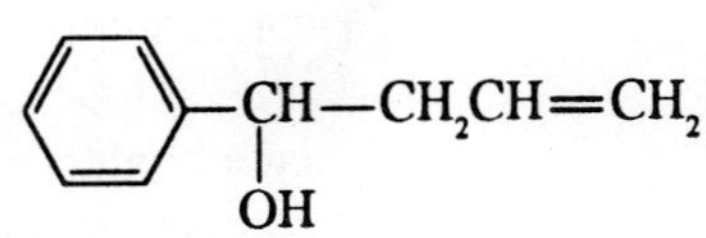

X n-$C_6H_{13}{-}CH_2OH$

Y $n\text{-}C_6H_{13}-CH_2OH$

Z $C_6H_5-\underset{OH}{CH}-C_6H_5$

AA CH_3-O-CH_2CN

BB $CCl_3(CH_2)_3CO_2H$

CC

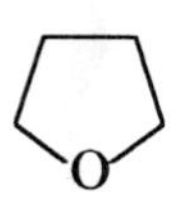

DD $\underset{OH}{CH_2}CH_2CH_2\underset{Cl}{CH_2} \longrightarrow \underset{Cl}{CH_2}CH_2CH_2\underset{Cl}{CH_2}$

EE

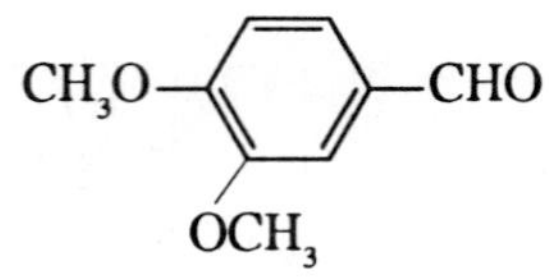

FF

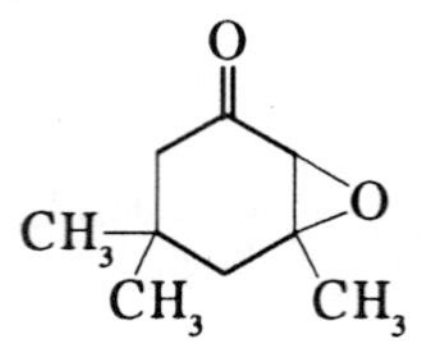

GG $(C_6H_5)_3C-\underset{O}{\overset{\|}{C}}-C_6H_5$ (via pinacol rearrangement)

HH $(CH_3)_2CH-COO^-Na^+ + CHI_3$ (iodoform test)

9.8 a. The compound to be synthesized is $CH_3\underset{C_6H_5}{CH}-\overset{OH}{CH}CH_3$. The Grignard reaction requires CH_3CHO and $CH_3\underset{C_6H_5}{CH}MgBr$

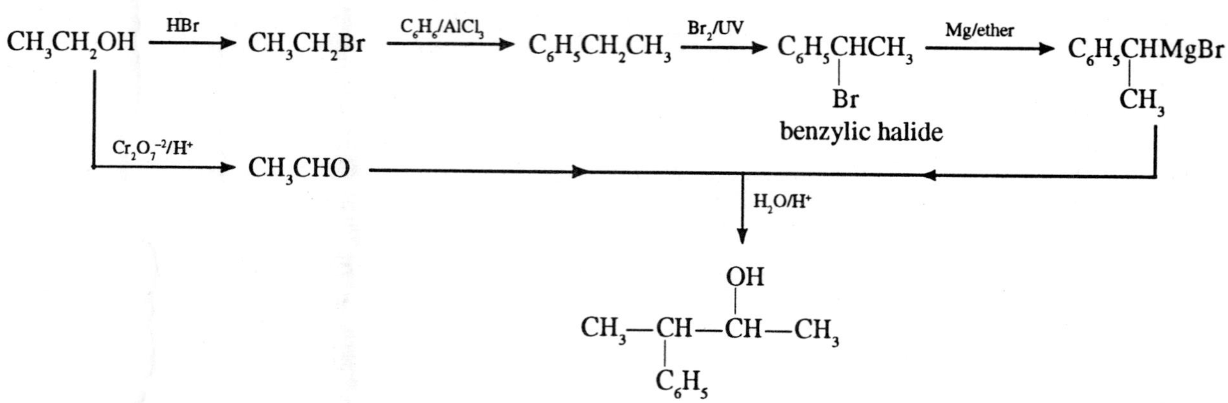

b.

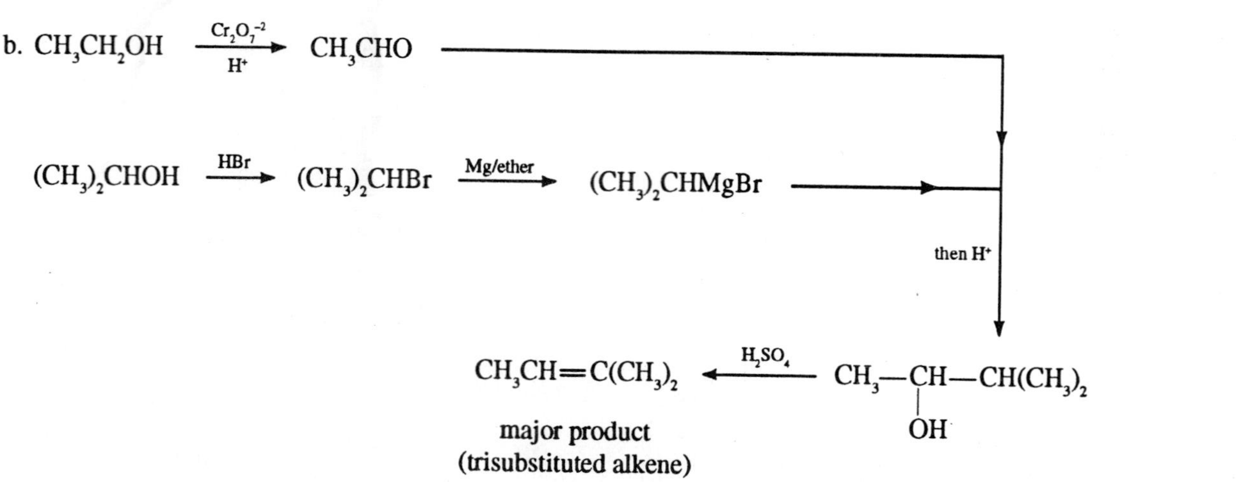

c.

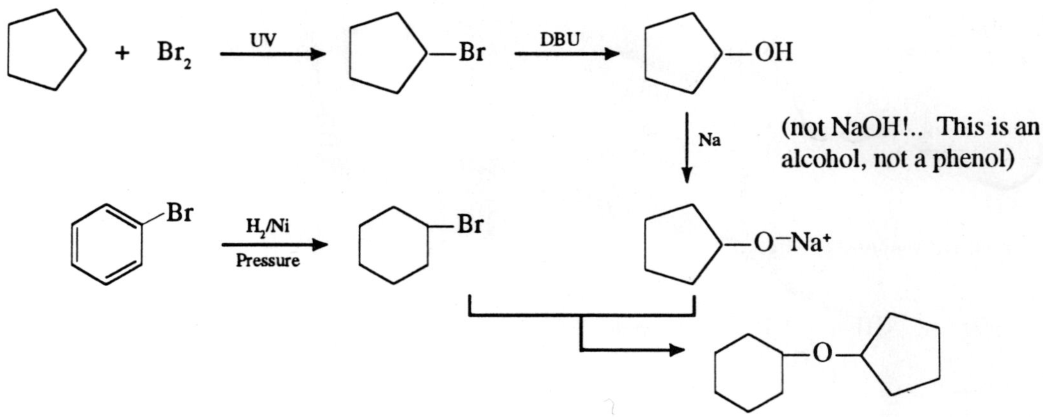

d. The compound has five carbons; therefore, propene must be cleaved. A two-carbon compound is produced by ozonolysis:

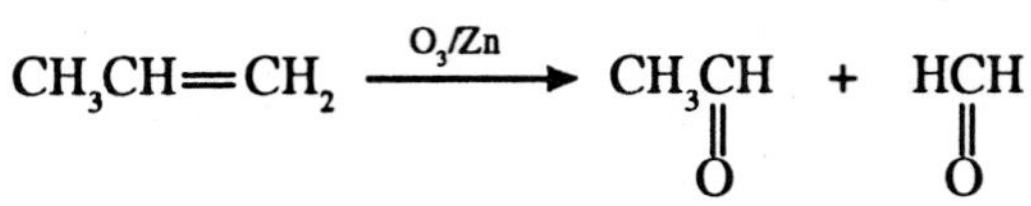

Then proceed with the Grignard reaction:

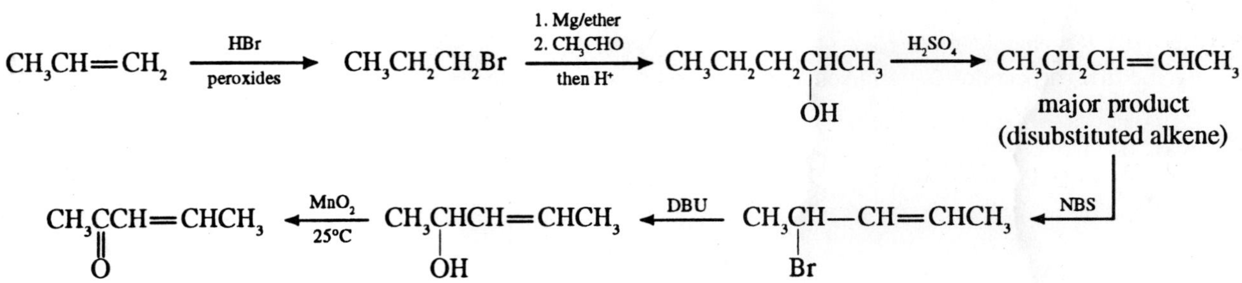

e. The product is

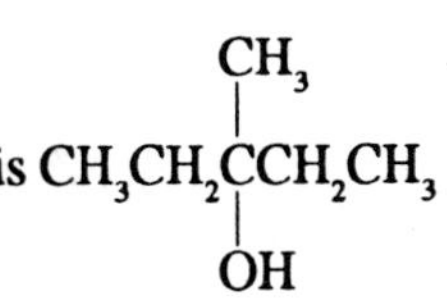

This is a six-carbon 3° alcohol which can be prepared via the Grignard reaction:

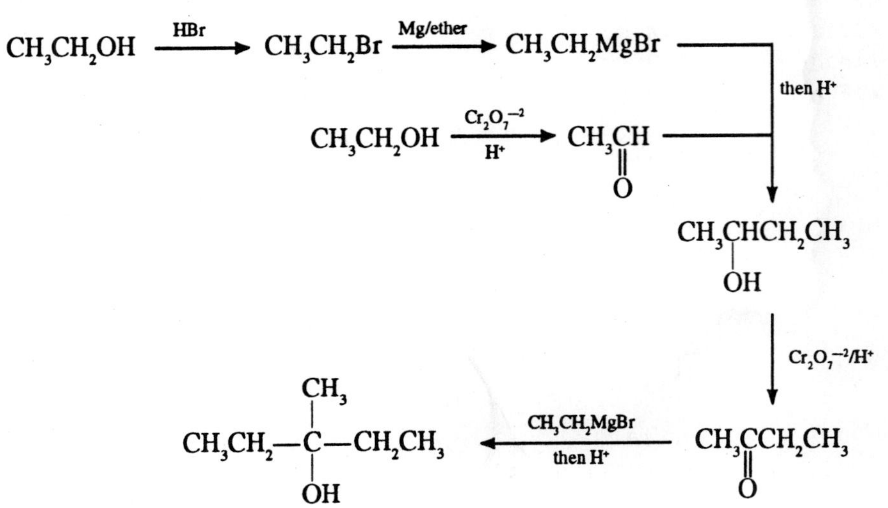

f. 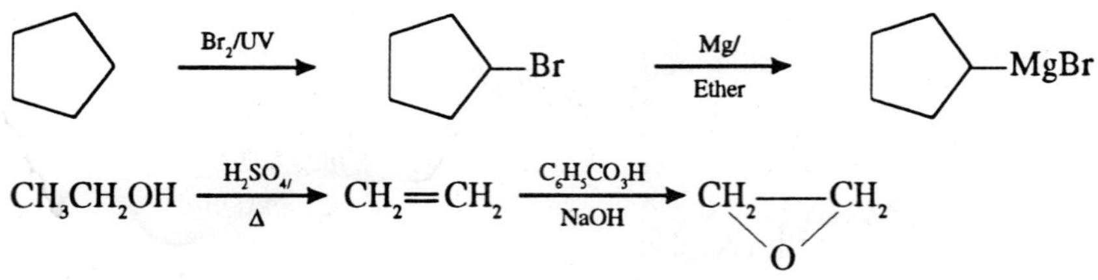

Reacting the above reagents:

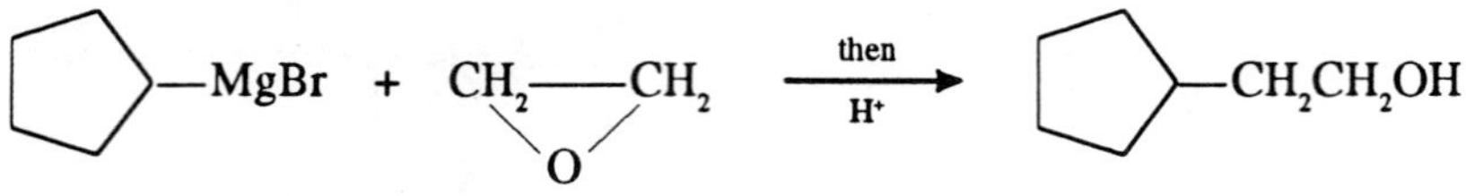

g. $C_6H_6 \xrightarrow[H_2SO_4]{HNO_3/} C_6H_5NO_2 \xrightarrow{Sn/HCl} C_6H_5NH_2 \xrightarrow{NaNO_2/HCl} C_6H_5N_2^+ \xrightarrow[heat]{H_2O/} C_6H_5OH \xrightarrow{NaOH} C_6H_5O^-Na^+$

$C_6H_5O^-Na^+ \xrightarrow{C_2H_5Br} C_6H_5OC_2H_5$

9.9 The reaction proceeds via protonation of the double bond and formation of the most stable carbocation:

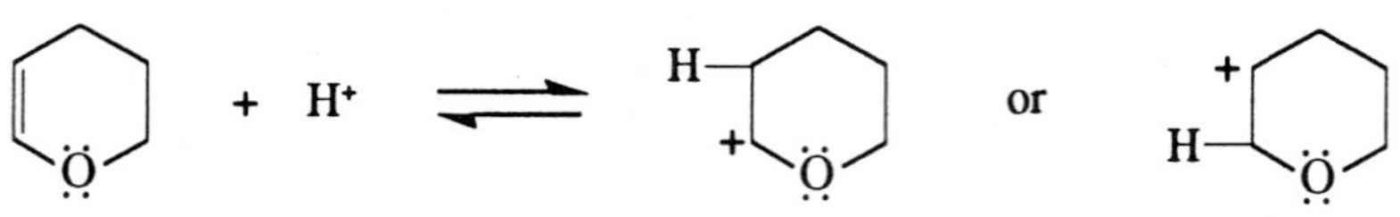

The first carbocation is stabilized by a very significant resonance structure:

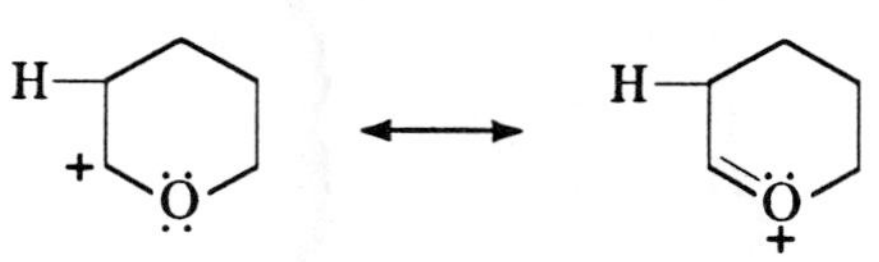

All elements in the latter structure obey the octet rule. In all other carbocation sructures, the carbons bearing the positive charge are electron deficient (they carry only 6 electrons).

Answers for Multiple Choice Test

9.1 a	**9.6** d	**9.11** b	**9.16** d
9.2 c	**9.7** c	**9.12** a	**9.17** d
9.3 e	**9.8** b	**9.13** a	**9.18** c
9.4 b	**9.9** b	**9.14** d	**9.19** d
9.5 b	**9.10** a	**9.15** b	**9.20** c

10 Carboxylic Acids and Derivatives

Questions

10.1 Name the following compounds:

a. $C_2H_5CO_2C_2H_5$

b. $(CH_3CO)_2O$

c. C_6H_5COCl

d. $CH_3COOCH_2CH_2Br$

e. $C_6H_5COOCH_3$

f. $C_6H_5CH_2COOH$

g. $CH_3COO^-NH_4^+$

h. $HCON(CH_3)_2$

i. $(CH_3)_2CHCON(CH_3)_2$

j. $CH_3CH_2CHCH_2COOH$
   $CH{=}CH_2$

k. $HO_2C—C{=}C—CO_2H$
   H H

l. $(CH_3CH_2CHCO)_2O$
   Br

m. 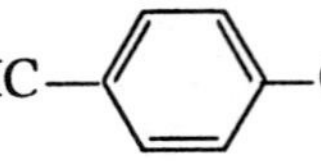

n. 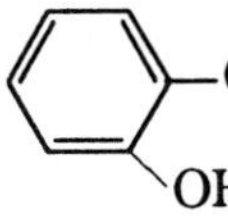

o. 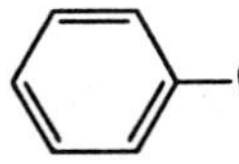

p. 

q. 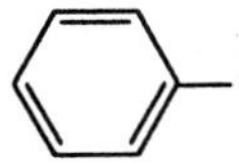

r. $C_2H_5CNHCH_2CH_2SH$
   O

s. 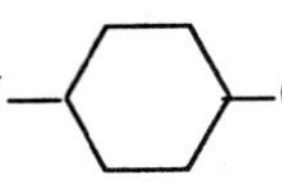

t. 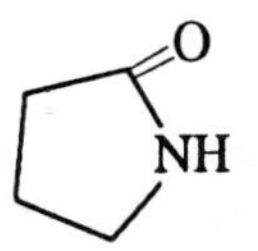

u. 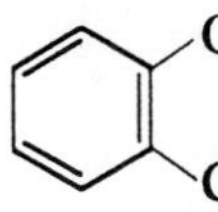

10.2 Give structures for the following compounds and then provide alternative names:

a. methyl acetate
b. N-methyl phenylacetamide
c. butanedioic acid
d. *trans*-2-butenoic acid
e. valeric anhydride
f. α-methoxy-n-butyric acid
g. *p*-bromophenyl formate
h. trifluoroacetic anhydride
i. propanoyl chloride
j. α-aminoisocaproic acid
k. calcium formate
l. N-(2-chloroethyl)-N-methylbenzamide
m. isovaleronitrile
n. N-bromo-N-phenyl acetamide

10.3 Match the following acids and their common names:

a. isophthalic acid
b. valeric acid
c. phthalic acid
d. *p*-toluic acid
e. maleic acid
f. succinic acid
g. capric acid
h. caproic acid
i. picric acid
j. oxalic acid
k. fumaric acid
l. terephthalic acid
m. adipic acid

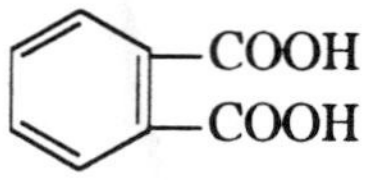

A

$HOOC(CH_2)_4COOH$

F

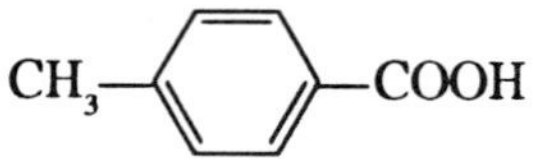

J

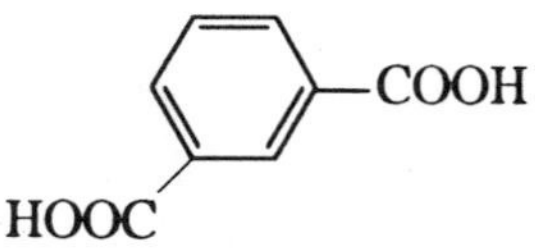

B

$CH_3(CH_2)_3COOH$

G

cis-HOOC—CH=CH—COOH

K

C

$CH_3(CH_2)_8COOH$

H

$HOOC(CH_2)_2COOH$

L

HOOC—COOH

D

$CH_3(CH_2)_4COOH$

I

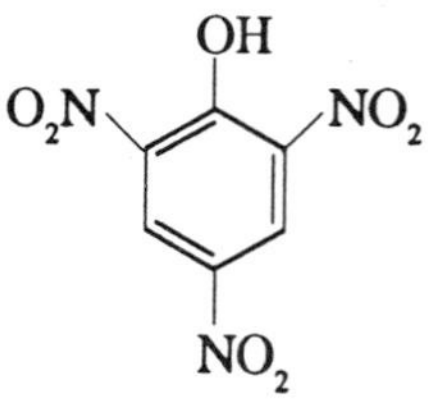

M

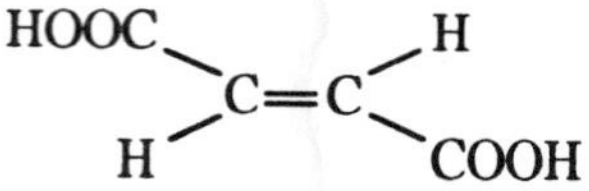

E

10.4 Explain why

a. amides are less reactive than esters toward nucleophiles.
b. acyl halides are more easily hydrolyzed than alkyl halides.
c. esters are generally synthesized via the acyl halide method rather than the acid-catalyzed carboxylic acid/alcohol method.
d. amides cannot be synthesized by the reaction of carboxylic acids with an amine (or ammonia) at room temperature.
e. hydrolysis of benzonitrile to produce benzoic acid yields higher yields in basic medium than in acidic medium.
f. $CH_3CONHCOCH_3$ is a stronger acid than $CH_3CONHCH_2CH_3$.
g. amides are not soluble in dilute HCl, but amines are.
h. the synthesis of an amide via an acyl halide requires a two-mole equivalent of the amine (or ammonia).

10.5 Identify the following compounds using the given clues:

a. C_7H_5OCl, yields HCl upon exposure to water
b. $C_8H_8O_3$, soluble in dilute $NaHCO_3$, optically active
c. $C_6H_{10}O_2$, optically active; hydrogenation yields $C_6H_{12}O_2$ (optically inactive)
d. C_9H_9N, optically inactive; yields 3-phenyl propanoic acid upon hydrolysis
e. $C_6H_{12}O_2$, insoluble in dilute NaOH; hydrolysis yields acetic acid and $C_4H_{10}O$ (positive iodoform)

10.6 Fill in the missing reagents or products in the reactions below:

a. 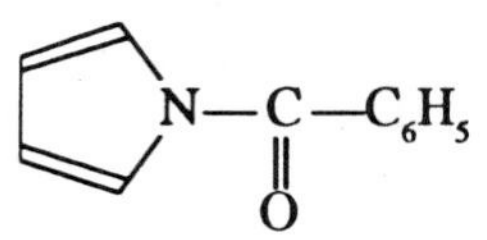 $\xrightarrow[\text{2. H}^+]{\text{1. LiAlH}_4}$ **A** + **B**

A: soluble in dilute HCl

B: positive Jones test, immediate reaction with Lucas

b. 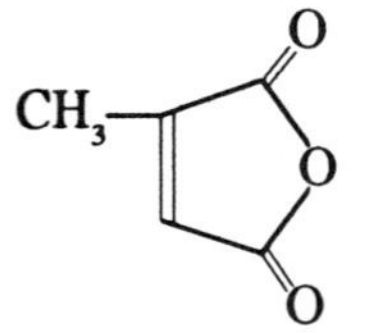 + H_2O $\xrightarrow[\text{heat}]{H^+}$ **C**

c.  + $COCl_2$ $\longrightarrow$ **D** + HCl

(1:1 molar ratio)

d. $2ClCH_2CO_2H$ + CH_3NH_2 $\xrightarrow[\text{2. acidification}]{\text{1. NaOH, 50°C}}$ **E** ($C_5H_9NO_4$)

e. $2C_4H_9MgBr$ + $HCO_2C_2H_5$ $\xrightarrow[H^+]{\text{then}}$ **F** (positive Lucas, 5 min)

f. CH_2CO_2H–CH_2CO_2H + P_2O_5 $\xrightarrow{-H_2O}$ **G**

g. CCl_3CO_2H + $SOCl_2$ $\xrightarrow[\text{(solvent)}]{\text{DMF}}$ **H** + SO_2 + HCl

h. $C_6H_5COOC(CH_3)_3$ + CH_3OH $\xrightarrow[\text{4 days}]{\text{reflux}}$ **I** + **J**

J: positive Lucas

i. $C_8H_{17}CH{=}CH(CH_2)_7COCl$ + $C_4H_9NH_2$ $\xrightarrow{\text{pyridine}}$ **K**

j. 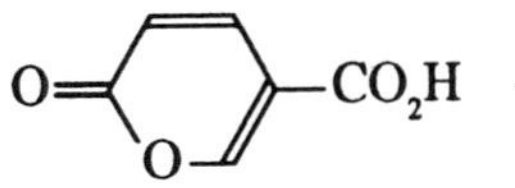 + CH_3OH $\xrightarrow{H^+}$ **L**

k. 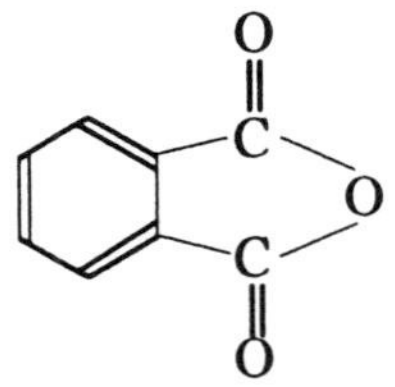 + $SOCl_2$ $\longrightarrow$ **M** ($C_8H_4O_2Cl_2$) + SO_2

l. $(COOH)_2$ + $Ca(OH)_2$ $\xrightarrow[\text{temperature}]{\text{room}}$ **N**

m. $C_6H_5CH_2CO_2H$ + **O** $\longrightarrow$ $C_6H_5CH_2COCl$

n. n-C_3H_7—$C(CH_3)(C_2H_5)Cl$ + Mg/ether $\xrightarrow[\text{2. H}^+/H_2O]{\text{1. CO}_2}$ **P**

o. 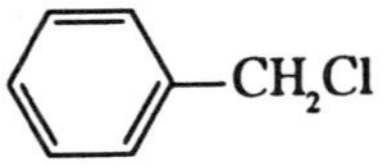 + $KMnO_4$/NaOH $\xrightarrow[H^+]{\text{then}}$ **Q**

p. 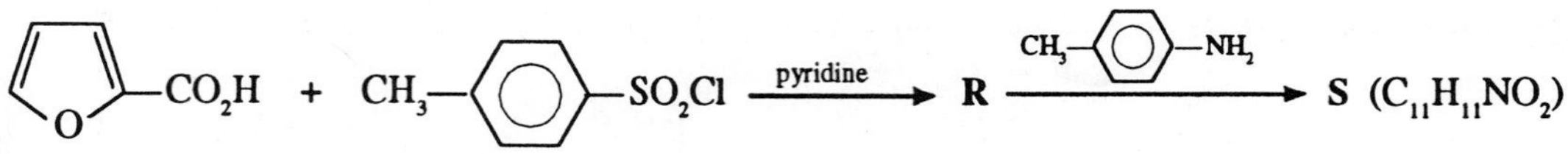

q. $n\text{-}C_6H_{13}CO_2H + NH_3 \xrightarrow{190°C} \mathbf{T}$

r. $(CO_2H)_2 + 2CH_3OH \xrightarrow[\text{heat}]{H^+} \mathbf{U}$

s. $C_9H_{19}CH{=}C(CH_3)CO_2CH_3 \xrightarrow[\text{ethanol}]{\text{KOH/}} \mathbf{V} + CH_3OH$

t. $(CH_3)_2CHC(=O){-}Cl \xrightarrow{NH_3} \mathbf{W} + HCl$

u. 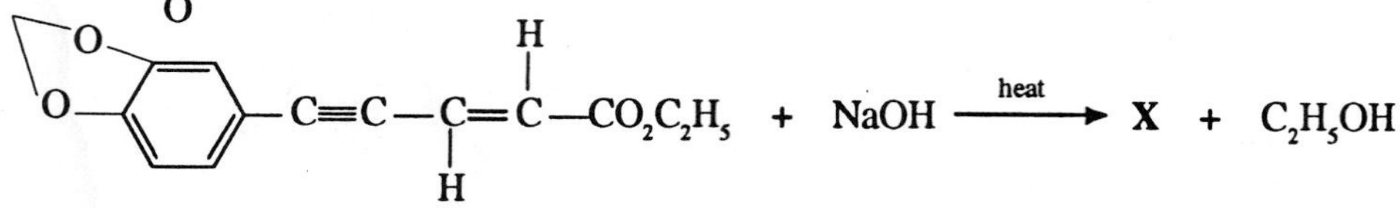

v. $C_6H_5C(NHC_6H_5){=}N{-}NH_2 + C_6H_5SOCl \text{ (1 mole)} \longrightarrow \mathbf{Y} + HCl$

w. 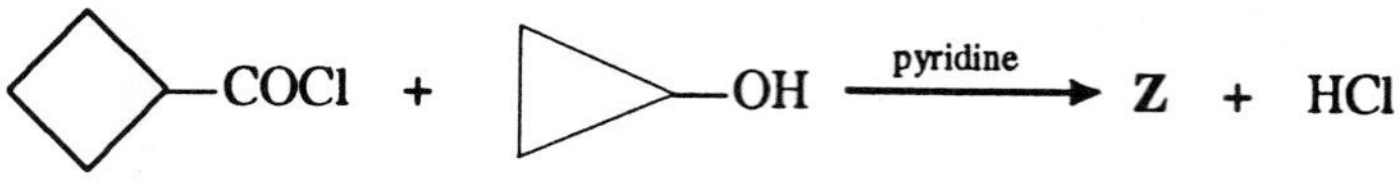

10.7 Fill in the compounds missing in the road maps below

a. 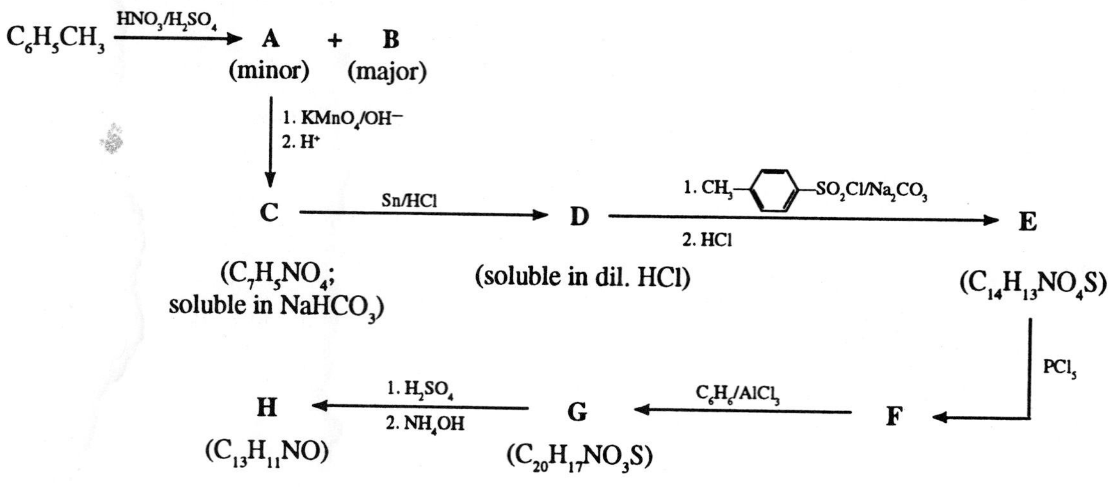

b. $C_6H_6 \xrightarrow[AlCl_3]{(CH_3CO)_2O} \mathbf{I} \xrightarrow[HCl]{\text{NaCN/}} \mathbf{J} \xrightarrow[H^+]{H_2O/} \mathbf{K} \xrightarrow[\text{2. HCl}]{\text{1. NaOH}} \mathbf{L}$

K: $(C_9H_{11}NO_2)$; L: (soluble in $NaHCO_3$)

c. $ClCH_2CO_2H \xrightarrow{Na_2CO_3} \mathbf{M} \xrightarrow{NaCN} \mathbf{N} \xrightarrow[\text{heat}]{NaOH/H_2O} \mathbf{O} \xrightarrow{Ca(OH)_2} \mathbf{P} \xrightarrow{HCl} \mathbf{Q}$

O: $(C_3H_2O_4Na_2)$

d. 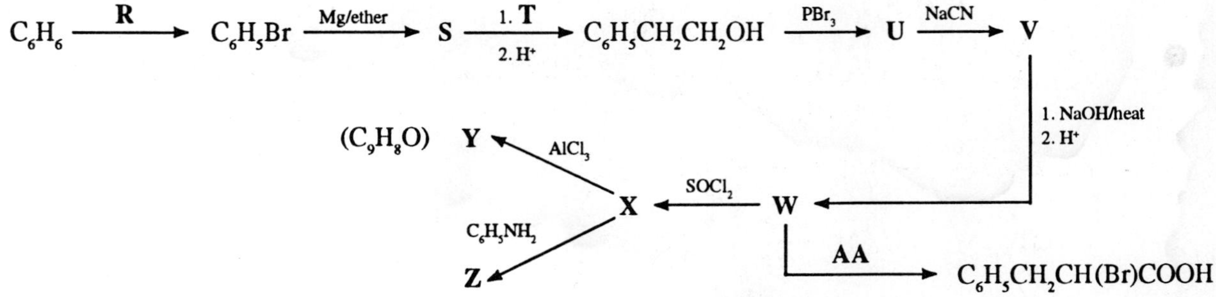

e. 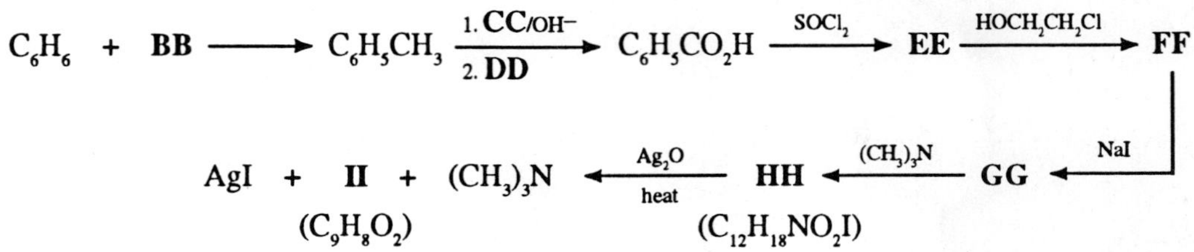

10.8 Suggest plausible mechanisms that would explain the formation of the indicated compounds:

a. $CH_3COOH + CH_3OH \xrightarrow{H^+} CH_3COOCH_3 + H_2O$

b. $(CH_2)_5$(—NH—C=O) + $H^+/H_2O \longrightarrow HOOC(CH_2)_5CONH_2$

c. $CH_3C{\equiv}N + OH^-/H_2O \longrightarrow CH_3COO^- + NH_3$

d. $CH_3C{\equiv}N + H^+/H_2O \longrightarrow CH_3COOH + NH_4^+$

e. 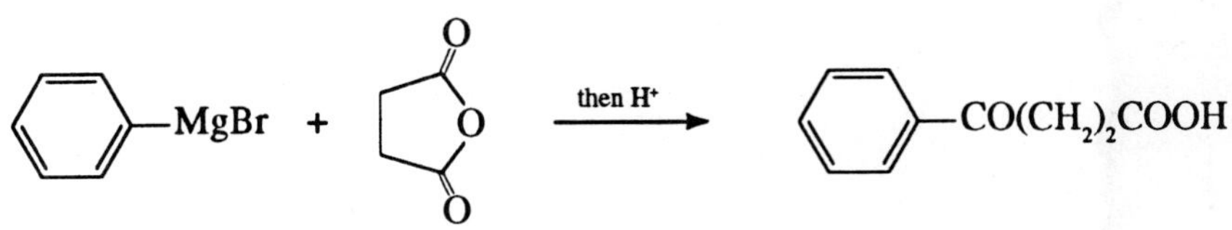

f. 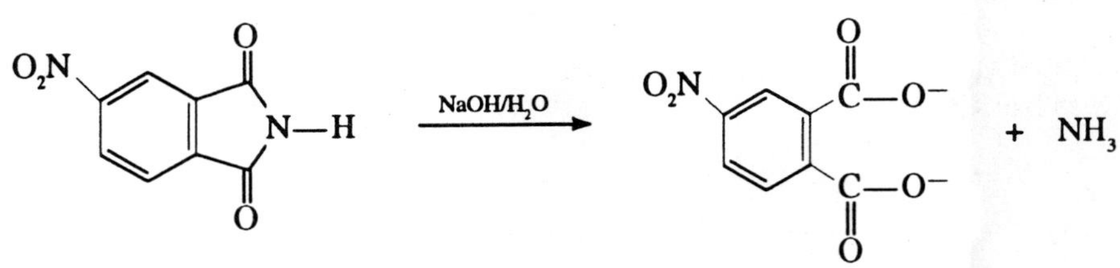

g. 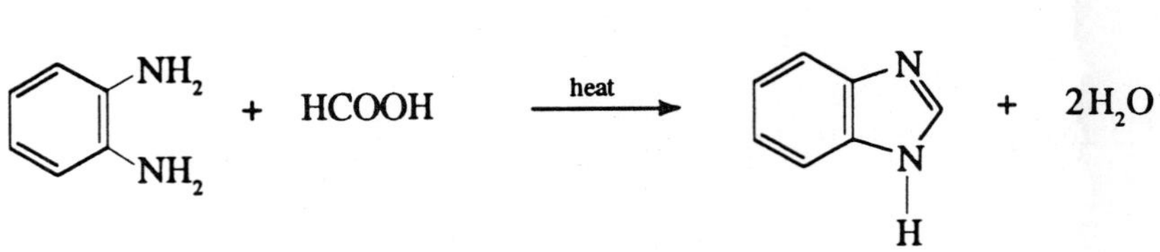

h. $CH_2{=}CHCH_2{-}N{=}C{=}S + H^+/H_2O \longrightarrow CH_2{=}CHCH_2NH_2$

i. $C_2H_5CH(OH)CH_3 + CH_3C{\equiv}N \xrightarrow{H^+/H_2O} CH_3C(=O)NHCH(CH_3)C_2H_5 + H_2O$

j. $HC{\equiv}CCH_2C(CH_3)(OC_2H_5)_2 \xrightarrow[2.\ (CH_3CO)_2O]{1.\ n\text{-}C_4H_9Li} CH_3C(=O){-}C{\equiv}C{-}CH_2C(CH_3)(OC_2H_5)_2$

k. 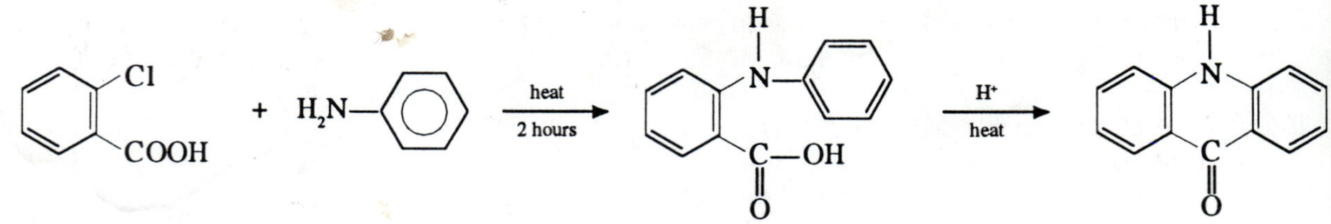

l. 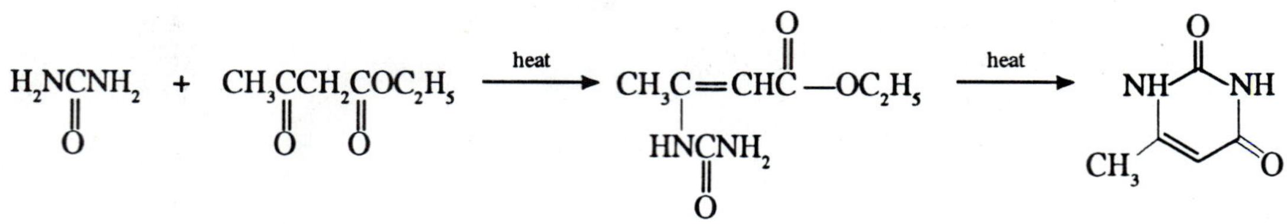

m. 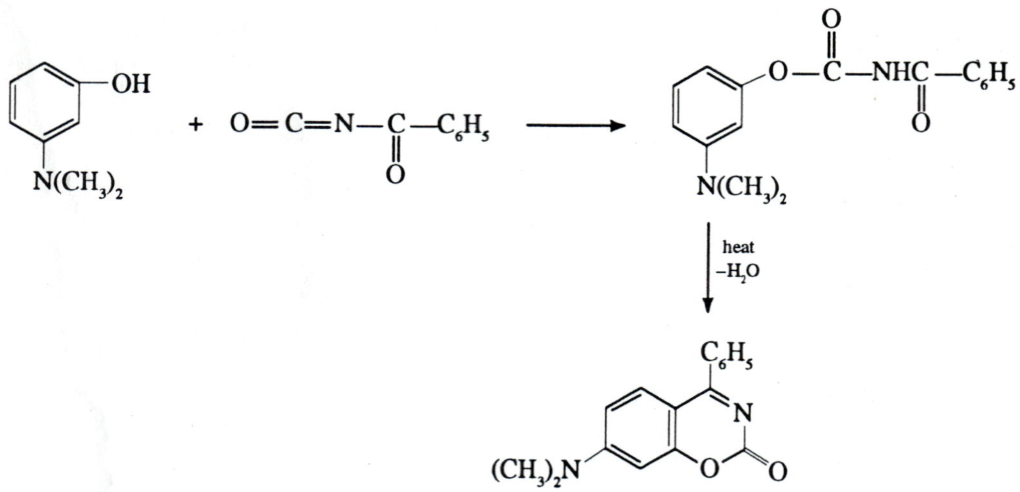

n. 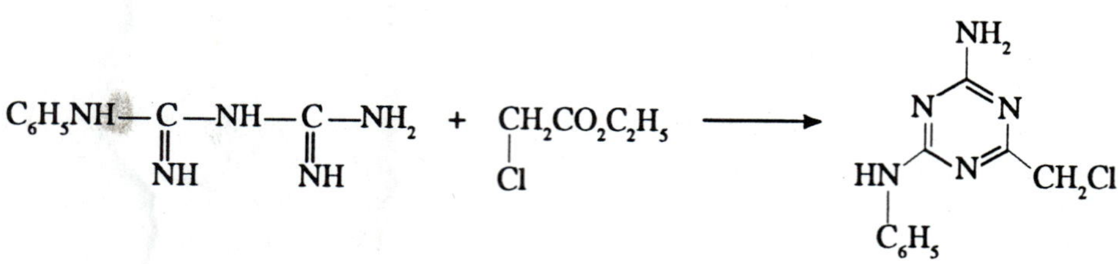

o. 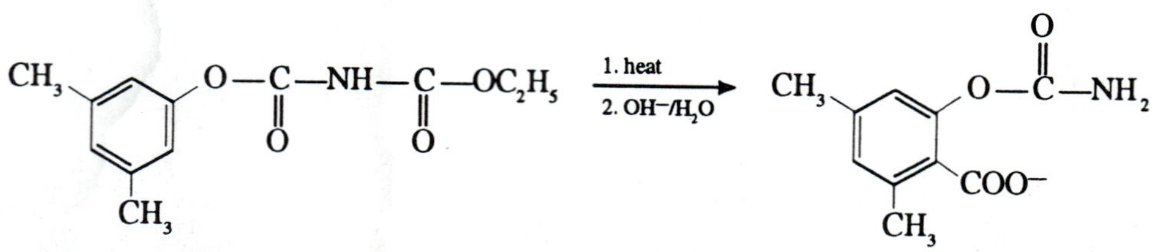

p. 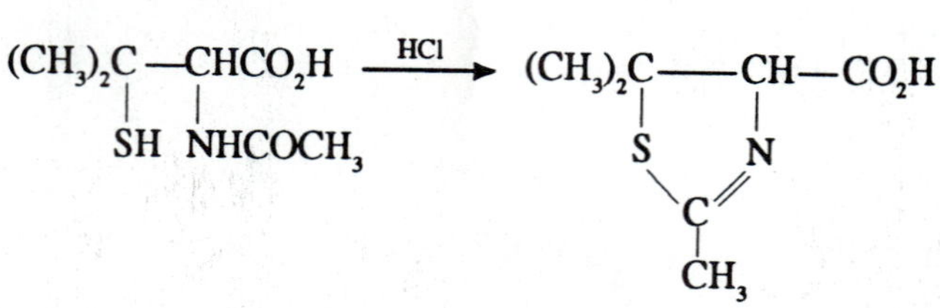

10.9 Arrange the acids below in order of decreasing acidity:

a. 2-bromobutanoic, butanoic, 3-bromobutanoic
b. 2-fluorobutanoic, 2-methylbutanoic, butanoic
c. *p*-nitrobenzoic, *p*-toluic, benzoic, *p*-chlorobenzoic
d. acetic, formic, chloroformic, propanoic
e. fluoroacetic, bromoacetic, iodoacetic, chloroacetic

10.10 Suggest synthetic routes that would lead to the preparation of $HOOC(CH_2)_9COOH$ from the indicated starting reagents.
a. $N{\equiv}C(CH_2)_9C{\equiv}N$ b. $OH(CH_2)_{10}COOH$ c. $HOOC(CH_2)_4CO(CH_2)_4COOH$

10.11 Suggest plausible syntheses of acetic acid from
a. methane (two ways)
b. ethane
c. propane
d. 2-butanol
e. methyl acetate
f. *n*-propylbenzene

10.12 0.11 g of an unknown monocarboxylic acid are neutralized by 12.5 mL of a 0.1 *N* NaOH solution.
a. Calculate the molecular weight of the compound,
b. Suggest a possible formula.

10.13 Exactly 1.05 g of an unknown carboxylic acid was titrated and neutralized by 25.0 mL of a 0.5 *N* NaOH solution. Elemental analysis showed 57.48% carbon and 3.63% hydrogen. When nitrated, the acid gave only one mononitro derivative. Suggest a plausible structure.

10.14 Suggest the simplest chemical test that would help distinguish between each of the following pairs of compounds. Describe what you would do and what you would see.

a. CH_3CH_2COOH and CH_3COOCH_3

b. 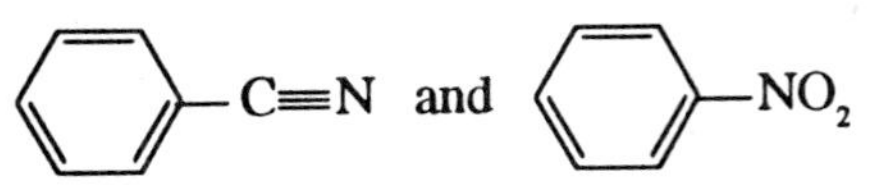

c. 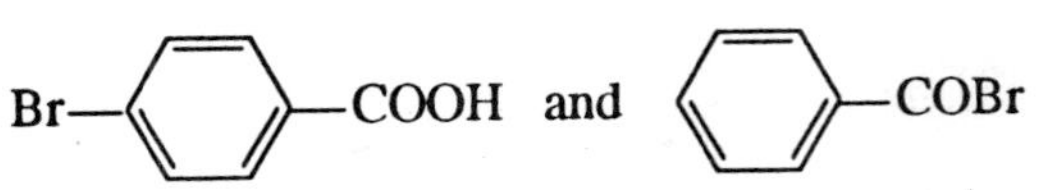

d. 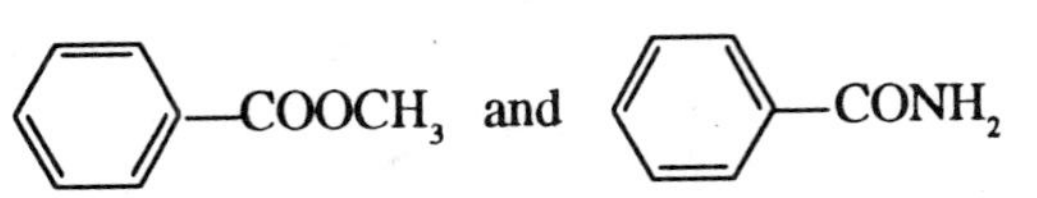

e. 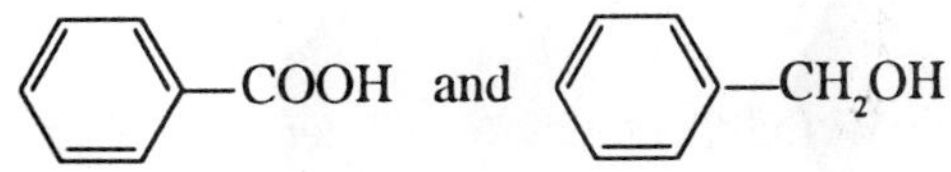

f. 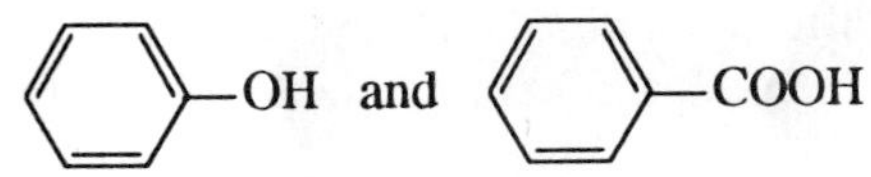

g. 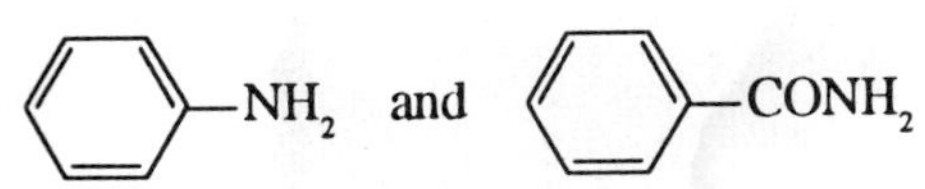

h. 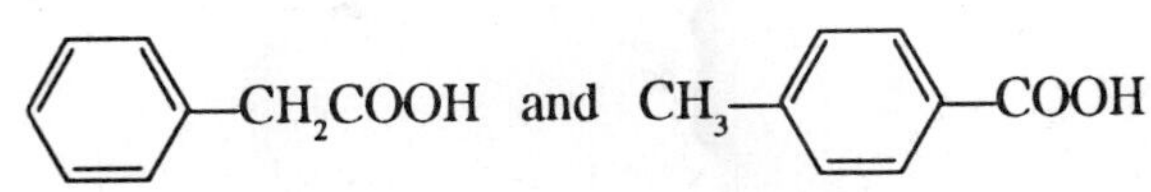

10.15 Outline efficient syntheses of the indicated compounds from the given reactants as the only organic sources
a. $C_6H_5CO^{18}OCH_3$ from $CH_3^{18}OH$ and benzene
b. $CH_3CH_2{}^{14}CH_2COOH$ from $^{14}CH_3OH$ and ethanol
c. isocaproic acid from isopropyl alcohol
d. *n*-butyric acid from ethanol
e. diphenyl ketone from benzene
f. methyl-α-phenyl acetate from benzene and methanol
g. N-methyl valeramide from alcohols of three carbons or less
h. *m*-nitrobenzamide from benzonitrile

10.16 The following are the IR spectra of two compounds of molecular formula $C_8H_8O_2$. What are their structures?

a.

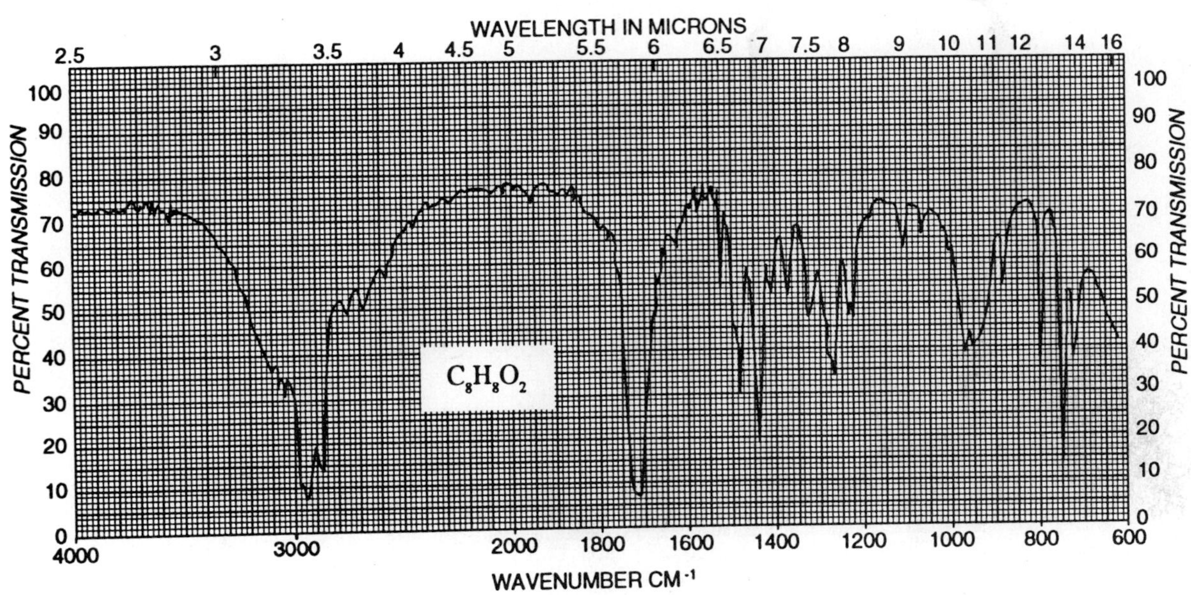

b.

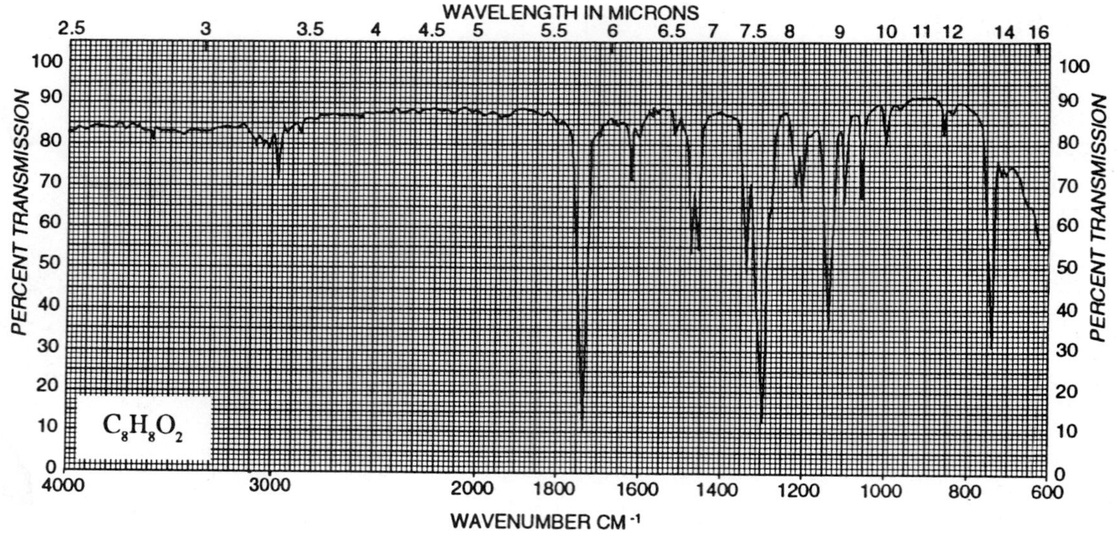

10.17 Compare the spectra of methyl benzoate ($C_8H_8O_2$) and methyl salicylate (methyl *o*-hydroxybenzoate, $C_8H_8O_3$) seen below. Explain

a. the differences in the carbonyl frequencies and

b. why the peak at 3300 cm^{-1} due to –OH is broad.

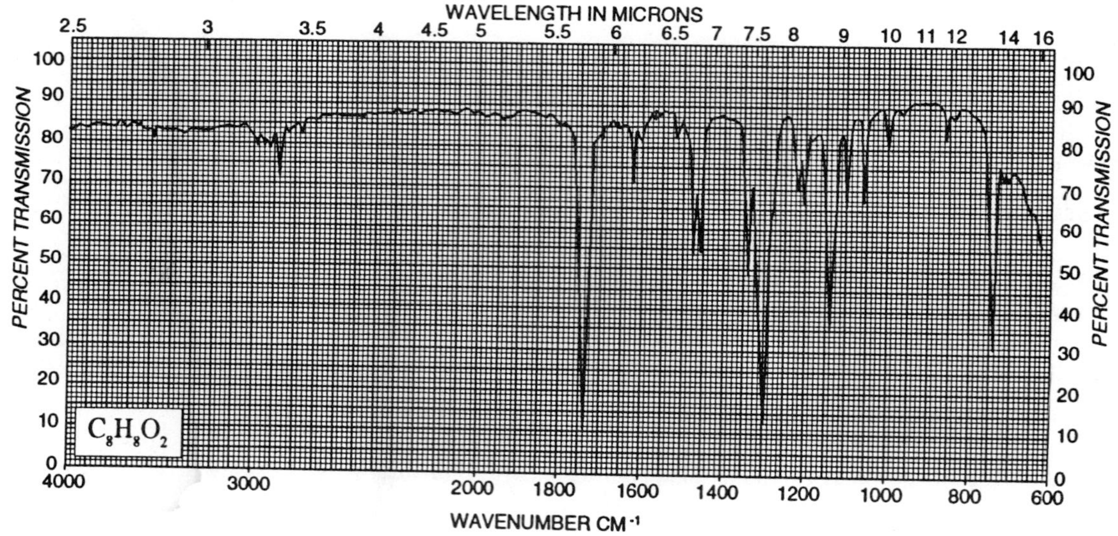

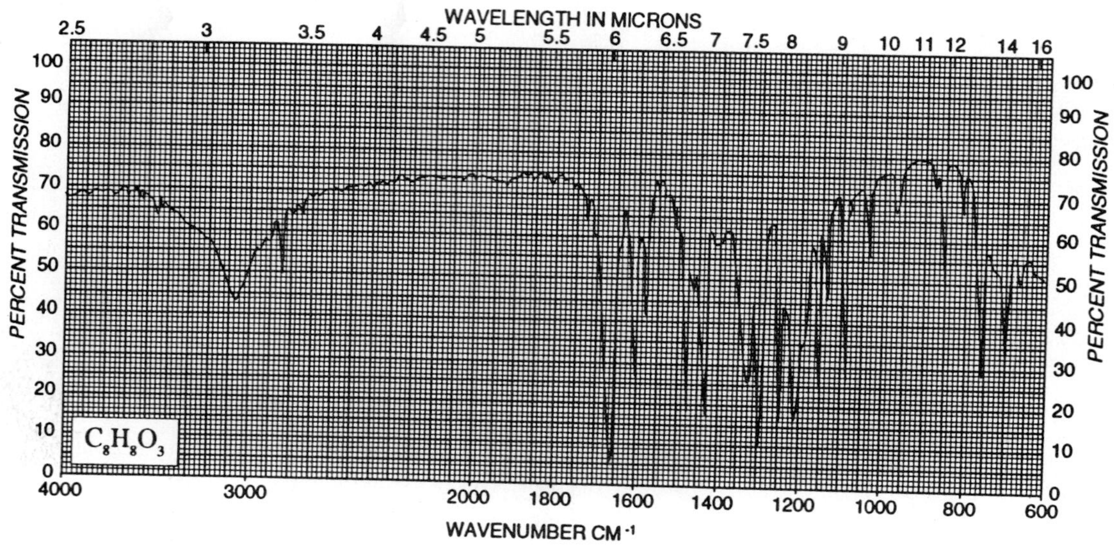

Multiple Choice Test

10.1 The number of carboxylic acids of molecular formula $C_5H_{10}O_2$ is

a. two   c. four
b. three   d. five

10.2 Rank the following in order of decreasing boiling point:

CH_3COOCH_3	$(CH_3)_2CHCH_2CH_3$	$CH_3CH_2CH(OH)CH_3$	CH_3CH_2COOH
A	**B**	**C**	**D**

a. A > B > C > D   c. B > D > A > C
b. A > D > C > B   d. D > C > A > B

10.3 A mixture of 1-pentanol and butanoic acid is dissolved in ether and shaken with a 10% aqueous sodium bicarbonate solution. Which is the correct distribution of the mixture components in each of the two phases?

	Ether solution	Aqueous bicarbonate solution
a.	1-pentanol	butanoic acid
b.	1-pentanol	sodium butanoate
c.	sodium butanoate	1-pentanol
d.	butanoic acid	pentanol

10.4 Rank the following compounds in decreasing acidity

A hydrocyanic acid ($pK_a = 9.31$)   **C** phenol ($pK_a = 9.89$)
B chloroacetic acid ($pK_a = 2.85$)   **D** lactic acid ($pK_a = 3.86$)

a. A > B > C > D   c. C > A > D > B
b. B > D > A > C   d. D > B > A > C

10.5 Which of the following alcohols yields a carboxylic acid when treated with hot $KMnO_4$?

$(CH_3)_2CHOH$	$(CH_3)_2CHCH_2OH$	$(CH_3)_3CCH_2OH$	$(CH_3)_3COH$
A	**B**	**C**	**D**

a. all but D   c. only B
b. B and C   d. A and B

10.6 Which are the hydrolysis products of the following compound

a. 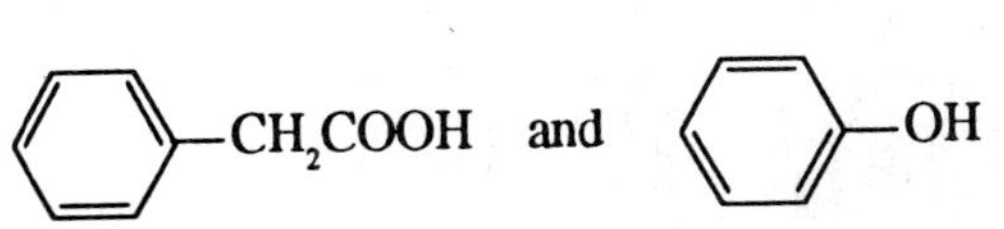

c. 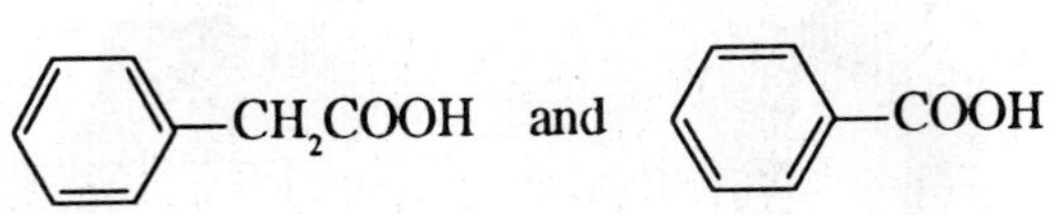

b. 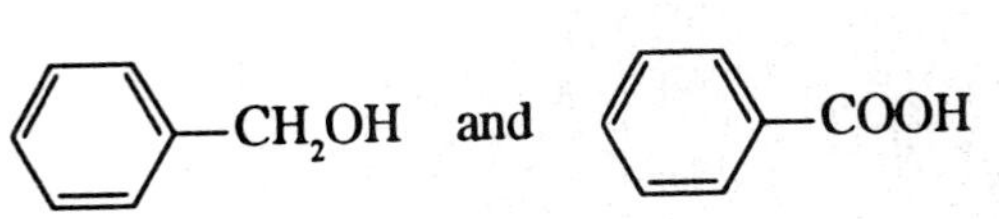

d. 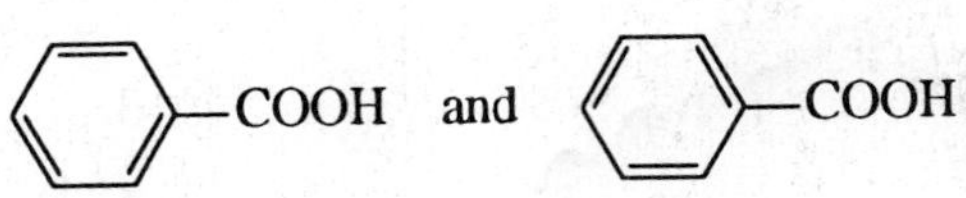

10.7 Which of the following is considered a soap?

a. $CH_3(CH_2)_{14}COOH$

b. $CH_3(CH_2)_{14}COOCH_2$
$CH_3(CH_2)_{14}COOCH$
$CH_3(CH_2)_{14}COOCH_2$

c. $CH_2O^-Na^+$
CHO^-Na^+
$CH_2O^-Na^+$

d. $CH_3(CH_2)_{14}COO^-Na^+$

10.8 Carboxylic acids are ________ acids than alcohols because ________.
a. stronger; the carboxylate anion is destabilized by resonance
b. stronger; the carboxylate anion is stabilized by resonance
c. weaker; the alkoxide anion is more basic because of the alkyl group's inductive effect
d. weaker; the carboxyl group is more stabilized by resonance than the carboxylate anion

10.9 The acid-catalyzed esterification of a carboxylic acid with an alcohol involves the attack of the
a. carboxylic acid –OH group, which displaces the protonated hydroxyl group of the alcohol.
b. alcohol –OH on the carboxylate anion.
c. alcohol –OH on the protonated carbonyl of the acid.
d. alcohol –OH on the neutral carbonyl of the acid.

10.10 *n*-Decanoic acid is ________ in water, ________ in dilute NaOH, ________ in $NaHCO_3$ and ________ in HCl.
a. insoluble; soluble; soluble; insoluble
b. soluble; soluble; soluble; soluble
c. insoluble; soluble; insoluble; soluble
d. soluble; soluble; soluble; insoluble

10.11 Reaction of acetic acid with ethyl magnesium bromide yields
a. $(CH_3CH_2)_2C{=}O$
b. $(CH_3CH_2)_2C(OH)CH_3$
c. C_2H_6
d. no reaction

10.12 What is the reagent for the conversion

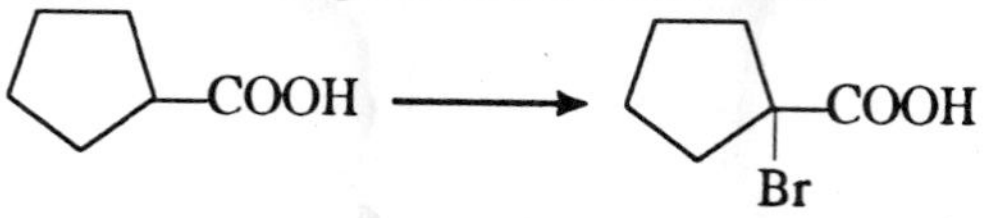

a. Br_2/P
b. Br_2/Fe
c. $SOBr_2$
d. $Br_2/ZnCl_2$

10.13 Which of the following compound is benzoyl chloride?

a. 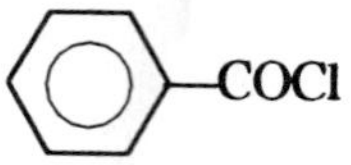

b. 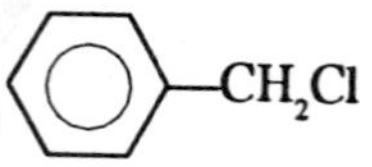

c. 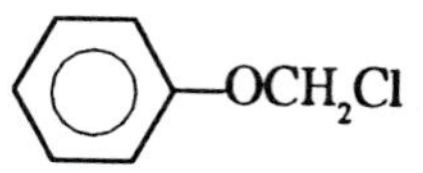

d. 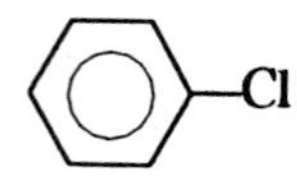

10.14 Which of the following is hydrolyzed the slowest by base?
a. $(CH_3CH_2CO)_2O$
b. CH_3CH_2COCl
c. $CH_3CH_2CONH_2$
d. $CH_3CH_2COOCH_3$

10.15 What is the product in this sequence?

$$(CH_3)_2CHCOOCH_3 \xrightarrow{LiAlH_4} \xrightarrow{PBr_3} \xrightarrow{NaCN} \xrightarrow{H_2O/H^+} \text{product}$$

a. $(CH_3)_2CHCONH_2$
b. $(CH_3)_2CHCH_2COOH$
c. $(CH_3)_2CHCH_2CH_2OH$
d. $(CH_3)_2CHCH_2COO^-NH_4^+$

10.16 The following compound is an example

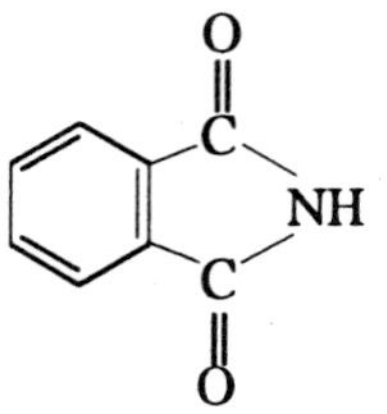

a. lactam
b. imine
c. imide
d. amide

10.17 Which of the following reactions does not produce an ester?

a. $CH_3CH_2OH + CH_3CH_2COOH/H^+/heat \longrightarrow$
b. $CH_3COCl + CH_3CH_2OH \longrightarrow$
c. $CH_3CH_2OH + H_3PO_4 \longrightarrow$
d. $CH_3CH_2O^-Na^+ + CH_3CH_2I \longrightarrow$

10.18 What is the name of this compound?

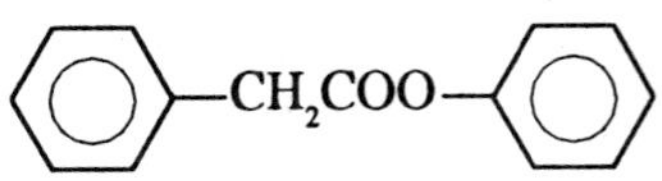

a. benzyl benzoate
b. phenyl benzoate
c. phenyl phenylacetate
d. benzyl phenolate

10.19 What is the correct descending order in acid strength of the following compounds?

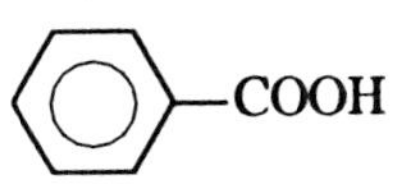

A

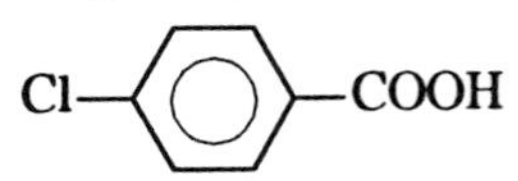

B

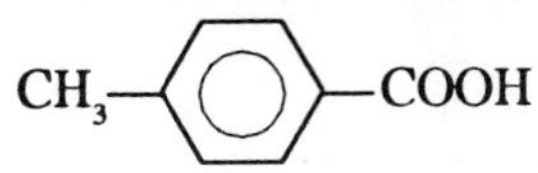

C

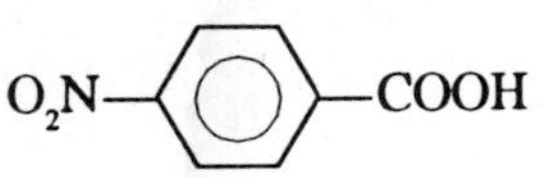

D

a. A > D > C > B
b. B > A > D > C
c. C > D > A > B
d. D > B > A > C

10.20 Which route does not give a high yield of a carboxylic acid?

a. 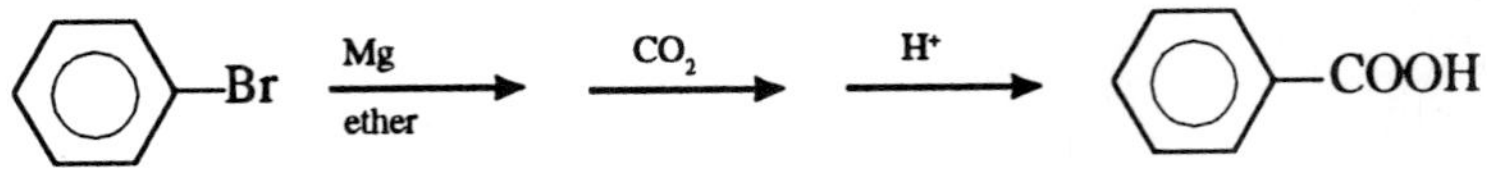

b. 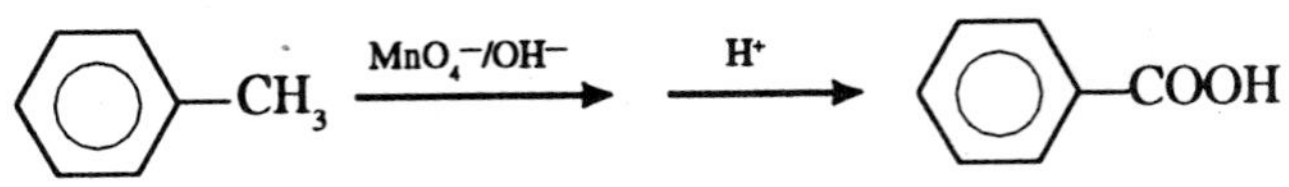

c. 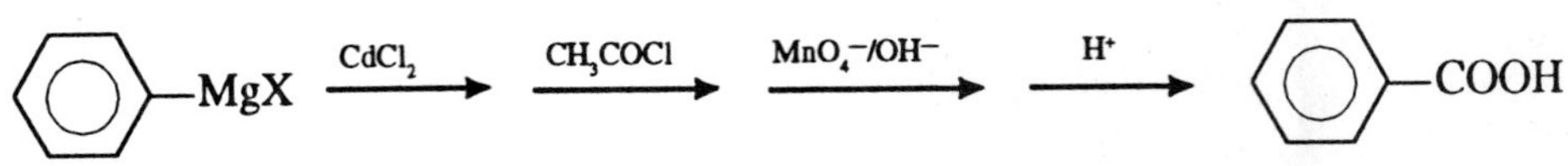

d. $(CH_3)_3CBr \xrightarrow[NaOH/heat]{NaCN/H_2O} \xrightarrow{H^+/H_2O} (CH_3)_3CCOOH$

Answers for Questions

10.1 a. ethyl propanoate, or ethyl propionate
b. ethanoic anhydride, or acetic anhydride
c. benzoyl chloride, or phenyl methanoyl chloride
d. β-bromoethyl acetate or 2-bromoethyl ethanoate
e. methyl benzoate
f. phenylacetic acid, or phenyl ethanoic acid
g. ammonium acetate or ammonium ethanoate
h. dimethylformamide, or dimethyl methanamide
i. N,N,2-trimethyl propanamide or N,N-dimethylisobutyramide or N,N,α-trimethyl propionamide

j. 3-ethyl-4-pentenoic acid, or β-vinyl valeric acid
k. *cis*-butenedioic acid, or maleic acid
l. α-bromo-*n*-butyric anhydride, or 2-bromobutanoic anhydride
m. *p*-formylbenzoic acid, or 4-methanoylphenyl methanoic acid or 4-methanoyl benzoic acid
n. 4-(2-hydroxyphenyl) butanoic acid, or δ-(*o*-hydroxyphenyl) butyric acid
o. phenyl acetamide, or phenyl ethanamide
p. N-*p*-nitrobenzylphenylacetamide, or N-(4-nitrophenylmethyl) phenyl ethanamide
q. benzonitrile, or cyanobenzene
r. N-(2-mercaptoethyl) propanamide, or N-(β-mercaptoethyl) propionamide
s. N-methyl-4-aminocyclohexanecarboxamide
t. δ-butyrolactam
u. aspirin, or acetylsalicylic acid

10.2

a. CH_3COOCH_3
methyl ethanoate

b. $C_6H_5CH_2CONHCH_3$
N-methyl phenylethanamide

c. $HOOCCH_2CH_2COOH$
succinic acid

d. 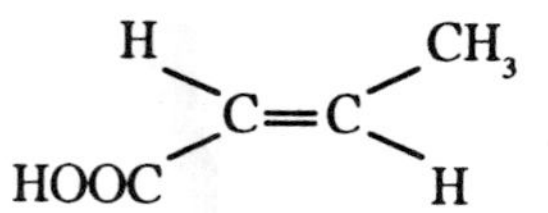
crotonic acid

e. $(CH_3CH_2CH_2CH_2CO)_2O$
pentanoic anhydride

f. $CH_3CH_2CH(OCH_3)COOH$
2-methoxybutanoic acid

g. 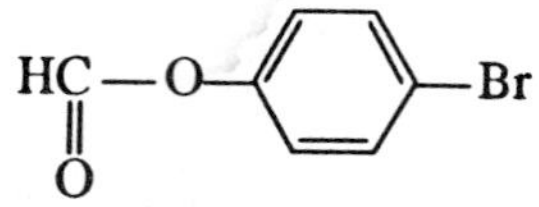
4-bromophenyl methanoate

h. $(CF_3CO)_2O$
trifluoroethanoic anhydride

i. CH_3CH_2COCl
propionyl chloride

j. $(CH_3)_2CHCH_2CH(NH_2)COOH$
2-amino-4-methylpentanoic acid

k. $(HCOO)_2Ca$
calcium methanoate

l. 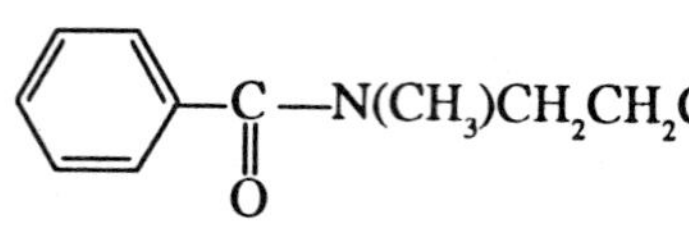
N-(β-chloroethyl)-N-methylbenzamide

m. $(CH_3)_2CHCH_2C{\equiv}N$
3-methyl-1-butanenitrile

n. 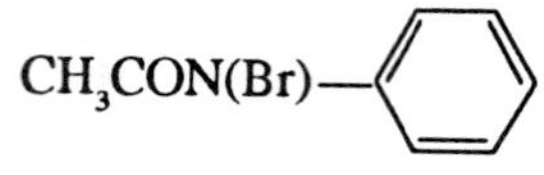
N-bromo-N-phenylethanamide

10.3

a. B	d. J	f. L	h. I	j. D	l. C
b. G	e. K	g. H	i. M	k. E	m. F
c. A					

10.4 a. Amides have a resonance form that leads to the delocalization of nitrogen's non-bonded pair of electrons. This leads to a partial loss of the electrophilic character of the carbonyl carbon:

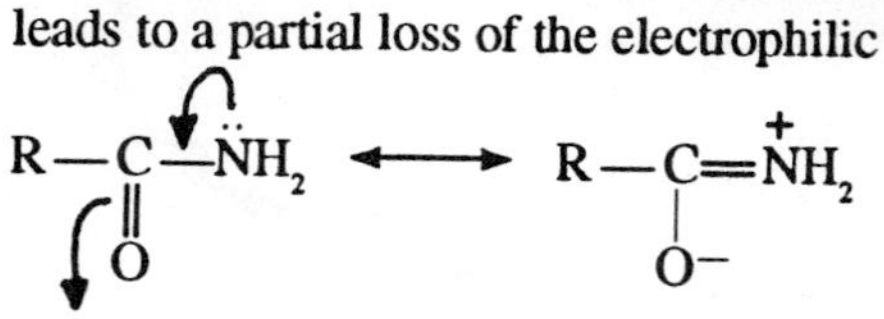

The dipolar resonance form is more significant in amides than in esters because nitrogen is less electronegative than oxygen. Moreover, the amino group ($-NH_2$) is one of the poorest leaving groups (strongest base). As a result, nucleophiles react much more sluggishly with amides.

b. First the carbonyl carbon is more polar because of the electronegative oxygen. Moreover, hydrolysis of an acyl halide leads to a tetravalent carbon intermediate, while hydrolysis of alkyl halides leads to a transition state that possesses a ten-electron, four-bonded carbon:

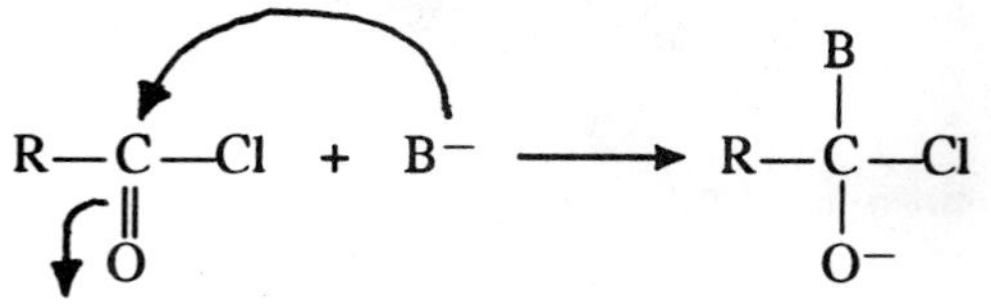

(intermediate; tetravalent carbon)

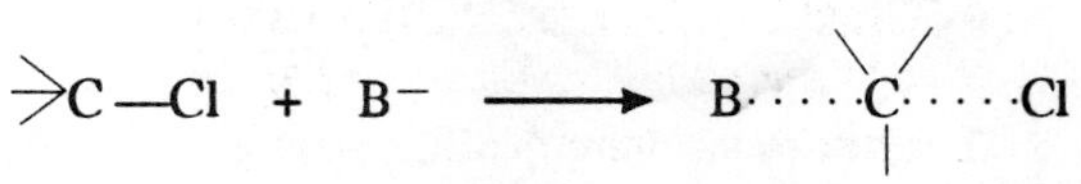

(transition state; 10-electron, 4-bonded carbon)
(Note: Dotted lines represent partial bonds)

c. The two synthetic methods are

Method 1. 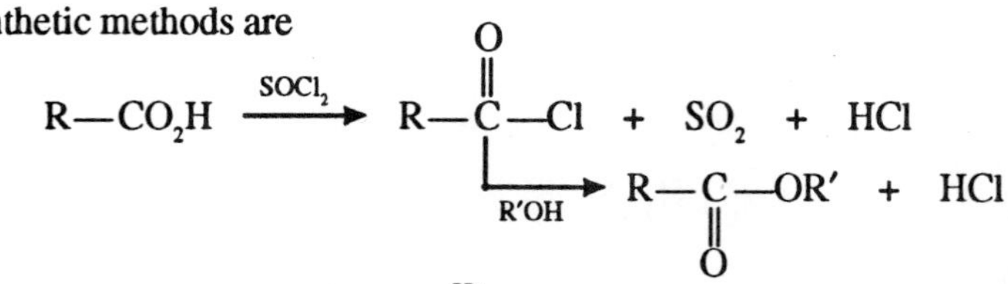

Method 2. $R{-}CO_2H + R'OH \xrightarrow{H^+} R{-}CO_2R' + H_2O$

In Method 1 SO_2 and HCl are formed; since they are gases at room temperature, each step is driven almost to completion according to Le Châtelier's principle, thus providing much higher yields.

d. The carboxylic acid and the amine or ammonia (which are bases) will give a salt and water:

$R{-}CO_2H + R'NH_2 \longrightarrow RCO_2^- NHR'^+ + H_2O.$

Heat is needed to remove water and thus drive the equilibrium toward the right.

e. The reactions are:

acidic medium: $R{-}C{\equiv}\ddot{N} + H_2O \xrightarrow{H^+} RCO_2^- NH_4^+ \xrightarrow{H^+} RCO_2H + NH_4^+$

basic medium: $R{-}C{\equiv}\ddot{N} + H_2O \xrightarrow{OH^-} RCO_2^- + NH_3$

The reaction in basic medium yields a gas (NH_3), which drives the reaction almost to completion according to Le Châtelier's principle.

f. Loss of the acidic hydrogen on the nitrogen yields an anion that is stabilized by three resonance structures in the first compound and only two in the second compound.

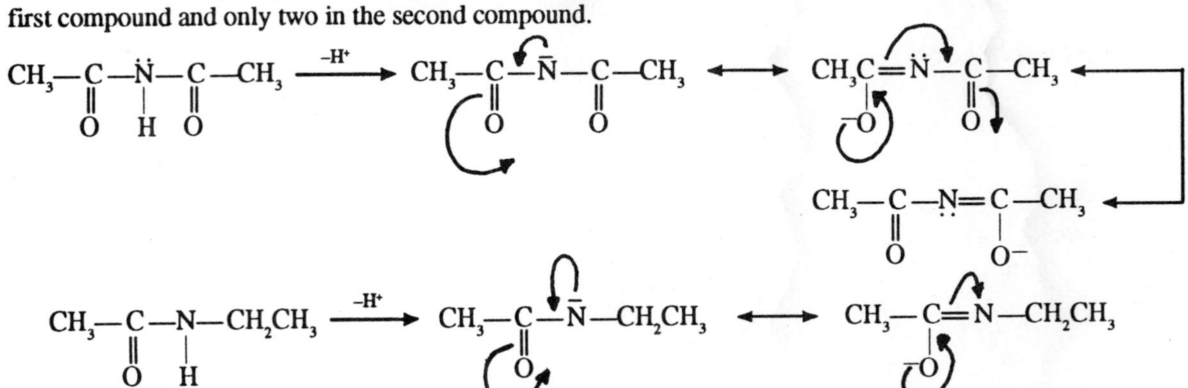

g. The electron pair on the amide nitrogen is involved in resonance delocalization, and therefore, is not as available as the amine nitrogen, whose non-bonded pair of electrons is localized:

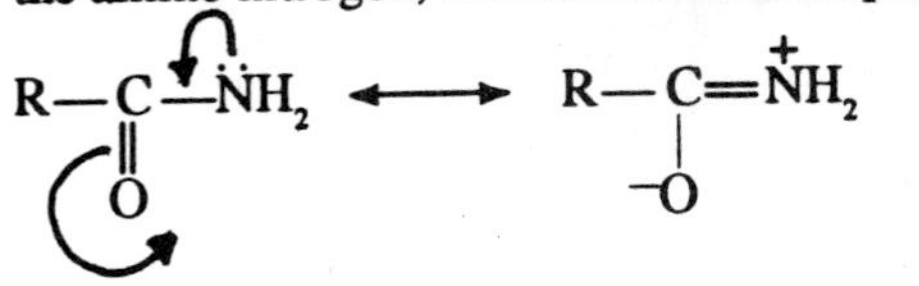

The equation is $R{-}\overset{\overset{\displaystyle O}{\|}}{C}{-}Cl$ + $R'NH_2$ $\longrightarrow$ $R{-}\overset{\overset{\displaystyle O}{\|}}{C}{-}NHR'$ + HCl

The formed HCl will react with $R'NH_2$ (a base)

$R'NH_2 + HCl \longrightarrow R'NH_3^+Cl^-$

Thus, the maximum possible yield is 50%. Using a 2-mole equivalent of the amine will double the amount of amine available and will thus increase the yield. Experimentally, a tertiary amine (such as pyridine) is used as a scavenger to trap the HCl.

.5 a. 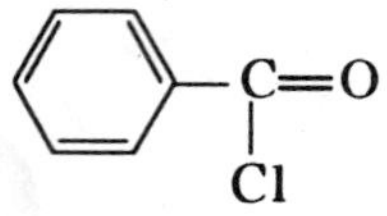

c. $CH_2{=}CH\overset{*}{C}HCOOH$ (with C_2H_5 on the starred carbon)

e. $CH_3COOCHCH_2CH_3$ (with CH_3 on the CH)

b. 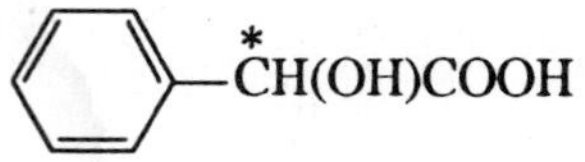

d. 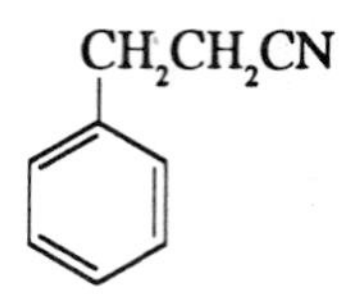

10.6 A 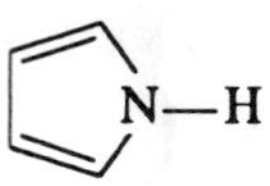 (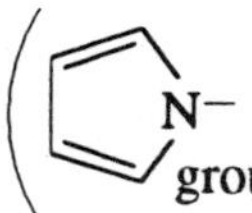 is an excellent leaving group because it is aromatic.)

I 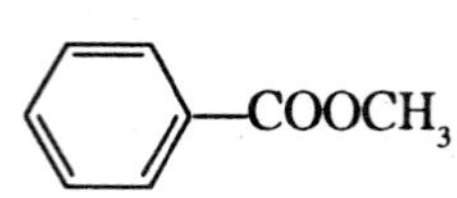

B 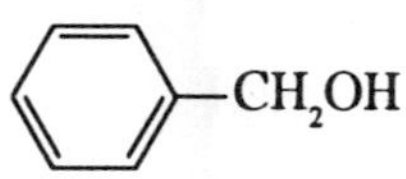

J $(CH_3)_3COH$

C 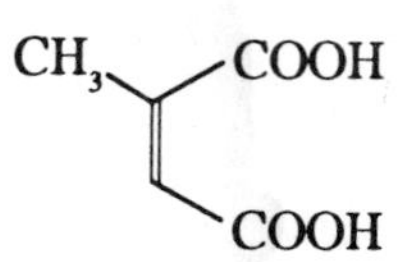

K $C_8H_{17}CH{=}CH(CH_2)_7\overset{\overset{\displaystyle O}{\|}}{C}NHC_4H_9$

D 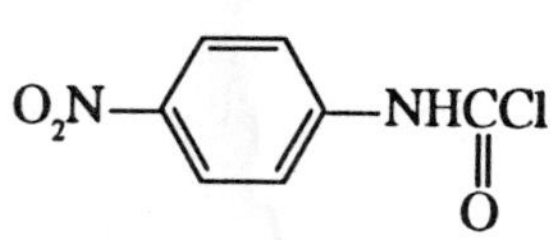

L 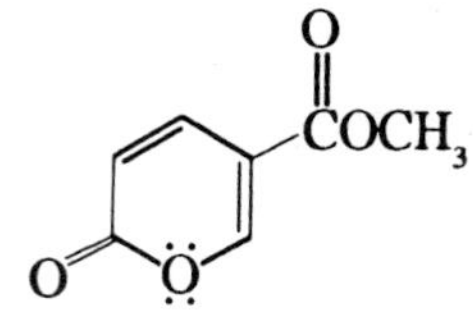

(*Note*: the lactone is aromatic and therefore does not easy undergo hydrolysis)

E $CH_3N(CH_2COOH)_2$

M 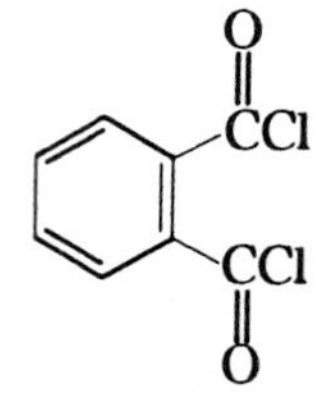

F $(C_4H_9)_2CHOH$

N $(COO^-)_2Ca^{+2}$

G 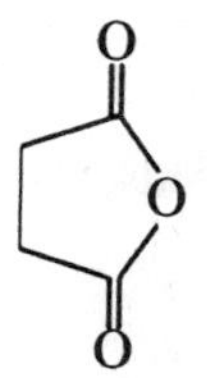

O PCl_3 or $SOCl_2$

H CCl_3COCl

P $n\text{-}C_3H_7{-}C(CH_3)(C_2H_5){-}COOH$

Q 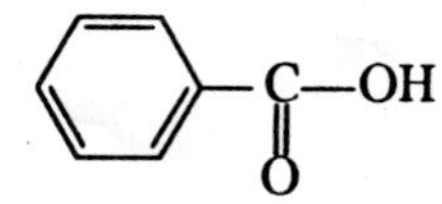

R 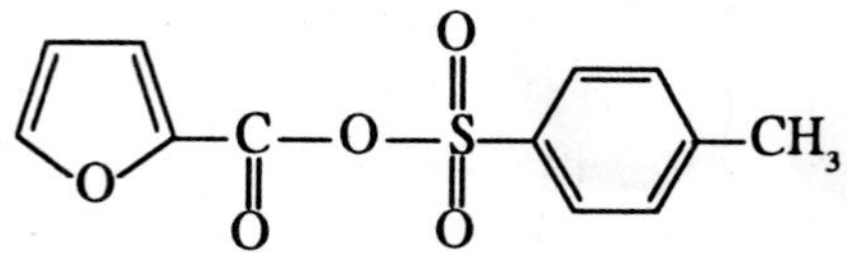

S 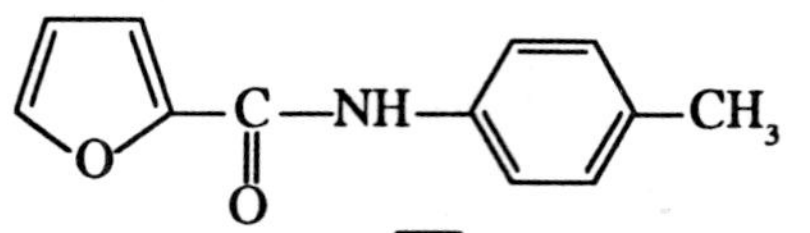

(the tosyl group, CH$_3$– 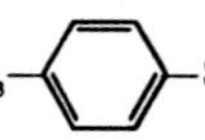 –SO$_3^-$, is an excellent leaving group)

T $n\text{-}C_6H_{13}\underset{\underset{O}{\|}}{C}NH_2$

U $(\text{-}\underset{\underset{O}{\|}}{C}OCH_3)_2$

V $C_9H_{19}CH{=}\underset{\underset{CH_3}{|}}{C}{-}COO^-K^+$

W $(CH_3)_2CH\underset{\underset{O}{\|}}{C}NH_2$

X 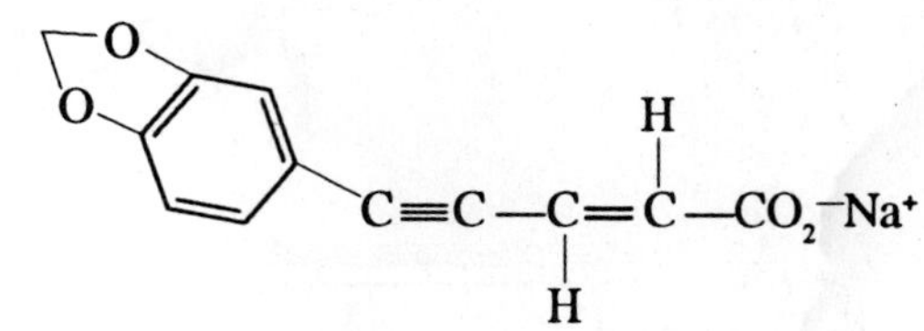

Y 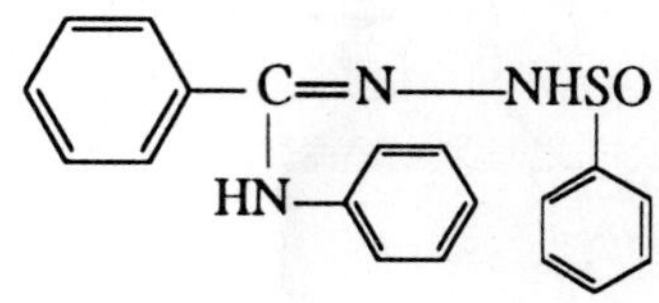

(the reacted nitrogen is the *most* basic)

Z 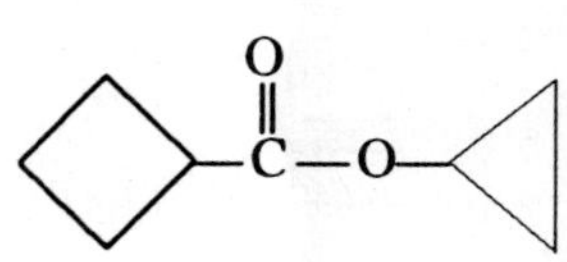

10.7 **A** 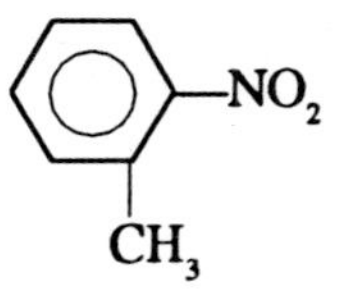

B 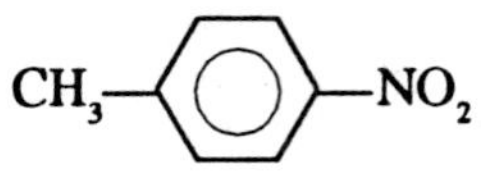

C 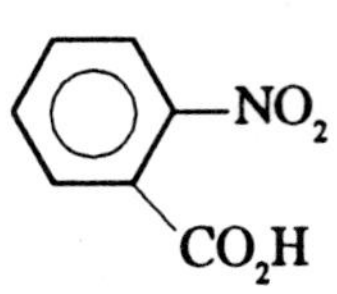

D 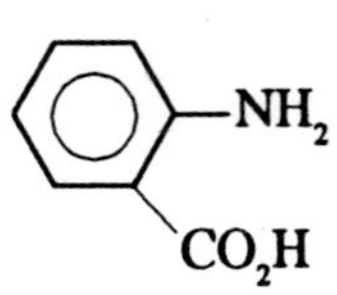

E 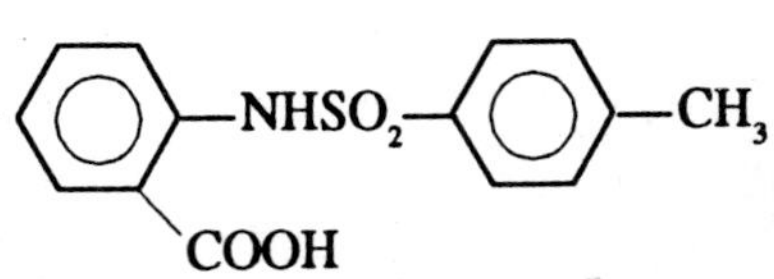

F 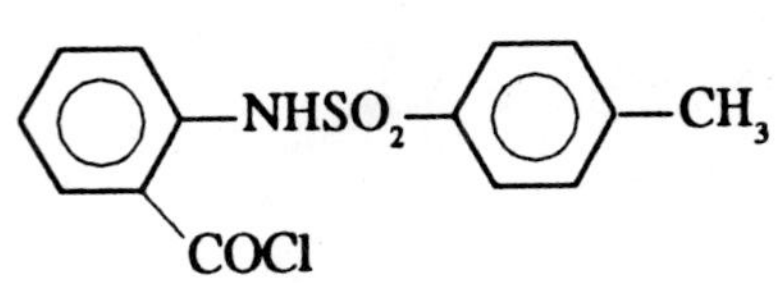

G 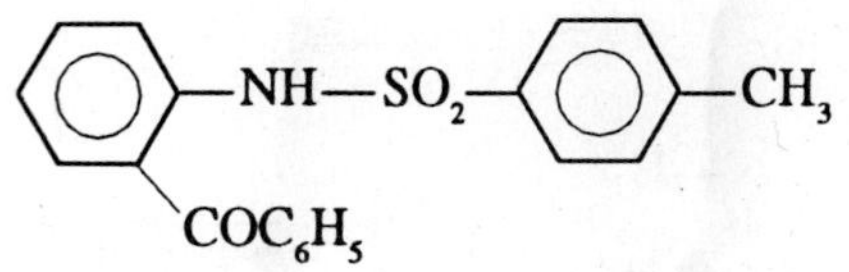

H 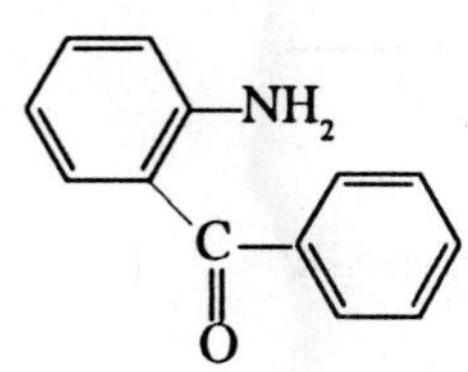

I 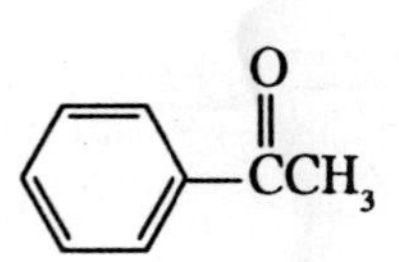

J 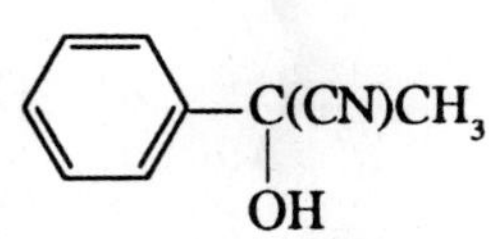

K 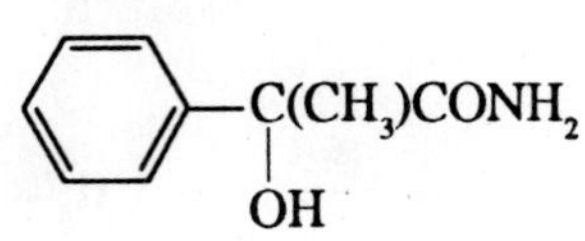

L 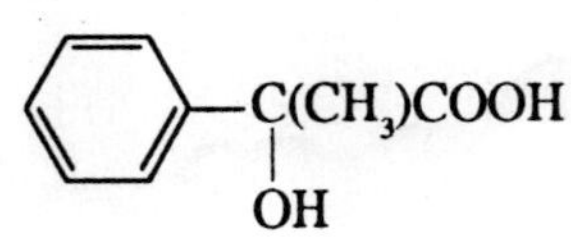

M	$ClCH_2CO_2Na$	**U**	$C_6H_5(CH_2)_2Br$	**AA**	Br_2/P
N	$NC—CH_2CO_2Na$	**V**	$C_6H_5CH_2CH_2CN$	**BB**	$CH_3Cl/AlCl_3$
O	$CH_2(CO_2Na)_2$	**W**	$C_6H_5CH_2CH_2COOH$	**CC**	$KMnO_4$
P	$CH_2(CO_2)_2Ca$	**X**	$C_6H_5CH_2CH_2COCl$	**DD**	H^+
Q	$CH_2(CO_2H)_2$	**Y**	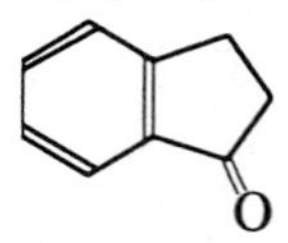	**EE**	C_6H_5COCl
R	Br_2/Fe or $FeBr_3$			**FF**	$C_6H_5COOCH_2CH_2Cl$
S	C_6H_5MgBr			**GG**	$C_6H_5CO_2CH_2CH_2I$
T	$CH_2—CH_2$ (bridged by O)			**HH**	$C_6H_5CO_2CH_2CH_2\overset{+}{N}(CH_3)_3I^-$
		Z	$C_6H_5CH_2CH_2CONHC_6H_5$	**II**	$C_6H_5CO_2CH{=}CH_2$

10.8 a.

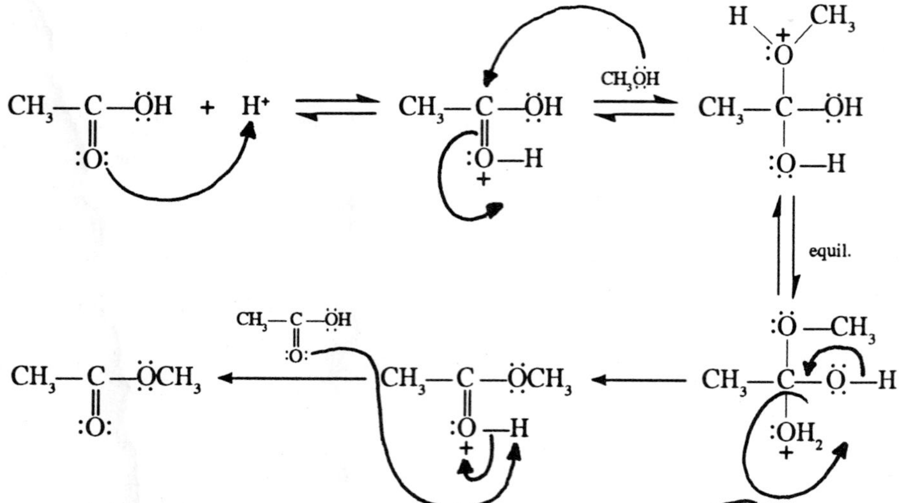

b.

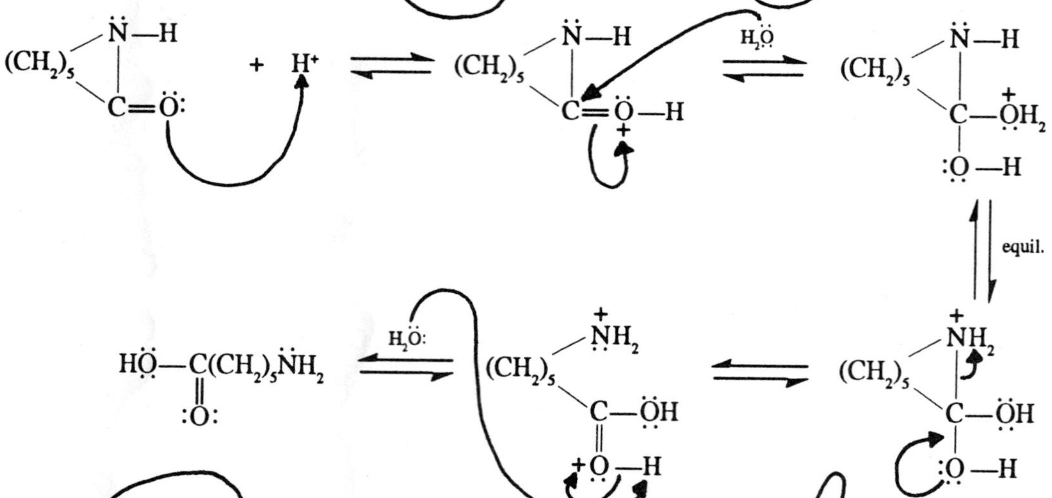

c.

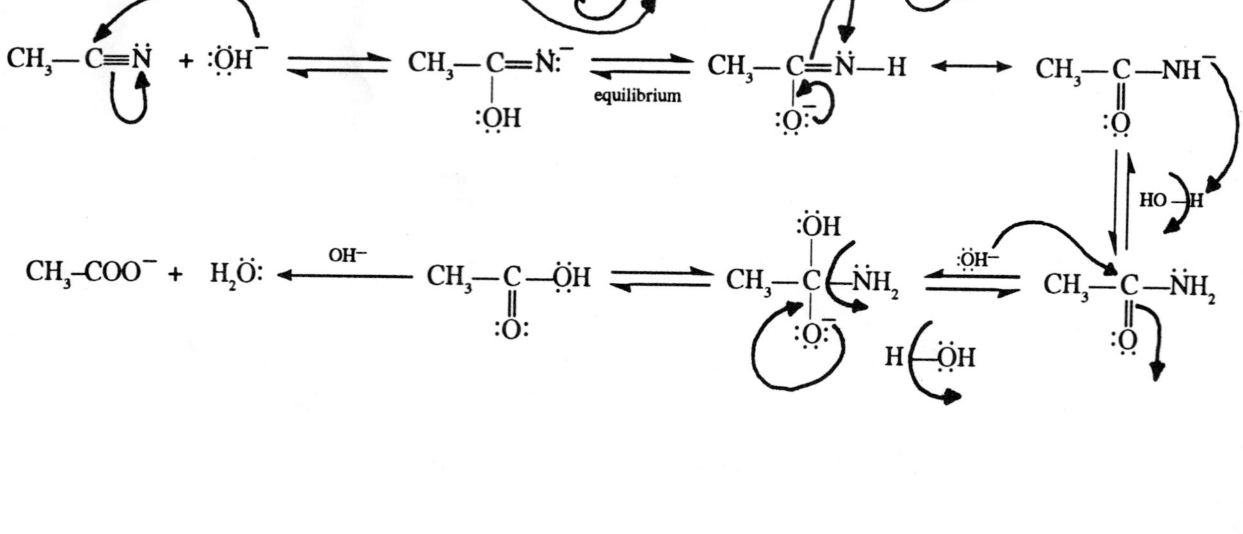

d.

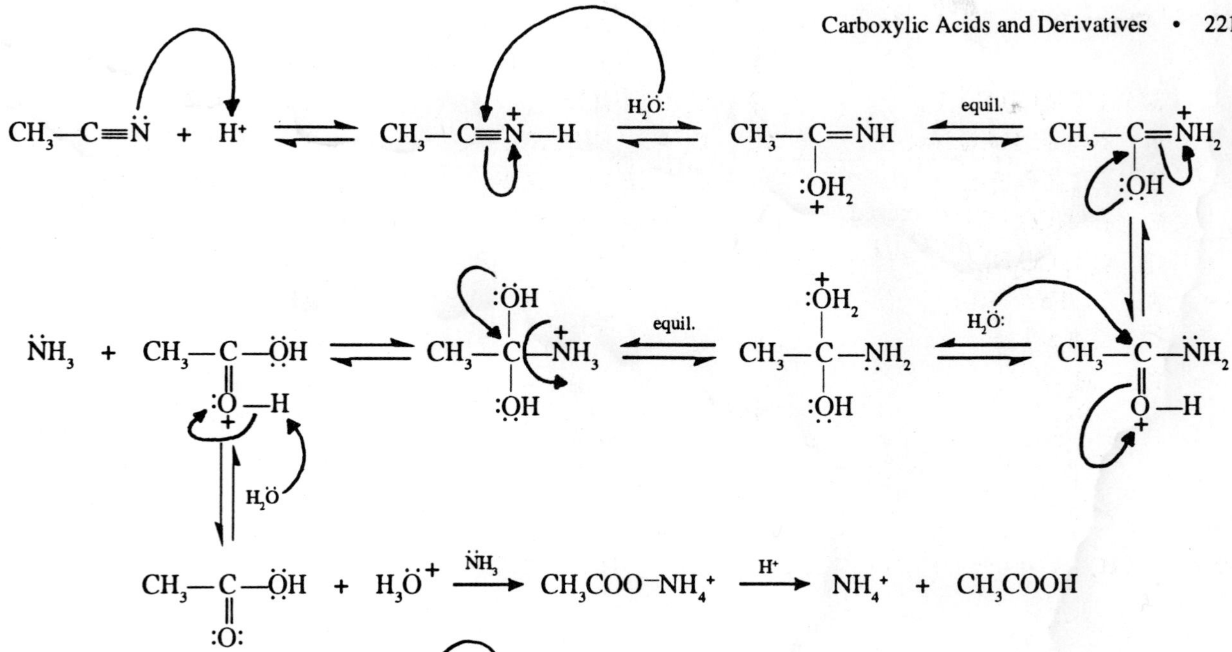

e.

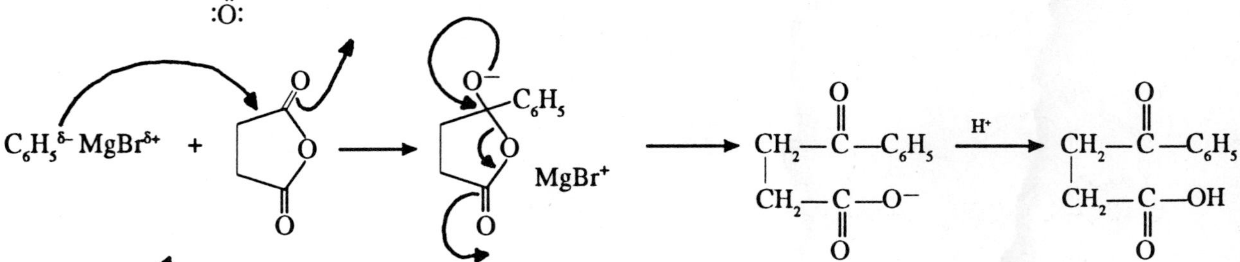

f.

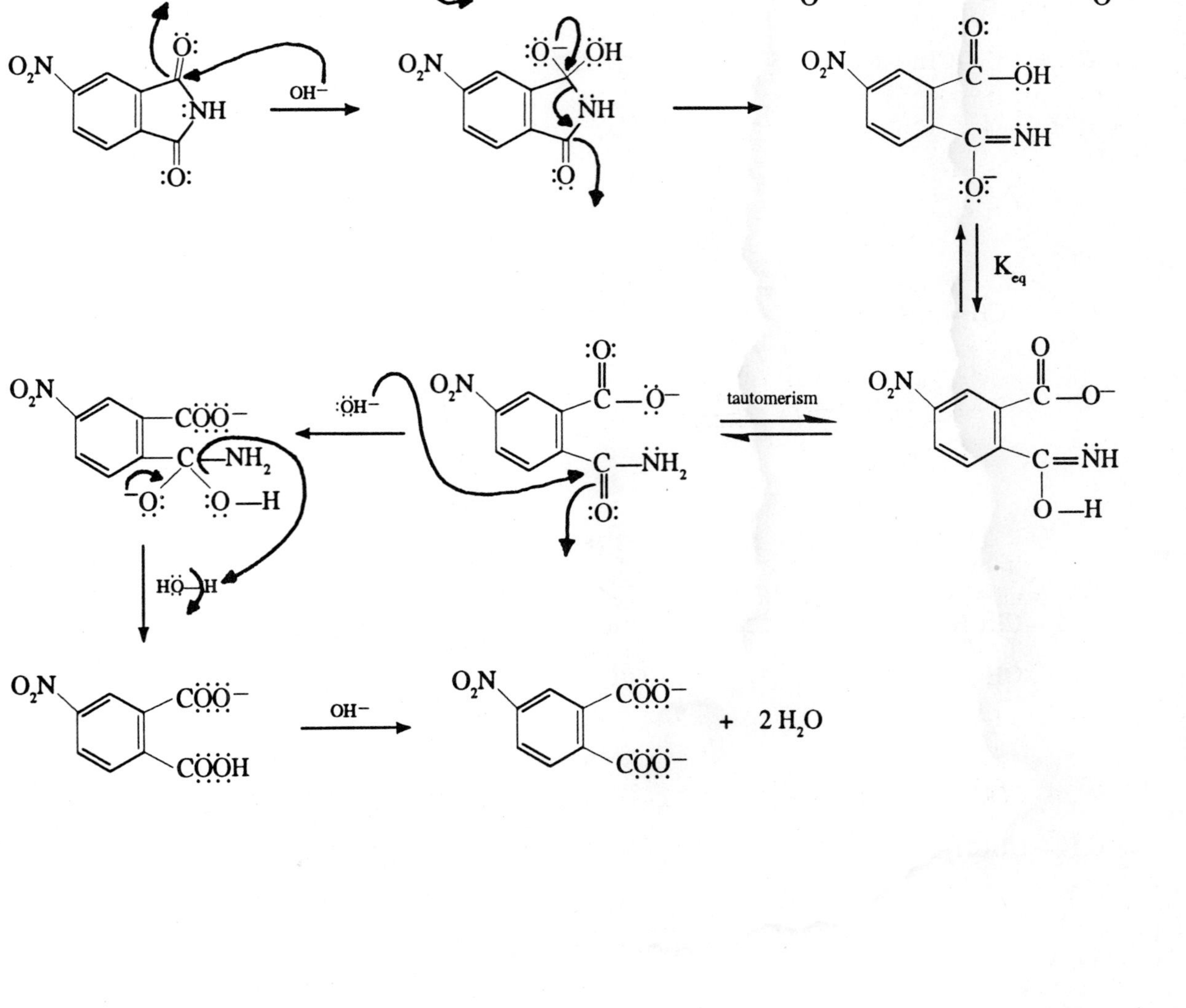

g.

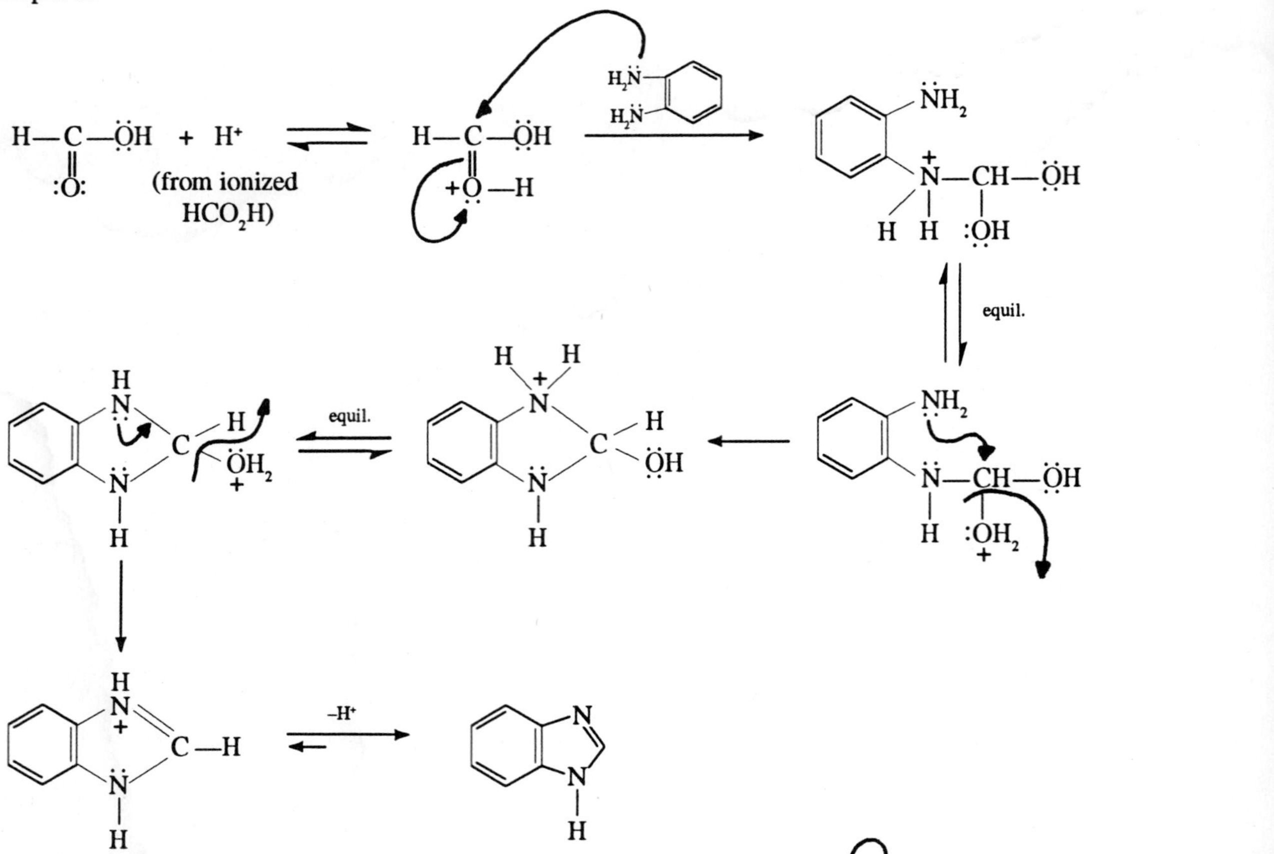
(from ionized HCO_2H)
equil.
equil.
$-H^+$

h.

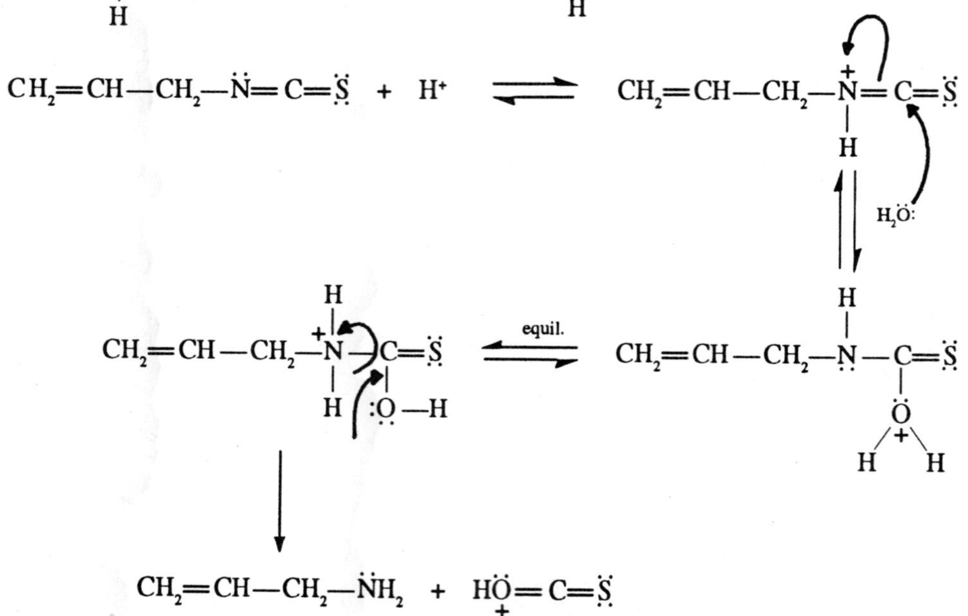
equil.

i.

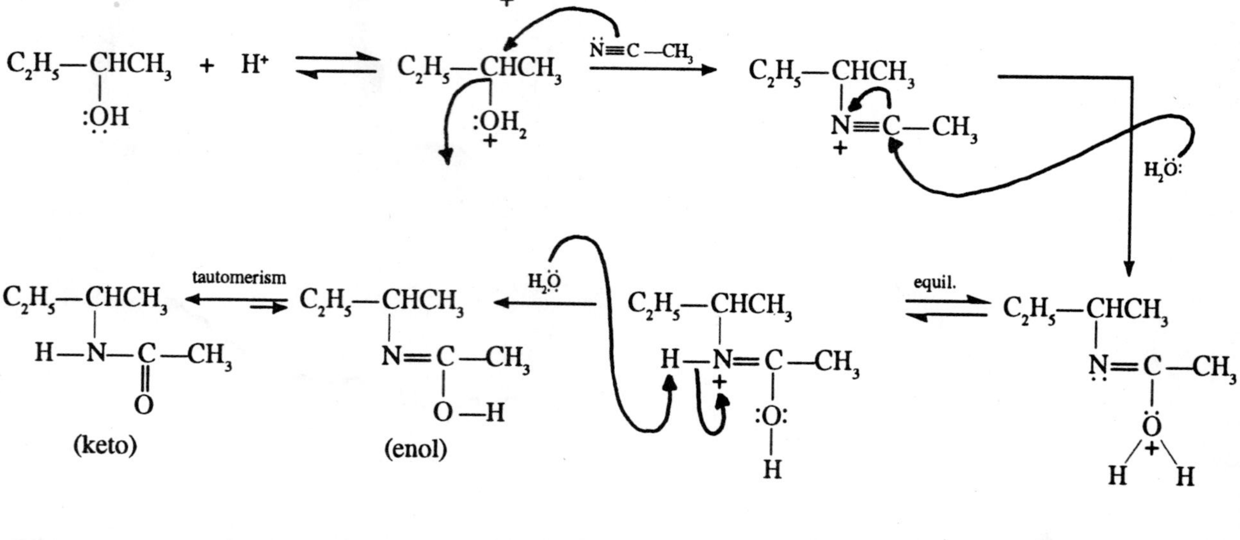
tautomerism
equil.
(keto)
(enol)

j.

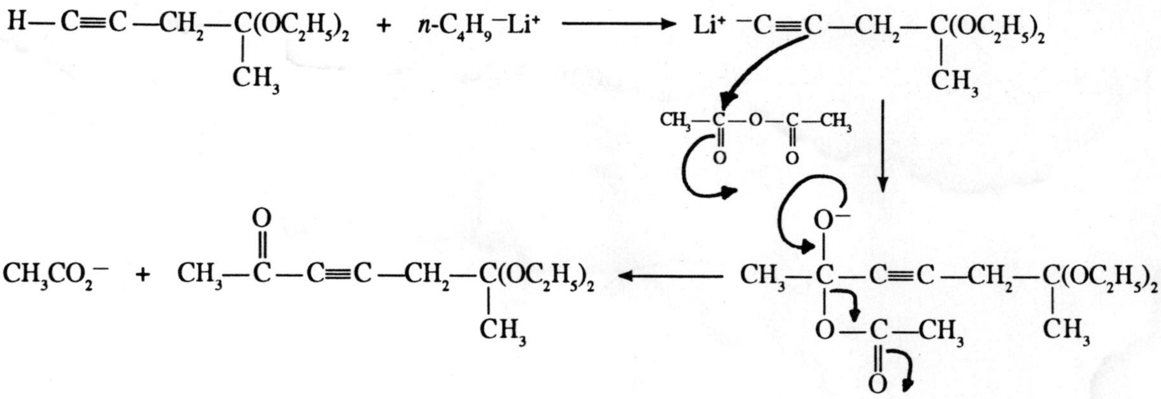

k.

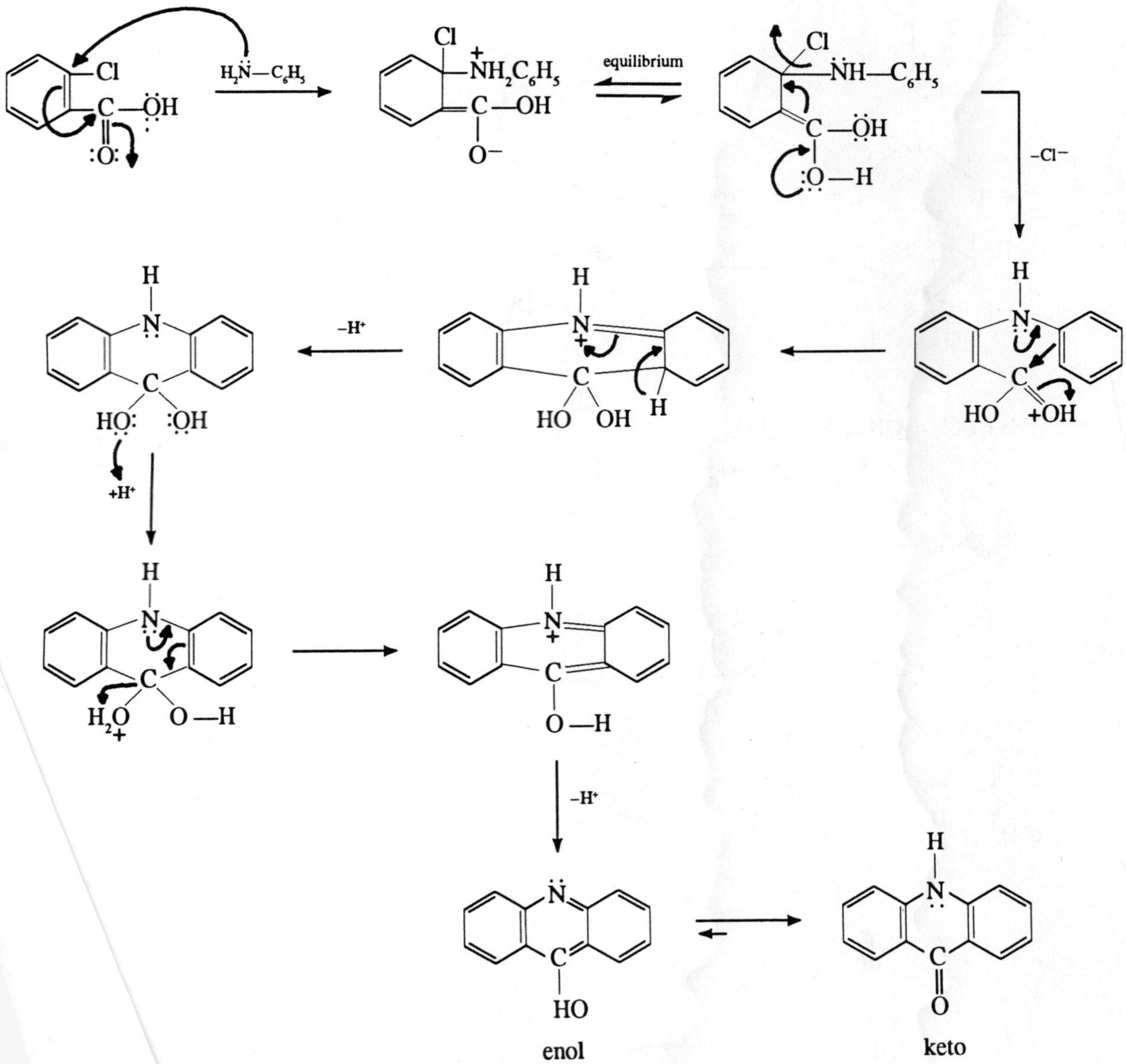

n.

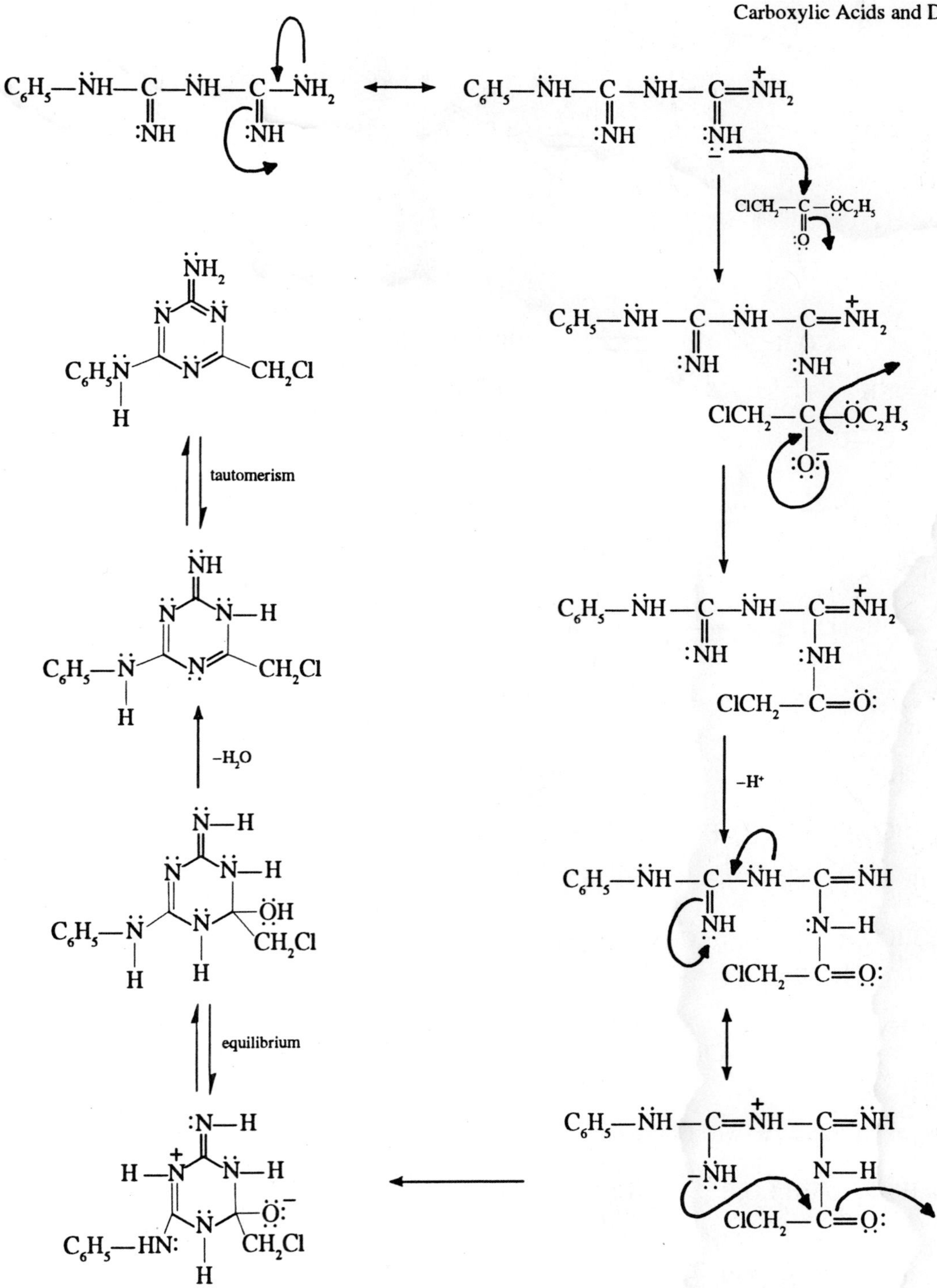

o.

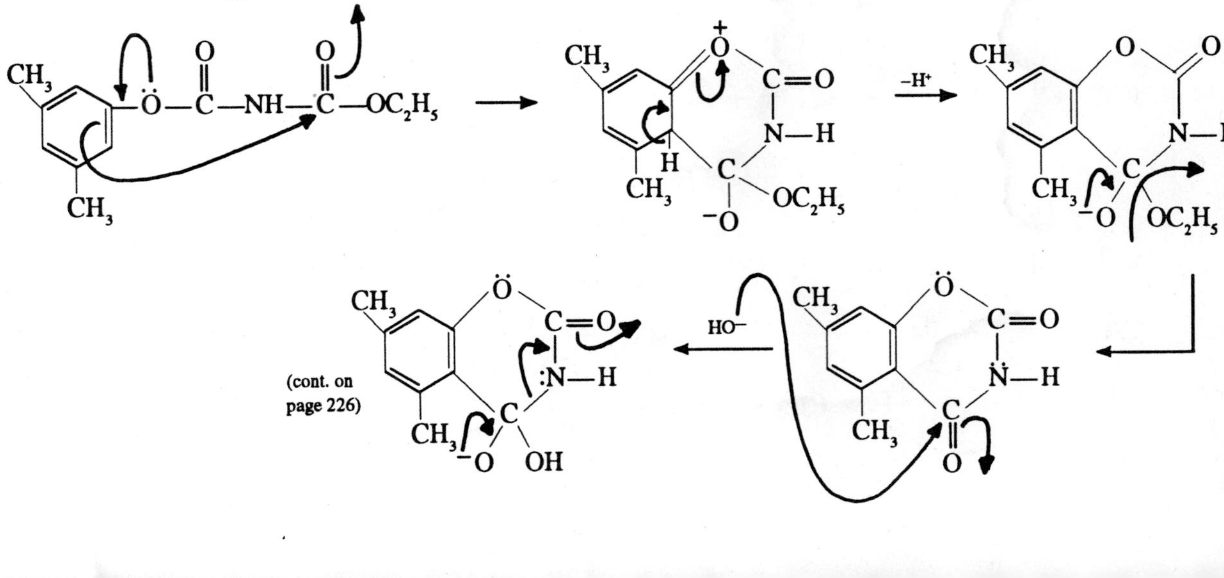

(cont. on page 226)

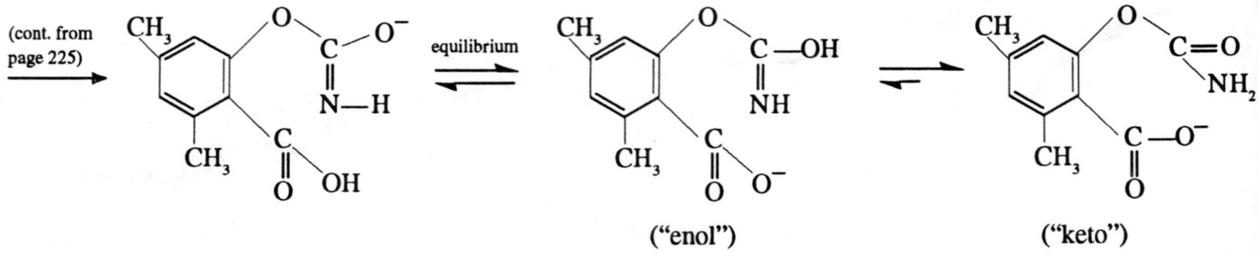

p.

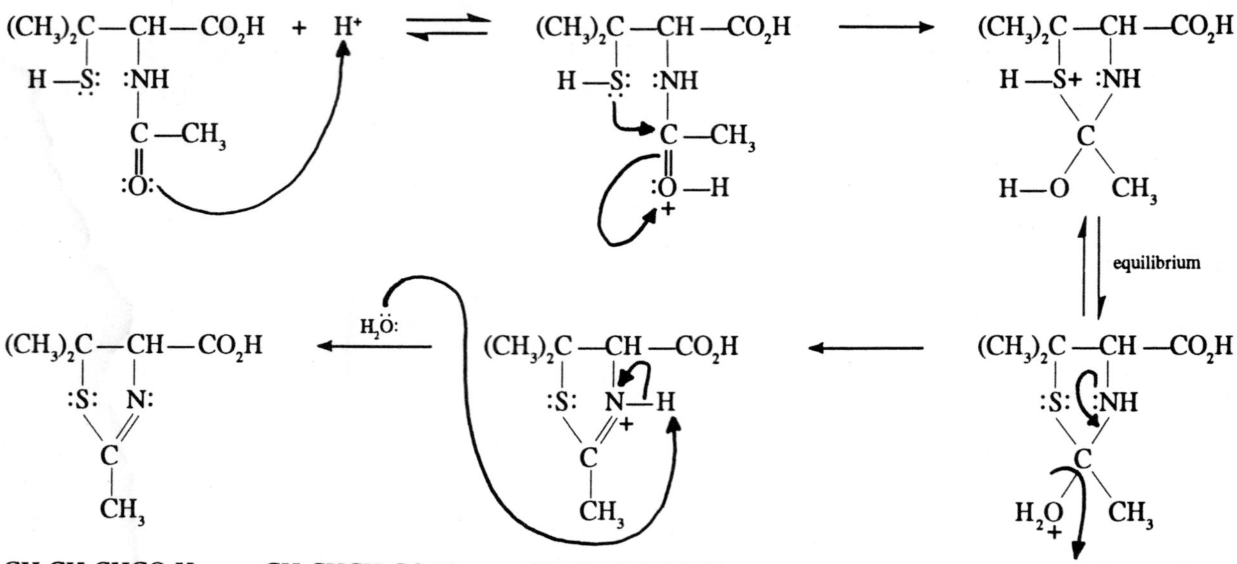

10.9 a. $CH_3CH_2\underset{\substack{|\\ Br}}{C}HCO_2H$ > $CH_3\underset{\substack{|\\ Br}}{C}HCH_2CO_2H$ > $CH_3CH_2CH_2CO_2H$

b. $CH_3CH_2\underset{\substack{|\\ F}}{C}HCO_2H$ > $CH_3CH_2CH_2CO_2H$ > $CH_3CH_2\underset{\substack{|\\ CH_3}}{C}HCO_2H$

c.

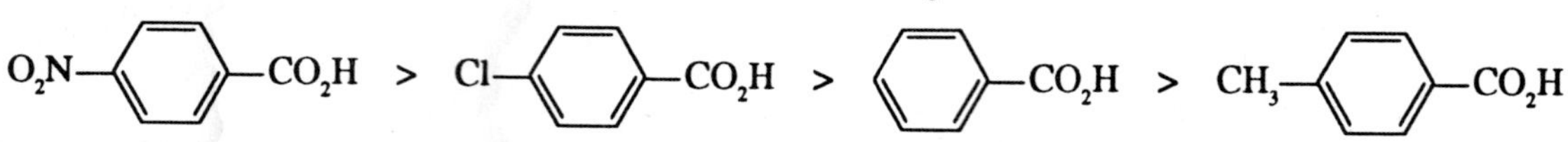

d. $ClCO_2H$ > HCO_2H > CH_3CO_2H > $CH_3CH_2CO_2H$

e. FCH_2CO_2H > $ClCH_2CO_2H$ > $BrCH_2CO_2H$ > ICH_2CO_2H

10.10 a. H^+/H_2O/heat
b. $Cr_2O_7^{-2}/H^+$ (or dil. $KMnO_4/OH^-$/heat, then acidification)
c. Zn(Hg)/HCl

10.11 a. $CH_4 + Br_2 \xrightarrow{UV} CH_3Br \xrightarrow{CN^-} CH_3CN \xrightarrow[2.\ H^+]{1.\ OH^-/heat} CH_3CO_2H$

$CH_3Br \xrightarrow[ether]{Mg} CH_3MgBr \xrightarrow[2.\ H^+]{1.\ CO_2} CH_3CO_2H$

b. $C_2H_6 + Br_2 \xrightarrow{UV} CH_3CH_2Br \xrightarrow{OH^-} CH_3CH_2OH \xrightarrow[2.\ H^+]{1.\ KMnO_4/OH^-} CH_3CO_2H$

c. $CH_3CH_2CH_3 + Br_2 \xrightarrow{UV} CH_3CH(Br)CH_3 \xrightarrow{alc.\ KOH} CH_3CH{=}CH_2 \xrightarrow[2.\ H^+]{1.\ KMnO_4/OH^-} CH_3CO_2H$

d. $CH_3\underset{\substack{|\\ OH}}{C}HCH_2CH_3 \xrightarrow[heat]{H_2SO_4} CH_3CH{=}CHCH_3 \xrightarrow[2.\ H^+]{1.\ KMnO_4/OH^-} CH_3CO_2H$

j.

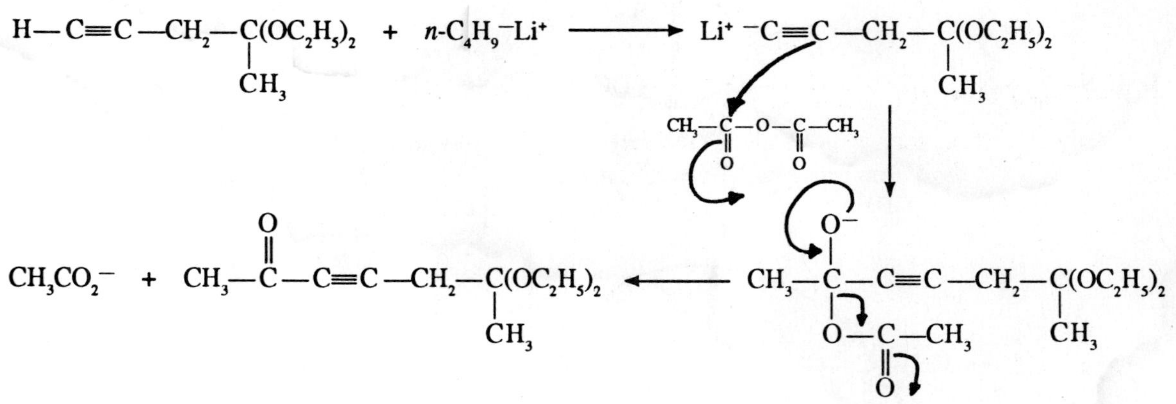

k.

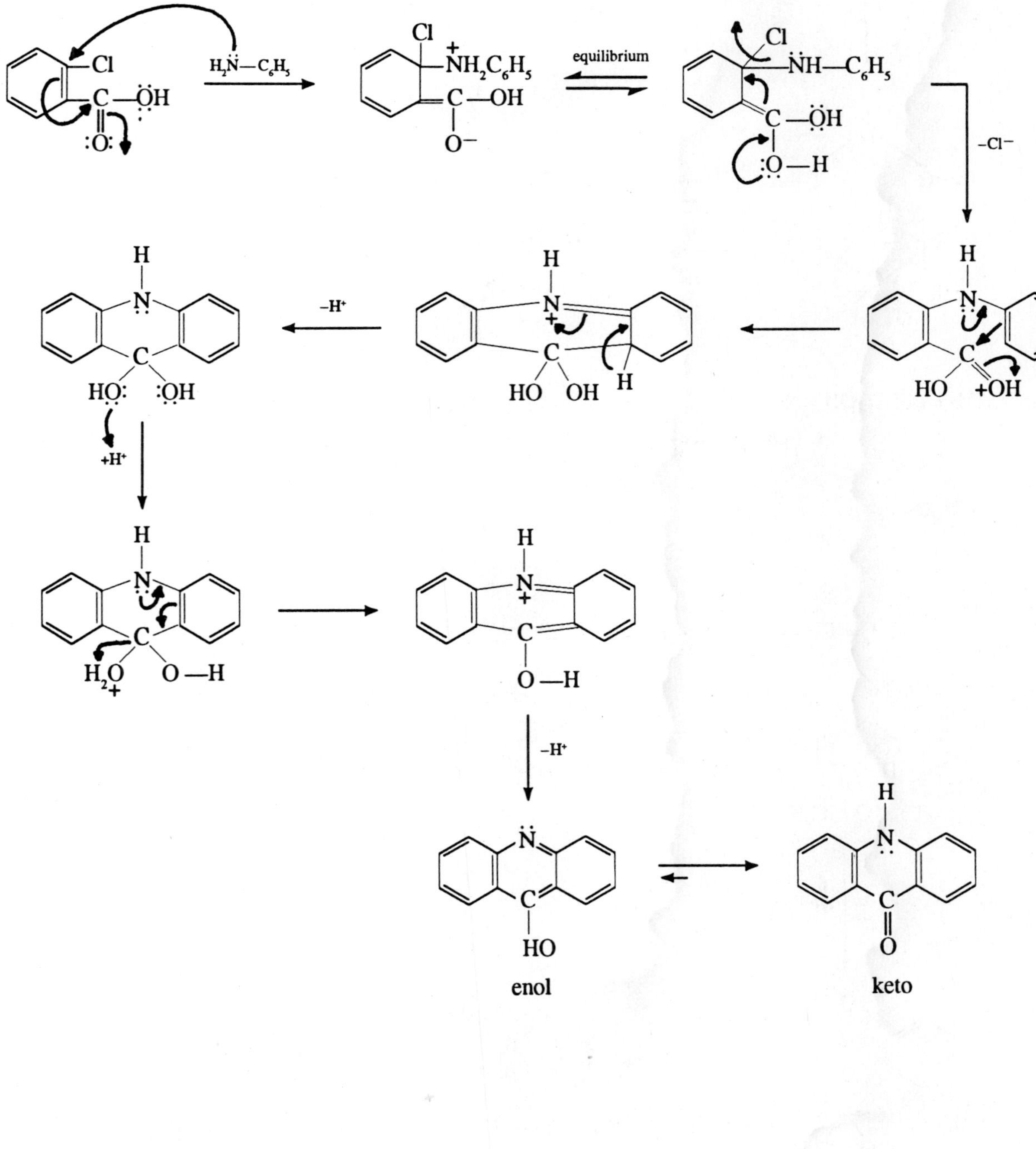

l.

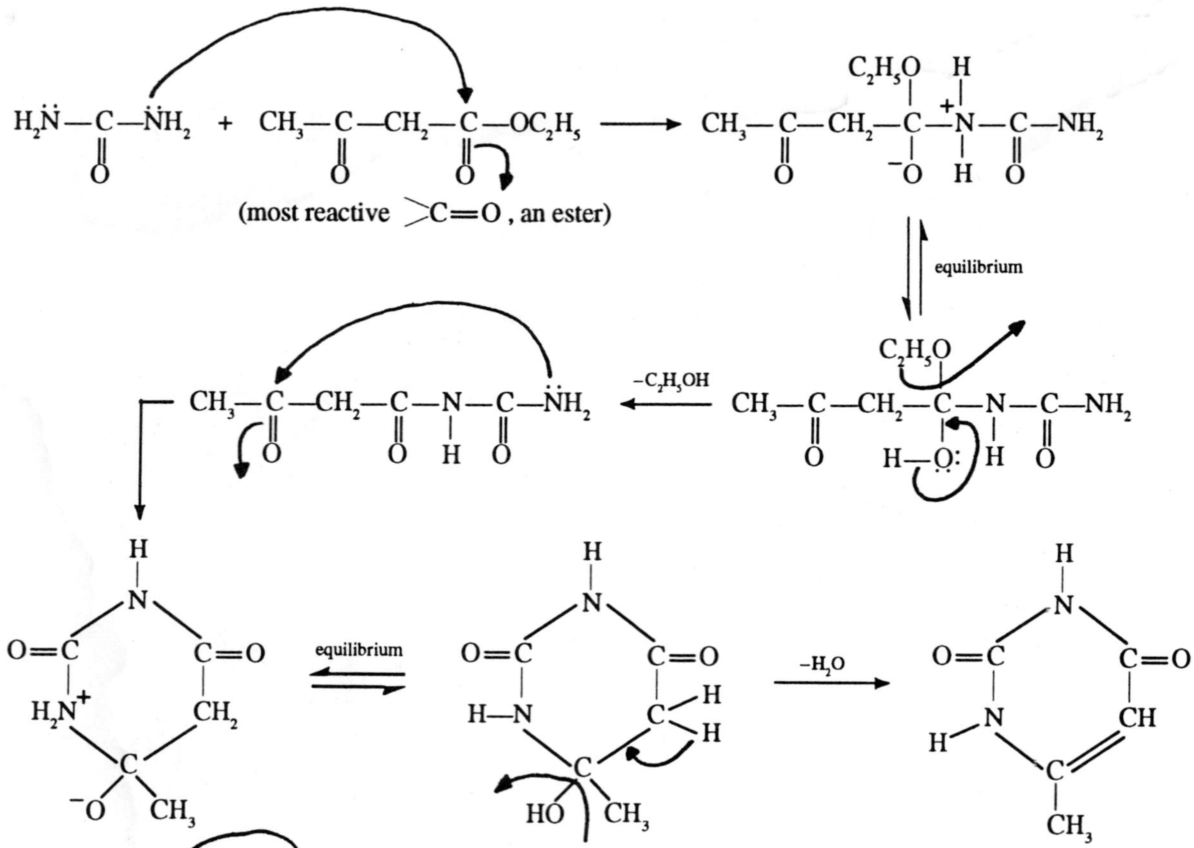
(most reactive >C=O, an ester)
equilibrium
$-C_2H_5OH$
equilibrium
$-H_2O$

m.

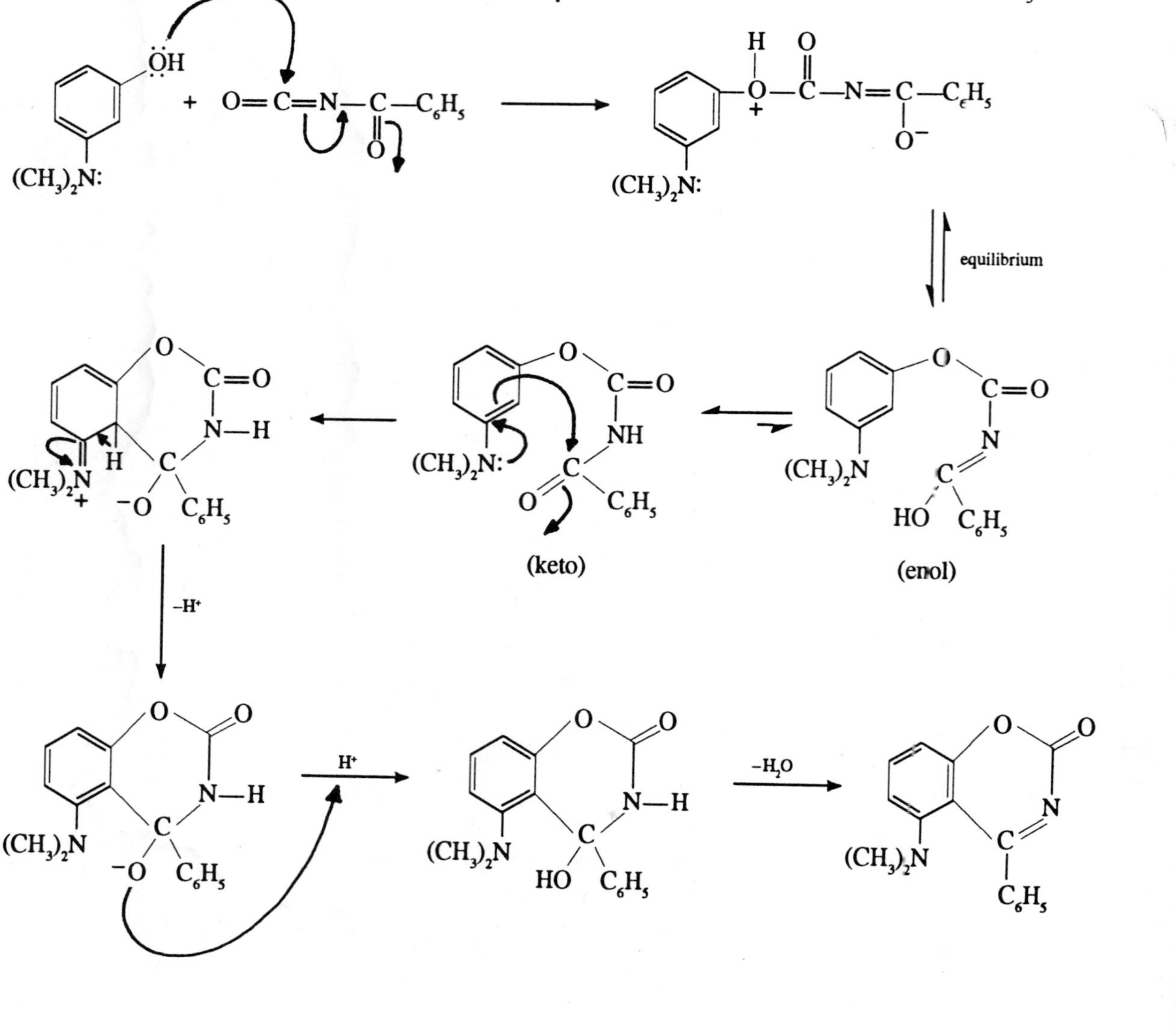
equilibrium
(keto)
(enol)
$-H^+$
H^+
$-H_2O$

e. $CH_3CO_2CH_3 \xrightarrow{OH^-/H_2O} CH_3CO_2^- \xrightarrow{H^+} CH_3CO_2H$

f. $C_6H_5CH_2CH_2CH_3 + Br_2 \xrightarrow{UV} C_6H_5CH(Br)CH_2CH_3 \xrightarrow{\text{alc. KOH}} C_6H_5CH{=}CHCH_3 \xrightarrow[2.\ H^+]{1.\ KMnO_4/OH^-} CH_3CO_2H + C_6H_5CO_2H$

10.12 At the end point of titration, the number of moles of acid neutralized equals the number of moles of base used up in the titration:

$\text{Mass}_{acid} / \text{Equivalent weight}_{acid} = \text{Volume}_{base} \times \text{Normality}_{base}$

Rearranging the equation and using the information given:

$\text{Equivalent weight}_{acid} = 0.11\text{g}/(0.0125 \text{ mL} \times 0.1 \text{ g/mL})$
$= 88$

Since the acid is monocarboxylic, the equivalent weight is equal to the molecular weight (88), and R—COOH is C_3H_7COOH. Possible structures are $CH_3CH_2CH_2COOH$ and $(CH_3)_2CHCOOH$.

10.13 Using the same equation as in 10.12,

$\text{Equivalent weight}_{acid} = 10.5 \text{ g}/(0.025 \text{ mL} \times 0.5 \text{ g/mL})$
$= 84$

The molecular weight of the acid is a multiple of 84, and can be determined by the number of available protons. First calculate the empirical formula:

Moles of carbon:	57.48/12.01	= 4.79
Moles of hydrogen:	3.63/1.00	= 3.60
Moles of oxygen:	100 – (57.48 + 3.63)/16.00	= 2.43

The empirical formula is $C_4H_3O_2$; the molecular formula is some multiple (i.e., $C_{4x}H_{3x}O_{4x}$). Since the acid yielded a mononitro derivative when nitrated, it is aromatic; therefore, the most reasonable molecular formula is $C_8H_6O_4$.

Because a single mononitro derivative is formed, the structure is 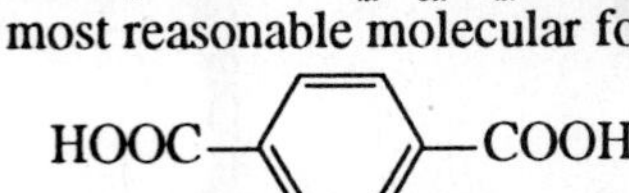

10.14 a. The first compound will be soluble in NaOH or $NaHCO_3$ while the second one will not. Note that litmus paper is *not* a chemical test.

b. Only the first compound will liberate NH_3 when treated with hot dilute NaOH.

c. Only the second compound will precipitate AgBr upon treatment with an aqueous $AgNO_3$ solution (aryl halides do not react with $AgNO_3$). $NaHCO_3$ will dissolve both compounds, because benzoyl bromide reacts readily with water: $C_6H_5COBr + NaHCO_3(aq) \longrightarrow C_6H_5CO_2Na + NaBr + H_2O + CO_2\uparrow$

d. The second compound will react with hot NaOH to yield NH_3; the first compound will not.

e. The first compound will react with $NaHCO_3$; the second one will not. CO_2 bubbles indicate a positive result.

f. Both compounds are soluble in dilute NaOH, but only the second compound is soluble in $NaHCO_3$ (bubble formation).

g. Only the first compound (an amine) is soluble in dilute HCl; the second one (an amide) is not.

h. Both compounds are carboxylic acids, and are therefore soluble in basic solution. Oxidation with hot $KMnO_4$ and further acidification yields benzoic acid (in the first case) and phthalic acid, a dicarboxylic acid (in the second case). The second derivative will then need twice as much base for complete neutralization.

10.15 a. 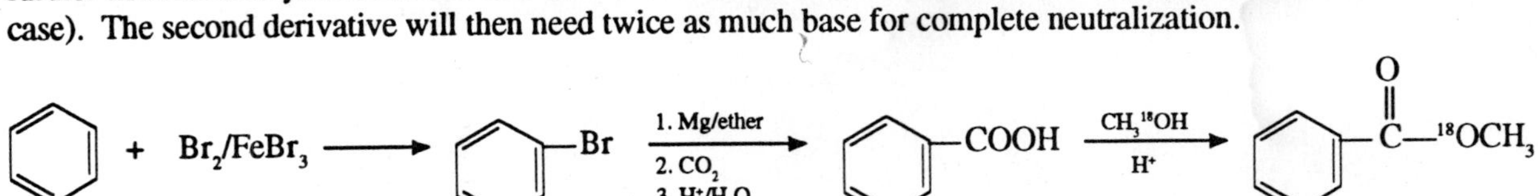

Note: The yield in the final step can be improved by the intermediate formation of 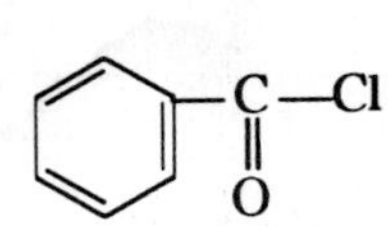 (via $SOCl_2$) and subsequent reaction with $CH_3{}^{18}OH$.

b.

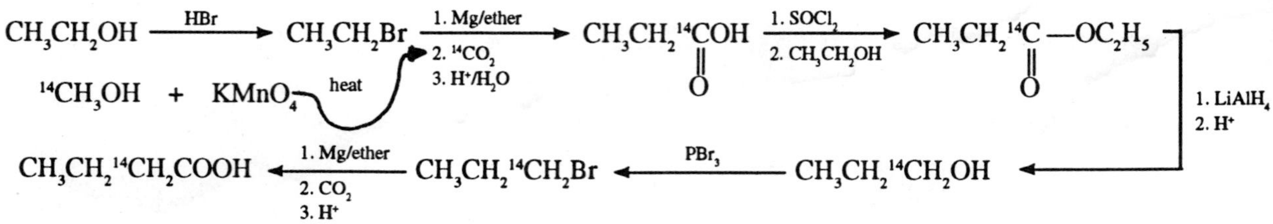

Note: The reduction $CH_3CH_2{}^{14}COH$ (C=O) ⟶ $CH_3CH_2{}^{14}CH_2OH$ can be accomplished (although with lower yields) by using excess $LiAlH_4$ (followed by H_2O/H^+) without the intermediate formation of the ester $CH_3CH_2{}^{14}COC_2H_5$ (C=O).

c.

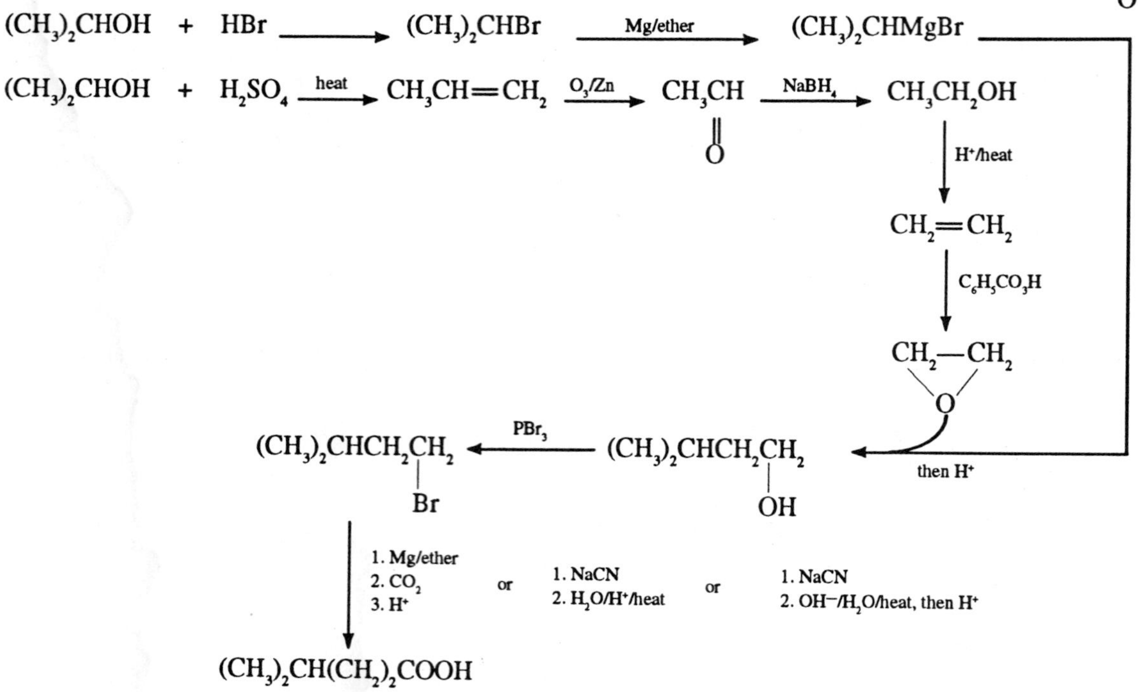

d.

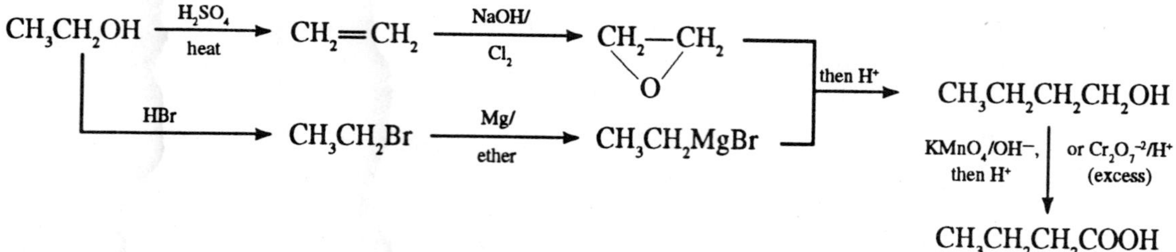

e.

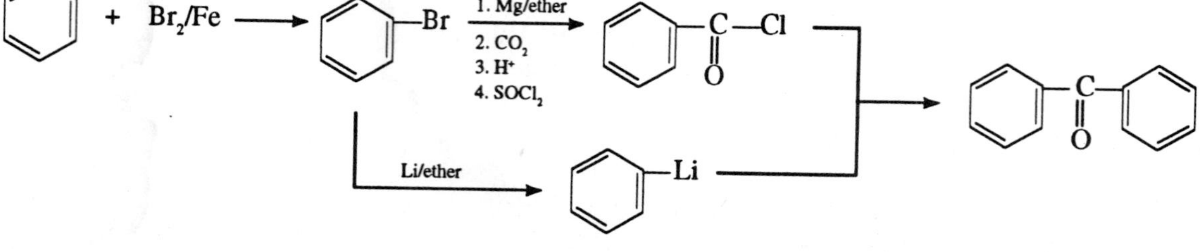

alternative method:

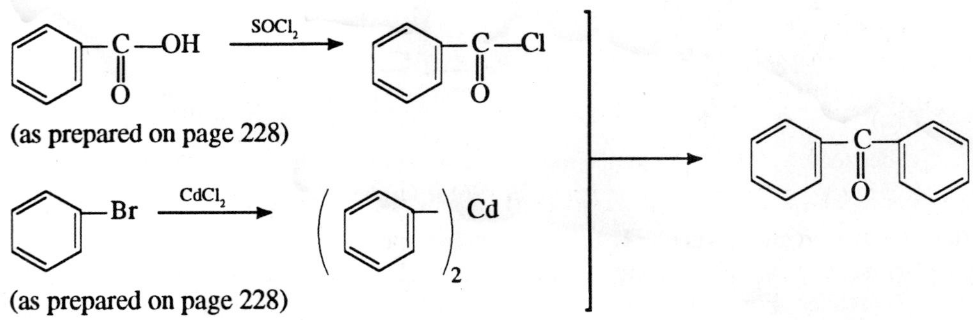

f.

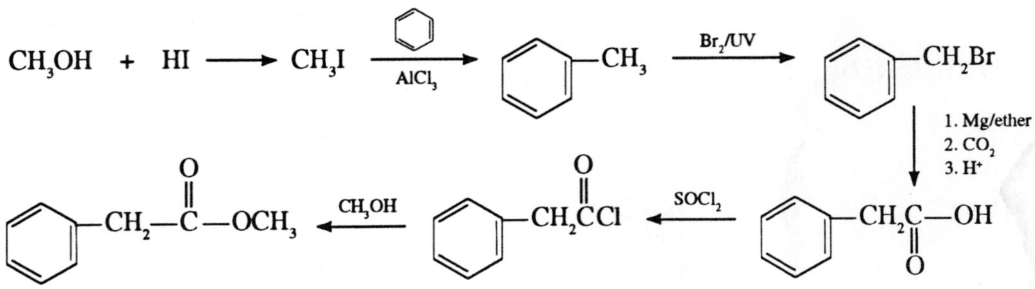

g.

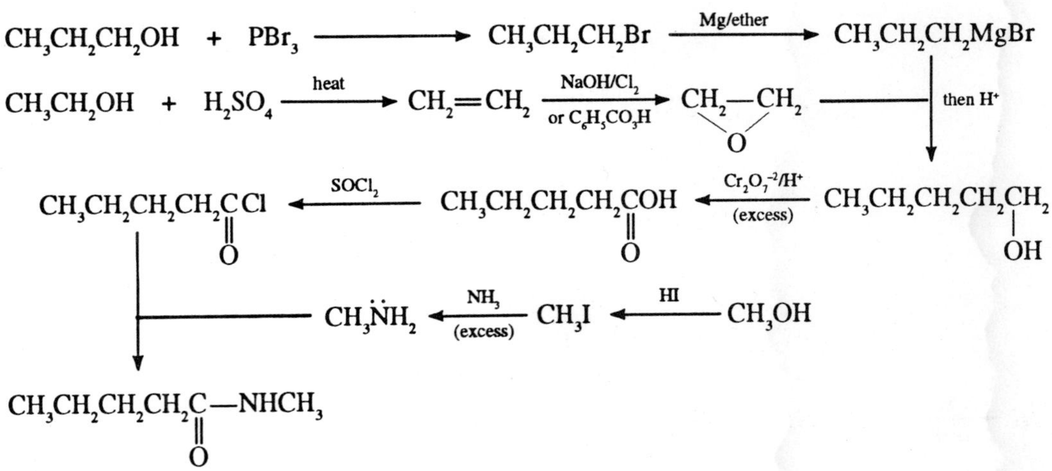

h.

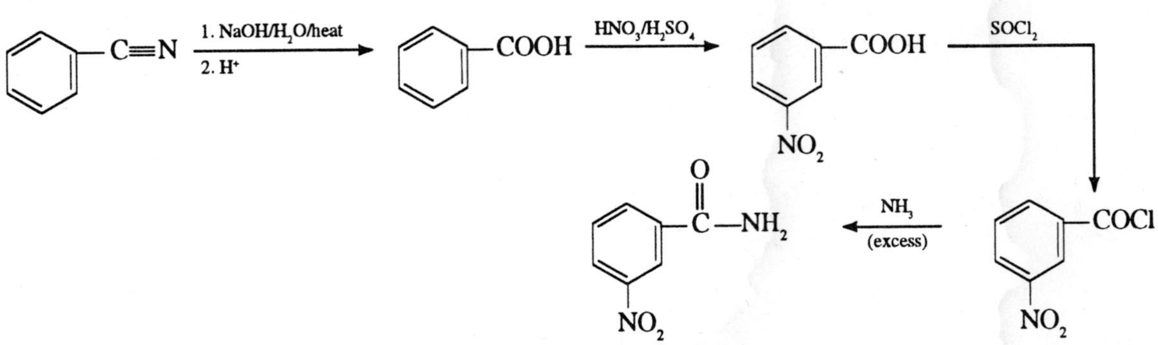

10.16 Both compounds have the same molecular formula $C_8H_8O_2$ which has five double bonds and/or rings. Both have peaks at 1600 and 1500 cm^{-1} (—C$\overset{..}{=}$C— aromatic) and both of them possess a >C=O (around 1700 cm^{-1}). Therefore, all five degrees of unsaturation are accounted for. The first compound has a broad peak at 3300–2500 cm^{-1} (—COOH) and should be a monosubstituted benzene ring (900–700 cm^{-1} pattern). The natural conclusion is that the first compound is C_6H_5—CH_2COOH. The second compound is an ester (>C=O at 1735 cm^{-1}), and has a monosubstituted benzene ring. Therefore, the compound is C_6H_5—$COOCH_3$ or C_6H_5—$OOCCH_3$. Further spectroscopic analysis is required to distinguish between the latter two compounds (e.g., ^{1}H NMR, ^{13}C NMR, or mass spectroscopy).

10.17 The structures are

methyl benzoate ($C_8H_8O_2$) and methyl *o*-hydroxybenzoate ($C_8H_8O_3$)

In the latter compound there is an intramolecular H-bonding that is responsible for the weakening of the C=O character. Consequently, the wavenumber is decreased considerably (1670 cm^{-1} vs. 1735 cm^{-1} in the case of methyl benzoate) and the band is broader. For the same reason, the –OH peak of methyl salicylate is broad because of the same reason (H-bonding).

Answers for Multiple Choice Test

10.1 c	**10.6** b	**10.11** c	**10.16** c
10.2 d	**10.7** d	**10.12** a	**10.17** d
10.3 b	**10.8** b	**10.13** a	**10.18** c
10.4 b	**10.9** c	**10.14** c	**10.19** d
10.5 b	**10.10** a	**10.15** b	**10.20** d

11 Aldehydes and Ketones

Questions

11.1 Name the following compounds:

a. CH_3CHO
b. $(CH_3)_3CCHO$
c. 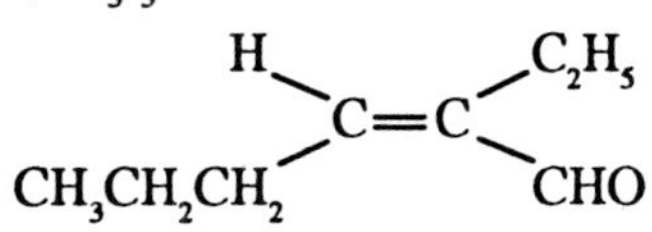
d. $(CH_3)_3CCOC(CH_3)_3$
e. $ClCH_2CH_2CHO$
f. $C_2H_5CO(CH_2)_2CH(CH_3)CHO$
g. $(CH_3)_3CCOCH_2CH_3$
h. $CH_2{=}CHCOCH_3$
i. $(CH_3)_2CHCOCH_3$
j. 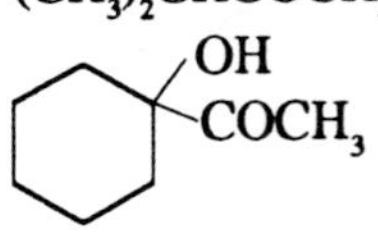
k. 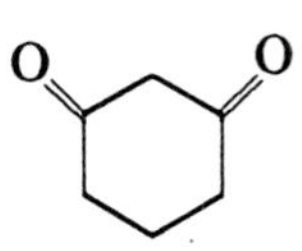
l. 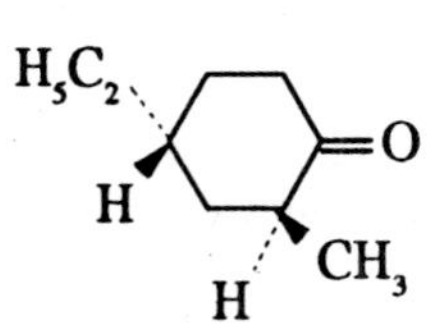
m. 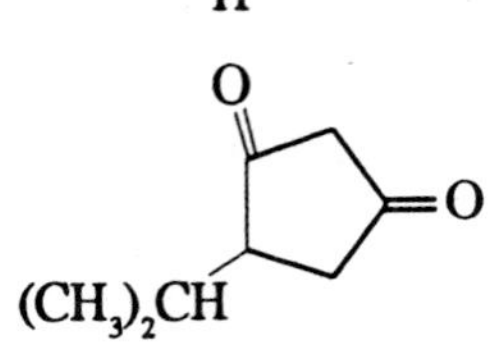
n. 
o. 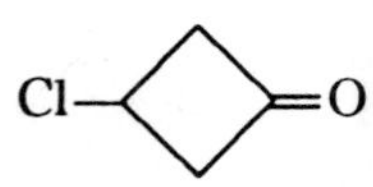
p. 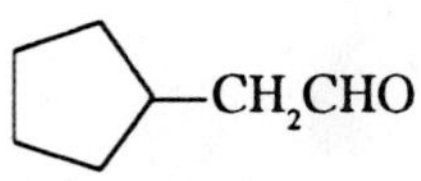
q. 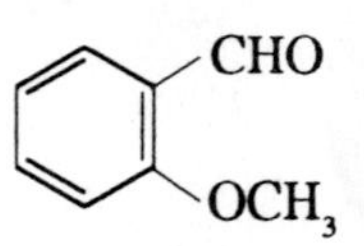
r. 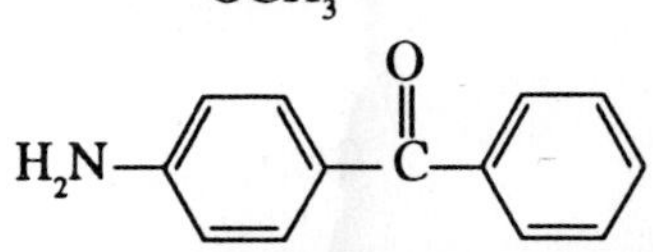
s. $CH_3COCH_2CH_2CH(CH_3)C(CH_3){=}CH_2$
t. $CH_3COCH_2CH{=}CH_2$
u. $C_6H_5C(CH_3)(OCH_3)_2$
v. $Cl_3CCH(OH)_2$
w. 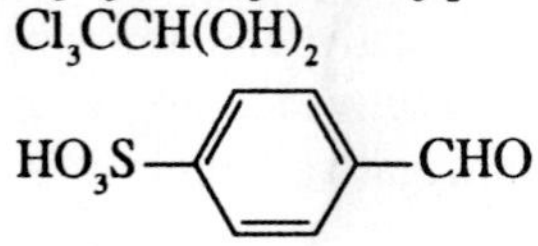

11.2 Give structures for the following

a. acetone
b. 4-chlorocyclohexanone
c. dibenzalacetone
d. pinacolone
e. isopropionaldehyde
f. ethanedial
g. 4-bromo-2-ethylcyclopentanone
h. 1,4-diacetylbenzene
i. acetone oxime
j. 2-propenal
k. 2-nitrobenzophenone
l. 3,3-dimethylcyclohexanone

m. 2,2,4-trichlorocyclopentanone
n. 6-bromo-2-oxocyclohexanecarbaldehyde
o. trans-benzalacetophenone
p. acetone semicarbazone
q. 2-methyl-4-oxohexanal
r. *m*-nitrobenzaldehyde
s. cyclohexanone oxime

11.3 Match the following compounds with the given family names:

a. imine
b. enamine
c. hydrazone
d. oxime
e. semicarbazone
f. 2,4-DNP
g. imide
h. amine

$(C_6H_5)_2C{=}NNH_2$
A

$(C_6H_5)_2C{=}NOH$
B

$(CH_3)_3CNH_2$
C

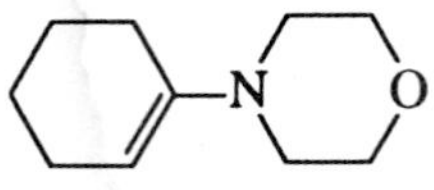

D

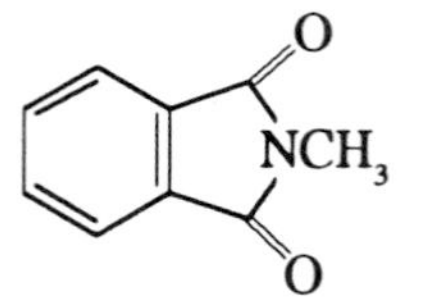

E

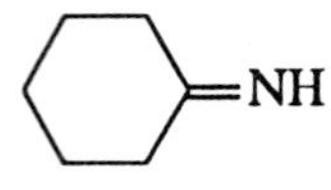

F

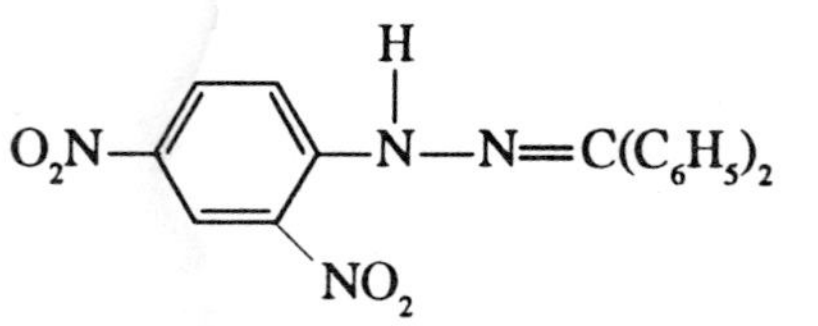

G

$H_2N-\overset{O}{\overset{\|}{C}}-NHN{=}C(C_6H_5)_2$
H

11.4 Identify the following as hemiacetals, hemiketals, acetals, ketals or ethers:

a. 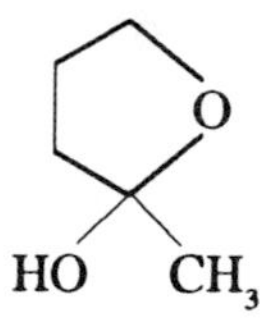

c. 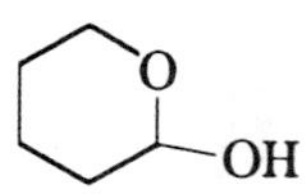

e. 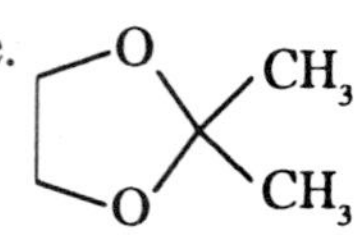

g. 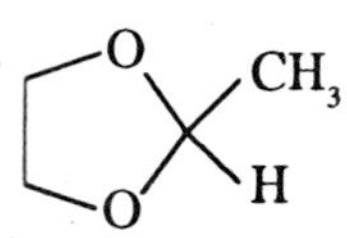

i. 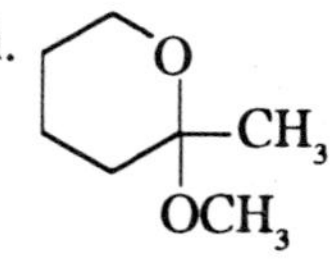

b. 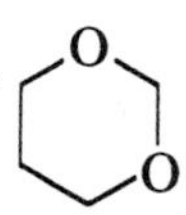

d. 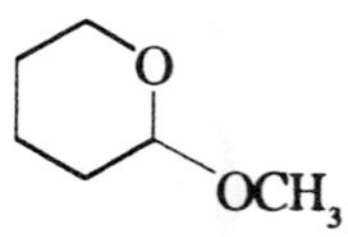

f. 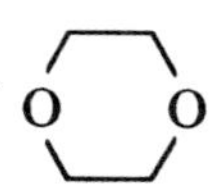

h. 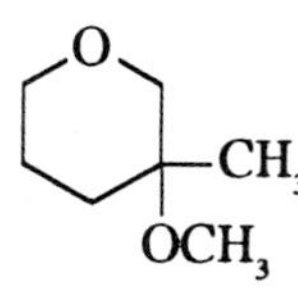

j. 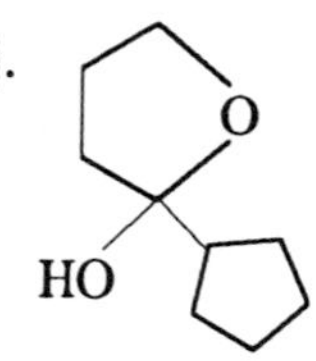

11.5 Identify the structures below from their molecular formulas, chemcial test results, and spectral data:

a. $C_9H_{10}O$; positive 2,4-DNP; negative iodoform; negative Tollens'.
b. $C_9H_{10}O$; positive 2,4-DNP; positive iodoform; negative Tollens'; hot $KMnO_4$ oxidation yields benzoic acid.
c. $C_9H_{10}O$; positive 2,4-DNP; negative Tollens'; IR peaks at 1690, 1600, 1503 cm^{-1}; hot $KMnO_4$ oxidation yields

d. $C_9H_{10}O$; positive 2,4-DNP; positive Tollens'; IR peaks at 2750, 1710, 1600, 1500 cm^{-1}; optically inactive; hot $KMnO_4$ oxidation yields benzoic acid.
e. $C_9H_{10}O$; positive 2,4-DNP; positive Tollens'; IR peaks at 1710, 1600, 1500 cm^{-1}; hot $KMnO_4$ oxidation yields phthalic acid.
f. $C_9H_{10}O$; positive 2,4-DNP; positive Tollens'; optically active.
g. $C_4H_{10}O_2$; no 3300 cm^{-1} peak in IR; acid hydrolysis yields acetaldehyde.
h. C_8H_8O; Clemmensen reduction yields compound that has only one mononitro derivative.
i. $C_9H_{10}O$; positive 2,4-DNP; hot $KMnO_4$ oxidation yields 1,3,5-benzene tricarboxylic acid.
j. $C_9H_{18}O$; positive 2,4-DNP; one ^{1}H NMR signal.

11.6 Fill in the missing products in the reactions below. If no reaction occurs, state so.

a. 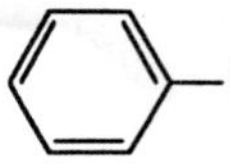 —CHO $\xrightarrow[\text{then } H^+]{\text{conc. NaOH}}$ A

b. 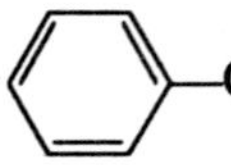 —CH_2OH $\xrightarrow[\text{CrO}_3\text{, pyridine}]{}$ B $\xrightarrow[\text{(excess)}]{Na_2Cr_2O_7/H^+}$ C

c. $CH_3CH(OH)CH_2CH_3 \xrightarrow{Na_2Cr_2O_7/H^+}$ D

d. $CH_3COCH_3 \xrightarrow[\text{then } H^+]{LiAlH_4}$ E

e. CH_3COOH + Zn (Hg)/HCl $\longrightarrow$ F

f. $CH_3COCH_2CH_3$ + $I_2/KI/OH^-$ $\longrightarrow$ G

g. $H_2C{=}O$ + CN^- $\xrightarrow{H^+}$ H $\xrightarrow[\text{heat}]{H_2O/H^+}$ I

h. 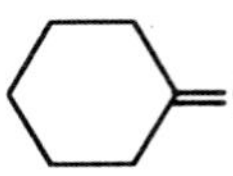 =O $\xrightarrow[\text{then } H^+]{CH_3C{\equiv}CMgBr}$ J

i. $C_6H_5COCH_3$ + $C_6H_5NH_2$ $\xrightarrow{H^+}$ K

j. 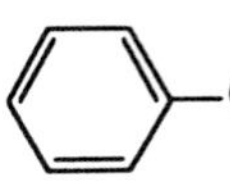 —CHO + $Ag(NH_3)_2^+$ $\longrightarrow$ L

k. $(C_6H_5)_2C{=}O$ + PCl_5 $\longrightarrow$ M ($C_{13}H_{10}Cl_2$)

l. CH_3CHO + $H_2NCONHNH_2$ $\xrightarrow{H^+}$ N

m. 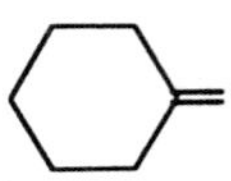 =O $\xrightarrow[\text{then } H^+/H_2O]{LiAlH_4}$ O

n. 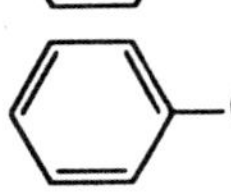 —CHO + $HS(CH_2)_2SH$ $\xrightarrow{H^+}$ P $\xrightarrow{\text{Raney nickel}}$ Q

o. $(CH_3CO)_2O$ + $C_6H_5CH_3$ $\xrightarrow{H_3PO_4}$ R

p. $HOOC(CH_2)_4CO(CH_2)_4COOH$ $\xrightarrow[\text{then KOH/HOCH}_2\text{CH}_2\text{OH}]{NH_2NH_2/H^+}$ S

q. $(CH_3)_2CO$ + $H_2C{=}O$ $\xrightarrow[\text{then } H^+]{\text{NaOH/ethanol}}$ T (C_4H_6O) $\xrightarrow{(CH_3)_2NH/HCl}$ U ($C_6H_{13}NO$)

r. 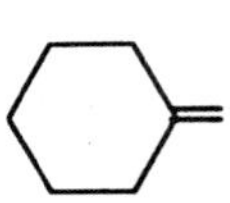 =O + $N{\equiv}CCH_2COOH$ $\xrightarrow{CH_3COO^-NH_4^+}$ V ($C_9H_{11}NO_2$)

s. $CH_3CH{=}CHCHO$ + HOCl $\longrightarrow$ W

t. 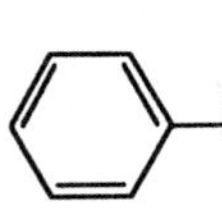 —NH_2 + $CH_2{=}CHC{\equiv}N$ $\xrightarrow[\text{catalyst}]{Cu^{+2}}$ X ($C_9H_{10}N_2$)

u. 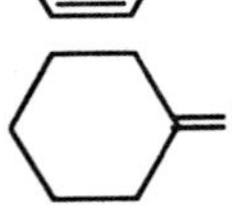 =O + Br_2/CH_3COOH $\longrightarrow$ Y

v. $H_2C{=}O$ + $NaOD/D_2O$ $\longrightarrow$ Z

w. $(CH_3)_2C(OH)C(OH)(CH_3)_2$ $\xrightarrow{H^+}$ AA

x. $CH_3COC_6H_5 + NH_2OH \xrightarrow{H^+} \textbf{BB} \xrightarrow[H_2O]{H^+/heat} \textbf{CC}\ (C_8H_9NO)$

y. $(CH_3)_3CCHO + I_2/KI/OH^- \longrightarrow \textbf{DD}$

z. $CH_2{=}CHCOCH_3 \xrightarrow[\text{then } H^+]{NaBH_4} \textbf{EE}$ (C_4H_8O; IR peaks at 3300, 1620, 1090 cm^{-1})

11.7 Suggest plausible mechanisms that would justify the formation of the indicated products:

a.

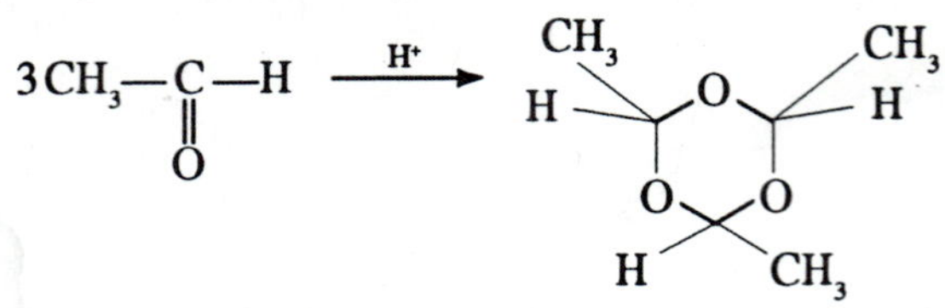

b. $(CH_3)_3C-\underset{\overset{\|}{O}}{C}-H \xrightarrow{\text{conc. NaOH}} (CH_3)_3C-CH_2OH + (CH_3)_3C-\underset{\overset{\|}{O}}{C}-O^-Na^+$

c. $C_6H_5-\underset{\overset{\|}{O}}{C}-H + NH_2NHC_6H_5 \xrightarrow{H^+} C_6H_5-CH{=}N-NHC_6H_5 + H_2O$

d. $(CH_3)_2\overset{+}{S}-O^- + CH_3CH_2O-\overset{\overset{O}{\|}}{\underset{\overset{\|}{O}}{S}}-C_6H_5 \xrightarrow{HCO_3^-} (CH_3)_2S + CH_3\overset{\overset{O}{\|}}{C}-H + C_6H_5-SO_3^-$

e. $CH_3-\underset{\overset{\|}{O}}{C}-CH_3 + NH_2NH_2 \xrightarrow[-H_2O]{+H^+} CH_3-\underset{\overset{\|}{N-NH_2}}{C}-CH_3 \xrightarrow[(CH_2OH)_2]{KOH/heat} CH_3CH_2CH_3 + N_2$

f.

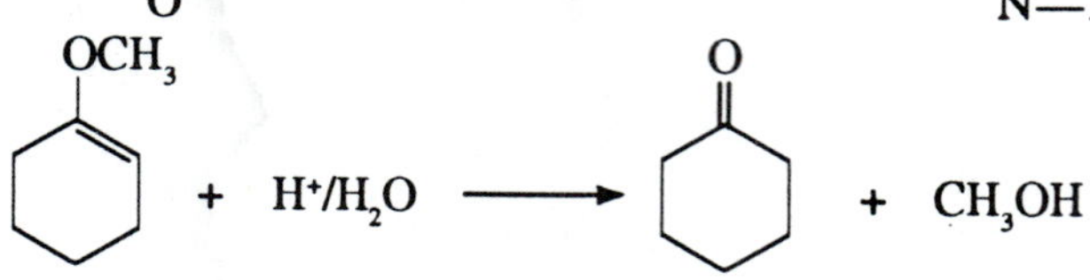

g. $CH_3-\underset{\overset{\|}{O}}{C}-CH_3 + 2\,C_6H_5-\underset{\overset{\|}{O}}{C}-H \xrightarrow[-2H_2O]{H^+/EtOH} C_6H_5-CH{=}CH-\underset{\overset{\|}{O}}{C}-CH{=}CH-C_6H_5$

h. $2C_6H_5-\underset{\overset{\|}{O}}{C}-H + CH_3\underset{\overset{\|}{O}}{C}CH_3 \xrightarrow[EtOH]{NaOH/} C_6H_5-CH{=}CH-\underset{\overset{\|}{O}}{C}-CH{=}CH-C_6H_5$

i.

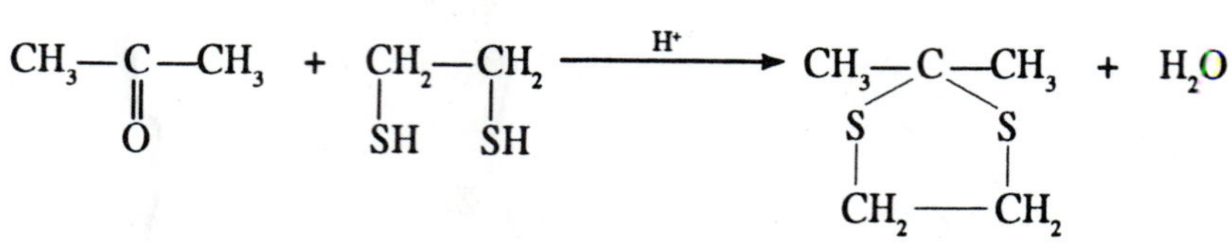

j. $CH_3-\underset{\overset{\|}{O}}{C}-CH_3 + NH_2OH \xrightarrow{HCl} CH_3-\underset{\overset{\|}{N-OH}}{C}-CH_3 + H_2O$

k.

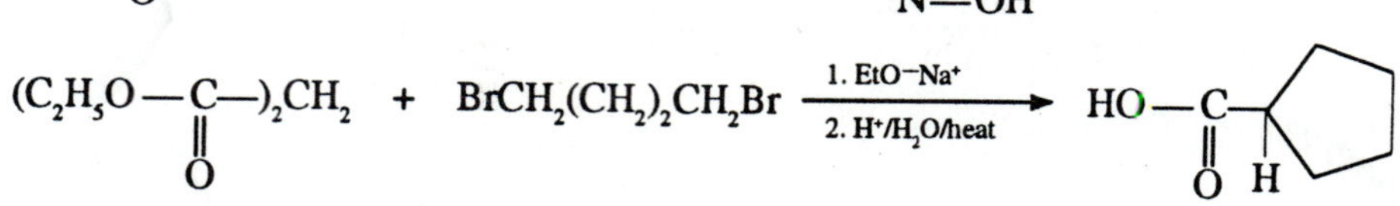

l. $C_6H_5-\underset{\overset{\|}{O}}{C}-H + CH_3C{\equiv}N \xrightarrow{CH_3COO^-/H_2O} C_6H_5CH{=}CH-CN$

m. 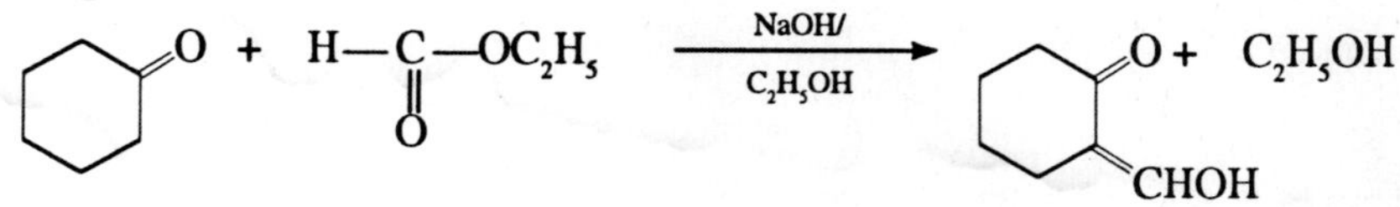

n. 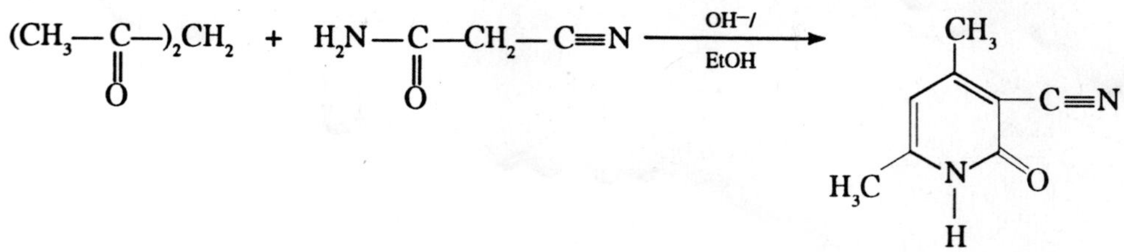

o. 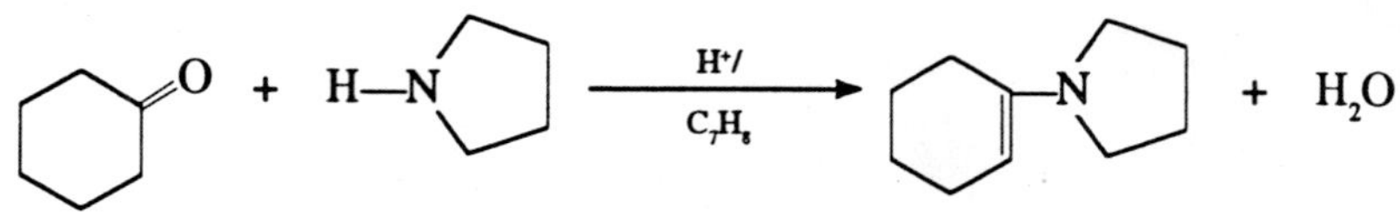

p. 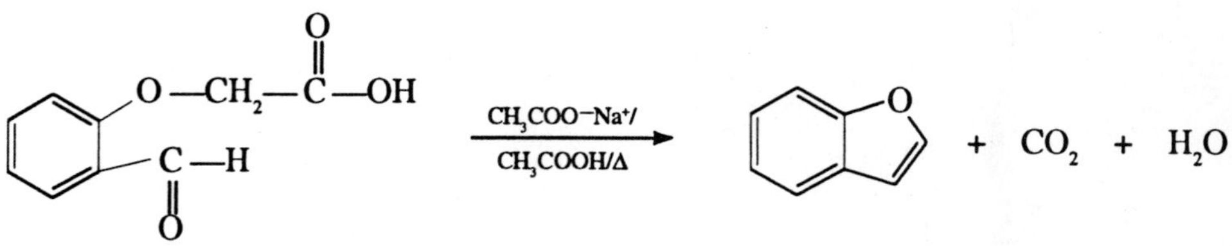

q. 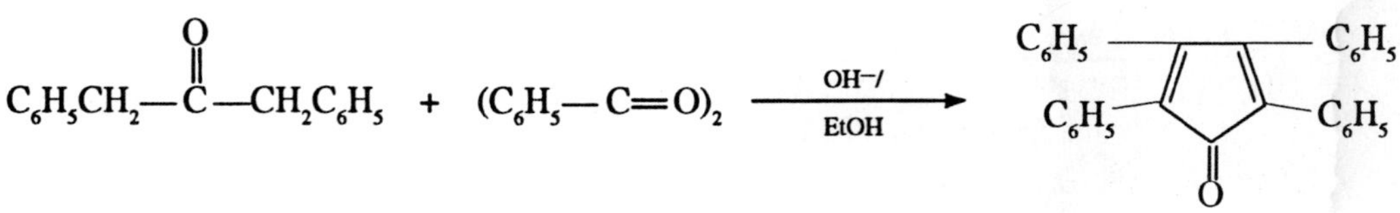

r. 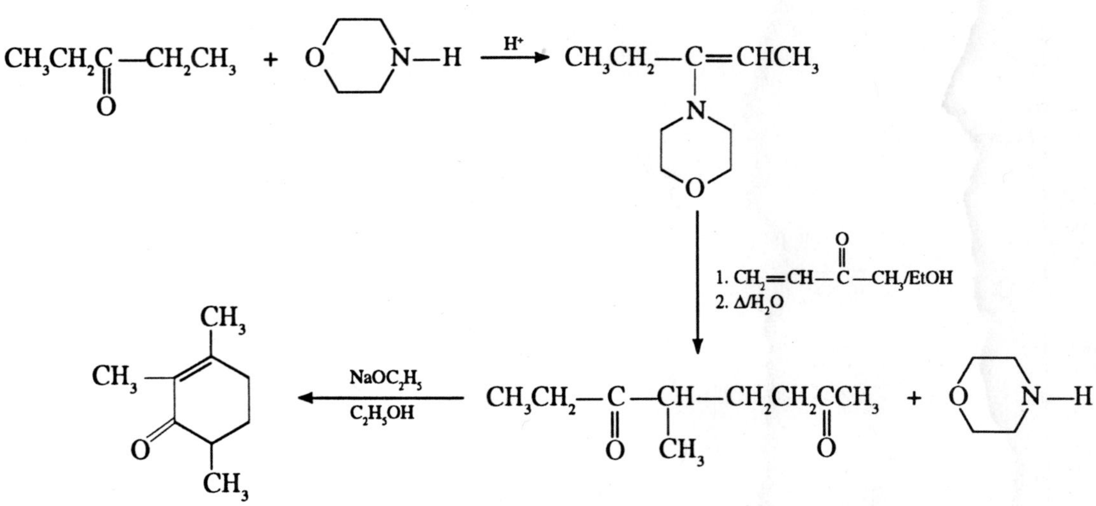

s. 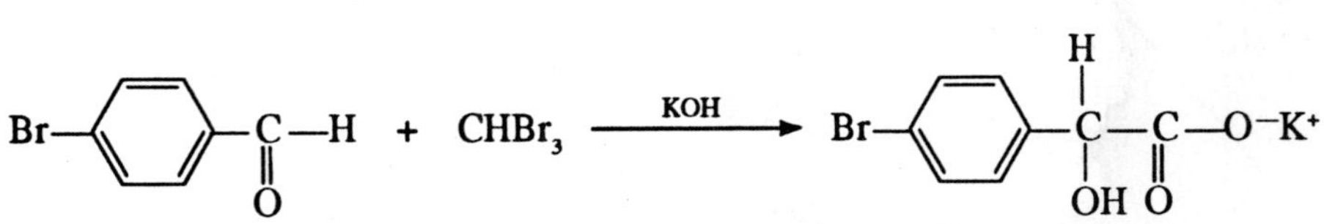

t. 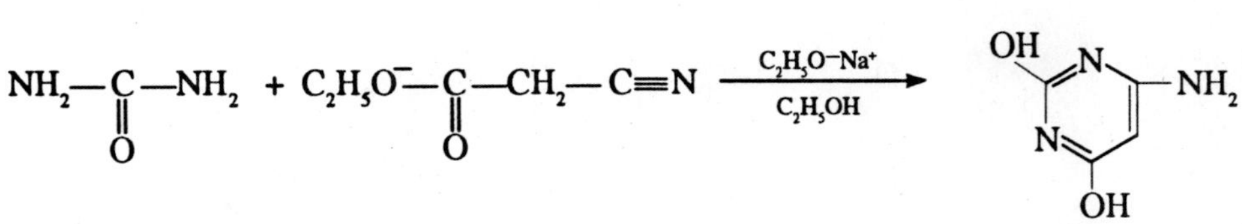

11.8 Identify the products and reagents in the synthetic routes below:

a.

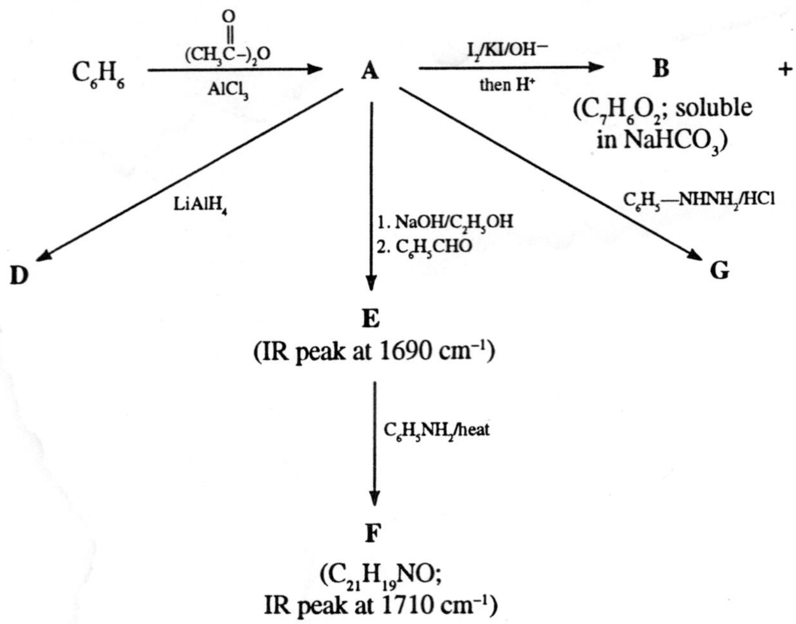

b.

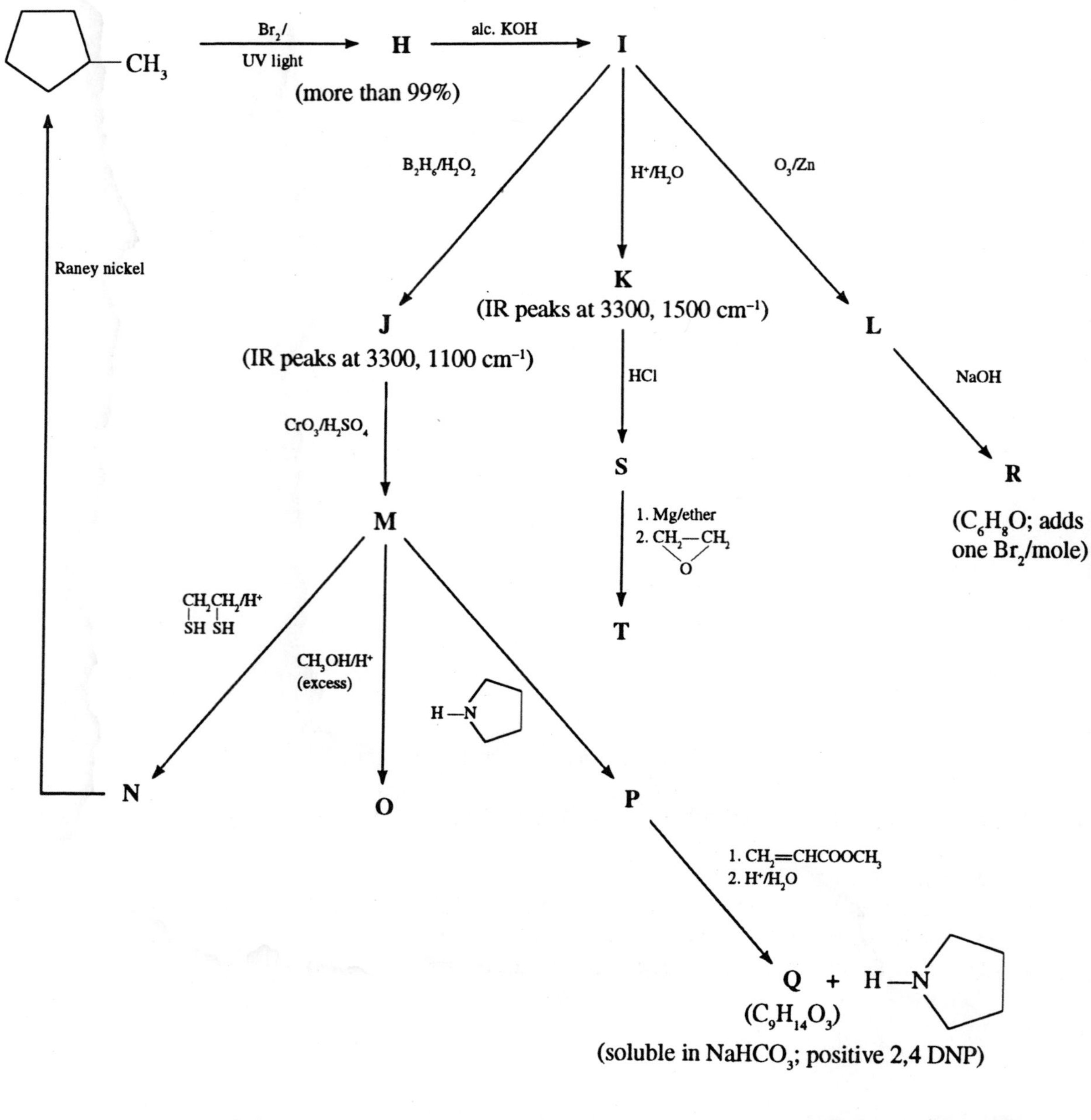

c.

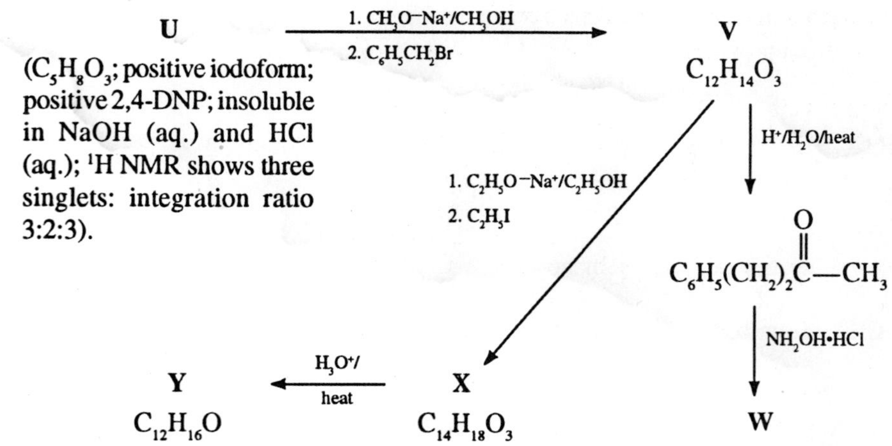

11.9 Give a reasonable explanation for the following observations:

a.

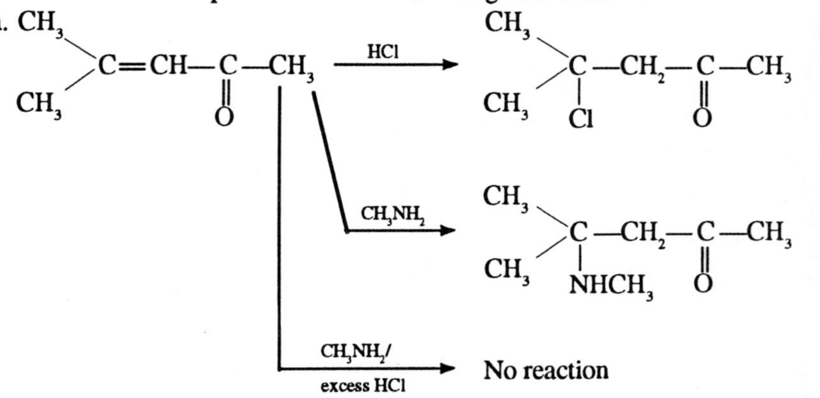

b. Compound **A** but not **B** is isolated when carbonyl compounds are reacted with semicarbazide hydrochloride:

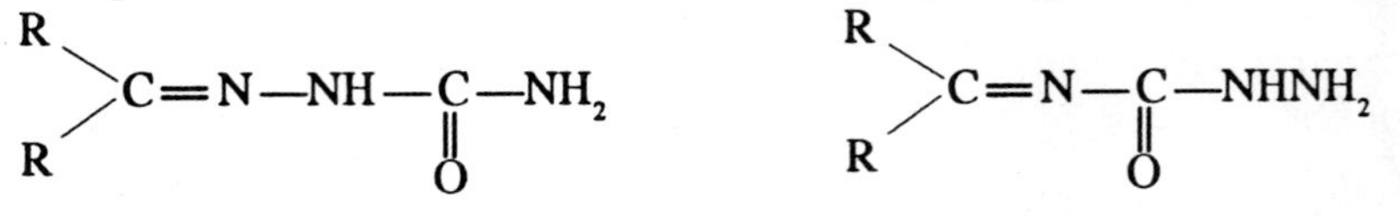

A **B**

c. The Clemmensen reduction cannot be performed for the following conversion:

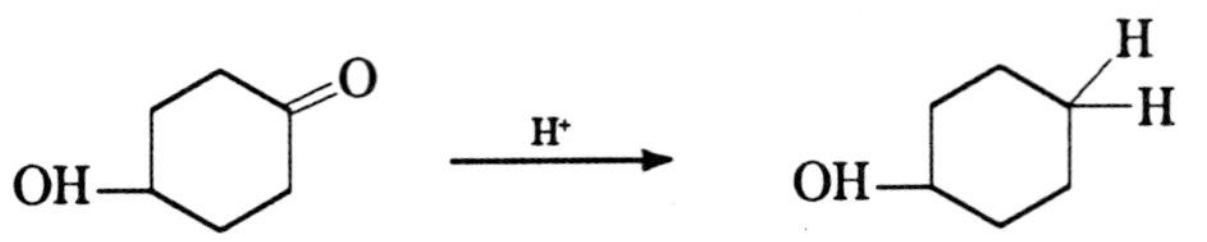

d. In the reaction

$$CH_3CH_2Br + (CH_3)_2\overset{+}{S}—O^- \longrightarrow CH_3—\underset{\overset{\|}{O}}{C}—H + (CH_3)_2S + HBr$$

the yields are considerably higher when sodium bicarbonate is present.

11.10 The base-catalyzed (NaOH/ethanol) reaction of 1 mole of acetone with 2 moles of benzaldehyde yields a yellow solid ($C_{17}H_{14}O$).

a. What is the structure of the product?

b. What will be the result when only 1 mole of benzaldehyde is added? When acetone is not added at all, while the concentration of base is high? When benzaldehyde is not added at all?

c. Why is the self-condensation of acetone unlikely in the presence of benzaldehyde?

11.11 Show how to synthesize the following from the indicated reactants

a. 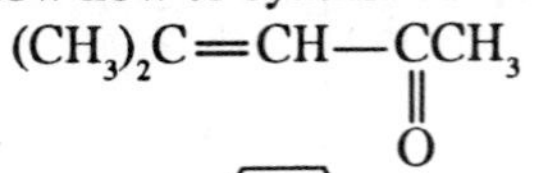 from propane

b. 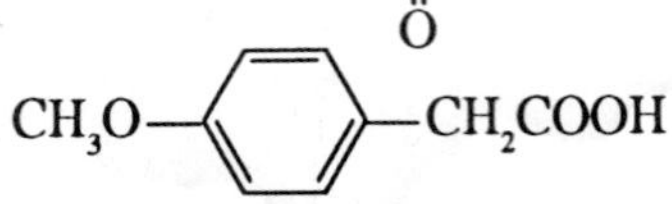 from benzene and methanol as the only organic sources.

c. 4-methyl-3-heptanone from alcohols that contain *three* or fewer carbons.

d. 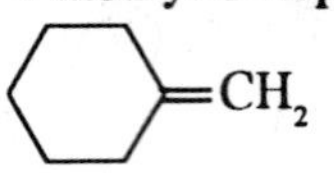 from cyclohexanol and iodomethane

e. 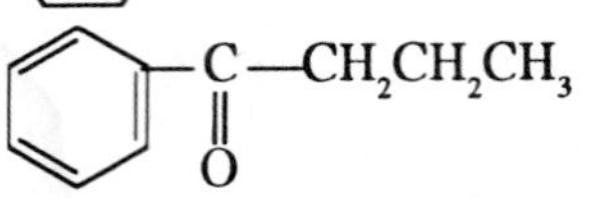 from benzene and any other needed reagents (four different methods)

f. 2-butenal from ethanol as your only organic source
g. 2-hydroxy-2-methylbutane from any alcohols of three carbons or less
h. 1,3-dihydroxybutane from ethanol as your only organic source
i. isohexane from acetone as your only organic source

11.12 Describe the results of the simplest chemical test (except melting point differences of ammonia derivatives [e.g., 2,4-DNP]) to distinguish between the following compounds of each pair:
a. 3-phenylpropanal and 1-phenyl-2-propanone
b. 3-phenylpropanal and 1-phenyl-2-butanone
c. benzyl alcohol and benzaldehyde
d. butanone and *n*-butyl methyl ether
e. *p*-methylacetophenone and benzyl methyl ketone
f. CH_3CH_2COOH and CH_3CH_2CHO
g. cyclohexyl methyl ketone and acetophenone
h. $CH_3CH_2COCH_3$ and $CH_3CH_2COOCH_3$
i. 2-cyclohexenone and 3-cyclohexenone

11.13 Suggest an IR frequency that would distinguish between the following pairs of compounds:
a. 1-phenyl-1-propanone and 1-phenyl-2-propanone
b. $C_6H_5OCH_2C{\equiv}CH$ and $C_6H_5COCH_3$
c. $C_6H_5CH_2CH_2CHO$ and $C_6H_5CH_2COCH_3$
d. *p*-methylacetophenone and acetophenone
e. acetophenone and benzophenone

11.14 Identify the compounds whose molecular formulas and ^{1}H NMR spectra are given below:
a. $C_{15}H_{14}O$; 2.1 ppm (2H, t); 2.4 ppm (2H, t); 7.9–7.2 ppm (10H)
b. $C_{16}H_{16}O$; 1.7 ppm (6H, s); 7.9–7.2 ppm (10H)
c. $C_5H_{10}O$; 1.4 ppm (9H, s); 9.8 ppm (1H, s)
d. $C_6H_{12}O$; 1.4 ppm (9H, s); 2.2 ppm (2H, d); 9.7 ppm (1H, t)

Multiple Choice Test

11.1 Which of the following combinations is correct? Aldehydes

A are more reactive than ketones toward nucleophiles.
B are more easily oxidized than ketones.
C are less reactive than ketones toward nucleophiles.
D are less easily oxidized than ketones.

a. A, B b. C, D c. A, D d. B,C

11.2 The IUPAC name of acetone is _______ .

a. ethanone b. dimethyl ketone c. propanone d. 2-butanone

11.3 The IUPAC name of the compound below is

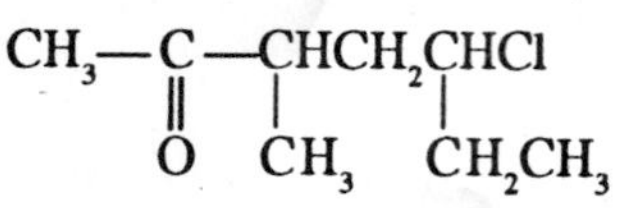

a. 1-chloro-1,3-dimethyl-4-pentanone
b. 5-chloro-3-methyl-2-heptanone
c. 5-chloro-3,5-dimethyl-2-hexanone
d. 3-chloro-5-methyl-6-heptanone

11.4 Which of the following will have the highest boiling point?
a. $CH_3(CH_2)_2CH_2OH$ b. $(C_2H_5)_2O$ c. $CH_3(CH_2)_2CHO$ d. $CH_3COC_2H_5$

11.5 What is the product of the reaction of $HOCH_2CHO$ with C_2H_5MgBr?
a. $C_2H_5CH(OH)CH_2OH$ b. $C_2H_5OCH_2CHO$ c. $C_2H_5CH(OH)CHO$ d. C_2H_6

11.6 Which reagent would not reduce benzophenone to diphenylmethane?
a. Zn(Hg)/HCl
b. NH_2NH_2/H^+, followed by KOH/ethylene glycol/heat
c. $HS(CH_2)_2SH/H^+$, followed by Raney nickel
d. CrO_3/pyridine

11.7 Which of the following reactions would lead to the formation of an aldehyde?

A $(CH_3)_2C{=}CHCH_3$ + O_3/Zn, H_2O
B $C_6H_5C{\equiv}CH$ + $H_2O/H^+/Hg^{+2}$
C $(CH_3)_2CHCH_2OH$ + MnO_2, 25°C
D C_6H_5COCl + $H_2/Pd/BaSO_4$
E $(CH_3)_2CHCH_2OH$ + CrO_3, pyridine
F $C_6H_5CHCl_2$ + H_2O

a. A, C, and E b. C and E c. all but B d. all but B and F

11.8 Which of the following reagents cannot be used to distinguish between butanal and 2-butanone
a. Ag_2O/OH^- b. $Cr_2O_7^{-2}/H^+$ c. Cu^{+2} d. $NaBH_4$

11.9 Which reagent would be used for the following transformation

$C_6H_5COCl \longrightarrow C_6H_5COCH_3$

a. CH_3MgI, then H^+ b. CH_3OH/H^+ c. $(CH_3)_2Cd$ d. CH_3Li

11.10 The reaction below produces a(n) _______ .

$C_6H_5CHO + CN^- \xrightarrow{H_2O/H^+}$

a. optically active compound
b. racemic pair
c. *meso* compound
d. a pair of diastereomers

11.11 Which compound would not form an imine with *p*-nitrobenzaldehyde?
a. NH_3 b. $CH_3CH_2NH_2$ c. $(CH_3)_2CHNH_2$ d. $(CH_3CH_2)_2NH$

11.12 Which of the following compounds would not undergo a Cannizzaro reaction?

a. $(CH_3)_3CCHO$ b. HCHO c. 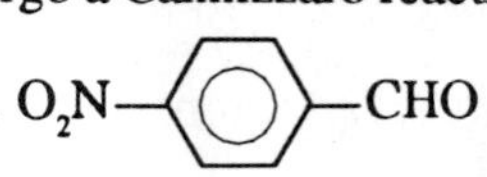

d. CH_3CHO

11.13 What is the four-membered intermediate in the Wittig reaction called?
a. betaine b. oxaphosphetane c. ylide d. phosphonium ion

11.14 During the first step of the base-catalyzed (OH^-) aldol condensation,
a. the carbonyl is protonated.
b. the HO^- ion attacks the carbonyl group.
c. the HO^- ion abstracts an α-hydrogen to form an enolate.
d. the carbonyl gets ionized.

11.15 Which are the most acidic hydrogens in the compound below?

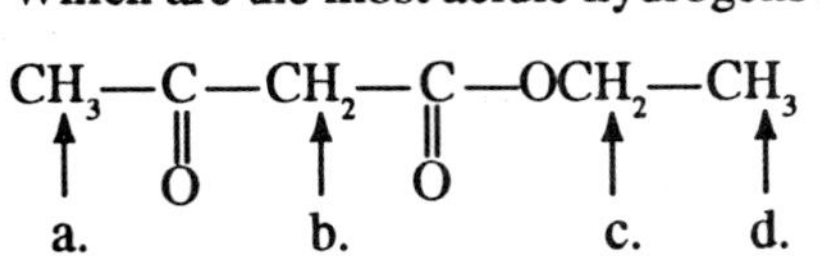

11.16 Decarboxylation is common in _______ .
a. β-oxoketones
b. β-ketocarboxylic acids
c. β-oxoarenes
d. β-hydroxyketones

11.17 At room temperature in the liquid phase $CH_3COCH_2COOCH_3$ is at equilibrium with _______ .
a. $H_2C{=}C(OH)CH_2COOCH_3$
b. $CH_3C(OH){=}CHCOOCH_3$
c. $CH_3COCH{=}C(OH)OCH_3$
d. $CH_3COCH_2C(OH){=}CH_2$

11.18 Which of the following species would not react with benzalacetophenone in a Michael addition?
a. $C_6H_5NH_2$ b. $CH_3O^-Na^+$ c. CH_3CH_2I d. NaOH

11.19 What is the product of reaction between benzalacetophenone and methylamine, CH_3NH_2?

a.

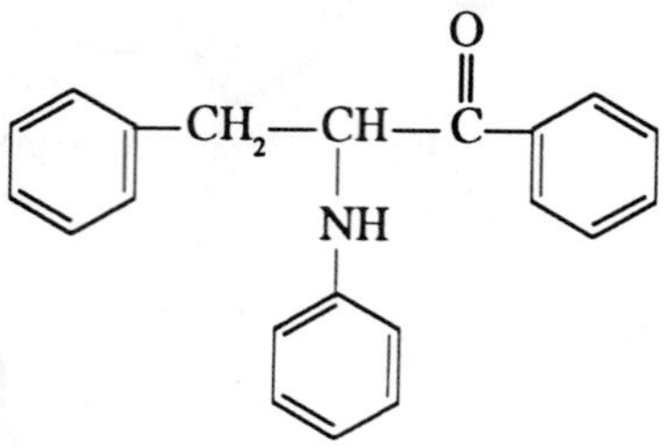

c.

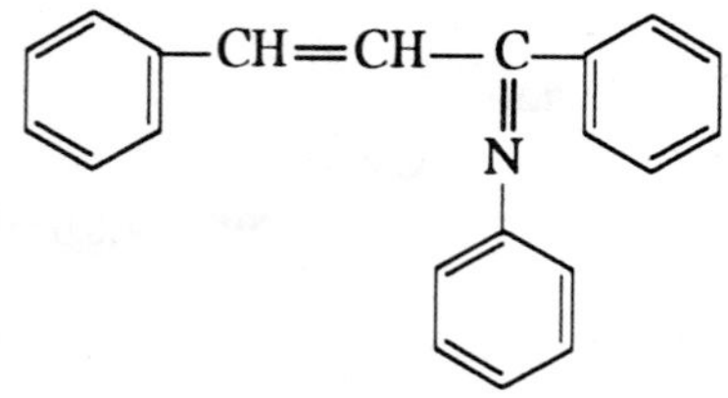

b.

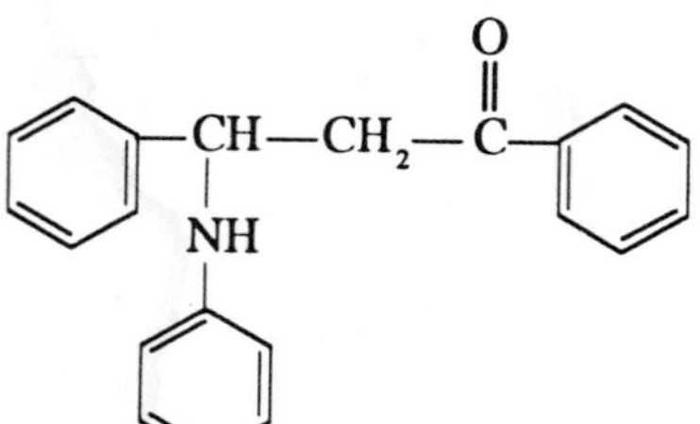

d.

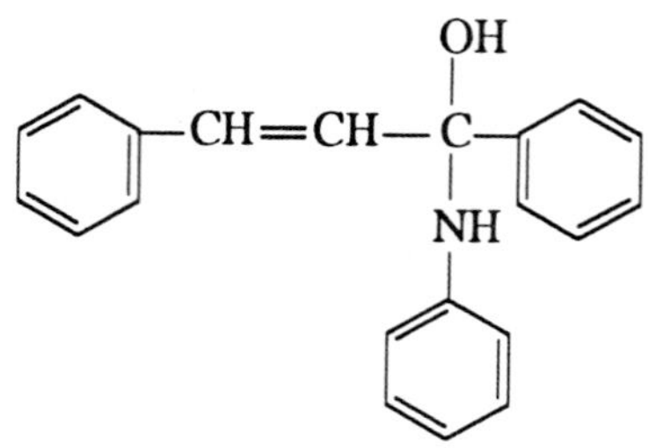

11.20 The number of exchangeable hydrogens in the ^{1}H NMR spectrum of a NaOD/D_2O solution of acetylacetone is ______.

a. O b. 2 c. 6 d. 8

Answers for Questions

11.1 a. ethanal or acetaldehyde
b. 2,2-dimethylpropanal, or trimethylacetaldehyde, or dimethylpropionaldehyde
c. (*E*)-2-ethyl-2-hexenal
d. tetramethyl-3-pentanone, or di-*t*-butyl ketone
e. 3-chloropropanal, or β-chloropropionaldehyde, but *not* 3-chloropropanal
f. 2-methyl-5-oxoheptanal
g. 2,2-dimethyl-3-pentanone, or *t*-butyl ethyl ketone
h. 3-buten-2-one, or methyl vinyl ketone
i. 3-methyl-2-butanone, or iospropyl methyl ketone
j. 1-acetylcyclohexanol, or 1-hydroxycyclohexyl methyl ketone
k. 1,3-cyclohexanedione
l. *trans*-4-ethyl-2-methylcyclohexanone, or (2*S*,4*R*)-2-methyl-4-ethylcyclohexanone
m. 4-isopropyl-1,3-cyclopentanedione
n. 1-chlorocyclopropanecarbaldehyde, or 1-chloro-1-formylcyclopropane
o. 3-chlorocyclobutanone
p. cyclopentylethanal, or cyclopentylacetaldehyde
q. *o*-methoxybenzaldehyde
r. 4-aminobenzophenone, or *p*-aminodiphenyl ketone
s. 5,6-dimethyl-6-hepten-2-one
t. 4-penten-2-one, or allyl methyl ketone
u. acetophenone dimethyl acetal, or 1,1-dimethoxy-1-phenyl ethane
v. chloral hydrate
w. 4-formylbenzenesulfonic acid

11.2 a. $(CH_3)_2C{=}O$

b. 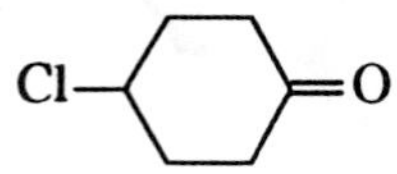

c. 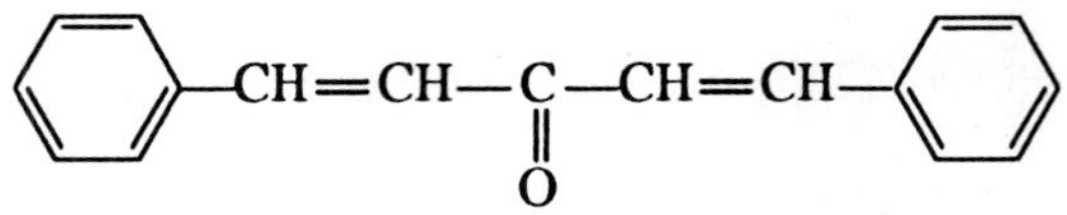

d. $(CH_3)_3C{-}\overset{\overset{\displaystyle O}{\|}}{C}{-}CH_3$

e. it cannot exist (! . . .)

f. OHC —CHO

g. 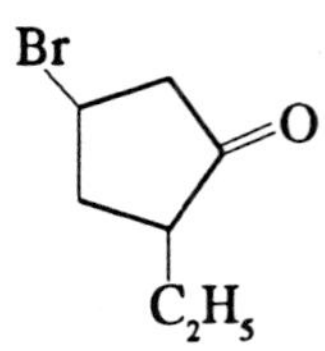

h. 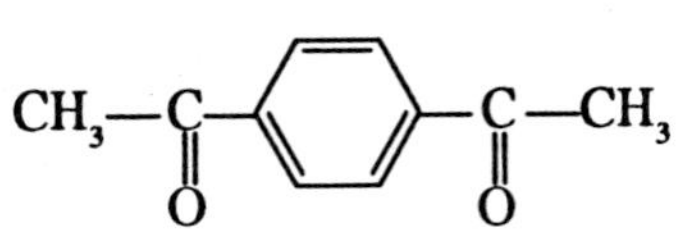

i. $(CH_3)_2C{=}NOH$

j. $CH_2{=}CHCHO$

k. 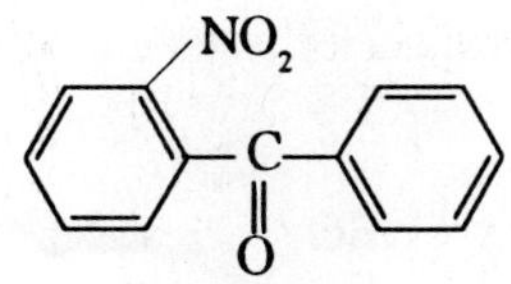

l. 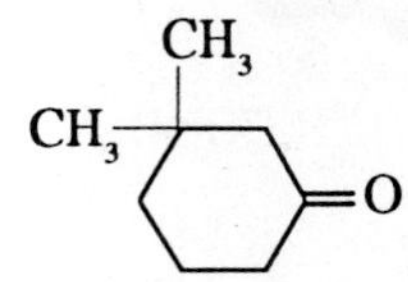

m. 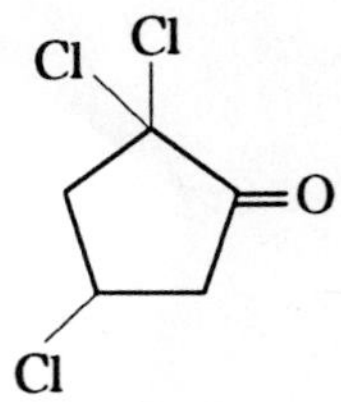

n. 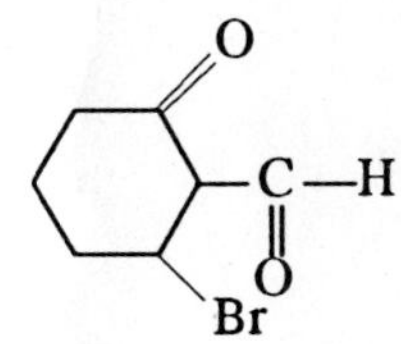

o. 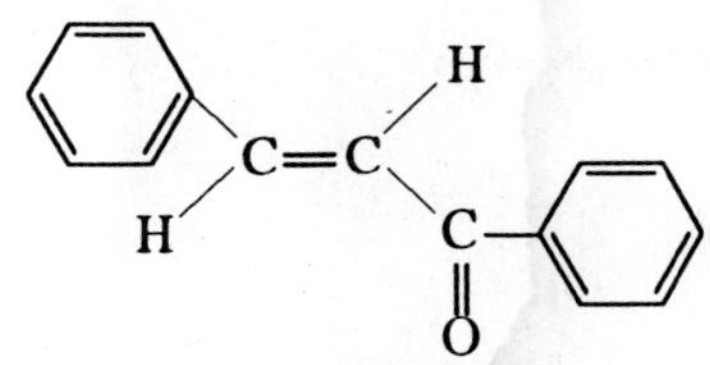

p. $(CH_3)_2C{=}N{-}NHCONH_2$

q. $C_2H_5\overset{\overset{\displaystyle O}{\|}}{C}CH_2\overset{\overset{\displaystyle CH_3}{|}}{C}H\underset{\underset{\displaystyle O}{\|}}{C}H$

r. 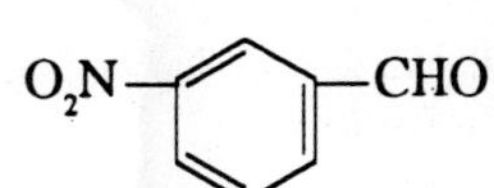

s. 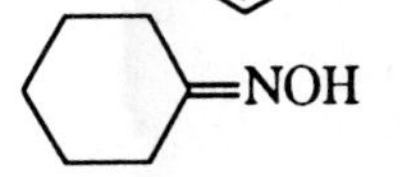

11.3 a. F b. D c. A d. B e. H f. G g. E h. C

11.4 Hemiacetals: c Hemiketals: a, j Acetals: b, d, g Ketals: e, i Ethers: f, h

11.5 a. $C_6H_5{-}\overset{\overset{\displaystyle O}{\|}}{C}{-}CH_2CH_3$

b. $C_6H_5CH_2{-}\underset{\underset{\displaystyle O}{\|}}{C}{-}CH_3$

c. 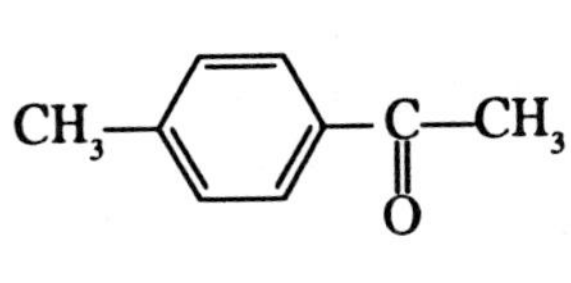

d. $C_6H_5CH_2CH_2CHO$

e. 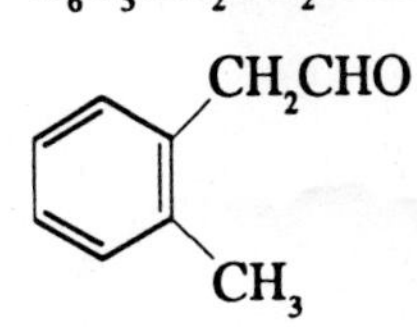

f. $C_6H_5CH(CH_3)CHO$

g. $CH_3CH(OCH_3)_2$

h. 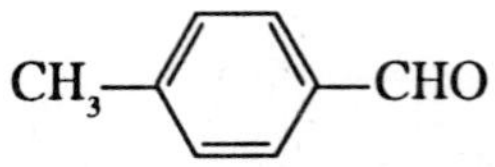

i. 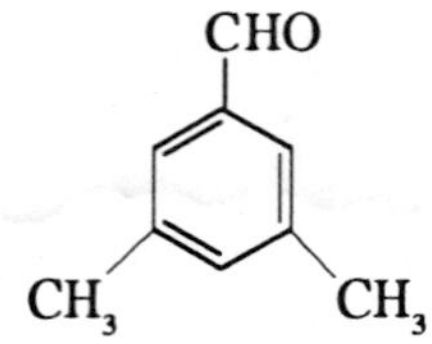

j. $[(CH_3)_3C]_2C{=}O$

11.6

A $C_6H_5CH_2OH$ + C_6H_5COOH

B C_6H_5CHO + Cr^{+3}

C C_6H_5COOH + Cr^{+3}

D $CH_3COCH_2CH_3$ + Cr^{+3}

E $(CH_3)_2CHOH$

F no reaction (CH_3COOH is an acid)

G CHI_3 + $C_2H_5COO^-$

H $N{\equiv}C{-}CH_2OH$

I $HOOCCH_2OH$

J 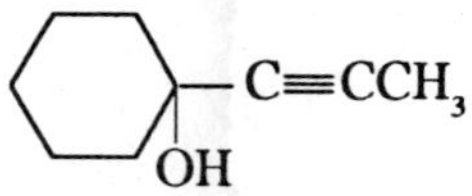

K $C_6H_5C({=}NC_6H_5)CH_3$ + H_2O

L 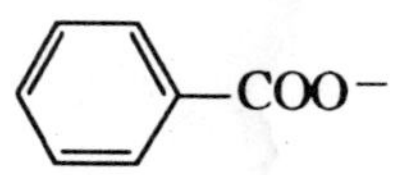 + Ag

M $(C_6H_5)_2CCl_2$

N $CH_3CH{=}N{-}NHCONH_2$ + H_2O

O 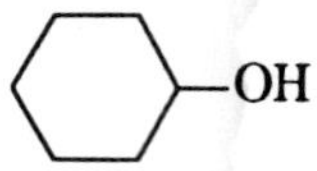

P 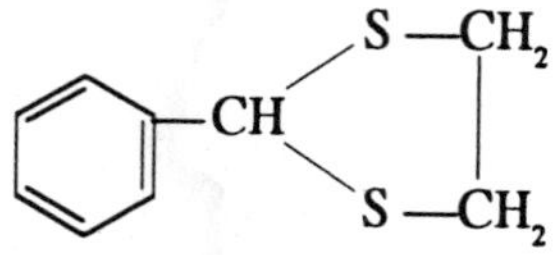 + H_2O

Q $C_6H_5CH_3$

R 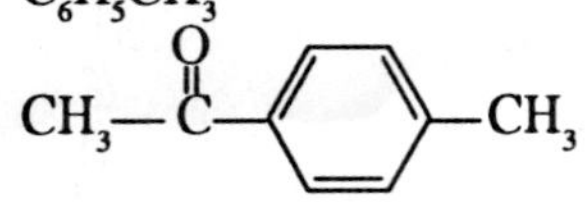 + CH_3COOH

S $HOOC(CH_2)_4CH_2(CH_2)_4COOH$ + N_2

T $CH_3COCH{=}CH_2$ + H_2O

U $CH_3COCH_2CH_2N(CH_3)_2$

V 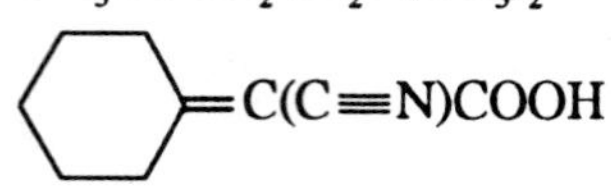 + H_2O

W $CH_3CH(OH)CH(Cl)CHO$

X $C_6H_5NH{-}CH_2CH_2{-}C{\equiv}N$

Y 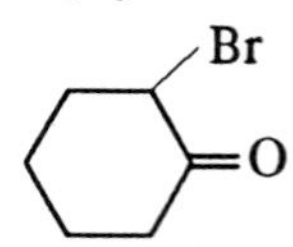

Z CH_3OD + HCOOD

AA $CH_3\overset{O}{\overset{\|}{C}}C(CH_3)_3$ (major) + $CH_2{=}C(CH_3)C(CH_3){=}CH_2$ (minor) + H_2O

BB $CH_3C({=}N{-}OH)C_6H_5$ (syn and anti) + H_2O

CC $CH_3\underset{\underset{O}{\|}}{C}NHC_6H_5$

DD no reaction

EE $CH_2{=}CHCH(OH)CH_3$

11.7 a.

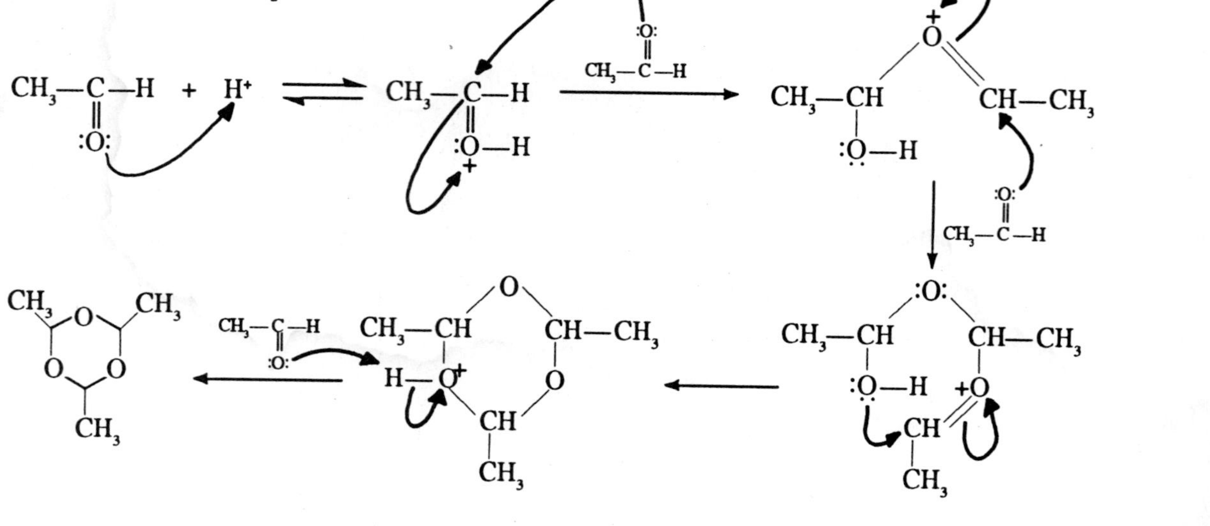

b. The products of this reaction are an alcohol and a carboxylic acid salt, and the reaction occurs in the presence of concentrated NaOH; together, these facts suggest the Cannizzaro reaction, which occurs with aldehydes that have no α-hydrogens.

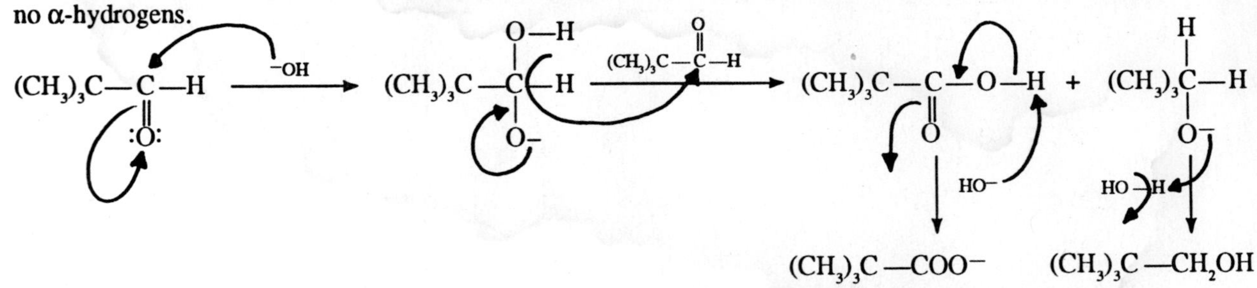

c. This is an acid-catalyzed reaction of a carbonyl compound with an ammonia derivative.

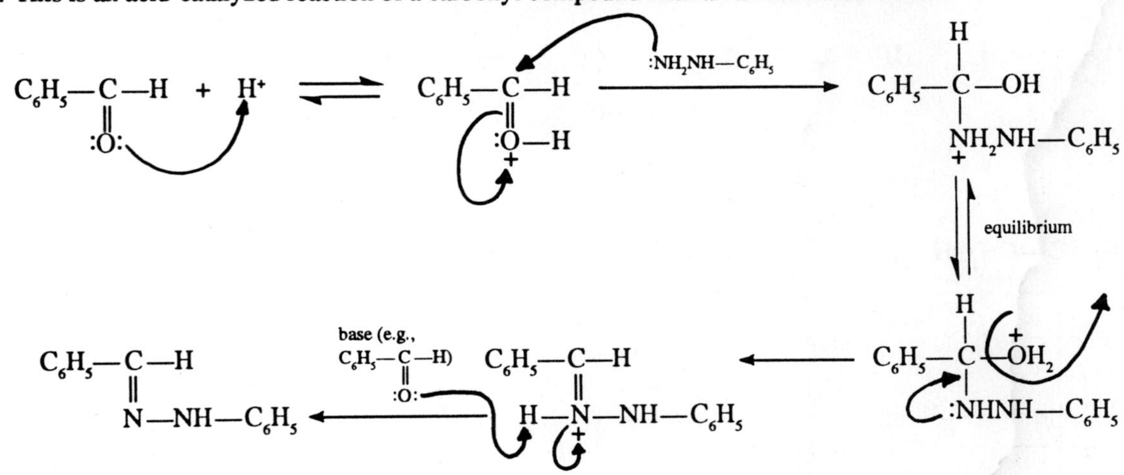

d. Comparing the reactants with the products, we see

1. the loss of dimethyl sulfoxide's oxygen and its transfer to the ethyl part of the sulfonate,
2. the loss of the $^{-}O{-}S(=O)_2{-}C_6H_5$ group.

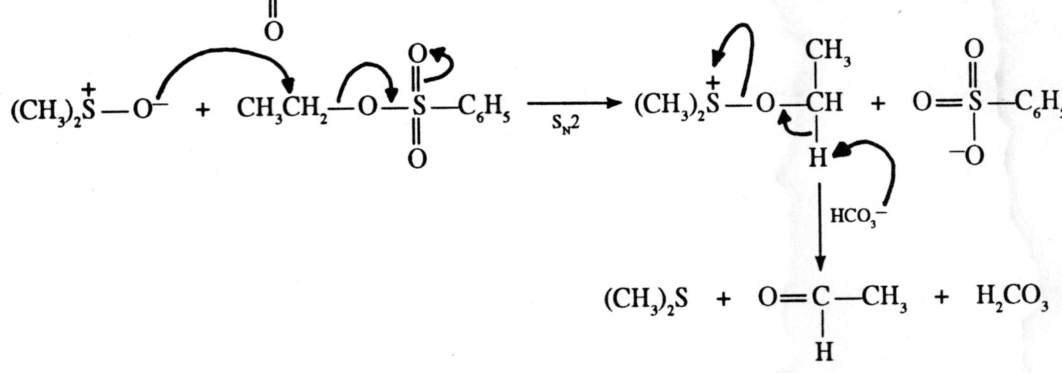

e. The mechanism of the first step is similar to that in part (c). The second step is a Wolff-Kishner reduction.

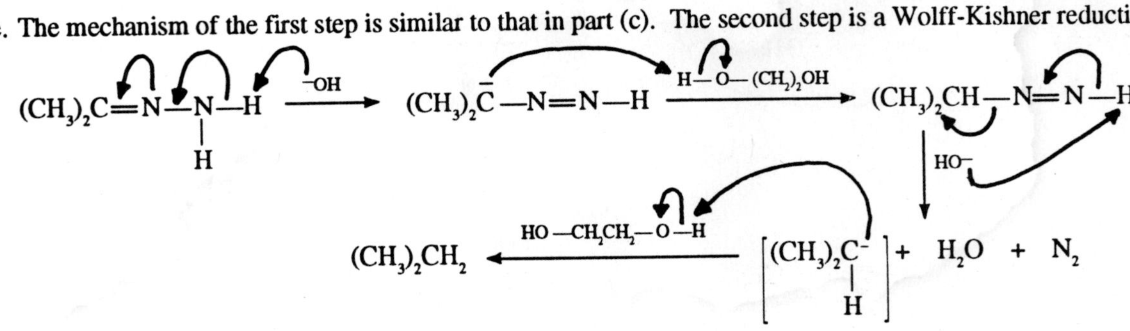

f.

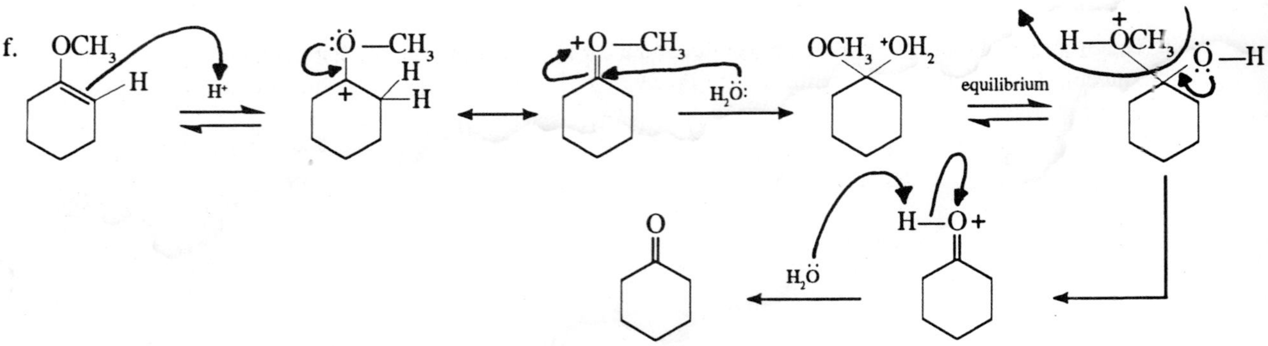

g. This is an acid-catalyzed aldol condensation.

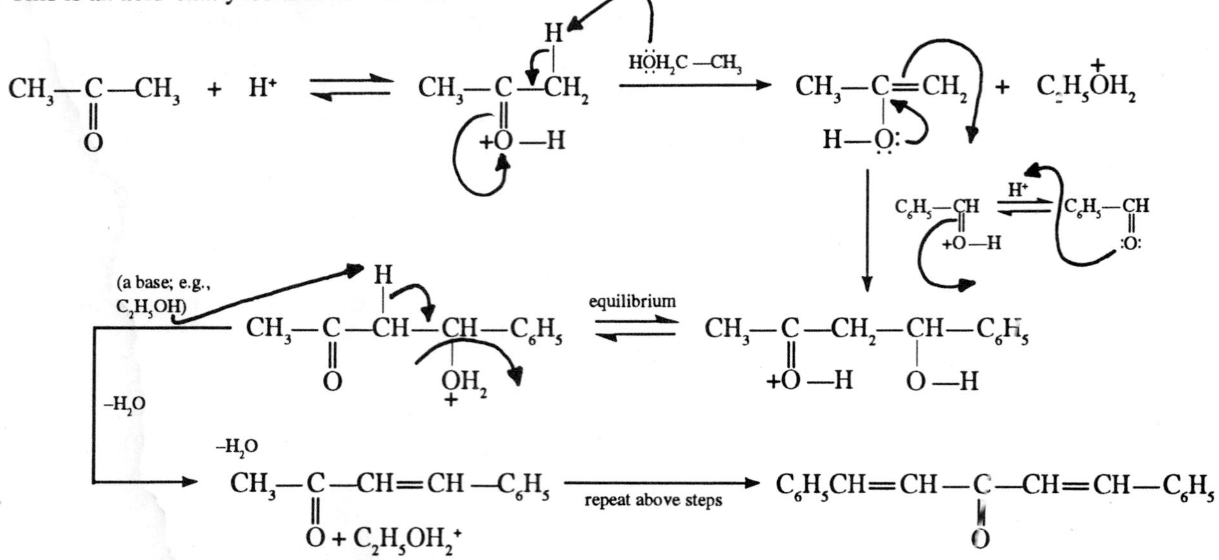

h. The product is a bis (α, β-unsaturated ketone), which suggests the aldol condensation (this time, base-catalyzed).

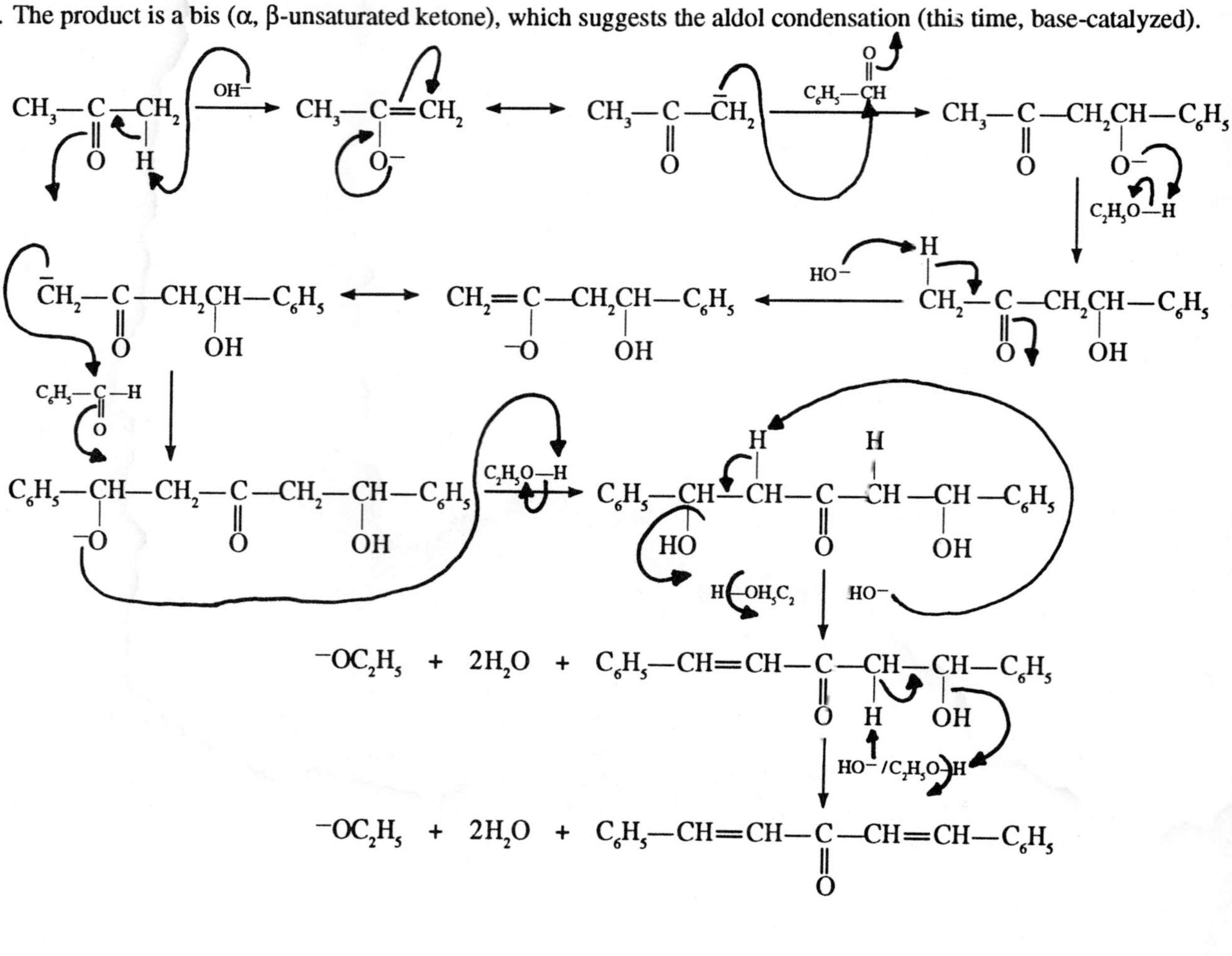

i. This is an acid-catalyzed thioketal formation (similar to formation of a regular ketal).

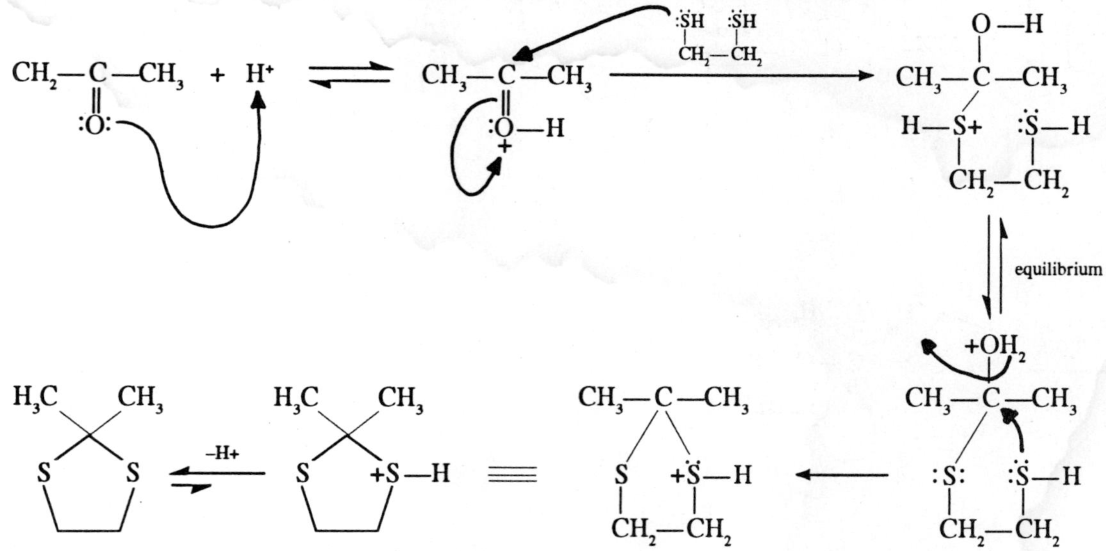

j. The reaction is similar to that in part (c).

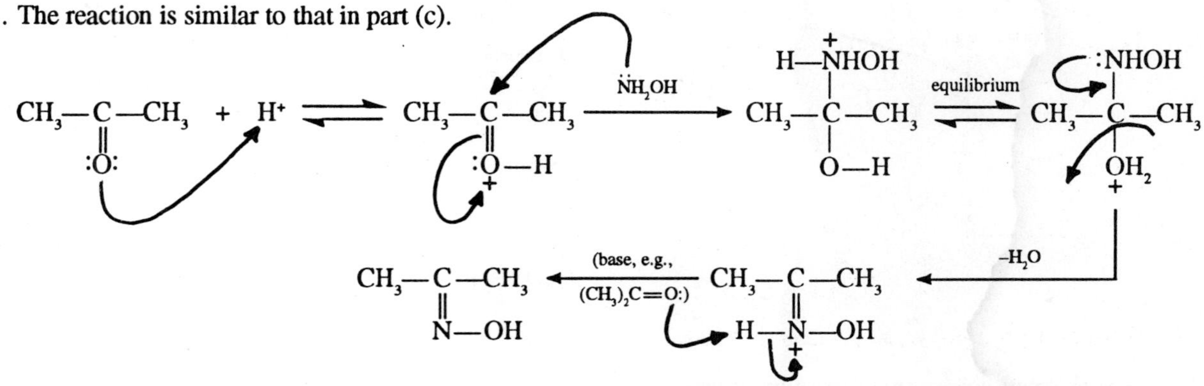

k. Comparison of the reactants and products suggests carbanion formation, followed by nucleophilic substitution, cylization, hydrolysis, and decarboxylation of a β-ketoacid.

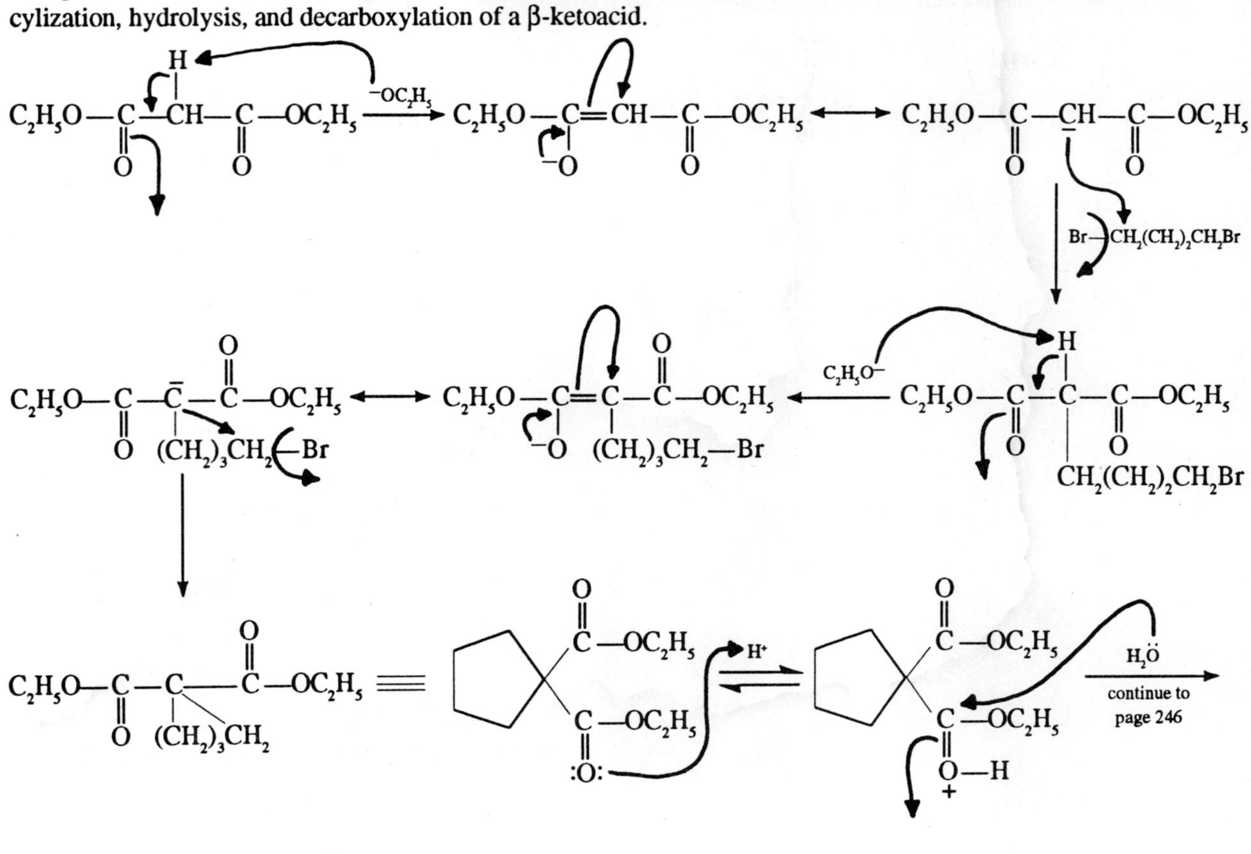

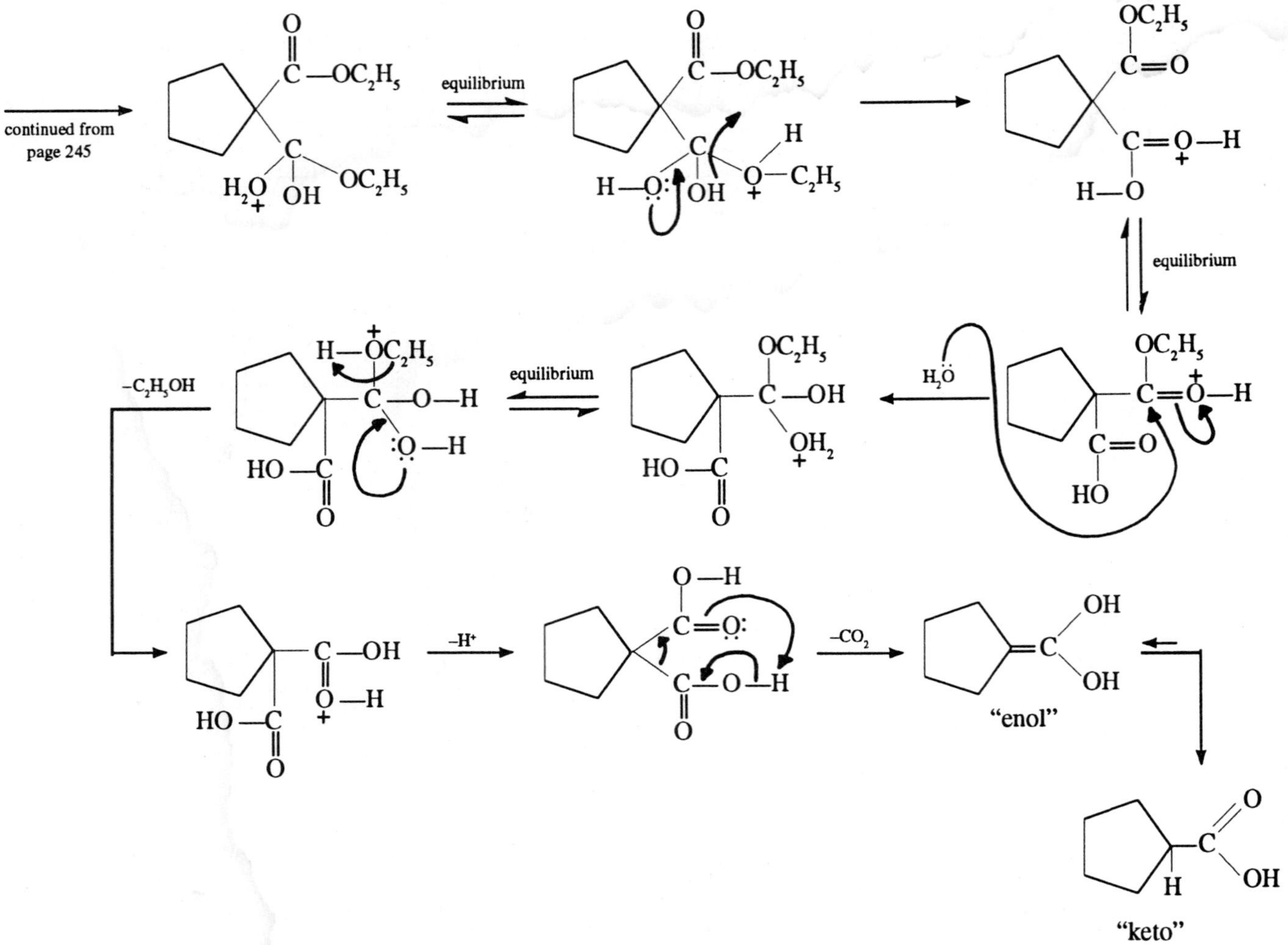

1. The products suggest the formation of a carbanion and an aldol-type condensation. The acetate acts as the base.

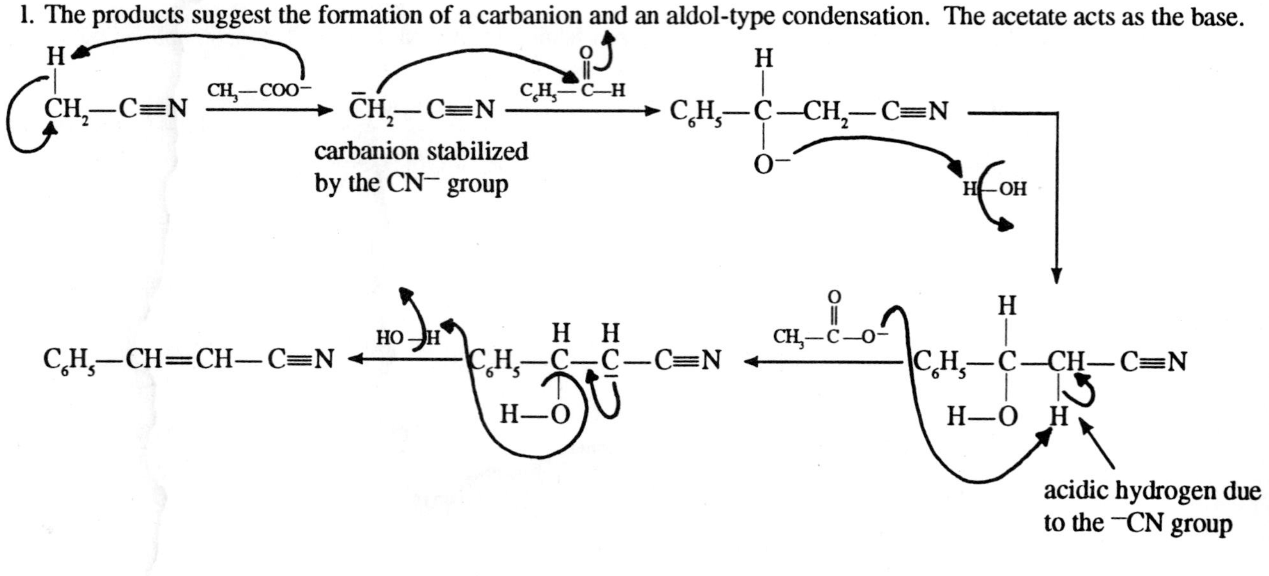

m. This is another base-catalyzed condensation that involves a carbanion formation and subsequent attack at the carbonyl.

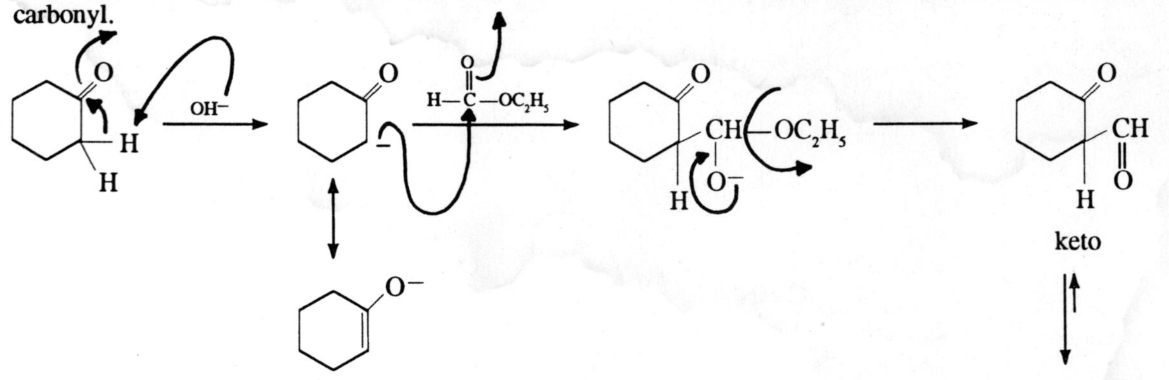

In this instance, the enol is more stable because of α, β-conjugation with the carbonyl carbon.

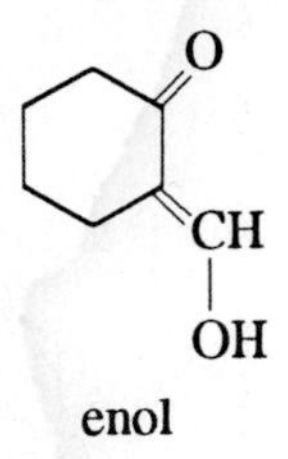

n. In the product

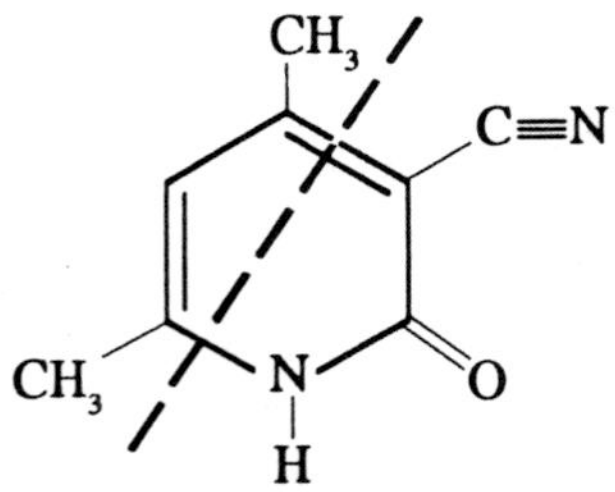

the dashed line suggests the way the two reagent molecules should be attached via carbanion formation.

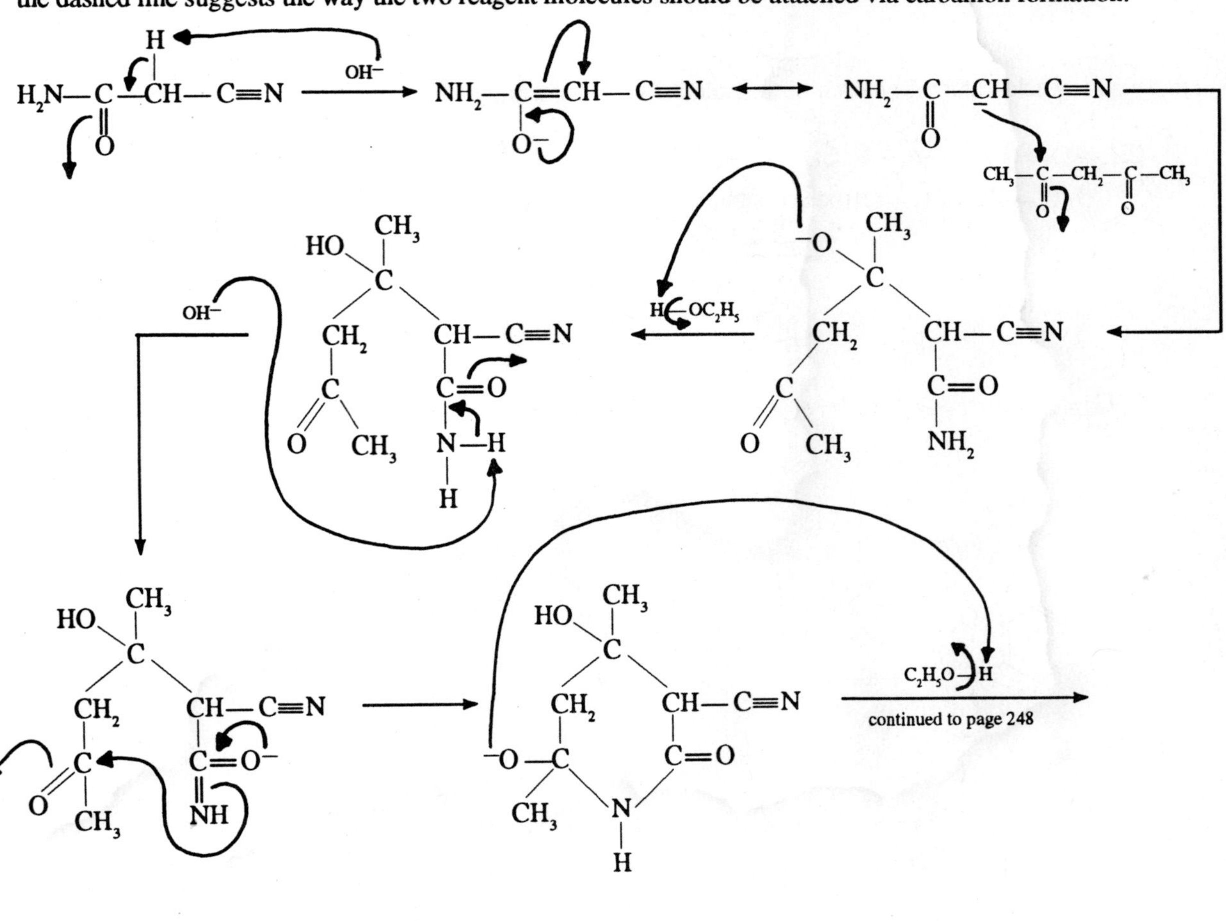

continued to page 248

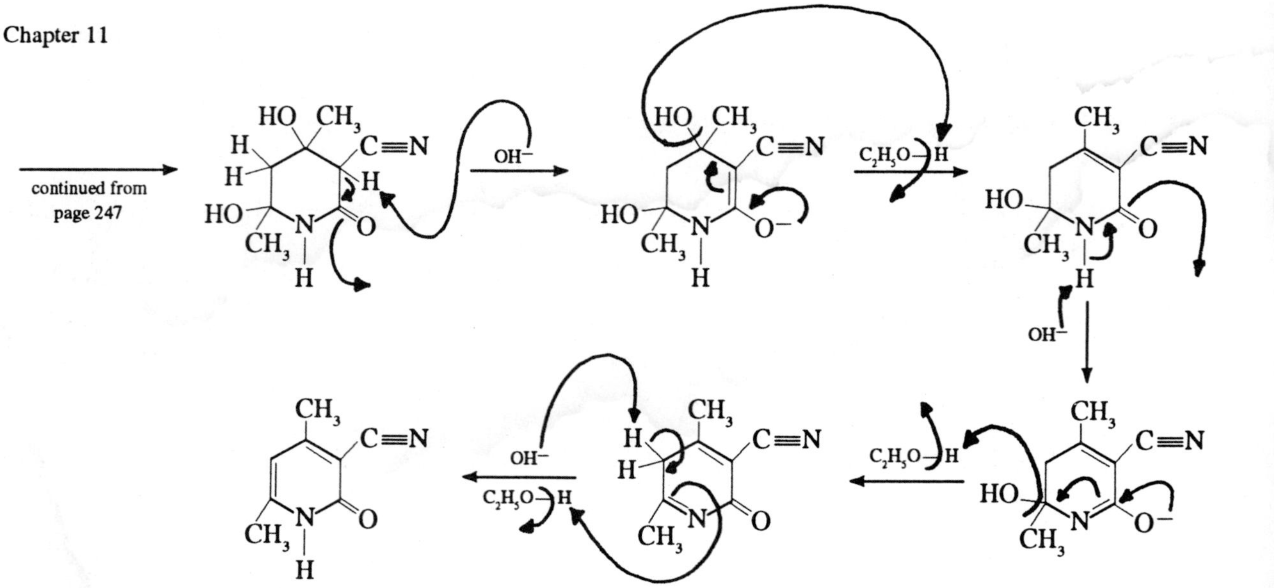

o. This is an enamine synthesis that is acid catalyzed (toluene is used to remove the water azeotropically).

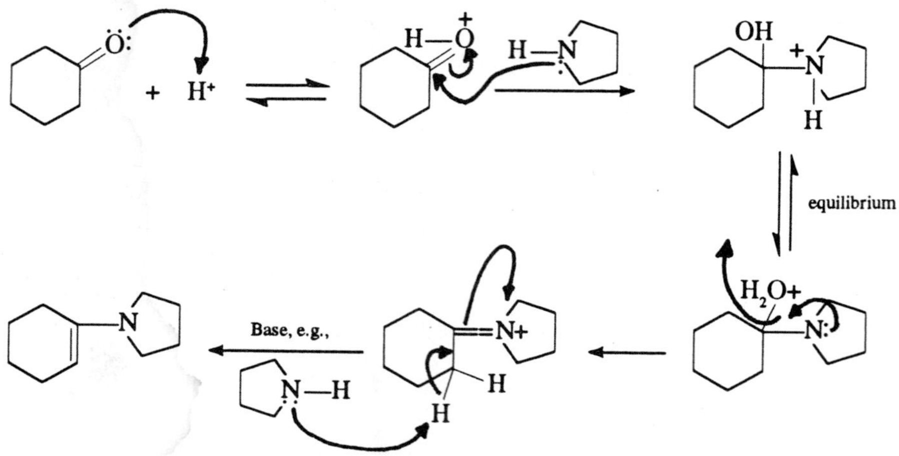

p. This is an acid-catalyzed intramolecular aldol-like condensation.

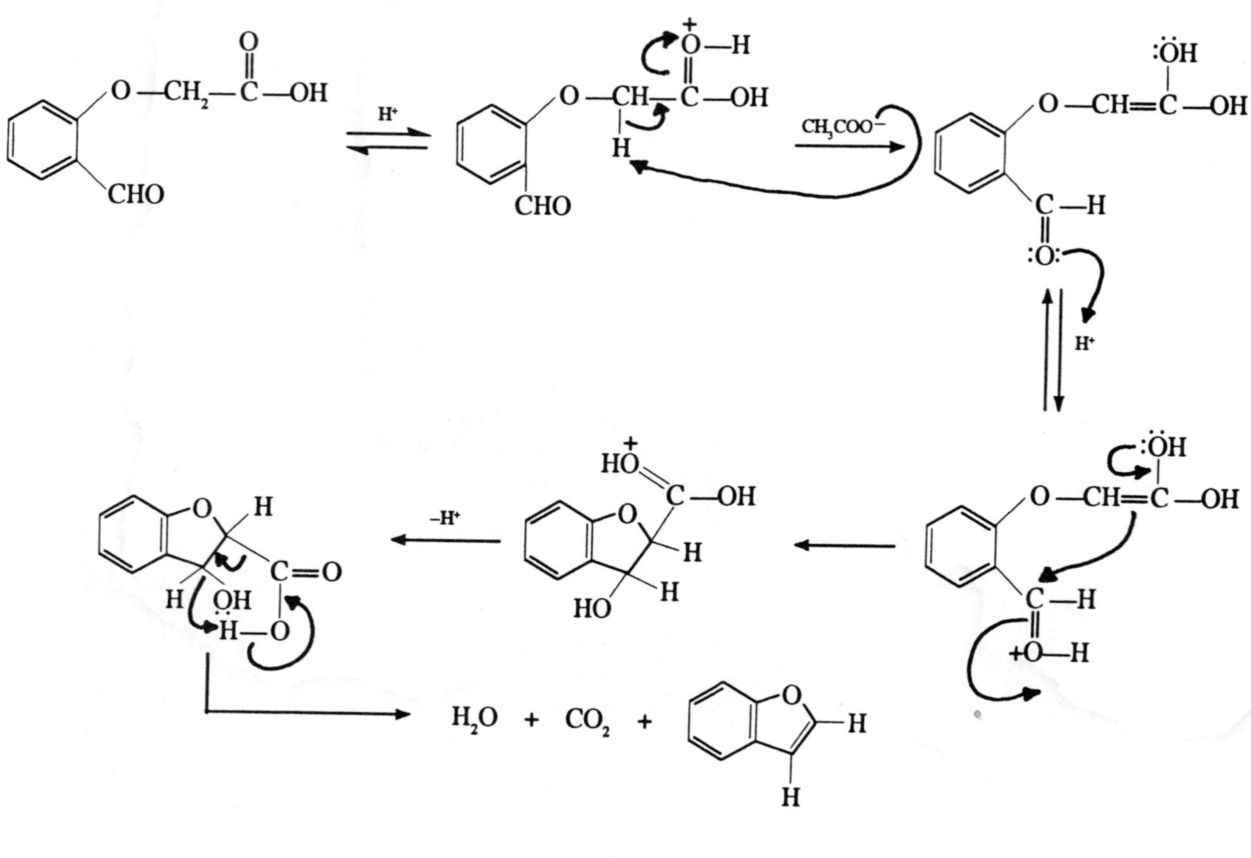

q. This is a double aldol condensation/dehydration reaction.

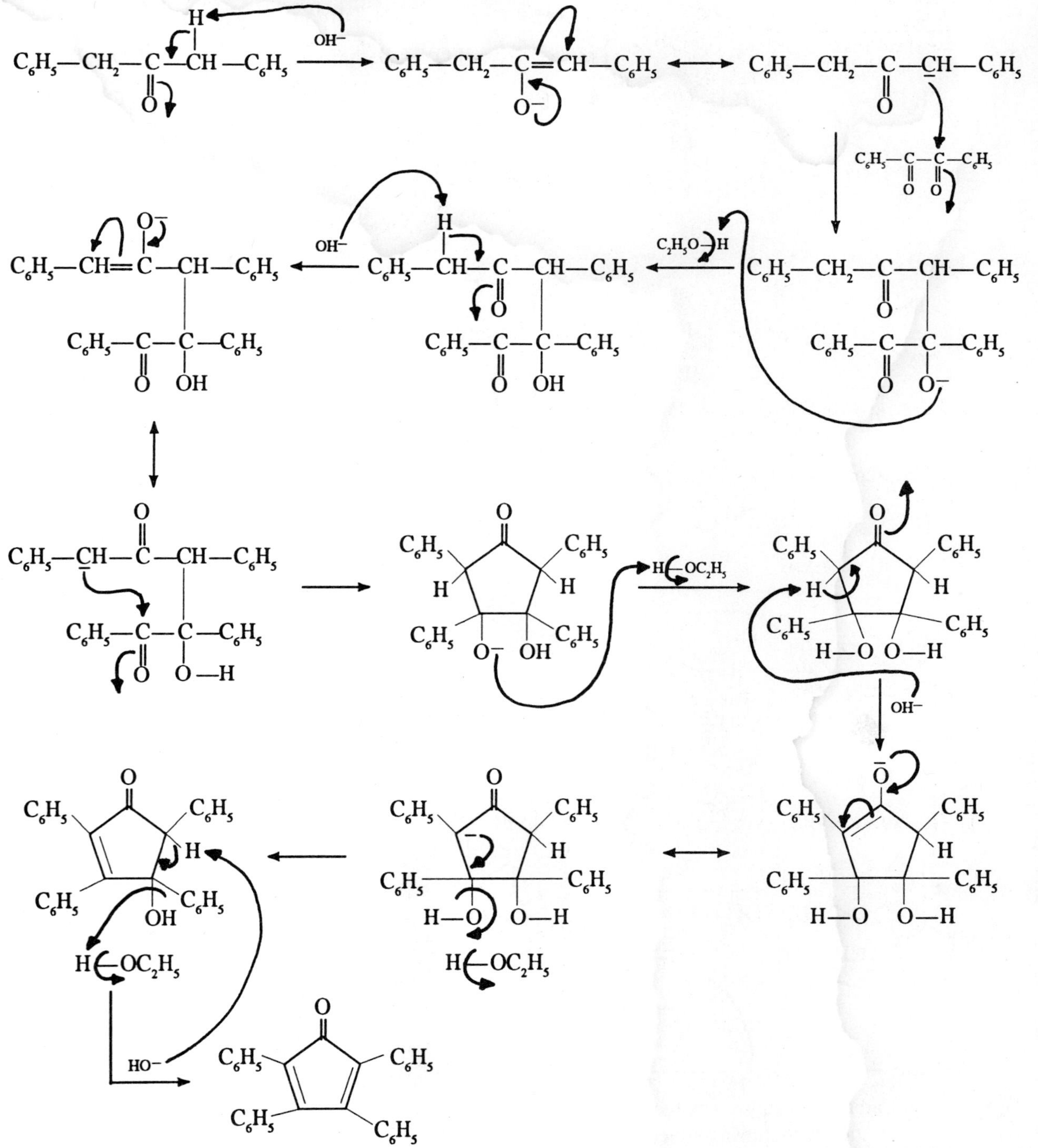

r. The first step is the synthesis of an enamine (similar to the mechanism in part (o). The second step is an enamine reaction with an α, β-unsaturated ketone (Michael-type addition), followed by hydrolysis. The third step is an intramolecular base-catalyzed aldol condensation, followed by dehydration.

Second step:

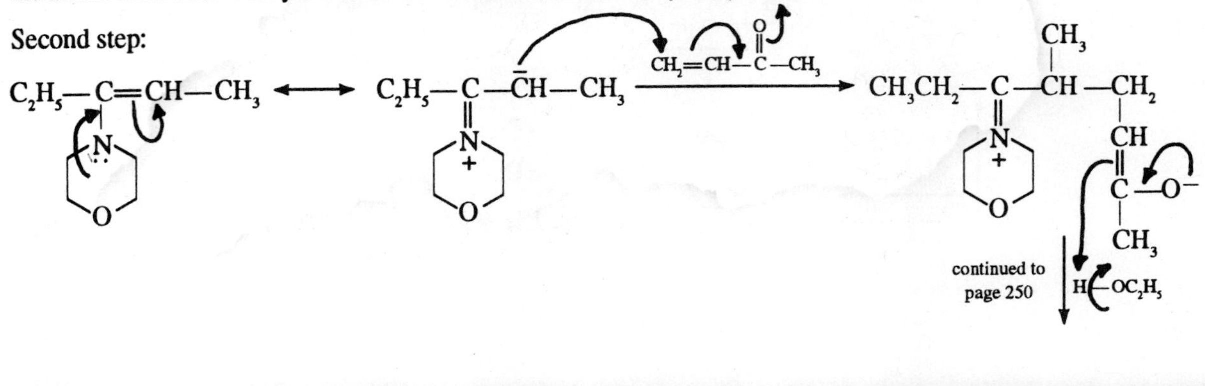

continued to page 250

continued from page 249

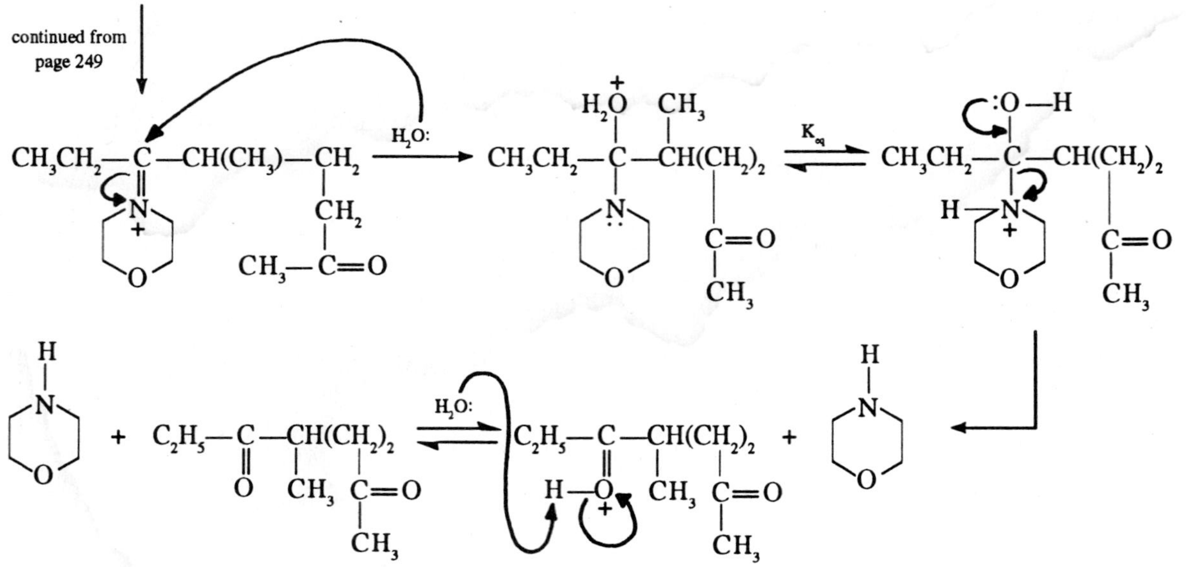

Third step:

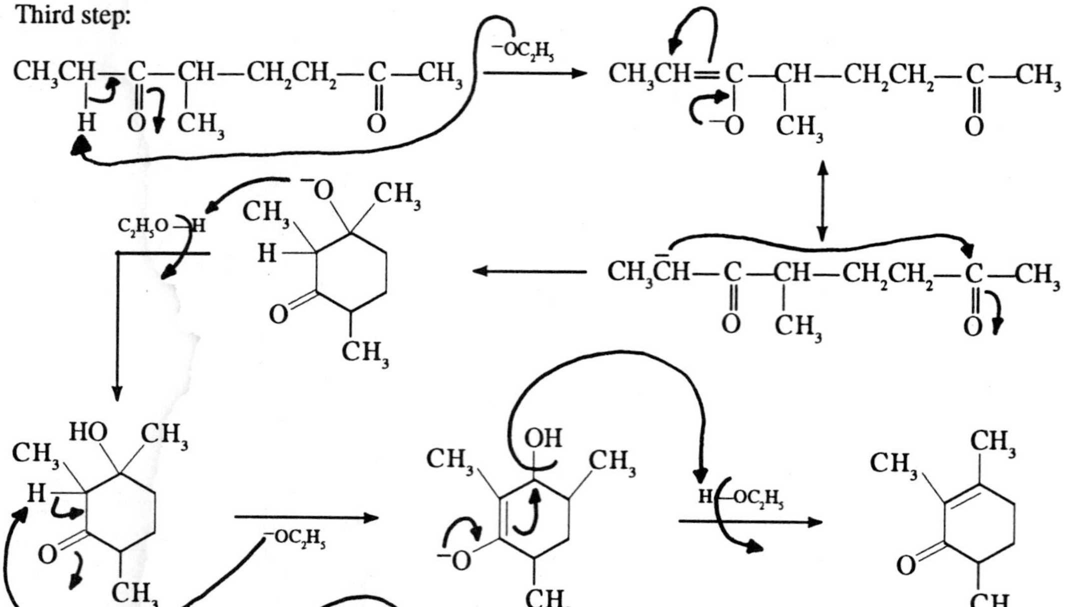

s.

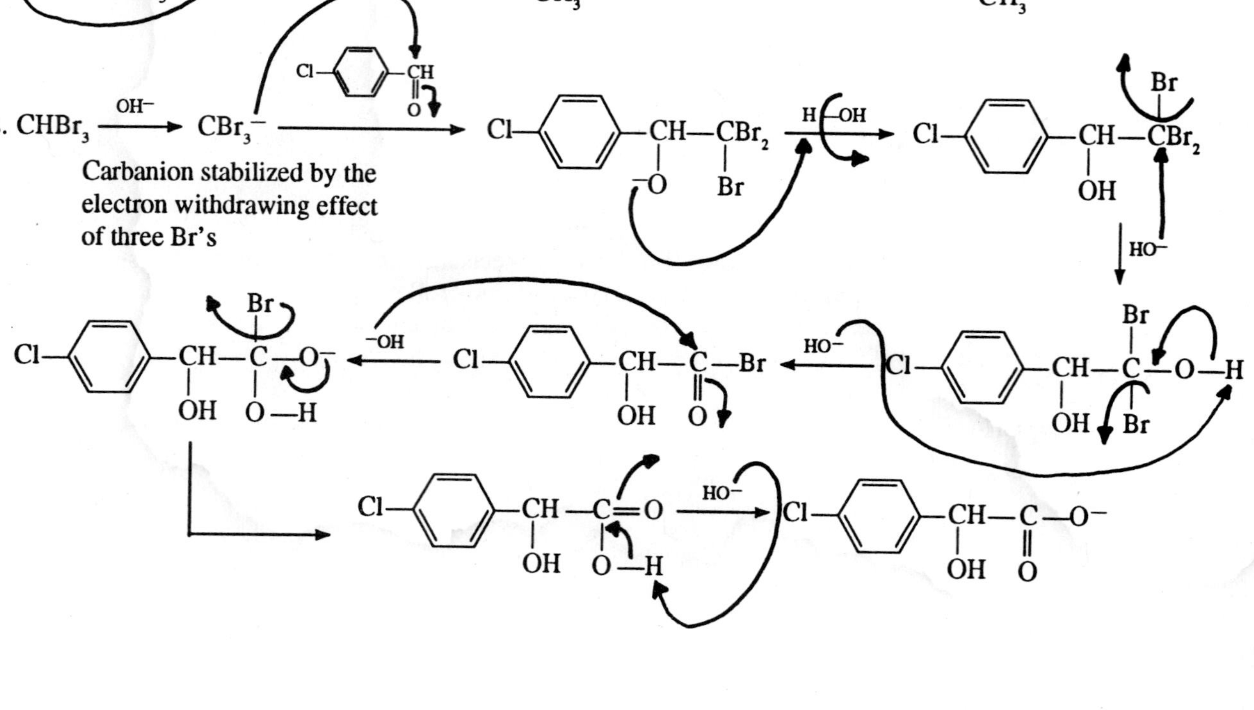

t. In the product

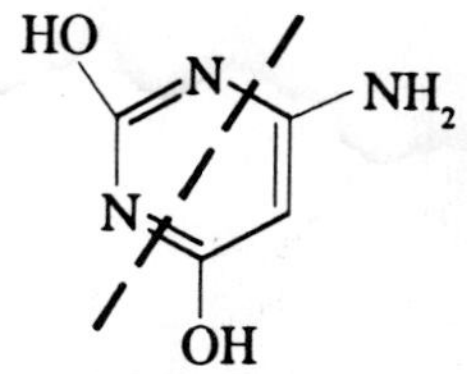

the dashed line suggests the way the two molecules should be attached to each other. Loss of the amide hydrogens in urea, $NH_2\underset{\underset{O}{\|}}{C}NH_2$, occurs because of the stability of the carbanion formed.

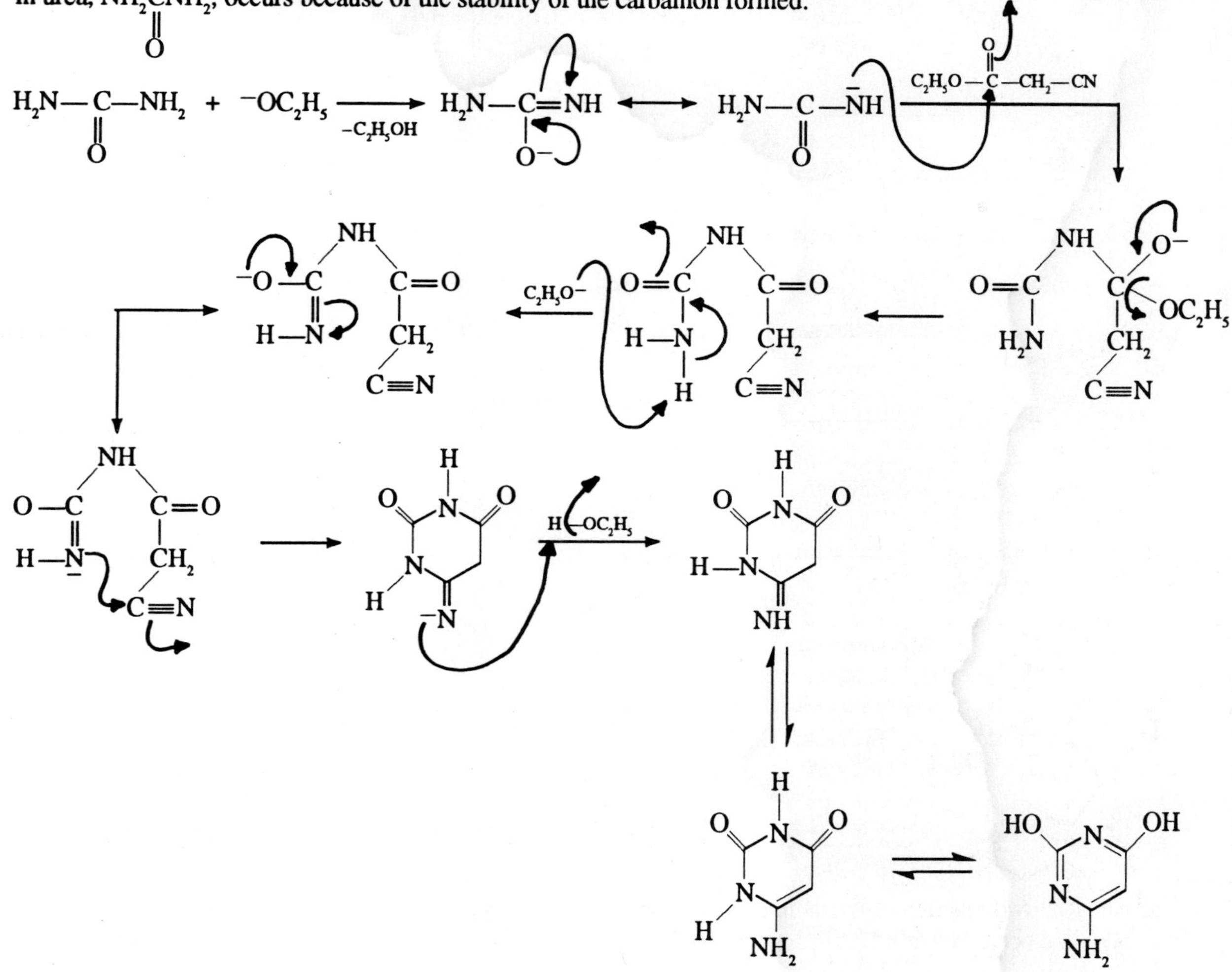

"keto" "enol"

11.8 **A** $C_6H_5-\underset{\underset{O}{\|}}{C}CH_3$

B $C_6H_5CO_2H$

C CHI_3

D $C_6H_5-\underset{\underset{OH}{|}}{CH}-CH_3$

E $C_6H_5-\underset{\underset{O}{\|}}{C}-CH=CH-C_6H_5$

F $C_6H_5-\underset{\underset{O}{\|}}{C}-CH_2-\underset{\underset{HN-C_6H_5}{|}}{CH}-C_6H_5$

G $C_6H_5-\underset{\underset{CH_3}{|}}{C}=N-NH-C_6H_5$

H

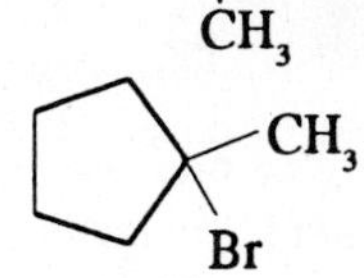

I

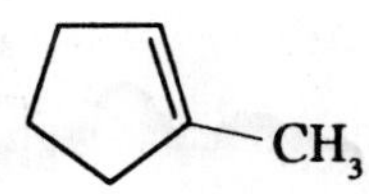

J
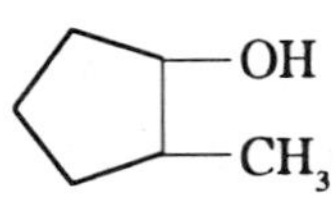

K
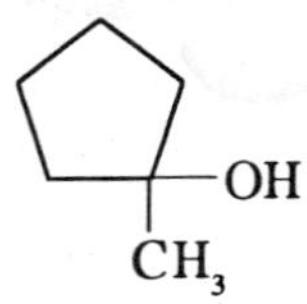

L
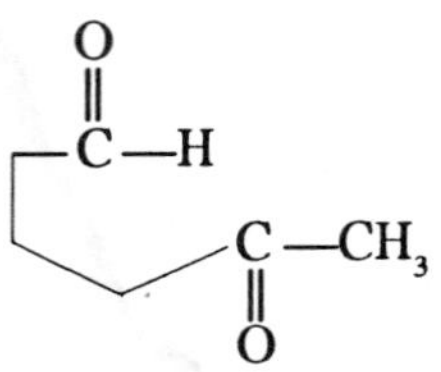

M
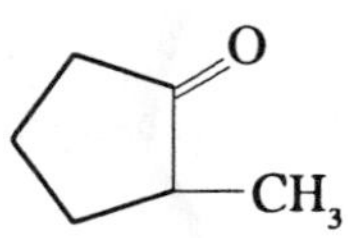

N
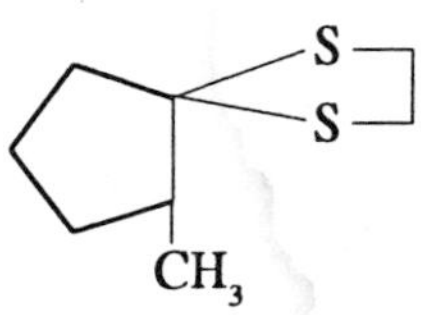

O
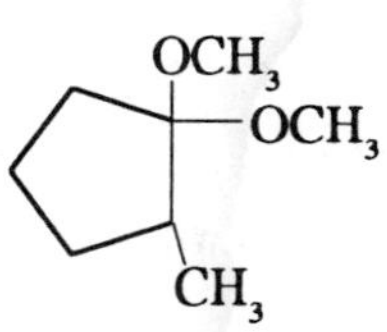

P
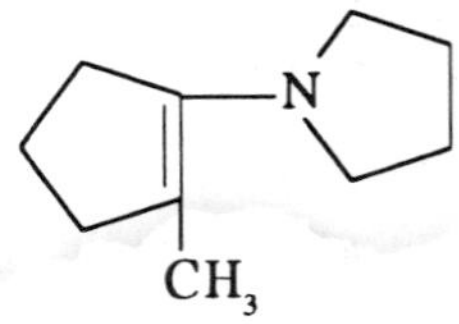

Q
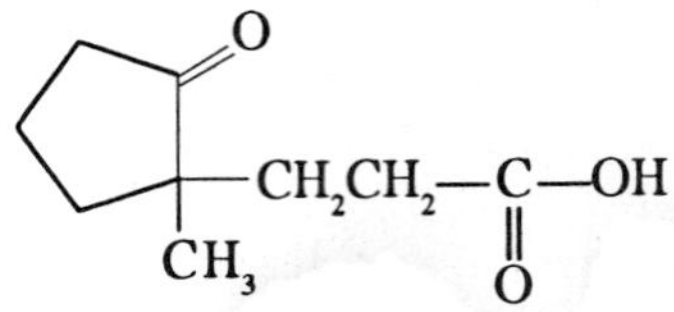

R
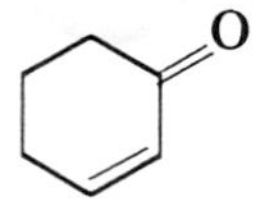

S
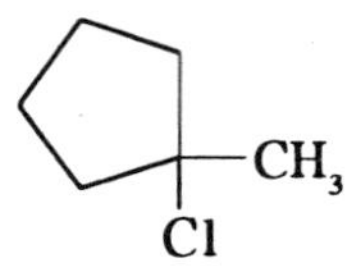

T
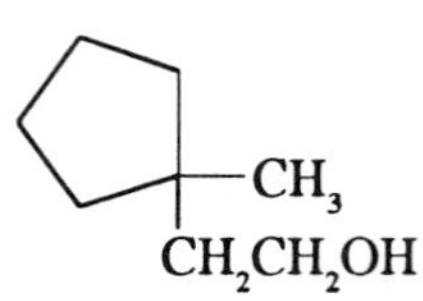

U
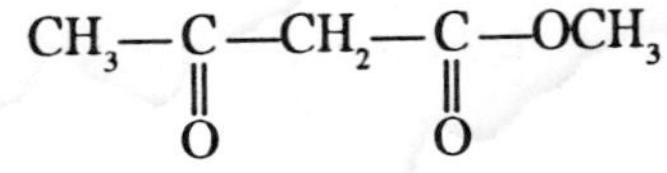

V
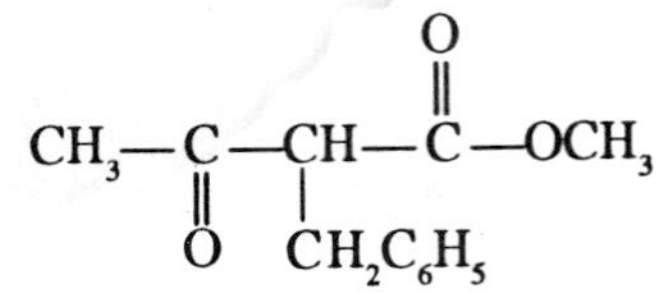

W
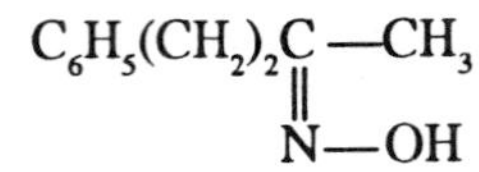

X
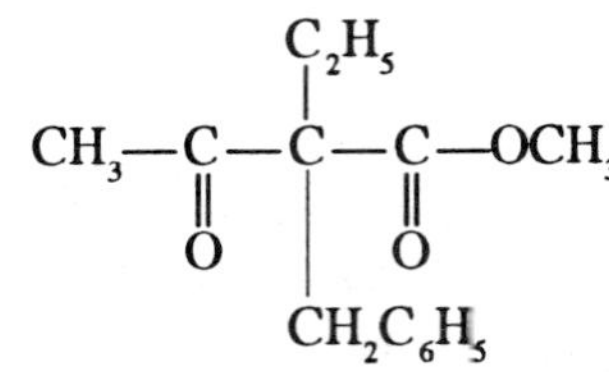

Y
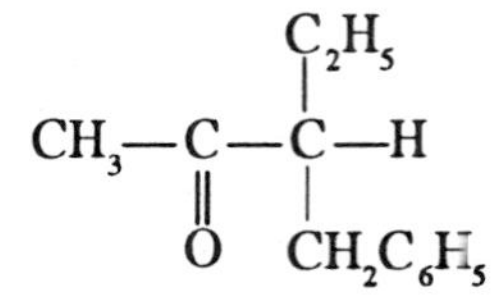

11.9 a. The first two reactions are examples of a 1,4-additions to an α, β-unsaturated ketone. However, the third reaction does not occur because HCl reacts with CH_3NH_2:

$$\underset{\text{(base)}}{CH_3NH_2} + \underset{\text{(acid)}}{HCl} \longrightarrow \underset{\text{(salt)}}{CH_3\overset{+}{N}H_3Cl^-}$$

Thus, nitrogen's unshared pair of electrons is unavailable for nucleophilic attack at the β-carbon.

b. In semicarbazide hydrochloride

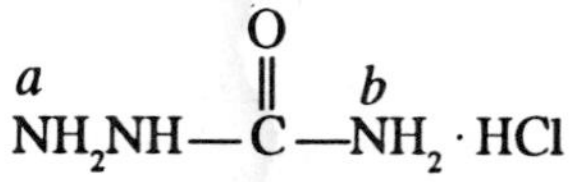

the nitrogen labelled (*a*) is amine-like, but the one labelled (*b*) is an amide nitrogen. Nitrogen (*a*) is more nucleophilic, because the electron pair on nitrogen (*b*) contributes to the resonance forms:

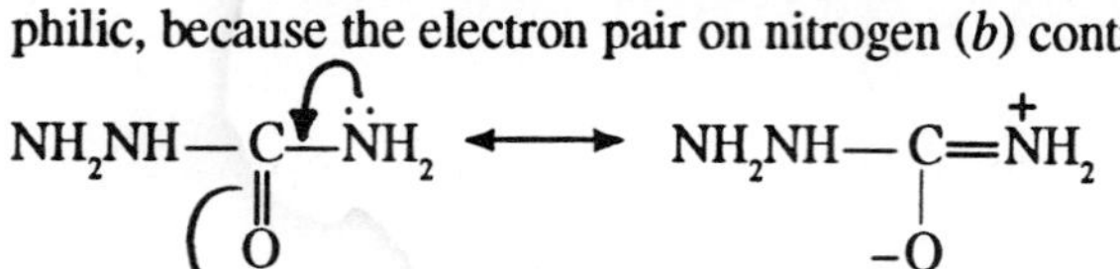

c. In Clemmensen reduction the carbonyl group reacts with Zn(Hg)/HCl, and HCl reacts easily with the reactant's hydroxyl group, producing an alkyl halide.

d. $NaHCO_3$ is a scavenger that neutralizes the hydrobromic acid produced. This affects the equilibrium by pulling the reaction to the right-hand side according to Le Châtelier's principle.

11.10 a. The aldol condensation product is dibenzalacetone; its structure is $C_6H_5{-}CH{=}CH{-}\underset{\underset{\displaystyle O}{\|}}{C}{-}CH{=}CH{-}C_6H_5$

b. With 1 mole of benzaldehyde, the product would be $C_6H_5CH{=}CHCOCH_3$ along with some unreacted acetone, and an equimolar amount of dibenzalacetone. With no acetone and *high* base concentration the products would be $C_6H_5COO^-$ + $C_6H_5CH_2O^-$, via the Cannizzaro reaction. With no benzaldehyde, the product would be $(CH_3)_2C(OH)CH_2COCH_3$ or (in high base concentration) $(CH_3)_2C{=}CHCOCH_3$.

c. Because the carbonyl group of benzaldehyde is more susceptible than that of acetone toward nucleophilic attack of the carbanion. Acetone's methyl groups have an electron-releasing (inductive) effect, which decreases the electrophilic properties of the carbonyl group and can sterically hinder the approach of the carbanion.

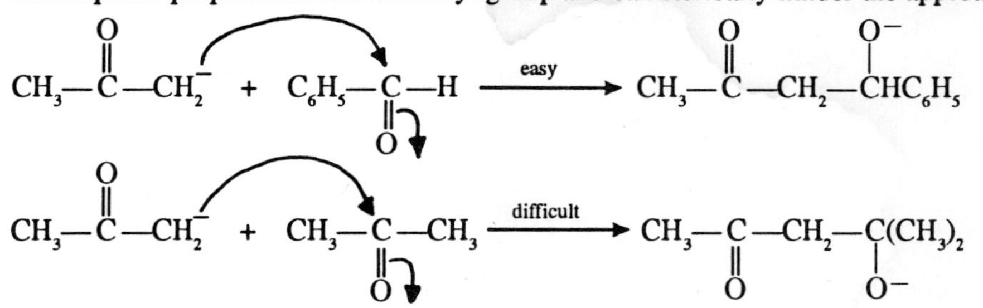

11.11 a. The structure of the product (an α, β-unsaturated ketone) suggests an aldol condensation and subsequent dehydration. Thus, working backwards:

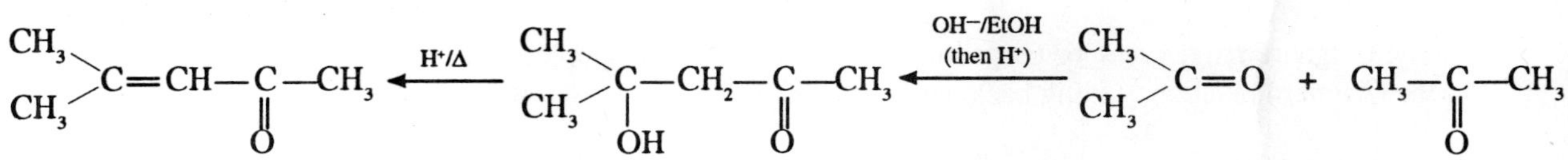

Acetone can be prepared from propane as follows:

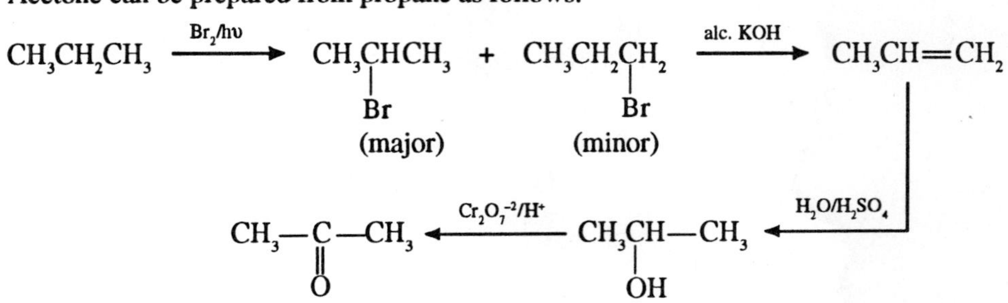

b. The structure of the product suggests the insertion of an ^-OH group and subsequent methylation, and insertion of a $-CH_3$ group and its conversion to $-CH_2\underset{\underset{\displaystyle O}{\|}}{C}{-}OH$

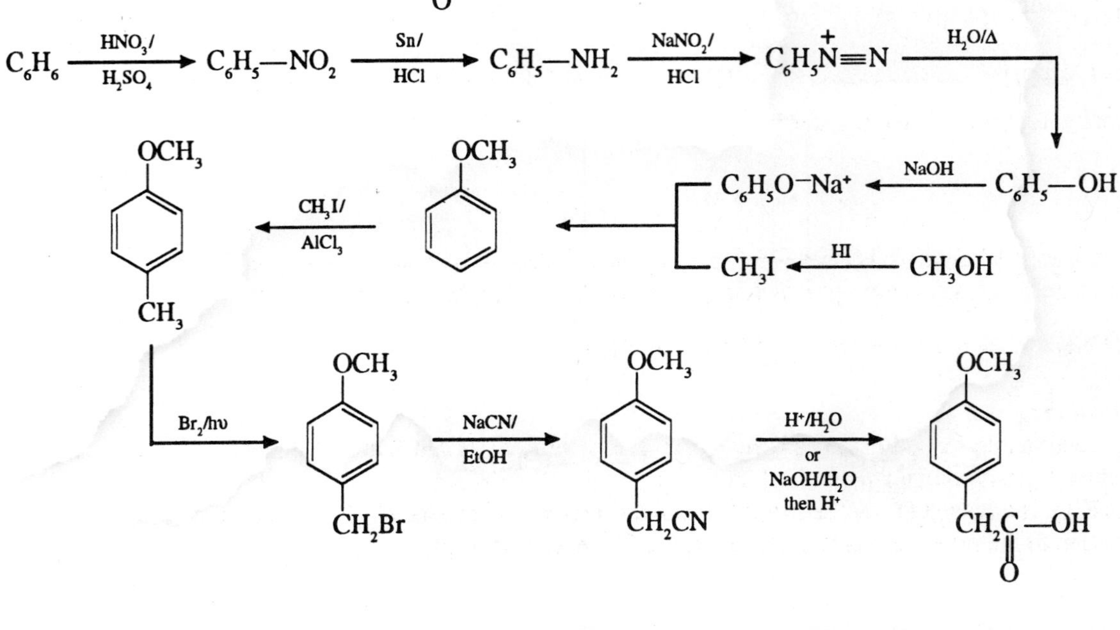

c. The product is $CH_3CH_2C(=O)—CH(CH_3)—CH_2CH_2CH_3$.

This can be prepared by oxidizing the corresponding alcohol, which itself can be synthesized via the Grignard reaction. Thus, going backwards,

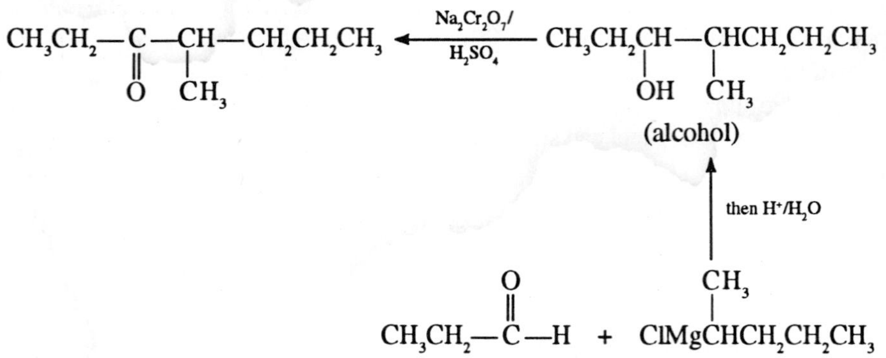

Propanal can be made by oxidation of propanol:

$$CH_3CH_2CH_2OH \xrightarrow[H_2SO_4]{Na_2Cr_2O_7} CH_3CH_2C(=O)—H$$

The Grignard reagent can be prepared starting with the corresponding alcohol, which can also be synthesized via the Grignard reaction. Going backwards again:

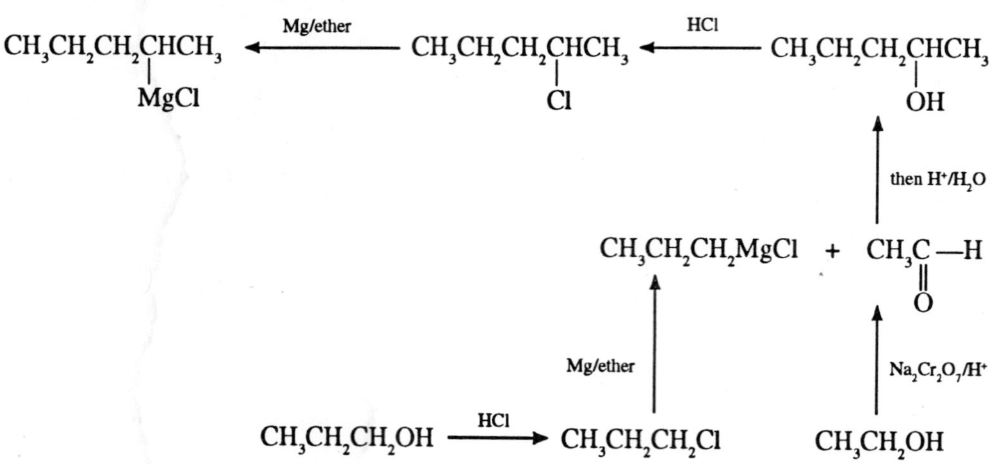

d. The structure of the product suggests the use of the Wittig reaction; going backwards

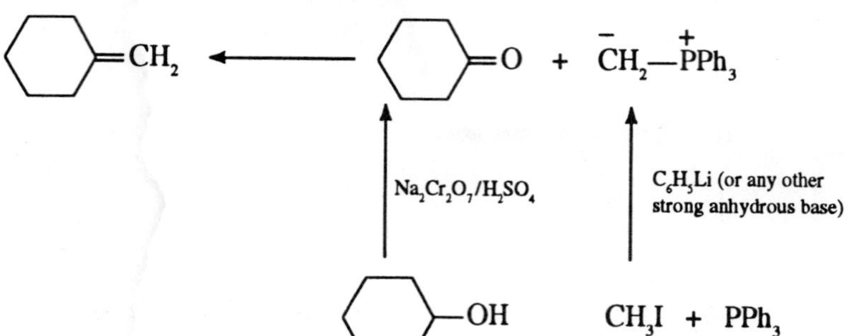

e. *Method 1:*

$$C_6H_6 + CH_3(CH_2)_2C(=O)—Cl \xrightarrow{AlCl_3} C_6H_5—C(=O)—(CH_2)_2CH_3$$

Method 2:

$$C_6H_6 + Br_2 \xrightarrow{Fe} C_6H_5Br \xrightarrow[\text{ether}]{Mg/} C_6H_5MgBr \xrightarrow{CdCl_2} (C_6H_5)_2Cd \xrightarrow{CH_3CH_2C(=O)-Cl} C_6H_5-C(=O)-(CH_2)_2CH_3$$

Method 3:

$$C_6H_6 + Br_2 \xrightarrow{Fe} C_6H_5-Br \xrightarrow[\text{ether}]{Mg/} C_6H_5MgBr \xrightarrow{CH_3(CH_2)_2C(=O)-H} C_6H_5CH(OH)(CH_2)_2CH_3$$

$$C_6H_5CH(OH)(CH_2)_2CH_3 \xrightarrow{Na_2Cr_2O_7/H_2SO_4} C_6H_5-C(=O)-(CH_2)_2CH_3$$

Method 4:

$$C_6H_6 + (CH_3(CH_2)_2-C(=O)-)_2O \xrightarrow{H_3PO_4} C_6H_5-C(=O)-(CH_2)_2CH_3$$

f. The product is $CH_3CH{=}CH-C(=O)-H$ (an α, β-unsaturated aldehyde), which suggests an aldol condensation and subsequent dehydration. Thus, going backwards:

$$CH_3CH{=}CH-C(=O)-H \xleftarrow[\text{then } H^+/\Delta]{OH^-/EtOH} 2\,CH_3C(=O)-H$$

Acetaldehyde can be prepared from ethanol via dichromate oxidation

$$CH_3CH_2OH \xrightarrow[H_2SO_4]{Na_2Cr_2O_7/} CH_3C(=O)-H$$

g. The product is $CH_3-C(CH_3)(OH)-CH_2CH_3$, a five-carbon, 3° alcohol that can be prepared by reacting a Grignard reagent with a ketone. Thus,

$$CH_3C(=O)CH_3 + CH_3CH_2MgBr \longrightarrow (CH_3)_2C(OH)-CH_2CH_3$$

Acetone can be prepared by oxidizing 2-propanol:

$$(CH_3)_2CHOH + Na_2Cr_2O_7/H^+ \longrightarrow (CH_3)_2C{=}O$$

The Grignard reagent can be prepared from ethanol:

$$CH_3CH_2OH \xrightarrow{HBr} CH_3CH_2Br \xrightarrow[\text{ether}]{Mg/} CH_3CH_2MgBr$$

h. The product is $CH_3CH(OH)-CH_2CH_2(OH)$

Ethanol can be converted to the corresponding aldehyde (acetaldehyde), and then treated with a base in an aldol condensation:

$$2\,CH_3CH_2OH \xrightarrow[H_2SO_4]{Na_2Cr_2O_7/} 2\,CH_3C(=O)-H \xrightarrow[\text{then } H^+]{OH^-/EtOH} CH_3-CH(OH)-CH_2-C(=O)-H$$

Reduction with $NaBH_4$ should yield the product.

i. The product is $CH_3-\underset{|}{CH}(CH_3)-CH_2CH_2CH_3$.

Since acetone, a three-carbon molecule, is the only starting material, the Grignard reaction with acetone appears to be the easiest method:

$$CH_3-\overset{O}{\overset{\|}{C}}-CH_3 + CH_3CH_2CH_2MgBr \xrightarrow[\text{then } H^+]{\text{ether}} CH_3-C(CH_3)(OH)-CH_2CH_2CH_3$$

The hydroxyl group can be removed by dehydration and hydrogenation:

$$CH_3-C(CH_3)(OH)-CH_2CH_2CH_3 \xrightarrow{H^+/\Delta} CH_3-C(CH_3){=}CHCH_2CH_3 \xrightarrow{H_2/Pt} CH_3-CH(CH_3)CH_2CH_2CH_3$$

The Grignard reagent can be prepared from acetone as follows:

$$CH_3-\overset{O}{\overset{\|}{C}}-CH_3 \xrightarrow[\text{then } H^+]{NaBH_4} CH_3CH(OH)CH_3 \xrightarrow{H^+/\Delta} CH_3CH{=}CH_2 \xrightarrow[\text{peroxides}]{HBr/} CH_3CH_2CH_2Br \xrightarrow[\text{ether}]{Mg/} CH_3CH_2CH_2MgBr$$

11.12 a. $C_6H_5-CH_2CH_2-\overset{O}{\overset{\|}{C}}-H$ will show a positive Tollens' test, reacting with $Ag(NH_3)_2^+$ to form a silver mirror, while $C_6H_5CH_2-\overset{O}{\overset{\|}{C}}-CH_3$ will not. Also, the second compound will show a positive iodoform test (I_2/KI/NaOH), producing CHI_3, which is a yellow ppt. The first compound will not, since it is not a methyl ketone or acetaldehyde.

b. $C_6H_5(CH_2)_2\overset{O}{\overset{\|}{C}}-H$ will give a positive Tollens' test (see part (a)), while $C_6H_5-\overset{O}{\overset{\|}{C}}-CH_2CH_3$ will not.

c. $C_6H_5-\overset{O}{\overset{\|}{C}}-H$ will give a positive 2,4-DNP test (an orange/red precipitate when treated with 2,4-dinitrophenylhydrazine/H_2SO_4) and will also give a positive Tollen's test (see part (a)). $C_6H_5CH_2OH$ will be the only compound to give a positive Lucas test ($HCl/ZnCl_2 \longrightarrow$ two layers). $C_6H_5CH_2OH$ will neither react with 2,4-DNP or with Tollen's reagent.

d. $CH_3-\overset{O}{\overset{\|}{C}}-CH_2CH_3$ will give a positive 2,4-DNP test (see part (c)), while $CH_3(CH_2)_3OCH_3$ will give a negative test (no orange/red precipitate). Also, the first compound will give a positive iodoform test (see part (a)) because it is a methyl ketone, while the second will not.

e. Both compounds are methyl ketones, and therefore show positive results for both the iodoform and the 2,4-DNP test. To distinguish between them, oxidize each with $KMnO_4$/NaOH, acidify, and titrate the formed acids.

$CH_3-C_6H_4-\overset{O}{\overset{\|}{C}}-CH_3$ will yield $HO-\overset{O}{\overset{\|}{C}}-C_6H_4-\overset{O}{\overset{\|}{C}}-OH$, which needs twice as much standardized base for neutralization as $C_6H_5-\overset{O}{\overset{\|}{C}}-OH$ (which is produced by oxidizing $C_6H_5-CH_2-\overset{O}{\overset{\|}{C}}-CH_3$).

f. $CH_3CH_2—\underset{\underset{O}{\|}}{C}—H$ is an aldehyde that shows positive results from the Tollens' test (see part (a)) and from the 2,4-DNP test (see part (c)). $CH_3CH_2\underset{\underset{O}{\|}}{C}—OH$ is a carboxylic acid that gives negative results with those tests, but dissolves in dilute NaOH or sodium bicarbonate.

g. Both compounds are methyl ketones and show positive results with the 2,4-DNP and the iodoform tests. The difference is that the first compound has an alicyclic ring, while the second one has an aromatic ring. Nitration of

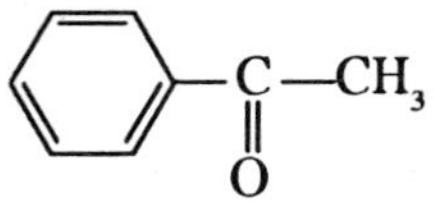

with conc. HNO_3/conc. H_2SO_4 will yield a yellow precipitate upon pouring the reaction mixture into an ice-water mixture.

h. The first compound is a methyl ketone that should give positive results with both the 2,4-DNP (see part (c)) and iodoform (see part (a)) tests. The second compound is an ester and does not react with either reagent.

i. The first compound (2-cyclohexenone) is an α, β-unsaturated ketone that should react with a nucleophile (e.g., aniline) in a Michael-type addition. The resulting product (a solid) should *not* decolorize a Br_2/CCl_4 solution, which indicates the loss of the double bond. On the other hand, the second compound (3-cyclohexenone) is *not* an α, β-unsaturated ketone, and therefore will not undergo Michael-type additions.

11.13 a. The first compound's $>C{=}O$ wavenumber will appear just below 1700 cm^{-1} (conjugated system), while that of the second compound should be around 1710–1720 cm^{-1}.

b. The first compound's triple bond should show up as a 2200 cm^{-1} peak, which will not appear for the second structure. The presence of the triple bond can be verified by an absorption peak at 3300 cm^{-1} due to the acetylenic hydrogen, $—C{\equiv}CH$.

c. The first compound (an aldehyde) shows peaks at 2850 and 2750 cm^{-1}, while the second compound shows a peak at 1435 cm^{-1} (CH_3).

d. Only the first compound shows a sharp, strong peak at 810 cm^{-1} (*p*-substitution).

e. Only the first compound shows a peak at 1435 cm^{-1} (CH_3).

11.14 a. $C_6H_5CH_2CH_2COC_6H_5$

b. $(CH_3)_2C(C_6H_5)COC_6H_5$

c. $(CH_3)_3C—CHO$

d. $(CH_3)_3CCH_2CHO$

Answers for Multiple Choice Test

11.1 a	11.6 d	11.11 d	11.16 b
11.2 c	11.7 c	11.12 d	11.17 b
11.3 b	11.8 d	11.13 b	11.18 c
11.4 a	11.9 c	11.14 c	11.19 b
11.5 d	11.10 b	11.15 b	11.20 d

12 Amines

Questions

12.1 Name the following amines and classify them as 1°, 2°, or 3° amines, or 4° ammonium salts:

a. $(CH_3)_2NH$

b. $CH_3CH_2CH(NH_2)CH_3$

c. H_2NCH_2COOH

d. $(C_2H_5)_4N^+BF_4^-$

e. $CH_2{=}CHCH_2CH_2NH_2$

f. $C_6H_5CH_2CH_2NH_2$

g. 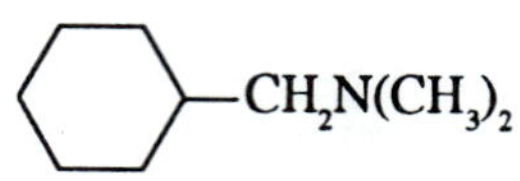

h. 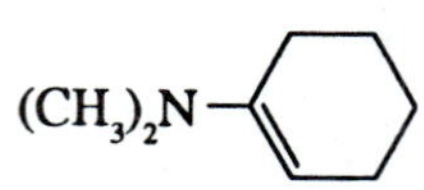

i. 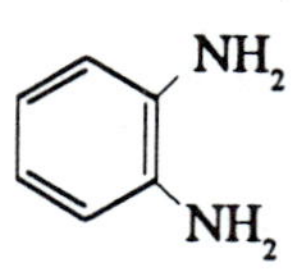

j. 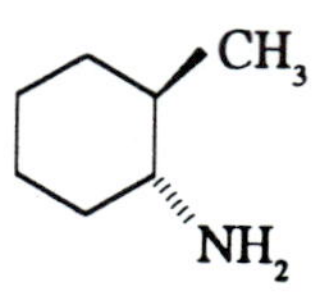

k. 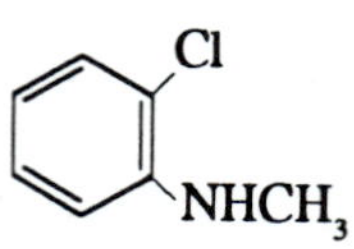

l. 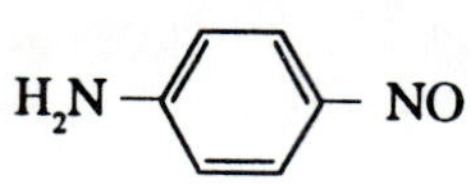

m. 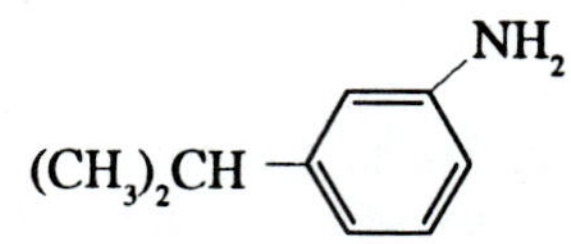

n. 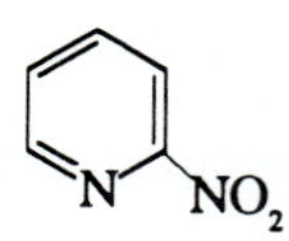

o. 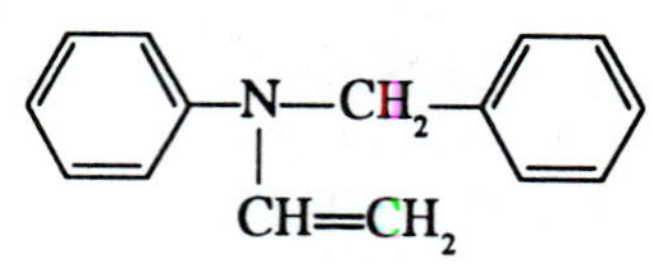

p. $(CH_3)_3CNH_2$

q. CH_3—N(C_2H_5)($CH_2CH_2CH_3$)

r. $H_2N(CH_2)_5NH_2$

12.2 Give condensed formulas for the following compounds and classify them as 1°,2°,3° amines or 4° ammonium salts:

a. diethylamine
b. benzylcyclopentylmethylammonium sulfate
c. aniline
d. cyclohexylamine
e. diphenylamine
f. 4-ethylpyridine
g. isobutylamine
h. 1,4-butanediamine
i. *p*-(N,N-dimethylamino)aniline
j. *p*-nitrobenzylamine
k. 1-phenyl-2-aminopropane
l. 3-hexylamine
m. *p*-allylaniline
n. N-ethyl-N,N-dimethylanilinium perchlorate
o. 4-amino-4-(1-propyl) heptane
p. 2,4,6-trinitrobenzylamine
q. cyclopropylethylmethylamine
r. 2-nitropyrrole

12.3 Identify the following as 1°, 2°, or 3° amines, 4° ammonium salts, amides, imines, enamines, lactams, imides, amino acids, hydrazones, or oximes.

a. 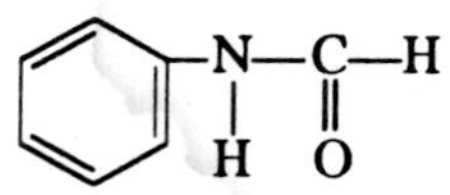

b. 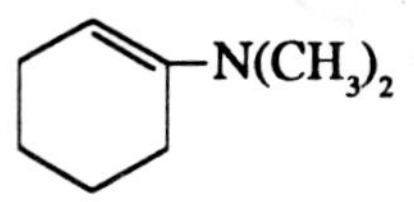

c. 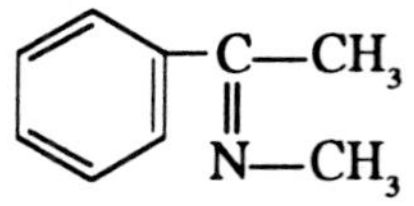

d. $H_2NCH_2CH_2COOH$

e. $(CH_3)_4N^+Cl^-$

f. 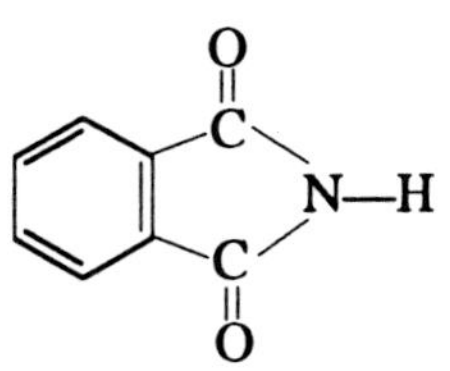

g. 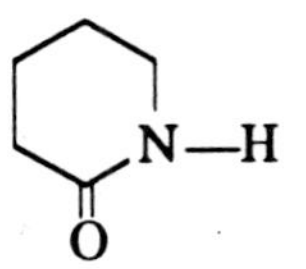

h. 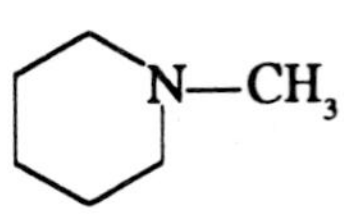

i. 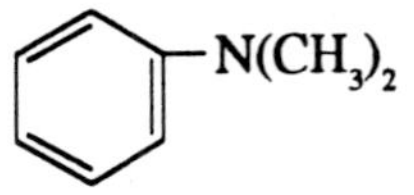

j. 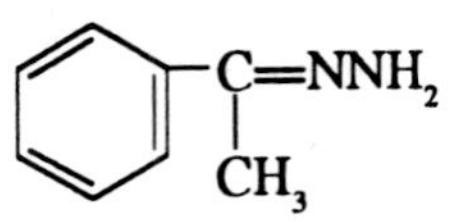

k. 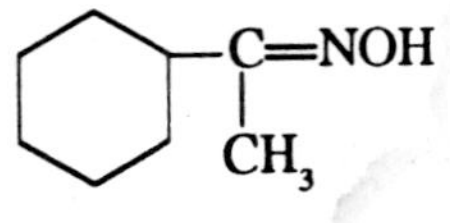

l. 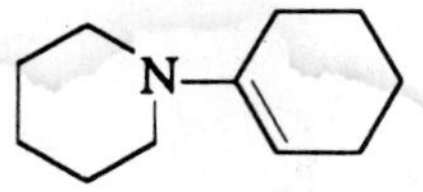

m. CH_3NH_2

n. $CH_3NHC_2H_5$

o. $(CH_3)_3C—N(CH_3)_2$

p. 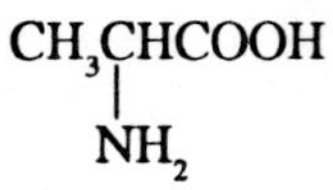

q. 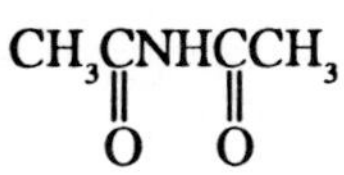

r. $(C_2H_5)_3\overset{+}{N}CH_3\ Cl^-$

s. 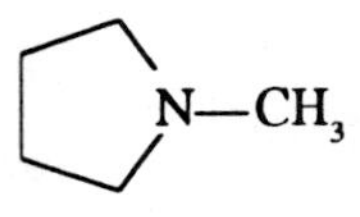

t. 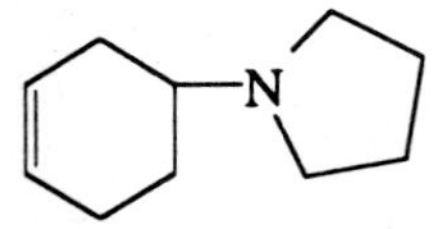

u. 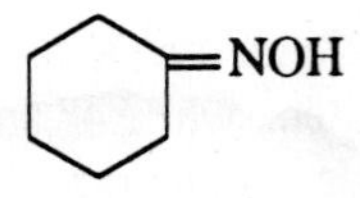

v. 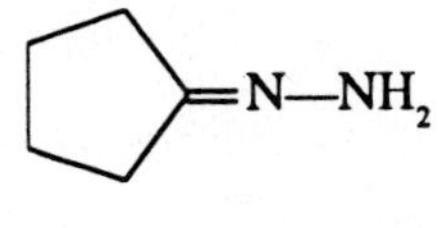

w. $(C_2H_5)_2NH$

x. 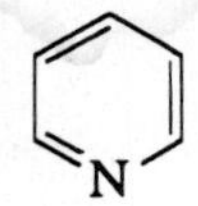

y. 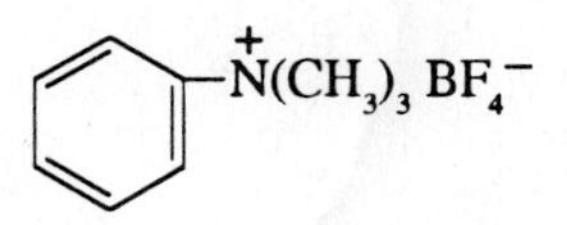

z. $(C_2H_5)_2C{=}NNH_2$

aa. 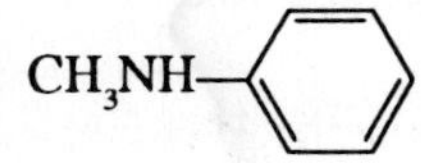

bb. 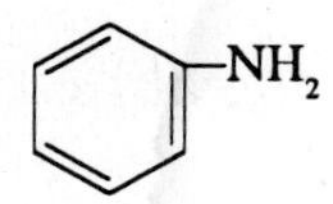

cc. 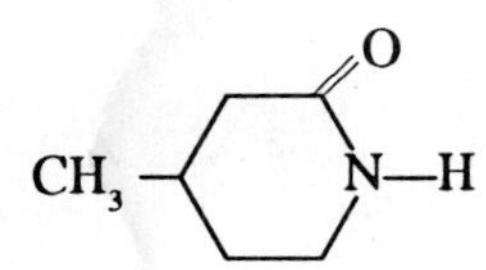

dd. 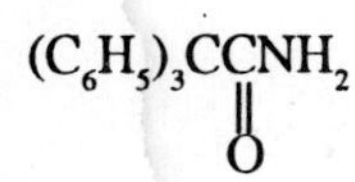

ee. 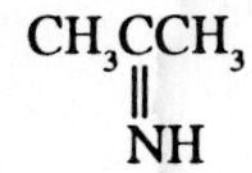

ff. 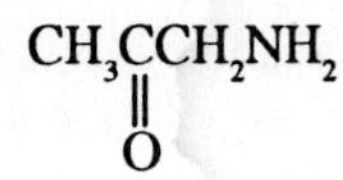

gg. 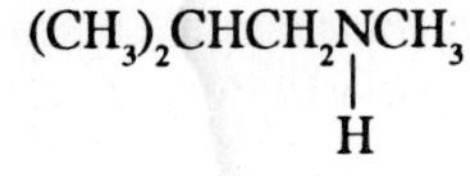

12.4 Identify the most basic nitrogen in the compounds below:

a. 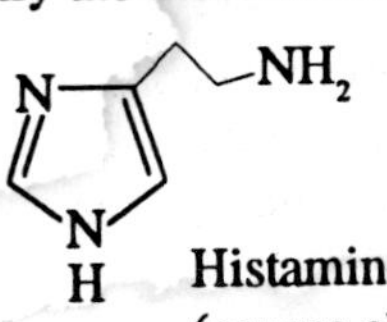

Histamine
(causes allergic reactions)

b. 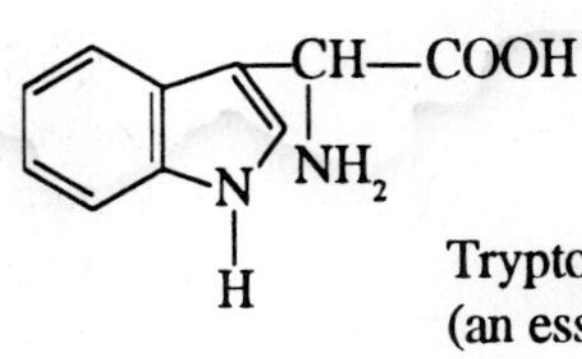

Tryptophan
(an essential amino acid)

c.

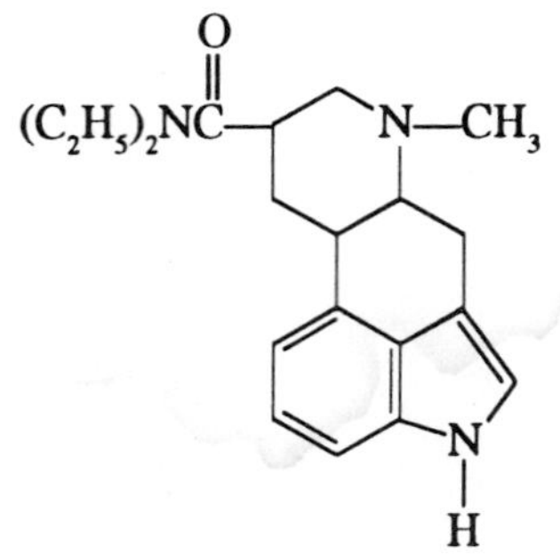

d. $H_2NCH_2CH_2CH_2CH_2CHCOOH$ (with NH_2 on the CH)

lysine
(an essential aminoacid)

e.

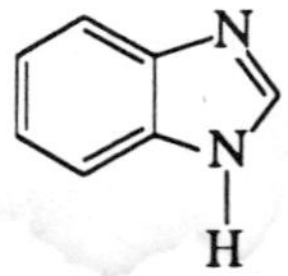

Benzimidazole

f.

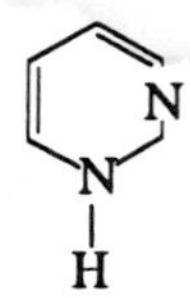

Imidazole

g. $H_2NCONHNH_2$
Semicarbazide

12.5 Select the more basic compound from each pair below. Justify your answer.

a. CH_3NH_2 and $(CH_3)_2NH$

b. $(C_2H_5)_2NH$ and $(C_2H_5)_3N$

c.

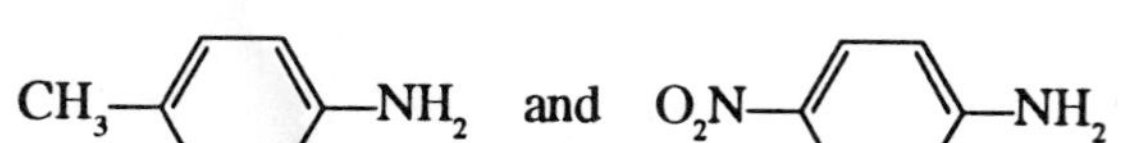

d. CH_3CHCH_2COOH (with NH_2 on C-3) and $CH_2CH_2CH_2COOH$ (with NH_2 on C-4)

e.

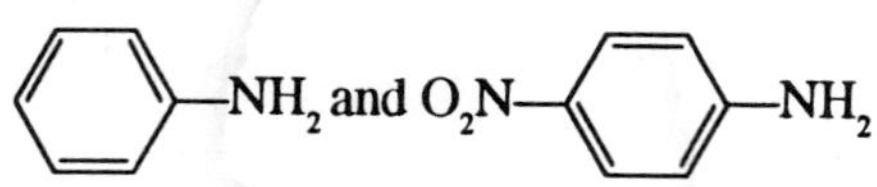

f.

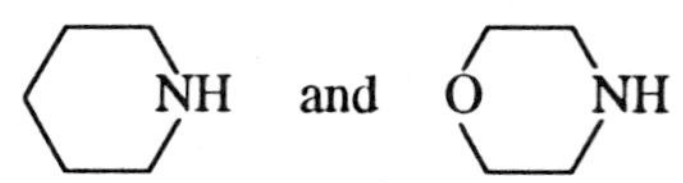

g.

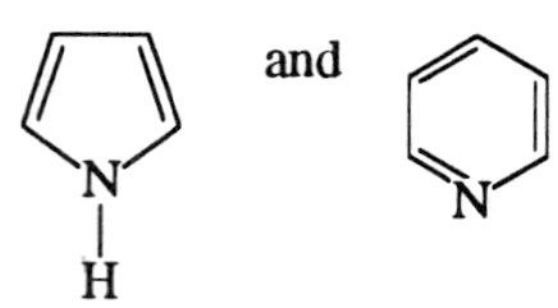

h.

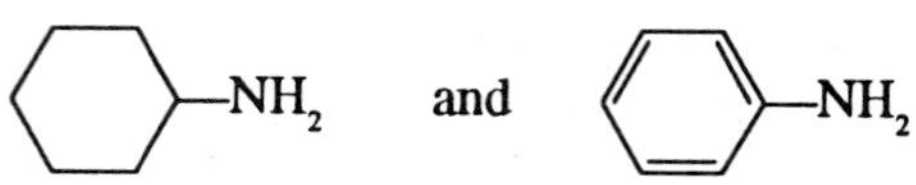

i.

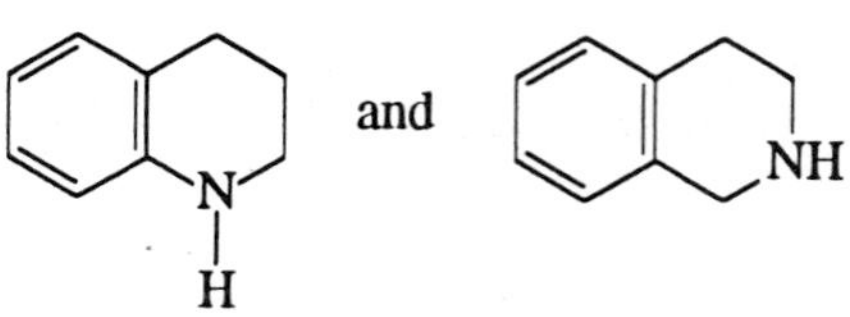

12.6 Explain why
a. when a primary amine reacts with acetic anhydride, the presence of an equimolar amount of pyridine leads to higher yields.
b. triethylamine is less soluble in water than ethylamine.
c. chiral amines cannot be isolated as enantiomers, while 4° ammonium salts can.
d. treatment of aniline with H_2SO_4/HNO_3 yields, among other products, all three mononitrated derivatives.

12.7 Suggest plausible mechanisms that would explain the formation of the indicated products. Use arrows to show the flow of electrons.

a.

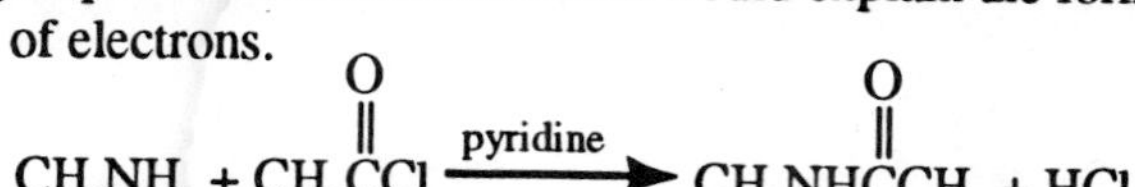

b.

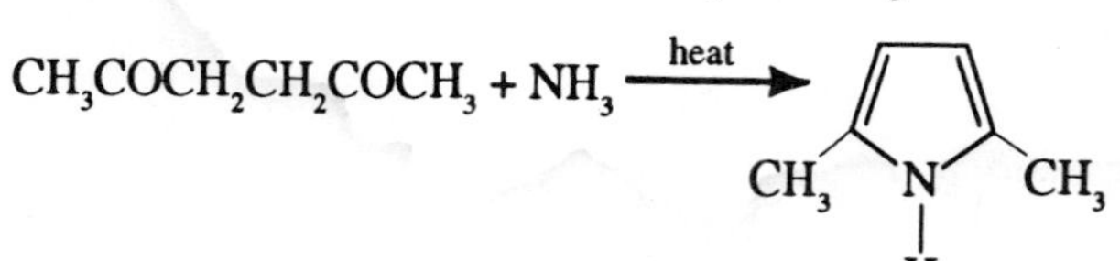

c. 

d. 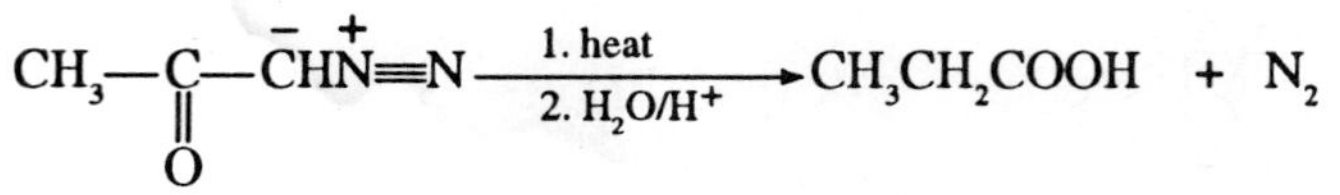

e. 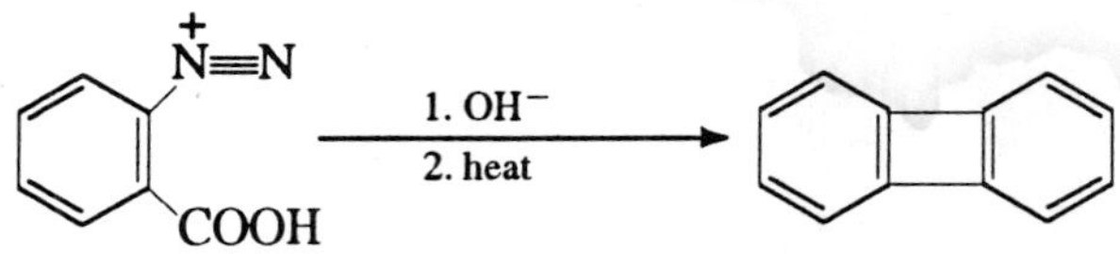

f. 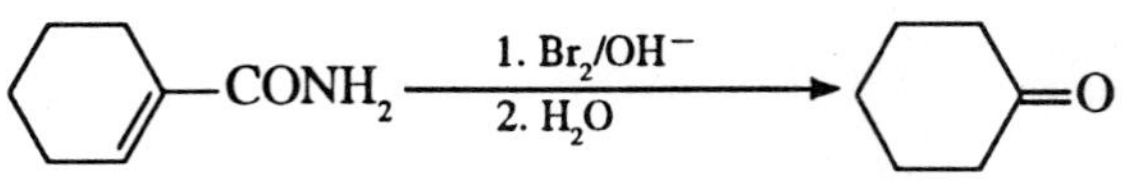

g. 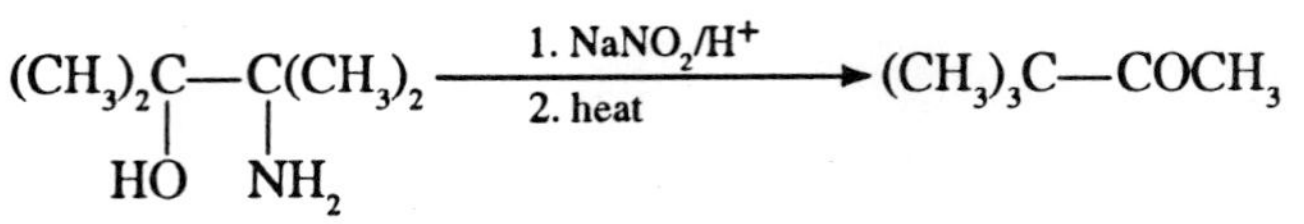

h. 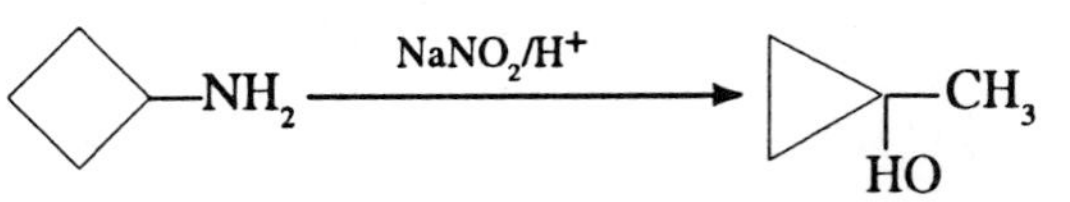

12.8 Fill in the missing reactants or products in the reactions below:

a. 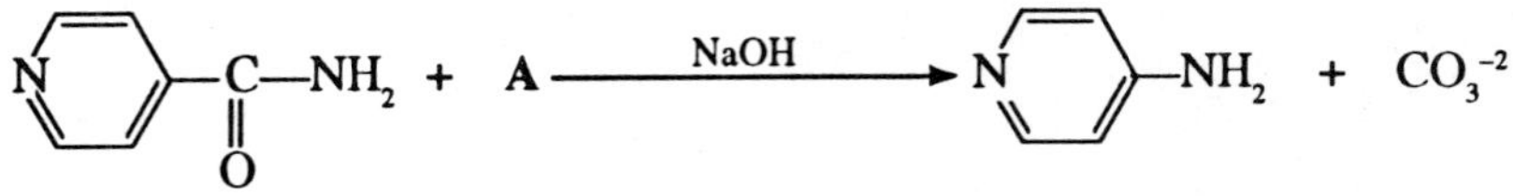

b. 

c. $C_6H_5\text{—}COCH_2COOC_2H_5 + C_6H_5NH_2 \xrightarrow{\text{heat}} \mathbf{C}\ (C_{15}H_{13}NO_2) + C_2H_5OH$

d. 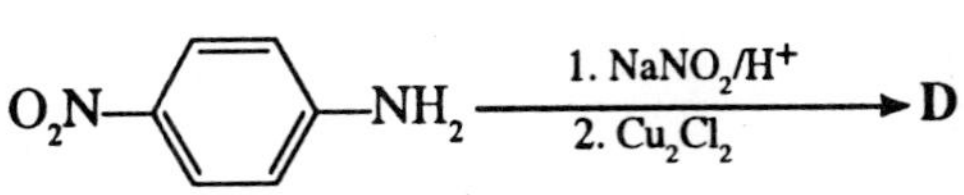

e. 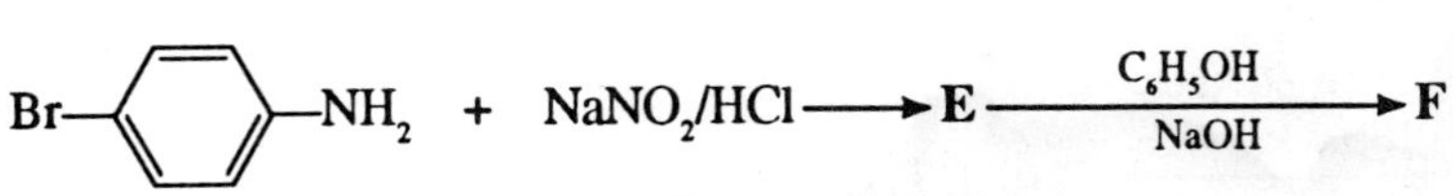

f. $H_2NCH_2COOH + (CH_3CO)_2O \longrightarrow \mathbf{G} + CH_3COOH$

g. 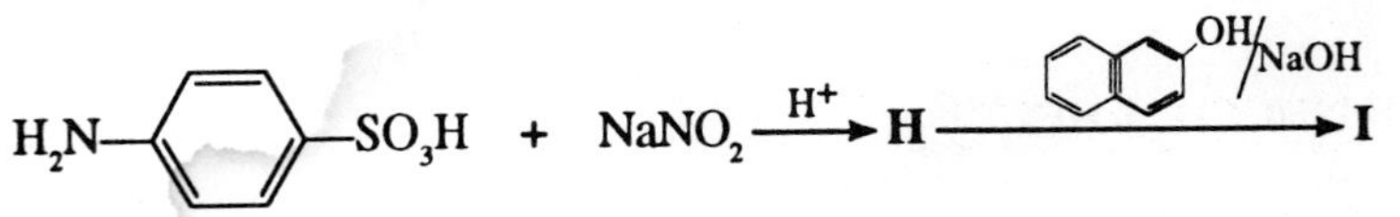

h. $HOCH_2CH_2NH_2 + HBr\ (\text{excess}) \longrightarrow \mathbf{J}\ (C_2H_7NBr_2)$

i. 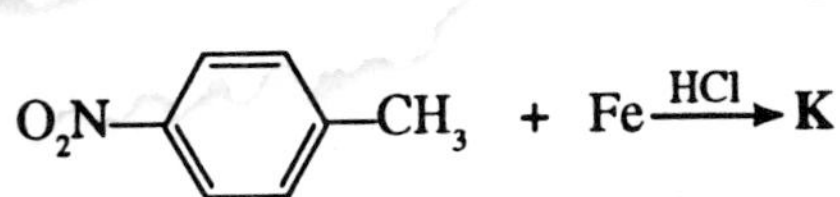

j. $[(CH_3)_2CH]_3N + NaNO_2 \xrightarrow{H^+} \mathbf{L}$

k. 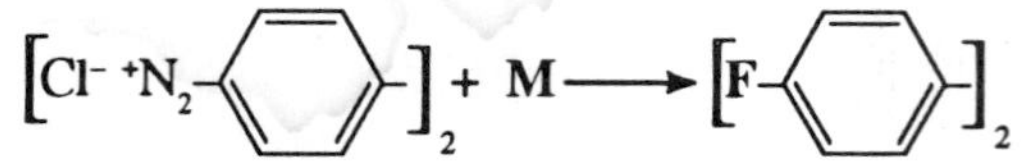

l. $(CH_3)_2NH{\cdot}HCl + NaNO_2 \xrightarrow[heat]{HCl} \mathbf{N}$

m. 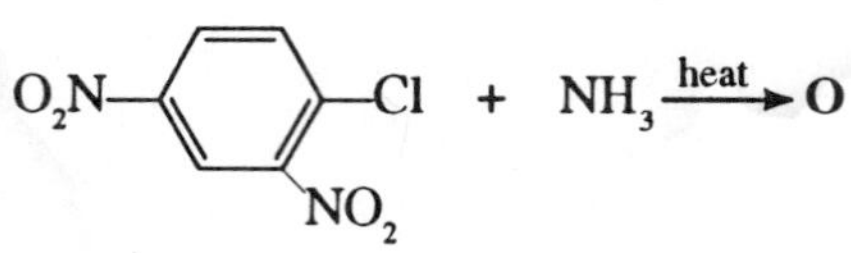

n. $ClCO_2C_2H_5 + \mathbf{P} \longrightarrow CH_3NHCO_2C_2H_5$

o. $C_6H_5N_2^+Cl^- + \mathbf{Q} \longrightarrow C_6H_5I$

p. $C_6H_5NHCH_3 + \mathbf{R} \longrightarrow C_6H_5N(NO)CH_3$

q. 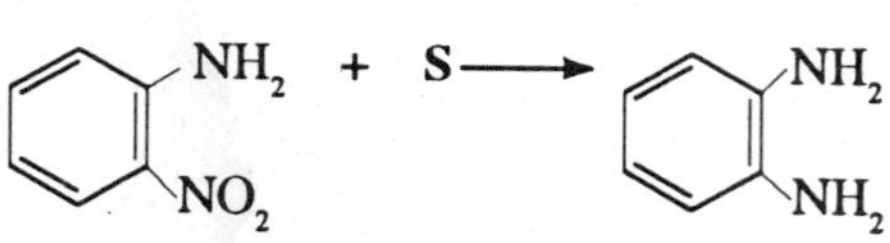

r. 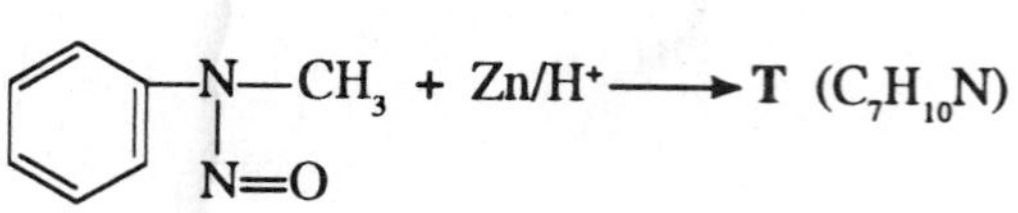

s. 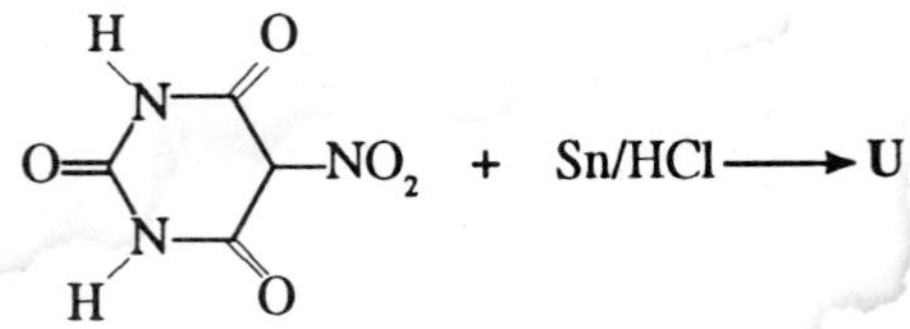

t. 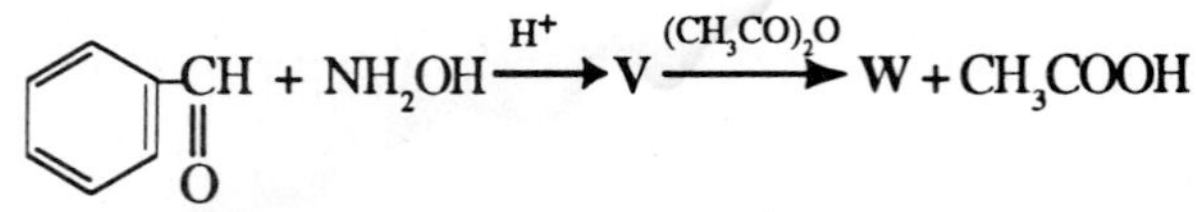

u. $C_6H_5N_2^+Cl^- + \mathbf{X} \longrightarrow C_6H_5C{\equiv}N$

v. $C_6H_5N(CH_3)_2 + HNO_2 \longrightarrow \mathbf{Y}$

w. $(n{-}C_4H_9)C(=O){-}N_3 \xrightarrow{heat} \mathbf{Z}$

x. 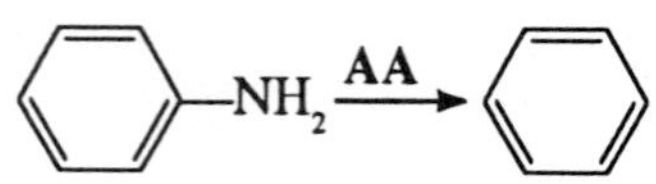

y. $\eta\text{-}C_4H_9NH_2 + NaNO_2/H^+ \longrightarrow \mathbf{BB} \xrightarrow[heat]{H_2O/} \mathbf{CC} + \mathbf{DD}$ (isomers of molecular formula $C_4H_{10}O$)

z. 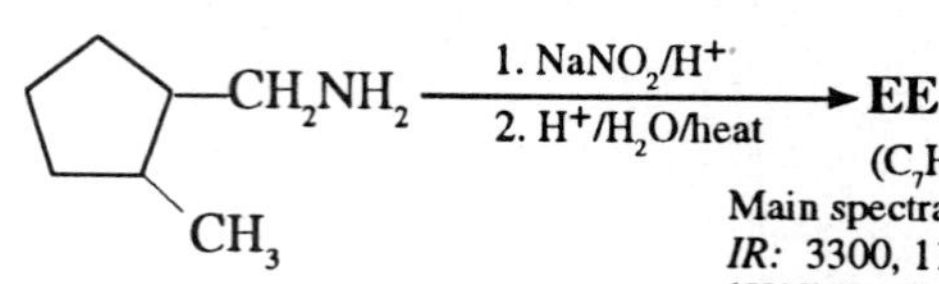

EE ($C_7H_{14}O$)
Main spectral features:
IR: 3300, 1100 cm^{-1}
^{1}HNMR: 3.5 ppm (1H, quartet)
1.2 ppm (3H, doublet)

12.9 Identify the compounds in the roadmap below using the spectra provided:

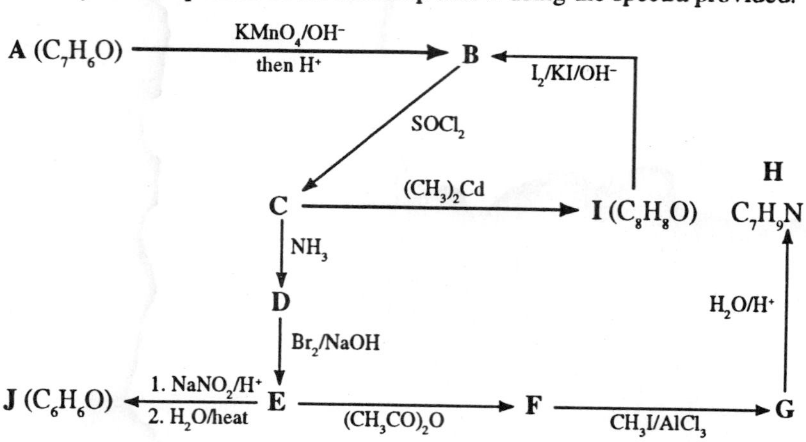

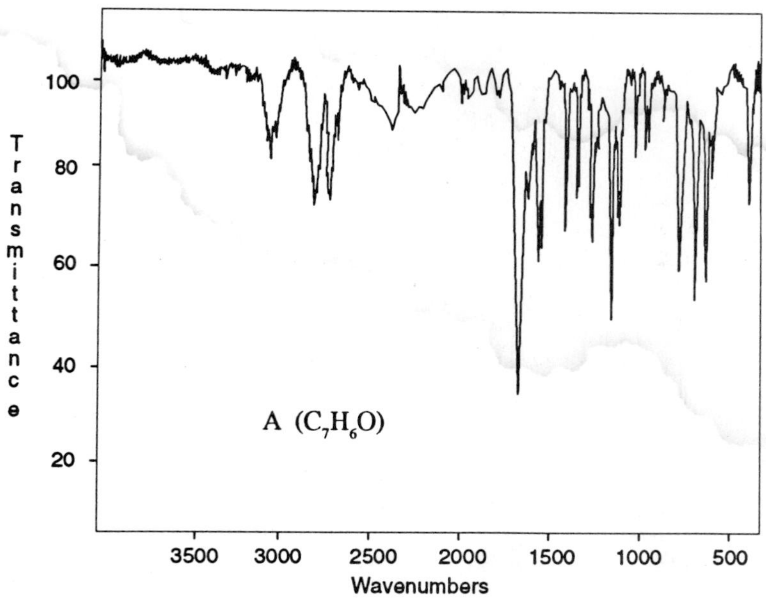

100
80
60
40
20
Transmittance
3500
3000
2500
2000
1500
1000
500
Wavenumbers
A (C_7H_6O)

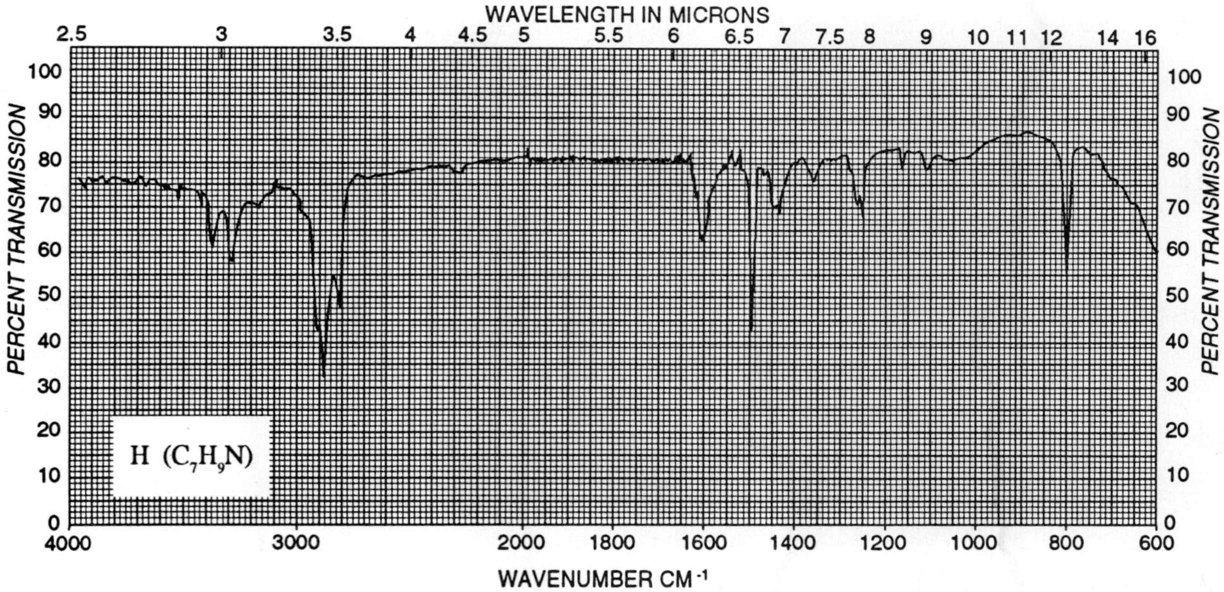

WAVELENGTH IN MICRONS
2.5
3
3.5
4
4.5
5
5.5
6
6.5
7
7.5
8
9
10
11
12
14
16
PERCENT TRANSMISSION
100
90
80
70
60
50
40
30
20
10
0
4000
3000
2000
1800
1600
1400
1200
1000
800
600
WAVENUMBER CM-1
H (C_7H_9N)

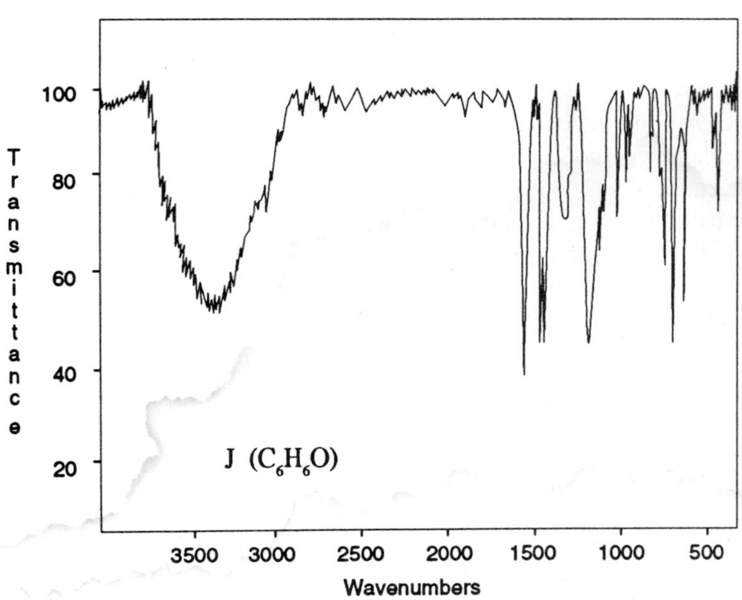

100
80
60
40
20
Transmittance
3500
3000
2500
2000
1500
1000
500
Wavenumbers
J (C_6H_6O)

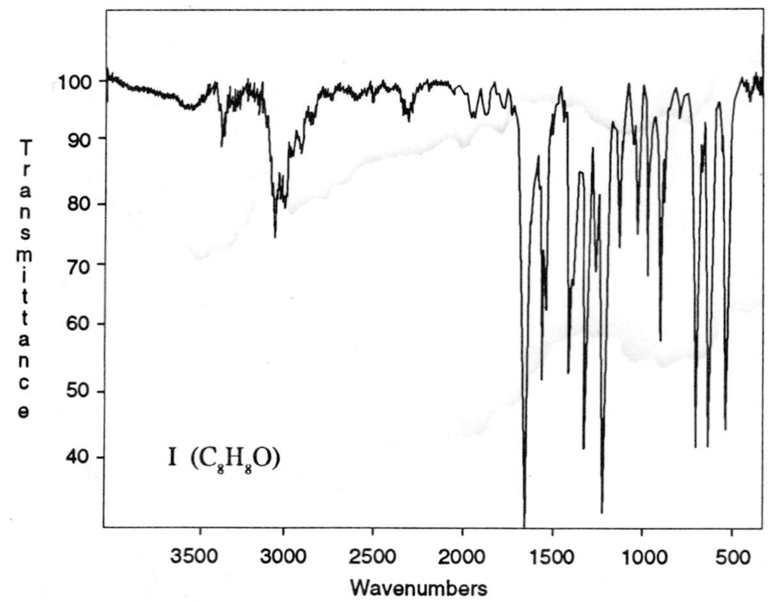

12.10 Suggest a synthesis of the indicated compounds from the given starting materials:

a. *m*-chlorotoluene from *m*-nitrotoluene
b. toluene from *p*-nitrotoluene
c. di-*n*-propylamine from 1-propanol
d. *p*-cyanotoluene from benzene
e. *p*-chlorobenzoic acid from nitrobenzene
f. *m*-bromoaniline from benzoic acid

12.11 Suggest chemical tests that would help distinguish between the following pairs of compounds. State what you do and what you see.

a. benzylamine and benzamide
b. cyclohexylamine and aniline
c. pyridine and benzene
d. N-ethylaniline and *p*-ethylaniline
e. diethylamine and triethylamine
f.

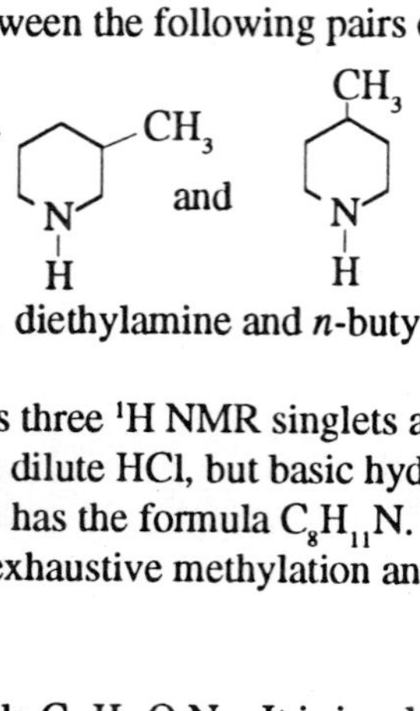

g. diethylamine and *n*-butylamine

12.12 Compound **A** with molecular formula $C_{10}H_{13}NO$ shows three ^{1}H NMR singlets at 3.4, 2.2, and 1.6 ppm (3H each), and a multiplet in the 7 ppm region (4H). **A** is insoluble in dilute HCl, but basic hydrolysis and subsequent acidification yields acetic acid and an amine salt, whose free base **B** has the formula $C_8H_{11}N$. High pressure hydrogenation yields **C** ($C_8H_{17}N$), which yields 3-methylcyclohexene upon exhaustive methylation and Ag_2O treatment. Identify **A**, **B**, and **C**.

12.13 Novocaine, a local anesthetic, has the molecular formula $C_{13}H_{20}O_2N_2$. It is insoluble in water and dilute NaOH, but is soluble in dilute HCl. Its IR spectrum shows two peaks in the 3500–3400 cm^{-1} region. When novocaine is boiled in aqueous NaOH, it slowly dissolves. This alkaline solution can be shaken with ether, and the layers separated. Acidification of the aqueous layer causes the precipitation of a white solid **A** ($C_7H_7O_2N$), but *excessive* acidification redissolves **A**. Evaporation of the ether layer leaves compound **B** ($C_6H_{15}ON$), which, when dissolved in water, turns red litmus paper blue. Treatment of **B** with acetic anhydride gives compound **C** ($C_8H_{17}O_2N$), which is soluble in dilute HCl, but insoluble in water and dilute base. **B** can be independently prepared by the action of diethylamine on ethylene oxide. What are the structures of **A**, **B**, and novocaine?

12.14 Identify the following compounds from the given spectral data:

	Molecular formula	Mass Spec	^{1}H NMR	IR
a.	C 65.7%, H 15.1%, N 19.2%	–	2.6 ppm (2H, t) 1.6 ppm (4H, m). 1.1 ppm (2H, s) 0.9 ppm (3H , t)	–
b.	$C_4H_{11}N$	–	2.8 ppm (1H, sextet) 1.8 ppm (2H, s) 1.5–0.8 ppm (8H, m)	–

c.	$C_4H_{11}N$	–	1.3 ppm (2H, s) 1.1 ppm (9H, s)	–
d.	$M^+ = 117$, peaks at 77, 91		7.2 ppm (5H) 3.5 ppm (2H, s)	3040, 2950, 2250, 1600, 1595, 785, 745 cm^{-1}
e.	$C_5H_7NO_2$	–	4.2 ppm (2H, q) 3.5 ppm (2H, s) 1.4 ppm (3H, t)	2980, 2260, 1740, 1380, 1200 cm^{-1}
f.	$C_{10}H_{13}NO_2$	–	9.5 ppm (1H, broad) 7.6–6.8 ppm (4H, doublet of doublets) 4.0 ppm (2H, q) 2.1 ppm (3H, s) 1.4 ppm (3H, t)	3450, 1650, 1610, 1500, 810 cm^{-1}

Multiple Choice Test

12.1 What is the correct sequence of the following compounds for decreasing boiling points?

CH_3CH_2COOH	$CH_3CH_2CH_2OH$	$CH_3CH_2NH_2$	$CH_3CH_2C{\equiv}CH$
A	**B**	**C**	**D**

a. A > B > C > D
b. A > C > B > D
c. B > A > C > D
d. A > B > D > C

12.2 Which is the weakest base?
a. N-methylaniline b. benzylamine c. aniline d. cyclohexylamine

12.3 Which of the following reactions does not yield an amine (assume that the medium is neutralized during the isolation step)?
a. $CH_3CH_2C{\equiv}N$ + 2 H_2/Pt
b. $C_6H_5NO_2$ + H_2O/H^+
c. $CH_3CH_2C{=}NH$ + H_2/Pt
d. $CH_3CONHCH_2CH_3$ + H_2O/H^+

12.4 In the synthesis of aniline from nitrobenzene, aniline can be separated from unreacted nitrobenzene by using an extraction procedure that involves ether and ______.
a. water b. aqueous HCl c. aqueous NaOH d. aqueous $NaHCO_3$

12.5 There are ____ amines of molecular formula $C_4H_{11}N$.
a. 5 b. 6 c. 7 d. 8

12.6 The number of 2° amines of molecular formula $C_4H_{11}N$ is ______.
a. 1 b. 2 c. 3 d. 4

12.7 The pH of a tetramethylammonium chloride solution
a. is less than 7 b. is exactly 7 c. is higher than 7

12.8 The reaction of butanamide with Br_2/NaOH yields ______.
a. 1-butanol b. *n*-butylamine c. 2-bromobutanamide d. 1-aminopropane

12.9 Which of the following compounds *can* be prepared via the Gabriel synthesis?

n-butylamine	aniline	N-methylpropylamine	trimethylamine
A	**B**	**C**	**D**

a. A only b. A and B c. all but C d. all of the above

12.10 Which compound would not produce an amine with $LiAlH_4$?
a. $C_2H_5CH{=}NOH$ b. CH_3CONH_2 c. $C_2H_5C{\equiv}N$ d. $C_6H_5NO_2$

12.11 Reaction of N,N-dimethylaniline with $NaNO_2$/HCl yields
a. benzene diazonium chloride
b. N-methyl-N-nitrosoaniline
c. *p*-nitroso-N,N-dimethylaniline
d. phenol

12.12 Reaction of compound A with $NaNO_2/H^+$ followed by water yields *p*-methylphenol. A is therefore _____.
a. N-methlaniline
b. *p*-methylaniline
c. (*p*-methylphenyl)methylamine
d. N,N-dimethylaniline

12.13 Which procedure removes the amino group from aniline?
a. diazotization and acidification with dilute H_2SO_4
b. reduction with $LiAlH_4$ or H_2/Ni
c. reduction with Sn/HCl
d. diazotization and further reaction with H_3PO_2

12.14 Which compound will not react with $C_6H_5N{\equiv}N^+Cl^-$ to yield an azo dye?
a. phenol b. benzene c. N,N-dimethylaniline d. 2-naphthol

12.15 Which compound reacts with benzenesulfonyl chloride to give a product that is insoluble in basic *and* acidic medium?
a. N-methylaniline b. pyridine c. aniline d. N,N-dimethylaniline

12.16 Which is the correct sequence of reagents required to prepare *p*-toluidine from benzene?
a. $CH_3I/AlCl_3$; HNO_3/H^+; Sn/HCl
b. HNO_3/H^+; $CH_3I/AlCl_3$; Sn/HCl
c. $Br_2/FeBr_3$; $CH_3I/AlCl_3$; NH_3
d. $CH_3I/AlCl_3$; NH_3; $Br_2/FeBr_3$

12.17 Which reagent would react with a diazonium salt, but will not liberate nitrogen gas?
a. HBF_4 b. CuCN c. KI d. $C_6H_5O^-Na^+$

12.18 Which method would produce N-methylbenzylamine?
a. $C_6H_5CH_2Cl$ + NH_2OH/HCl, followed by H_2/Pt
b. C_6H_5CHO + CH_3NH_2, followed by H_2/Ni
c. $C_6H_5COCH_3$ + NH_3/H^+, followed by H_2/Ni
d. C_6H_6 + $(CH_3CO)_2O/H^+$, followed by NH_2OH/HCl

12.19 Exhaustive methylation and Ag_2O treatment of methyl (2-methylcyclohexyl) amine yields _____-methylcyclohexene.
a. 1 b. 2 c. 3 d. 4

12.20 A compound ($C_8H_{11}N$) gives the following ^{1}H NMR spectrum: 1.3 ppm (3H , d); 1.7 ppm (2H, s); 4.1 (1H, q); 7.3 (5H, m). What is the compound?
a. benzylmethylamine b. 2-phenylaminoethane c. N-ethylaniline d. 1-phenylaminoethane

Answers for Questions

12.1
a. dimethylamine; 2°
b. *sec*-butylamine or 2-aminobutane; 1°
c. aminoacetic acid or glycine; 1°
d. tetraethylammonium tetrafluoroborate; 4°
e. 3-butenylamine; 1°
f. 2-phenylethanamine or β-phenylethylamine; 1°
g. (cyclohexylmethyl) dimethylamine; 3°
h. 1-(N,N-dimethylamino) cyclohexene; 3°
i. 1,2-diaminobenzene or o–phenylenediamine; 1°
j. *trans*-(2-methylcyclohexyl) amine; 1°
k. *o*-chloro-N-methylaniline; 2°
l. *p*-nitrosoaniline; 1°
m. *m*-isopropylaniline; 1°
n. 2-nitropyridine; 3°
o. benzylphenylvinylamine; 3°
p. *t*-butylamine; 1°
q. ethylmethyl-n-propyl amine; 3°
r. 1,5-diaminopentane, or 1,5-pentanediamine, or pentamethylenediamine; 1°

12.2 a. $(C_2H_5)_2NH$, 2°

b. 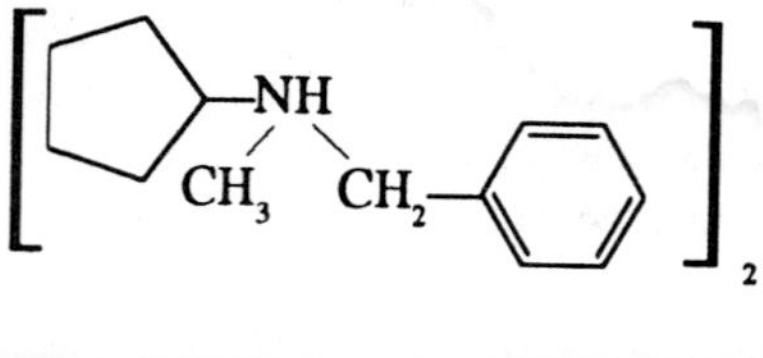

SO_4, 4°

c. 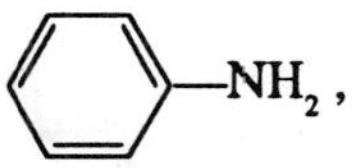 , 1°

d. 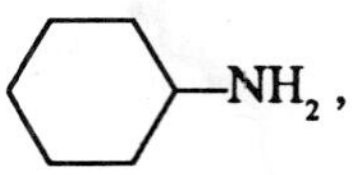 , 1°

e. $(C_6H_5)_2NH$, 2°

f. 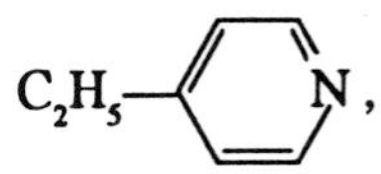 , 3°

g. $(CH_3)_2CHCH_2NH_2$, 1°

h. $NH_2(CH_2)_4NH_2$, 1°

i. 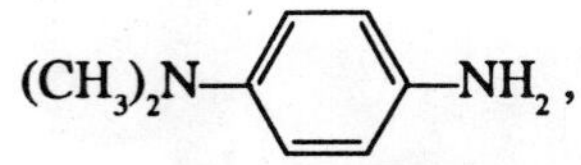 , 1°

j. 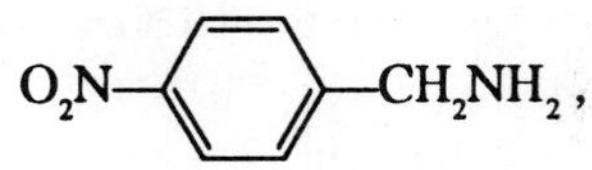 , 1°

k. 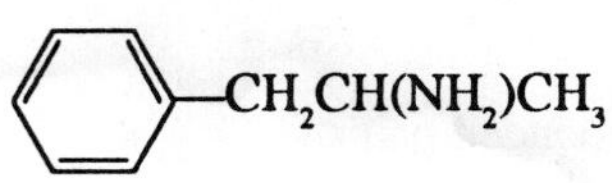 , 1°

l. $CH_3CH_2CH(NH_2)CH_2CH_2CH_3$, 1°

m. 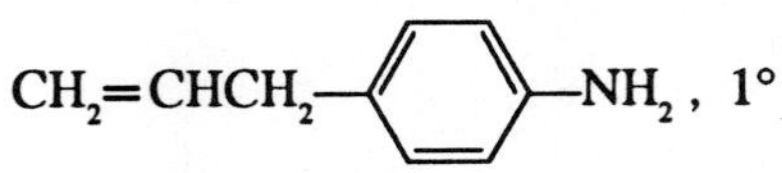 , 1°

n. $[(CH_3)_2N(C_2H_5)(C_6H_5)]^+ClO_4^-$, 4°

o. $(CH_3CH_2CH_2)_3CNH_2$, 1°

p. 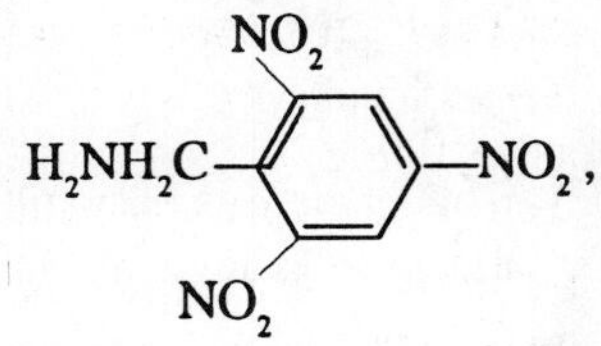 , 1°

q. 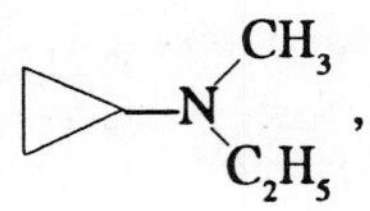 , 3°

r. 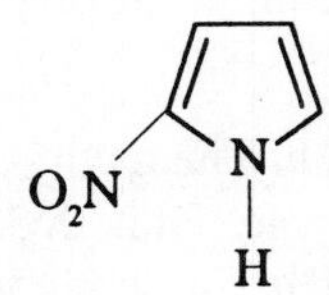 , 2°

12.3 *1° Amines:* m, bb, ff; *2° Amines:* n, w, aa, gg; *3° Amines:* h, i, o, s, t, x; *4° Ammonium salts:* e, r, y; *Amides:* a, dd; *Amino acids:* d, p; *Imines:* c, ee; *Enamines:* b, l; *Imides:* f, q; *Lactams:* g, cc; *Hydrazones:* j, v, z; *Oximes:* k, u.

Note: Structures (b) and (l) are (and react like) *enamines*; they are also 3° amines. Structures (i) and (x) are 3° amines that technically can be classified as enamines, although they do not react as enamines.

12.4 a.

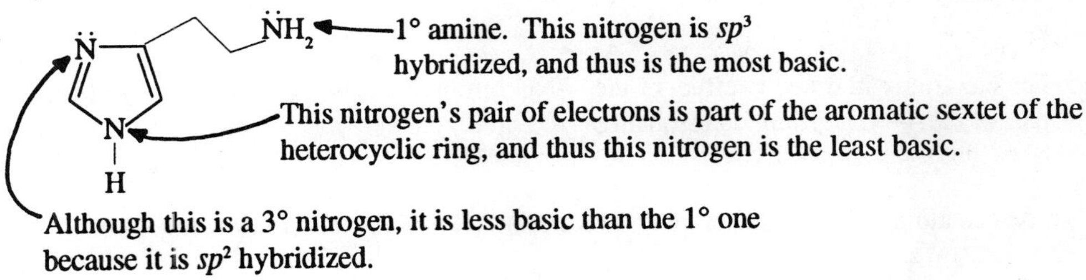

b.

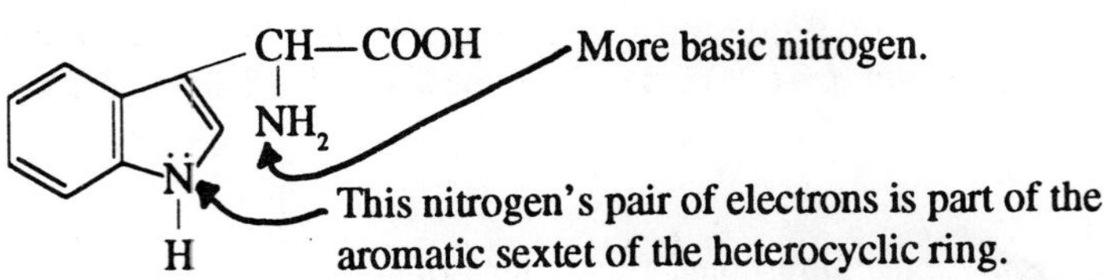

c.

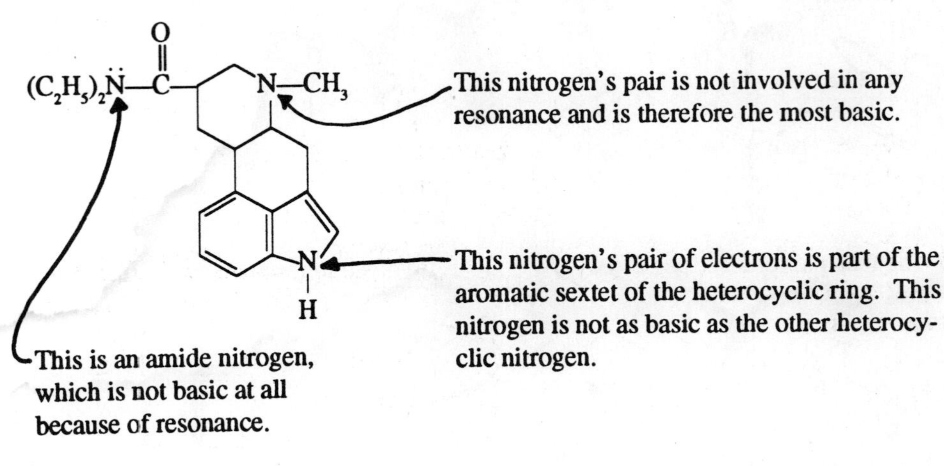

d. $H_2NCH_2CH_2CH_2CH_2CH—COOH$ (with NH_2 on the CH)

Most basic nitrogen: the terminal H_2N. The NH_2 on the CH: This nitrogen is less basic because of the electron-withdrawing effect of the —COOH group.

e. 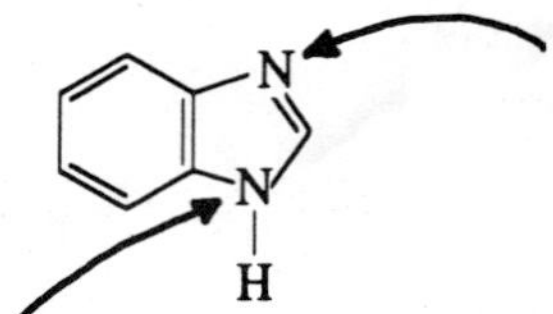

This is the more basic nitrogen. The non-bonded electron pair is not involved in the aromatic sextet.

This is the less basic nitrogen. The nonbonded pair of electrons is involved in the aromatic sextet.

f. Same as for part (e).

g. $H_2N—C(=O)—NH—NH_2$ — terminal NH_2: Amine nitrogen; most basic.

Both of these nitrogens are amide nitrogens and are not basic, since their non-bonded electrons are involved in resonance with the carbonyl.

12.5 a. $(CH_3)_2NH$: The methyl groups' electron-donating effect (via induction)intensifies the basicity of the nonbonded electron pair. Dimethylamine has two methyl groups, while methylamine has one.

b. $(C_2H_5)_2NH$: Steric hindrance in 3° amines reduces the availability of nitrogen's nonbonded electron pair.

c. 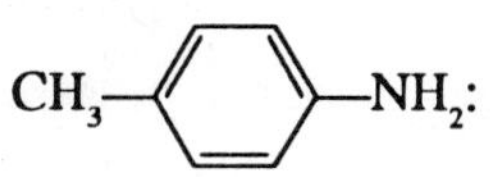: The electron-donating effect of the methyl group increases the availability of the nitrogen's electron pair.

d. $H_2NCH_2CH_2CH_2COOH$: The electron-withdrawing —COOH group is further away from the amino group than it is on the other compound.

e. 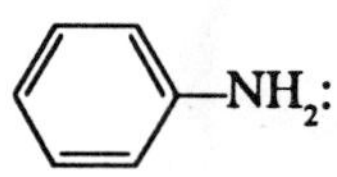: The electron-withdrawing effect of the $—NO_2$ group decreases the availability of the electron pair of the $—NH_2$ group via resonance.

f. 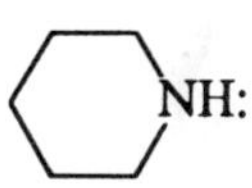: The oxygen atom of the other compound (morpholine) is electron-withdrawing.

g. 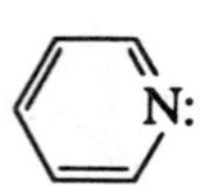: The electron pair on the other compound (pyrrole) is involved in ring aromaticity.

h. 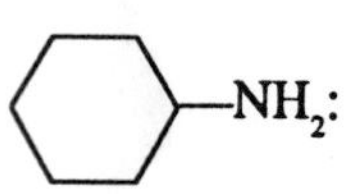: Aniline's electron pair is involved in resonance with the aromatic ring (unlike that of cyclohexylamine).

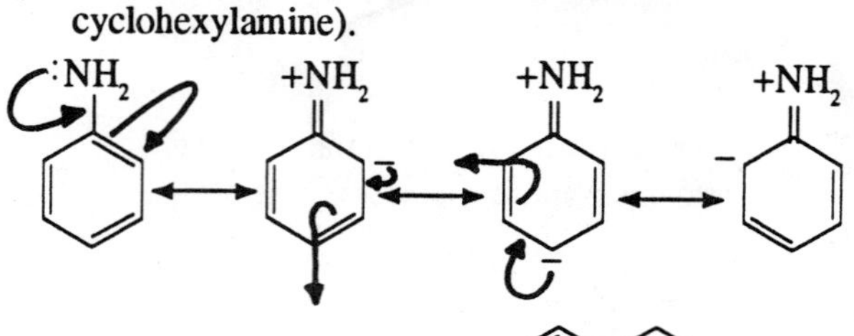

i. 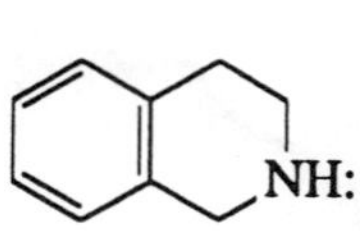: The lone pair of electrons in 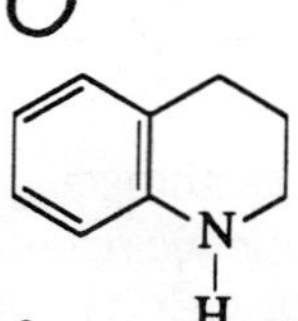 is involved in resonance with the aromatic ring.

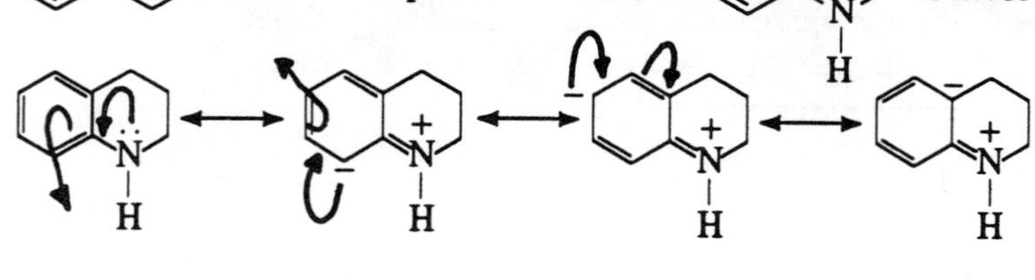

12.6 a. The reaction is: $R{-}\ddot{N}H_2 + CH_3{-}\underset{\|}{C}(=O){-}O{-}C(=O){-}CH_3 \longrightarrow R{-}N(H){-}C(=O)CH_3 + CH_3COOH$

(1° amine) (acetic anhydride) (amide) (acetic acid)

The weak acid formed (acetic acid) will react with the amine in an acid-base reaction:

$R{-}NH_2 + CH_3COOH \longrightarrow RNH_3^{+}CH_3COO^{-}$

(base) (acid) (salt)

The added pyridine acts as a scavenger and traps the acetic acid that is formed. As a result, the amine is free to react with the acetic anhydride, thus increasing the yield:

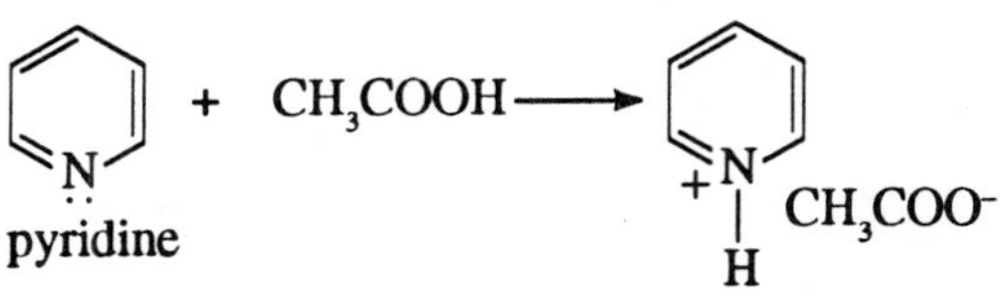

b. Hydrogen bonding is responsible for the solubility of several organic compounds in water. 1° and 2° amines can form such bonds. 3° amines cannot bond as effectively.

c. The barrier of inversion between the two pyramidal (*tri*valent) forms of the two enentiomeric amines is very low to the extent that it makes their isolation extremely difficult. This is not the case with quaternary ammonium salts, in which a *tetra*valent nitrogen is involved.

d. The amino, —NH$_2$, group is *o-/p*-directing. However, amines react with acids to give salts.

$C_6H_5NH_2 + H^+ \longrightarrow C_6H_5\overset{+}{N}H_3$

The protonated amino group, —$\underset{+}{N}H_3$, is deactivating to the aromatic ring by induction; thus it is *m*-directing.

12.7 a.

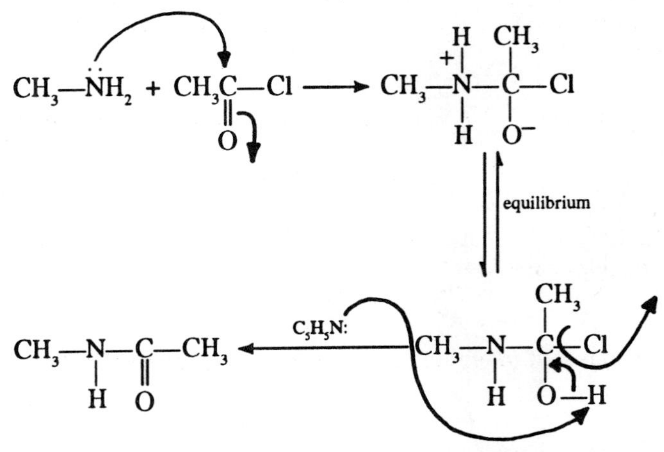

b. This is a condensation reaction between the basic nitrogen of NH$_3$ and the two carbonyl groups:

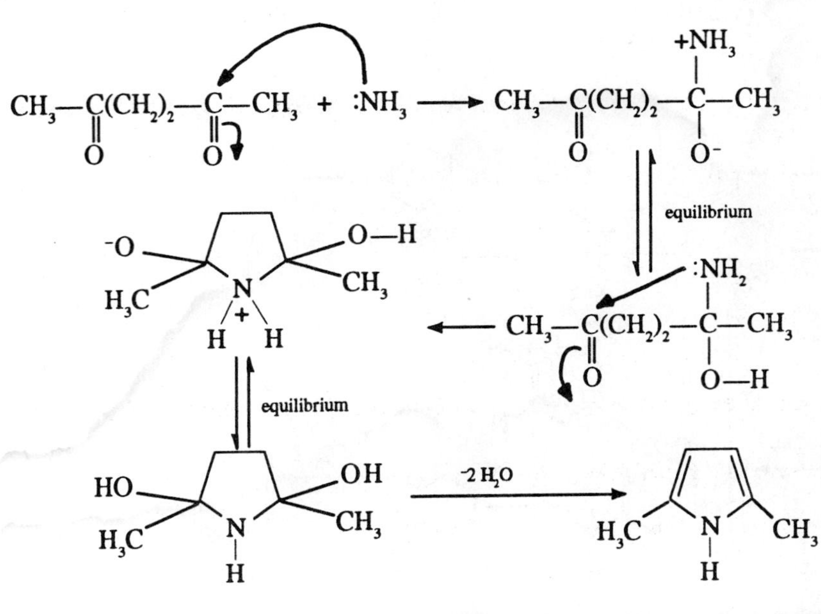

c. This is similar to the condensation of part (b), but this time it is acid-catalyzed.

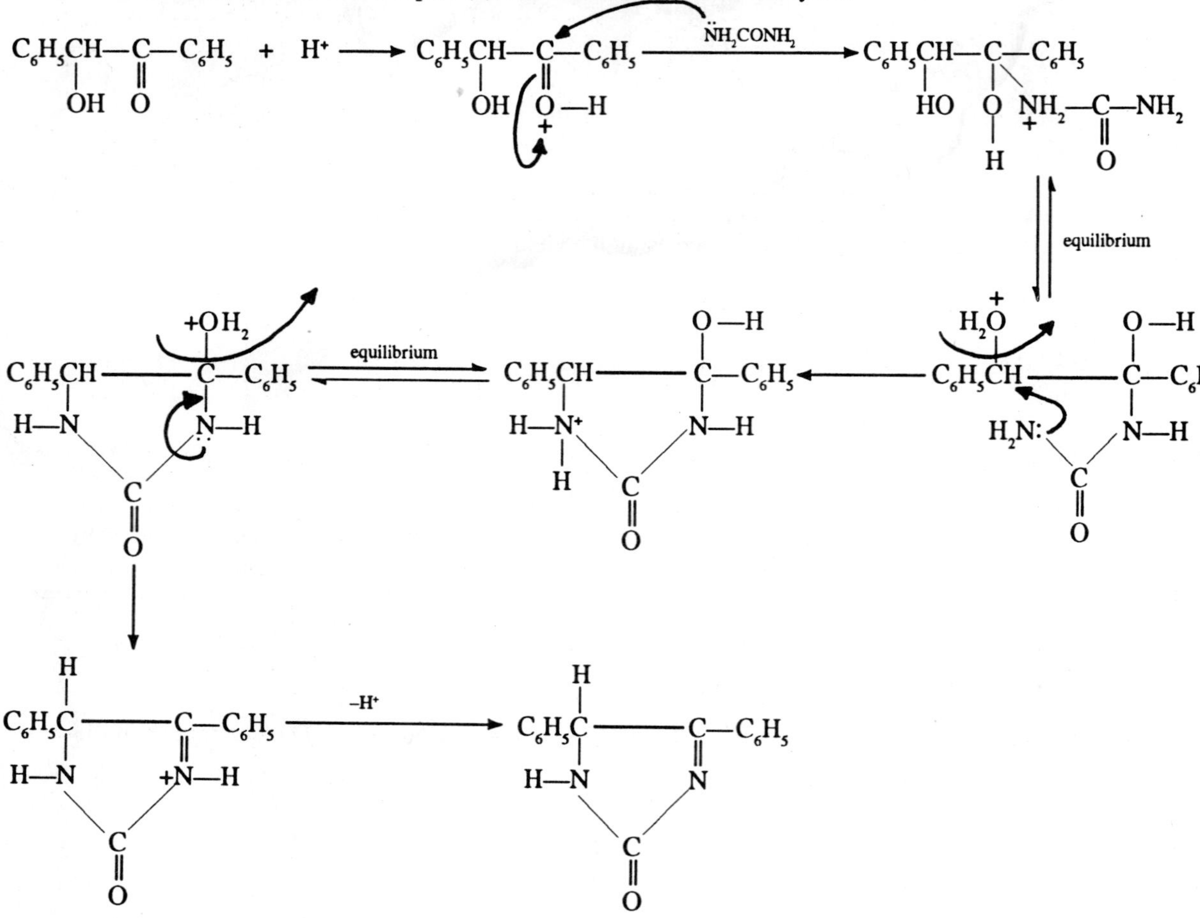

d. This reaction is similar to the Hoffman degradation of amides. The leaving group in this case is N_2 instead of Br^-.

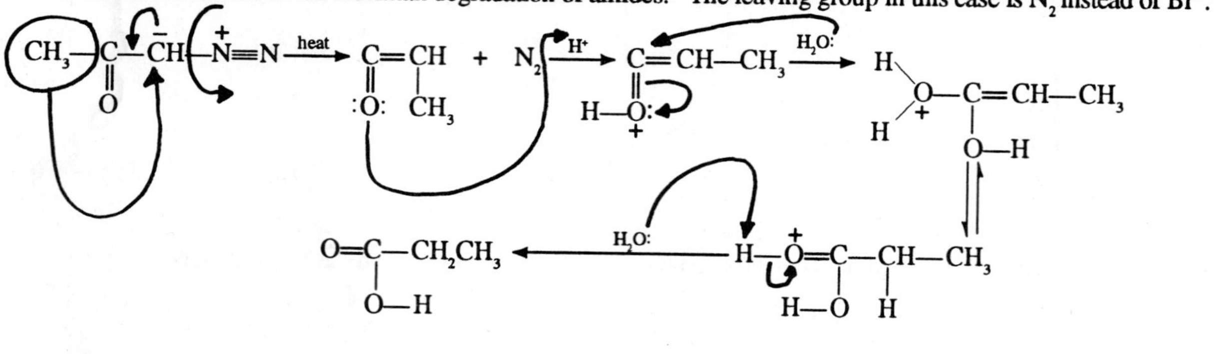

e.

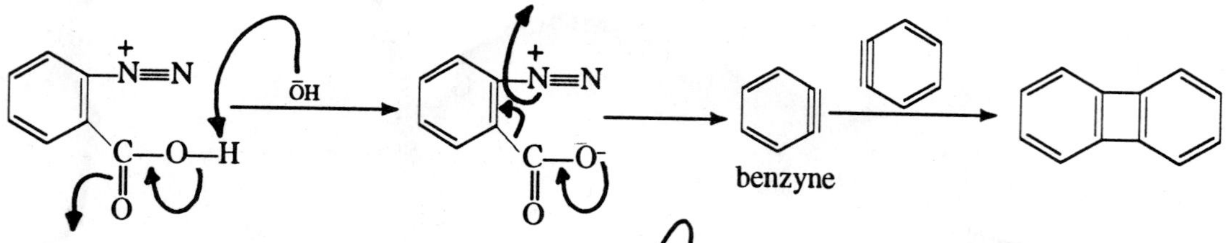

f.

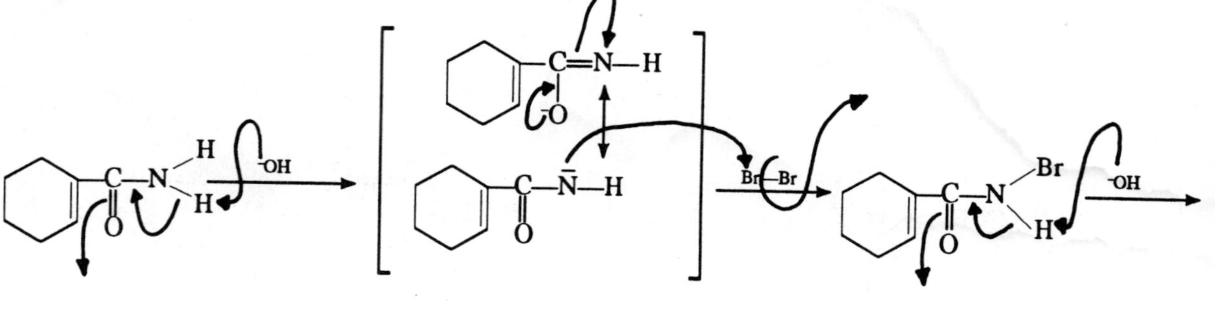

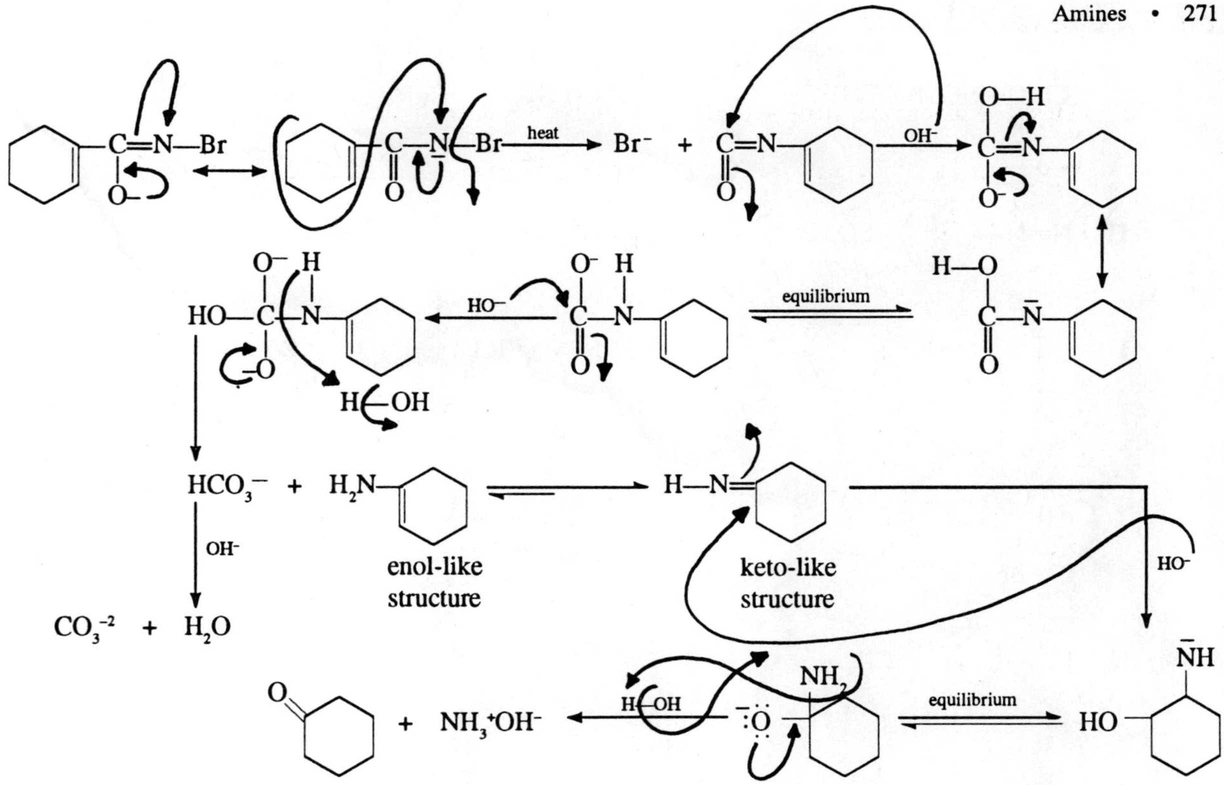

g.

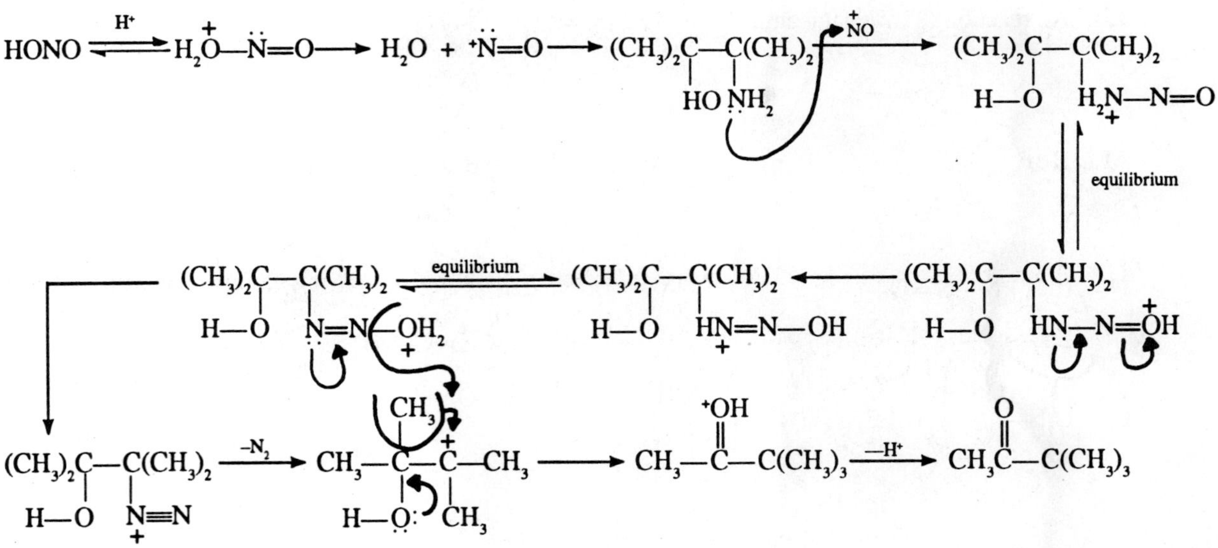

h.

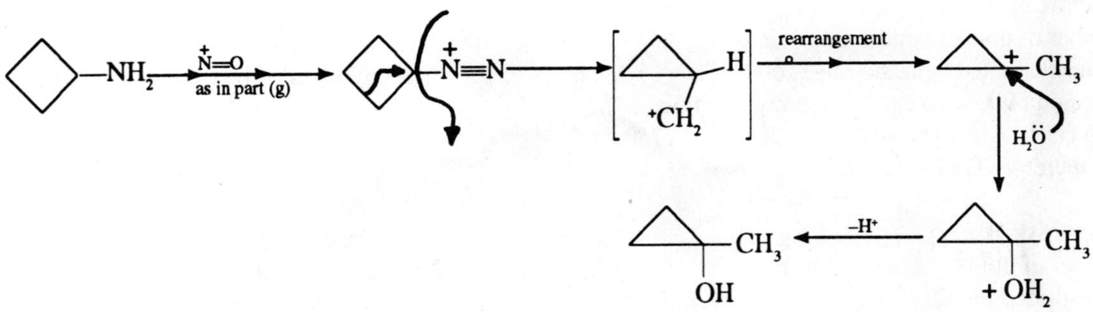

12.8 A Br_2 C $C_6H_5COCH_2CONHC_6H_5$ E

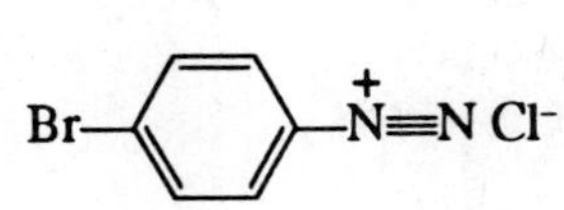

B $(C_6H_5)_3COH$ D

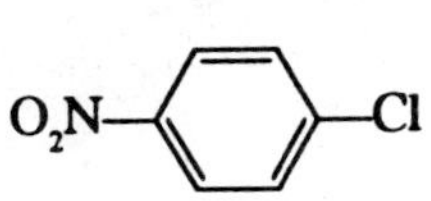

F

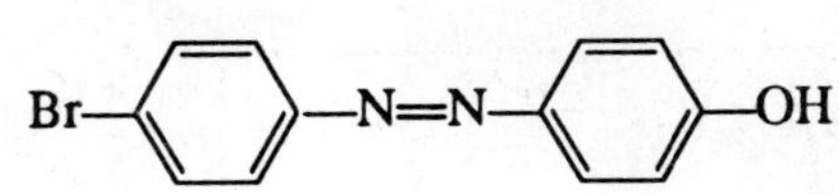

G $CH_3C(=O)$—NHCH$_2$COOH

H 

I 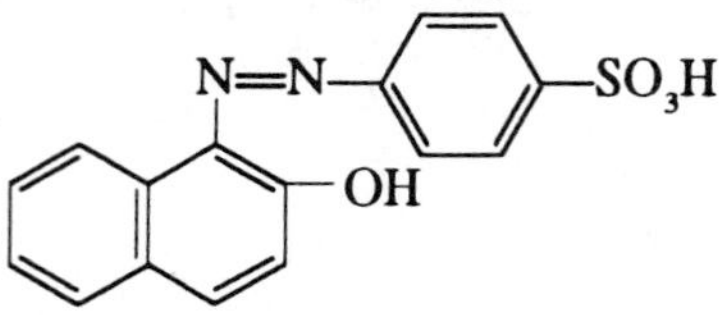

J $CH_3CH(Br)$—$NH_3^+Br^-$

K 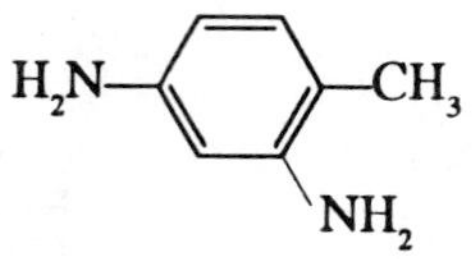

L No reaction (3° aliphatic amine)

M HBF_4

N $(CH_3)_2N$—N=O

O 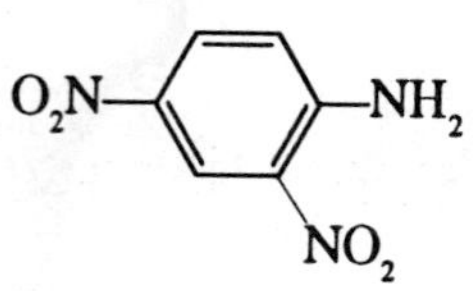

P CH_3NH_2

Q KI

R $NaNO_2/HCl$

S Fe/HCl; or Sn/HCl; or H_2, Pd/C

T 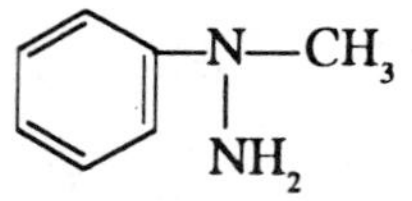

U 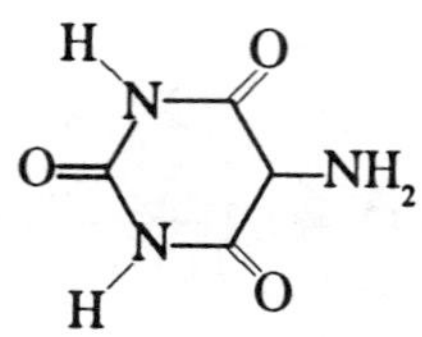

V 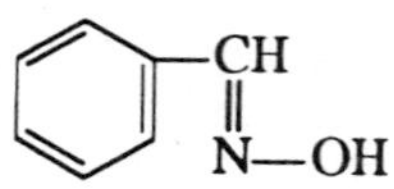

W 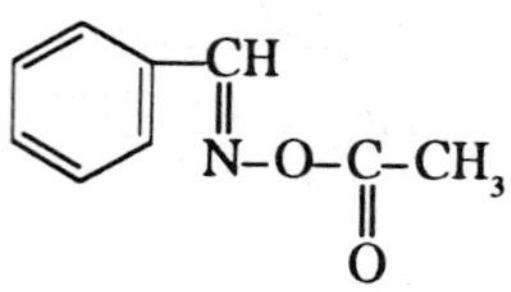

X CuCN

Y 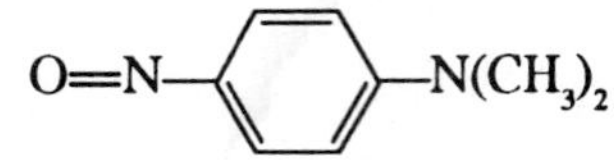

Z (n—C_4H_9)—N=C=O + N_2

AA 1. $NaNO_2/HCl$
2. H_3PO_2

BB (n—C_4H_9)—$\overset{+}{N}\equiv N$

CC n—C_4H_9OH

DD $CH_3CH_2CH(OH)CH_3$

EE 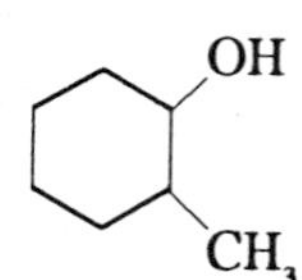

12.9 Structure of **A**:
Number of double bonds and/or rings: 5.
Monosubstituted benzene ring: 1600, 1580, 1470 cm^{-1} and region 900–650 cm^{-1}. Carbonyl (conjugated): 1695 cm^{-1}. Aldehyde —C—H: 2830 and 2750 cm^{-1}.
A is therefore C_6H_5—CHO.

Structure of **H**:
Number of double bonds and/or rings: 4.
p-substituted benzene ring: 1610, 1520 cm^{-1} and region 900–650 cm^{-1}. Amino (—NH_2) group: doublet at 3410 and 3320 cm^{-1}. Methyl group: 1435 cm^{-1}.
H is therefore *p*-CH_3—C_6H_4—NH_2.

Structure of **J**:
Number of double bonds and/or rings: 4.
Monosubstituted benzene ring: 1610, 1490 cm^{-1} and region 90–650 cm^{-1}. Hydroxyl peak: 3250 cm^{-1}.
J is therefore C_6H_5—OH.

Structure of **I**:
Number of double bonds and/or rings: 5.
Monosubstituted benzene ring: 1600, 1470 cm^{-1} and region 900–650 cm^{-1}). Conjugated carbonyl: 1700 cm^{-1}. Methyl group: 1435 cm^{-1}.
I is therefore C_6H_5—C(=O)—CH_3.

Structures of compounds **A** through **J**:

A $C_6H_5{-}CHO$

B $C_6H_5{-}COOH$

C $C_6H_5{-}COCl$

D $C_6H_5{-}CONH_2$

E $C_6H_5{-}NH_2$

F $C_6H_5NH{-}\overset{O}{\overset{\|}{C}}{-}CH_3$

G 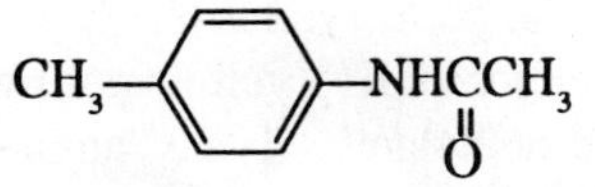

H 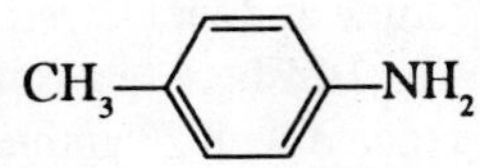

I $CH_3{-}\underset{\underset{O}{\|}}{C}{-}C_6H_5$

J $C_6H_5{-}OH$

12.10 a. 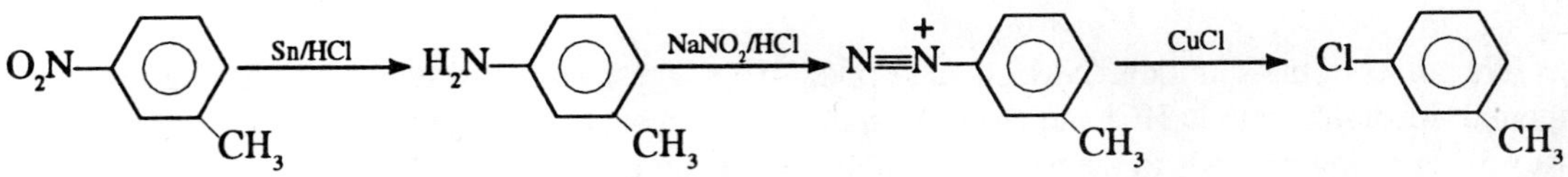

b. 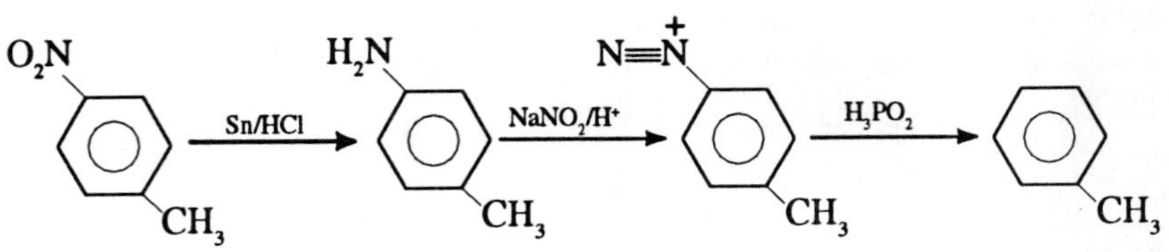

c. 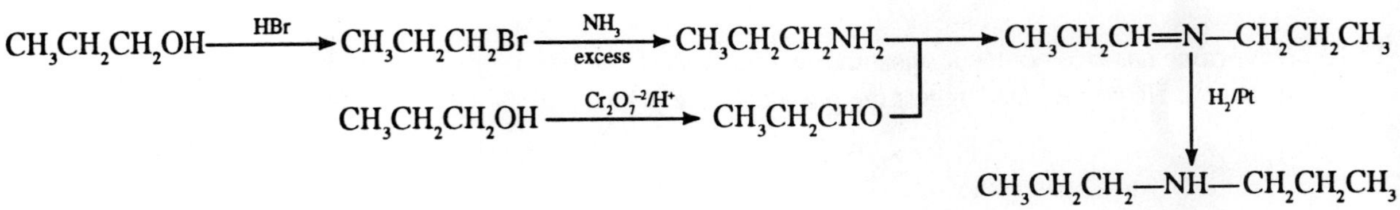

d. 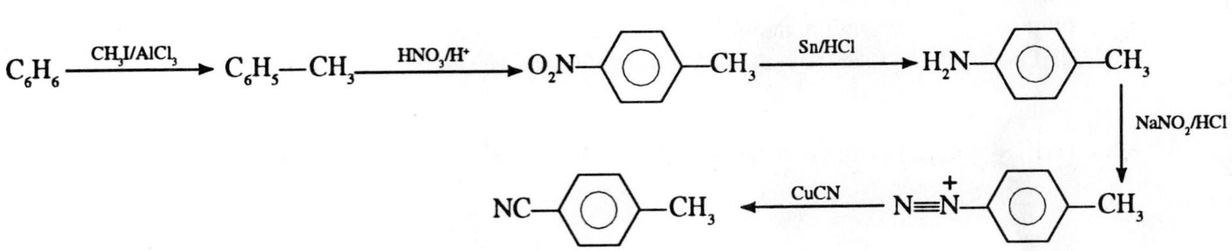

e. 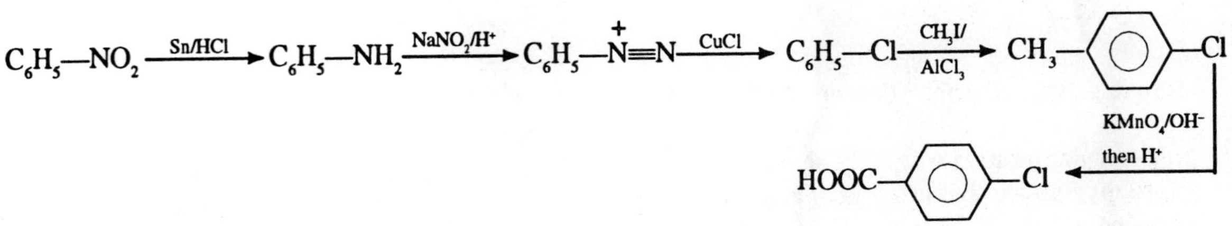

f. 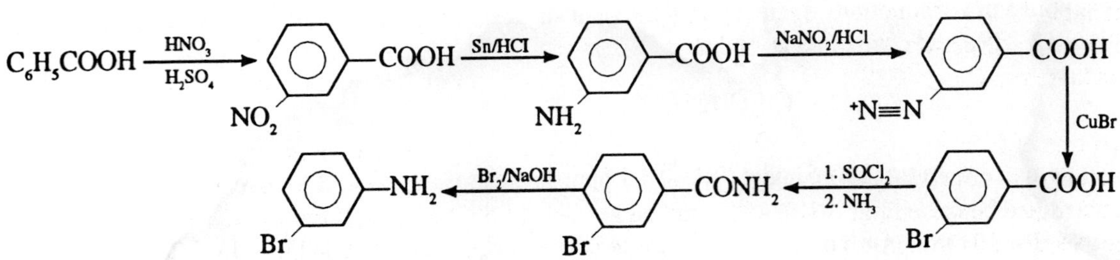

12.11 a. Benzylamine is a 1° amine, but benzamide is an amide—therefore, a neutral compound. Thus, the former is soluble in dilute acid; the latter is not.

b. Aniline is an aromatic (1°) amine and yields an azo dye when treated with $NaNO_2/HCl$ and a basic solution of β-naphthol. Cyclohexylamine produces a mixture of alkenes and alcohols when reacted with $NaNO_2/HCl$.

c. Pyridine is an amine and should be soluble in dilute acid, while benzene is a neutral compound and therefore, is, insoluble in any acid.

d. *p*-Ethylaniline is a 1° aromatic amine and should yield an azo dye (see part [b]). N-ethylaniline is a 2° aromatic amine and yields yellow N-nitroso compounds upon treatment with $NaNO_2$.
e. The Hinsberg test can distinguish between a 2° amine (diethylamine) and a 3° amine (triethylamine). When reacted with 2° amines, benzenesulfonyl chloride yields an insoluble benzenesulfonamide that appears as a precipitate. 3° amines do not react with benzenesulfonylchloride.
f. Exhaustive methylation of these compounds followed by Hoffman elimination produces dienes that give different products upon ozonolysis. Thus, the first compound yields CH_3COCH_2CHO, while the second one yields $CH_3CH(CHO)_2$. Only the first product produces CHI_3 (yellow ppt.) when reacted with I_2/KI/NaOH.
g. Both compounds are aliphatic amines, but diethylamine is a 2° amine that when treated with benzenesulfonyl chloride, yields a precipitate that is insoluble in both excess acid and base. *n*-Butylamine yields a benzene sulfonamide that is soluble in base.

12.12 **A** has five points of unsaturation. NMR data (7 ppm, 4H) suggest a disubstituted benzene ring. **A** is a neutral compound (insoluble both in HCl and NaOH), and its nitrogen is probably part of an amide. This is proven by the hydrolysis reaction that leads to the formation of acetic acid and amine **B** ($C_8H_{11}N$). Hydrogenation yields an aliphatic amine (the benzene ring is converted to a cyclohexane ring), which upon exhaustive methylation and Ag_2O treatment gives 3-methylcyclohexene. The only amine that yields 3-methylcyclohexene as the only product upon such treatment is

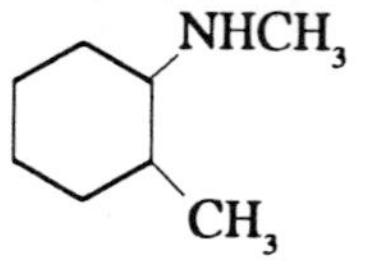

Thus, **B** is

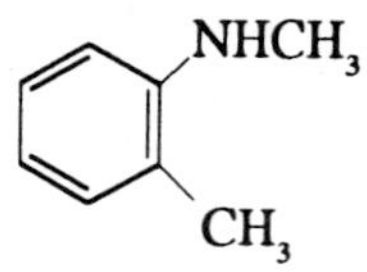

and **A** is

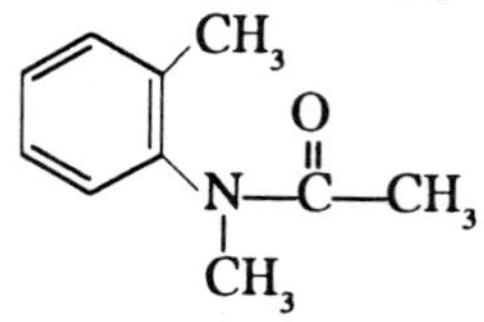

12.13 Novocaine has five points of unsaturation and at least one of its nitrogens is an amine nitrogen (since the compound is soluble in HCl). IR data suggest the presence of a primary amine ($R\text{-}NH_2$). Novocaine is hydrolyzed to **A** and **B**:

$$\underset{\text{Novocaine}}{(C_{13}H_{20}O_2N_2)} + H_2O \longrightarrow \underset{\mathbf{A}}{(C_7H_7O_2N)} + \underset{\mathbf{B}}{(C_6H_{15}ON)}$$

B has no points of unsaturation, and it is an amine, since it turns the red litmus paper to blue. Its formula must be $(CH_3CH_2)_2N\text{—}CH_2CH_2OH$, the reaction product of $(CH_3CH_2)_2NH$ +

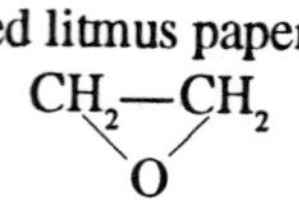

A has both a —COOH (soluble in NaOH) and $\text{—}NH_2$ (soluble in HCl) groups. Its structure is

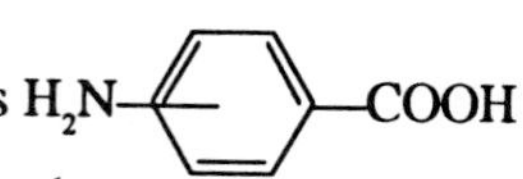

(*o*—, *m*—, or *p*—)

Thus, novocaine has the formula $(C_2H_5)_2NCH_2CH_2OOC$

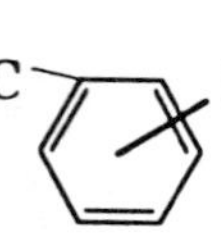

NH_2 (*o*—, *m*—, or *p*—).

More (e.g., spectroscopic) data are needed for the complete elucidation of the structure.

12.14 a. Molecular formula is $C_4H_{11}N$. No points of unsaturation. Triplet (2H) at 2.6 ppm is due to the $CH_2\text{—}\mathbf{CH_2}\text{—}NH_2$ group. Singlet (2H) at 1.1 ppm is due to $\text{—}NH_2$. Triplet (3H) at 0.9 ppm suggests $\mathbf{CH_3}\text{—}CH_2$. Thus, the compound is $CH_3CH_2CH_2CH_2NH_2$
b. No points of unsaturation. Sextet (1H) at 2.8 ppm suggests $CH_3\text{—}\underset{\displaystyle N}{\underset{|}{\mathbf{CH}}}\text{—}CH_2\text{—}$. Singlet (2H) at 1.8 ppm is due to $\text{—}NH_2$. Thus, the compound is $\underset{CH_3CHCH_2CH_3}{\overset{NH_2}{|}}$.
c. No points of unsaturation. Singlet (9H) at 1.1 ppm suggests $(CH_3)_3C\text{—}$. Left over is $\text{—}NH_2$, thus this compound is $(CH_3)_3C\text{—}NH_2$.
d. Mass spectroscopy peaks at 77 and 91, suggest $C_6H_5\text{—}CH_2\text{—}$. Leftover (117 – 91 =) 26 amu is due to C≡N (which is supported by the IR peak at 2250 cm^{-1}). The compound is $C_6H_5CH_2C{\equiv}N$.
e. Three points of unsaturation. IR suggests —C≡N (2260 cm^{-1}) and > C=O (ester, 1740 cm^{-1}). NMR peaks at 1.4 ppm (3H) and 4.2 ppm (2H) suggest the $CH_3CH_2O\text{—}$ moiety. Thus, this compound is $CH_3CH_2OOC\text{—}CH_2C{\equiv}N$.
f. Five points of unsaturation. NMR signal (7.6–6.8 ppm, 4H) suggests a *p*-disubstituted aromatic ring (supported by IR peak at 810 cm^{-1}). NMR signals at 1.4 ppm (3H, t) and 4.0 ppm (2H, q) suggest $CH_3CH_2O\text{—}$. Singlet at 2.1 ppm (3H) indicates $CH_3\text{—}CO\text{—}$. Only an amide carbonyl is possible, since the IR peak is pretty low (1650 cm^{-1}). This amide is

N-monosubstituted (1H in NMR at 9.5 ppm; IR peak at 3450 cm^{-1}). The compound is composed, therefore, of the following groups: C_6H_4 (*p*-substituted), CH_3CH_2—O— and CH_3—CO—NH—. Its formula, therefore, is

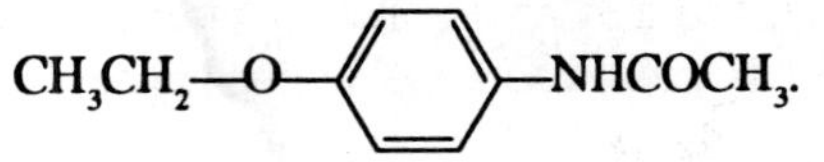

Answers for Multiple Choice Test

12.1 a	**12.6** c	**12.11** c	**12.16** a
12.2 c	**12.7** a	**12.12** b	**12.17** d
12.3 b	**12.8** d	**12.13** d	**12.18** b
12.4 b	**12.9** a	**12.14** b	**12.19** c
12.5 d	**12.10** d	**12.15** a	**12.20** d

13 Fats, Carbohydrates, and Amino Acids

Questions

Lipids

13.1 A general test for the presence of fats consists of heating the substance with a dehydrating agent, such as $KHSO_4$. Explain.

13.2 Fat molecules tend to clump together when mixed with water. Why?

13.3 State the factors responsible for the rancidity of fats.

13.4 Define ketone bodies.

13.5 Describe the meaning of *β-oxidation* of fatty acids.

13.6 Define and compare the following

a. fats and oils
b. soaps and detergents
c. fats and waxes
d. ABS and LAS detergents

13.7 Define the term *saponification number*. How many pounds of NaOH are required for the saponification of 5 pounds of fat with a saponification number of 200.0?

13.8 Given the following iodine numbers, state which is the most saturated fat or oil.

Fat or Oil	**Iodine Number** (g of I_2 per 100 g of fat)
Butter	30
Lard	90
Corn oil	120
Soybean oil	140

13.9 What is the general formula of phospholipids?

13.10 A 4.36 g sample of a lipid (MW=218) is saponified with excess NaOH, and the excess base is back-titrated with HCl. Thus, 60.0 mL of 1.50 M NaOH is added, and the excess is neutralized with 75.0 mL of 0.40 M HCl. The only acid found is acetic acid. What is the structure of the lipid?

13.11 Match the following vitamins with their scientific names and the diseases associated with their deficiency.

Vitamin	Scientific Name	Disease and Consequences
A	Thiamine	Beriberi
B_1	Retinol	Scurvy
B_{12}	Ascorbic acid	Night blindness
C	Cobalamin coenzymes	Hemolysis of red blood cells
D_2	α-Tocopherol	Rickets (bone-crippling disease)
E	Phylloquinone	Decreased ability of blood to clot
K_1	Calciferol	Pernicious anemia

Carbohydrates

13.12 Answer true or false for each of the statements below.

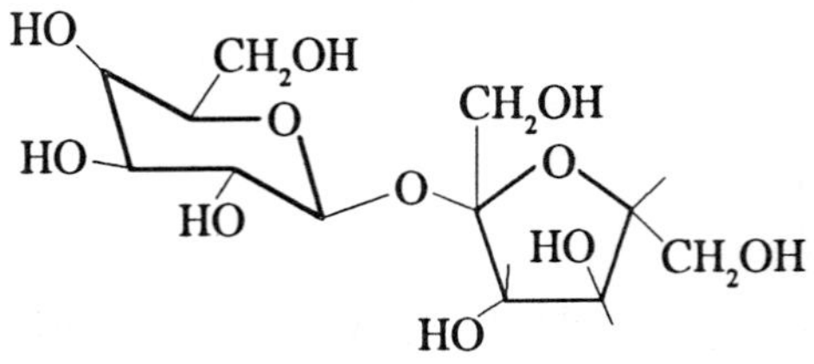

a. will form a single monosaccharide on hydrolysis.
b. undergoes mutarotation.
c. exists in two anomeric forms.
d. forms an osazone.
e. will not react with Benedict's solution.

13.13 Write the structures of the expected products in the following sequence of reactions:

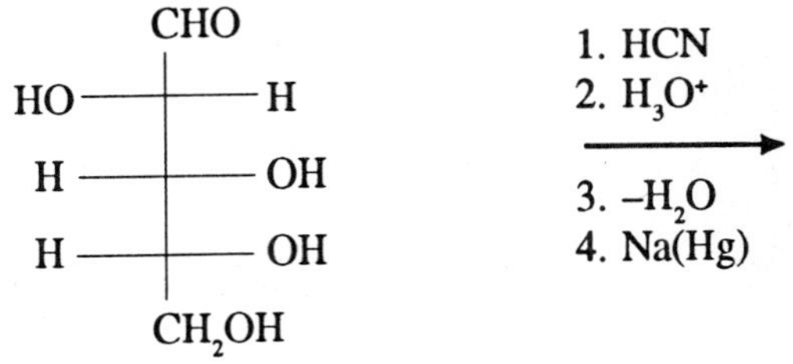

The products are related. They are ______ .
a. *meso* compounds
b. enantiomers
c. anomers
d. epimers
e. ketones

13.14 (+)-Glucose, (+)-mannose, and (–)-fructose all give the same osazone. What does this prove about the structures of these sugars?

13.15 When the amylose protion of starch is fully methylated and then hydrolyzed by enzymes that are specific for the α-linkage, 2,3,6-tri-O-methyl-D-glucose is obtained along with about 0.5% of 2,3,4,6-tetra-O-methyl-D-glucose. What information about the structure of amylose does this suggest?

13.16 Distinguish between α-glucose and β-glucose.

13.17 If glucose is in the hemiacetal form, why does it undergo reactions that aldehydes undergo?

13.18 Distinguish between starch and cellulose.

13.19 During the metabolism of glucose, glucose-6-phosphate is converted to fructose-6-phosphate and then to fructose-1,6-diphosphate. In the presence of the enzyme aldolase, fructose-1,6-diphosphate is converted into dihydroxyacetone phosphate and glyceraldehyde-3-phosphate. What common organic reaction does this last reaction resemble?

13.20 Suggest a structure for the disaccharide maltose on the basis of the following data:
a. Maltose is a reducing sugar, it mutarotates, and it forms an osazone.
b. It is cleaved into two moles of D-(+)-glucose by the enzyme maltase, which is specific for α-glucosides.
c. When oxidized, maltose forms a monocarboxylic acid, maltobionic acid.
d. When maltobionic acid is methylated and then hydrolyzed, 2,3,4,6-tetra-O-methyl glucose and 2,3,5,6-tetra-O-methylgluconic acid are isolated.

13.21 Define:
a. fructofuranose
b. mutarotation
c. glycogen
d. glycopyranose
e. reducing sugar
f. insulin
g. diabetes mellitus
h. glycogenesis
i. glycogenolysis
j. photosynthesis

13.22 Identify the configuration of carbohydrates **A** through **F**.

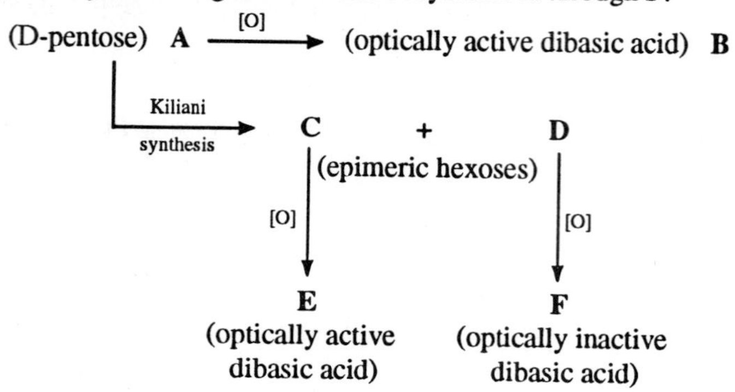

13.23 The anaerobic breakdown of glucose leads to the formation of lactic acid (glycolysis) and can be expressed by the overall equilibrium

$$C_6H_{12}O_6 \ + \ 2P_i \ + \ 2ADP \longrightarrow 2CH_3CH(OH)CO_2H \ + \ 2ATP \ + \ 2H_2O$$

where P_i = inorganic phosphate
Note that two moles of ATP are formed from ADP + P_i in the process. The overall equation can be separated into two steps.

Step 1. $C_6H_{12}O_6 \longrightarrow 2CH_3CH(OH)CO_2H$ $\quad \Delta G° = -47.0$ kcal/mole

Step 2. $2P_i \ + \ 2ADP \longrightarrow 2ATP \ + \ 2H_2O$ $\quad \Delta G° = +14.6$ kcal/mole

a. What is the overall free energy change for the reaction?
b. Calculate the K_{eq} value. Considering this value is the breakdown of glucose a reversible or an irreversible process?

Amino acids and Proteins

13.24 Define:
a. alpha amino acid
b. basic amino acid
c. zwitterion
d. isoelectric point
e. pK of an acid
f. heterocycloic amino acid
g. peptide linkage
h. polypeptide
i. alpha helix
j. 1° structure
k. 2° structure
l. 3° structure
m. 4° structure
n. Biuret test
o. denaturation
p. salting out
q. amphoteric compound

13.25 Match each item in the first column with the correct choice in the second column:

a. prosthetic group	________	1. disulfide crosslink
b. 2° structure	________	2. conjugated protein
c. electrophoresis	________	3. alpha helix
d. isoelectric point	________	4. color test
e. basic amino acid	________	5. end group analysis
f. cystine	________	6. glycine
g. ninhydrin	________	7. minimum solubility
h. 2,4-DNFB	________	8. fractionation of proteins
i. only optically inactive amino acid	________	9. tyrosine
	________	10. lysine

13.26 a. If 5 mL of 0.10 M $H_3\overset{+}{N}—CH_2CO_2H$ ($K_i = 4.6 \times 10^{-3}$) is added to 5 mL of 0.05 M NaOH, what will be the pH of the resulting solution?

b. What ratio of $[H_3\overset{+}{N}—CH_2COO^-] / [H_3\overset{+}{N}—CH_2COOH]$ is required to prepare a solution with a pH of 6?

13.27 For amino acid lysine $H_2N—(CH_2)_4CH(NH_2)—CO_2H$ $pK_1 = 2.18$, $pK_2 = 8.95$ and $pK_3 = 10.53$. What is the pI value?

13.28 A mixture with a pH=6 of alanine, valine, aspartic acid, and lysine was subjected to electrophoresis. On the basis of the following pK values,
a. which amino acid migrated toward the anode?
b. which amino acid migrated toward the cathode?
c. which amino acid remained stationary?

Amino acid	pK_1	pK_2	pK_3
Valine	2.3	9.3	
Alanine	2.3	9.3	
Aspartic acid	1.88	3.65	9.60
Lysine	2.18	8.95	10.53

13.29 What is the net charge (0, +1, or –1) of alanine, aspartic acid, and lysine at:
a. pH = 1 b. pH = 6 c. pH = 12

13.30 a. Write the structure of the tripeptide Ala-Leu-Tyr
b. Write the structure of the same tripeptide after reaction with 2,4-dinitrofluorobenzene and indicate the structures of the products obtained after partial hydrolysis.

13.31 Hydrolysis of a polypeptide with an approximate molecular weight of 1,000 gave four amino acids: Gly, Ala, Phe, and Val. Tests to determine the identity of the C-terminal and N-terminal units were negative. When partially hydrolyzed, the following products were found: Phe-Ala, Ala-Val, Gly-Phe, and Ala-Val-Gly. What is the likely structure of the polypeptide?

13.32 Complete the following reactions:

a.

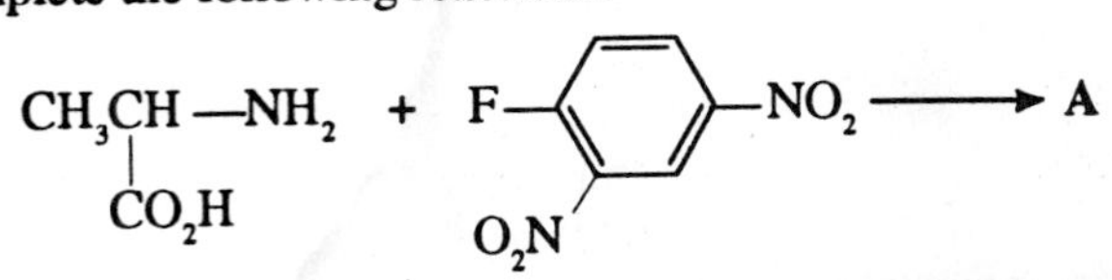

b.

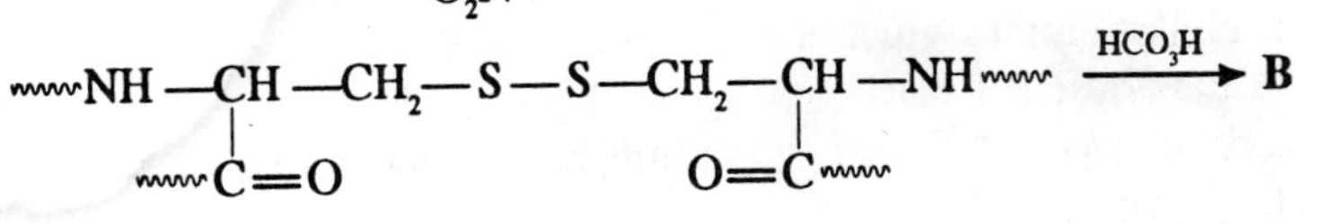

13.33 Match the reagents or techniques below with the proper protein chemistry task.

a. $(NH_4)_2SO_4$ _____ 1. Cleavage of carboxyl side of methionine residue
b. trypsin _____ 2. Cleavage of disulfide bonds
c. HCO_3H _____ 3. Salting out of proteins
d. CNBr _____ 4. Determination of amino-terminal residue of a peptide
e. 2,4-DNFB _____ 5. Cleavage of carboxyl side of lysine and arginine residues
f. $CuSO_4$/base _____ 6. Measurement of amino acid's mobility in an electric field
g. ninhydrin _____ 7. Colorimetric reaction for the qualitative and quantitative determination of proteins
h. electrophoresis _____ 8. Detection of amino acids

Multiple Choice Test

13.1 Which projection(s) correspond(s) to D-glyceraldehyde?

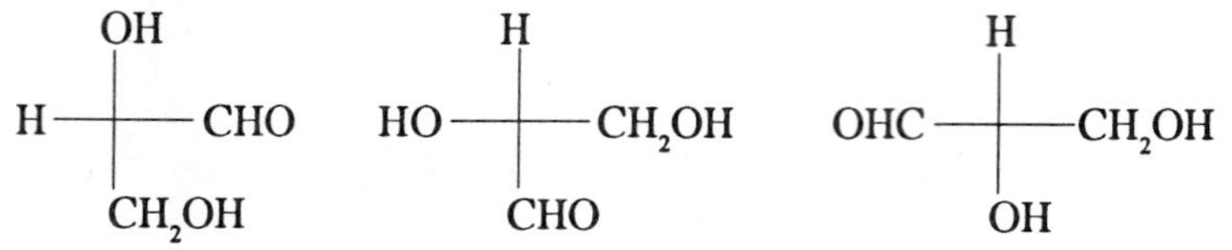

A **B** **C**

a. A b. B c. C d. A and B

13.2 Glucose and mannose are _______ .

a. enantiomers
b. anomers
c. epimers
d. structural isomers

13.3 Which of the following matches the definition of aldohexose?

A is a monosaccharide
B contains a —CHO group
C is a disaccharide
D is a nonreducing sugar

a. A and B b. B and C c. A and D d. all but D

13.4 The existence of α-glucose and β-glucose is indicative of

a. the existence of two conformational isomers of glucose.
b. the existence of two enantiomers of glucose.
c. the formation of a cyclic hemiacetal of glucose.
d. the presence of a keto/aldehyde equilibrium in glucose.

13.5 The relationship between α-glucose and β-glucose is that they are _______ .

a. tautomers
b. epimers
c. anomers
d. enantiomers

13.6 Which of the following is correct about sucrose?

a. It is a reducing sugar.
b. It undergoes mutarotation in water.
c. It hydrolyzes to fructose and glucose.
d. It reacts with Fehling's and/or Benedict's reagent(s).

13.7 The main difference between amylose and cellulose is

a. the glucose units in amylose are linked α-1-4, while in cellulose they are linked β-1-4.
b. amylose is a D-glucose polymer, while cellulose is an L-glucose polymer.
c. cellulose is a cross-linked polymer, while amylose is linear.
d. amylose is a 1,4-linked polymer, while cellulose is 1,6-linked.

13.8 An aqueous solution of glucose behaves as an aldehyde because
a. glucose is actually a cyclic aldehyde.
b. it is hydrolyzed by water to the free aldehyde.
c. it is a ketone, but is in equilibrium with the aldehyde form.
d. its predominant form (the cyclic hemiacetal) is in equilibrium with the free aldehyde form.

13.9 The fact that sucrose is a nonreducing sugar, proves that the _______ unit of sucrose is linked to the _______ unit via a _______ linkage.
a. glucose, fructose, C_5 to C_6
b. glucose, fructose, C_1 to C_2
c. fructose, glucose, C_1 to C_2
d. fructose, glucose, C_5 to C_6

13.10 Periodate oxidation of an aldopentose yields _______ molecules of HCOOH, _______ molecules of HCHO, and _______ molecules of CO_2.
a. 4, 1, 0
b. 2, 1, 2
c. 3, 1, 1
d. 3, 2, 0

13.11 Epimeric aldohexoses give the same
a. cyanohydrin when treated with HCN.
b. osazones.
c. aldaric acid when oxidized with HNO_3.
d. both (a) and (b)

13.12 Which statement is false about naturally occuring amino acids?
a. They belong to the D-family.
b. They have high melting points.
c. In neutral aqueous solution, they exist as zwitterions.
d. They have characteristic isoelectric points.

13.13 Amino acids are chromatographically detected by means of _______ .
a. 2,4-DNP
b. cyanogen bromide
c. ninhydrin
d. $AgNO_3$

13.14 Which of the following represents a peptide bond?
a. —CO—NH—CO—
b. —CO—NH—
c. —NH—CO—NH—
d. —N=C(OH)—NH—

13.15 The correct matching of isoelectric points and amino acids is

1. alanine
2. lysine
3. aspartic acid

A 10.5
B 3.0
C 6.0

a. 1-A, 2-B, 3-C
b. 1-B, 2-C, 3-A
c. 1-C, 2-B, 3-A
d. 1-C, 2-A, 3-B

13.16 Match the following names and reagents.

1. Edman
2. Sanger
3. Merrifield

A solid state synthesis
B polypeptide degradation using C_6H_5-NCS
C reaction of aminoacids with 2,4-DNFB

a. 1-C, 2-B, 3-A
b. 1-A, 2-B, 3-C
c. 1-C, 2-A, 3-B
d. 1-B, 2-C, 3-A

13.17 Which of the following is true about the solid state synthesis?

A It can be automated.
B It cannot produce the S—S linkage.
C It is applicable only to the synthesis of small peptides.
D The product in each step does not need to be purified.
E The yield in each step is virtually 100%.

a. A, C, D b. A, D, E c. A, B, D d. all but C

13.18 Match the following aminoacids with the correct facts

1. glycine
2. cysteine
3. cystine
4. tyrosine

A the only amino acid carrying a phenol group
B the only amino acid carrying a thiol group
C the only amino acid carrying the disulfide bond
D the only optically inactive aminoacid

a. 1-D, 2-C, 3-B, 4-A
b. 1-D, 2-B, 3-C, 4-A
c. 1-D, 2-A, 3-B, 4-C
d. 1-A, 2-C, 3-B, 4-D

13.19 The amide linkage in proteins has restricted rotation because

a. the carbonyl is a planar group.
b. the linkage is stabilized by the resonance structure 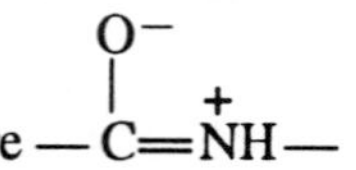

c. the linkage is hydrogen bonded to other amide bonds.
c. all of the above.

13.20 Complete hydrolysis of a nucleic acid yields

A a mixture of heterocyclic bases.
B a mixture of amino acids.
C an aldohexose.
D phosphate ion, P_i.
E an aldopentose.

a. A, C, and D
b. A, D, and E
c. A, B, C, and D
d. A, B, D, and E

Answers

Lipids

13.1 When fats are heated, glycerol is liberated; glycerol, in turn, is then dehydrated, forming the pungent aldehyde *acrolein*.

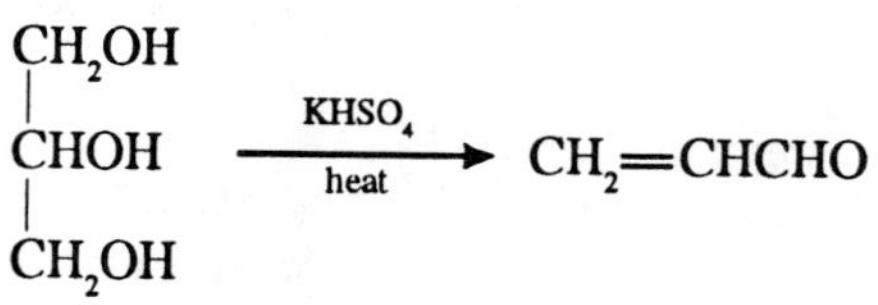

glycerol acrolein

13.2 Fat molecules contain both hydrophilic and hydrophobic segments. The molecules' hydrophobic parts tend to associate within a "clump" and exclude water, while their hydrophilic parts associate farther away from the central hydrophobic part of the "clump" and form its exterior, which is in contact with the surrounding water.

13.3 Fats can undergo hydrolytic cleavage, liberating fatty acids, some of which, notably butyric acid, have a rancid odor and taste. Heat, moisture, microorganisms, and enzymes contribute to this process. Rancidity, however, is more often due to oxidation—particularly oxidation of unsaturated fatty acids at allylic positions. The resultant peroxides degrade to aldehydes, ketones, and acids.

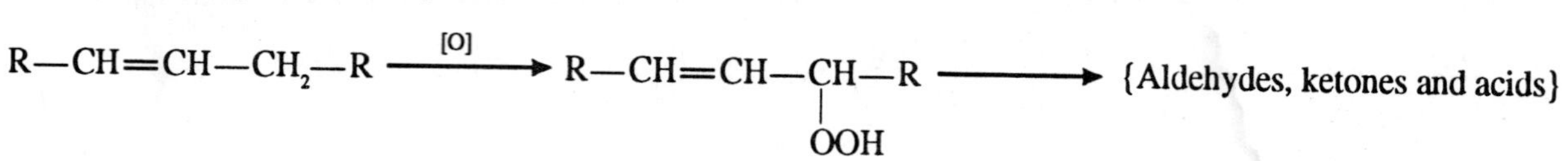

peroxide

Light, heat, and moist air promote oxidation to peroxides. Oxidation of oils that accumulate on the surface of the skin is promoted by the warmth of the body (especially in the presence of oxygen and water). The resulting compounds are responsible for the odor associated with perspiration.

13.4 When carbohydrate metabolism is impaired (for instance, in untreated diabetics or in people who are severely restricting their calorie intake in an attempt to "diet"), fatty acid metabolism is accelerated. Under normal circumstances, fatty acids are oxidized to acetyl-CoA, which is then further metabolized. With increased fatty acid metabolism comes an abundance of acetyl-CoA, which is condensed to acetoacetic acid by the liver. Acetoacetic acid is a precursor of both acetone and β-hydroxybutyric acid. Acetoacetic acid, acetone, and β-hydroxybutyric acid are called *ketone bodies,* and the condition in which they accumulate in the blood is known as *ketosis.* When untreated, ketosis can lead to acidosis, coma, and death.

13.5 Fatty acids are broken down two-carbons at a time. Each two-carbon unit forms a molecule of acetyl-CoA, which then enters the citric acid cycle. The breakdown proceeds by oxidation at the β-carbon atom of the acids in the presence of enzymes and cofactors:

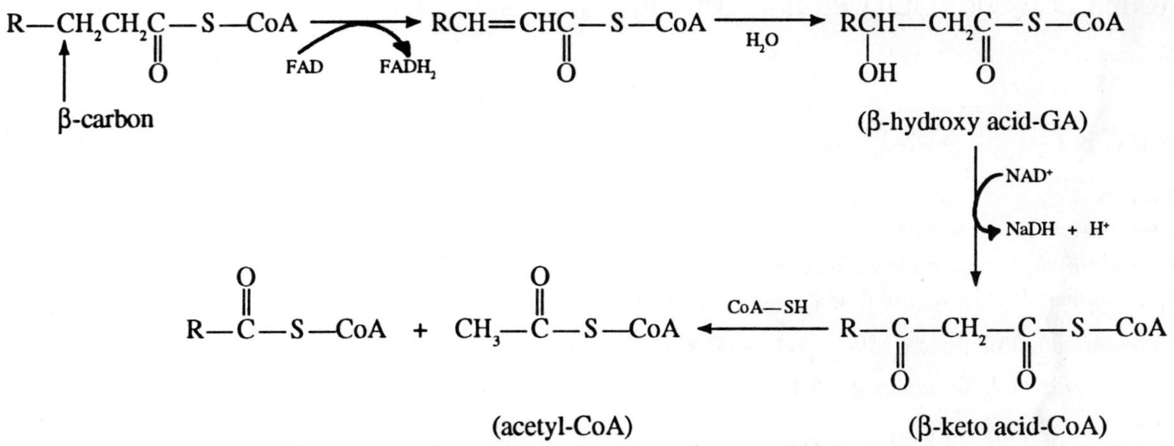

CoA—SH = Coenzyme A
FAD = Flavin Adenine Dinucleotide
NAD^+ = Nicotinamide Adenine Dinucleotide

The addition of water in the second step is a Michael-like addition, while the last step in this sequence is a retro-Claisen condensation.

13.6 a. At room temperature, fats are solid, mostly saturated-carboxylate triesters of glycerol (1,2,3-propanetriol). They are also called *triglycerols* or *triglycerides. Oils* are liquids at room temperature and have similar structure; however their carboxylic acids are mostly unsaturated (unlike those of fats). The acids found in fats and oils beginning with butyric acid contain an even number of carbon atoms.

$CH_2OOC(CH_2)_{16}CH_3$	$CH_2OOC(CH_2)_7CH{=}CH(CH_2)_7CH_3$
$CHOOC(CH_2)_{16}CH_3$	$CHOOC(CH_2)_7CH{=}CH(CH_2)_7CH_3$
$CH_2OOC(CH_2)_{16}CH_3$	$CH_2OOC(CH_2)_7CH{=}CH(CH_2)_7CH_3$
tristearin (a saturated triglyceride)	triolein (an unsaturated triglyceride)

b. *Soaps* are the salts of fatty acids, such as the sodium salt, the potassium salt. *Detergents* are synthetic salts of sulfonic acids containing long organic chains.

$n\text{-}C_{17}H_{35}COO^-Na^+$

sodium stearate
(a soap)

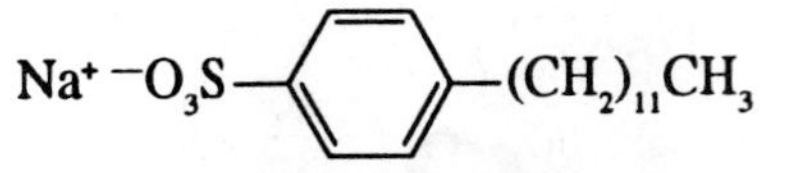

sodium *p*-dodecylbenzenesulfonate
(a detergent)

c. A *fat* is the triester of a fatty acid and glycerol, while a *wax* is a monoester of a fatty acid and a long chain alcohol.

n-$C_{25}H_{51}COO$-n-$C_{32}H_{65}$

wax

$CH_2OOC(CH_2)_{16}CH_3$
|
$CHOOC(CH_2)_{16}CH_3$
|
$CH_2OOC(CH_2)_{16}CH_3$

fat

d. The *ABS* (alkylbenzenesulfonate) detergents have branched alkyl groups and are not biodegradable. The *LAS* (linear alkylsulfonate) detergents do not have branched chains and are biodegradable.

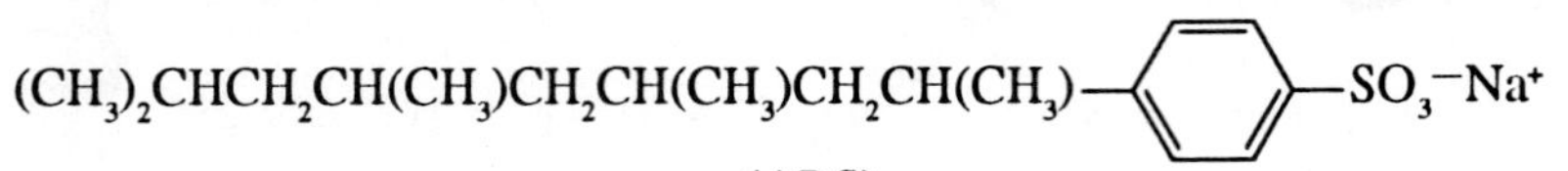

(ABS)

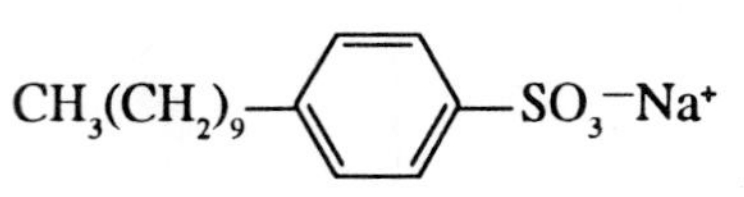

(LAS)

13.7 The *saponification number* is defined as the number of milligrams of KOH required to saponify one gram of fat. Thus, # moles of KOH required to saponify 1 g of fat is

the saponification number ÷ MW_{KOH} = 0.2000 g ÷ 56 g/mole = 0.0036 moles.
5 lb × 454 g/lb x 0.0036 moles/g = 8.17 moles of KOH required for 5 lbs of fat.

The number of moles of KOH necessary is the same as the number of moles of NaOH necessary. Thus, the weight of NaOH needed is

8.17 moles × 40 $g/mole_{NaOH}$ = 327 g (or 327 g ÷ 454 g/lb = 0.72 lb.

13.8 The most saturated fat or oil is the one with the least degree of unsaturation (lowest iodine number). Thus, butter is the most saturated and soybean oil the most unsaturated.

13.9 Phospholipids (such as lecithin and cephalin) are mixed esters of glycerols with a fatty acid, phosphoric acid, and an amino alcohol (choline, ethanolamine, serine).

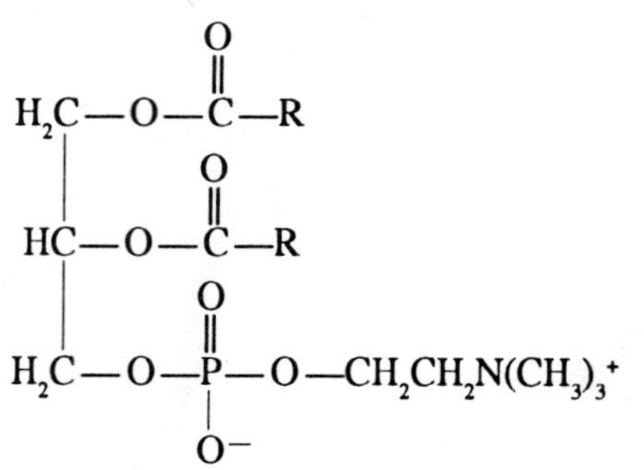

lecithin

13.10 The overall procedure is summarized as follows:

lipid + NaOH ⟶ acetate + alcohol (+ excess NaOH)

The excess NaOH solution is removed and titrated with HCl

NaOH + HCl ⟶ NaCl + H_2O

mmoles of lipid: (4.36 g ÷ 218 g/mole) = 0.020 moles = 20 mmoles.

mmoles of NaOH added: M × V = 1.5 moles/L × 0.060 mL = 0.090 moles.

mmoles of HCl needed = mmoles of excess NaOH: $M \times V = 0.4$ moles/L $\times$ 0.075 L = 0.030 moles = 30 mmoles.

mmoles of NaOH needed for the hydrolysis of the fat: 90 mmoles – 30 mmoles = 60 mmoles.

Thus, 20 mmoles of lipid requires 60 mmoles of NaOH for saponification. Consequently, the lipid has three ester groups with acetate as the acid part (i.e., it is a triglyceride). Therefore, of the 218 amu of the lipid $3 \times (15+12+32) =$ 177 amu are due to acetate. The 41 amu left over corresponds to C_3H_5 (glycerol) and the ester is, therefore,

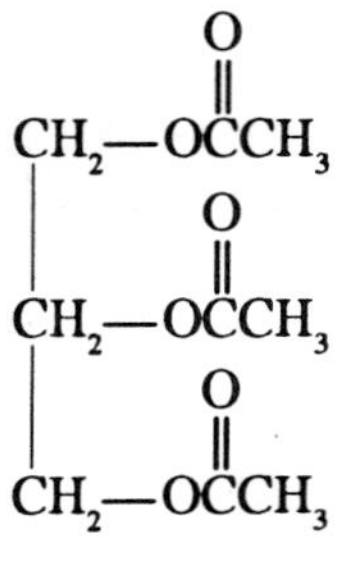

13.11 Vitamin A, retinol, night blindness
Vitamin B_1, thiamine, beriberi
B_{12}, cobalamin coenzymes, pernicious anemia
C, ascorbic acid, scurvy
D_2, calciferol, rickets
E, α-tocopherol, hemolysis of red blood cells
K_1, phylloquinone, decreased ability of blood to clot

Carbohydrates

13.12 Only (e) is true.

13.13

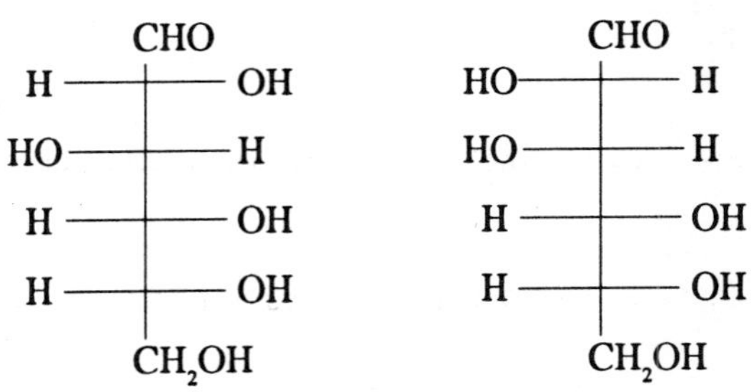

d. They are epimers

13.14 All possess the same configuration at carbons 3, 4, and 5.

13.15

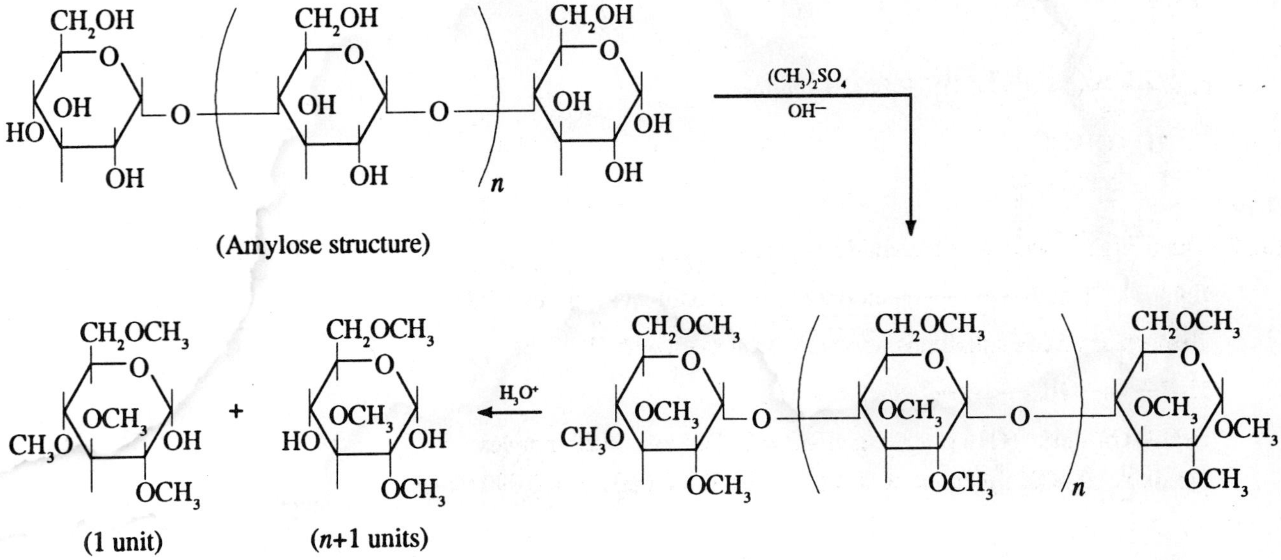

(Amylose structure)

(1 unit) (n+1 units)

Amylose is composed of α-glucose monomer units joined through the 1,4 position. The small amount (0.5%) of 2,3,4,6-tetra-O-methyl-O-glucose comes from the terminal unit of each biopolymer chain. The results indicate a polymer chain with little branching. An approximate molecular weight can be calculated from the data as follows:

$$\frac{0.5 \text{ units}}{99.5 \text{ units}} = \frac{1 \text{ unit}}{n \text{ units}}$$

$n = 199$ units

Since each glucose unit's molecular weight (minus water) is 162, the molecular weight of amylose is about $199 \times 162 = 32{,}238$.

13.16 α-glucose and β-glucose are cyclic hemiacetals that differ in configuration only at position 1. Such diastereomers are called *anomers*.

13.17 In solution, the cyclic hemiacetal forms are in equilibrium through the open aldehyde form:

α-D-glucose ⇌ open aldehyde from ⇌ β-D-gluose

The drift in optical rotation of a freshly prepared solution of either α-D-glucose or β-D-glucose to an equilibrium value is called mutarotation. The presence of a small amount (approximately 0.2%) of the free aldehyde is the reason for typical aldehyde reaction.

13.18 Starch and cellulose differ in the stereochemistry of the glycosidic linkage. Starch is a polymer of α-glucose and the 1,4-linkage between units is an α-glycosidic linkage. Cellulose, on the other hand, is a polymer of β-glucose with β-glycosidic linkages.

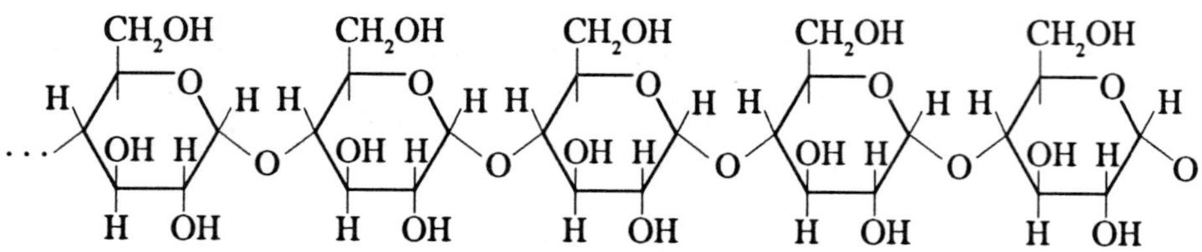

Starch

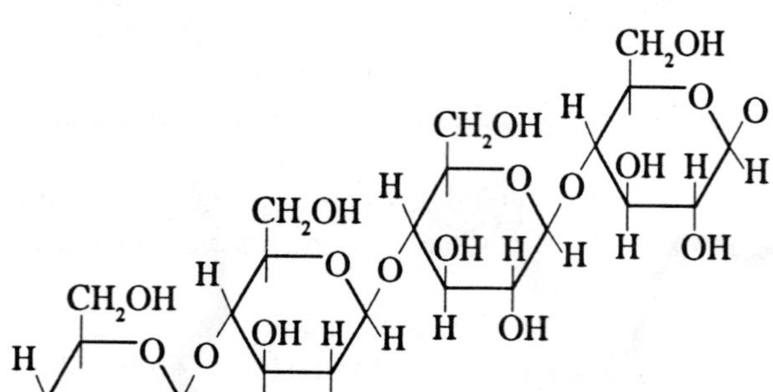

Cellulose

13.19

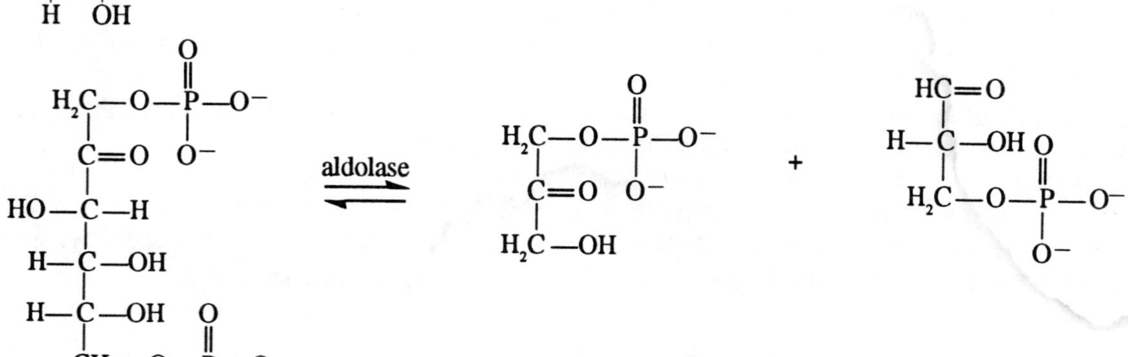

This resembles a retro-aldol reaction

13.20

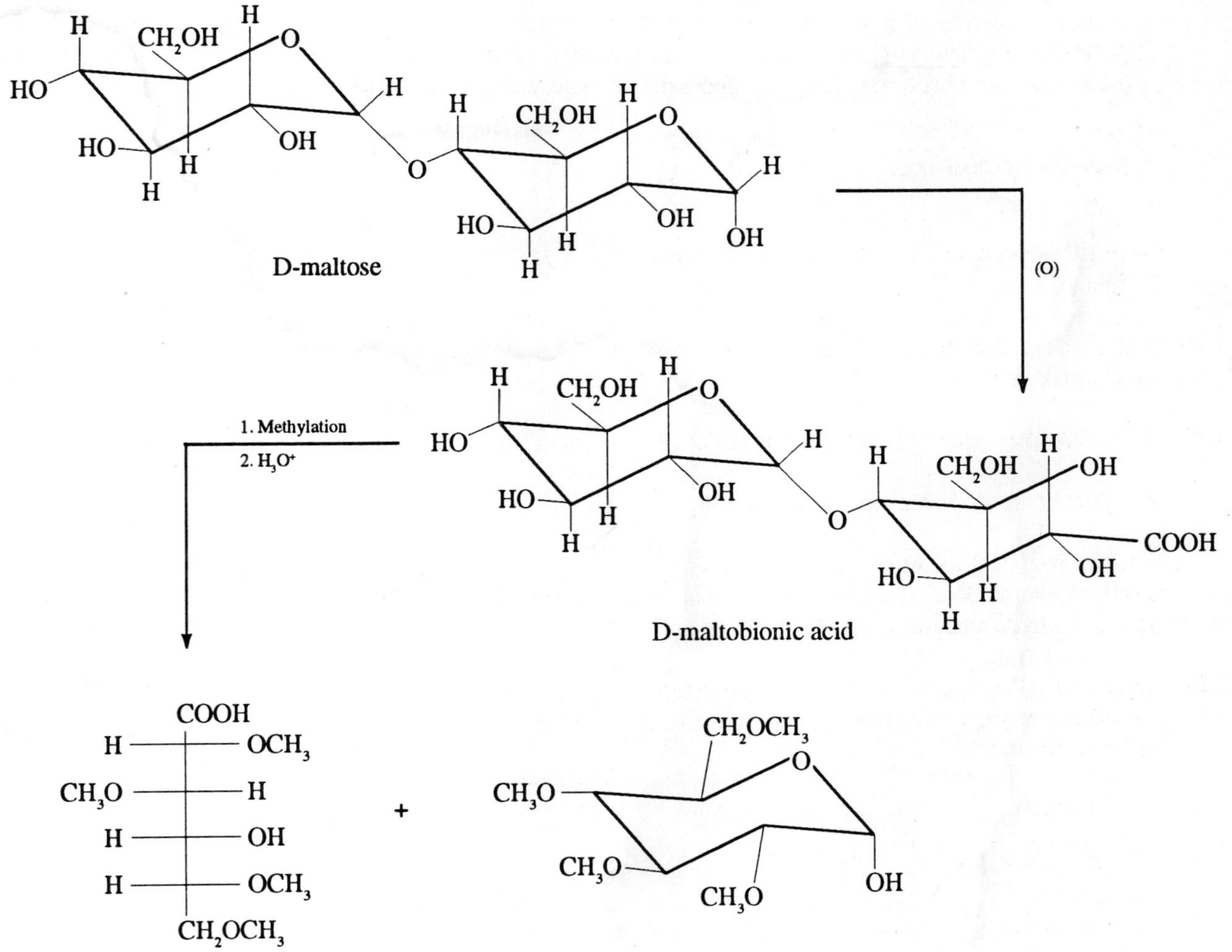

A 1,4 linkage is suggested by the fact that 2,3,5,6-tetra-O-methyl gluconic acid has not been methylated in position 4, and that oxidation of maltose results in the formation of a monocarboxylic acid. Therefore, the glycosidic linkage must be from position 1 of the nonreducing glucose unit to position 4 of the reducing glucose unit.

13.21 a. Fructofuranose: the structure of fructose represented as a cyclic compound in the form of the heterocyclic ring *furan*.

b. Mutarotation: the interconversion of pure α- and β-cyclic forms of simple sugars to give an equilibrium mixture of the two forms. The phenomenon occurs because of the ability of the two anomers to interconvert via the intermediate open-chain form.

c. Glycogen: a polysaccharide that is the storage form of carbohydrates in the animal body.

d. Glucopyranose: the structure of glucose represented as a cyclic compound in the form of the heterocyclic 6-membered ring *pyran*.

e. Reducing sugar: a sugar that contains free or masked aldehyde or ketone groups (in the form of acetal or ketals) capable of reducing metal ions; for example, $Ag^+ \rightarrow Ag^0$ or $Cu^{+2} \rightarrow Cu^{+1}$.

f. Insulin: a hormone produced by the pancreas, which functions to maintain the proper blood sugar level in the normal metabolism of glucose.

g. Diabetes mellitus: an impairment of carbohydrate oxidation ordinarily caused by the lack of insulin and characterized by high blood sugar levels and ketone bodies in blood and urine.

h. Glycogenesis: the synthesis of glycogen from glucose by liver and muscle tissue.

i. Glycogenolysis: the breakdown of glycogen to glucose, which is then liberated into the bloodstream.

j. Photosynthesis: the formation of carbohydrates in cells of green plants from carbon dioxide and water in the presence of sunlight and chorophyll.

13.22

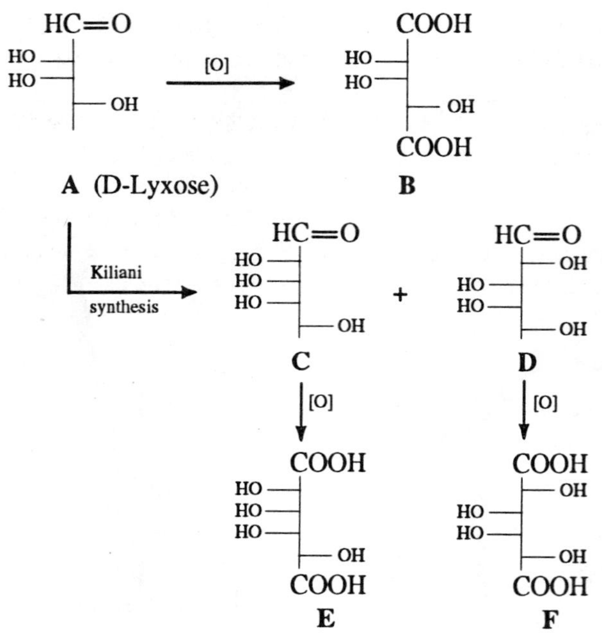

13.23 a. $\Delta G^{\circ}_{\text{overall}} = (-47.0) + 14.6 = -32.4$ kcal/mole

b. $\Delta G = -RT\ln K_{eq} = -2.303\,RT\,(\log_{10} k_{eq})$

$$\log_{10} K_{eq} = \frac{\Delta G}{(-2.303)RT} = \frac{-32{,}400}{(-2.303)(1.98)(298)} = +23.84$$

Therefore, $K_{eq} = 7 \times 10^{23}$. Thus, is the reaction is irreversible.

Amino acids and Proteins

13.24 a. Alpha amino acid: a carboxylic acid with an amino group in the 2- (or α-) position. The general formula is R—CH(NH_2)—COOH.

b. Basic amino acid: an amino acid with more basic goups (–NH_2) than acidic groups (–COOH). Such an example is lysine (lys) H_2N—$CH_2CH_2CH_2CH_2CH(NH_2)COOH$.

c. Zwitterion: a charge-separated form (usually a dipolar ion) of an otherwise covalent compound. For example, α-amino acids (i.e., HOOC—CH(R)NH_2) exist in the dry, solid state as dipolar ions, and the carboxyl group, —COOH, is present as the carboxylate anion, —COO^-, and the amine group, —NH_2, as the ammonium group, —NH_3^+. Thus, H_2N—CH(R)COOH (covalent form) ^+H_3N—CH(R)COO^- (zwitterion (dipolar ion) form)

d. Isoelectric point: The pH at which there is no net migration of an amino acid or protein under the influence of an electric field. At this pH, the amino acid or protein molecule contains an equal number of positive and negative charges.

e. pK of an acid: The pH at which there are equal concentrations of a weak acid HA and its anion A^-. It is mathematically equal to ($-\log_{10}K$), where K is the dissociation constant of the acid.

f. Heterocyclic amino acid: an amino acid with a heterocyclic ring in its structure, such as histidine and tryptophan.

g. Peptide linkage: an amide linkage between the carboxyl group of an amino acid and the amino group of another that results in the elimination of one molecule of water.

h. Polypeptide: Any peptide with more than 10 amino acid residues.

i. Alpha helix: a chain of amino acid units wound into a spiral that retains this configuration by hydrogen bonding.

j. Primary structure: the sequence of amino acids and location of disulfide bridges (if present) in a polypeptide or protein.

k. Secondary structure: the three-dimensional repeat structure of a peptide or protein from one end of the chain to the other. This includes, the α-helix and the β-pleated sheet.

l. Tertiary structure: the three-dimensional shape that results from the further bending and folding of the α-helix.

m. Quaternary structure: the way the polypeptide chains are packed together. The term is applicable only to compounds that contain more than one polypeptide chain.

I

Appendix
Important Spectroscopic Data

^{1}H NMR Spectroscopy

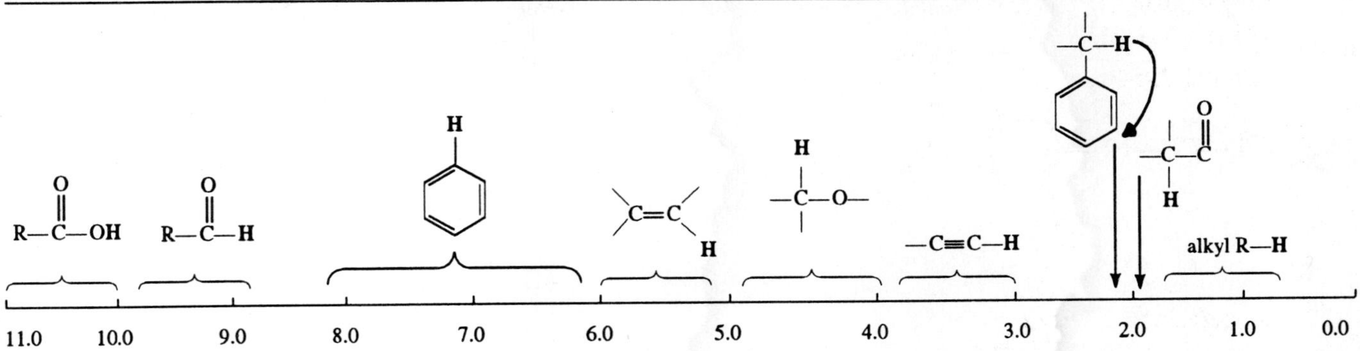

Mass Spectroscopy

Isotope Relative Abundance

	P^+		$P^+ + 1$		$P^+ + 2$	
Element	Isotope	Relative Abundance	Isotope	Relative Abundance	Isotope	Relative Abundance
Hydrogen	^{1}H	100.000	^{2}H	0.015		
Carbon	^{12}C	100.000	^{13}C	1.122		
Nitrogen	^{14}N	100.000	^{15}N	0.037		
Oxygen	^{16}O	100.000	^{17}O	0.371	^{18}O	0.205
Fluorine	^{19}F	100.000				
Sulfur	^{32}S	100.000	^{33}S	0.080	^{34}S	4.44
Chlorine	^{35}Cl	100.000			^{37}Cl	32.40
Bromine	^{79}Br	100.00			^{81}Br	97.86

Relative Intensities of Isotope Peaks for Bromine and Chlorine Combinations[1]

		Br_0	Br_1	Br_2
Cl_0	$P^+ + 2$	—	98.0	196.0
	$P^+ + 4$	—	—	96.1
Cl_1	$P^+ + 2$	32.5	130.6	228.0
	$P^+ + 4$	—	31.9	159.0
	$P^+ + 6$	—	—	31.2
Cl_2	$P^+ + 2$	65.0	163.0	261.1
	$P^+ + 4$	10.6	74.4	234.2
	$P^+ + 6$	—	10.4	83.3
	$P^+ + 8$	—	—	10.2

[1]Values are approximate. $P^+ = 100\%$

Infrared Spectroscopy

Hydrocarbons

Group	IR range	Group	IR range
—C≡C—	2260–2190 cm^{-1}	**—C≡C—H**	3300 cm^{-1}
—C=C—	1650–1620 cm^{-1}	**—C=C—H**	3100 cm^{-1}
—C⋯C— (aromatic)	1600 and 1500 cm^{-1}	**—C⋯C—H** (aromatic)	just above 3000 cm^{-1}
—C—C— (aliphatic)	below 1000 cm^{-1}	**—C—C—H** (aliphatic)	just below 3000 cm^{-1}
cis—RCH=CHR′	675-730 cm^{-1} (variable)	trans—RCH=CHR′	965–975
—CH_3	1435 cm^{-1}	$(CH_3)_2CH$— (iso)	1380 cm^{-1}, doublet

Carbonyl group in various compounds: 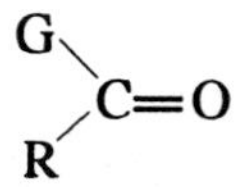

G	Family	Formula	IR range
Halogen	Acyl halide	R—C(=O)—X X=Cl, Br, I	1840 cm^{-1}
—OOCR	Anhydride	R—C(=O)—O—C(=O)—R	1800 cm^{-1}
—OR′	Ester	R—C(=O)—OR′	1740 cm^{-1}
—OH	Carboxylic acid	R—C(=O)—OH	1710–1720 cm^{-1}
—R	Ketone	R—C(=O)—R′	1710–1720 cm^{-1}
—H	Aldehyde	R—C(=O)—H	1710–1720 cm^{-1}
—NH_2, —NHR′, —NR'_2	Amide	R—C(=O)—NH_2, R—C(=O)—NHR′, R—C(=O)—NR'_2	1650 cm^{-1}
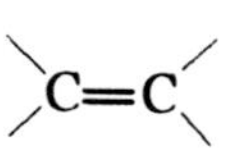	α, β-unsaturated	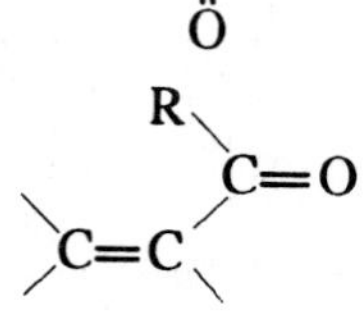	1695–1680 cm^{-1}

Other

—O—H	3300 cm^{-1} (for R—COOH, broad peak covers the range from 3300 to 2500 cm^{-1})	
—C—O—	(RCH_2—OH, 1° alcohol) (R_2CH—OH, 2° alcohol) (R_3C—OH, 3° alcohol)	1050 cm^{-1} 1100 cm^{-1} 1150 cm^{-1}
—NO_2		1550 and 1345 cm^{-1}
—N—H	(R—NH_2, 1° amine) (R_2NH, 2° amine)	3500 and 3400 cm^{-1} 3350 cm^{-1}

Infrared Aromatic Substitution Patterns[1]

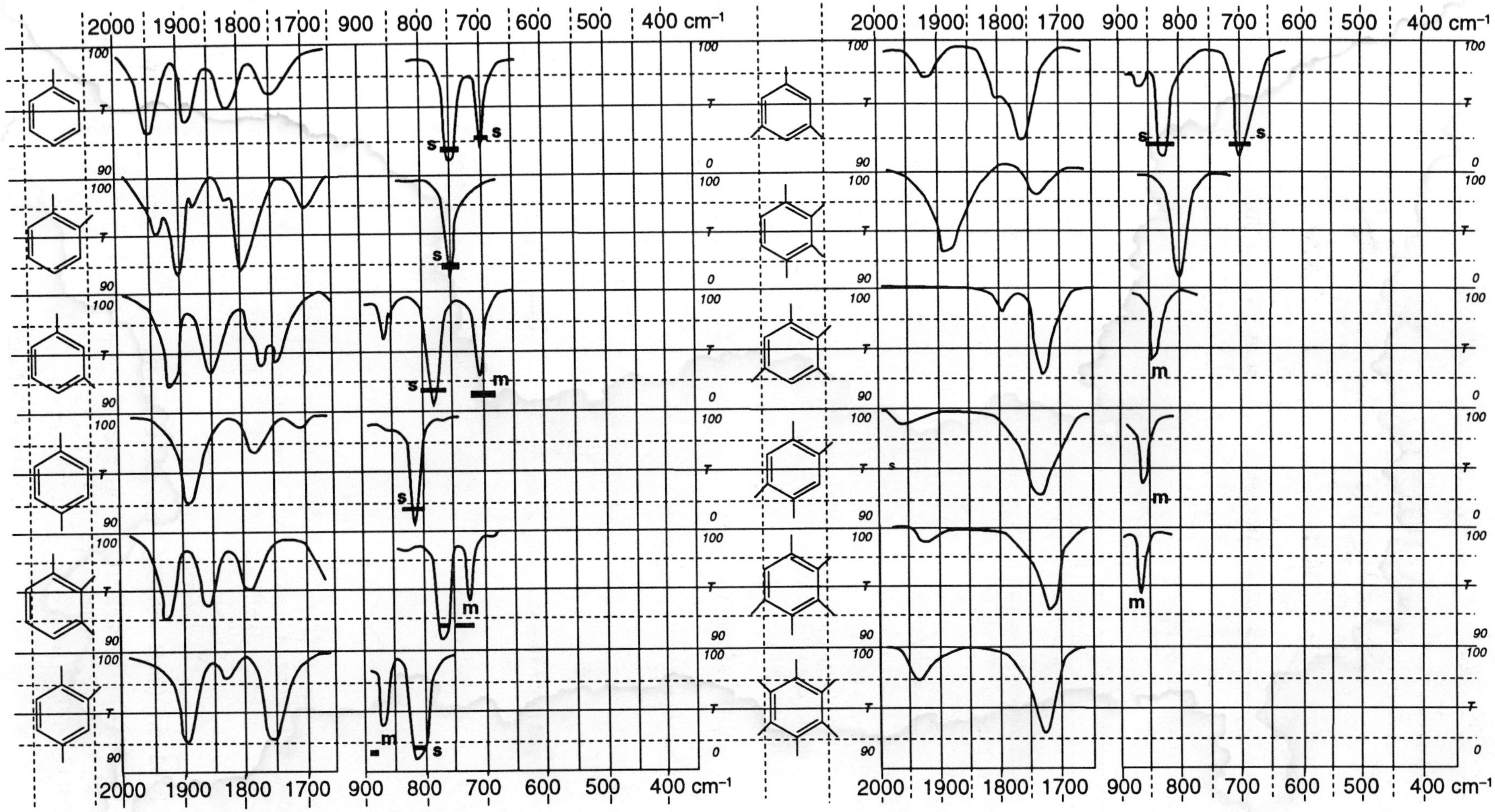

[1]With permission of Professor Charles F. Hammer, Chemistry Department, Georgetown University, Washington, D.C. 20057.

II

Appendix: Absurd Mechanisms

Shown below are several incorrect mechanistic steps, in addition to some reactions that yield improbable products. The reasons for their unlikely occurrence are described.

a.

Loss of a hydride (H^-) ion without the participation of an oxidizing agent normally does not occur, because H^- is a very poor leaving group. An exception is the (usually intermolecular) H^- transfer in the Cannizzaro reaction, in which an aldehyde that does not have an α-hydrogen disproportionates to a 1° alcohol and a carboxylic acid. Moreover, the above electron movement leads to loss

of a hydrogen molecule (H_2) and the formation of a carbocation 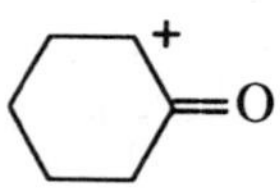 which is very unstable due to the presence of the carbonyl group next to the positively charged carbon.

b.

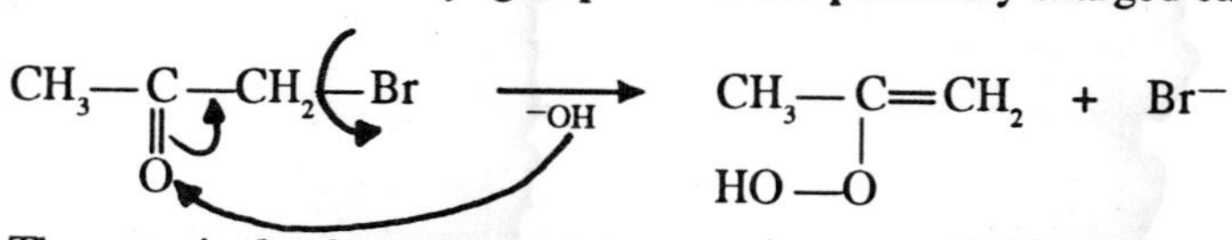

The negatively charged hydroxide ion cannot attack the electronegative oxygen of the carbonyl group.

c.

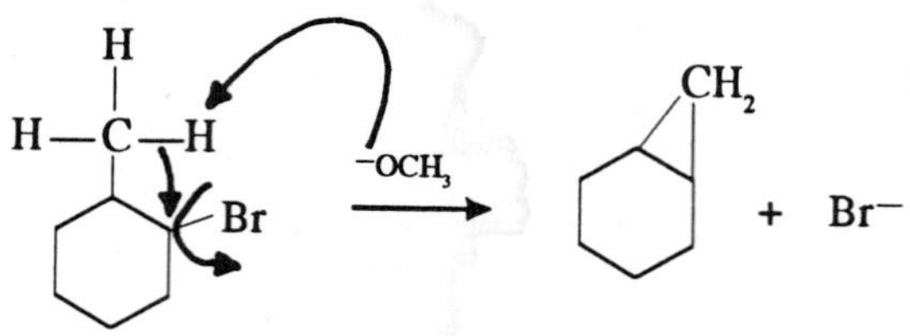

It is unlikely that the hydrogen will be lost, because it is not acidic enough. In other words, it is not bound to a carbon that is attached to an electronegative group (e.g., a carbonyl group).

d.

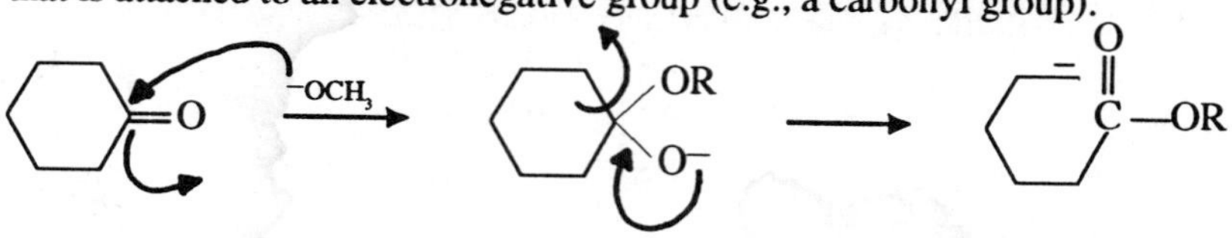

In the second step, the alkyl chain is a very poor leaving group since there is no electronegative group directly attached to it.

e. 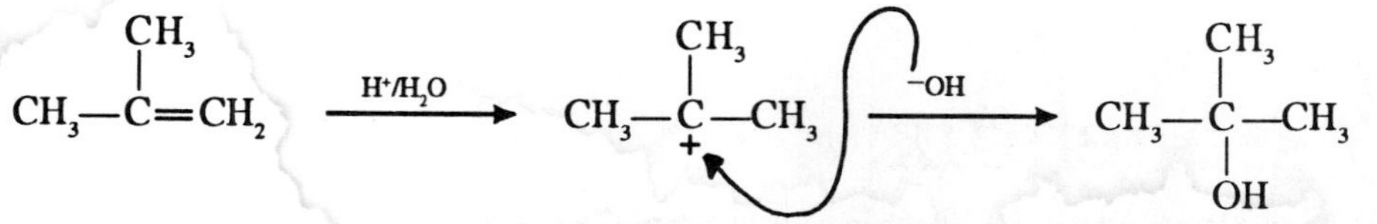

The concentration of the hydroxide (HO⁻) anion is practically zero in acid medium. The nucleophile in this case is the water molecule.

f. 

The starred (*) carbon that is β-to the carbonyl is already tetravalent (it carries two hydrogens), and this mechanism would give that same carbon, in the product, ten electrons (five bonds) around it.

g. 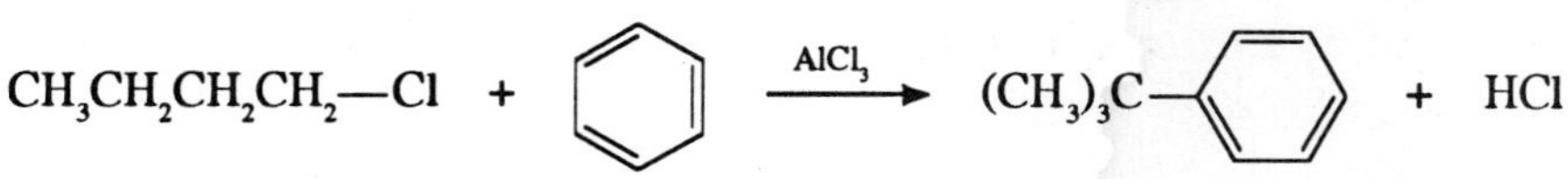

This is an example of the Friedel-Crafts alkylation. Such a product would result from the formation of the *t*-butyl carbocation through the following sequence:

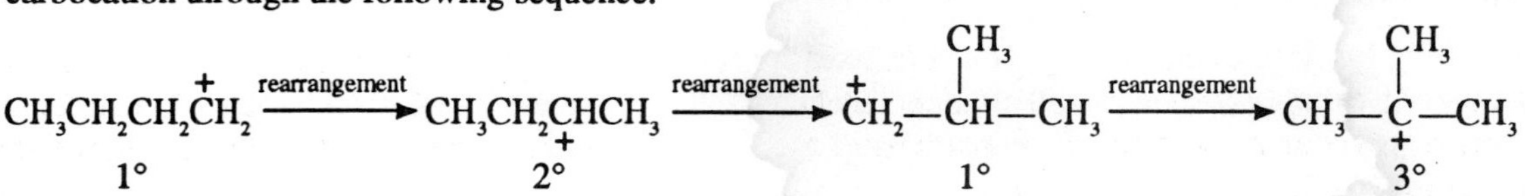

This sequence is unlikely since in order to form the *t*-butyl carbocation, the intermediate will have to go through the unlikely route of 2° ⟶ 1° ⟶ 3° carbocation rearrangement.

h. 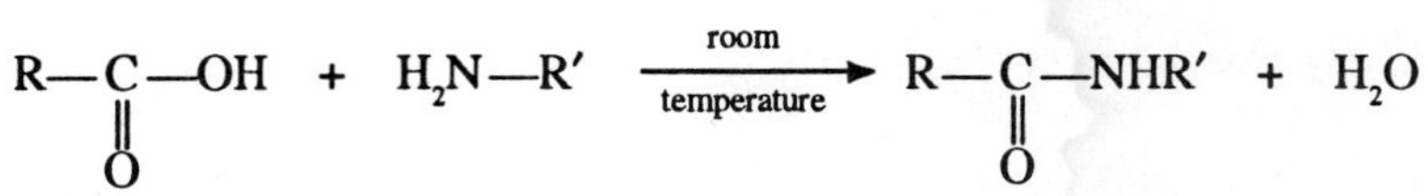

carboxylic acid (an acid) amine (a base)

This reaction involves an acid and a base, and the expected product is the salt $R{-}C(=O){-}O^- \; R'NH_3^+$

i. 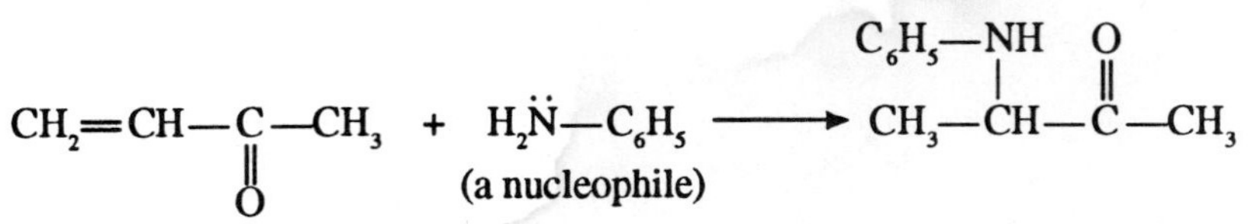

α, β-unsaturated ketones have the following resonance form: $\overset{+}{C}H_2{-}CH{=}C(O^-){-}CH_3$

Thus, nucleophiles will react at the β-position, which, in this case, leads to the formation of $C_6H_5NH{-}CH_2CH_2{-}C(=O){-}CH_3$.